2021 UNIFORM PLUMBING CODE®

AN AMERICAN NATIONAL STANDARD
IAPMO/ANSI UPC 1 – 2021

IAPMO®

Twenty-Ninth Edition
First Printing, February 2020
Second Printing, January 2021
Third Printing, May 2021

Published by the International Association of Plumbing and Mechanical Officials
4755 E. Philadelphia Street • Ontario, CA 91761-2816 – USA
Main Phone: (909) 472-4100 • Main Fax: (909) 472-4150

Important Notices and Disclaimers

The 2021 edition of the *Uniform Plumbing Code* is developed through a consensus standards development process approved by the American National Standards Institute. This process brings together volunteers representing varied viewpoints and interests to achieve consensus on plumbing issues. While the International Association of Plumbing and Mechanical Officials (IAPMO) administers the process and establishes rules to promote fairness in the development of consensus, it does not independently test, evaluate, or verify the accuracy of any information or the soundness of any judgments contained in its codes and standards.

IAPMO disclaims liability for any personal injury, property, or other damages of any nature whatsoever, whether special, indirect, consequential, or compensatory, directly or indirectly resulting from the publication, use of, or reliance on this document. IAPMO also makes no guarantee or warranty as to the accuracy or completeness of any information published herein.

In issuing and making this document available, IAPMO is not undertaking to render professional or other services for or on behalf of any person or entity. Nor is IAPMO undertaking to perform any duty owed by any person or entity to someone else. Anyone using this document should rely on his or her own independent judgment or, as appropriate, seek the advice of a competent professional in determining the exercise of reasonable care in any given circumstances.

Updating IAPMO Codes

Users of IAPMO codes should be aware that IAPMO codes may be amended from time to time through the issuance of Tentative Interim Amendments or corrected by Errata. IAPMO codes consist of the current edition of the document together with any Tentative Interim Amendment and any Errata in effect.

In order to determine whether an IAPMO code has been amended through the issuance of Tentative Interim Amendments or corrected by Errata, please visit the IAPMO Group codes information pages on IAPMO's website (www.iapmo.org). The codes information pages provide a list of IAPMO codes with up-to-date, specific information including any issued Tentative Interim Amendments and Errata.

To access the codes information pages for a specific code, go to http://codes.iapmo.org to select from the list of IAPMO codes. For Tentative Interim Amendments, go to the standard council decisions. For Errata, select the archived revision information.

Origin and Development

The advantages of a statewide adopted *Uniform Plumbing Code* are recognized throughout the industry. Disorder in the industry because of widely divergent plumbing practices and the use of many different, often conflicting, plumbing codes by local jurisdictions influenced the Western Plumbing Officials Association (now the International Association of Plumbing and Mechanical Officials [IAPMO]) to form a committee. This committee of plumbing inspectors, master and journeyman plumbers, and plumbing engineers, backed by public utility companies and the plumbing industry to create a basic plumbing document for general use. The product of this effort, the first edition of the *Uniform Plumbing Code*® (UPC®) was adopted by IAPMO in 1945. The widespread use of this code over the past seven decades by jurisdictions throughout the United States and internationally is testament to its merit.

Publishing of the 2003 *Uniform Plumbing Code* was a significant milestone because it was the first time in the history of the United States a plumbing code was developed through a true consensus process. The 2021 edition represents the most current approaches in the plumbing field and is the seventh edition developed under the ANSI consensus process. Contributions to the content of this code consists of diverse interests as consumers, enforcing authorities, installers/maintainers, labor, manufacturers, research/standards/ testing laboratories, special experts, and users.

The *Uniform Plumbing Code* provides consumers with safe and sanitary plumbing systems while, at the same time, allowing latitude for innovation and new technologies. The public at large is invited and encouraged to take part in IAPMO's open consensus code development process. This code is updated every three years. The *Uniform Plumbing Code* is dedicated to all those who, in working to achieve "the ultimate plumbing code," have unselfishly devoted their time, effort, and personal funds to create and maintain this, the finest plumbing code in existence today.

The *Uniform Plumbing Code* updates every three years in revision cycles that begin twice each year that takes two years to complete.

Each revision cycle advances according to a published schedule that includes final dates for all major events and contains four basic steps as follows:
1. Public and Committee Proposal Stage;
2. Comment Stage;
3. Association Technical Meeting;
4. Council Appeals and Issuance of Code.

IAPMO develops "full consensus" codes built on a foundation of maximum participation and agreement by a broad range of interests. This philosophy has led to producing technically sound codes that promote health and safety, yet do not stifle design or development.

It is important to stress that the process remains committed to the principles of consensus code development where consensus Technical Committees and Correlating Committees revise codes. The public and membership is offered multiple opportunities to debate, provide input and raise concerns through Amending Motions at the annual Assembly Consideration Session. Anyone may submit an appeal related to the issuance of a document through the IAPMO Standards Council.

The 2021 *Uniform Plumbing Code* is supported by the Mechanical Contractors Association of America (MCAA), the Plumbing-Heating-Cooling Contractors National Association (PHCC-NA), the United Association (UA), and the World Plumbing Council (WPC). The presence of these logos, while reflecting support, does not imply any ownership of the copyright to the UPC, which is held exclusively by IAPMO. Further, the logos of these associations indicate the support of IAPMO's open consensus process being used to develop IAPMO's codes and standards.

The addresses of the organizations are as follows:

ASSE – 18927 Hickory Creek Drive, Suite 220 • Mokena, IL 60448 • (708) 995-3019
MCAA – 1385 Piccard Drive • Rockville, MD 20850 • (301) 869-5800
PHCC-NA – PO Box 6808 • Falls Church, VA 22040-6808 • (800) 533-7694
UA – Three Park Place • Annapolis, MD 21401 • (410) 269-2000
WPC – World Plumbing Council Secretariat, Auf der Mauer 11 • Postfach CH 8021 • Zurich, Switzerland • www.WorldPlumbing.org

Adoption

The *Uniform Plumbing Code* is available for adoption and use by jurisdictions in the United States and Internationally. Its use within a governmental jurisdiction is accomplished through adoption by reference in accordance with applicable jurisdictional laws. At adoption, jurisdictions should insert the applicable information in bracketed words in the sample ordinance. The sample legislation for adoption of the *Uniform Plumbing Code* on page xiii provides key components, regulations and resolutions.

Revision Markings

Solid vertical lines in the margins indicate a technical change from the requirements of the 2018 edition. An arrow (⟵) in the margin indicates where an entire section, paragraph, exception, figure, or table has been deleted, or an item in a list of items or a table has been deleted.

A double right angle («) in the margin indicates that the text or a table has been relocated within the code. The table found on page xv points out the relocations in the 2021 edition of the *Uniform Plumbing Code*.

TIA TIA indicates that the revision is the result of a Tentative Interim Amendment.
TIA
TIA For further information on tentative interim amendments see Section 5 of the IAPMO
TIA Regulations Governing Committee Projects available at http://codes.iapmo.org/

A reference in brackets [ ] following a section or paragraph indicates material that has been extracted from another document. A reference in brackets { } following a section or paragraph indicates material that has been extracted from another document and has been modified further by the Technical Committee. This reprinted material is not the complete and official position of the source document on the referenced subject that is represented by the standard in its entirety. Material contained in this document that is taken or extracted from NFPA standards is used with permission of the National Fire Protection Association. This material is not the complete and official position of the NFPA on the reference subject, which is represented solely by the relevant standard in its entirety. NFPA standards can be accessed at www.nfpa.org. In addition, this extracted material may include revisions or modifications developed through IAPMO's standards development process. Therefore, NFPA disclaims responsibility for the content of this Code.

Text that is extracted pursuant to IAPMO's Extract Guidelines, but outside of the regular revision process is denoted with the use of the source document in the margin. This text is not fully processed by IAPMO in accordance with ANSI's public announcement consensus requirements for an American National Standard (ANS) nor approved by ANSI's Board of Standards Review. The next revision cycle processes such text in accordance with those requirements.

FORMAT OF THE UNIFORM PLUMBING CODE

The format of the *Uniform Plumbing Code* (UPC) arranges each chapter in accordance with a specific subject matter. However, Chapter 3 is dedicated to general requirements that are applicable to every chapter. The subject matters are divided as follows:

CHAPTERS	SUBJECTS
1	Administration
2	Definitions
3	General Regulations
4	Plumbing Fixtures and Fixture Fittings
5	Water Heaters
6	Water Supply and Distribution
7	Sanitary Drainage
8	Indirect Wastes
9	Vents
10	Traps and Interceptors
11	Storm Drainage
12	Fuel Gas Piping
13	Health Care Facilities and Medical Gas and Medical Vacuum Systems
14	Firestop Protection
15	Alternate Water Sources for Nonpotable Applications
16	Nonpotable Rainwater Catchment Systems
17	Referenced Standards
Appendix A	Recommended Rules for Sizing the Water Supply System
Appendix B	Explanatory Notes on Combination Waste and Vent Systems
Appendix C	Alternate Plumbing Systems
Appendix D	Sizing Storm Water Drainage Systems
Appendix E	Manufactured/Mobile Home Parks and Recreational Vehicle Parks
Appendix F	Firefighter Breathing Air Replenishment Systems
Appendix G	Sizing of Venting Systems
Appendix H	Private Sewage Disposal Systems
Appendix I	Installation Standards
Appendix J	Combination of Indoor and Outdoor Combustion and Ventilation Opening Design
Appendix K	Potable Rainwater Catchment Systems
Appendix L	Sustainable Practices
Appendix M	Peak Water Demand Calculator
Appendix N	Impact of Water Temperature on the Potential for Scalding and Legionella Growth

The following is a summary of the scope and intent of the provisions addressed within the chapters and appendices of the *Uniform Plumbing Code*:

Chapter 1 Administration.

Chapter 1 regulates the application, enforcement, and administration of subsequent requirements of the code. As well as establishing the scope of the code, this chapter is concerned with enforcing the requirements contained in the body of the code. A plumbing code, as with any other code, is intended to be adopted as a legally enforceable document to safeguard health, safety, property and public welfare. The code cannot be effective without satisfactory provisions for its administration and enforcement. The Authority Having Jurisdiction is to review the proposed and completed work and to decide whether a plumbing system conforms to the code requirements. As a public servant, the Authority Having Jurisdiction enforces the code in an unbiased, proper manner. The design professional is responsible for the design of a safe plumbing system. The contractor is responsible for installing the system in accordance with the plans.

Chapter 2 Definitions.

To maintain consistency and encourage the use of common terminology, Chapter 2 establishes definitions to provide clarity of terms and promote the use of a common language throughout the code. Understanding definitions within the context of their application enables greater collaboration, efficiency, standardization and interpretation in applying and enforcing terms used throughout the code. Codes are technical documents, and every term can impact the meaning of the code text. Terms not defined have a normally accepted meaning.

Chapter 3 General Regulations.

Chapter 3 regulates the general requirements, not specific to other chapters, for installing plumbing systems. Many regulations are not specific plumbing requirements, but relate to the overall plumbing system. This chapter contains safety requirements for installing plumbing and also contains nonplumbing requirements for identifying pipe, pipe fittings, traps, fixtures, materials and devices used in plumbing systems. Listing method of approval, based on applicable nationally recognized standards, for the safe and proper installation of plumbing systems is essential to ensure protection of public health, safety, and welfare. The safety requirements provide protection for piping, material, and structures, with provisions for installation practices, removing stress and strain of the pipe, sleeving, and hanger support. The building's structural stability is protected by the regulations for cutting and notching of structural members.

Chapter 4 Plumbing Fixtures and Fixture Fittings.

This chapter regulates the minimum number of plumbing fixtures of a specific type and quality for each building. The fixtures must be properly installed to be usable by the individuals occupying the building. The quality and design of every fixture must conform to the applicable referenced standard. Compliance with this chapter will result in a building or structure having acceptable plumbing fixtures for the sanitary, hygienic, cleaning, washing and food preparation needs of the occupants.

Chapter 5 Water Heaters.

Chapter 5 regulates the design, approval, installation, and safety devices of fuel burning and other types of water heaters with the combustion air requirements for ventilation and dilution of flue gases for appliances installed in buildings. This chapter does not apply to direct-vent appliances. In addition, this chapter regulates the design, construction, installation, maintenance of chimneys, vents and their connections to fuel burning appliances. Methods to supply combustion air may be supplied from an indoor air supply, outdoor air supply, a combination of indoor and outdoor air supply, mechanical air supply, or an engineered system. Combustion air provisions are based on the number of openings and the total opening size required based on the total energy input rating of the appliance. Acceptable air supply for combustion and ventilation is necessary for proper operation of fuel burning appliances. A shortage of combustion air can result in incomplete combustion and production of poisonous gases, such as carbon monoxide or appliance overheating. Ventilation air provides cooling for the appliance casing and internal controls. Inadequate ventilation of the space in which an appliance is installed can result in increased surrounding temperatures that stress the appliance itself or other appliances in the vicinity.

Chapter 6 Water Supply and Distribution.

Chapter 6 regulates the design, material and installation of water supply and distribution systems, including residential fire sprinklers. The water supply and distribution system is designed to achieve the correct water pressure and flow rates and avoid cross connections. For fixtures to perform properly, an acceptable supply of potable water is essential to their operation and use. Cross connections and backflow are ranked as the highest priority because of the long history of recognized health risks posed by cross connections, outbreaks, or cases of waterborne disease. Piping materials and components are evaluated for their possible effect on the potable water with which they are in contact. The intent is to control the potential adverse health effects produced by indirect additives, products, and materials that come in contact with potable water. When selecting materials for water supply and distribution systems, consider water pressure, water temperature, compatibility with the water supply, durability, support, and sustainability.

In addition, this chapter regulates the design, location, materials, and installation of multipurpose and stand-alone sprinkler systems that do not include the use of antifreeze. Where systems are installed as a portion of the water distribution system under the requirements of this chapter and are not provided with a fire department connection, backflow protection for the water supply system is not required.

Chapter 7 Sanitary Drainage.

This chapter regulates the design and installation of sanitary drainage systems to ensure they will work as intended. Drainage piping should not be oversized nor undersized, and constructed of approved materials to guard against fouling, deposit of solids, clogging, and with cleanouts so arranged that the pipes may be readily cleaned. The purpose of the sanitary drainage system is to remove effluent discharged from plumbing fixtures and other equipment to an approved point of disposal, such as a public sanitary system or private sewage disposal system.

The basics of a sanitary drainage system include public and private sewage disposal; selection of materials; installation of the building drain and sewer; joining methods for pipe and fittings; drainage fixture units for sizing the drainage system; sumps and ejectors; vent sizing and length of vents; and testing.

Chapter 8 Indirect Wastes.

Chapter 8 regulates indirect waste connections that are required for plumbing fixtures and plumbing appliances dealing with food preparation, dishwashing, potable liquids, and similar equipment. An indirect connection prevents sewage from backing up into a fixture or appliance, thus providing protection against potential health hazards. The waste pipe discharges through an air gap or air break into a waste receptor or standpipe. The protection in the form of an air gap is necessary when the contamination is a potential health hazard or cross connection with the potable water system. Where there is no possibility of contaminating the potable water (nonpotable discharge), the indirect waste pipe may connect in the form of an air break. This method is often preferred to prevent splashing. In addition, health care facilities and special wastes must be protected from contamination that may result from the connection to the drainage system. The waste must be treated to prevent any damage to the piping or sewage treatment process. Waste receptors are sized and designed to prevent splashing and allow for peak discharge conditions.

Chapter 9 Vents.

Chapter 9 regulates the material, design, and installation of vents. A vent system is a pipe or pipes installed in a drainage system that provide a flow of air to and from the system to ventilate it, provide a circulation of air to eliminate trap siphonage, and reduce back-pressure and vacuum surge. In addition, vents provide the rapid and silent flow of waste without exposing occupants of the building to any sewer gases. Proper installation of vents is crucial, as a telltale sign that there is a problem in the drain and vent system is related to the elevation of the horizontal portion of the venting. Venting is not limited to sanitary drainage systems. Venting methods are applicable to other drainage systems such as those for chemical waste, graywater waste, and clear water waste. Sizing the venting system is directly tied to the design of the drainage system. For example, the velocities in the drainage system and its peak flow rates affect the diameters in the venting system. Where the vertical distance between a fixture outlet and trap is excessive, velocities in the entire drainage system will be greater than those in the vent sizing table. All venting methods in this chapter are categorized as either dry vents or wet vents. Vent stacks, stack vents, branch vents, island vents, relief vents, and individual vents are dry vents. Wet vents (horizontal or vertical), circuit vents, combination drain and vents are versions of "wet venting" in which the vent is wetted by drainage flow.

Chapter 10 Traps and Interceptors.

Chapter 10 regulates the material, design, and installation of traps, interceptors, and separators. Traps are required on drainage type plumbing fixtures and must be self-scouring without interior partitions. Interceptors, on the other hand, are designed to control what goes down a drain. Interceptors are used to keep harmful substances from entering the sanitary drainage system, such as grease, sand, oil and other materials. The retained materials need periodic removal to maintain efficiency and function of the separating device. The capacity of an interceptor is based on retention and flow rate. There are many types of interceptors that are used at beauty salons, hospitals, meat, fish or foul packaging, refineries, repair garages, gas stations, car washing facilities, various plants, factories, and processing sites. The designer of the building is responsible for locating interceptors with the expectation for the frequency of maintenance, ease of cleaning and floor space for equipment.

Chapter 11 Storm Drainage.

Chapter 11 regulates the removal of stormwater from roofs, yards, paved areas, and similar areas. The objective of storm drainage systems is to provide a conduit or channel through which runoff will be carried from a point of collection to a point of disposal; this protects the property and the public from the uncontrolled flow of runoff and ensures that drains and inlets are adequately sized to receive the volume of runoff that flows to the drains. For the purpose of system design, it's necessary to specify the duration of a selected storm. All methods used to determine volumes and peak flow use historical data. Drain location must be coordinated with the architectural design of the building. When selecting the type of roof drain to use, the roof construction and its thickness, along with the intended use of the roof, are required. Where the roof perimeter extends above the roof in such a manner that water is entrapped and causes ponding, or if any portion of the roof is designed so water can pond, secondary drainage is required. Where secondary drainage is required, scuppers, or a secondary system of roof drains and pipes, are installed to prevent the accumulation of excessive rainwater.

Chapter 12 Fuel Gas Piping.

Chapter 12 regulates the installation of gas piping in a building, structure or within the property lines of buildings up to 5 psi. Gas piping systems must supply the minimum volume of gas required by each gas appliance to perform their proper operation under working conditions without exceeding the maximum pressure specified by each manufacturer. Because of the hazards associated with fuel gas, it is important to ensure the gas system has been inspected and tested, and that it is safe to turn on the gas supply to the building.

Chapter 13 Health Care Facilities and Medical Gas and Medical Vacuum Systems.

Chapter 13 regulates the installation, inspection, testing, maintenance, performance, and safe practices for medical gas and vacuum systems located in health care facilities. This chapter addresses the installation and maintenance of health care fixtures, devices, and equipment. The purpose of medical gas and medical vacuum systems is to provide safe and sufficient flows at required pressures to the medical gas outlet or vacuum inlet terminals. System design and layout should allow convenient access by the medical staff to outlet and inlet terminals, valves, and equipment during patient care or emergencies, as safety is of primary concern.

Chapter 14 Firestop Protection.

Chapter 14 regulates piping penetrations of fire-resistance-rated walls, partitions, floors, floor and ceiling assemblies, roof and ceiling assemblies, or shaft enclosures through firestopping. To firestop is to create a physical barrier that impedes the spread of smoke, gases, and flames from one compartment in the building design to the next. The firestop is seen as a part that is essential to protecting the lives of people who live or work in the structure, increasing the chances of not succumbing to smoke or gases before they are able to evacuate the building. Fireproofing of this type helps to restore the fire-resistant properties of the building materials before the openings were created as part of the construction process.

Chapter 15 Alternate Water Sources for Nonpotable Applications.

Chapter 15 regulates gray water sources, reclaimed (recycled) water sources and on-site treated nonpotable water systems. Water sources include subsurface irrigation, subsoil irrigation, and mulch basin systems. Subsoil water irrigation provides a means to disperse shallow drip irrigation lines and mulch basins that collect and spread water in various applications. The reclaimed water provisions to on-site nonpotable water systems include gray water and other nonpotable water sources that are used for on-site applications. Water reuse is integral to sustainable water management because it allows water to remain in the environment and be preserved for future use while meeting the water requirements of the present. Water reuse reduces energy use by removing added potable water treatment, offsetting water demands, and providing water for energy production.

Chapter 16 Nonpotable Rainwater Catchment Systems.

Chapter 16 regulates nonpotable rainwater catchment systems that include irrigation; toilet and urinal flushing with proper treatment; provisions where permits are required; maintenance of alternate water sources; and minimum water quality. This chapter provides guidance on how to optimize rainwater use while ensuring there is a decrease of risk to consumers from poor design, installation, and maintenance. Rainwater harvesting is the process of capturing, channeling, and storing water runoff for later use. Most systems are constructed of three principal components: the catchment area, the collection device, and the conveyance system.

Chapter 17 Referenced Standards.

Chapter 17 provides two comprehensive tables with referenced standards. The standards listed in Table 1701.1 are applied as indicated in the applicable reference section(s). A list of additional approved standards, publications, practices, and guides that are not referenced in specific sections appear in Table 1701.2.

Referenced standards set forth specific details of accepted practices, materials specifications, or test methods in many specialized applications. Standards provide an efficient method of conveying complex information and specifications on the performance requirements for materials, products, systems, application, and installation. The manner and purpose for a standard's use and, in turn, code compliance, must be definitive in all references to the standard. If the standard is intended to be a requirement for judging code compliance, the code must state its intent for use. The standard should adequately address a defined need and at the same time specify the minimum performance requirements, technical characteristics and methods of testing, and required test results.

The referenced standards tables are organized in a manner that makes it easy to find specific standards in alphabetical order, and by acronym of the publishing agency of the standard. The tables list the title of the standard, the edition, and any addenda. Contact information for each publishing agency is provided at the end of the chapter.

Appendix A Recommended Rules for Sizing the Water Supply System.

Appendix A provides a method of sizing the water supply and distribution system that provides precise calculations to establish the proper pressures and flow to the system's fixtures. The goal of sizing the system is to deliver an acceptable volume of water to the most hydraulically remote fixture during minimum pressure and maximum flow conditions; provide satisfactory water pressure to the most hydraulically remote fixture during minimum pressure and maximum flow conditions; and to prevent excessive water velocity during maximum flow conditions.

Appendix B Explanatory Notes on Combination Waste and Vent Systems.

Appendix B contains general guidelines for the design and installation of combination waste and vent systems. These systems are designed for waste piping and are purposely oversized to serve as both a waste and vent pipe to avoid excessive pneumatic effects at fixture drains.

Appendix C Alternate Plumbing Systems.

The intent of this appendix is to provide clarification of procedures for the design and approval of engineered plumbing systems, alternate materials, and equipment that are not specifically covered in other parts of the code. Alternative methods are allowed to be used where approved by the authority having jurisdiction. Approval of alternatives is based on a demonstration showing that the method or material used is at least equivalent in strength, deflection, and capacity to that provided by the prescriptive methods and materials.

Appendix D Sizing Storm Water Drainage Systems.

Appendix D provides general guidelines for the sizing of stormwater drainage systems. There are two pieces of information that must always be a given. They are the roof size and the rate of rainfall for the area.

Appendix E Manufactured/Mobile Home Parks and Recreational Vehicle Parks.

The provisions of this appendix apply to the plumbing and drainage systems of mobile home and recreational vehicle parks. These provisions also apply to the use, maintenance, and installation for supplying fuel gas, water, electricity, and disposal of sewage from accessory buildings or structures, and building components.

Appendix F Firefighter Breathing Air Replenishment Systems.

Appendix F provides guidance on installing firefighter breathing air replenishment systems. System components include outside fire department connection panel, interior air fill panel or station, interconnected piping distribution system and pressure monitoring switch. Fire departments access the system through an outside connection panel and are able to pump air into the system. The firefighters inside the structure access the system at fill stations that are found throughout the building. The piping distribution system is made from stainless tubing or other approved materials. It delivers compressed air to the building interior air fill stations and interior air fill panels. The tubing also acts as a conduit in the interior of the building between the outside connection panel and the air storage system. If the system becomes over-pressurized, the air monitoring system also acts as a pressure relief. A system isolation valve is placed alongside each interior air fill station and interior air fill panel to isolate the system.

Appendix G Sizing of Venting Systems.

Appendix G provides added information on the sizing of gas vents. This appendix is useful to the end user for the proper sizing of venting systems. A series of examples are given that show how to use the tables and other requirements of Chapter 5.

Appendix H Private Sewage Disposal Systems.

Appendix H provides general guidelines for the materials, design, and installation of new or existing private sewage disposal systems. Where a building cannot be served by a public sewer system, the building site must be provided with a system for treating the waste water generated from the use of plumbing fixtures in the building. The appendix addresses site evaluations, materials, soil absorption systems, holding tanks, cesspools and on-site waste-water treatment systems. Private sewage disposal systems must be designed based on the soil conditions, constructed using approved materials, and installed according to prescribed dimensions.

Appendix I Installation Standards.

Installation Standards (IS) are industry standards that specify procedures for installation of plumbing and mechanical products, and the inspection methods and procedures employed for the examination of such systems. These comprehensive standards add detailed information and guidance where a manufacturer or product installation instructions may not fully address. They give the user confidence with options from industry experts for installing plumbing products and assemblies in a safe and professional manner.

Appendix J Combination of Indoor and Outdoor Combustion and Ventilation Opening Design.

Appendix J provides an example of how to determine the required combination of indoor and outdoor combustion air opening sizes for appliances. The combustion air example also provides a table that contains the required volume of a space per the appliance Btu/h input that is based on the standard method.

Appendix K Potable Rainwater Catchment Systems.

Potable rainwater catchment system is defined as a system that uses the principal of collecting and using rain from a rooftop or other man-made, aboveground collection surface. This appendix applies to new rainwater catchment installations, as well as changes, additions, maintenance, and repairs to existing installations. Rainwater harvesting is the practice of collecting the water produced during rainfall events before it has a chance to run off into a river or stream or soak into the ground and become groundwater.

Appendix L Sustainable Practices.

This appendix provides a comprehensive set of technically sound provisions that encourage sustainable practices and works toward improving the design and construction of plumbing systems that result in a positive long-term environmental impact. Environmental sustainability is important because it involves natural resources that human beings need for economic or manufactured capital. Their sustainability is defined by their reliance on infinitely available resources that are naturally occurring, constant, and free to access.

Appendix M Peak Water Demand Calculator.

This appendix provides a method for estimating the demand load for the building water supply and principal branches for single- and multi-family dwellings with water-conserving plumbing fixtures, fixture fittings, and appliances.

Appendix N Impact Of Water Temperature On The Potential For Scalding And Legionella Growth

Legionnaires' disease is a serious issue that is a threat to many citizens. It is known that the higher the temperature, the lower the risk for Legionella growth, but the higher the temperature, the higher the possibility for scalding. This appendix addresses Legionella growth potential and scald potential within specified temperature ranges. It contains definitions for terms used to describe water temperatures such as cold, tepid, warm, tempered hot, and disinfecting hot. Since plumbing systems operate within various temperature ranges, the public has a document that identifies such temperature ranges and the "potential" dangers within these temperature ranges.

SAMPLE LEGISLATION FOR ADOPTION OF THE UNIFORM PLUMBING CODE

The Uniform Codes are designed to be adopted by jurisdictions through an ordinance. Jurisdictions wishing to adopt the 2021 *Uniform Plumbing Code* as an enforceable regulation governing plumbing systems by reference should ensure the legal basis under which adoption and implementation are included in the ordinance.

The following sample ordinance is a guide for drafting an ordinance for adoption that addresses key components regulations and resolutions.

ORDINANCE NO.

An ordinance of the [JURISDICTION] adopting the 2021 edition of the *Uniform Plumbing Code*, regulating and controlling the design, construction, quality of materials, erection, installation, alteration, repair, location, relocation, replacement, addition to, use or maintenance of plumbing systems in the [JURISDICTION]; providing for the issuance of permits and collection of fees therefor; repealing Ordinance No. of the [JURISDICTION] and all other ordinances and parts of the ordinances in conflict therewith.

The [GOVERNING BODY] of the [JURISDICTION] does ordain as follows:

Section 1 Codes Adopted by Reference. That certain documents, three (3) copies of which are on file in the office of the [JURISDICTION'S KEEPER OF RECORDS] and the [JURISDICTION], being marked and designated as the 2021 *Uniform Plumbing Code*, including Appendix Chapters [FILL IN THE APPENDIX CHAPTERS BEING ADOPTED], as published by the International Association of Plumbing and Mechanical Officials, be and is hereby adopted as the Code of the [JURISDICTION], in the State of [STATE NAME] regulating and controlling the design, construction, quality of materials, erection, installation, alteration, repair, location, relocation, replacement, addition to, use or maintenance of plumbing systems as herein provided; providing for the issuance of permits and collection of fees therefor; and each and all of the regulations, provisions, penalties, conditions and terms of such 2021 *Uniform Plumbing Code* on file in the office of the [JURISDICTION] are hereby referred to, adopted, and made a part hereof, as if fully set out in this ordinance.

Section 2 Modifications. The following sections are hereby revised:
Section 101.1. Insert: [NAME OF JURISDICTION]
Section 104.5. Insert: [APPROPRIATE FEE SCHEDULE]

Section 3 Conflicting Ordinances Repealed. That Ordinance No. of [JURISDICTION] entitled
[TITLE OF THE ORDINANCE OR ORDINANCES IN EFFECT AT THE PRESENT TIME SO THAT THEY WILL BE REPEALED BY MENTION] and all other ordinances or parts of ordinances in conflict herewith are hereby repealed.

Section 4 Preemption. [JURISDICTION] hereby fully occupies and preempts the entire field of regulation of design, construction, quality of materials, erection, installation, alteration, repair, location, relocation, replacement, addition to, use or maintenance of plumbing systems; and provision for the issuance of permits and collection of fees therefor; within the boundaries of [JURISDICTION]. [AS APPROPRIATE] Cities, towns, and counties or other municipalities may enact only those laws and ordinances relating to this field as specifically authorized by state law and consistent with this ordinance. Local laws and ordinances that are inconsistent with, more restrictive than, or exceed the requirements of [ORDINANCE NO.] shall not be enacted and are hereby expressly preempted and repealed, regardless of the nature of the code, charter, or home rule status of such city, town, county, or municipality.

Section 5 Severability. That if any section, subsection, sentence, clause or phrase of this ordinance is, for any reason, held to be unconstitutional, such decision shall not affect the validity of the remaining portions of this ordinance. The [GOVERNING BODY] hereby declares that it would have passed this ordinance, and each section, subsection, clause or phrase thereof, irrespective of the fact that any one or more sections, subsections, sentences, clauses and phrases be declared unconstitutional.

Section 6 Legal Notice. That the [JURISDICTION'S KEEPER OF RECORDS] is hereby ordered and directed to cause this ordinance to be published. (An additional provision may be required to direct the number of times the ordinance is to be published and to specify that it is to be in a newspaper in general circulation. Posting may also be required.)

Section 7 Violations and Penalties. [INCORPORATE PENALTIES FOR VIOLATIONS]

Section 8 Effective Date. That this ordinance and the rules, regulations, provisions, requirements, orders and matters established and adopted hereby shall take effect and be in full force and effect [TIME PERIOD] from and after the date of its final passage and adoption.

COMMITTEE ON UNIFORM PLUMBING CODE

These lists represent the membership at the time the Committee was balloted on the final text of this edition. Since that time, changes in the membership may have occurred.

IAPMO Standards Council
Bill Erickson, Chairman
CJ Erickson Plumbing Company [U]

JT Baca, State of New Mexico [E]
Carl Crimmins, MN State Pipe Trades, Retired [SE]
James Majerowicz, Plumbers JAC LU 130 [L]
Rich Prospal, Prospal Consulting Services [C]
Linden Raimer, Raimer Consulting Services, LLC [U]
Ron Rice, City of St. Paul, Retired [C]
Robert "Bud" Riestenberg, Piping Systems, Inc. [U]

Nonvoting

Gabriella M. Davis, Secretary
International Association of Plumbing and Mechanical Officials

Hugo Aguilar, Recording Secretary
International Association of Plumbing and Mechanical Officials

IAPMO Uniform Plumbing Code Technical Committee
Dan Daniels, *Chairman*
Self [C]

Bob Adler, City of San Jose [E]
Sarah Aguilar McCabe Plumbing [I/M]
Julius Ballanco JB Engineering & Code Consulting [SE]
Domenico Barbato, City of Los Angeles Department of Building & Safety [E]
DJ Berger National ITC Corporation [U]
Raymond Boyd, United Association [L]
Jeremy Brown, NSF International [R/S/T]
Richard Church, Plastic Pipe & Fittings Association (PPFA) [M]
Pennie Feehan Copper Development Association (CDA) [M]
Dave Mann, CA State Pipe Trades [L]
Rick Moreno Astro Mechanical/CPMCA [I/M]
John Nielsen State of Idaho – Division of Building Safety [E]
Phil Ribbs PHR Consultants [SE]
Arnold Rodio Pace Setter Plumbing [I/M]
Robert Sewell UA Local 159 [L]
Matt Sigler Plumbing Manufacturers International (PMI) [M]
Billy Smith ASPE [SE]
Larry Soskin ACE Duraflo [M]
Jim Stack, Plumbing Heating Cooling Contractors (PHCC) [I/M]
Don Taylor Dittmann Plumbing [U]

Alternates

Gary Bonetti Jr, Local 342 [L]
Terry Burger, NSF International [R/S/T]
Michael Cudahy, Plastic Pipe & Fittings Association (PPFA) [M]
Shawn Hargis, City of Los Angeles Department of Building & Safety [E]
David Ledda, City/County of San Francisco [E]
James Pavesic, United Association [L]
Chuck White, Plumbing Heating Cooling Contractors (PHCC) [I/M]

Nonvoting

Heath Dehn, National Fire Protection Association (NFPA) [R/S/T]
Enrique Gonzalez, IAPMO Staff Liaison
Doug Marian, ASSE [R/S/T]
Laura Moreno, National Fire Protection Association (NFPA) [R/S/T]
David Straub, Ex-Officio IAPMO [E]

COMMITTEE MEMBERSHIP CLASSIFICATION ABBREVIATIONS

These classifications apply to Technical Committee members and represent their principal interest in the activity of a committee.

M *Manufacturer:* A representative of a maker or marketer of a product, assembly or system, or portion thereof, that is affected by the standard.
U *User:* A representative of an entity that is subject to the provisions of the standard or that voluntarily uses the standard.
I/M *Installer/Maintainer:* A representative of an entity that is in the business of installing or maintaining a product, assembly or system affected by the standard.
L *Labor:* A labor representative or employee concerned with safety in the workplace.
R/S/T *Research/Standards/Testing Laboratory:* A representative of an independent research organization; an organization that develops codes, standards or other similar documents; or an independent testing laboratory.
E *Enforcing Authority:* A representative or an agency or an organization that promulgates and/or enforces standards.
C *Consumer:* A person who is, or represents, the ultimate purchaser of a product, system, or service affected by the standard, but who is not included in the *User* classification.
SE *Special Expert:* A person not representing any of the previous classifications, but who has special expertise in the scope of the standard or portion thereof.

2021 Location	2018 Location
209.0 General Anesthesia and Levels of Sedation/Analgesia	214.0 Levels of Sedation
209.0 Deep Sedation/Analgesia	214.0 Deep Sedation/Analgesia
209.0 General Anesthesia	214.0 General Anesthesia
209.0 Minimal Sedation (Anxiolysis)	214.0 Minimal Sedation (Anxiolysis)
209.0 Moderate Sedation/ Analgesia (Conscious Sedation)	214.0 Moderate Sedation/ Analgesia (Conscious Sedation)
210.0 Health Care Facility's Governing Body	209.0 Governing Body
216.0 Nonwater Urinal with Drain Cleansing Action	223.0 Urinal, Hybrid
509.3.4	509.3.4.1
509.6.1(2) – 509.6.1(7)	509.6.1.1 – 509.6.1.6
509.6.1.1	509.6.1.7
509.7.4(1) – 509.7.4(3)	509.7.4.1 – 509.7.4.3
609.9 – 609.12.2	609.8 – 609.11.2
Table 702.1(1)	Table 702.2(1)
Table 702.2	Table 702.2(2)
911.3	911.4
911.3.1	911.4.1
911.3.2	911.4.2
1208.6.4.2 – 1208.6.4.5	1208.6.4.1 – 1208.6.4.4
1208.6.7.1 – 1208.6.7.3	1208.6.7(1) – 1208.6.7(3)
1208.6.9 – 1208.6.10	1208.6.10 – 1208.6.11
Table 1208.6.9.2	Table 1208.6.10.2
1208.6.10.1	1208.6.11.2
1208.6.10.4 – 1208.6.13.5	1208.6.11.3 – 1208.6.14.5
1208.8.6, 1208.8.7	1208.8.7, 1208.8.8
1208.10.6	1208.8.6
1208.11 – 1208.15.1	1208.10 – 1208.14.1
1210.3	1210.2
1210.3.1	1208.6.9
1210.3.2 – 1210.8.2	1210.2.1 – 1210.7.2
Table 1210.3.5.1	Table 1210.2.4.1
1211.4 – 1211.7.1	1211.3 – 1211.6.1
1301.4, 1301.5	1301.5, 1301.6
1302.1.2, 1302.1.3	1302.1.1, 1302.1.2
1304.2 – 1304.5	1304.5 – 1304.6.3
1306.0 – 1306.3	1319.0 – 1319.3
1306.4	1317.4

2021 Location	2018 Location
1307.0	1312.0
1307.1	1301.4
1307.3, 1307.4	1312.1, 1312.2
1308.0 – 1308.2	1315.0 – 1315.2
1308.3	1312.3
1310.0, 1310.1	1313.0, 1313.1
1310.3 – 1310.3.2	1313.2 – 1313.2.2
1310.4	1313.7
1310.5	1313.7.2
1311.1, 1311.2	1313.5, 1313.5.1
1311.3 – 1311.6	1313.6 – 1313.6.3
1312.0 – 1312.2	1314.0 – 1314.2
1312.3	1314.4
1313.1 – 1313.5	1314.5 – 1314.5.4
1314.0	1311.0
1314.2	1311.1.1
1314.3	1311.1.3
1314.4	1311.1.2
1314.5	1311.1
1314.6 – 1314.10	1311.2 – 1311.6
1314.10.3	1311.6.3
1314.11	1311.7
1314.12, 1314.12.1	1311.9, 1311.9.1
1315.0 – 1315.3	1316.0 – 1316.2.1
1315.15	1316.3
1316.1, 1316.2	1315.2.2, 1315.3
1317.1.2	1317.1.1
1318.0 – 1318.4	1308.0 – 1308.4
1318.6	1310.1
1319.1	1308.5
1320.0 – 1320.2.1	1309.0 – 1309.3
1321.1 – 1321.5	1309.4 – 1309.4.4
1321.7 – 1321.12.4	1309.5 – 1309.10.4
1322.1	1309.16
1322.1.1	1309.16.2
1322.1.3	1309.16.3
1322.1.5, 1322.2	1309.16.4, 1309.17
1322.2.1	1309.16.1

2021 Location	2018 Location
1322.3 – 1322.5	1309.11 – 1309.13
1322.6	1309.15
1322.6.1	1309.14
1322.7	1309.18
1323.0	1310.0
1323.1 – 1323.1.6	1310.2 – 1310.2.6
Table 1323.1.4(1)	Table 1310.2.1(2)
Table 1323.1.4(2)	Table 1310.2.1(3)
Table 1323.1.4(3)	Table 1310.2.1(4)
Table 1323.1.4(4)	Table 1310.2.1(5)
Table 1323.1.4(5)	Table 1310.2.1(6)
Table 1323.1.4(6)	Table 1310.2.1(7)
Table 1323.1.5(1)	Table 1310.2.1(1)
Table 1323.1.5(2)	Table 1310.2.2(1)
1323.2 – 1323.2.2	1310.3 – 1310.3.2
1323.3, 1323.3.1	1310.4, 1310.4.1
1323.3.2	1310.4.3
1323.3.3	1310.4.2
1323.4 – 1323.4.5	1310.5 – 1310.5.5
Table 1323.4.4	Table 1310.5.4(1)
1323.5	1310.3.3
1323.5.1 – 1323.7.2	1310.6 – 1310.8.2
1323.8	1304.2
1323.9, 1323.9.1	1310.9, 1310.9.1
1323.10 – 1323.10.3	1306.1 – 1306.3
1323.11 – 1323.11.6	1307.1 – 1307.7
1323.12 – 1323.13.2	1310.10 – 1310.11.2
1323.14 – 1323.14.6	1311.10 – 1311.10.6
1323.15	1316.4
1324.0 – 1324.2.1	1318.0 – 1318.3
1324.3	1318.4.1
1324.4	1318.4
1324.5 – 1324.5.1.2	1318.5 – 1318.5.3
1324.5.2 – 1324.5.2.6	1318.8 – 1318.8.6
1324.5.3 – 1324.5.3.2	1318.12 – 1318.12.2
1324.5.4 – 1324.5.5.5	1318.9 – 1318.10.5
1324.5.5.6	1318.10.7
1324.5.5.7	1318.10.6

2021 Location	2018 Location
1324.5.7 – 1324.5.7.2	1318.16 – 1318.16.2
1324.5.8	1309.8.9(4)
1324.5.9 – 1324.5.9.2	1309.8.9(1) – 1309.8.9(3)
1324.5.9.3, 1324.5.9.4	1309.8.9(5), 1309.8.9(6)
1324.5.10.1 – 1324.5.11	1318.14.3 – 1318.15
Table 1324.5.11	Table 1318.15
1325.1	1304.3
1325.4	1313.3
1325.6	1314.2.1
1325.10	1317.2
1326.1	1304.4
1327.2	1311.8
1327.3	1317.3
Table 1327.6	Table 1310.5.4(2)
1327.7 – 1327.7.6	1318.11 – 1318.11(6)
1327.8 – 1327.8.2	1318.14 – 1318.14.2
1602.9	1602.9.2
1603.0 – 1603.2	1602.9, 1602.9.1
Figure 1603.2	Figure 1602.9.1
1603.3	1603.2
1603.4 – 1603.5	1602.9.3 – 1602.9.6
Table 1603.5	Table 1602.9.6
1603.6	1603.0, 1603.1
1603.7 – 1603.20	1603.3 – 1603.16
Table K 104.3.1	Table K 104.3(1)
Table K 104.3.3	Table K 104.3(2)
K 104.4	K 106.0, K 106.1
K 104.4.1, K 104.4.2	K 104.4, K104.5
K 104.4.3	K 106.2
K 104.5	K 104.6
K 106.0, K 106.1	K 106.3
K 106.2 – K 106.5	K 106.4 – K 106.7
L 402.8 – L 402.10	L 402.7 – L 402.9
L 404.8 – L404.11	L 404.7 – L 404.10
L 411.14	L 411.12

TABLE OF CONTENTS

CHAPTER 1	**ADMINISTRATION**	**1**
101.0	General	1
101.1	Title	1
101.2	Scope	1
101.3	Purpose	1
101.4	Unconstitutional	1
101.5	Validity	1
102.0	Applicability	1
102.1	Conflicts Between Codes	1
102.2	Existing Installations	1
102.3	Maintenance	1
102.4	Additions, Alterations, Renovations, or Repairs	1
102.5	Health and Safety	1
102.6	Changes in Building Occupancy	1
102.7	Moved Structures	1
102.8	Appendices	1
103.0	Duties and Powers of the Authority Having Jurisdiction	2
103.1	General	2
103.2	Liability	2
103.3	Applications and Permits	2
103.4	Right of Entry	2
104.0	Permits	2
104.1	Permits Required	2
104.2	Exempt Work	2
104.3	Application for Permit	2
104.4	Permit Issuance	3
104.5	Fees	4
105.0	Inspections and Testing	4
105.1	General	4
105.2	Required Inspections	4
105.3	Testing of Systems	5
105.4	Connection to Service Utilities	5
106.0	Violations and Penalties	5
106.1	General	5
106.2	Notices of Correction or Violation	5
106.3	Penalties	5
106.4	Stop Orders	6
106.5	Authority to Disconnect Utilities in Emergencies	6
106.6	Authority to Condemn	6
107.0	Board of Appeals	6
107.1	General	6
107.2	Limitations of Authority	6
Table 104.5	Plumbing Permit Fees	7
CHAPTER 2	**DEFINITIONS**	**9**
201.0	General	9
201.1	Applicability	9
202.0	Definition of Terms	9
202.1	General	9
CHAPTER 3	**GENERAL REGULATIONS**	**21**
301.0	General	21
301.1	Applicability	21
301.2	Minimum Standards	21
301.3	Alternate Materials and Methods of Construction Equivalency	21
301.4	Flood Hazard Areas	21
301.5	Alternative Engineered Design	22
302.0	Iron Pipe Size (IPS) Pipe	22
302.1	General	22
303.0	Disposal of Liquid Waste	22
303.1	General	22
304.0	Connections to Plumbing System Required	22
304.1	General	22
305.0	Damage to Drainage System or Public Sewer	22
305.1	Unlawful Practices	22
306.0	Industrial Wastes	22
306.1	Detrimental Wastes	22
306.2	Safe Discharge	22
307.0	Location	22
307.1	System	22
307.2	Ownership	23
308.0	Improper Location	23
308.1	General	23
309.0	Workmanship	23
309.1	Engineering Practices	23
309.2	Concealing Imperfections	23
309.3	Burred Ends	23
309.4	Installation Practices	23
309.5	Sound Transmission	23
309.6	Dead Legs	23
310.0	Prohibited Fittings and Practices	23
310.1	Fittings	23

310.2 Drainage and Vent Piping 23
310.3 Waste Connection 23
310.4 Use of Vent and Waste Pipes 23
310.5 Obstruction of Flow 23
310.6 Dissimilar Metals 23
310.7 Direction of Flow 23
310.8 Screwed Fittings 23
311.0 Independent Systems 23
311.1 General 23
312.0 Protection of Piping, Materials, and Structures 23
312.1 General 23
312.2 Installation 23
312.3 Building Sewer and Drainage Piping 24
312.4 Corrosion, Erosion, and Mechanical Damage 24
312.5 Protectively Coated Pipe 24
312.6 Freezing Protection 24
312.7 Fire-Resistant Construction 24
312.8 Waterproofing of Openings 24
312.9 Steel Nail Plates 24
312.10 Sleeves 24
312.11 Structural Members 24
312.12 Rodentproofing 24
313.0 Hangers and Supports 24
313.1 General 24
313.2 Material 24
313.3 Suspended Piping 24
313.4 Alignment 24
313.5 Underground Installation 24
313.6 Hanger Rod Sizes 24
Table 313.6 Hanger Rod Sizes 24
313.7 Gas Piping 24
314.0 Trenching, Excavation, and Backfill 24
314.1 Trenches 24
314.2 Tunneling and Driving 25
314.3 Open Trenches 25
314.4 Excavations 25
315.0 Joints and Connections 25
315.1 Unions 25
315.2 Prohibited Joints and Connections 25
316.0 Increasers and Reducers 25
316.1 General 25
317.0 Food-Handling Establishments 25
317.1 General 25
318.0 Test Gauges 25
318.1 General 25
318.2 Pressure Tests (10 psi or less) ...25
318.3 Pressure Tests (greater than 10 psi to 100 psi) 25
318.4 Pressure Tests (exceeding 100 psi) 25
318.5 Pressure Range 26
319.0 Medical Gas and Vacuum Systems 26
319.1 General 26
320.0 Rehabilitation of Piping Systems ..26
320.1 General 26
Table 313.3 Hangers and Supports 26

CHAPTER 4 PLUMBING FIXTURES AND FIXTURE FITTINGS 27

401.0 General 27
401.1 Applicability 27
401.2 Quality of Fixtures 27
402.0 Installation 27
402.1 Cleaning 27
402.2 Joints 27
402.3 Securing Fixtures 27
402.4 Wall-Hung Fixtures 27
402.5 Setting 27
402.6 Flanged Fixture Connections 27
402.7 Supply Fittings 27
402.8 Installation 27
402.9 Design and Installation of Plumbing Fixtures 28
402.10 Slip Joint Connections 28
402.11 Future Fixtures 28
403.0 Accessible Plumbing Facilities 28
403.1 General 28
403.2 Fixtures and Fixture Fittings for Persons with Disabilities 28
403.3 Exposed Pipes and Surfaces 28
404.0 Waste Fittings and Overflows 28
404.1 Waste Fittings 28
404.2 Overflows 28
405.0 Prohibited Fixtures 28
405.1 Prohibited Water Closets 28
405.2 Prohibited Urinals 28
405.3 Miscellaneous Fixtures 28
406.0 Special Fixtures and Specialties ...28
406.1 Water and Waste Connections 28
406.2 Special Use Sinks 28
406.3 Special Use Fixtures 28
406.4 Zinc Alloy Components 28

407.0 Lavatories .28
407.1 Application .28
407.2 Water Consumption .28
407.3 Limitation of Hot Water Temperature for Public Lavatories .29
407.4 Transient Public Lavatories .29
407.5 Waste Outlet .29
407.6 Overflow .29
408.0 Showers .29
408.1 Application .29
408.2 Water Consumption .29
408.3 Individual Shower and Tub-Shower Combination Control Valves .29
408.4 Waste Outlet .29
408.5 Finished Curb or Threshold .29
408.6 Shower Compartments .29
408.7 Lining for Showers and Receptors .30
408.8 Public Shower Floors .30
408.9 Location of Valves and Heads .30
408.10 Water Supply Riser .30
409.0 Bathtubs and Whirlpool Bathtubs .30
409.1 Application .30
409.2 Waste Outlet .31
409.3 Overflow .31
409.4 Limitation of Hot Water Temperature in Bathtubs and Whirlpool Bathtubs .31
409.5 Backflow Protection .31
409.6 Installation and Access .31
410.0 Bidets .31
410.1 Application .31
410.2 Backflow Protection .31
410.3 Limitation of Water Temperature in Bidets .31
411.0 Water Closets .31
411.1 Application .31
411.2 Water Consumption .31
411.3 Water Closet Seats .31
411.4 Personal Hygiene Devices .31
412.0 Urinals .31
412.1 Application .31
412.2 Backflow Protection .32
413.0 Flushing Devices .32
413.1 Where Required .32
413.2 Flushometer Valves .32
413.3 Flush Tanks .32
413.4 Water Supply for Flush Tanks .32
413.5 Overflows in Flush Tanks .32
414.0 Dishwashing Machines .32
414.1 Application .32
414.2 Backflow Protection .32
414.3 Drainage Connection .32
415.0 Drinking Fountains .32
415.1 Application .32
415.2 Drinking Fountain Alternatives .32
415.3 Drainage Connection .32
415.4 Location .32
416.0 Emergency Eyewash and Shower Equipment .32
416.1 Application .32
416.2 Water Supply .32
416.3 Installation .32
416.4 Location .32
416.5 Drain .33
417.0 Faucets and Fixture Fittings .33
417.1 Application .33
417.2 Deck Mounted Bath/Shower Valves .33
417.3 Handheld Showers .33
417.4 Faucets and Fixture Fittings with Hose Connected Outlets .33
417.5 Separate Controls for Hot and Cold Water .33
417.6 Low-Pressure Water Dispenser .33
418.0 Floor Drains .33
418.1 Application .33
418.2 Strainer .33
418.3 Location of Floor Drains .33
418.4 Food Storage Areas .33
418.5 Floor Slope .33
419.0 Food Waste Disposers .33
419.1 Application .33
419.2 Drainage Connection .33
419.3 Water Supply .33
420.0 Sinks .33
420.1 Application .33
420.2 Water Consumption .33
420.3 Pre-Rinse Spray Valve .33
420.4 Waste Outlet .34
421.0 Floor Sinks .34
421.1 Application .34
421.2 Strainers .34
422.0 Minimum Number of Required Fixtures .34
422.1 Fixture Count .34

422.2 Separate Facilities .34
422.3 Fixture Requirements for Special Occupancies .34
422.4 Toilet Facilities Serving Employees and Customers .34
422.5 Toilet Facilities for Workers .34
Table 422.1 Minimum Plumbing Facilities .35

CHAPTER 5 WATER HEATERS .41
501.0 General .41
501.1 Applicability .41
Table 501.1(1) Water Heaters .41
502.0 Permits .41
502.1 General .41
503.0 Inspection .41
503.1 Inspection of Chimneys or Vents .41
503.2 Final Water Heater Inspection .41
504.0 Water Heater Requirements .41
504.1 Location .41
504.2 Vent .41
Table 501.1(2) First Hour Rating .41
504.3 Clearance .42
504.4 Pressure-Limiting Devices .42
504.5 Temperature-Limiting Devices .42
504.6 Temperature, Pressure, and Vacuum Relief Devices .42
505.0 Oil-Burning and Other Water Heaters .42
505.1 Water Heaters .42
505.2 Safety Devices .42
505.3 Oil-Fired Water Heaters .42
505.4 Indirect-Fired Water Heaters .42
506.0 Air for Combustion and Ventilation .42
506.1 General .42
506.2 Indoor Combustion Air .42
506.3 Indoor Opening Size and Location .43
506.4 Outdoor Combustion Air .43
506.5 Combination Indoor and Outdoor Combustion Air .45
506.6 Engineered Installations .45
506.7 Mechanical Combustion Air Supply .45
506.8 Louvers, Grilles, and Screens .45
506.9 Combustion Air Ducts .45
507.0 Appliance and Equipment Installation Requirements .46
507.1 Dielectric Insulator .46
507.2 Seismic Provisions .46
507.3 Appliance Support .46
507.4 Ground Support .46
507.5 Drainage Pan .46
507.6 Added or Converted Equipment or Appliances .46
507.7 Types of Gases .46
507.8 Safety Shutoff Devices for Unlisted LP-Gas Appliance Used Indoors .46
507.9 Use of Air or Oxygen Under Pressure .46
507.10 Protection of Gas Appliances from Fumes or Gases other than Products of Combustion .46
507.11 Process Air .46
507.12 Flammable Vapors .46
507.13 Installation in Residential Garages .46
507.14 Installation in Commercial Garages .47
507.15 Installation in Aircraft Hangars .47
507.16 Venting of Flue Gases .47
507.17 Extra Device or Attachment .47
507.18 Adequate Capacity of Piping .47
507.19 Avoiding Strain on Gas Piping .47
507.20 Gas Appliance Pressure Regulators .47
507.21 Venting of Gas Appliance Pressure Regulators .47
507.22 Bleed Lines for Diaphragm-Type Valves .47
507.23 Combination of Appliances and Equipment .47
507.24 Installation Instructions .47
507.25 Protection of Outdoor Appliances .48
507.26 Accessibility for Service .48
507.27 Clearance to Combustible Materials .48
508.0 Appliances on Roofs .48
508.1 General .48
508.2 Installation of Appliances on Roofs .48
508.3 Appliances on Roofs .48
508.4 Appliances in Attics and Under-Floor Spaces .48
509.0 Venting of Appliances .49
509.1 Listing .49
509.2 Connection to Venting Systems .49

509.3	Minimum Safe Performance	49
509.4	Type of Venting System to be Used	50
Table 509.4	Type of Venting System to be Used	50
509.5	Masonry, Metal, and Factory-Built Chimneys	50
509.6	Gas Vents	52
Table 509.6.1	Roof Slope Heights	53
509.7	Single-Wall Metal Pipe	55
Table 509.7.3.4(1)	Clearances for Connectors	55
Table 509.7.3.4(2)	Reduction of Clearances with Specified Forms of Protection	57
509.8	Through-the-Wall Vent Termination	58
Table 509.8.2	Through-the-Wall Direct-Vent Termination Clearances	58
509.9	Condensation Drain	59
509.10	Vent Connectors for Category I Appliances	59
Table 509.10.1.3	Minimum Thickness for Galvanized Steel Vent Connectors for Low-Heat Appliances	59
Table 509.10.1.4	Minimum Thickness for Steel Vent Connectors for Medium-Heat Appliances	60
509.11	Vent Connectors for Category II, Category III, and Category IV Appliances	61
509.12	Draft Hoods and Draft Controls	61
509.13	Manually Operated Dampers	62
509.14	Automatically Operated Vent Dampers	62
509.15	Obstructions	62
510.0	Sizing of Category I Venting Systems	62
510.1	Single Appliance Vent Table 510.1.2(1) through Table 510.1.2(6)	62
510.2	Multiple Appliance Vent Table 510.2(1) through Table 510.2(9)	63
Table 510.2.1	Vent Connector Maximum Length	64
Table 510.1.2(1)	Type B Double-Wall Gas Vent	68
Table 510.1.2(2)	Type B Double-Wall Gas Vent	71
Table 510.1.2(3)	Masonry Chimney	73
Table 510.1.2(4)	Masonry Chimney	75
Table 510.1.2(5)	Single-Wall Metal Pipe or Type B Asbestos-Cement Vent	77
Table 510.1.2(6)	Exterior Masonry Chimney	78
Table 510.2(1)	Type B Double-Wall Vent	79
Table 510.2(2)	Type B Double-Wall Vent	83
Table 510.2(3)	Masonry Chimney	85
Table 510.2(4)	Masonry Chimney	87
Table 510.2(5)	Single-Wall Metal Pipe or Type B Asbestos-Cement Vent	89
Table 510.2(6)	Exterior Masonry Chimney	89
Table 510.2(7)	Exterior Masonry Chimney	90
Table 510.2(8)	Exterior Masonry Chimney	91
Table 510.2(9)	Exterior Masonry Chimney	92
CHAPTER 6	**WATER SUPPLY AND DISTRIBUTION**	**93**
601.0	General	93
601.1	Applicability	93
601.2	Water Supply and Flushing	93
601.3	Identification of a Potable and Nonpotable Water System	93
Table 601.3.2	Minimum Length of Color Field and Size of Letters	93
602.0	Unlawful Connections	93
602.1	Prohibited Installation	93
602.2	Cross-Contamination	94
602.3	Backflow Prevention	94
602.4	Approval by Authority	94
603.0	Cross-Connection Control	94
603.1	General	94
603.2	Approval of Devices or Assemblies	94
603.3	Backflow Prevention Devices, Assemblies, and Methods	94
Table 603.2	Backflow Prevention Devices, Assemblies, and Methods	95
Table 603.3.1	Minimum Air Gaps for Water Distribution	97
603.4	General Requirements	97
603.5	Specific Requirements	98
604.0	Materials	100
604.1	Pipe, Tube, and Fittings	100
604.2	Lead Content	100
604.3	Copper or Copper Alloy Tube	100
604.4	Hard-Drawn Copper or Copper Alloy Tubing	100
604.5	Flexible Connectors	100
604.6	Cast-Iron Fittings	100
604.7	Malleable Iron Fittings	100
604.8	Previously Used Piping and Tubing	100
604.9	Epoxy Coating	100
604.10	Plastic Materials	100

Table 604.1	Materials for Building Supply and Water Distribution Piping and Fittings	101
604.11	Solder	101
604.12	Flexible Corrugated Connectors	101
604.13	Water Heater Connectors	101
605.0	Joints and Connections	102
605.1	Copper or Copper Alloy Pipe, Tubing, and Joints	102
605.2	CPVC Plastic Pipe and Joints	102
605.3	CPVC/AL/CPVC Plastic Pipe and Joints	103
605.4	Ductile Iron Pipe and Joints	103
605.5	Galvanized Steel Pipe and Joints	103
605.6	PE Plastic Pipe/Tubing and Joints	103
605.7	PE-AL-PE Plastic Pipe/Tubing and Joints	104
605.8	PE-RT	104
605.9	PEX Plastic Tubing and Joints	104
605.10	PEX-AL-PEX Plastic Tubing and Joints	104
605.11	Polypropylene (PP) Piping and Joints	105
605.12	PVC Plastic Pipe and Joints	105
605.13	Stainless Steel Pipe and Joints	105
605.14	Slip Joints	105
605.15	Dielectric Unions	105
605.16	Joints Between Various Materials	105
606.0	Valves	106
606.1	General	106
606.2	Fullway Valve	106
606.3	Multidwelling Units	106
606.4	Multiple Openings	106
606.5	Control Valve	106
606.6	Accessible	106
606.7	Multiple Fixtures	106
606.8	Check Valve Required	106
606.9	Leak Detection Devices	106
607.0	Potable Water Supply Tanks	106
607.1	General	106
607.2	Potable Water Tanks	106
607.3	Venting	106
607.4	Overflow	106
607.5	Valves	106
608.0	Water Pressure, Pressure Regulators, Pressure Relief Valves, and Vacuum Relief Valves	106
608.1	Inadequate Water Pressure	106
608.2	Excessive Water Pressure	106
608.3	Expansion Tanks, and Combination Temperature and Pressure-Relief Valves	107
608.4	Pressure Relief Valves	107
608.5	Discharge Piping	107
608.6	Water-Heating Devices	107
608.7	Vacuum Relief Valves	107
609.0	Installation, Testing, Unions, and Location	107
609.1	Installation	107
609.2	Trenches	107
609.3	Under Concrete Slab	107
609.4	Testing	108
609.5	Unions	108
609.6	Location	108
609.7	Abutting Lot	108
609.8	Pumps	108
609.9	Low-Pressure Cutoff Required on Booster Pumps for Water Distribution Systems	108
609.10	Disinfection of Potable Water System	108
609.11	Water Hammer	108
609.12	Pipe Insulation	108
610.0	Size of Potable Water Piping	109
610.1	Size	109
610.2	Pressure Loss	109
610.3	Quantity of Water	109
610.4	Sizing Water Supply and Distribution Systems	109
610.5	Sizing per Appendices A and C	109
610.6	Friction and Pressure Loss	109
610.7	Conditions for Using Table 610.4	109
610.8	Size of Meter and Building Supply Pipe Using Table 610.4	109
610.9	Size of Branches	109
610.10	Sizing for Flushometer Valves	109
Table 610.10	Flushometer Fixture Units for Water Sizing Using Table 610.3	110
610.11	Sizing Systems for Flushometer Tanks	110
610.12	Sizing for Velocity	110
610.13	Exceptions	110
Table 610.3	Water Supply Fixture Units (WSFU) and Minimum Fixture Branch Pipe Sizes	111

Table 610.4	Fixture Unit Table for Determining Water Pipe and Meter Sizes	112
611.0	Drinking Water Treatment Units	113
611.1	Application	113
611.2	Air Gap Discharge	113
611.3	Connection Tubing	113
611.4	Sizing of Residential Softeners	113
Table 611.4	Sizing of Residential Water Softeners	113
612.0	Residential Fire Sprinkler Systems	113
612.1	Where Required	113
612.2	Types of Systems	113
612.3	Sprinklers	113
Table 611.1	Drinking Water Treatment Units	113
Table 612.3.3.1	Locations Where Intermediate Temperature Sprinklers are Required	114
Table 612.3.6	Minimum Separation From Obstruction	115
612.4	Sprinkler Piping System	116
612.5	Sprinkler Piping Design	116
Table 612.5.3.2(1)	Water Service Pressure Loss (PL_{ws})	117
Table 612.5.3.2(2)	Minimum Water Meter Pressure Loss (PL_m)	118
Table 612.5.3.2(3)	Elevation Loss (PL_e)	118
612.6	Instructions and Signs	118
612.7	Inspection and Testing	119
Table 612.5.3.2(4)	Allowable Pipe Length for ¾ Inch Type M Copper Water Tubing	120
Table 612.5.3.2(5)	Allowable Pipe Length for 1 Inch Type M Copper Water Tubing	121
Table 612.5.3.2(6)	Allowable Pipe Length for ¾ Inch IPS CPVC Pipe	122
Table 612.5.3.2(7)	Allowable Pipe Length for 1 Inch IPS CPVC Pipe	123
Table 612.5.3.2(8)	Allowable Pipe Length for ¾ Inch PEX Tubing	124
Table 612.5.3.2(9)	Allowable Pipe Length for 1 Inch PEX Tubing	125
CHAPTER 7	**SANITARY DRAINAGE**	**127**
PART I	**Drainage Systems**	127
701.0	General	127
701.1	Applicability	127
701.2	Drainage Piping	127
701.3	Drainage Fittings	127
701.4	Continuous Wastes	127
701.5	Lead	127
701.6	Caulking Ferrules	127
701.7	Soldering Bushings	127
Table 701.6	Caulking Ferrules	127
Table 701.7	Soldering Bushings	127
702.0	Fixture Unit Equivalents	127
702.1	Trap Size	127
Table 701.2	Materials for Drain, Waste, Vent Pipe and Fittings	128
Table 702.1	Drainage Fixture Unit Values (DFU)	129
Table 702.1(1)	Maximum Drainage Fixture Units for a Trap and Trap Arm	130
702.2	Intermittent Flow	130
Table 702.2	Discharge Capacity in Gallons Per Minute for Intermittent Flow Only	130
702.3	Continuous Flow	130
703.0	Size of Drainage Piping	130
703.1	Minimum Size	130
703.2	Maximum Number of Fixture Units	130
703.3	Sizing per Appendix C	130
704.0	Fixture Connections (Drainage)	130
704.1	Inlet Fittings	130
704.2	Single Vertical Drainage Pipe	130
704.3	Commercial Sinks	130
705.0	Joints and Connections	130
705.1	ABS and ABS Co-Extruded Plastic Pipe and Joints	130
705.2	Cast-Iron Pipe and Joints	130
Table 703.2	Maximum Unit Loading and Maximum Length of Drainage and Vent Piping	131
705.3	Copper or Copper Alloy Pipe (DWV) and Joints	131
705.4	Galvanized Steel Pipe and Joints	132
705.5	Polyethylene (PE) Sewer Pipe	132
705.6	PVC and PVC Co-Extruded Plastic Pipe and Joining Methods	132
705.7	Stainless Steel Pipe and Joints	132
705.8	Vitrified Clay Pipe and Joints	133
705.9	Special Joints	133
705.10	Joints Between Various Materials	133
706.0	Changes in Direction of Drainage Flow	133

706.1	Approved Fittings	133
706.2	Horizontal to Vertical	133
706.3	Horizontal to Horizontal	133
706.4	Vertical to Horizontal	133
707.0	Cleanouts	133
707.1	Plug	133
Table 707.1	Cleanouts	134
707.2	Approved	134
Table 707.2	Cleanout Materials for Drain, Waste, and Vent	134
707.3	Watertight and Gastight	134
707.4	Location	134
707.5	Cleaning	134
707.6	Extension	134
707.7	Interceptor	134
707.8	Access	134
707.9	Clearance	134
707.10	Fittings	134
707.11	Pressure Drainage Systems	134
707.12	Countersunk Cleanout Plugs	134
707.13	Hubless Blind Plugs	135
707.14	Trap Arms	135
708.0	Grade of Horizontal Drainage Piping	135
708.1	General	135
709.0	Gravity Drainage Required	135
709.1	General	135
710.0	Drainage of Fixtures Located Below the Next Upstream Manhole or Below the Main Sewer Level	135
710.1	Backflow Protection	135
710.2	Sewage Discharge	135
710.3	Sewage Ejector and Pumps	135
710.4	Discharge Line	135
710.5	Size of Building Drains and Sewers	135
710.6	Backwater Valves	135
710.7	Drainage and Venting Systems	135
710.8	Sump and Receiving Tank Construction	135
710.9	Alarm	135
710.10	Sump and Receiving Tank Covers and Vents	136
710.11	Air Tanks	136
710.12	Grinder Pump Ejector	136
710.13	Macerating Toilet Systems and Pumped Waste Systems	136
711.0	Suds Relief	136
711.1	General	136
712.0	Testing	136
712.1	Media	136
712.2	Water Test	136
712.3	Air Test	136
PART II	**Building Sewers**	137
713.0	Sewer Required	137
713.1	Where Required	137
713.2	Private Sewage Disposal System	137
713.3	Public Sewer	137
713.4	Public Sewer Availability	137
713.5	Permit	137
713.6	Lot	137
713.7	Installation	137
714.0	Damage to Public Sewer or Private Sewage Disposal System	137
714.1	Unlawful Practices	137
714.2	Prohibited Water Discharge	137
714.3	Prohibited Sewer Connection	137
714.4	Commercial Food Waste Disposer	137
714.5	Tanks	137
715.0	Building Sewer Materials	137
715.1	Materials	137
715.2	Joining Methods and Materials	137
715.3	Existing Sewers	137
716.0	Markings	137
716.1	General	137
717.0	Size of Building Sewers	137
717.1	General	137
Table 717.1	Maximum/Minimum Fixture Unit Loading on Building Sewer Piping	138
718.0	Grade, Support, and Protection of Building Sewers	138
718.1	Slope	138
718.2	Support	138
718.3	Protection from Damage	138
719.0	Cleanouts.	138
719.1	Locations	138
719.2	No Additional Cleanouts	138
719.3	Building Sewer Cleanouts	138
719.4	Cleaning	138
719.5	Access	138
719.6	Manholes	138
720.0	Sewer and Water Pipes	138
720.1	General	138

721.0	Location	138
721.1	Building Sewer	138
Table 721.1	Minimum Horizontal Distance Required from Building Sewer	139
721.2	Abutting Lot	139
722.0	Abandoned Sewers and Sewage Disposal Facilities	139
722.1	Building (House) Sewer	139
722.2	Cesspools, Septic Tanks, and Seepage Pits	139
722.3	Filling	139
722.4	Ownership	139
722.5	Disposal Facilities	139
723.0	Building Sewer Test	139
723.1	General	139
CHAPTER 8	**INDIRECT WASTES**	**141**
801.0	General	141
801.1	Applicability	141
801.2	Air Gap or Air Break Required	141
801.3	Food and Beverage Handling Establishments	141
801.4	Bar and Fountain Sink Traps	141
801.5	Connections from Water Distribution System	141
801.6	Sterilizers	141
801.7	Drip or Drainage Outlets	141
802.0	Approvals	141
802.1	General	141
803.0	Indirect Waste Piping	141
803.1	Materials	141
803.2	Copper and Copper Alloys	141
803.3	Pipe Size and Length	141
804.0	Indirect Waste Receptors	142
804.1	Standpipe Receptors	142
805.0	Pressure Drainage Connections	142
805.1	General	142
806.0	Sterile Equipment	142
806.1	General	142
807.0	Appliances	142
807.1	Non-Classed Apparatus	142
807.2	Undiluted Condensate Waste	142
807.3	Domestic Dishwashing Machine	142
808.0	Cooling Water	142
808.1	General	142
809.0	Drinking Fountains	142
809.1	General	142
810.0	Steam and Hot Water Drainage Condensers and Sumps	142
810.1	High-Temperature Discharge	142
Table 810.1	Pipe Connections in Blowoff Condensers and Sumps	143
810.2	Sumps, Condensers, and Intercepting Tanks	143
810.3	Cleaning	143
810.4	Strainers	143
811.0	Chemical Wastes	143
811.1	Pretreatment	143
811.2	Waste and Vent Pipes	143
811.3	Joining Materials	143
811.4	Access	143
811.5	Permanent Record	143
811.6	Chemical Vent	143
811.7	Discharge	143
811.8	Diluted Chemicals	143
812.0	Clear Water Wastes	143
812.1	General	143
813.0	Swimming Pools	143
813.1	General	143
814.0	Condensate Waste and Control	143
814.1	Condensate Disposal	143
814.2	Condensate Control	144
814.3	Condensate Waste Pipe Material and Sizing	144
Table 814.3	Minimum Condensate Pipe Size	144
814.4	Appliance Condensate Drains	144
814.5	Point of Discharge	144
814.6	Condensate Waste from Air-Conditioning Coils	144
814.7	Plastic Fittings	144
CHAPTER 9	**VENTS**	**145**
901.0	General	145
901.1	Applicability	145
901.2	Vents Required	145
901.3	Trap Seal Protection	145
902.0	Vents Not Required	145
902.1	Interceptor	145
902.2	Bars, Soda Fountains, and Counter	145
903.0	Materials	145
903.1	Applicable Standards	145
903.2	Use of Copper or Copper Alloy Tubing	145
903.3	Changes in Direction	145
904.0	Size of Vents	145

904.1	Size	145
904.2	Length	145
905.0	Vent Pipe Grades and Connections	145
905.1	Grade	145
905.2	Horizontal Drainage Pipe	146
905.3	Vent Pipe Rise	146
905.4	Roof Termination	146
905.5	Location of Opening	146
905.6	Common Vertical Pipe	146
906.0	Vent Termination	146
906.1	Roof Termination	146
906.2	Clearance	146
906.3	Use of Roof	146
906.4	Outdoor Installations	146
906.5	Joints	146
906.6	Lead	146
906.7	Frost or Snow Closure	146
907.0	Vent Stacks and Relief Vents	146
907.1	Drainage Stack	146
907.2	Yoke Vent	146
908.0	Wet Venting	146
908.1	Vertical Wet Venting	146
908.2	Horizontal Wet Venting for a Bathroom Group	146
909.0	Special Venting for Island Fixtures	147
909.1	General	147
910.0	Combination Waste and Vent Systems	147
910.1	Where Permitted	147
910.2	Approval	147
910.3	Vents	147
910.4	Size	147
910.5	Vertical Waste Pipe	147
910.6	Cleanouts	147
910.7	Fixtures	147
911.0	Circuit Venting	147
911.1	Circuit Vent Permitted	147
911.2	Circuit Vent Size and Connection	148
911.3	Relief Vent	148
911.4	Slope and Size of Horizontal Branch	148
911.5	Additional Fixtures	148
912.0	Engineered Vent System	148
912.1	General	148
912.2	Minimum Requirements	148
CHAPTER 10	**TRAPS AND INTERCEPTORS**	**149**
1001.0	General	149
1001.1	Applicability	149
1001.2	Where Required	149
1002.0	Traps Protected by Vent Pipes	149
1002.1	Vent Pipes	149
1002.2	Fixture Traps	149
Table 1002.2	Horizontal Lengths of Trap Arms	149
1002.3	Change of Direction	149
1002.4	Vent Pipe Opening	149
1003.0	Traps — Described	149
1003.1	General Requirements	149
1003.2	Slip Joint Fittings	149
1003.3	Size	149
1004.0	Traps	149
1004.1	Prohibited	149
1004.2	Movable Parts	149
1005.0	Trap Seals	150
1005.1	General	150
1006.0	Floor Drain Traps	150
1006.1	General	150
1007.0	Trap Seal Protection	150
1007.1	General	150
1007.2	Trap Seal Primers	150
1008.0	Building Traps	150
1008.1	General	150
1009.0	Interceptors (Clarifiers) and Separators	150
1009.1	Where Required	150
1009.2	Approval	150
1009.3	Design	150
1009.4	Relief Vent	150
1009.5	Location	150
1009.6	Maintenance of Interceptors	150
1009.7	Discharge	150
1010.0	Slaughterhouses, Packing Establishments, etc	150
1010.1	General	150
1011.0	Minimum Requirements for Auto Wash Racks	150
1011.1	General	150
1012.0	Commercial and Industrial Laundries	150
1012.1	General	150
1013.0	Bottling Establishments	150
1013.1	General	150
1014.0	Grease Interceptors	150

1014.1	General	150
1014.2	Hydromechanical Grease Interceptors	151
Table 1014.2.1	Hydromechanical Grease Interceptor Sizing Using Gravity Flow Rates	151
1014.3	Gravity Grease Interceptors	152
Table 1014.3.6	Gravity Grease Interceptor Sizing	152
1015.0	FOG (Fats, Oils, and Greases) Disposal System	153
1015.1	Purpose	153
1015.2	Components, Materials, and Equipment	153
1015.3	Sizing and Installation	153
1015.4	Performance	153
1016.0	Sand Interceptors	153
1016.1	Discharge	153
1016.2	Authority Having Jurisdiction	153
1016.3	Construction and Size	153
1016.4	Separate Use	153
1017.0	Oil and Flammable Liquid Interceptors	153
1017.1	Interceptors Required	153
1017.2	Design of Interceptors	153
CHAPTER 11	**STORM DRAINAGE**	**155**
1101.0	General	155
1101.1	Applicability	155
1101.2	Where Required	155
1101.3	Storm Water Drainage to Sanitary Sewer Prohibited	155
1101.4	Material Uses	155
Table 1101.4.6	Materials for Subsoil Drain Pipe and Fittings	155
1101.5	Expansion Joints Required	155
1101.6	Subsoil Drains	155
1101.7	Building Subdrains	156
1101.8	Areaway Drains	156
1101.9	Window Areaway Drains	156
1101.10	Filling Stations and Motor Vehicle Washing Establishments	156
1101.11	Paved Areas	156
1101.12	Roof Drainage	156
1101.13	Cleanouts	157
1101.14	Rainwater Sumps	157
1101.15	Traps on Storm Drains and Leaders	157
1101.16	Leaders, Conductors, and Connections	157
1102.0	Roof Drains	157
1102.1	Applications	157
1102.2	Dome Strainers Required	157
1102.3	Roof Drain Flashings	157
1103.0	Size of Leaders, Conductors, and Storm Drains	157
1103.1	Vertical Conductors and Leaders	157
1103.2	Size of Horizontal Storm Drains and Sewers	157
1103.3	Size of Roof Gutters	157
1103.4	Side Walls Draining onto a Roof	157
Table 1103.1	Sizing Roof Drains, Leaders, and Vertical Rainwater Piping	158
1104.0	Values for Continuous Flow	158
1104.1	General	158
1105.0	Controlled-Flow Roof Drainage	158
1105.1	Application	158
Table 1105.1(1)	Controlled-Flow Maximum Roof Water Depth	159
Table 1105.1(2)	Distance of Scupper Bottoms Above Roof	159
1105.2	Setback Roofs	159
1106.0	Engineered Storm Drainage System	159
1106.1	General	159
1106.2	Siphonic Roof Drainage Systems	159
1106.3	Siphonic Roof Drains	159
1107.0	Testing	159
1107.1	Testing Required	159
1107.2	Methods of Testing Storm Drainage Systems	159
Table 1103.2	Sizing of Horizontal Rainwater Piping	160
Table 1103.3	Size of Gutters	161
CHAPTER 12	**FUEL GAS PIPING**	**163**
1201.0	General	163
1201.1	Applicability	163
1202.0	Coverage of Piping System	163
1202.1	General	163
1202.2	Piping System Requirements	163
1202.3	Applications	163
1203.0	Inspection	163
1203.1	Inspection Notification	163
1203.2	Excavation	163
1203.3	Type of Inspections	163

1203.4 Inspection Waived 164
1204.0 Certificate of Inspection 164
1204.1 Issuance .164
1204.2 Gas Supplier 164
1204.3 Unlawful .164
1205.0 Authority to Render Gas Service 164
1205.1 Authorized Personnel 164
1205.2 Outlets 164
1206.0 Authority to Disconnect 164
1206.1 Disconnection 164
1206.2 Notice 164
1206.3 Capped Outlets 164
1207.0 Temporary Use of Gas 164
1207.1 General .164
1208.0 Gas Piping System Design, Materials, and Components 164
1208.1 Installation of Piping System 164
1208.2 Provision for Location of Point of Delivery 164
1208.3 Interconnections Between Gas Piping Systems 164
1208.4 Sizing of Gas Piping Systems . . .164
Table 1208.4.1 Approximate Gas Input for Typical Appliances 165
1208.5 Maximum Operating Pressure in Buildings 165
1208.6 Acceptable Piping Materials and Joining Methods 165
Table 1208.6.9.2 Specifications for Threading Metallic Pipe 167
1208.7 Gas Meters 168
1208.8 Gas Pressure Regulators 169
1208.9 Overpressure Protection Required .169
1208.10 Overpressure Protection Devices 169
1208.11 Pressure Limitation Requirements 170
1208.12 Backpressure Protection 170
1208.13 Low-Pressure Protection 170
1208.14 Shutoff Valves 171
1208.15 Expansion and Flexibility 171
1209.0 Excess Flow Valve 171
1209.1 General .171
1210.0 Gas Piping Installation 171
1210.1 Piping Underground 171
1210.2 CSST Piping Systems 172
1210.3 Installation of Aboveground Piping 172
Table 1210.3.5.1 Support of Piping 173
1210.4 Concealed Piping in Buildings . . .173
1210.5 Piping in Vertical Chases 173
1210.6 Gas Pipe Turns 174
1210.7 Drips and Sediment Traps 174
1210.8 Outlets 174
1210.9 Manual Gas Shutoff Valves 174
1210.10 Prohibited Devices 175
1210.11 Systems Containing Gas-Air Mixtures Outside the Flammable Range 175
1210.12 Systems Containing Flammable Gas-Air Mixtures 175
1211.0 Electrical Bonding and Grounding 176
1211.1 Pipe and Tubing Other than CSST 176
1211.2 Bonding of CSST Gas Piping 176
1211.3 Arc-Resistant Jacketed CSST . . .176
1211.4 Prohibited Use 176
1211.5 Lightning Protection System 176
1211.6 Electrical Circuits 176
1211.7 Electrical Connections 177
1212.0 Appliance and Equipment Connections to Building Piping . . .177
1212.1 Connecting Appliances and Equipment 177
1212.2 Suspended Low-Intensity Infrared Tube Heaters 177
1212.3 Use of Nonmetallic Gas Hose Connectors 177
1212.4 Injection (Bunsen) Burners 177
1212.5 Connection of Portable and Mobile Industrial Gas Appliances 177
1212.6 Appliance Shutoff Valves and Connections 178
1212.7 Quick-Disconnect Devices 178
1212.8 Gas Convenience Outlets 178
1212.9 Sediment Trap 178
1212.10 Installation of Piping 178
1212.11 Liquefied Petroleum Gas Facilities and Piping 178
1213.0 Pressure Testing, Inspection, and Purging 178
1213.1 Piping Installations 178
1213.2 Test Preparation 179
1213.3 Test Pressure 179
1213.4 Detection of Leaks and Defects . .179
1213.5 Piping System Leak Test 179
1213.6 Purging Requirements 179

Table 1213.6.1 Size and Length of Piping 180
1214.0 Required Gas Supply 181
1214.1 General 181
1214.2 Volume 181
1214.3 Gas Appliances 181
1214.4 Size of Piping Outlets 181
1215.0 Required Gas Piping Size 181
1215.1 Pipe Sizing Methods 181
1215.2 Sizing of Gas Piping Systems . . .181
1215.3 Sizing Equations 181
Table 1215.3 *Cr* and *Y* for Natural Gas and Undiluted Propane at Standard Conditions 182
1215.4 Sizing of Piping Sections 182
1215.5 Engineering Methods 182
1215.6 Variable Gas Pressure 182
Table 1215.2(1) Schedule 40 Metallic Pipe 184
Table 1215.2(2) Schedule 40 Metallic Pipe 185
Table 1215.2(3) Schedule 40 Metallic Pipe 186
Table 1215.2(4) Schedule 40 Metallic Pipe 187
Table 1215.2(5) Schedule 40 Metallic Pipe 188
Table 1215.2(6) Schedule 40 Metallic Pipe 189
Table 1215.2(7) Semi-Rigid Copper Tubing 190
Table 1215.2(8) Semi-Rigid Copper Tubing 191
Table 1215.2(9) Semi-Rigid Copper Tubing 192
Table 1215.2(10) Semi-Rigid Copper Tubing 193
Table 1215.2(11) Semi-Rigid Copper Tubing 194
Table 1215.2(12) Semi-Rigid Copper Tubing 195
Table 1215.2(13) Semi-Rigid Copper Tubing 196
Table 1215.2(14) Corrugated Stainless Steel Tubing (CSST) 197
Table 1215.2(15) Corrugated Stainless Steel Tubing (CSST) 198
Table 1215.2(16) Corrugated Stainless Steel Tubing (CSST) 199
Table 1215.2(17) Corrugated Stainless Steel Tubing (CSST) 200
Table 1215.2(18) Corrugated Stainless Steel Tubing (CSST) 201
Table 1215.2(19) Polyethylene Plastic Pipe 202
Table 1215.2(20) Polyethylene Plastic Pipe 203
Table 1215.2(21) Polyethylene Plastic Pipe 204
Table 1215.2(22) Polyethylene Plastic Tubing 205
Table 1215.2(23) Polyethylene Plastic Tubing 205
Table 1215.2(24) Schedule 40 Metallic Pipe 206
Table 1215.2(25) Schedule 40 Metallic Pipe 207
Table 1215.2(26) Schedule 40 Metallic Pipe 208
Table 1215.2(27) Schedule 40 Metallic Pipe 209
Table 1215.2(28) Semi-Rigid Copper Tubing 210
Table 1215.2(29) Semi-Rigid Copper Tubing 211
Table 1215.2(30) Semi-Rigid Copper Tubing 212
Table 1215.2(31) Corrugated Stainless Steel Tubing (CSST) 213
Table 1215.2(32) Corrugated Stainless Steel Tubing (CSST) 214
Table 1215.2(33) Corrugated Stainless Steel Tubing (CSST) 215
Table 1215.2(34) Polyethylene Plastic Pipe 216
Table 1215.2(35) Polyethylene Plastic Pipe 217
Table 1215.2(36) Polyethylene Plastic Tubing 218

CHAPTER 13 HEALTH CARE FACILITIES AND MEDICAL GAS AND MEDICAL VACUUM SYSTEMS 219

PART I General Requirements 219
1301.0 General Requirements 219
1301.1 Applicability 219
1301.2 Where Not Applicable 219
1301.3 Conflict of Requirements 219
1301.4 Where Required 219
1301.5 Existing Systems 219
1302.0 Design Requirements 219
1302.1 Risk Categories 219
1302.2 Patient Care Spaces 219
1302.3 Anesthesia 219
1302.4 Wet Procedure Locations 219
1303.0 Health Care Facilities 219
1303.1 Drinking Fountain Control Valves 219
1303.2 Psychiatric Patient Rooms 219
1303.3 Locations for Ice Storage 220
1303.4 Sterilizers and Bedpan Steamers 220
1303.5 Aspirators 220
1303.6 Drains 220
1303.7 Clinical Sinks 220
1303.8 Water Supply for Hospitals 220
1303.9 Work Performed in Occupied Healthcare Facilities 220
1304.0 Medical Gas and Medical Vacuum Piping Systems 220
1304.1 General 220
1304.2 Certification of Systems 220
1304.3 Construction Documents 220
1304.4 Extent of Work 220
1304.5 Record 220
1305.0 System Performance 220
1305.1 Required Operating Pressures . . .220
1305.2 Minimum Flow Rates 220

1305.3 Minimum Station Outlets and Inlets 220
1306.0 System Certification 220
1306.1 Certification 220
1306.2 Certification Tests 220
Table 1305.1 Standard Designation Colors and Operating Pressures for Gas and Vacuum Systems 221
1306.3 Report Items 221
1306.4 Components 221
PART II **Category 1 Piped Gas and Vacuum Systems** 221
1307.0 Central Supply Systems 221
1307.1 Terms 221
Table 1305.2 Minimum Flow Rates 221
1307.2 Nature of Hazards of Gas and Vacuum Systems 221
1307.3 Permitted Locations for Medical Gases 221
Table 1305.3 Minimum Outlets and Inlets Per Station 222
1307.4 Materials 222
1308.0 Pressure-Regulating Equipment . .222
1308.1 Where Required 222
1308.2 Pressure Relief Valves 222
1308.3 Pressure-Relief Valve Requirements 223
1309.0 Oxygen Concentrator Supply Units 223
1309.1 Oxygen Requirements 223
1309.2 Particulate Size 223
1309.3 Suitability 223
1309.4 Compatible Materials 223
1309.5 Oxygen Concentrator Components 223
1309.6 Supply Air Quality 223
1309.7 Electrical Components 223
1309.8 Vent Valve 223
1309.9 Valved Sample Port 223
1309.10 Suitable Filter 223
1309.11 Check Valve 223
1309.12 Outlet Valve 223
1309.13 Oxygen Concentration Monitor . . .223
1310.0 Category 1 Medical Air Central Supply Systems 224
1310.1 Quality of Medical Air 224
1310.2 Uses of Medical Air 224
1310.3 Medical Air Compressors 224
1310.4 Medical Air Receivers 224
1310.5 Valves 224
1311.0 Compressor Intake 224
1311.1 Air Sources 224
1311.2 Medical Air Compressor Source .224
1311.3 Air Intakes 224
1311.4 Location .224
1311.5 Separate Compressors 225
1311.6 Screening 225
1312.0 Medical Surgical Vacuum Central Supply Systems 225
1312.1 General 225
1312.2 Medical-Surgical Vacuum Sources 225
1312.3 Vacuum Receivers 225
1312.4 Vacuum Filtration 225
1312.5 Piping Arrangement and Redundancies 225
1312.6 Piping Serviceability 226
1312.7 Shutoff Valve 226
1313.0 Medical-Surgical Vacuum Exhaust .226
1313.1 Vacuum Source Exhausts 226
1313.2 Location .226
1313.3 Screening 226
1313.4 Dips and Loops 226
1313.5 Multiple Pumps 226
1314.0 Valves 226
1314.1 Gas and Vacuum Shutoff Valves 226
1314.2 Security 226
1314.3 Labeled 226
1314.4 Accessibility 226
1314.5 Valve Types 226
Table 1314.5 Positive Pressure Gases 226
1314.6 Source Valves 227
1314.7 Main Line Valve 227
1314.8 Riser Valves 227
1314.9 Service Valves 227
1314.10 Zone Valves 227
1314.11 In-Line Shutoff Valves 227
1314.12 Valves for Future Connections . . .227
1315.0 Station Outlets and Inlets 227
1315.1 General 227
1315.2 Required Valves 227
1315.3 Secondary Valve 227
1315.4 Identification 227
1315.5 Threaded Outlets/Fittings 227
1315.6 Gas-Specific Station Outlet/Inlet 227

1315.7 Common Parts 227
1315.8 Marking of Components 228
1315.9 Components Not Specific to a Vacuum 228
1315.10 Factory-Installed Copper Inlet Tubes 228
1315.11 Factory-Installed Copper Outlet Tubes 228
1315.12 Protection from Damage 228
1315.13 Multiple Wall Outlets/Inlets 228
1315.14 Nonstandard Operation Pressures 228
1315.15 Post Installation 228
1316.0 Pressure and Vacuum Indicator Locations 228
1316.1 Isolation 228
1316.2 Pressure and Vacuum Indicator Locations 228
1317.0 Warning Systems 228
1317.1 Category 1 228
1318.0 Piping Materials for Field-Installed Positive Pressure Medical Gas Systems 229
1318.1 General 229
1318.2 Cleaning 229
1318.3 Delivery 229
1318.4 Tubes for Medical Gas Systems .229
1318.5 Manufacturer Markings 229
1318.6 Documentation 229
1319.0 Piping Materials for Field-Installed Medical-Surgical Vacuum Systems 229
1319.1 Tubes for Medical Vacuum Systems 229
1320.0 Joints and Connections 229
1320.1 General 229
1320.2 Changes in Direction 229
1321.0 Brazed Joints 229
1321.1 Brazed Joints and Fittings 229
1321.2 Tube Joints 230
1321.3 Filler Metals 230
1321.4 Copper-to-Copper Joints 230
1321.5 Accessible 230
1321.6 Purging 230
1321.7 Tube Ends 230
1321.8 Cleaning Procedures 230
1321.9 Brazing Dissimilar Metals 230
1321.10 Nitrogen Purge 230
1321.11 Assembling and Heating Brazed Joints 231
1321.12 Inspection of Brazed Joints 231
1322.0 Welded Joints 231
1322.1 Welded Joints Procedure 231
1322.2 Welding for Stainless Tube 232
1322.3 Memory Metal Fittings 232
1322.4 Axially Swaged Fittings 232
1322.5 Threaded Fittings 232
1322.6 Other Types of Fittings 232
1322.7 Prohibited Joints 232
1323.0 Installation of Piping and Equipment 232
1323.1 Required Pipe Sizing 232
Table 1323.1.5(1) System Sizing – Flow Requirements for Station Outlets and Inlets1 233
Table 1323.1.5(2) Maximum Permitted Pressure Loss In Medical Gas and Medical Vacuum Systems 233
1323.2 Pipe Protection 233
1323.3 Location of Piping 233
1323.4 Pipe Support 233
Table 1323.1.4(1) Pressure Loss for Medical Air 234
Table 1323.1.4(2) Pressure Loss for Nitrogen 234
Table 1323.1.4(3) Pressure Loss for Nitrous Oxide and Carbon Dioxide 235
Table 1323.1.4(4) Pressure Loss for Oxygen 235
Table 1323.1.4(5) Pressure Loss for Vacuum 236
Table 1323.1.4(6) Pressure Loss for Vacuum (Category 3) 236
1323.5 Frost Protection 237
Table 1323.4.4 Maximum Pipe Support Spacing 237
1323.6 Connectors 238
1323.7 Prohibited System Interconnections 238
1323.8 Manufacturer's Instructions 238
1323.9 Changes in System Use 238
1323.10 Qualifications of Installers 238
1323.11 Qualification of Brazing Procedures and Brazing 238
1323.12 Breaching or Penetrating Medical Gas Piping 239
1323.13 Labeling, Identification and Operating Pressure 239
1323.14 Identification of Shutoff Valves . . .239
1323.15 Identification 240
1324.0 Performance Criteria and Testing Category 1 (Gases, Medical Surgical Vacuum) 240
1324.1 Where Required 240
1324.2 Breached Systems 240

1324.3	Test Gas	240
1324.4	Initial Piping Blowdown	240
1324.5	Initial Pressure Tests– Medical Gas and Vacuum Systems	240
Table 1324.5.11	Gas Concentrations	242
PART III	**Category 2 Piped Gas and Vacuum Systems**	242
1325.0	Category 2 Piped Gas and Vacuum Systems	242
1325.1	General	242
1325.2	Nature of Hazards of Gas and Vacuum Systems	243
1325.3	Central Supply Systems	243
1325.4	Category 2 Medical Air Supply Systems	243
1325.5	Oxygen Concentrators	243
1325.6	Category 2 Medical-Surgical Vacuum	243
1325.7	Valves	243
1325.8	Station Outlets and Inlets	243
1325.9	Pressure and Vacuum Indicators	243
1325.10	Warning Systems (Category 2)	243
1325.11	Category 2 Distribution	243
1325.12	Labeling and Identification	243
1325.13	Performance Criteria and Testing — Category 2 (Gas, Medical–Surgical and Vacuum)	243
PART IV	**Category 3 Piped Gas and Vacuum Systems**	243
1326.0	Category 3 Piped Gas and Vacuum Systems	243
1326.1	General	243
1326.2	Nature of Hazards of Gas and Vacuum Systems	243
1326.3	Medical Air Supply Systems	243
1326.4	Oxygen Central Supply Systems Using Concentrators	243
1326.5	Medical–Surgical Vacuum	243
1326.6	Valves	243
1326.7	Station Outlets and Inlets	243
1326.8	Pressure and Vacuum Indicators	243
1326.9	Warning Systems	243
1326.10	Distribution	244
1326.11	Labeling and Identification	244
PART V	**Dental Gas and Vacuum Systems**	244
1327.0	Dental Gas and Vacuum Systems	244
1327.1	General	244
1327.2	Emergency Shutoff Valves (Oxygen and Nitrous Oxide)	244
1327.3	Warning Systems (Oxygen and Nitrous Oxide)	244
1327.4	Initial Pressure Test	244
1327.5	Maximum Copper Tube Support Spacing	244
1327.6	Maximum Plastic Pipe Support Spacing	244
Table 1327.5	Maximum Copper Tube Support Spacing	245
Table 1327.6	Maximum Plastic Pipe Support Spacing	245
1327.7	Standing Pressure Tests for Oxygen and Nitrous Oxide Piping	245
1327.8	Verifier Operational Pressure Test	245
CHAPTER 14	**FIRESTOP PROTECTION**	**247**
1401.0	General	247
1401.1	Applicability	247
1402.0	Construction Documents	247
1402.1	Penetrations	247
1403.0	Installation	247
1403.1	Materials	247
1404.0	Combustible Piping Installations	247
1404.1	General Requirements	247
1404.2	Fire-Resistance Rating	247
1404.3	Firestop Systems	247
1404.4	Connections	247
1404.5	Insulation and Coverings	247
1404.6	Sleeves	247
1405.0	Noncombustible Piping Installations	247
1405.1	General Requirements	247
1405.2	Fire-Resistance Rating	247
1405.3	Firestop Systems	247
1405.4	Connections	247
1405.5	Unshielded Couplings	248
1405.6	Sleeves	248
1405.7	Insulation and Coverings	248
1406.0	Required Inspection	248
1406.1	General	248
1406.2	Examination	248
1406.3	Penetrations	248
1406.4	Field Installations	248

CHAPTER 15 ALTERNATE WATER SOURCES FOR NONPOTABLE APPLICATIONS 249

1501.0 General 249
1501.1 Applicability 249
1501.2 System Design 249
1501.3 Permit 249
1501.4 Component Identification 249
1501.5 Maintenance and Inspection 249
Table 1501.5 Minimum Alternate Water Source Testing, Inspection, and Maintenance Frequency 249
1501.6 Operation and Maintenance Manual 249
1501.7 Minimum Water Quality Requirements 250
1501.8 Material Compatibility 250
1501.9 Commercial, Industrial, and Institutional Restroom Signs 250
1501.10 System Controls 250
1502.0 Inspection and Testing 250
1502.1 General 250
1502.2 Supply System Inspection and Test 250
1502.3 Annual Cross-Connection Inspection and Testing 250
1502.4 Separation Requirements 251
1502.5 Abandonment 251
1502.6 Sizing 251
1503.0 Gray Water Systems 251
1503.1 General 251
1503.2 System Requirements 251
1503.3 Connections to Potable and Reclaimed (Recycled) Water Systems 251
1503.4 Location 252
Table 1503.4 Location of Gray Water System ...252
1503.5 Plot Plan Submission 252
1503.6 Prohibited Location 252
1503.7 Drawings and Specifications 252
1503.8 Procedure for Estimating Gray Water Discharge 252
1503.9 Gray Water System Components 253
1504.0 Subsurface Irrigation System Zones 253
1504.1 General 253
1504.2 Required Area of Subsurface Irrigation Fields, Subsoil Irrigation Fields, and Mulch Basins 253
1504.3 Determination of Maximum Absorption Capacity 254
1504.4 Groundwater Level 254
Table 1504.2 Design of Six Typical Soils 254
1504.5 Subsurface and Subsoil Irrigation Field Design and Construction 254
Table 1504.5.5 Subsurface Irrigation Design Criteria for Six Typical Soils 254
1504.6 Mulch Basin Design and Construction 254
1504.7 Subsoil Irrigation Field 255
Table 1504.7.3 Subsoil Irrigation Field Construction 255
1504.8 Gray Water System Color and Marking Information 255
1504.9 Other Collection and Distribution Systems 255
1504.10 Testing 255
1504.11 Maintenance 255
1505.0 Reclaimed (Recycled) Water Systems 255
1505.1 General 255
1505.2 Permit 255
1505.3 System Changes 255
1505.4 Connections to Potable or Reclaimed (Recycled) Water Systems 255
1505.5 Initial Cross-Connection Test 256
1505.6 Reclaimed (Recycled) Water System Materials 256
1505.7 Reclaimed (Recycled) Water System Color and Marking Information 256
1505.8 Valves 256
1505.9 Hose Bibbs 256
1505.10 Required Appurtenances 256
1505.11 Same Trench as Potable Water Pipes 256
1505.12 Signs 256
1505.13 Inspection and Testing 256
1506.0 On-Site Treated Nonpotable Water Systems 256
1506.1 General 256
1506.2 Plumbing Plan Submission 256
1506.3 System Changes 256
1506.4 Connections to Potable or Reclaimed (Recycled) Water Systems 256
1506.5 Initial Cross-Connection Test 256
1506.6 On-Site Treated Nonpotable Water System Materials 256

1506.7	On-Site Treated Nonpotable Water Devices and Systems	256
1506.8	On-Site Treated Nonpotable Water System Color and Marking Information	257
1506.9	Design and Installation	257
1506.10	Valves	257
1506.11	Signs	257
1506.12	Inspection and Testing	257
CHAPTER 16	**NONPOTABLE RAINWATER CATCHMENT SYSTEMS**	**259**
1601.0	General	259
1601.1	Applicability	259
1601.2	System Design	259
1601.3	Permit	259
1601.4	Component Identification	259
1601.5	Maintenance and Inspection	259
Table 1601.5	Minimum Alternate Water Source Testing, Inspection, and Maintenance Frequency	259
1601.6	Operation and Maintenance Manual	260
1601.7	Minimum Water Quality Requirements	260
1601.8	Material Compatibility	260
1601.9	System Controls	260
1601.10	Separation Requirements	260
1601.11	Abandonment	260
1601.12	Sizing	260
1602.0	Nonpotable Rainwater Catchment Systems	260
1602.1	General	260
1602.2	Plumbing Plan Submission	260
1602.3	System Changes	260
1602.4	Connections to Potable or Reclaimed (Recycled) Water Systems	260
1602.5	Initial Cross-Connection Test	260
1602.6	Sizing	260
1602.7	Rainwater Catchment System Materials	261
1602.8	Rainwater Catchment System Color and Marking Information	261
1602.9	Deactivation and Drainage for Cross-Connection Test	261
1603.0	Design and Installation	261
1603.1	Rainwater Catchment Systems	261
1603.2	Outside Hose Bibbs	261
1603.3	Construction	261
1603.4	Rainwater Catchment Collection Surfaces	261
1603.5	Minimum Water Quality	261
1603.6	Rainwater Storage Tanks	261
1603.7	Location	261
1603.8	Above Grade	261
1603.9	Below Grade	261
Table 1603.5	Minimum Water Quality	262
1603.10	Drainage and Overflow	262
1603.11	Opening and Access Protection	262
1603.12	Marking	262
1603.13	Storage Tank Venting	262
1603.14	Pumps	262
1603.15	Roof Drains	262
1603.16	Water Quality Devices and Equipment	262
1603.17	Freeze Protection	262
1603.18	Debris Removal	262
1603.19	Required Filters	262
1603.20	Roof Gutters	263
1604.0	Signs	263
1604.1	General	263
1604.2	Commercial, Industrial, and Institutional Restroom Signs	263
1604.3	Equipment Room Signs	263
1605.0	Inspection and Testing	263
1605.1	General	263
1605.2	Supply System Inspection and Test	263
1605.3	Annual Cross-Connection Inspection and Testing	263
CHAPTER 17	**REFERENCED STANDARDS**	**265**
1701.0	General	265
1701.1	Standards	265
Table 1701.1	Referenced Standards	265
1701.2	Standards, Publications, Practices, and Guides	279
Table 1701.2	Standards, Publications, Practices, and Guides	279
APPENDICES	**TABLE OF CONTENTS**	**287**
Appendix A	Recommended Rules for Sizing the Water Supply System	289
Appendix B	Explanatory Notes on Combination Waste and Vent Systems	303
Appendix C	Alternate Plumbing Systems	305

Appendix D	Sizing Storm Water Drainage Systems	313
Appendix E	Manufactured/Mobile Home Parks and Recreational Vehicle Parks	321
Appendix F	Firefighter Breathing Air Replenishment Systems	331
Appendix G	Sizing of Venting Systems	335
Appendix H	Private Sewage Disposal Systems	345
Appendix I	Installation Standards	355
Appendix J	Combination of Indoor and Outdoor Combustion and Ventilation Opening Design	393
Appendix K	Potable Rainwater Catchment Systems	395
Appendix L	Sustainable Practices	399
Appendix M	Peak Water Demand Calculator	415
Appendix N	Impact Of Water Temperature On The Potential for Scalding and Legionella Growth	421

USEFUL TABLES **423**

INDEX **433**

CHAPTER 1

ADMINISTRATION

101.0 General.

101.1 Title. This document shall be known as the "Uniform Plumbing Code,'' may be cited as such, and will be referred to herein as "this code."

101.2 Scope. The provisions of this code shall apply to the erection, installation, alteration, repair, relocation, replacement, addition to, use, or maintenance of plumbing systems within this jurisdiction.

101.3 Purpose. This code is an ordinance providing minimum requirements and standards for the protection of the public health, safety, and welfare.

101.4 Unconstitutional. Where a section, subsection, sentence, clause, or phrase of this code is, for a reason, held to be unconstitutional, such decision shall not affect the validity of the remaining portions of this code. The legislative body hereby declares that it would have passed this code, and each section, subsection, sentence, clause, or phrase thereof, irrespective of the fact that one or more sections, subsections, sentences, clauses, and phrases are declared unconstitutional.

101.5 Validity. Where a provision of this code, or the application thereof to a person or circumstance, is held invalid, the remainder of the code, or the application of such provision to other persons or circumstances, shall not be affected thereby.

102.0 Applicability.

102.1 Conflicts Between Codes. Where the requirements within the jurisdiction of this plumbing code conflict with the requirements of the mechanical code, this code shall prevail. In instances where this code, applicable standards, or the manufacturer's installation instructions conflict, the more stringent provisions shall prevail. Where there is a conflict between a general requirement and a specific requirement, the specific requirement shall prevail.

102.2 Existing Installations. Plumbing systems lawfully in existence at the time of the adoption of this code shall be permitted to have their use, maintenance, or repair continued where the use, maintenance, or repair is in accordance with the original design and location and no hazard to life, health, or property has been created by such plumbing system.

102.3 Maintenance. The plumbing and drainage system, both existing and new, of a premise under the Authority Having Jurisdiction shall be maintained in a sanitary and safe operating condition. Devices or safeguards required by this code shall be maintained in accordance with the code edition under which installed.

The owner or the owner's designated agent shall be responsible for maintenance of plumbing systems. To determine compliance with this subsection, the Authority Having Jurisdiction shall be permitted to cause a plumbing system to be reinspected.

102.4 Additions, Alterations, Renovations, or Repairs. Additions, alterations, renovations or repairs shall conform to that required for a new system without requiring the existing plumbing system to be in accordance with the requirements of this code. Additions, alterations, renovations, or repairs shall not cause an existing system to become unsafe, insanitary, or overloaded.

Additions, alterations, renovations, or repairs to existing plumbing installations shall comply with the provisions for new construction unless such deviations are found to be necessary and are first approved by the Authority Having Jurisdiction.

102.4.1 Building Sewers and Drains. Existing building sewers and building drains shall be permitted to be used in connection with new buildings or new plumbing and drainage work where they are found on examination and test to be in accordance with the requirements governing new work, and the proper Authority Having Jurisdiction shall notify the owner to make changes necessary to be in accordance with this code. No building, or part thereof, shall be erected or placed over a part of a drainage system that is constructed of materials other than those approved elsewhere in this code for use under or within a building.

102.4.2 Openings. Openings into a drainage or vent system, excepting those openings to which plumbing fixtures are properly connected or which constitute vent terminals, shall be permanently plugged or capped in an approved manner, using the appropriate materials in accordance with this code.

102.5 Health and Safety. Where compliance with the provisions of this code fails to eliminate or alleviate a nuisance, or other dangerous or insanitary condition that involves health or safety hazards, the owner or the owner's agent shall install such additional plumbing and drainage facilities or shall make such repairs or alterations as ordered by the Authority Having Jurisdiction.

102.6 Changes in Building Occupancy. Plumbing systems that are a part of a building or structure undergoing a change in use or occupancy, as defined in the building code, shall be in accordance with the requirements of this code that are applicable to the new use or occupancy.

102.7 Moved Structures. Parts of the plumbing system of a building or part thereof that is moved from one foundation to another, or from one location to another, shall be in accordance with the provisions of this code for new installations and completely tested as prescribed elsewhere in this section for new work, except that walls or floors need not be removed during such test where other equivalent means of inspection acceptable to the Authority Having Jurisdiction are provided.

102.8 Appendices. The provisions in the appendices are intended to supplement the requirements of this code and shall not be considered part of this code unless formally adopted as such.

103.0 Duties and Powers of the Authority Having Jurisdiction.

103.1 General. The Authority Having Jurisdiction shall be the Authority duly appointed to enforce this code. For such purposes, the Authority Having Jurisdiction shall have the powers of a law enforcement officer. The Authority Having Jurisdiction shall have the power to render interpretations of this code and to adopt and enforce rules and regulations supplemental to this code as deemed necessary in order to clarify the application of the provisions of this code. Such interpretations, rules, and regulations shall comply with the intent and purpose of this code.

In accordance with the prescribed procedures and with the approval of the appointing authority, the Authority Having Jurisdiction shall be permitted to appoint a such number of technical officers, inspectors, and other employees as shall be authorized from time to time. The Authority Having Jurisdiction shall be permitted to deputize such inspectors or employees as necessary to carry out the functions of the code enforcement agency.

The Authority Having Jurisdiction shall be permitted to request the assistance and cooperation of other officials of this jurisdiction so far as required in the discharge of the duties in accordance with this code or other pertinent law or ordinance.

103.2 Liability. The Authority Having Jurisdiction charged with the enforcement of this code, acting in good faith and without malice in the discharge of the Authority Having Jurisdiction's duties, shall not thereby be rendered personally liable for damage that accrues to persons or property as a result of an act or by reason of an act or omission in the discharge of duties. A suit brought against the Authority Having Jurisdiction or employee because of such act or omission performed in the enforcement of provisions of this code shall be defended by legal counsel provided by this jurisdiction until final termination of such proceedings.

103.3 Applications and Permits. The Authority Having Jurisdiction shall be permitted to require the submission of plans, specifications, drawings, and such other information in accordance with the Authority Having Jurisdiction, prior to the commencement of, and at a time during the progress of, work regulated by this code.

The issuance of a permit upon construction documents shall not prevent the Authority Having Jurisdiction from thereafter requiring the correction of errors in said construction documents or from preventing construction operations being carried on thereunder where in violation of this code or of other pertinent ordinance or from revoking a certificate of approval where issued in error.

103.3.1 Licensing. Provision for licensing shall be determined by the Authority Having Jurisdiction.

103.4 Right of Entry. Where it is necessary to make an inspection to enforce the provisions of this code, or where the Authority Having Jurisdiction has reasonable cause to believe that there exists in a building or upon premises a condition or violation of this code that makes the building or premises unsafe, insanitary, dangerous, or hazardous, the Authority Having Jurisdiction shall be permitted to enter the building or premises at reasonable times to inspect or to perform the duties imposed by the Authority Having Jurisdiction by this code, provided that where such building or premises is occupied, the Authority Having Jurisdiction shall present credentials to the occupant and request entry. Where such building or premises is unoccupied, the Authority Having Jurisdiction shall first make a reasonable effort to locate the owner or other person having charge or control of the building or premises and request entry. Where entry is refused, the Authority Having Jurisdiction has recourse to every remedy provided by law to secure entry.

Where the Authority Having Jurisdiction shall have first obtained an inspection warrant or other remedy provided by law to secure entry, no owner, occupant, or person having charge, care or control of a building or premises shall fail or neglect, after a request is made as herein provided, to promptly permit entry herein by the Authority Having Jurisdiction for the purpose of inspection and examination pursuant to this code.

104.0 Permits.

104.1 Permits Required. It shall be unlawful for a person, firm, or corporation to make an installation, alteration, repair, replacement, or remodel a plumbing system regulated by this code except as permitted in Section 104.2, or to cause the same to be done without first obtaining a separate plumbing permit for each separate building or structure.

104.2 Exempt Work. A permit shall not be required for the following:

(1) The stopping of leaks in drains, soil, waste, or vent pipe, provided, however, that a trap, drainpipe, soil, waste, or vent pipe become defective, and it becomes necessary to remove and replace the same with new material, the same shall be considered as new work and a permit shall be procured and inspection made as provided in this code.

(2) The clearing of stoppages, including the removal and reinstallation of water closets, or the repairing of leaks in pipes, valves, or fixtures, provided such repairs do not involve or require the replacement or rearrangement of valves, pipes, or fixtures.

Exemption from the permit requirements of this code shall not be deemed to grant authorization for work to be done in violation of the provisions of the code or other laws or ordinances of this jurisdiction.

104.3 Application for Permit. To obtain a permit, the applicant shall first file an application therefore in writing on a form furnished by the Authority Having Jurisdiction for that purpose. Such application shall:

(1) Identify and describe the work to be covered by the permit for which application is made.

(2) Describe the land upon which the proposed work is to be done by legal description, street address, or similar description that will readily identify and locate the proposed building or work.

(3) Indicate the use or occupancy for which the proposed work is intended.

(4) Be accompanied by construction documents in accordance with Section 104.3.1.

(5) Be signed by the permittee or the permittee's authorized agent. The Authority Having Jurisdiction shall be permitted to require evidence to indicate such authority.

(6) Give such other data and information in accordance with the Authority Having Jurisdiction.

104.3.1 Construction Documents. Construction documents, engineering calculations, diagrams, and other data shall be submitted in two or more sets with each application for a permit. The construction documents, computations, and specifications shall be prepared by, and the plumbing designed by, a registered design professional. Construction documents shall be drawn to scale with clarity to identify that the intended work to be performed is in accordance with the code.

Exception: The Authority Having Jurisdiction shall be permitted to waive the submission of construction documents, calculations, or other data where the Authority Having Jurisdiction finds that the nature of the work applied for is such that reviewing of construction documents is not necessary to obtain compliance with the code.

104.3.2 Plan Review Fees. Where a plan or other data is required to be submitted in accordance with Section 104.3.1, a plan review fee shall be paid at the time of submitting construction documents for review.

The plan review fees for plumbing work shall be determined and adopted by this jurisdiction.

The plan review fees specified in this subsection are separate fees from the permit fees specified in Section 104.5.

Where plans are incomplete or changed so as to require additional review, a fee shall be charged at the rate shown in Table 104.5.

104.3.3 Time Limitation of Application. Applications for which no permit is issued within 180 days following the date of application shall expire by limitation, plans and other data submitted for review thereafter, shall be returned to the applicant or destroyed by the Authority Having Jurisdiction. The Authority Having Jurisdiction shall be permitted to exceed the time for action by the applicant for a period not to exceed 180 days upon request by the applicant showing that circumstances beyond the control of the applicant have prevented the action from being taken. No application shall be extended more than once. In order to renew action on an application after expiration, the applicant shall resubmit plans and pay a new plan review fee.

104.4 Permit Issuance. The application, construction documents, and other data filed by an applicant for a permit shall be reviewed by the Authority Having Jurisdiction. Such plans shall be permitted to be reviewed by other departments of this jurisdiction to verify compliance with applicable laws under their jurisdiction. Where the Authority Having Jurisdiction finds that the work described in an application for permit and the plans, specifications, and other data filed therewith are in accordance with the requirements of the code and other pertinent laws and ordinances and that the fees specified in Section 104.5 have been paid, the Authority Having Jurisdiction shall issue a permit therefore to the applicant.

104.4.1 Approved Plans or Construction Documents. Where the Authority Having Jurisdiction issues the permit where plans are required, the Authority Having Jurisdiction shall endorse in writing or stamp the construction documents "APPROVED." Such approved construction documents shall not be changed, modified, or altered without authorization from the Authority Having Jurisdiction, and the work shall be done in accordance with approved plans.

The Authority Having Jurisdiction shall be permitted to issue a permit for the construction of a part of a plumbing system before the entire construction documents for the whole system have been submitted or approved, provided adequate information and detailed statements have been filed in accordance with the pertinent requirements of this code. The holder of such permit shall be permitted to proceed at the holder's risk without assurance that the permit for the entire building, structure, or plumbing system will be granted.

104.4.2 Validity of Permit. The issuance of a permit or approval of construction documents shall not be construed to be a permit for, or an approval of, a violation of the provisions of this code or other ordinance of the jurisdiction. No permit presuming to give authority to violate or cancel the provisions of this code shall be valid.

The issuance of a permit based upon plans, specifications, or other data shall not prevent the Authority Having Jurisdiction from thereafter requiring the correction of errors in said plans, specifications, and other data or from preventing building operations being carried on thereunder where in violation of this code or of other ordinances of this jurisdiction.

104.4.3 Expiration. A permit issued by the Authority Having Jurisdiction under the provisions of this code shall expire by limitation and become null and void where the work authorized by such permit is not commenced within 180 days from the date of such permit, or where the work authorized by such permit is suspended or abandoned at a time after the work is commenced for a period of 180 days. Before such work is recommenced, a new permit shall first be obtained to do so, and the fee, therefore, shall be one-half the amount required for a new permit for such work, provided no changes have been made or will be made in the original construction documents for such work, and provided further that such suspensions or abandonment has not exceeded 1 year.

104.4.4 Extensions. A permittee holding an unexpired permit shall be permitted to apply for an extension of the time within which work shall be permitted to commence under that permit where the permittee is unable to commence work within the time required by this section. The Authority Having Jurisdiction shall be permitted to extend the time for action by the permittee for a period not exceeding 180 days upon written request by the permittee showing that circumstances beyond the control of the permittee have prevented the action from being taken. No permit shall be extended more than once. In order to renew action on a permit after expiration, the permittee shall pay a new full permit fee.

104.4.5 Suspension or Revocation. The Authority Having Jurisdiction shall be permitted to, in writing, suspend or revoke a permit issued under the provisions of this code where the permit is issued in error or on the basis of incorrect information supplied or in violation of other ordinance or regulation of the jurisdiction.

104.4.6 Retention of Plans. One set of approved construction documents and computations shall be retained by the Authority Having Jurisdiction until final approval of the work covered therein.

One set of approved construction documents, computations, and manufacturer's installation instructions shall be returned to the applicant and said set shall be kept on the site of the building or work at times during which the work authorized thereby is in progress.

104.5 Fees. Fees shall be assessed in accordance with the provisions of this section and as set forth in the fee schedule, Table 104.5. The fees are to be determined and adopted by this jurisdiction.

104.5.1 Work Commencing Before Permit Issuance. Where work for which a permit is required by this code has been commenced without first obtaining said permit, a special investigation shall be made before a permit is issued for such work.

104.5.2 Investigation Fees. An investigation fee, in addition to the permit fee, shall be collected whether a permit is then or subsequently issued. The investigation fee shall be equal to the amount of the permit fee that is required by this code if a permit were to be issued. The payment of such investigation fee shall not exempt a person from compliance with other provisions of this code, nor from a penalty prescribed by law.

104.5.3 Fee Refunds. The Authority Having Jurisdiction shall be permitted to authorize the refunding of a fee as follows:

(1) The amount paid hereunder that was erroneously paid or collected.

(2) Refunding of not more than a percentage, as determined by this jurisdiction where no work has been done under a permit issued in accordance with this code.

The Authority Having Jurisdiction shall not authorize the refunding of a fee paid except upon written application filed by the original permittee not to exceed 180 days after the date of fee payment.

105.0 Inspections and Testing.

105.1 General. Plumbing systems for which a permit is required by this code shall be inspected by the Authority Having Jurisdiction.

No plumbing system or portion thereof shall be covered, concealed, or put into use until inspected and approved as prescribed in this code. Neither the Authority Having Jurisdiction nor the jurisdiction shall be liable for expense entailed in the removal or replacement of material required to permit inspection. Plumbing systems regulated by this code shall not be connected to the water, the energy fuel supply, or the sewer system until authorized by the Authority Having Jurisdiction.

105.2 Required Inspections. New plumbing work and such portions of existing systems as affected by new work, or changes, shall be inspected by the Authority Having Jurisdiction to ensure compliance with the requirements of this code and to ensure that the installation and construction of the plumbing system are in accordance with approved plans. The Authority Having Jurisdiction shall make the following inspections and other such inspections as necessary. The permittee or the permittee's authorized agent shall be responsible for the scheduling of such inspections as follows:

(1) The underground inspection shall be made after trenches or ditches are excavated and bedded, piping installed, and before backfill is put in place.

(2) Rough-in inspection shall be made prior to the installation of wall or ceiling membranes.

(3) Final inspection shall be made upon completion of the installation.

105.2.1 Uncovering. Where a drainage or plumbing system, building sewer, private sewage disposal system, or part thereof, which is installed, altered, or repaired, is covered or concealed before being inspected, tested, and approved as prescribed in this code, it shall be uncovered for inspection after notice to uncover the work has been issued to the responsible person by the Authority Having Jurisdiction.

The requirements of this section shall not be considered to prohibit the operation of plumbing installed to replace existing equipment or fixtures serving an occupied portion of the building in the event a request for inspection of such equipment or fixture has been filed with the Authority Having Jurisdiction not more than 72 hours after such replacement work is completed, and before a portion of such plumbing system is concealed by a permanent portion of the building.

105.2.1.1 Water Supply System. No water supply system or portion thereof shall be covered or concealed until it first has been tested, inspected, and approved.

105.2.1.2 Covering or Using. No plumbing or drainage system, building sewer, private sewer disposal system, or part thereof, shall be covered, concealed, or put into use until it has been tested, inspected, and accepted as prescribed in this code.

105.2.2 Other Inspections. In addition to the inspections required by this code, the Authority Having Jurisdiction shall be permitted to require other inspections to ascertain compliance with the provisions of this code and other laws that are enforced by the Authority Having Jurisdiction.

105.2.3 Inspection Requests. It shall be the duty of the person doing the work authorized by a permit to notify the Authority Having Jurisdiction that such work is ready for inspection. The Authority Having Jurisdiction shall be permitted to require that a request for inspection be filed not less than 1 working day before such inspection is desired. Such request shall be permitted to be made in writing or by telephone, at the option of the Authority Having Jurisdiction.

It shall be the duty of the person requesting inspections in accordance with this code to provide access to and means for inspection of such work.

105.2.4 Advance Notice. It shall be the duty of the person doing the work authorized by the permit to notify the Authority Having Jurisdiction, orally or in writing that said work is ready for inspection. Such notification shall be given not less than 24 hours before the work is to be inspected.

105.2.5 Responsibility. It shall be the duty of the holder of a permit to make sure that the work will stand the test prescribed before giving the notification.

The equipment, material, and labor necessary for inspection or tests shall be furnished by the person to whom the permit is issued or by whom inspection is requested.

105.2.6 Reinspections. A reinspection fee shall be permitted to be assessed for each inspection or reinspection where such portion of work for which inspection is called is not complete or where required corrections have not been made.

This provision shall not be interpreted as requiring reinspection fees the first time a job is rejected for failure to be in accordance with the requirements of this code, but as controlling the practice of calling for inspections before the job is ready for inspection or reinspection.

Reinspection fees shall be permitted to be assessed where the approved plans are not readily available to the inspector, for failure to provide access on the date for which the inspection is requested, or for deviating from plans requiring the approval of the Authority Having Jurisdiction.

To obtain reinspection, the applicant shall file an application therefore in writing upon a form furnished for that purpose and pay the reinspection fee in accordance with Table 104.5.

In instances where reinspection fees have been assessed, no additional inspection of the work will be performed until the required fees have been paid.

105.3 Testing of Systems. Plumbing systems shall be tested and approved in accordance with this code or the Authority Having Jurisdiction. Tests shall be conducted in the presence of the Authority Having Jurisdiction or the Authority Having Jurisdiction's duly appointed representative.

No test or inspection shall be required where a plumbing system, or part thereof, is set up for exhibition purposes and has no connection with a water or drainage system. In cases where it would be impractical to provide the required water or air tests, or for minor installations and repairs, the Authority Having Jurisdiction shall be permitted to make such inspection as deemed advisable in order to be assured that the work has been performed in accordance with the intent of this code. Joints and connections in the plumbing system shall be gastight and watertight for the pressures required by the test.

105.3.1 Defective Systems. An air test shall be used in testing the sanitary condition of the drainage or plumbing system of building premises where there is reason to believe that it has become defective. In buildings or premises condemned by the Authority Having Jurisdiction because of an insanitary condition of the plumbing system, or part thereof, the alterations in such system shall be in accordance with the requirements of this code.

105.3.2 Retesting. Where the Authority Having Jurisdiction finds that the work will not pass the test, necessary corrections shall be made, and the work shall be resubmitted for test or inspection.

105.3.3 Approval. Where prescribed tests and inspections indicate that the work is in accordance with this code, a certificate of approval shall be issued by the Authority Having Jurisdiction to the permittee on demand.

105.4 Connection to Service Utilities. No person shall make connections from a source of energy or fuel to a plumbing system or equipment regulated by this code and for which a permit is required until approved by the Authority Having Jurisdiction. No person shall make connection from a water-supply line nor shall connect to a sewer system regulated by this code and for which a permit is required until approved by the Authority Having Jurisdiction. The Authority Having Jurisdiction shall be permitted to authorize temporary connection of the plumbing equipment to the source of energy or fuel for the purpose of testing the equipment.

106.0 Violations and Penalties.

106.1 General. It shall be unlawful for a person, firm, or corporation to erect, construct, enlarge, alter, repair, move, improve, remove, convert, demolish, equip, use, or maintain plumbing or permit the same to be done in violation of this code.

106.2 Notices of Correction or Violation. Notices of correction or violation shall be written by the Authority Having Jurisdiction and shall be permitted to be posted at the site of the work or mailed or delivered to the permittee or their authorized representative.

Refusal, failure, or neglect to comply with such notice or order within 10 days of receipt thereof, shall be considered a violation of this code and shall be subject to the penalties set forth by the governing laws of the jurisdiction.

106.3 Penalties. A person, firm, or corporation violating a provision of this code shall be deemed guilty of a misdemeanor, and upon conviction thereof, shall be punishable by a fine, imprisonment, or both set forth by the governing laws of the jurisdiction. Each separate day or portion thereof, during which a violation of this code occurs or continues, shall be deemed to constitute a separate offense.

106.4 Stop Orders. Where work is being done contrary to the provisions of this code, the Authority Having Jurisdiction shall be permitted to order the work stopped by notice in writing served on persons engaged in the doing or causing such work to be done, and such persons shall forthwith stop work until authorized by the Authority Having Jurisdiction to proceed with the work.

106.5 Authority to Disconnect Utilities in Emergencies. The Authority Having Jurisdiction shall have the authority to disconnect a plumbing system to a building, structure, or equipment regulated by this code in case of emergency where necessary to eliminate an immediate hazard to life or property.

106.6 Authority to Condemn. Where the Authority Having Jurisdiction ascertains that a plumbing system or portion thereof, regulated by this code, has become hazardous to life, health, or property, or has become insanitary, the Authority Having Jurisdiction shall order in writing that such plumbing either be removed or placed in a safe or sanitary condition. The order shall fix a reasonable time limit for compliance. No person shall use or maintain defective plumbing after receiving such notice.

Where such plumbing system is to be disconnected, written notice shall be given. In cases of immediate danger to life or property, such disconnection shall be permitted to be made immediately without such notice.

107.0 Board of Appeals.

107.1 General. In order to hear and decide appeals of orders, decisions, or determinations made by the Authority Having Jurisdiction relative to the application and interpretations of this code, there shall be and is hereby created a Board of Appeals consisting of members who are qualified by experience and training to pass upon matters pertaining to plumbing design, construction, and maintenance and the public health aspects of plumbing systems and who are not employees of the jurisdiction. The Authority Having Jurisdiction shall be an ex-officio member and shall act as secretary to said board but shall have no vote upon a matter before the board. The Board of Appeals shall be appointed by the governing body and shall hold office at its pleasure. The board shall adopt rules of procedure for conducting its business and shall render decisions and findings in writing to the appellant with a duplicate copy to the Authority Having Jurisdiction.

107.2 Limitations of Authority. The Board of Appeals shall have no authority relative to interpretation of the administrative provisions of this code, nor shall the board be empowered to waive requirements of this code.

TABLE 104.5
PLUMBING PERMIT FEES

Permit Issuance

1. For issuing each permit .... *______
2. For issuing each supplemental permit .... *______

Unit Fee Schedule (in addition to Item 1 and Item 2 above)

1. For each plumbing fixture on one trap or a set of fixtures on one trap (including water, drainage piping, and backflow protection therefore) .... *______
2. For each building sewer and each trailer park sewer .... *______
3. Rainwater systems – per drain (inside building) .... *______
4. For each cesspool (where permitted) .... *______
5. For each private sewage disposal system .... *______
6. For each water heater, vent, or both .... *______
7. For each gas piping system of one to five outlets .... *______
8. For each additional gas piping system outlet, per outlet .... *______
9. For each industrial waste pretreatment interceptor, including its trap and vent, except kitchen-type grease interceptors functioning as fixture traps .... *______
10. For each installation, alteration, or repair of water piping, water treating equipment, or both .... *______
11. For each repair or alteration of drainage or vent piping, each fixture .... *______
12. For each lawn sprinkler system on one meter including backflow protection devices therefore .... *______
13. For atmospheric-type vacuum breakers not referenced in Item 12:
    - One to 5 .... *______
    - Over 5, each .... *______
14. For each backflow protective device other than atmospheric-type vacuum breakers:
    - Two inches (50 mm) in diameter and smaller .... *______
    - Over 2 inches (50 mm) in diameter .... *______
15. For each gray water system .... *______
16. For initial installation and testing of a reclaimed water system .... *______
17. For each annual cross-connection testing of a reclaimed water system (excluding initial test) .... *______
18. For each medical gas piping system serving one to five inlet(s)/outlet(s) for a specific gas .... *______
19. For each additional medical gas inlet(s)/outlet(s) .... *______

Other Inspections and Fees

1. Inspections outside of normal business hours .... *______
2. Reinspection fee .... *______
3. Inspections for which no fee is specifically indicated .... *______
4. Additional plan review required by changes, additions, or revisions to approved plans (minimum charge – ½ hour) .... *______

For SI units: 1 inch = 25 mm

* Jurisdiction will indicate their fees here

CHAPTER 2
DEFINITIONS

201.0 General.

201.1 Applicability. For the purpose of this code, the following terms have the meanings indicated in this chapter.

No attempt is made to define ordinary words, which are used in accordance with their established dictionary meanings, except where a word has been used loosely, and it is necessary to define its meaning as used in this code to avoid misunderstanding.

202.0 Definition of Terms.

202.1 General. The definitions of terms are arranged alphabetically according to the first word of the term.

203.0 – A –

ABS. Acrylonitrile-butadiene-styrene.

Accepted Engineering Practice. That which conforms to technical or scientific-based principles, tests, or standards that are accepted by the engineering profession.

Accessible. Where applied to a fixture, connection, appliance, or equipment, "accessible" means having access thereto, but which first may require the removal of an access panel, door, or similar obstruction.

Accessible, Readily. Having a direct access without the necessity of removing a panel, door, or similar obstruction.

Air Break. A physical separation which may be a low inlet into the indirect waste receptor from the fixture, appliance, or device indirectly connected.

Air Gap, Drainage. The unobstructed vertical distance through the free atmosphere between the lowest opening from a pipe, plumbing fixture, appliance, or appurtenance conveying waste to the flood-level rim of the receptor.

Air Gap, Water Distribution. The unobstructed vertical distance through the free atmosphere between the lowest opening from a pipe or faucet conveying potable water to the flood-level rim of a tank, vat, or fixture.

Alternate Water Source. Nonpotable source of water that includes but not limited to gray water, on-site treated nonpotable water, rainwater, and reclaimed (recycled) water.

Anchors. See Supports.

Appliance. A device that utilizes an energy source to produce light, heat, power, refrigeration, air conditioning, or compressed fuel gas. This definition also shall include a vented decorative appliance.

> **Appliance, Low-Heat.** A fuel-burning appliance that produces a continuous flue gas temperature, at the point of entrance to the flue, of not more than 1000°F (538°C).
>
> **Appliance, Medium-Heat.** A fuel-burning appliance that produces a continuous flue gas temperature, at the point of entrance to the flue, of more than 1000°F (538°C) and less than 2000°F (1093°C).

Appliance Categorized Vent Diameter/Area. The minimum vent diameter/area permissible for Category I appliances to maintain a nonpositive vent static pressure when tested in accordance with nationally recognized standards. [NFPA 54:3.3.6]

Appliance Fuel Connector. An assembly of listed semirigid or flexible tubing and fittings to carry fuel between a fuel-piping outlet and a fuel-burning appliance.

Approved. Acceptable to the Authority Having Jurisdiction.

Approved Testing Agency. An organization primarily established for purposes of testing to approved standards and approved by the Authority Having Jurisdiction.

Area Drain. A receptor designed to collect surface or storm water from an open area.

Aspirator. A fitting or device supplied with water or other fluid under positive pressure that passes through an integral orifice or constriction, causing a vacuum.

Authority Having Jurisdiction. The organization, office, or individual responsible for enforcing the requirements of a code or standard, or for approving equipment, materials, installations, or procedures. The Authority Having Jurisdiction shall be a federal, state, local, or other regional department or an individual such as a plumbing official, mechanical official, labor department official, health department official, building official, or others having statutory authority. In the absence of statutory authority, the Authority Having Jurisdiction may be some other responsible party. This definition shall include the Authority Having Jurisdiction's duly authorized representative.

204.0 – B –

Backflow. The flow of water or other liquids, mixtures, or substances into the distributing pipes of a potable supply of water from sources other than its intended source. See Backpressure Backflow and Backsiphonage.

Backflow Connection. An arrangement whereby backflow can occur.

Backflow Preventer. A backflow prevention device, an assembly, or another method to prevent backflow into the potable water system.

Backpressure Backflow. Backflow due to an increased pressure above the supply pressure, which may be due to pumps, boilers, gravity, or other sources of pressure.

Backsiphonage. The flowing back of used, contaminated, or polluted water from a plumbing fixture or vessel into a water supply pipe due to a pressure less than atmospheric in such pipe. See Backflow.

Backwater Valve. A device installed in a drainage system to prevent reverse flow.

Bathroom. A room equipped with a shower, bathtub, or combination bath/shower.

Bathroom, Half. A room equipped with only a water closet and lavatory.

Bathroom Group. Any combination of fixtures, not to exceed one water closet, two lavatories, either one bathtub or one combination bath/shower, and one shower, and may include a bidet and an emergency floor drain.

Battery of Fixtures. A group of two or more similar, adjacent fixtures that discharge into a common horizontal waste or soil branch.

Bedpan Steamer. A fixture that is used to sterilize bedpans by way of steam.

Boiler Blowoff. An outlet on a boiler to permit emptying or discharge of sediment.

Bonding Conductor or Jumper. A reliable conductor to ensure the required electrical conductivity between metal parts required to be electrically connected. [NFPA 70:100 (Part I)]

Bottle Filling Station. A plumbing fixture connected to the potable water distribution system and sanitary drainage system that is designed and intended for filling personal use drinking water bottles or containers not less than 10 inches (254 mm) in height. Such fixtures can be separate from or integral to a drinking fountain and can incorporate a water filter and a cooling system for chilling the drinking water.

Branch. A part of the piping system other than a main, riser, or stack.

Branch, Fixture. See Fixture Branch.

Branch, Horizontal. See Horizontal Branch.

Branch Vent. A vent connecting one or more individual vents with a vent stack or stack vent.

Building. A structure built, erected, and framed of component structural parts designed for the housing, shelter, enclosure, or support of persons, animals, or property of any kind.

Building Drain. That part of the lowest piping of a drainage system that receives the discharge from soil, waste, and other drainage pipes inside the walls of the building and conveys it to the building sewer beginning 2 feet (610 mm) outside the building wall.

Building Drain (Sanitary). A building drain that conveys sewage only.

Building Drain (Storm). A building drain that conveys storm water or another drainage, but no sewage.

Building Sewer. That part of the horizontal piping of a drainage system that extends from the end of the building drain and that receives the discharge of the building drain and conveys it to a public sewer, private sewer, private sewage disposal system, or another point of disposal.

Building Sewer (Combined). A building sewer that conveys both sewage and storm water or other drainage.

Building Sewer (Sanitary). A building sewer that conveys sewage only.

Building Sewer (Storm). A building sewer that conveys storm water or another drainage, but no sewage.

Building Subdrain. That portion of a drainage system that does not drain by gravity into the building sewer.

Building Supply. The pipe is carrying potable water from the water meter or another source of water supply to a building or other point of use or distribution on the lot.

205.0 – C –

Category 1. Activities, systems, or equipment whose failure is likely to cause major injury or death to patients, staff, or visitors. [NFPA 99:3.3.158.1]

Category 2. Activities, systems, or equipment whose failure is likely to cause minor injury to patients, staff, or visitors. [NFPA 99:3.3.158.2]

Category 3. Activities, systems, or equipment whose failure is not likely to cause injury to patients, staff, or visitors, but can cause discomfort. [NFPA 99:3.3.158.3]

Category 3 Vacuum System. A Category 3 vacuum distribution system that can be either a wet system designed to remove liquids, air-gas, or solids from the treated area; or a dry system designed to trap liquid and solids before the service inlet and to accommodate air-gas only through the service inlet. [NFPA 99:3.3.21]

Category 4. Activities, systems, or equipment whose failure would have no impact on patient care. [NFPA 99:3.3.158.4]

Certified Backflow Assembly Tester. A person who has shown competence to test and maintain backflow assemblies to the satisfaction of the Authority Having Jurisdiction.

Cesspool. A lined excavation in the ground that receives the discharge of a drainage system or part thereof, so designed as to retain the organic matter and solids discharging therein but permitting the liquids to seep through the bottom and sides.

Chemical Waste. See Special Wastes.

Chimney. One or more passageways, vertical or nearly so, for conveying flue or vent gases to the outdoors. [NFPA 54:3.3.18]

Chimney, Factory-Built. A chimney composed of listed factory-built components assembled in accordance with the manufacturer's installation instructions to form the completed chimney. [NFPA 54:3.3.18.2]

Chimney, Masonry. A field-constructed chimney of solid masonry units, bricks, stones, listed masonry chimney units, or reinforced Portland cement concrete, lined with suitable chimney flue liners. [NFPA 54:3.3.18.3]

Chimney, Metal. A chimney constructed of metal with a minimum thickness not less than 0.127 inches (3.23 mm) (No. 10 manufacturer's standard gauge) steel sheet.

Chimney Classifications:

Chimney, High-Heat Appliance-Type. A factory-built, masonry, or metal chimney suitable for removing the products of combustion from fuel-burning high-heat appliances producing combustion gases in excess of 2000°F (1093°C), measured at the appliance flue outlet.

Chimney, Low-Heat Appliance-Type. A factory-built, masonry, or metal chimney suitable for removing the products of combustion from fuel-burning low-heat appliances producing combustion gases not in excess of 1000°F (538°C) under normal operating conditions, but capable of producing combustion gases of 1400°F (760°C) during intermittent forced firing for periods up to one hour. Temperatures are measured at the appliance flue outlet.

Chimney, Medium-Heat Appliance-Type. A factory-built, masonry, or metal chimney suitable for removing the products of combustion from fuel-burning medium-heat appliances producing combustion gases, not in excess of 2000°F (1093°C), measured at the appliance flue outlet.

Chimney, Residential Appliance-Type. A factory-built or masonry chimney suitable for removing products of combustion from residential-type appliances producing combustion gases, not in excess of 1000°F (538°C), measured at the appliance flue outlet. Factory-built Type HT chimneys have high-temperature thermal shock resistance.

Circuit Vent. The vent that connects to a horizontal drainage branch and vents two traps to a maximum of eight traps connected into a battery of fixtures.

Clarifier. See Interceptor (Clarifier).

Clear Water Waste. Cooling water and condensate drainage from refrigeration and air-conditioning equipment; cooled condensate from steam heating systems, and cooled boiler blowdown water.

Clinical Sink. A fixture that has the same flushing and cleansing characteristics of a water closet that is used to receive the wastes from a bedpan. Also, known as a bedpan washer.

Coastal High Hazard Areas. An area within the flood hazard area that is subject to high-velocity wave action, and shown on a Flood Insurance Rate Map or other flood hazard map as Zone V, VO, VE or V1-30.

Code. A standard that is an extensive compilation of provisions covering broad subject matter or that is suitable for adoption into law independently of other codes and standards.

Combination Temperature and Pressure-Relief Valve. A relief valve that actuates when a set temperature, pressure, or both is reached. Also, known as a T&P Valve.

Combination Thermostatic/Pressure Balancing Valve. A mixing valve that senses outlet temperature and incoming hot and cold water pressure and compensates for fluctuations in incoming hot and cold water temperatures, pressures, or both to stabilize outlet temperatures.

Combination Waste and Vent System. A specially designed system of waste piping embodying the horizontal wet venting of one or more sinks or floor drains using a common waste and vent pipe adequately sized to provide free movement of air above the flow line of the drain.

Combined Building Sewer. See Building Sewer (Combined).

Combustible Material. A material that, in the form in which it is used and under the conditions anticipated, will ignite and burn; a material that does not meet the definition of noncombustible. [NFPA 54:3.3.64.1]

Commercial Modular System. A drinking water treatment unit system consisting of multiple components attached to a manifold, produced specifically for food service applications, and not intended for use in residential applications.

Common. That part of a plumbing system that is so designed and installed as to serve more than one appliance, fixture, building, or system.

Condensate. The liquid phase produced by condensation of a gas or vapor.

Conductor. A pipe inside the building that conveys storm water from the roof to a storm drain, combined building sewer, or other approved point of disposal.

Confined Space. A room or space having a volume less than 50 cubic feet per 1000 British thermal units per hour (Btu/h) (4.83 m^3/kW) of the aggregate input rating of all fuel-burning appliances installed in that space.

Construction Documents. Plans, specifications, written, graphic, and pictorial documents prepared or assembled for describing the design, location, and physical characteristics of the elements of a project necessary for obtaining a permit.

Contamination. An impairment of the quality of the potable water that creates an actual hazard to the public health through poisoning or the spread of disease by sewage, industrial fluids, or waste. Also, defined as High Hazard.

Continuous Vent. A vertical vent that is a continuation of the drain to which it connects.

Continuous Waste. A drain is connecting the compartments of a set of fixtures to a trap or connecting other permitted fixtures to a common trap.

Copper Alloy. A homogenous mixture of two or more metals in which copper is the primary component, such as brass and bronze.

CPVC. Chlorinated Polyvinyl Chloride.

Critical Care Area. See Patient Care Space, Category 1.

Critical Level. The critical level (C-L or C/L) marking on a backflow prevention device or vacuum breaker is a point conforming to approved standards and established by the testing laboratory (usually stamped on the device by the manufacturer) that determines the minimum elevation above the flood-level rim of the fixture or receptor served at which the device may be installed. Where a backflow prevention device does not bear a critical level marking, the bottom of the vacuum breaker, combination valve, or the bottom of such approved device shall constitute the critical level.

Cross-Connection. A connection or arrangement, physical or otherwise, between a potable water supply system and a plumbing fixture or a tank, receptor, equipment, or device, through which it may be possible for nonpotable, used, unclean, polluted, and contaminated water, or other substances to enter into a part of such potable water system under any condition.

206.0 – D –

Dead Leg. A section of potable water pipe which contains water that has no flow or does not circulate.

Debris Excluder. A device installed on the rainwater catchment conveyance system to prevent the accumulation of leaves, needles, or other debris in the system.

Department Having Jurisdiction. The Authority Having Jurisdiction, including any other law enforcement agency affected by a provision of this code, whether such agency is specifically named or not.

Design Flood Elevation. The elevation of the "design flood," including wave height, relative to the datum specified on the community's legally designated flood hazard map. In areas designated as Zone AO, the design flood elevation is the elevation of the highest existing grade of the building's perimeter plus the depth number (in feet) specified on the flood hazard map. In areas designated as Zone AO where a depth number is not specified on the map, the depth number is taken as being equal to 2 feet (610 mm).

Developed Length. The length along the centerline of a pipe and fittings.

Diameter. Unless specifically stated, "diameter" is the nominal diameter as designated commercially.

Direct-Vent Appliances. Appliances that are constructed and installed so that all air for combustion is derived directly from the outdoors and all flue gases are discharged to the outdoors. [NFPA 54:3.3.5.3]

Domestic Sewage. The liquid and water-borne wastes derived from the ordinary living processes, free from industrial wastes, and of such character as to permit satisfactory disposal, without special treatment, into the public sewer or by means of a private sewage disposal system.

Downspout. The rain leader from the roof to the building storm drain, combined building sewer, or other means of disposal located outside of the building. See Conductor and Leader.

Drain. A pipe that carries waste or waterborne wastes in a building drainage system.

Drainage System. Includes all the piping within public or private premises that conveys sewage, storm water, or other liquid wastes to a legal point of disposal, but does not include the mains of a public sewer system or a public sewage treatment or disposal plant.

Drinking Fountain. A plumbing fixture connected to the potable water distribution system and sanitary drainage system that provides drinking water in a flowing stream so that the user can consume water directly from the fixture without the use of accessories. Drinking fountains should also incorporate a bottle filling station and can incorporate a water filter and a cooling system for chilling the drinking water.

Dry Vent. A vent that does not receive the discharge of any sewage or waste.

Durham System. Soil or waste system in which all piping is threaded pipe, tubing, or other such rigid construction, using recessed drainage fittings to correspond to the types of piping.

207.0 – E –

Effective Ground-Fault Current Path. An intentionally constructed, low impedance electrically conductive path designed and intended to carry current under ground-fault conditions from the point of a ground fault on a wiring system to the electrical supply source and that facilitates the operation of the overcurrent protective device or ground-fault detectors. [NFPA 54:3.3.34]

Effective Opening. The minimum cross-sectional area at the point of water supply discharge measured or expressed in terms of (1) diameter of a circle or (2) where the opening is not circular, the diameter of a circle of equivalent cross-sectional area. (This applies to an air gap).

Emergency Floor Drain. A floor drain that does not receive discharge from any fixture drain or indirect waste pipe, and serves to protect from damage where accidental spills, leaks or fixture backups occur.

Essentially Nontoxic Transfer Fluid. Essentially nontoxic at practically nontoxic, Toxicity Rating Class 1 (reference "Clinical Toxicology of Commercial Products" by Gosselin, Smith, Hodge, & Braddock).

Exam Room Sink. A sink used in the patient exam room of a medical or dental office with a primary purpose of the washing of hands.

Excess Flow Valve (EFV). A valve designed to activate when the fuel gas passing through it exceeds a prescribed flow rate. [NFPA 54:3.3.99.3]

Existing Work. A plumbing system or any part thereof that has been installed prior to the effective date of this code.

Expansion Joint. A fitting or arrangement of pipe and fittings that permit the contraction and expansion of a piping system.

Expansion Tank. A vessel used to protect potable water systems from excessive pressure.

208.0 – F –

F Rating. The time period that the penetration firestop system limits the spread of fire through the penetration, where tested in accordance with ASTM E814 or UL 1479.

Fixture Branch. A water supply pipe between the fixture supply pipe and the water distribution pipe.

Fixture Drain. The drain from the trap of a fixture to the junction of that drain with any other drain pipe.

Fixture Fitting. A device that controls and guides the flow of water.

Fixture Supply. A water supply pipe is connecting the fixture with the fixture branch.

Fixture Unit. A quantity in terms of which the load-producing effects on the plumbing system of different kinds of plumbing fixtures are expressed on some arbitrarily chosen scale.

Flammable Vapor or Fumes. The concentration of flammable constituents in the air that exceeds 25 percent of its lower flammability limit (LFL).

Flood Hazard Area. The greater of the following two areas:

(1) The area within a floodplain subject to a 1 percent or greater chance of flooding in any given year.

(2) The area designated as a flood hazard area on a community's flood hazard map, or otherwise legally designated.

Flood Level. See Flooded.

Flood-Level Rim. The top edge of a receptor from which water overflows.

Flooded. A fixture is flooded where the liquid therein rises to the flood-level rim.

Flue Collar. That portion of an appliance designed for the attachment of a draft hood, vent connector, or venting system. [NFPA 54:3.3.44]

Flush Tank. A tank located above or integral with water closets, urinals, or similar fixtures for the purpose of flushing the usable portion of the fixture.

Flush Valve. A valve located at the bottom of a tank for flushing water closets and similar fixtures.

Flushometer Tank. A tank integrated within an air accumulator vessel that is designed to discharge a predetermined quantity of water to fixtures for flushing purposes.

Flushometer Valve. A valve that discharges a predetermined quantity of water to fixtures for flushing purposes and is actuated by direct water pressure.

FOG Disposal System. A grease interceptor that reduces nonpetroleum fats, oils, and grease (FOG) in the effluent by separation, mass, and volume reduction.

Fuel Gas. Natural, manufactured liquefied petroleum, or a mixture of these.

209.0 – G –

Gang or Group Shower. Two or more showers in a common area.

Gas Piping. An installation of pipe, valves, or fittings that are used to convey fuel gas, installed on a premise or in a building, but shall not include:

(1) A portion of the service piping.

(2) An approved piping connection 6 feet (1829 mm) or less in length between an existing gas outlet and a gas appliance in the same room with the outlet.

Gas Piping System. An arrangement of gas piping or regulators after the point of delivery and each arrangement of gas piping serving a building, structure, or premises, whether individually metered or not.

General Anesthesia and Levels of Sedation/Analgesia.

Deep Sedation/Analgesia. A drug-induced depression of consciousness during which patients cannot be easily aroused but respond purposefully following repeated or painful stimulation. The ability to independently maintain ventilatory function may be impaired. Patients may require assistance in maintaining a patent airway, and spontaneous ventilation may be inadequate. Cardiovascular function is usually maintained. [NFPA 99:3.3.66.2]

General Anesthesia. A drug-induced loss of consciousness during which patients are not arousable, even by painful stimulation. The ability to independently maintain ventilatory function is often impaired. Patients often require assistance in maintaining a patent airway, and positive pressure ventilation may be required because of depressed spontaneous ventilation or drug-induced depression of neuromuscular function. Cardiovascular function may be impaired. [NFPA 99:3.3.66.1]

Minimal Sedation (Anxiolysis). A drug-induced state during which patients respond normally to verbal commands. Although cognitive function and coordination may be impaired, ventilatory and cardiovascular functions are unaffected. [NFPA 99:3.3.66.4]

Moderate Sedation/Analgesia (Conscious Sedation). A drug-induced depression of consciousness during which patients respond purposefully to verbal commands, either alone or accompanied by light tactile stimulation. No interventions are required to maintain a patient airway, and spontaneous ventilation is adequate. Cardiovascular function is usually maintained. [NFPA 99:3.3.66.3]

Grade. The slope or fall of a line of pipe in reference to a horizontal plane. In drainage, it is usually expressed as the fall in a fraction of an inch (mm) or percentage slope per foot (meter) length of pipe.

Gravity Grease Interceptor. A plumbing appurtenance or appliance that is installed in a sanitary drainage system to intercept nonpetroleum fats, oils, and greases (FOG) from a wastewater discharge and is identified by volume, 30 minute retention time, baffle(s), not less than two compartments, a total volume of not less than 300 gallons (1135 L), and gravity separation. [These interceptors comply with the requirements of Chapter 10 or are designed by a registered design professional.] Gravity grease interceptors are generally installed outside.

Gray Water. Untreated wastewater that has not come into contact with toilet waste, kitchen sink waste, dishwasher waste or similarly contaminated sources. Gray water includes wastewater from bathtubs, showers, lavatories, clothes washers, and laundry tubs. Also, known as grey water, graywater, and greywater.

Gray Water Diverter Valve. A valve that directs gray water to the sanitary drainage system or a subsurface irrigation system.

Grease Interceptor. A plumbing appurtenance or appliance that is installed in a sanitary drainage system to intercept nonpetroleum fats, oil, and greases (FOG) from a wastewater discharge.

Grease Removal Device (GRD). A hydromechanical grease interceptor that automatically, mechanically removes non-petroleum fats, oils and grease (FOG) from the interceptor, the control of which are either automatic or manually initiated.

Grounding Electrode. A conducting object through which a direct connection to earth is established. [NFPA 70:100 (Part I)]

Group Wash Fixture. A lavatory that allows more than one person to utilize the fixture at the same time. The fixture has one or more drains and one or more faucets.

210.0 – H –

Hangers. See Supports.

»| **Health Care Facility's Governing Body.** The person or persons who have the overall legal responsibility for the operation of a health care facility. [NFPA 99:3.3.72]

Heat-Fusion Weld Joints. A joint used in some thermoplastic systems to connect the pipe to fittings or pipe lengths directly to one another (butt-fusion). This method of joining pipe to fittings includes socket-fusion, electro-fusion, and saddle-fusion. This method of welding involves the application of heat and pressure to the components, allowing them to fuse together forming a bond between the pipe and fitting.

High Hazard. See Contamination.

Horizontal Branch. A drain pipe extending laterally from soil or waste stack or building drain with or without vertical sections or branches, which receives the discharge from one or more fixture drains and conducts it to the soil or waste stack or the building drain.

Horizontal Pipe. A pipe or fitting that is installed in a horizontal position or which makes an angle of less than 45 degrees (0.79 rad) with the horizontal.

Hot Water. Water at a temperature exceeding or equal to 120°F (49°C).

House Drain. See Building Drain.

House Sewer. See Building Sewer.

Hydromechanical Grease Interceptor. A plumbing appurtenance or appliance that is installed in a sanitary drainage system to intercept nonpetroleum fats, oil, and grease (FOG) from a wastewater discharge and is identified by flow rate, and separation and retention efficiency. The design incorporates air entrainment, hydromechanical separation, interior baffling, or barriers in combination or separately, and one of the following:

(1) External flow control, with an air intake (vent), directly connected.

(2) External flow control, without air intake (vent), directly connected.

(3) Without external flow control, directly connected.

(4) Without external flow control, indirectly connected.

These interceptors comply with the requirements of Table 1014.2.1. Hydromechanical grease interceptors are generally installed inside.

211.0 – I –

Indirect-Fired Water Heater. A water heater consisting of a storage tank equipped with an internal or external heat exchanger used to transfer heat from an external source to heat potable water. The storage tank either contains heated potable water or water supplied from an external source, such as a boiler.

Indirect Waste Pipe. A pipe that does not connect directly to the drainage system but conveys liquid wastes by discharging into a plumbing fixture, interceptor, or receptacle that is directly connected to the drainage system.

Individual Vent. A pipe installed to vent a fixture trap, and that connects with the vent system above the fixture served or terminates in the open air.

Industrial Waste. Liquid or water-borne waste from industrial or commercial processes, except domestic sewage.

Insanitary. A condition that is contrary to sanitary principles or is injurious to health.

Conditions to which "insanitary" shall apply include the following:

(1) A trap that does not maintain a proper trap seal.

(2) An opening in a drainage system, except where lawful that is not provided with an approved liquid-sealed trap.

(3) A plumbing fixture or other waste discharging receptor or device that is not supplied with water sufficient to flush and maintain the fixture or receptor in a clean condition.

(4) A defective fixture, trap, pipe, or fitting.

(5) A trap, except where in this code exempted, directly connected to a drainage system, the seal of which is not protected against siphonage and backpressure by a vent pipe.

(6) A connection, cross-connection, construction, or condition, temporary or permanent that would permit or make possible by any means whatsoever for an unapproved foreign matter to enter a water distribution system used for domestic purposes.

(7) The preceding enumeration of conditions to which the term "insanitary" shall apply, shall not preclude the application of that term to conditions that are, in fact, insanitary.

Interceptor (Clarifier). A device designed and installed to separate and retain deleterious, hazardous, or undesirable matter from normal wastes and permit normal sewage or liquid wastes to discharge into the disposal terminal by gravity.

Invert. The lowest portion of the inside of a horizontal pipe.

212.0 – J –

Joint, Brazed. A joint obtained by joining of metal parts with alloys that melt at temperatures exceeding 840°F (449°C), but less than the melting temperature of the parts to be joined.

Joint, Compression. A multipiece joint with cup-shaped threaded nuts that, when tightened, compress tapered sleeves so that they form a tight joint on the periphery of the tubing they connect.

Joint, Flanged. One made by bolting together a pair of flanged ends.

Joint, Flared. A metal-to-metal compression joint in which a conical spread is made on the end of a tube that is compressed by a flare nut against a mating flare.

Joint, Mechanical. The general form for gas-tight or liquid-tight joints obtained by the joining of parts through a positive holding mechanical construction.

Joint, Press-Connect. A permanent mechanical joint incorporating an elastomeric seal or an elastomeric seal and corrosion resistant grip ring. The joint is made with a pressing tool and jaw or ring that complies with the manufacturer's installation instructions.

Joint, Soldered. A joint obtained by the joining of metal parts with metallic mixtures or alloys that melt at a temperature up to and including 840°F (449°C).

Joint, Welded. A gastight joint obtained by the joining of metal parts in the plastic molten state.

213.0 – K –

No definitions.

214.0 – L –

Labeled. Equipment or materials bearing a label of a listing agency (accredited conformity assessment body). See Listed (third-party certified).

Lavatories in Sets. Two or three lavatories that are served by one trap.

Leader. An exterior vertical drainage pipe for conveying storm water from roof or gutter drains. See Downspout.

Liquefied Petroleum Gas (LP-Gas) Facilities. Liquefied petroleum gas (LP-Gas) facilities include tanks, containers, container valves, regulating equipment, meters, appurtenances, or any combination thereof for the storage and supply of liquefied petroleum gas for a building, structure, or premises.

Liquid Waste. The discharge from a fixture, appliance, or appurtenance in connection with a plumbing system that does not receive fecal matter.

Listed (Third-Party Certified). Equipment or materials included in a list published by a listing agency (accredited conformity assessment body) that maintains periodic inspection of current production of listed equipment or materials and whose listing states either that the equipment or material complies with approved standards or has been tested and found suitable for use in a specified manner.

Listing Agency. An agency accredited by an independent and authoritative conformity assessment body to operate a material and product listing and labeling (certification) system and that are accepted by the Authority Having Jurisdiction, which is in the business of listing or labeling. The system includes initial and ongoing product testing, a periodic inspection on current production of listed (certified) products, and that makes available a published report of such listing in which specific information is included that the material or product is in accordance with applicable standards and found safe for use in a specific manner.

Lot. A single or individual parcel or area of land legally recorded or validated by other means acceptable to the Authority Having Jurisdiction on which is situated a building or which is the site of any work regulated by this code, together with the yards, courts, and unoccupied spaces legally required for the building or works, and that is owned by or is in the lawful possession of the owner of the building or works.

Low Hazard. See Pollution.

Low-Pressure Water Dispenser. A terminal fitting located downstream of a pressure reducing valve that dispenses drinking hot water above 71°C (160°F) or cold water or both at a pressure of 105 kPa (15 psi) or less.

215.0 – M –

Macerating Toilet System. A system comprised of a sump with macerating pump and with connections for a water closet and other plumbing fixtures, which is designed to accept, grind and pump wastes to an approved point of discharge.

Main. The principal artery of a system of continuous piping to which branches may be connected.

Main Sewer. See Public Sewer.

Main Vent. The principal artery of the venting system to which vent branches may be connected.

May. A permissive term.

Medical Air. For the purposes of this code, medical air is air supplied from cylinders, bulk containers, or medical air compressors, or reconstituted from oxygen USP and oil-free, dry nitrogen NF. [NFPA 99:3.3.96]

Medical Gas. A patient medical gas or medical support gas. (See also Patient Medical Gas and Medical Support Gas) [NFPA 99:3.3.104]

> **Manifold.** A device for connecting the outlets of one or more gas cylinders to the central piping system for that specific gas. [NFPA 99:3.3.99]
>
> **Medical Gas System.** An assembly of equipment and piping for the distribution of nonflammable medical gases such as oxygen, nitrous oxide, compressed air, carbon dioxide, and helium. [NFPA 99:3.3.105]
>
> **Medical Support Gas.** Nitrogen or instrument air used for any medical support purpose (e.g., to remove excess moisture from instruments before further processing, or to operate medical-surgical tools, air-driven booms, pendants, or similar applications) and, if appropriate to the procedures, used in laboratories and are not respired as part of any treatment. Medical support gas falls under the general requirements for medical gases. [NFPA 99:3.3.107]

Medical-Surgical Vacuum. A method used to provide a source of drainage, aspiration, and suction in order to remove body fluids from patients. [NFPA 99:3.3.108]

Medical-Surgical Vacuum System. An assembly of central vacuum-producing equipment and a network of piping for patient suction in medical, medical-surgical, and waste anesthetic gas disposal (WAGD) applications. [NFPA 99:3.3.109]

Mobile Home Park Sewer. That part of the horizontal piping of a drainage system that begins 2 feet (610 mm) downstream from the last mobile home site and conveys it to a public sewer, private sewer, private sewage disposal system, or other point of disposal.

Mulch. Organic materials, such as wood chips and fines, tree bark chips, and pine needles that are used in a mulch basin to conceal gray water outlets and permit the infiltration of gray water.

Mulch Basin. A subsurface catchment area for gray water that is filled with mulch and of sufficient depth and volume to prevent ponding, surfacing, or runoff.

216.0 – N –

Nitrogen, NF. Nitrogen complying as a minimum, with nitrogen NF. [NFPA 99:3.3.109.1]

Nonwater Urinal with Drain Cleansing Action. A nonwater urinal that conveys waste into the drainage system without the use of water for flushing and automatically performs a drain-cleansing action after a predetermined amount of time.

Nuisance. Includes, but is not limited to:

(1) A public nuisance known at common law or in equity jurisprudence.

(2) Where work regulated by this code is dangerous to human life or is detrimental to health and property.

(3) Inadequate or unsafe water supply or sewage disposal system.

217.0 – O –

Offset. A combination of elbows or bends in a line of piping that brings one section of the pipe out of line but into a line parallel with the other section.

Oil Interceptor. See Interceptor (Clarifier).

On-Site Treated Nonpotable Water. Nonpotable water, including gray water that has been collected, treated, and intended to be used on-site and is suitable for direct beneficial use.

218.0 – P –

Patient Care Space. Any space of a health care facility wherein patients are intended to be examined or treated. [NFPA 99:3.3.136]

Category 1 Space. Space in which failure of equipment or a system is likely to cause major injury or death of patients, staff, or visitors. [NFPA 99:3.3.136.1]

Category 2 Space. Space in which failure of equipment or a system is likely to cause minor injury to patients, staff, or visitors. [NFPA 99:3.3.136.2]

Category 3 Space. Space in which the failure of equipment or a system is not likely to cause injury to patients, staff, or visitors but can cause discomfort. [NFPA 99:3.3.136.3]

Category 4 Space. Space in which failure of equipment or a system is not likely to have a physical impact on patient care. [NFPA 99:3.3.136.4]

Patient Medical Gas. Piped gases such as oxygen, nitrous oxide, helium, carbon dioxide, and medical air that are used in the application of human respiration and the calibration of medical devices used for human respiration. [NFPA 99:3.3.142]

PB. Polybutylene.

PE. Polyethylene.

PE-AL-PE. Polyethylene-aluminum-polyethylene.

PE-RT. Polyethylene of raised temperature.

Penetration Firestop System. A specific assemblage of field-assembled materials, or a factory-made device, which has been tested to a standard test method and, where installed properly on penetrating piping materials, is capable of maintaining the fire- resistance rating of assemblies penetrated.

Person. A natural person, his heirs, executor, administrators, or assigns and shall also include a firm, corporation, municipal or quasi-municipal corporation, or governmental agency. The singular includes the plural, male includes female.

PEX. Cross-linked polyethylene.

PEX-AL-PEX. Cross-linked polyethylene–aluminum-cross-linked polyethylene.

Pipe. A cylindrical conduit or conductor is conforming to the dimensions commonly known as "pipe size."

Plumbing. The business, trade, or work having to do with the installation, removal, alteration, or repair of plumbing systems or parts thereof.

Plumbing Appliance. A special class of device or equipment that is intended to perform a special plumbing function. Its operation, control, or both may be dependent upon one or more energized components, such as motors, controls, heating elements, or pressure- or temperature-sensing elements. Such device or equipment may operate automatically through one or more of the following actions: a time cycle, a temperature range, a pressure range, a measured volume or weight; or the device or equipment may be manually adjusted or controlled by the user or operator.

Plumbing Appurtenance. A manufactured device, a prefabricated assembly, or an on-the-job assembly of component parts that is an adjunct to the basic piping system and plumbing fixtures. An appurtenance demands no additional water supply, nor does it add a discharge load to a fixture or the drainage system. It performs some useful function in the operation, maintenance, servicing, economy, or safety of the plumbing system.

Plumbing Fixture. An approved type installed receptacle, device or appliance that is supplied with water or that receives liquid or liquid-borne wastes and discharges such wastes into the drainage system to which it may be directly or indirectly connected. Industrial or commercial tanks, vats, and similar processing equipment are not plumbing fixtures, but may be connected to or discharged into approved traps or plumbing fixtures where and as otherwise provided for elsewhere in this code.

Plumbing Official. See Authority Having Jurisdiction.

Plumbing System. Includes all potable water, alternate water sources, building supply, and distribution pipes; all plumbing fixtures and traps; all drainage and vent pipes; and all building drains and building sewers, including their respective joints and connections, devices, receptors, and appurtenances within the property lines of the premises and shall include potable water piping, potable water treating or using equipment, medical gas and medical vacuum systems, liquid and fuel gas piping, and water heaters and vents for same.

Plumbing Vent. A pipe provided to ventilate a plumbing system, to prevent trap siphonage and backpressure, or to equalize the air pressure within the drainage system.

Plumbing Vent System. A pipe or pipes installed to provide a flow of air to or from a drainage system or to provide a circulation of air within such system to protect trap seals from siphonage and backpressure.

Point-of-Entry, Water Treatment Unit. A device serving the water distribution system of a building for the purposes of altering, modifying, adding, or removing minerals, chemicals, contaminants, and suspended solids in the water.

Point-of-Use, Water Treatment Unit. A device serving a single atmospheric outlet such as a faucet for the purposes of altering, modifying, adding, or removing any minerals, chemicals, contaminants, and suspended solids in water.

Pollution. An impairment of the quality of the potable water to the degree that does not create a hazard to the public health but which does adversely and unreasonably affect the aesthetic qualities of such potable water for domestic use. Also, defined as "Low Hazard."

Potable Water. Water that is satisfactory for drinking, culinary, and domestic purposes and that meets the requirements of the Health Authority Having Jurisdiction.

PP. Polypropylene.

Pre-fabricated Shower Enclosure. A factory-assembled watertight structure with enclosing walls, a drain, and door or open access way.

Pressure. The normal force exerted by a homogeneous liquid or gas, per unit of area, on the wall of the container.

Residual Pressure. The pressure available at the fixture or water outlet after allowance is made for pressure drop due to friction loss, head, meter, and other losses in the system during maximum demand periods.

Static Pressure. The pressure is existing without any flow.

Pressure-Balancing Valve. A mixing valve that senses incoming hot and cold water pressures and compensates for fluctuations in either to stabilize outlet temperature.

Pressure-Lock-Type Connection. A mechanical connection that depends on an internal retention device to prevent pipe or tubing separation. The connection is made by inserting the pipe or tubing into the fitting to a prescribed depth.

Private or Private Use. Applies to plumbing fixtures in residences and apartments, to private bathrooms in hotels, hospitals, and health care facilities, and to restrooms in commercial establishments where the fixtures are intended for the use of a family or an individual.

Private Sewage Disposal System. A septic tank with the effluent discharging into a subsurface disposal field, into one or more seepage pits, or into a combination of subsurface disposal field and seepage pit or of such other facilities as may be permitted under the procedures set forth elsewhere in this code.

Private Sewer. A building sewer that receives the discharge from more than one building drain and conveys it to a public sewer, private sewage disposal system, or another point of disposal.

Proportioning System for Medical Air USP. A central supply that produces medical air (USP) reconstituted from oxygen USP and nitrogen NF by means of a mixer or blender. [NFPA 99:3.3.102.1]

Public or Public Use. Applies to plumbing fixtures that are not defined as private or private use.

Public Sewer. A common sewer directly controlled by public authority.

Public Water System. A system for the provision to the public of water for human consumption through pipes or other constructed conveyances, if such system has at least fifteen service connections or regularly serves an average of twenty-five individuals daily for at least 60 days per year.

Push Fit Fitting. A mechanical fitting where the connection is assembled by pushing the tube or pipe into the fitting and is sealed with an o-ring.

PVC. Polyvinyl Chloride.

PVDF. Polyvinylidene Fluoride.

219.0 – Q –

Quick-Disconnect Device. A hand-operated device that provides a means for connecting and disconnecting a hose to a water supply, and that is equipped with a means to shut off the water supply when the device is disconnected.

Quick-Disconnect Device (Fuel Gas). A hand-operated device that provides a means for connecting and disconnecting an appliance or an appliance connector to a gas supply, and that is equipped with an automatic means to shut off the gas supply when the device is disconnected. [NFPA 54:3.3.28.3]

220.0 – R –

Rainwater. Natural precipitation that has not been contaminated by use.

Rainwater Catchment System. A system that utilizes the principal of collecting, storing, and using rainwater from a rooftop or other manmade, aboveground collection surface. Also, known as a rainwater harvesting system.

Rainwater Storage Tank. The central component of the rainwater catchment system. Also, known as a cistern or rain barrel.

Receptor. An approved plumbing fixture or device of such material, shape, and capacity as to adequately receive the discharge from indirect waste pipes, so constructed and located as to be readily cleaned.

Reclaimed Water. Nonpotable water provided by a water/wastewater utility that, as a result of tertiary treatment of domestic wastewater, meets requirements of the public health Authority Having Jurisdiction for its intended uses.

Registered Design Professional. An individual who is registered or licensed by the laws of the state to perform such design work in the jurisdiction.

Regulating Equipment. Includes valves and controls used in a plumbing system that is required to be accessible or readily accessible.

Relief Vent. A vent, the primary function of which is to provide circulation of air between drainage and vent systems or to act as an auxiliary vent on a specially designed system.

Remote Outlet. Where used for sizing water piping, it is the furthest outlet dimension, measuring from the meter, either the developed length of the cold-water piping or through the water heater to the furthest outlet on the hot-water piping.

Rim. See Flood-Level Rim.

Riser. A water supply pipe that extends vertically one full story or more to convey water to branches or fixtures.

Roof Drain. A drain installed to receive water collecting on the surface of a roof and to discharge it into a leader, downspout, or conductor.

Roof Washer. A device or method for removal of sediment and debris from a collection surface by diverting initial rainfall from entry into the cistern(s). Also, known as a first flush device.

Roughing-In. The installation of all parts of the plumbing system that can be completed prior to the installation of fixtures. This includes drainage, water supply, gas piping, vent piping, and the necessary fixture supports.

221.0 – S –

Sand Interceptor. See Interceptor (Clarifier).

Scavenging. Evacuation of exhaled mixtures of oxygen and nitrous oxide. [NFPA 99:3.3.159]

SDR. An abbreviation for "standard dimensional ratio," which is the specific ratio of the average specified outside diameter to the minimum wall thickness for outside controlled diameter plastic pipe.

Seam, Welded. See Joint, Welded.

Seepage Pit. A lined excavation in the ground which receives the discharge of a septic tank so designed as to permit the effluent from the septic tank to seep through its bottom and sides.

Septic Tank. A watertight receptacle that receives the discharge of a drainage system or part thereof, designed and constructed so as to retain solids, digest organic matter through a period of detention, and allow the liquids to discharge into the soil outside of the tank through a system of open joint piping or a seepage pit meeting the requirements of this code.

Service Piping. The piping and equipment between the street gas main and the gas piping system inlet that is installed by, and is under the control and maintenance of, the serving gas supplier.

Sewage. Liquid waste containing animal or vegetable matter in suspension or solution and that may include liquids containing chemicals in solution.

Sewage Ejector. A device for lifting sewage by entraining it on a high-velocity jet stream, air, or water.

Sewage Pump. A permanently installed mechanical device, other than an ejector, for removing sewage or liquid waste from a sump.

Shall. Indicates a mandatory requirement.

Shielded Coupling. An approved elastomeric sealing gasket with an approved outer shield and a tightening mechanism.

Shock Arrester. See Water Hammer Arrester.

Should. Indicates a recommendation or that which is advised but not required.

Single-Family Dwelling. A building designed to be used as a home by the owner of such building, which shall be the only dwelling located on a parcel of ground with the usual accessory buildings.

Size and Type of Tubing. See Diameter.

Slip Joint. An adjustable tubing connection, consisting of a compression nut, a friction ring, and a compression washer, designed to fit a threaded adapter fitting or a standard taper pipe thread.

Slope. See Grade.

Soil Pipe. A pipe that conveys the discharge of water closets, urinals, clinical sinks, or fixtures having similar functions of collection and removal of domestic sewage, with or without the discharge from other fixtures to the building drain or building sewer.

Special Wastes. Wastes that require some special method of handling, such as the use of indirect waste piping and receptors, corrosion-resistant piping, sand, oil or grease interceptors, condensers, or other pretreatment facilities.

Stack. The vertical main of a system of soil, waste, or vent piping extending through one or more stories.

Stack Vent. The extension of soil or waste stacks above the highest horizontal drain connected to the stack.

Standard. A document, the main text of which contains only mandatory provisions using the word "shall" to indicate requirements and which is in a form generally suitable for mandatory reference by another standard or code or for adoption into law. Nonmandatory provisions shall be located in an appendix, footnote, or fine print note and are not to be considered a part of the requirements of a standard.

Standard Cubic Feet per Minute (SCFM). Volumetric flow rate of gas in units of standard cubic feet per minute. [NFPA 99:3.3.168]

Station Inlet. An inlet point in a piped medical/surgical vacuum distribution system at which the user makes connections and disconnections. [NFPA 99:3.3.169]

Station Outlet. An outlet point in a piped medical gas distribution system at which the user makes connections and disconnections. [NFPA 99:3.3.170]

Sterilizer. A piece of equipment that disinfects instruments and equipment by way of heat.

Storm Drain. See Building Drain (Storm).

Storm Sewer. A sewer used for conveying rainwater, surface water, condensate, cooling water, or similar liquid wastes.

Subsoil Drain. A drain that collects subsurface or seepage water and conveys it to a place of disposal.

Subsoil Irrigation Field. Gray water irrigation field installed in a trench within the layer of soil below the topsoil. This system is typically used for irrigation of deep rooted plants.

Subsurface Irrigation Field. Gray water irrigation field installed below finished grade within the topsoil.

Sump. An approved tank or pit that receives sewage or liquid waste and which is located below the normal grade of the gravity system and which must be emptied by mechanical means.

Supports. Supports, hangers, and anchors are devices for properly supporting and securing pipe, fixtures, and equipment.

Surge Tank. A reservoir to modify the fluctuation in flow rates to allow for uniform distribution of gray water to the points of irrigation.

222.0 – T –

T Rating. The time period that the penetration firestop system, including the penetrating item, limits the maximum temperature rise of 325°F (181°C) above its initial temperature through the penetration on the nonfire side, where tested in accordance with ASTM E814 or UL 1479.

Tailpiece. The pipe or tubing that connects the outlet of a plumbing fixture to a trap.

Thermostatic (Temperature Control) Valve. A mixing valve that senses outlet temperature and compensates for fluctuations in incoming hot or cold water temperatures.

Toilet Facility. A room or space containing not less than one lavatory and one water closet.

Transition Gas Riser. A listed or approved section or sections of pipe and fittings used to convey fuel gas and installed in a gas piping system to provide a transition from belowground to aboveground.

Trap. A fitting or device so designed and constructed as to provide, where properly vented, a liquid seal that will prevent the back passage of air without materially affecting the flow of sewage or wastewater through it.

Trap Arm. Those portions of a fixture drain between a trap and the vent.

Trap Primer. A device and system of piping that maintains a water seal in a remote trap.

Trap Seal. The vertical distance between the crown weir and the top dip of the trap.

> **Crown Weir (Trap Weir).** The lowest point in the cross-section of the horizontal waterway at the exit of the trap.
>
> **Top Dip (of the trap).** The highest point in the internal cross-section of the trap at the lowest part of the bend (inverted siphon). By contrast, the bottom dip is the lowest point in the internal cross-section.

223.0 – U –

Unsanitary. See Insanitary.

User Outlet. See Station Outlet.

224.0 – V –

Vacuum. A pressure less than that exerted by the atmosphere.

Vacuum Breaker. See Backflow Preventer.

Vacuum Relief Valve. A device that prevents excessive vacuum in a pressure vessel.

Vacuum System-Level 1. A system consisting of central vacuum-producing equipment with pressure and operating controls, shutoff valves, alarm warning systems, gauges, and a network of piping extending to and terminating with suitable station inlets at locations where patient suction could be required.

Valve, Isolation. A valve that isolates one piece of equipment from another.

Valve, Pressure-Relief. A pressure-actuated valve held closed by a spring or other means and designed automatically to relieve pressure in excess of its setting.

Valve, Riser. A valve at the base of a vertical riser that isolates that riser.

Valve, Service. A valve is serving horizontal piping extending from a riser to a station outlet or inlet.

Valve, Source. A single valve at the source that controls a number of units that makes up the source.

Valve, Zone. A valve that controls the gas or vacuum to a particular area.

Vent. See Plumbing Vent; Dry Vent; Wet Vent.

Vent Connector, Gas. That portion of a gas venting system that connects a listed gas appliance to a gas vent and is installed within the space or area in which the appliance is located.

Vent Offset. An arrangement of two or more fittings and pipe installed for the purpose of locating a vertical section of vent pipe in a different but parallel plane with respect to an adjacent section of vertical vent pipe. [NFPA 54:3.3.102]

Vent Pipe. See Plumbing Vent.

Vent Stack. The vertical vent pipe installed primarily for the purpose of providing circulation of air to and from any part of the drainage system.

Vent System. See Plumbing Vent System.

Vented Flow Control Device. A device installed upstream from the hydromechanical grease interceptor having an orifice that controls the rate of flow through the interceptor, and an air intake (vent) downstream from the orifice, which allows air to be drawn into the flow stream.

Venting System. A continuous open passageway from the flue collar or draft hood of an appliance to the outdoors for the purpose of removing flue or vent gases. [NFPA 54:3.3.95.7]

> **Vent, Gases.** A listed factory-made vent pipe and vent fittings for conveying flue gases to the outdoors.

Type B Gas Vent. A factory-made gas vent listed by a nationally recognized testing agency for venting listed or approved appliances equipped to burn only gas.

Type BW Gas Vent. A factory-made gas vent listed by a nationally recognized testing agency for venting listed or approved gas-fired vented wall furnaces.

Type L Gas Vent. A venting system consisting of listed vent piping and fittings for use with oil-burning appliances listed for use with Type L or with listed gas appliances.

Vertical Pipe. A pipe or fitting that is installed in a vertical position or that makes an angle of not more than 45 degrees (0.79 rad) with the vertical.

225.0 – W –

Wall-Hung Water Closet. A water closet installed in such a way that no part of the water closet touches the floor.

Waste. See Liquid Waste and Industrial Waste.

Waste Pipe. A pipe that conveys only liquid waste, free of fecal matter.

Water-Conditioning or Treating Device. A device that conditions or treats a water supply to change its chemical content or remove suspended solids by filtration.

Water Distribution Pipe. In a building or premises, a pipe that conveys potable water from the building supply pipe to the plumbing fixtures and other water outlets.

Water Hammer Arrester. A device designed to provide protection against hydraulic shock in the building water supply system.

Water Heater or Hot Water Heating Boiler. An appliance designed primarily to supply hot water for domestic or commercial purposes and equipped with automatic controls limiting water temperature to a maximum of 210°F (99°C).

Water Main (Street Main). A water supply pipe for public or community use.

Water Supply System. The building supply pipe, the water distribution pipes, and the necessary connecting pipes, fittings, control valves, backflow prevention devices, and all appurtenances carrying or supplying potable water in or adjacent to the building or premises.

Water/Wastewater Utility. A public or private entity which may treat, deliver or do both functions to reclaimed (recycled) water, potable water, or both to wholesale or retail customers.

Welder, Pipe. A person who specializes in the welding of pipes and holds a valid certificate of competency from a recognized testing laboratory, based on the requirements of the ASME Boiler and Pressure Vessels code, Section IX.

Wet Procedure Locations. The area in a patient care space where a procedure is performed that is normally subject to wet conditions while patients are present, including standing fluids on the floor or drenching of the work area, either of which condition is intimate to the patient or staff. [NFPA 99:3.3.183]

Wet Vent. A vent that also serves as a drain.

Whirlpool Bathtub. A bathtub fixture equipped and fitted with a circulating piping system designed to accept, circulate, and discharge bathtub water upon each use.

226.0 – X –

No definitions.

227.0 – Y –

Yoke Vent. A pipe connecting upward from soil or waste stack to a vent stack to prevent pressure changes in the stacks.

228.0 – Z –

No definitions.

CHAPTER 3
GENERAL REGULATIONS

301.0 General.

301.1 Applicability. This chapter shall govern the general requirements, not specific to other chapters, for the installation of plumbing systems.

301.2 Minimum Standards. Pipe, pipe fittings, traps, fixtures, material, and devices used in a plumbing system shall be listed (third-party certified) by a listing agency (accredited conformity assessment body) as complying with the approved applicable recognized standards referenced in this code, and shall be free from defects. Unless otherwise provided for in this code, materials, fixtures, or devices used or entering into the construction of plumbing systems, or parts thereof shall be submitted to the Authority Having Jurisdiction for approval prior to being installed.

301.2.1 Marking. Each length of pipe and each pipe fitting, trap, fixture, material, and device used in a plumbing system shall have cast, stamped, or indelibly marked on it any markings required by the applicable referenced standards and listing agency, and the manufacturer's mark or name, which shall readily identify the manufacturer to the end user of the product. Where required by the approved standard that applies, the product shall be marked with the weight and the quality of the product. Materials and devices used or entering into the construction of plumbing and drainage systems, or parts thereof shall be marked and identified in a manner satisfactory to the Authority Having Jurisdiction. Such marking shall be done by the manufacturer. Field markings shall not be acceptable.

Exception: Markings shall not be required on nipples created from cutting and threading of approved pipe.

301.2.2 Standards. Standards listed or referred to in this chapter or other chapters cover materials that will conform to the requirements of this code, where used in accordance with the limitations imposed in this or other chapters thereof and their listing. Where a standard covers materials of various grades, weights, quality, or configurations, the portion of the listed standard that is applicable shall be used. Design and materials for special conditions or materials not provided for herein shall be permitted to be used by special permission of the Authority Having Jurisdiction after the Authority Having Jurisdiction has been satisfied as to their adequacy. A list of plumbing standards that appear in specific sections of this code is referenced in Table 1701.1. Standards referenced in Table 1701.1 shall be applied as indicated in the applicable referenced section. A list of additional approved standards, publications, practices, and guides that are not referenced in specific sections of this code appear in Table 1701.2. An IAPMO Installation Standard is referenced in Appendix I for the convenience of the users of this code. It is not considered as a part of this code unless formally adopted as such by the Authority Having Jurisdiction.

301.2.3 Plastic Pipe, Plastic Pipe Fittings, and Components. Plastic pipe, plastic pipe fittings, and components other than those for gas shall comply with NSF 14.

301.2.4 Cast-Iron Soil Pipe, Fittings, and Hubless Couplings. Cast-iron soil pipe, fittings, and hubless couplings shall be third party certified in accordance with ASTM C1277 and CISPI 310 for couplings and ASTM A888, ASTM A74, and CISPI 301 for pipes and fittings.

301.2.5 Existing Buildings. In existing buildings or premises in which plumbing installations are to be altered, repaired, or renovated, the Authority Having Jurisdiction has discretionary powers to permit deviation from the provisions of this code, provided that such proposal to deviate is first submitted for proper determination in order that health and safety requirements, as they pertain to plumbing, shall be observed.

301.3 Alternate Materials and Methods of Construction Equivalency. Nothing in this code is intended to prevent the use of systems, methods, or devices of equivalent or superior quality, strength, fire resistance, effectiveness, durability, and safety over those prescribed by this code. Technical documentation shall be submitted to the Authority Having Jurisdiction to demonstrate equivalency prior to installation. The Authority Having Jurisdiction shall have the authority to approve or disapprove the system, method, or device for the intended purpose.

However, the exercise of this discretionary approval by the Authority Having Jurisdiction shall have no effect beyond the jurisdictional boundaries of said Authority Having Jurisdiction. An alternate material or method of construction so approved shall not be considered as in accordance with the requirements, intent, or both of this code for a purpose other than that granted by the Authority Having Jurisdiction where the submitted data does not prove equivalency.

301.3.1 Testing. The Authority Having Jurisdiction shall have the authority to require tests, as proof of equivalency.

301.3.1.1 Tests. Tests shall be made in accordance with approved or applicable standards, by an approved testing agency at the expense of the applicant. In the absence of such standards, the Authority Having Jurisdiction shall have the authority to specify the test procedure.

301.3.1.2 Request by Authority Having Jurisdiction. The Authority Having Jurisdiction shall have the authority to require tests to be made or repeated where there is reason to believe that a material or device no longer is in accordance with the requirements on which its approval was based.

301.4 Flood Hazard Areas. Plumbing systems shall be located above the elevation in accordance with the building code for utilities and attendant equipment or the elevation of the lowest floor, whichever is higher.

Exception: Plumbing systems shall be permitted to be located below the elevation in accordance with the building code for utilities and attendant equipment or the elevation of the lowest floor, whichever is higher, provided that the systems are designed and installed to prevent water from entering or accumulating within their components, and the systems are constructed to resist hydrostatic and hydrodynamic loads and stresses, including the effects of buoyancy, during the occurrence of flooding to such elevation.

301.4.1 Coastal High Hazard Areas. Plumbing systems in buildings located in coastal high hazard areas shall be in accordance with the requirements of Section 301.4, and plumbing systems, pipes, and fixtures shall not be mounted on or penetrate through walls that are intended to breakaway under flood loads in accordance with the building code.

301.5 Alternative Engineered Design. An alternative engineered design shall comply with the intent of the provisions of this code and shall provide an equivalent level of quality, strength, effectiveness, fire resistance, durability, and safety. Material, equipment, or components shall be designed and installed in accordance with the manufacturer's installation instructions.

301.5.1 Permit Application. The registered design professional shall indicate on the design documents that the plumbing system, or parts thereof, is an alternative engineered design so that it is noted on the construction permit application. The permit and permanent permit records shall indicate that an alternative engineered design was part of the approved installation.

301.5.2 Technical Data. The registered design professional shall submit sufficient technical data to substantiate the proposed alternative engineered design and to prove that the performance meets the intent of this code.

301.5.3 Design Documents. The registered design professional shall provide two complete sets of signed and sealed design documents for the alternative engineered design for submittal to the Authority Having Jurisdiction. The design documents shall include floor plans and a riser diagram of the work. Where appropriate, the design documents shall indicate the direction of flow, pipe sizes, grade of horizontal piping, loading, and location of fixtures and appliances.

301.5.4 Design Approval. An approval of an alternative engineered design shall be at the discretion of the Authority Having Jurisdiction. The exercise of this discretionary approval by the Authority Having Jurisdiction shall have no effect beyond the jurisdictional boundaries of said Authority Having Jurisdiction. An alternative engineered design so approved shall not be considered as in accordance with the requirements, intent, or both of this code for a purpose other than that granted by the Authority Having Jurisdiction.

301.5.5 Design Review. The Authority Having Jurisdiction shall have the authority to require testing of the alternative engineered design in accordance with Section 301.3.1, including the authority to require an independent review of the design documents by a registered design professional selected by the Authority Having Jurisdiction and at the expense of the applicant.

301.5.6 Inspection and Testing. The alternative engineered design shall be tested and inspected in accordance with the submitted testing and inspection plan and the requirements of this code.

302.0 Iron Pipe Size (IPS) Pipe.

302.1 General. Iron, steel, copper, and copper alloy pipe shall be standard-weight iron pipe size (IPS) pipe.

303.0 Disposal of Liquid Waste.

303.1 General. It shall be unlawful for a person to cause, suffer, or permit the disposal of sewage, human excrement, or other liquid wastes, in a place or manner, except through and by means of an approved drainage system, installed and maintained in accordance with the provisions of this code.

304.0 Connections to Plumbing System Required.

304.1 General. Plumbing fixtures, drains, appurtenances, and appliances, used to receive or discharge liquid wastes or sewage, shall be connected properly to the drainage system of the building or premises, in accordance with the requirements of this code.

305.0 Damage to Drainage System or Public Sewer.

305.1 Unlawful Practices. It shall be unlawful for a person to deposit, by any means whatsoever, into a plumbing fixture, floor drain, interceptor, sump, receptor, or device, which is connected to a drainage system, public sewer, private sewer, septic tank, or cesspool, any ashes; cinders; solids; rags; inflammable, poisonous, or explosive liquids or gases; oils; grease; or any other thing whatsoever that is capable of causing damage to the drainage system or public sewer.

306.0 Industrial Wastes.

306.1 Detrimental Wastes. Wastes detrimental to the public sewer system or detrimental to the functioning of the sewage treatment plant shall be treated and disposed of as found necessary and directed by the Authority Having Jurisdiction.

306.2 Safe Discharge. Sewage or other waste from a plumbing system that is capable of being deleterious to surface or subsurface waters shall not be discharged into the ground or a waterway unless it has first been rendered safe by some acceptable form of treatment in accordance with the Authority Having Jurisdiction.

307.0 Location.

307.1 System. Except as otherwise provided in this code, no plumbing system, drainage system, building sewer, private sewage disposal system, or parts thereof shall be located in a lot other than the lot that is the site of the building, structure, or premises served by such facilities.

307.2 Ownership. No subdivision, sale, or transfer of ownership of existing property shall be made in such manner that the area, clearance, and access requirements of this code are decreased.

308.0 Improper Location.

308.1 General. Piping, fixtures, or equipment shall not be so located as to interfere with the normal use thereof or with the normal operation and use of windows, doors, or other required facilities.

309.0 Workmanship.

309.1 Engineering Practices. Design, construction, and workmanship shall be in accordance with accepted engineering practices and shall be of such character as to secure the results sought to be obtained by this code.

309.2 Concealing Imperfections. It is unlawful to conceal cracks, holes, or other imperfections in materials by welding, brazing, or soldering or by using therein or thereon paint, wax, tar, solvent cement, or other leak-sealing or repair agent.

309.3 Burred Ends. Burred ends of pipe and tubing shall be reamed to the full bore of the pipe or tube, and chips shall be removed.

309.4 Installation Practices. Plumbing systems shall be installed in a workmanlike manner which is in accordance with this code, applicable standards, and the manufacturer's installation instructions. All materials shall be installed so as not to adversely affect the systems and equipment or the structure of the building, and in compliance with all laws and other provisions of this code. All plumbing systems shall be in accordance with construction documents approved by the Authority Having Jurisdiction.

309.5 Sound Transmission. Plumbing piping systems shall be designed and installed in conformance with sound limitations as required in the building code.

309.6 Dead Legs. Dead legs shall have a method of flushing.

310.0 Prohibited Fittings and Practices.

310.1 Fittings. No double hub fitting, single or double tee branch, single or double tapped tee branch, side inlet quarter bend, running thread, band, or saddle shall be used as a drainage fitting, except that a double hub sanitary tapped tee shall be permitted to be used on a vertical line as a fixture connection.

310.2 Drainage and Vent Piping. No drainage or vent piping shall be drilled and tapped for the purpose of making connections thereto, and no cast-iron soil pipe shall be threaded.

310.3 Waste Connection. No waste connection shall be made to a closet bend or stub of a water closet or similar fixture.

310.4 Use of Vent and Waste Pipes. Except as hereinafter provided in Section 908.0 through Section 911.0, no vent pipe shall be used as a soil or waste pipe, nor shall a soil or waste pipe be used as a vent. Also, single-stack drainage and venting systems with unvented branch lines are prohibited.

310.5 Obstruction of Flow. No fitting, fixture and piping connection, appliance, device, or method of installation that obstructs or retards the flow of water, wastes, sewage, or air in the drainage or venting systems, in an amount exceeding the normal frictional resistance to flow, shall be used unless it is indicated as acceptable in this code or is approved in accordance with Section 301.2 of this code. The enlargement of a 3 inch (80 mm) closet bend or stub to 4 inches (100 mm) shall not be considered an obstruction.

310.6 Dissimilar Metals. Except for necessary valves, where intermembering or mixing of dissimilar metals occurs, the point of connection shall be confined to exposed or accessible locations.

310.7 Direction of Flow. Valves, pipes, and fittings shall be installed in correct relationship to the direction of flow.

310.8 Screwed Fittings. Screwed fittings shall be ABS, cast-iron, copper, copper alloy, malleable iron, PVC, steel, or other approved materials. Threads shall be tapped out of solid metal or molded in solid ABS or PVC.

311.0 Independent Systems.

311.1 General. The drainage system of each new building and new work installed in an existing building shall be separate and independent from that of any other building, and, where available, every building shall have an independent connection with a public or private sewer.

Exception: Where one building stands in the rear of another building on an interior lot, and no private sewer is available or can be constructed to the rear building through an adjoining court, yard, or driveway, the building drain from the front building shall be permitted to be extended to the rear building.

312.0 Protection of Piping, Materials, and Structures.

312.1 General. Piping passing under or through walls shall be protected from breakage. Piping passing through or under cinders or other corrosive materials shall be protected from external corrosion in an approved manner. Approved provisions shall be made for expansion of hot water piping. Voids around piping passing through concrete floors on the ground shall be sealed.

312.2 Installation. Piping in connection with a plumbing system shall be so installed that piping or connections will not be subject to undue strains or stresses, and provisions shall be made for expansion, contraction, and structural settlement. No plumbing piping shall be directly embedded in concrete or masonry. No structural member shall be seriously weakened or impaired by cutting, notching, or otherwise, as defined in the building code.

312.3 Building Sewer and Drainage Piping. No building sewer or other drainage piping or part thereof, constructed of materials other than those approved for use under or within a building, shall be installed under or within 2 feet (610 mm) of a building or structure, or less than 1 foot (305 mm) below the surface of the ground.

312.4 Corrosion, Erosion, and Mechanical Damage. Piping subject to corrosion, erosion, or mechanical damage shall be protected in an approved manner.

312.5 Protectively Coated Pipe. Protectively coated pipe or tubing shall be inspected and tested, and a visible void, damage, or imperfection to the pipe coating shall be repaired in an approved manner.

312.6 Freezing Protection. No water, soil, or waste pipe shall be installed or permitted outside of a building, in attics or crawl spaces, or in an exterior wall unless, where necessary, adequate provision is made to protect such pipe from freezing.

312.7 Fire-Resistant Construction. Piping penetrations of fire-resistance-rated walls, partitions, floors, floor/ceiling assemblies, roof/ceiling assemblies, or shaft enclosures shall be protected in accordance with the requirements of the building code and Chapter 14, "Firestop Protection."

312.8 Waterproofing of Openings. Joints at the roof around pipes, ducts, or other appurtenances shall be made watertight by the use of lead, copper, galvanized iron, or other approved flashings or flashing material. Exterior wall openings shall be made watertight. Counterflashing shall not restrict the required internal cross-sectional area of the vent.

312.9 Steel Nail Plates. Plastic and copper or copper alloy piping penetrating framing members to within 1 inch (25.4 mm) of the exposed framing shall be protected by steel nail plates not less than No. 18 gauge (0.0478 inches) (1.2 mm) in thickness. The steel nail plate shall extend along the framing member not less than 1½ inches (38 mm) beyond the outside diameter of the pipe or tubing.

Exception: See Section 1210.4.3.

312.10 Sleeves. Sleeves shall be provided to protect piping through concrete and masonry walls, and concrete floors.

Exception: Sleeves shall not be required where openings are drilled or bored.

312.10.1 Building Loads. Piping through concrete or masonry walls shall not be subject to a load from building construction.

312.10.2 Exterior Walls. In exterior walls, annular space between sleeves and pipes shall be sealed and made watertight, as approved by the Authority Having Jurisdiction. A penetration through fire-resistive construction shall be in accordance with Section 312.7.

312.10.3 Firewalls. A pipe sleeve through a firewall shall have space around the pipe completely sealed with an approved fire-resistive material in accordance with other codes.

312.11 Structural Members. A structural member weakened or impaired by cutting, notching, or otherwise shall be reinforced, repaired, or replaced so as to be left in a safe structural condition in accordance with the requirements of the building code.

312.12 Rodentproofing. Strainer plates on drain inlets shall be designed and installed so that no opening exceeds ½ of an inch (12.7 mm) in the least dimension.

312.12.1 Meter Boxes. Meter boxes shall be constructed in such a manner as to restrict rodents or vermin from entering a building by following the service pipes from the box into the building.

312.12.2 Metal Collars. In or on buildings where openings have been made in walls, floors, or ceilings for the passage of pipes, such openings shall be closed and protected by the installation of approved metal collars securely fastened to the adjoining structure.

312.12.3 Tub Waste Openings. Tub waste openings in framed construction to crawl spaces at or below the first floor shall be protected by the installation of approved metal collars or metal screen securely fastened to the adjoining structure with no opening exceeding ½ of an inch (12.7 mm) in the least dimension.

313.0 Hangers and Supports.

313.1 General. Piping, fixtures, appliances, and appurtenances shall be supported in accordance with this code, the manufacturer's installation instructions, and in accordance with the Authority Having Jurisdiction.

313.2 Material. Hangers and anchors shall be of sufficient strength to support the weight of the pipe and its contents. Piping shall be isolated from incompatible materials.

313.3 Suspended Piping. Suspended piping shall be supported at intervals not to exceed those shown in Table 313.3.

313.4 Alignment. Piping shall be supported in such a manner as to maintain its alignment and prevent sagging.

313.5 Underground Installation. Piping in the ground shall be laid on a firm bed for its entire length; where other support is otherwise provided, it shall be approved in accordance with Section 301.2.

313.6 Hanger Rod Sizes. Hanger rod sizes shall be not smaller than those shown in Table 313.6.

TABLE 313.6
HANGER ROD SIZES

PIPE AND TUBE SIZE (inches)	ROD SIZE (inches)
½ – 4	⅜
5 – 8	½
10 – 12	⅝

For SI units: 1 inch = 25.4 mm

313.7 Gas Piping. Gas piping shall be supported by metal straps or hooks at intervals not to exceed those shown in Table 1210.3.5.1.

314.0 Trenching, Excavation, and Backfill.

314.1 Trenches. Trenches deeper than the footing of a building or structure, and paralleling the same, shall be located not less than 45 degrees (0.79 rad) from the bottom exterior edge of the footing, or as approved in accordance with Section 301.0.

314.2 Tunneling and Driving. Tunneling and driving shall be permitted to be done in yards, courts, or driveways of a building site. Where sufficient depth is available to permit, tunnels shall be permitted to be used between open-cut trenches.

Tunnels shall have a clear height of 2 feet (610 mm) above the pipe and shall be limited in length to one-half the depth of the trench, with a maximum length of 8 feet (2438 mm). Where pipes are driven, the drive pipe shall be not less than one size larger than the pipe to be laid.

314.3 Open Trenches. Excavations required to be made for the installation of a building drainage system or part thereof, within the walls of a building, shall be open trench work and shall be kept open until the piping has been inspected, tested, and accepted.

314.4 Excavations. Excavations shall be completely backfilled as soon after inspection as practicable. Precaution shall be taken to ensure compactness of backfill around piping without damage to such piping. Trenches shall be backfilled in thin layers to 12 inches (305 mm) above the top of the piping with clean earth, which shall not contain stones, boulders, cinder fill, frozen earth, construction debris, or other materials that will damage or break the piping or cause corrosive action. Mechanical devices such as bulldozers, graders, etc., shall be permitted to be then used to complete backfill to grade. Fill shall be properly compacted. Precautions shall be taken to ensure permanent stability for pipe laid in filled or made ground.

Underground thermoplastic pipe and fittings for sewers and other gravity flow applications shall be installed in accordance with this code and Section 314.4.1.

314.4.1 Installation of Thermoplastic Pipe and Fittings. Trench width for thermoplastic sewer pipe shall be not less than 1.25 times the outside diameter of the piping plus 12 inches (305 mm) or the outside diameter of the piping plus not less than 16 inches (406 mm). Thermoplastic piping shall be bedded in not less than 4 inches (102 mm) of granular fill supporting the piping. The backfill for thermoplastic piping shall be compacted along the sides of the piping in 6 inch (152 mm) layers and continue to not less than 12 inches (305 mm) above the piping. Compaction shall be not less than an 85 percent standard proctor density.

315.0 Joints and Connections.

315.1 Unions. Approved unions shall be permitted to be used in drainage piping where accessibly located in the trap seal or between a fixture and its trap; in the vent system, except underground or in wet vents; at any point in the water supply system; and in gas piping as permitted by Section 1212.5.1.

315.2 Prohibited Joints and Connections. A fitting or connection that has an enlargement, chamber, or recess with a ledge, shoulder, or reduction of pipe area that offers an obstruction to flow through the drain shall be prohibited.

316.0 Increasers and Reducers.

316.1 General. Where different sizes of pipes and fittings are to be connected, the proper size increasers or reducers or reducing fittings shall be used between the two sizes. Copper alloy or cast-iron body cleanouts shall not be used as a reducer or adapter from cast-iron drainage pipe to iron pipe size (IPS) pipe.

317.0 Food-Handling Establishments.

317.1 General. Food or drink shall not be stored, prepared, or displayed beneath soil or drain pipes unless those areas are protected against leakage or condensation from such pipes reaching the food or drink as described below. Where building design requires that soil or drain pipes be located over such areas, the installation shall be made with the least possible number of joints and shall be installed to connect to the nearest adequately sized vertical stack with the provisions as follows:

(1) Openings through floors over such areas shall be sealed watertight to the floor construction.

(2) Floor and shower drains installed above such areas shall be equipped with integral seepage pans.

(3) Soil or drain pipes shall be of an approved material as listed in Chapter 17 and Section 701.2. Materials shall comply with established standards. Cleanouts shall be extended through the floor construction above.

(4) Piping subject to operation at temperatures that will form condensation on the exterior of the pipe shall be thermally insulated.

(5) Where pipes are installed in ceilings above such areas, the ceiling shall be of the removable type or shall be provided with access panels to form a ready access for inspection of piping.

318.0 Test Gauges.

318.1 General. Tests in accordance with this code, which are performed utilizing dial gauges, shall be limited to gauges having the following pressure graduations or incrementations.

318.2 Pressure Tests (10 psi or less). Required pressure tests of 10 pounds-force per square inch (psi) (69 kPa) or less shall be performed with gauges of 0.10 psi (0.69 kPa) incrementation or less.

318.3 Pressure Tests (greater than 10 psi to 100 psi). Required pressure tests exceeding 10 psi (69 kPa) but less than or equal to 100 psi (689 kPa) shall be performed with gauges of 1 psi (7 kPa) incrementation or less.

318.4 Pressure Tests (exceeding 100 psi). Required pressure tests exceeding 100 psi (689 kPa) shall be performed with gauges incremented for 2 percent or less of the required test pressure.

318.5 Pressure Range. Test gauges shall have a pressure range not exceeding twice the test pressure applied.

319.0 Medical Gas and Vacuum Systems.

319.1 General. Such piping shall be in accordance with the requirements of Chapter 13. The Authority Having Jurisdiction shall require evidence of the competency of the installers and verifiers.

320.0 Rehabilitation of Piping Systems.

320.1 General. Where pressure piping systems are rehabilitated using an epoxy lining system, it shall be in accordance with ASTM F2831.

TABLE 313.3
HANGERS AND SUPPORTS

MATERIALS	TYPES OF JOINTS	HORIZONTAL	VERTICAL
Cast	Lead and Oakum	5 feet, except 10 feet where 10 foot lengths are installed[1, 2, 3]	Base and each floor, not to exceed 15 feet
	Compression Gasket	Every other joint, unless over 4 feet then support each joint[1, 2, 3]	Base and each floor, not to exceed 15 feet
Cast-Iron Hubless	Shielded Coupling	Every other joint, unless over 4 feet then support each joint[1,2,3,4]	Base and each floor, not to exceed 15 feet
Copper & Copper Alloys	Soldered, Brazed, Threaded, or Mechanical	1½ inches and smaller, 6 feet; 2 inches and larger, 10 feet	Each floor, not to exceed 10 feet[5]
Steel Pipe for Water or DWV	Threaded or Welded	¾ inch and smaller, 10 feet; 1 inch and larger, 12 feet	Every other floor, not to exceed 25 feet[5]
Steel Pipe for Gas	Threaded or Welded	½ inch, 6 feet; ¾ inch and 1 inch, 8 feet; 1¼ inches and larger, 10 feet	½ inch, 6 feet; ¾ inch and 1 inch, 8 feet; 1¼ inches every floor level
Schedule 40 PVC and ABS DWV	Solvent Cemented	All sizes, 4 feet; allow for expansion every 30 feet[3]	Base and each floor; provide mid-story guides; provide for expansion every 30 feet
CPVC	Solvent Cemented	1 inch and smaller, 3 feet; 1¼ inches and larger, 4 feet	Base and each floor; provide mid-story guides
CPVC-AL-CPVC	Solvent Cemented	½ inch, 5 feet; ¾ inch, 65 inches; 1 inch, 6 feet	Base and each floor; provide mid-story guides
Lead	Wiped or Burned	Continuous Support	Not to exceed 4 feet
Steel	Mechanical	In accordance with standards acceptable to the Authority Having Jurisdiction	
PEX	Cold Expansion, Insert and Compression	1 inch and smaller, 32 inches; 1¼ inches and larger, 4 feet	Base and each floor; provide mid-story guides
PEX-AL-PEX	Metal Insert and Metal Compression	½ inch ¾ inch 1 inch } All sizes 98 inches	Base and each floor; provide mid-story guides
PE-AL-PE	Metal Insert and Metal Compression	½ inch ¾ inch 1 inch } All sizes 98 inches	Base and each floor; provide mid-story guides
PE-RT	Insert and Compression	1 inch and smaller, 32 inches; 1¼ inches and larger, 4 feet	Base and each floor; provide mid-story guides
Polypropylene (PP)	Fusion weld (socket, butt, saddle, electrofusion), threaded (metal threads only), or mechanical	1 inch and smaller, 32 inches; 1¼ inches and larger, 4 feet	Base and each floor; provide mid-story guides

For SI units: 1 inch = 25.4 mm, 1 foot = 304.8 mm

Notes:

[1] Support adjacent to joint, not to exceed 18 inches (457 mm).

[2] Brace not to exceed 40 foot (12 192 mm) intervals to prevent horizontal movement.

[3] Support at each horizontal branch connection.

[4] Hangers shall not be placed on the coupling.

[5] Vertical water lines shall be permitted to be supported in accordance with recognized engineering principles with regard to expansion and contraction, where first approved by the Authority Having Jurisdiction.

CHAPTER 4

PLUMBING FIXTURES AND FIXTURE FITTINGS

401.0 General.

401.1 Applicability. This chapter shall govern the materials and installation of plumbing fixtures, including faucets and fixture fittings, and the minimum number of plumbing fixtures required based on occupancy.

401.2 Quality of Fixtures. Plumbing fixtures shall be constructed of dense, durable, non-absorbent materials and shall have smooth, impervious surfaces, free from unnecessary concealed fouling surfaces.

402.0 Installation.

402.1 Cleaning. Plumbing fixtures shall be installed in a manner to afford easy access for repairs and cleaning. Pipes from fixtures shall be run to the nearest wall.

402.2 Joints. Where a fixture comes in contact with the wall or floor, the joint between the fixture and the wall or floor shall be made watertight.

402.3 Securing Fixtures. Floor-outlet or floor-mounted fixtures shall be rigidly secured to the drainage connection and to the floor, where so designed, by screws or bolts of copper, copper alloy, or other equally corrosion-resistant material.

402.4 Wall-Hung Fixtures. Wall-hung fixtures shall be rigidly supported by metal supporting members so that no strain is transmitted to the connections. Floor-affixed supports for off-the-floor plumbing fixtures for public use shall comply with ASME A112.6.1M. Framing-affixed supports for off-the-floor water closets with concealed tanks shall comply with ASME A112.6.2. Flush tanks and similar appurtenances shall be secured by approved non-corrosive screws or bolts.

402.5 Setting. Fixtures shall be set level and in proper alignment with reference to adjacent walls. No water closet or bidet shall be set closer than 15 inches (381 mm) from its center to a side wall or obstruction or closer than 30 inches (762 mm) center to center to a similar fixture. The clear space in front of a water closet, lavatory, or bidet shall be not less than 24 inches (610 mm). No urinal shall be set closer than 12 inches (305 mm) from its center to a side wall or partition or closer than 24 inches (610 mm) center to center.

Exception: The installation of paper dispensers or accessibility grab bars shall not be considered obstructions.

402.6 Flanged Fixture Connections. Fixture connections between drainage pipes and water closets, floor outlet service sinks and urinals shall be made using an approved copper alloy, hard lead, ABS, PVC, or iron flanges caulked, soldered, solvent cemented; rubber compression gaskets; or screwed to the drainage pipe. The connection shall be bolted with an approved gasket, washer, or setting compound between the fixture and the connection. The bottom of the flange shall be set on an approved firm base.

Wall-mounted water closet fixtures shall be securely bolted to an approved carrier fitting. The connecting pipe between the carrier fitting and the fixture shall be an approved material and designed to accommodate an adequately sized gasket. Gasket material shall be neoprene, felt, or similar approved types.

402.6.1 Closet Rings (Closet Flanges). Closet rings (closet flanges) for water closets or similar fixtures shall be of an approved type and shall be copper alloy, copper, hard lead, cast-iron, galvanized malleable iron, ABS, PVC, or other approved materials. Each such closet ring (closet flange) shall be approximately 7 inches (178 mm) in diameter and, where installed, shall, together with the soil pipe, present a 1½ inch (38 mm) wide flange or face to receive the fixture gasket or closet seal.

Caulked-on closet rings (closet flanges) shall be not less than ¼ of an inch (6.4 mm) thick and not less than 2 inches (51 mm) in overall depth.

Closet rings (closet flanges) shall be burned or soldered to lead bends or stubs, shall be caulked to cast-iron soil pipe, shall be solvent cemented to ABS and PVC, and shall be screwed or fastened in an approved manner to other materials.

Closet bends or stubs shall be cut-off to present a smooth surface even with the top of the closet ring before the rough inspection is called.

Closet rings (closet flanges) shall be adequately designed and secured to support fixtures connected thereto.

402.6.2 Securing Closet Flanges. Closet screws, bolts, washers, and similar fasteners shall be of copper alloy, copper, or other listed equally corrosion-resistant materials. Screws and bolts shall be of a size and number to properly support the fixture installed.

402.6.3 Securing Floor-Mounted, Back-Outlet Water Closet Bowls. Floor-mounted, back-outlet water closet bowls shall be set level with an angle of 90 degrees (1.57 rad) between the floor and wall at the centerline of the fixture outlet. The floor and wall shall have a flat mounting surface not less than 5 inches (127 mm) to the right and left of the fixture outlet centerline. The fixture shall be secured to the wall outlet flange or drainage connection and the floor by corrosion-resistant screws or bolts. The closet flange shall be secured to a firm base.

Where floor-mounted, back-outlet water closets are used, the soil pipe shall be not less than 3 inches (80 mm) in diameter. Offset, eccentric, or reducing closet flanges shall not be used.

402.7 Supply Fittings. The supply lines and fittings for every plumbing fixture shall be so installed as to prevent backflow in accordance with Chapter 6.

402.8 Installation. Fixtures shall be installed in accordance with the manufacturer's installation instructions.

402.9 Design and Installation of Plumbing Fixtures. Plumbing fixtures shall be installed in accordance with the manufacturer's installation instructions. The means of backflow prevention shall not be compromised by the designated fixture fitting mounting surface.

402.10 Slip Joint Connections. Fixtures having concealed slip joint connections shall be provided with an access panel or utility space not less than 12 inches (305 mm) in its least dimension and so arranged without obstructions as to make such connections accessible for inspection and repair.

402.11 Future Fixtures. Where provisions are made for the future installation of fixtures, those provided for shall be considered in determining the required sizes of the drain and water supply piping. Construction for future installations shall be terminated with a plugged fitting or fittings. Where the plugged fitting is at the point where the trap of a fixture is installed, the plumbing system for such fixture shall be complete and be in accordance with the plumbing requirements of this code.

403.0 Accessible Plumbing Facilities.

403.1 General. Where accessible facilities are required in applicable building regulations, the facilities shall be installed in accordance with those regulations.

403.2 Fixtures and Fixture Fittings for Persons with Disabilities. Plumbing fixtures and fixture fittings for persons with disabilities shall be in accordance with ICC A117.1 and the applicable standards referenced in Chapter 4.

403.3 Exposed Pipes and Surfaces. Water supply and drain pipes under accessible lavatories and sinks shall be insulated or otherwise be configured to protect against contact. Protectors, insulators, or both shall comply with ASME A112.18.9 or ASTM C1822.

404.0 Waste Fittings and Overflows.

404.1 Waste Fittings. Waste fittings shall comply with ASME A112.18.2/CSA B125.2, ASTM F409 or Table 701.2 for aboveground drainage piping and fittings.

404.2 Overflows. Where a fixture is provided with an overflow, the waste shall be so arranged that the standing water in the fixture shall not rise in the overflow where the stopper is closed or remain in the overflow where the fixture is empty. The overflow pipe from a fixture shall be connected to the house or inlet side of the fixture trap, except that overflow on flush tanks shall be permitted to discharge into the water closets or urinals served by them, but it shall be unlawful to connect such overflows with any other part of the drainage system.

405.0 Prohibited Fixtures.

405.1 Prohibited Water Closets. Water closets having an invisible seal or an unventilated space or having walls which are not thoroughly washed at each discharge shall be prohibited. A water closet that might permit siphonage of the contents of the bowl back into the tank shall be prohibited.

405.2 Prohibited Urinals. Trough urinals and urinals with an invisible seal shall be prohibited.

405.3 Miscellaneous Fixtures. Fixed wooden, or tile wash trays or sinks for domestic use shall not be installed in a building designed or used for human habitation. No sheet metal-lined wooden bathtub shall be installed or reconnected. No dry or chemical closet (toilet) shall be installed in a building used for human habitation unless first approved by the Health Officer.

406.0 Special Fixtures and Specialties.

406.1 Water and Waste Connections. Baptisteries, ornamental and lily ponds, aquaria, ornamental fountain basins, and similar fixtures and specialties requiring water, waste connections, or both shall be submitted for approval to the Authority Having Jurisdiction prior to installation.

406.2 Special Use Sinks. Restaurant kitchen and other special use sinks shall be permitted to be made of approved-type bonderized and galvanized sheet steel of not less than No. 16 U.S. gauge (0.0635 inches) (1.6 mm). Sheet-metal plumbing fixtures shall be adequately designed, constructed, and braced in an approved manner to accomplish their intended purpose.

406.3 Special Use Fixtures. Special use fixtures shall be made of one of the following:

(1) Soapstone

(2) Chemical stoneware

(3) Copper-based alloy

(4) Nickel-based alloy

(5) Corrosion-resistant steel

(6) Other materials suited for the intended use of the fixture

406.4 Zinc Alloy Components. Zinc alloy components shall comply with applicable nationally recognized standards and shall be used in accordance with their listing.

407.0 Lavatories.

407.1 Application. Lavatories shall comply with ASME A112.19.1/CSA B45.2, ASME A112.19.2/CSA B45.1, ASME A112.19.3/CSA B45.4, ASME A112.19.12, CSA B45.5/IAPMO Z124, CSA B45.8/IAPMO Z403, CSA B45.11/IAPMO Z401 or CSA B45.12/IAPMO Z402. Group wash fixtures shall comply with the requirements of Section 401.2. Every 20 inches (508 mm) of rim space of a group wash fixture shall be considered as one lavatory for determining the number of lavatories required in accordance with Table 422.1. Lavatory assemblies with automatic soap dispensers, faucets, or hand dryers shall comply with IAPMO IGC 127.

407.2 Water Consumption. The maximum water flow rate of faucets shall comply with Section 407.2.1 and Section 407.2.2.

407.2.1 Maximum Flow Rate. The maximum flow rate for public lavatory faucets shall not exceed 0.5 gpm at 60 psi (1.9 L/m at 414 kPa) and 2.2 gpm at 60 psi (8.3 L/m at 414 kPa) for private lavatory faucets.

407.2.2 Metering Faucets. Metered faucets shall deliver a maximum of 0.25 gallons (0.95 L) per metering cycle.

407.3 Limitation of Hot Water Temperature for Public Lavatories. Hot water delivered from public-use lavatories shall be limited to a maximum temperature of 120°F (49°C). The maximum temperature shall be regulated by one of the following means:

(1) A limiting device conforming to either ASSE 1070/ASME A112.1070/CSA B125.70, or

(2) A water heater conforming to ASSE 1084.

407.4 Transient Public Lavatories. Self-closing or metering faucets shall be installed on lavatories intended to serve the transient public, such as those in, but not limited to service stations, train stations, airports, restaurants, and convention halls.

407.5 Waste Outlet. Lavatories shall have a waste outlet and fixture tailpiece not less than 1¼ inches (32 mm) in diameter. Continuous wastes and fixture tailpieces shall be constructed from the materials specified in Section 701.4. Waste outlets shall be provided with an approved stopper or strainer.

407.6 Overflow. Where overflows are provided, they shall be installed in accordance with Section 404.2.

408.0 Showers.

408.1 Application. Manufactured shower receptors and shower bases shall comply with ASME A112.19.1/CSA B45.2, ASME A112.19.2/CSA B45.1, ASME A112.19.3/CSAB45.4, CSA B45.12/IAPMO Z402, or CSA B45.5/IAPMO Z124. Prefabricated shower enclosures shall comply with IAPMO IGC 154.

408.2 Water Consumption. Showerheads shall have a maximum flow rate of not more than 2.5 gpm at 80 psi (9.5 L/m at 552 kPa).

408.3 Individual Shower and Tub-Shower Combination Control Valves. Showers and tub-shower combinations shall be provided with individual control valves of the pressure balance, thermostatic, or combination pressure balance/thermostatic mixing valve type that provide scald and thermal shock protection for the rated flow rate of the installed showerhead. These valves shall be installed at the point of use and comply with ASSE 1016/ASME A112.1016/CSA B125.16 or ASME A112.18.1/CSA B125.1.

408.3.1 Gang Showers. Where gang showers are supplied with a single temperature-controlled water supply pipe, it shall be controlled by a mixing valve that complies with ASSE 1069.

408.3.2 Temperature Limiting. The maximum water temperature discharging from an individual showerhead shall be limited to 120°F (49°C) by one of the following methods:

(1) A shower or tub/shower combination valve conforming to ASSE 1016/ASME A112.1016/CSA B125.16 where either:

  (a) The valve is field-adjusted to the required maximum temperature, or

  (b) The handle position, stop, or temperature limiting control is set in accordance with the manufacturer's instructions to the required maximum temperature.

(2) For gang showers supplied by a single water supply pipe, a mixing valve that conforms to ASSE 1069 that is field-adjusted to the required maximum temperature.

(3) A limiting device conforming to either ASSE 1070/ASME A112.1070/CSA B125.70 or CSA B125.3.

(4) A water heater conforming to ASSE 1084.

(5) A temperature actuated flow reduction device conforming to ASSE 1062.

408.4 Waste Outlet. Showers shall have a waste outlet and fixture tailpiece not less than 2 inches (50 mm) in diameter. Fixture tailpieces shall be constructed from the materials specified in Section 701.2 for drainage piping. Strainers serving shower drains shall have a waterway at least equivalent to the area of the tailpiece.

408.5 Finished Curb or Threshold. Where a shower receptor has a finished dam, curb, or threshold, it shall be not less than 1 inch (25.4 mm) lower than the sides and back of such receptor. In no case, shall a dam or threshold be less than 2 inches (51 mm) or exceeding 9 inches (229 mm) in depth where measured from the top of the dam or threshold to the top of the drain. Each such receptor shall be provided with an integral nailing flange to be located where the receptor meets the vertical surface of the finished interior of the shower compartment. The flange shall be watertight and extend vertically not less than 1 inch (25.4 mm) above the top of the sides of the receptor. The finished floor of the receptor shall slope uniformly from the sides towards the drain not less than ⅛ inch per foot (10.4 mm/m), nor more than ½ inch per foot (41.6 mm/m).

Thresholds shall be of sufficient width to accommodate a minimum 22 inch (559 mm) door. Shower doors shall open so as to maintain not less than a 22 inch (559 mm) unobstructed opening for egress. Where there is a shower without a threshold, the floor space within the same room shall be considered a wet location and shall comply with the requirements of the building, residential, and electrical codes.

Exceptions:

(1) Showers in accordance with Section 403.2.

(2) A cast-iron shower receptor flange shall be not less than 0.3 of an inch (7.62 mm) in height.

(3) For flanges not used as a means of securing, the sealing flange shall be not less than 0.3 of an inch (7.62 mm) in height.

408.6 Shower Compartments. Shower compartments, regardless of shape, shall have a minimum finished interior of 1024 square inches (0.6606 m^2) and shall also be capable of encompassing a 30 inch (762 mm) circle. The minimum required area and dimensions shall be measured at a height equal to the top of the threshold and a point tangent to its centerline. The area and dimensions shall be maintained to a point of not less than 70 inches (1778 mm) above the shower drain outlet with no protrusions other than the fixture valve or valves, showerheads, soap dishes, shelves, and safety grab bars, or rails. Fold-down seats in accessible shower stalls shall be permitted to protrude into the 30 inch (762 mm) circle.

Exceptions:

(1) Showers that are designed to be in accordance with ICC A117.1.

(2) The minimum required area and dimension shall not apply for a shower receptor having overall dimensions of not less than 30 inches (762 mm) in width and 60 inches (1524 mm) in length.

408.7 Lining for Showers and Receptors. Shower receptors built on-site shall be watertight and shall be constructed from approved-type dense, nonabsorbent, and noncorrosive materials. Each such receptor shall be adequately reinforced, shall be provided with an approved flanged floor drain designed to make a watertight joint on the floor, and shall have smooth, impervious, and durable surfaces.

Shower receptors shall have the subfloor and rough side of walls to a height of not less than 3 inches (76 mm) above the top of the finished dam or threshold shall be first lined with sheet plastic, lead, or copper, or shall be lined with other durable and watertight materials. Showers that are provided with a built in place, permanent seat or seating area that is located within the shower enclosure, shall be first lined with sheet plastic, lead, copper, or shall be lined with other durable and watertight materials that extend not less than 3 inches (76 mm) above horizontal surfaces of the seat or the seating area.

Lining materials shall be pitched ¼ inch per foot (20.8 mm/m) to weep holes in the subdrain of a smooth and solidly formed subbase. Such lining materials shall extend upward on the rough jambs of the shower opening to a point not less than 3 inches (76 mm) above the horizontal surfaces of the seat or the seating area, the top of the finished dam or threshold and shall extend outward over the top of the permanent seat, permanent seating area, or rough threshold and be turned over and fastened on the outside face of both the permanent seat, permanent seating area, or rough threshold and the jambs.

Nonmetallic shower subpans or linings shall be permitted to be built up on the job site of not less than three layers of standard grade 15 pound (6.8 kg) asphalt impregnated roofing felt. The bottom layer shall be fitted to the formed subbase and each succeeding layer thoroughly hot-mopped to that below. Corners shall be carefully fitted and shall be made strong and watertight by folding or lapping, and each corner shall be reinforced with suitable webbing hot-mopped in place.

Folds, laps, and reinforcing webbing shall extend not less than 4 inches (102 mm) in all directions from the corner, and webbing shall be of approved type and mesh, producing a tensile strength of not less than 50 pounds per square foot (lb/ft^2) (244 kg/m^2) in either direction. Nonmetallic shower subpans or linings shall be permitted to consist of multilayers of other approved equivalent materials suitably reinforced and carefully fitted in place on the job site as elsewhere required in this section.

Linings shall be properly recessed and fastened to the approved backing so as not to occupy the space required for the wall covering, and shall not be nailed or perforated at a point that is less than 1 inch (25.4 mm) above the finished dam or threshold. An approved type subdrain shall be installed with a shower subpan or lining. Each such subdrain shall be of the type that sets flush with the subbase and shall be equipped with a clamping ring or other device to make a tight connection between the lining and the drain. The subdrain shall have weep holes into the waste line. The weep holes located in the subdrain clamping ring shall be protected from clogging.

408.7.1 PVC Sheets. Plasticized polyvinyl chloride (PVC) sheets shall conform to ASTM D4551. Sheets shall be joined by solvent cementing in accordance with the manufacturer's installation instructions.

408.7.2 Chlorinated Polyethylene (CPE) Sheets. Nonplasticized chlorinated polyethylene sheets shall conform to ASTM D4068. The liner shall be joined in accordance with the manufacturer's installation instructions.

408.7.3 Sheet Lead. Sheet lead shall weigh not less than 4 lb/ft^2 (19.5 kg/m^2) and shall be coated with an asphalt paint or other approved coating. The lead sheet shall be insulated from conducting substances, other than the connecting drain, by 15 pound (6.8 kg) asphalt felt or an equivalent. Sheet lead shall be joined by burning.

408.7.4 Sheet Copper. Sheet copper shall comply with ASTM B152 and shall weigh not less than 12 ounces per square foot (oz/ft^2) (3.7 kg/m^2) or No. 24 B & S Gauge (0.02 inches) (0.51 mm). The copper sheet shall be insulated from conducting substances, other than the connecting drain, by 15 pound (6.8 kg) asphalt felt or an equivalent. Sheet copper shall be joined by brazing or soldering.

408.7.5 Tests for Shower Receptors. Shower receptors shall be tested for watertightness by filling with water to the level of the rough threshold. The test plug shall be so placed that both upper and under sides of the subpan shall be subjected to the test at the point where it is clamped to the drain.

408.8 Public Shower Floors. Floors of public shower rooms shall have a nonskid surface and shall be drained in such a manner that wastewater from one bather shall not pass over areas occupied by other bathers. Gutters in public or gang shower rooms shall have rounded corners for easy cleaning and shall be sloped not less than 2 percent toward drains. Drains in gutters shall be spaced at a maximum of 8 feet (2438 mm) from sidewalls nor more than 16 feet (4877 mm) apart.

408.9 Location of Valves and Heads. Control valves and showerheads shall be located on the sidewall of shower compartments or otherwise arranged so that the showerhead does not discharge directly at the entrance to the compartment so that the bather can adjust the valves before stepping into the shower spray.

408.10 Water Supply Riser. A water supply riser from the shower valve to the showerhead outlet, whether exposed or not, shall be securely attached to the structure.

409.0 Bathtubs and Whirlpool Bathtubs.

409.1 Application. Bathtubs shall comply with ASME A112.19.1/CSA B45.2, ASME A112.19.2/CSA B45.1,

ASME A112.19.3/CSA B45.4, CSA B45.5/IAPMO Z124, or CSA B45.12/IAPMO Z402. Whirlpool bathtubs shall comply with ASME A112.19.7/CSA B45.10. Pressure sealed doors within a bathtub or whirlpool bathtub enclosure shall comply with ASME A112.19.15.

409.2 Waste Outlet. Bathtubs and whirlpool bathtubs shall have a waste outlet and fixture tailpiece not less than 1½ inches (40 mm) in diameter. Fixture tailpieces shall be constructed from the materials specified in Section 701.2 for drainage piping. Waste outlets shall be provided with an approved stopper or strainer.

409.3 Overflow. Where overflows are provided, they shall be installed in accordance with Section 404.2.

409.4 Limitation of Hot Water Temperature in Bathtubs and Whirlpool Bathtubs. The maximum hot water temperature discharging from the bathtub and whirlpool bathtub filler shall be limited to 120°F (49°C). The maximum temperature shall be regulated by one of the following means:

(1) A limiting device conforming to either ASSE 1070/ASME A112.1070/CSA B125.70 or CSA B125.3.

(2) A water heater conforming to ASSE 1084.

409.5 Backflow Protection. The water supply to a bathtub and whirlpool bathtub filler valve shall be protected by an air gap or in accordance with Section 417.0.

409.6 Installation and Access. Bathtubs and whirlpool bathtubs shall be installed in accordance with the manufacturer's installation instructions. Access openings shall be of a size and opening to permit the removal and replacement of the circulation pump.

Whirlpool pump access located in the crawl space shall be located not more than 20 feet (6096 mm) from an access door, trap door, or crawl hole.

The circulation pump shall be located above the crown weir of the trap.

The pump and the circulation piping shall be self-draining to minimize water retention. Suction fittings on whirlpool bathtubs shall comply with ASME A112.19.7/CSA B45.10.

409.6.1 Flexible PVC Hoses and Tubing. Flexible PVC hoses and tubing intended to be used on whirlpool bathtub water circulation systems or pneumatic systems shall comply with IAPMO Z1033.

410.0 Bidets.

410.1 Application. Bidets shall comply with ASME A112.19.2/CSA B45.1 or ASME A112.19.3/CSA B45.4.

410.2 Backflow Protection. The water supply to the bidet shall be protected by an air gap or in accordance with Section 603.3.2, Section 603.3.5, or Section 603.3.6.

410.3 Limitation of Water Temperature in Bidets. The maximum hot water temperature discharging from a bidet shall be limited to 110°F (43°C). The maximum temperature shall be regulated by one of the following means:

(1) A limiting device conforming to either ASSE 1070/ASME A112.1070/CSA B125.70 or CSA B125.3.

(2) A water heater conforming to ASSE 1084.

411.0 Water Closets.

411.1 Application. Water closets shall comply with ASME A112.19.2/CSA B45.1, ASME A112.19.3/CSA B45.4, or CSA B45.5/IAPMO Z124. Water closet bowls for public use shall be of the elongated type. In nurseries, schools, and other similar places where plumbing fixtures are provided for the use of children less than 6 years of age, water closets shall be of a size and height suitable for children's use.

411.2 Water Consumption. Water closets shall have a maximum consumption not to exceed 1.6 gallons (6.0 Lpf) of water per flush.

411.2.1 Dual Flush Water Closets. Dual flush water closets shall comply with ASME A112.19.14. The effective flush volume for dual flush water closets shall be defined as the composite, average flush volume of two reduced flushes and one full flush.

411.2.2 Flushometer Valve Activated Water Closets. Flushometer valve activated water closets shall have a maximum flush volume of 1.6 gallons (6.0 Lpf) of water per flush.

411.3 Water Closet Seats. Water closet seats shall be properly sized for the water closet bowl type, and shall be of smooth, non-absorbent material. Seats, for public use, shall be of the elongated type and either of the open front type or have an automatic seat cover dispenser. Plastic seats shall comply with IAPMO Z124.5.

411.4 Personal Hygiene Devices. Water closets with integral personal hygiene devices shall comply with ASME A112.4.2/CSA B45.16.

412.0 Urinals.

412.1 Application. Urinals shall comply with ASME A112.19.2/CSA B45.1, ASME A112.19.19, or CSA B45.5/IAPMO Z124. Urinals shall have an average water consumption not to exceed 1 gallon (3.8 Lpf) of water per flush.

412.1.1 Nonwater Urinals. Nonwater urinals shall have a liquid barrier sealant to maintain a trap seal. Nonwater urinals shall permit the uninhibited flow of waste through the urinal to the sanitary drainage system. Nonwater urinals shall be cleaned and maintained in accordance with the manufacturer's instructions after installation. Where nonwater urinals are installed, not less than one water supplied fixture rated at not less than 1 water supply fixture unit (WSFU) shall be installed upstream on the same drain line to facilitate drain line flow and rinsing. Where nonwater urinals are installed, they shall have a water distribution line rough-in to each individual urinal location to allow for the installation of an approved backflow prevention device in the event of a retrofit.

412.1.2 Nonwater Urinals with Drain Cleansing Action. Nonwater urinals with drain cleansing action shall comply with ASME A112.19.19 and shall be cleaned, maintained and installed in accordance with the manufacturer's installation instructions.

412.2 Backflow Protection. A water supply to a urinal shall be protected by an approved-type vacuum breaker or other approved backflow prevention device in accordance with Section 603.5.

413.0 Flushing Devices.

413.1 Where Required. Each water closet, urinal, clinical sink, or other plumbing fixture that depends on trap siphonage to discharge its waste contents shall be provided with a flushometer valve, flushometer tank, or flush tank designed and installed so as to supply water in sufficient quantity and rate of flow to flush the contents of the fixture to which it is connected, to cleanse the fixture, and to refill the fixture trap, without excessive water use. Flushing devices shall comply with the antisiphon requirements in accordance with Section 603.5.

413.2 Flushometer Valves. Flushometer valves and flushometer tanks shall comply with ASSE 1037/ASME A112.1037/CSA B125.37, and shall be installed in accordance with Section 603.5.1. No manually controlled flushometer valve shall be used to flush more than one urinal, and each such urinal flushometer valve shall be an approved, self-closing type discharging a predetermined quantity of water. Flushometers shall be installed so that they will be accessible for repair. Flushometer valves shall not be used where the water pressure is insufficient to operate them properly. Where the valve is operated, it shall complete the cycle of operation automatically, opening fully, and closing positively under the line water pressure. Each flushometer shall be provided with a means for regulating the flow through it.

413.3 Flush Tanks. Flush tanks for manual flushing shall be equipped with a flush valve that complies with ASME A112.19.5/CSA B45.15 and an antisiphon fill valve (ballcock) that complies with ASSE 1002/ASME A112.1002/CSA B125.12 and installed in accordance with Section 603.5.2.

413.4 Water Supply for Flush Tanks. An adequate quantity of water shall be provided to flush and clean the fixture served. The water supply for flushing tanks and flushometer tanks equipped for manual flushing shall be controlled by a float valve or other automatic device designed to refill the tank after each discharge and to shut completely off the water flow to the tank where the tank is filled to operational capacity. Provision shall be made to automatically supply water to the fixture to refill the trap seal after each flushing.

413.5 Overflows in Flush Tanks. Flush tanks shall be provided with overflows discharging into the water closet or urinal connected thereto. Overflows supplied as original parts with the fixture shall be of sufficient size to prevent tank flooding at the maximum rate at which the tank is supplied with water under normal operating conditions and where installed in accordance with the manufacturer's installation instructions.

414.0 Dishwashing Machines.

414.1 Application. Domestic dishwashing machines shall comply with UL 749. Commercial dishwashing machines shall comply with NSF 3 and UL 921.

414.2 Backflow Protection. The water supply connection to a commercial dishwashing machine shall be protected by an air gap or a backflow prevention device in accordance with Section 603.3.2, Section 603.3.5, Section 603.3.6, or that complies with ASSE 1004.

414.3 Drainage Connection. Domestic dishwashing machines shall discharge indirectly through an air gap fitting in accordance with Section 807.3 into a waste receptor, a wye branch fitting on the tailpiece of a kitchen sink, or dishwasher connection of a food waste disposer. Commercial dishwashing machines shall discharge indirectly through an air break or direct connection. The indirect discharge for commercial dishwashing machines shall be in accordance with Section 807.1, and the direct discharge shall be in accordance with Section 704.3.

415.0 Drinking Fountains.

415.1 Application. Drinking fountains shall be self-closing and comply with ASME A112.19.1/CSA B45.2, ASME A112.19.2/CSA B45.1, or ASME A112.19.3/CSA B45.4. Drinking fountains shall also comply with NSF 61. Permanently installed electric water coolers shall also comply with UL 399.

415.2 Drinking Fountain Alternatives. Where food is consumed indoors, water stations shall be permitted to be substituted for drinking fountains. Bottle filling stations shall be permitted to be substituted for drinking fountains up to 50 percent of the requirements for drinking fountains. Drinking fountains shall not be required for an occupant load of 30 or less.

415.3 Drainage Connection. Drinking fountains shall be permitted to discharge directly into the drainage system or indirectly through an air break in accordance with Section 809.1.

415.4 Location. Drinking fountains shall not be installed in toilet rooms.

416.0 Emergency Eyewash and Shower Equipment.

416.1 Application. Emergency eyewash and shower equipment shall comply with ISEA Z358.1.

416.2 Water Supply. Emergency eyewash and shower equipment shall not be limited in the water supply flow rates. Where hot and cold water is supplied to an emergency shower or eyewash station, the temperature of the water supply shall be controlled by a temperature actuated mixing valve complying with ASSE 1071. Where water is supplied directly to an emergency shower or eyewash station from a water heater, the water heater shall comply with ASSE 1085. The flow rate, discharge pattern, and temperature of flushing fluids shall be provided in accordance with ISEA Z358.1.

416.3 Installation. Emergency eyewash and shower equipment shall be installed in accordance with the manufacturer's installation instructions.

416.4 Location. Emergency eyewash and shower equipment shall be located on the same level as the hazard and accessible for immediate use. The path of travel shall be free of obstructions and shall be clearly identified with signage.

416.5 Drain. A drain shall not be required for emergency eyewash or shower equipment. Where a drain is provided, the discharge shall be in accordance with Section 811.0.

417.0 Faucets and Fixture Fittings.

417.1 Application. Faucets and fixture fittings shall comply with ASME A112.18.1/CSA B125.1. Fixture fittings covered under the scope of NSF 61 shall comply with the requirements of NSF 61.

417.2 Deck Mounted Bath/Shower Valves. Deck mounted bath/shower transfer valves with integral backflow protection shall comply with ASME A112.18.1/CSA B125.1. This shall include handheld showers, and other bathing appliances mounted on the deck of bathtubs or other bathing appliances that incorporate a hose or pull out feature.

417.3 Handheld Showers. Handheld showers shall comply with ASME A112.18.1/CSA B125.1. Handheld showers with integral backflow protection shall comply with ASME A112.18.1/CSA B125.1 or shall have a backflow prevention device that complies with ASME A112.18.3 or ASSE 1014.

417.4 Faucets and Fixture Fittings with Hose Connected Outlets. Faucets and fixture fittings with pull out spout shall comply with ASME A112.18.1/CSA B125.1. Faucets and fixture fittings with pull out spouts with integral backflow protection shall comply with ASME A112.18.1/CSA B125.1 or shall have a backflow preventer device that complies with ASME A112.18.3.

417.5 Separate Controls for Hot and Cold Water. Where two separate handles control the hot and cold water, the left-hand control of the faucet where facing the fixture fitting outlet shall control the hot water. Faucets and diverters shall be connected to the water distribution system so that hot water corresponds to the left side of the fixture fitting.

Single-handle mixing valves installed in showers and tub-shower combinations shall have the flow of hot water corresponding to the markings on the fixture fitting.

417.6 Low-Pressure Water Dispenser. Beverage faucets shall comply with ASME A112.18.1/CSA B125.1. Low-pressure water dispensers that dispense electrically heated water and have a reservoir vented to the atmosphere shall comply with ASSE 1023. Electric devices that heat water shall comply with UL 499.

418.0 Floor Drains.

418.1 Application. Floor drains shall comply with ASME A112.3.1, ASME A112.6.3, or CSA B79.

418.2 Strainer. Floor drains shall be considered plumbing fixtures and each such drain shall be provided with an approved-type strainer having a waterway equivalent to the area of the tailpiece. Floor drains shall be of an approved type and shall provide a watertight joint on the floor.

418.3 Location of Floor Drains. Floor drains shall be installed in the following areas:

(1) Toilet rooms containing two or more water closets or a combination of one water closet and one urinal, except in a dwelling unit.

(2) Commercial kitchens and in accordance with Section 704.3.

(3) Laundry rooms in commercial buildings and common laundry facilities in multi-family dwelling buildings.

(4) Boiler rooms.

418.4 Food Storage Areas. Where drains are provided in storerooms, walk-in freezers, walk-in coolers, refrigerated equipment, or other locations where food is stored, such drains shall have indirect waste piping. Separate waste pipes shall be run from each food storage area, each with an indirect connection to the building sanitary drainage system. Traps shall be provided in accordance with Section 801.3.2 of this code and shall be vented.

Indirect drains shall be permitted to be located in freezers or other spaces where freezing temperatures are maintained, provided that traps, where supplied, shall be located where the seal will not freeze. Otherwise, the floor of the freezer shall be sloped to a floor drain located outside of the storage compartment.

418.5 Floor Slope. Floors shall be sloped to floor drains.

419.0 Food Waste Disposers.

419.1 Application. Food waste disposal units shall comply with UL 430. Residential food waste disposers shall also comply with ASSE 1008.

419.2 Drainage Connection. Approved wye or other directional-type branch fittings shall be installed in continuous wastes connecting or receiving the discharge from a food waste disposer. No dishwasher drain shall be connected to a sink tailpiece, continuous waste, or trap on the discharge side of a food waste disposer.

419.3 Water Supply. A cold water supply shall be provided for food waste disposers. Such connection to the water supply shall be protected by an air gap or backflow prevention device in accordance with Section 603.2.

420.0 Sinks.

420.1 Application. Sinks shall comply with ASME A112.19.1/CSA B45.2, ASME A112.19.2/CSA B45.1, ASME A112.19.3/CSA B45.4, CSA B45.5/IAPMO Z124, CSA B45.8/IAPMO Z403, or CSA B45.12/IAPMO Z402. Moveable sink systems shall comply with ASME A112.19.12. Sink assemblies with automatic soap dispensers, faucets, or hand dryers shall comply with IAPMO IGC 127.

420.2 Water Consumption. Sink faucets shall have a maximum flow rate of not more than 2.2 gpm at 60 psi (8.3 L/m at 414 kPa).

Exceptions:

(1) Clinical sinks

(2) Laundry trays

(3) Service sinks

420.3 Pre-Rinse Spray Valve. Commercial food service pre-rinse spray valves shall have a maximum flow rate of 1.6

gallons per minute (gpm) at 60 pounds-force per square inch (psi) (6.0 L/m at 414 kPa) and shall be equipped with an integral automatic shutoff.

420.4 Waste Outlet. Kitchen and laundry sinks shall have a waste outlet and fixture tailpiece not less than 1½ inches (40 mm) in diameter. Service sinks shall have a waste outlet and fixture tailpiece not less than 2 inches (50 mm) in diameter. Fixture tailpieces shall be constructed from the materials specified in Section 701.2 for drainage piping. Waste outlets shall be provided with an approved strainer.

421.0 Floor Sinks.

421.1 Application. Floor sinks shall comply with ASME A112.6.7.

421.2 Strainers. The waste outlet of a floor sink shall be provided with an approved strainer or grate that is removable and accessible.

422.0 Minimum Number of Required Fixtures.

422.1 Fixture Count. Plumbing fixtures shall be provided for the type of building occupancy and in the minimum number shown in Table 422.1. The total occupant load and occupancy classification shall be determined in accordance with the building code. Occupancy classification not shown in Table 422.1 shall be considered separately by the Authority Having Jurisdiction.

The minimum number of fixtures shall be calculated at 50 percent male and 50 percent female based on the total occupant load. Where information submitted indicates a difference in the distribution of the sexes such information shall be used to determine the number of fixtures for each sex. Once the occupancy load and occupancy are determined, Table 422.1 shall be applied to determine the minimum number of plumbing fixtures required. Where applying the fixture ratios in Table 422.1 results in fractional numbers, such numbers shall be rounded to the next whole number. For multiple occupancies, fractional numbers shall be first summed and then rounded to the next whole number.

422.1.1 Family or Assisted-Use Toilet and Bathing Facilities. Where family or assisted-use toilet and bathing rooms are required, in applicable building regulations, the facilities shall be installed in accordance with those regulations.

422.2 Separate Facilities. Separate toilet facilities shall be provided for each sex.

Exceptions:

(1) Residential installations.

(2) In occupancies with a total occupant load of 10 or less, including customers and employees, one toilet facility, designed for use by no more than one person at a time, shall be permitted for use by both sexes.

(3) In business and mercantile occupancies with a total occupant load of 50 or less including customers and employees, one toilet facility, designed for use by no more than one person at a time, shall be permitted for use by both sexes.

422.2.1 Single Use Toilet Facilities. Single use toilet facilities and family or assisted use toilet facilities shall be identified with signage indicating use by either sex.

422.2.2 Family or Assisted-Use Toilet Facilities. Where a separate toilet facility is required for each sex, and each toilet facility is required to have only one water closet, two family or assisted-use toilet facilities shall be permitted in place of the required separate toilet facilities.

422.3 Fixture Requirements for Special Occupancies. Additional fixtures shall be permitted to be required where unusual environmental conditions or referenced activities are encountered. In food preparation areas, fixture requirements shall be permitted to be dictated by health codes.

422.4 Toilet Facilities Serving Employees and Customers. Each building or structure shall be provided with toilet facilities for employees and customers. Requirements for customers and employees shall be permitted to be met with a single set of restrooms accessible to both groups.

Required toilet facilities for employees and customers located in shopping malls or centers shall be permitted to be met by providing a centrally located toilet facility accessible to several stores. The maximum travel distance from entry to any store to the toilet facility shall not exceed 300 feet (91 440 mm).

Required toilet facilities for employees and customers in other than shopping malls or centers shall have a maximum travel distance not to exceed 500 feet (152 m).

422.4.1 Access to Toilet Facilities. In multi-story buildings, accessibility to the required toilet facilities shall not exceed one vertical story. Access to the required toilet facilities for customers shall not pass through areas designated as for employee use only such as kitchens, food preparation areas, storage rooms, closets, or similar spaces. Toilet facilities accessible only to private offices shall not be counted to determine compliance with this section.

422.5 Toilet Facilities for Workers. Toilet facilities shall be provided and maintained in a sanitary condition for the use of workers during construction.

TABLE 422.1
MINIMUM PLUMBING FACILITIES[1]

Each building shall be provided with sanitary facilities, including provisions for persons with disabilities as prescribed by the Department Having Jurisdiction. Table 422.1 applies to new buildings, additions to a building, and changes of occupancy or type in an existing building resulting in increased occupant load.

TYPE OF OCCUPANCY[2]	WATER CLOSETS (FIXTURES PER PERSON)[3]		URINALS (FIXTURES PER PERSON)[4]	LAVATORIES (FIXTURES PER PERSON)[5]		BATHTUBS OR SHOWERS (FIXTURES PER PERSON)	DRINKING FOUNTAINS/ FACILITIES (FIXTURES PER PERSON)	OTHER
A-1 Assembly occupancy (fixed or permanent seating)- theaters, concert halls, and auditoriums	Male 1: 1-100 2: 101-200 3: 201-400	Female 1: 1-25 2: 26-50 3: 51-100 4: 101-200 6: 201-300 8: 301-400	Male 1: 1-200 2: 201-300 3: 301-400 4: 401-600	Male 1: 1-200 2: 201-400 3: 401-600 4: 601-750	Female 1: 1-100 2: 101-200 4: 201-300 5: 301-500 6: 501-750	—	1: 1-250 2: 251-500 3: 501-750	1 service sink or laundry tray
	Over 400, add 1 fixture for each additional 500 males and 1 fixture for each additional 125 females.		Over 600, add 1 fixture for each additional 300 males.	Over 750, add 1 fixture for each additional 250 males and 1 fixture for each additional 200 females.			Over 750, add 1 fixture for each additional 500 persons.	
A-2 Assembly occupancy- restaurants, pubs, lounges, nightclubs and banquet halls	Male 1: 1-50 2: 51-150 3: 151-300 4: 301-400	Female 1: 1-25 2: 26-50 3: 51-100 4: 101-200 6: 201-300 8: 301-400	Male 1: 1-200 2: 201-300 3: 301-400 4: 401-6000	Male 1: 1-150 2: 151-200 3: 201-400	Female 1: 1-150 2: 151-200 4: 201-400	—	1: 1-250 2: 251-500 3: 501-750	1 service sink or laundry tray
	Over 400, add 1 fixture for each additional 250 males and 1 fixture for each 125 females.		Over 600, add 1 fixture for each additional 300 males.	Over 400, add 1 fixture for each additional 250 males and 1 fixture for each additional 200 females			Over 750, add 1 fixture for each additional 500 persons.	
A-3 Assembly occupancy (typical without fixed or permanent seating)- arcades, places of worship, museums, libraries, lecture halls, gymnasiums (without spectator seating), indoor pools (without spectator seating)	Male 1: 1-100 2: 101-200 3: 201-400	Female 1: 1-25 2: 26-50 3: 51-100 4: 101-200 6: 201-300 8: 301-400	Male 1: 1-100 2: 101-200 3: 201-400 4: 401-600	Male 1: 1-200 2: 201-400 3: 401-600 4: 601-750	Female 1: 1-100 2: 101-200 4: 201-300 5: 301-500 6: 501-750	—	1: 1-250 2: 251-500 3: 501-750	1 service sink or laundry tray
	Over 400, add 1 fixture for each additional 500 males and 1 fixture for each additional 125 females.		Over 600, add 1 fixture for each additional 300 males.	Over 750, add 1 fixture for each additional 250 males and 1 fixture for each additional 200 females.			Over 750, add 1 fixture for each additional 500 persons.	
A-4 Assembly occupancy (indoor activities or sporting events with spectator seating)- swimming pools, skating rinks, arenas, and gymnasiums	Male 1: 1-100 2: 101-200 3: 201-400	Female 1: 1-25 2: 26-50 3: 51-100 4: 101-200 6: 201-300 8: 301-400	Male 1: 1-100 2: 101-200 3: 201-400 4: 401-600	Male 1: 1-200 2: 201-400 3: 401-750	Female 1: 1-100 2: 101-200 4: 201-300 5: 301-500 6: 501-750	—	1: 1-250 2: 251-500 3: 501-750	1 service sink or laundry tray
	Over 400, add 1 fixture for each additional 500 males and 1 fixture for each additional 125 females.		Over 600, add 1 fixture for each additional 300 males.	Over 750, add 1 fixture for each additional 250 males and 1 fixture for each additional 200 females.			Over 750, add 1 fixture for each additional 500 persons.	

TABLE 422.1
MINIMUM PLUMBING FACILITIES[1] (continued)

TYPE OF OCCUPANCY[2]	WATER CLOSETS (FIXTURES PER PERSON)[3]		URINALS (FIXTURES PER PERSON)[4]	LAVATORIES (FIXTURES PER PERSON)[5]		BATHTUBS OR SHOWERS (FIXTURES PER PERSON)	DRINKING FOUNTAINS/ FACILITIES (FIXTURES PER PERSON)	OTHER
A-5 Assembly occupancy (outdoor activities or sporting events)- amusement parks, grandstands and stadiums	Male 1: 1-100 2: 101-200 3: 201-400	Female 1: 1-25 2: 26-50 3: 51-100 4: 101-200 6: 201-300 8: 301-400	Male 1: 1-100 2: 101-200 3: 201-400 4: 401-600	Male 1: 1-200 2: 201-400 3: 401-750	Female 1: 1-100 2: 101-200 4: 201-300 5: 301-500 6: 501-750	—	1: 1-250 2: 251-500 3: 501-750	1 service sink or laundry tray
	Over 400, add 1 fixture for each additional 500 males and 1 fixture for each additional 125 females.		Over 600, add 1 fixture for each additional 300 males.	Over 750, add 1 fixture for each additional 250 males and 1 fixture for each additional 200 females.			Over 750, add 1 fixture for each additional 500 persons.	
B Business occupancy (office, professional or service type transactions)- banks, vet clinics, hospitals, car wash, banks, beauty salons, ambulatory health care facilities, laundries and dry cleaning, educational institutions (above high school), or training facilities not located within school, post offices and printing shops	Male 1: 1-50 2: 51-100 3: 101-200 4: 201-400	Female 1: 1-15 2: 16-30 3: 31-50 4: 51-100 8: 101-200 11: 201-400	Male 1: 1-100 2: 101-200 3: 201-400 4: 401-600	Male 1: 1-75 2: 76-150 3: 151-200 4: 201-300 5: 301-400	Female 1: 1-50 2: 51-100 3: 101-150 4: 151-200 5: 201-300 6: 301-400	—	1 per 150	1 service sink or laundry tray
	Over 400, add 1 fixture for each additional 500 males and 1 fixture for each additional 150 females.		Over 600, add 1 fixture for each additional 300 males.	Over 400, add 1 fixture for each additional 250 males and 1 fixture for each additional 200 females.				
E Educational occupancy- private or public schools	Male 1 per 50	Female 1 per 30	Male 1 per 100	Male 1 per 40	Female 1 per 40	—	1 per 150	1 service sink or laundry tray
F1, F2 Factory or Industrial occupancy-fabricating or assembly work	Male 1: 1-50 2: 51-75 3: 76-100	Female 1: 1-50 2: 51-75 3: 76-100	—	Male 1: 1-50 2: 51-75 3: 76-100	Female 1: 1-50 2: 51-75 3: 76-100	1 shower for each 15 persons exposed to excessive heat or to skin contamination with poisonous, infectious or irritating material	1: 1-250 2: 251-500 3: 501-750	1 service sink or laundry tray
	Over 100, add 1 fixture for each additional 40 persons.		—	Over 100, add 1 fixture for each additional 40 persons.			Over 750, add 1 fixture for each additional 500 persons.	
I-1 Institutional occupancy (houses more than 16 persons on a 24-hour basis)- substance abuse centers, assisted living, group homes, or residential facilities	Male 1 per 15	Female 1 per 15	—	Male 1 per 15	Female 1 per 15	1 per 8	1 per 150	1 service sink or laundry tray

TABLE 422.1
MINIMUM PLUMBING FACILITIES[1] (continued)

TYPE OF OCCUPANCY[2]		WATER CLOSETS (FIXTURES PER PERSON)[3]	URINALS (FIXTURES PER PERSON)[4]	LAVATORIES (FIXTURES PER PERSON)[5]	BATHTUBS OR SHOWERS (FIXTURES PER PERSON)	DRINKING FOUNTAINS/ FACILITIES (FIXTURES PER PERSON)	OTHER
I-2 Institutional occupancy-medical, psychiatric, surgical or nursing homes	Hospitals and nursing homes-individual rooms and ward room	1 per room	—	1 per room	1 per room	1 per 150	1 service sink or laundry tray
		1 per 8 patients	—	1 per 10 patients	1 per 20 patients		
	Hospital Waiting or Visitor Rooms	1 per room	—	1 per room	—	1 per room	—
	Employee Use	Male 1: 1-15 2: 16-35 3: 36-55 Female 1: 1-15 3: 16-35 4: 36-55 Over 55, add 1 fixture for each additional 40 persons.	—	Male 1 per 40 Female 1 per 40	—	—	—
I-3 Institutional occupancy (houses more than 5 people)	Prisons	1 per cell	—	1 per cell	1 per 20	1 per cell block/floor	—
	Correctional facilities or juvenile center	1 per 8	—	1 per 10	1 per 8	1 per floor	1 service sink or laundry tray
	Employee Use	Male 1: 1-15 2: 16-35 3: 36-55 Female 1: 1-15 3: 16-35 4: 36-55 Over 55, add 1 fixture for each additional 40 persons.	—	Male 1 per 40 Female 1 per 40	—	1 per 150	—
I-4 Institutional occupancy (any age that receives care for less than 24 hours)		Male 1: 1-15 2: 16-35 3: 36-55 Female 1: 1-15 3: 16-35 4: 36-55 Over 55, add 1 fixture for each additional 40 persons.	—	Male 1 per 40 Female 1 per 40	—	1 per 150	1 service sink or laundry tray
M Mercantile occupancy (the sale of merchandise and accessible to the public)		Male 1: 1-100 2: 101-200 3: 201-400 Female 1: 1-100 2: 101-200 4: 201-300 6: 301-400	Male 0: 1-200 1: 201-400	Male 1: 1-200 2: 201-400 Female 1: 1-200 2: 201-300 3: 301-400	—	1: 1-250 2: 251-500 3: 501-750	1 service sink or laundry tray
		Over 400, add 1 fixture for each additional 500 males and 1 fixture for each 200 females.	Over 400, add 1 fixture for each additional 500 males.	Over 400, add 1 fixture for each additional 500 males and 1 fixture for each 400 females.	—	Over 750, add 1 fixture for each additional 500 persons.	—
R-1 Residential occupancy (minimal stay)-hotels, motels, bed and breakfast homes		1 per sleeping room	—	1 per sleeping room	1 per sleeping room	—	1 service sink or laundry tray

TABLE 422.1
MINIMUM PLUMBING FACILITIES[1] (continued)

TYPE OF OCCUPANCY[2]		WATER CLOSETS (FIXTURES PER PERSON)[3]		URINALS (FIXTURES PER PERSON)[4]	LAVATORIES (FIXTURES PER PERSON)[5]		BATHTUBS OR SHOWERS (FIXTURES PER PERSON)	DRINKING FOUNTAINS/ FACILITIES (FIXTURES PER PERSON)	OTHER
R-2 Residential occupancy (long-term or permanent)	Dormitories	Male 1 per 10	Female 1 per 8	Male 1 per 25	Male 1 per 12	Female 1 per 12	1 per 8	1 per 150	1 service sink or laundry tray
		Add 1 fixture for each additional 25 males and 1 fixture for each additional 20 females.		Over 150, add 1 fixture for each additional 50 males.	Add 1 fixture for each additional 20 males and 1 fixture for each additional 15 females.				
	Employee Use	Male 1: 1-15 2: 16-35 3: 36-55	Female 1: 1-15 3: 16-35 4: 36-55	—	Male 1 per 40	Female 1 per 40	—	—	
		Over 55, add 1 fixture for each additional 40 persons							
	Apartment house/unit	1 per apartment		—	1 per apartment		1 per apartment	—	1 kitchen sink per apartment. 1 laundry tray or 1 automatic clothes washer connection per unit or 1 laundry tray or 1 automatic clothes washer connection for each 12 units
R-3 Residential occupancy (long-term or permanent in nature) for more than 5 but does not exceed 16 occupants)		Male 1 per 10	Female 1 per 8	—	Male 1 per 12	Female 1 per 12	1 per 8	1 per 150	1 service sink or laundry tray
		Add 1 fixture for each additional 25 males and 1 fixture for each additional 20 females.			Add 1 fixture for each additional 20 males and 1 fixture for each additional 15 females.				
R-3 Residential occupancy (one and two family dwellings)		1 per one and two family dwelling		—	1 per one and two family dwelling		1 per one and two family dwelling	—	1 kitchen sink and 1 automatic clothes washer connection per one and two family dwelling
R-4 Residential occupancy (residential care or assisted living)		Male 1 per 10	Female 1 per 8	—	Male 1 per 12	Female 1 per 12	1 per 8	1 per 150	1 service sink or laundry tray
		Add 1 fixture for each additional 25 males and 1 fixture for each additional 20 females.			Add 1 fixture for each additional 20 males and 1 fixture for each additional 15 females.				

TABLE 422.1
MINIMUM PLUMBING FACILITIES[1] (continued)

TYPE OF OCCUPANCY[2]	WATER CLOSETS (FIXTURES PER PERSON)[3]		URINALS (FIXTURES PER PERSON)[4]	LAVATORIES (FIXTURES PER PERSON)[5]		BATHTUBS OR SHOWERS (FIXTURES PER PERSON)	DRINKING FOUNTAINS/ FACILITIES (FIXTURES PER PERSON)	OTHER
S-1, S-2 Storage occupancy-storage of goods, warehouse, aircraft hanger, food products, appliances	Male 1: 1-100 2: 101-200 3: 201-400	Female 1: 1-100 2: 101-200 3: 201-400	—	Male 1: 1-200 2: 201-400 3: 401-750	Female 1: 1-200 2: 201-400 3: 401-750	—	1: 1-250 2: 251-500 3: 501-750	1 service sink or laundry tray
	Over 400, add 1 fixture for each additional 500 males and 1 fixture for each additional 150 females.			Over 750, add 1 fixture for each additional 500 persons.			Over 750, add 1 fixture for each additional 500 persons.	

Notes:

1 The figures shown are based upon one fixture being the minimum required for the number of persons indicated or any fraction thereof.

2 A restaurant is defined as a business that sells food to be consumed on the premises.
a. The number of occupants for a drive-in restaurant shall be considered as equal to the number of parking stalls.
b. Hand-washing facilities shall be available in the kitchen for employees.

3 The total number of required water closets for females shall be not less than the total number of required water closets and urinals for males.

4 For each urinal added in excess of the minimum required, one water closet shall be permitted to be deducted. The number of water closets shall not be reduced to less than two-thirds of the minimum requirement.

5 Metering or self-closing faucets shall be installed on lavatories intended to serve the transient public.

CHAPTER 5

WATER HEATERS

501.0 General.

501.1 Applicability. The regulations of this chapter shall govern the construction, location, and installation of fuel-burning and other types of water heaters heating potable water, together with chimneys, vents, and their connectors. The minimum capacity for storage water heaters shall be in accordance with the first-hour rating listed in Table 501.1(2). No water heater shall be hereinafter installed that does not comply with the manufacturer's installation instructions and the type and model of each size thereof approved by the Authority Having Jurisdiction. A list of accepted water heater appliance standards is referenced in Table 501.1(1). Listed appliances shall be installed in accordance with the manufacturer's installation instructions. Unlisted water heaters shall be permitted in accordance with Section 504.3.2.

TABLE 501.1(1)
WATER HEATERS

TYPE	STANDARD
Electric, Household Storage	UL 174
Oil-Fired Storage Tank	UL 732
Gas-Fired, 75 000 Btu/h or less, Storage	CSA Z21.10.1
Gas-Fired, Above 75 000 Btu/h, Storage and Instantaneous	CSA Z21.10.3
Electric, Commercial Storage	UL 1453
Solid Fuel-Fired	UL 2523
Electric Instantaneous	UL 499

For SI units: 1000 British thermal units per hour = 0.293 kW

502.0 Permits.

502.1 General. It shall be unlawful for a person to install, remove, or replace or cause to be installed, removed, or replaced a water heater without first obtaining a permit from the Authority Having Jurisdiction to do so.

503.0 Inspection.

503.1 Inspection of Chimneys or Vents. This inspection shall be made after chimneys, vents, or parts thereof, authorized by the permit, have been installed and before such vent or part thereof has been covered or concealed.

503.2 Final Water Heater Inspection. This inspection shall be made after work authorized by the permit has been installed. The Authority Having Jurisdiction will make such inspection as deemed necessary to be assured that the work has been installed in accordance with the intent of this code. No appliance or part thereof shall be covered or concealed until the same has been inspected and approved by the Authority Having Jurisdiction.

504.0 Water Heater Requirements.

504.1 Location. Water heater installations in bedrooms and bathrooms shall comply with one of the following [NFPA 54:10.27.1]:

(1) Water heater shall be of the direct-vent type. [NFPA 54: 10.27.1(2)]

(2) Fuel-burning water heaters shall be permitted to be installed in a closet located in the bedroom or bathroom provided the closet is equipped with a listed, gasketed door assembly and a listed self-closing device. The self-closing door assembly shall meet the requirements of Section 504.1.1. The door assembly shall be installed with a threshold and bottom door seal and shall meet the requirements of Section 504.1.2. Combustion air for such installations shall be obtained from the outdoors in accordance with Section 506.4. The closet shall be for the exclusive use of the water heater.

504.1.1 Self-Closing Doors. Self-closing doors shall swing easily and freely, and shall be equipped with a self-closing device to cause the door to close and latch each time it is opened. The closing mechanism shall not have a hold-open feature.

504.1.2 Gasketing. Gasketing on gasketed doors or frames shall be furnished in accordance with the published listings of the door, frame, or gasketing material manufacturer.

Exception: Where acceptable to the Authority Having Jurisdiction, gasketing of non-combustible or limited-combustible material shall be permitted to be applied to the frame, provided closing and latching of the door are not inhibited.

504.2 Vent. Water heaters of other than the direct-vent type shall be located as close as practical to the chimney or gas vent.

TABLE 501.1(2)
FIRST HOUR RATING[1]

Number of Bathrooms	1 to 1.5			2 to 2.5				3 to 3.5			
Number of Bedrooms	1	2	3	2	3	4	5	3	4	5	6
First Hour Rating,[2] Gallons	38	49	49	49	62	62	74	62	74	74	74

For SI units: 1 gallon = 3.785 L

Notes:

[1] The first-hour rating is found on the "Energy Guide" label.

[2] Solar water heaters shall be sized to meet the appropriate first-hour rating as shown in the table.

504.3 Clearance. The clearance requirements for water heaters shall comply with Section 504.3.1 or Section 504.3.2.

504.3.1 Listed Water Heaters. The clearances shall not be such as to interfere with combustion air, draft hood clearance and relief, and accessibility for servicing. Listed water heaters shall be installed in accordance with their listings and the manufacturer's installation instructions.

504.3.2 Unlisted Water Heaters. Unlisted water heaters shall be installed with a clearance of 12 inches (305 mm) on all sides and rear. Combustible floors under unlisted water heaters shall be protected in an approved manner. [NFPA 54:10.27.2.2]

504.4 Pressure-Limiting Devices. A water heater installation shall be provided with overpressure protection using an approved, listed device installed in accordance with the terms of its listing and the manufacturer's installation instructions.

504.5 Temperature-Limiting Devices. A water heater installation or a hot water storage vessel installation shall be provided with overtemperature protection by means of an approved, listed device installed in accordance with the terms of its listing and the manufacturer's installation instructions.

504.6 Temperature, Pressure, and Vacuum Relief Devices. Temperature, pressure, and vacuum relief devices or combinations thereof, and automatic gas shutoff devices shall be installed in accordance with the terms of their listings and the manufacturer's installation instructions. A shutoff valve shall not be placed between the relief valve and the water heater or on discharge pipes between such valves and the atmosphere. The hourly British thermal units (Btu) (kW•h) discharge capacity or the rated steam relief capacity of the device shall be not less than the input rating of the water heater. Discharge piping shall be installed in accordance with Section 608.5.

505.0 Oil-Burning and Other Water Heaters.

505.1 Water Heaters. Water heaters deriving heat from fuels or types of energy other than gas shall comply with the standards referenced in Table 501.1(1), Section 505.3, or Section 505.4. Vents or chimneys for such appliances shall be of approved types. An adequate supply of air for combustion and for adequate ventilation of heater rooms or compartments shall be provided. Each such appliance shall be installed in a location approved by the Authority Having Jurisdiction and local and state fire-prevention agencies.

505.2 Safety Devices. Storage-type water heaters and hot water boilers deriving heat from fuels or types of energy other than gas, shall be provided with, in addition to the primary temperature controls, an over-temperature safety protection device that complies with and is installed in accordance with nationally recognized applicable standards for such devices and a combination temperature and pressure-relief valve.

505.3 Oil-Fired Water Heaters. Oil-fired water heaters shall be installed in accordance with NFPA 31.

505.4 Indirect-Fired Water Heaters. Indirect-fired water heaters shall be in accordance with the applicable sections of the ASME Boiler and Pressure Vessel Code or shall comply with one of the other applicable standards shown in Table 501.1(1). Each water heater shall bear a label in accordance with ASME requirements, or an approved testing agency, certifying and attesting that such an appliance has been tested, inspected and meets the requirements of the applicable standards or code.

505.4.1 Single-Wall Heat Exchanger. An indirect-fired water heater that incorporates a single-wall heat exchanger shall be in accordance with the following requirements:

(1) The heat transfer medium shall be either potable water or contain fluids recognized as safe by the Food and Drug Administration (FDA) as food grade.

(2) Bear a label with the word "Caution," followed by the following statements:

   (a) The heat-transfer medium shall be potable water or other nontoxic fluid recognized as safe by the FDA.

   (b) The maximum operating pressure of the heat exchanger shall not exceed the maximum operating pressure of the potable water supply.

(3) The word "Caution" and the statements in letters shall have an uppercase height of not less than 0.120 of an inch (3.048 mm). The vertical spacing between lines of type shall be not less than 0.046 of an inch (1.168 mm). Lowercase letters shall be compatible with the uppercase letter size specification.

506.0 Air for Combustion and Ventilation.

506.1 General. Air for combustion, ventilation, and dilution of flue gases for appliances installed in buildings shall be obtained by application of one of the methods covered in Section 506.2 through Section 506.7.3. Where the requirements of Section 506.2 are not met, outdoor air shall be introduced in accordance with methods covered in Section 506.4 through Section 506.7.3.

Exception: This provision shall not apply to direct-vent appliances. [NFPA 54:9.3.1.1]

506.1.1 Other Types of Appliances. Appliances of other than natural draft design, appliances not designated as Category I vented appliances, and appliances equipped with power burners shall be provided with combustion, ventilation, and dilution air in accordance with the appliance manufacturer's instructions. [NFPA 54:9.3.1.2]

506.1.2 Draft Hood and Regulators. Where used, a draft hood or a barometric draft regulator shall be installed in the same room or enclosure as the appliance served so as to prevent any difference in pressure between the hood or regulator and the combustion air supply. [NFPA 54:9.3.1.4]

506.1.3 Makeup Air. Where exhaust fans, clothes dryers, and kitchen ventilation systems interfere with the operation of appliances, makeup air shall be provided. [NFPA 54:9.3.1.5]

506.2 Indoor Combustion Air. The required volume of indoor air shall be determined in accordance with the method in Section 506.2.1 or Section 506.2.2 except that where the air infiltration rate is known to be less than 0.40 ACH (air change per hour), the method in Section 506.2.2 shall be used. The total required volume shall be the sum of the required volume calculated for all appliances located within the space. Rooms communicating directly with the space in which the appliances are installed through openings not furnished with doors, and through combustion air openings sized and located in accordance with Section 506.3, are considered a part of the required volume. [NFPA 54:9.3.2]

506.2.1 Standard Method. The minimum required volume shall be 50 cubic feet per 1000 British thermal units per hour (Btu/h) (4.83 m^3/kW). [NFPA 54:9.3.2.1]

506.2.2 Known Air Infiltration Rate Method. Where the air infiltration rate of a structure is known, the minimum required volume shall be determined as follows [NFPA 54:9.3.2.2]:

(1) For appliances other than fan-assisted, calculate using the following Equation 506.2.2(1). [NFPA 54:9.3.2.2(1)]

[Equation 506.2.2(1)]

$$\text{Required Volume}_{other} \geq \frac{21\text{ ft}^3}{ACH}\left(\frac{I_{other}}{1000\text{ Btu/h}}\right)$$

(2) For fan-assisted appliances, calculate using the following Equation 506.2.2(2). [NFPA 54:9.3.2.2(2)]

[Equation 506.2.2(2)]

$$\text{Required Volume}_{fan} \geq \frac{15\text{ ft}^3}{ACH}\left(\frac{I_{fan}}{1000\text{ Btu/h}}\right)$$

Where:

I_{other} = All appliances other than fan-assisted input in (Btu/h)

I_{fan} = Fan-assisted appliance input in (Btu/h)

ACH = Air change per hour (percent of volume of space exchanged per hour, expressed as a decimal)

For SI units: 1 cubic foot = 0.0283 m^3, 1000 British thermal units per hour = 0.293 kW

(3) For purposes of these calculations, an infiltration rate greater than 0.60 ACH shall not be used in the equations in Section 506.2.2(1) and Section 506.2.2(2). [NFPA 54:9.3.2.2(3)]

506.3 Indoor Opening Size and Location. Openings used to connect indoor spaces shall be sized and located in accordance with the following:

(1) Combining spaces on the same story. Each opening shall have a minimum free area of 1 square inch per 1000 Btu/h (0.002 m^2/kW) of the total input rating of all appliances in the space but not less than 100 square inches (0.065 m^2). One permanent opening shall commence within 12 inches (305 mm) of the top of the enclosure and one permanent opening shall commence within 12 inches (305 mm) of the bottom of the enclosure (see Figure 506.3). The minimum dimension of air openings shall not be less than 3 inches (76 mm).

(2) Combining spaces in different stories. The volumes of spaces in different stories shall be considered as communicating spaces where such spaces are connected by one or more permanent openings in doors or floors having a total minimum free area of 2 square inches per 1000 Btu/h (0.004 m^2/kW) of total input rating of all appliances. [NFPA 54:9.3.2.3]

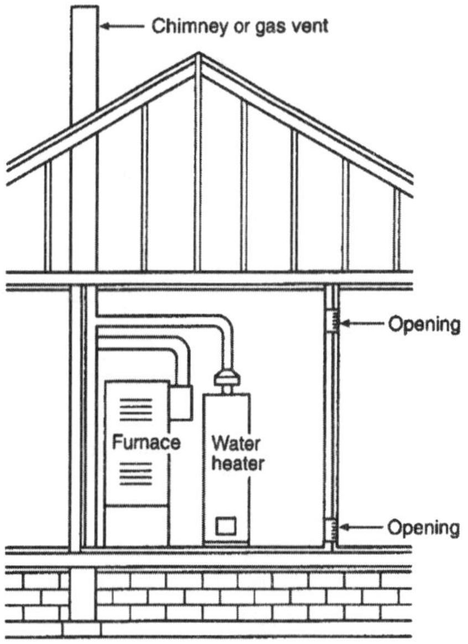

FIGURE 506.3
ALL COMBUSTION AIR FROM ADJACENT INDOOR SPACES THROUGH INDOOR COMBUSTION AIR OPENINGS
[NFPA 54: FIGURE A.9.3.2.3(1)]

506.4 Outdoor Combustion Air. Outdoor combustion air shall be provided through opening(s) to the outdoors in accordance with the methods in Section 506.4.1 or Section 506.4.2. The minimum dimension of air openings shall not be less than 3 inches (76 mm). [NFPA 54:9.3.3]

506.4.1 Two Permanent Openings Method. Two permanent openings, one commencing within 12 inches (305 mm) of the top of the enclosure and one commencing within 12 inches (305 mm) of the bottom of the enclosure, shall be provided. The openings shall communicate directly, or by ducts, with the outdoors or spaces that freely communicate with the outdoors, as follows:

(1) Where directly communicating with the outdoors or where communicating to the outdoors through vertical ducts, each opening shall have a minimum free area of 1 square inch per 4000 Btu/h (0.0005 m^2/kW) of total input rating of all appliances in the enclosure. [See Figure 506.4.1(1) and Figure 506.4.1(2)]

(2) Where communicating with the outdoors through horizontal ducts, each opening shall have a minimum free area of 1 square inch per 2000 Btu/h (0.001 m^2/kW) of total input rating of all appliances in the enclosure. [NFPA 54:9.3.3.1] [See Figure 506.4.1(3)]

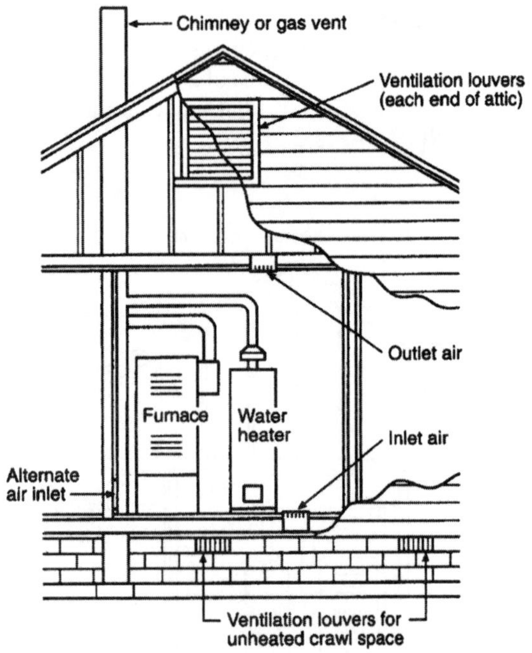

FIGURE 506.4.1(1)
ALL COMBUSTION AIR FROM OUTDOORS – INLET AIR FROM VENTILATED CRAWL SPACE AND OUTLET AIR TO VENTILATED ATTIC
[NFPA 54: FIGURE A.9.3.3.1(1)(a)]

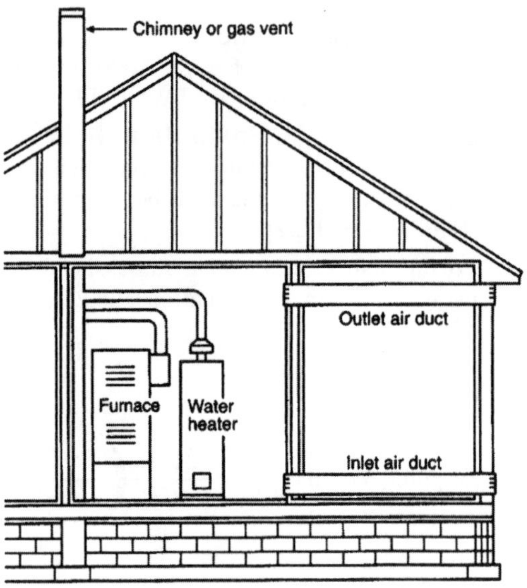

FIGURE 506.4.1(3)
ALL COMBUSTION AIR FROM OUTDOORS THROUGH HORIZONTAL DUCTS
[NFPA 54: FIGURE A.9.3.3.1(2)]

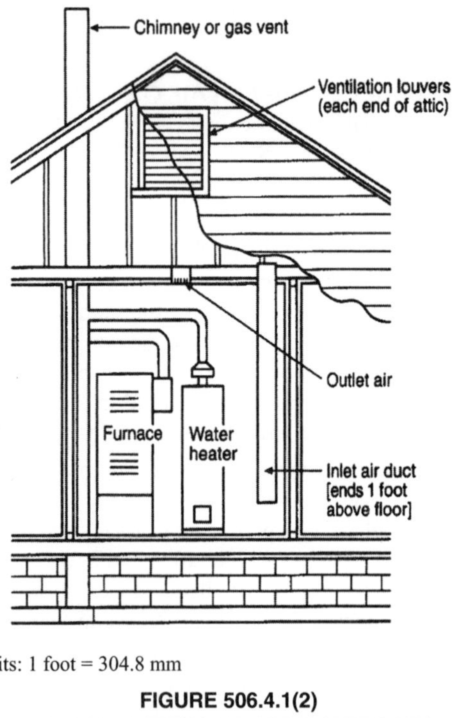

For SI units: 1 foot = 304.8 mm

FIGURE 506.4.1(2)
ALL COMBUSTION AIR FROM OUTDOORS THROUGH VENTILATED ATTIC
[NFPA 54: FIGURE A.9.3.3.1(1)(b)]

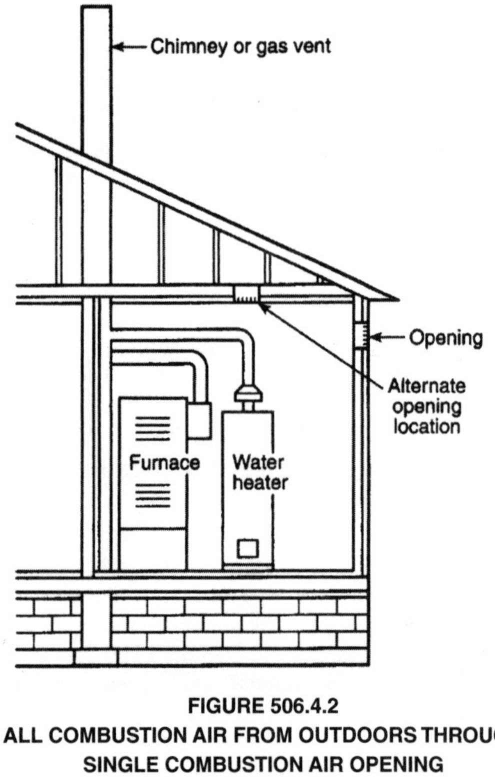

FIGURE 506.4.2
ALL COMBUSTION AIR FROM OUTDOORS THROUGH SINGLE COMBUSTION AIR OPENING
[NFPA 54: FIGURE A.9.3.3.2]

506.4.2 One Permanent Opening Method. One permanent opening, commencing within 12 inches (305 mm) of the top of the enclosure, shall be provided. The appliance shall have clearances of at least 1 inch (25.4 mm) from the sides and back and 6 inches (152 mm) from the front of the appliance. The opening shall directly communicate with the outdoors or shall communicate through a vertical or horizontal duct to the outdoors or spaces that freely communicate with the outdoors (see Figure 506.4.2) and shall have a minimum free area of the following:

(1) One square inch per 3000 Btu/h (0.0007 m²/kW) of the total input rating of all appliances located in the enclosure.

(2) Not less than the sum of the areas of all vent connectors in the space. [NFPA 54:9.3.3.2]

506.5 Combination Indoor and Outdoor Combustion Air. The use of a combination of indoor and outdoor combustion air shall be in accordance with Section 506.5.1 through Section 506.5.3. [NFPA 54:9.3.4] (See Appendix J for example calculations)

506.5.1 Indoor Openings. Where used, openings connecting the interior spaces shall comply with Section 506.3. [NFPA 54:9.3.4(1)]

506.5.2 Outdoor Opening(s) Location. Outdoor opening(s) shall be located in accordance with Section 506.4. [NFPA 54:9.3.4(2)]

506.5.3 Outdoor Opening(s) Size. The outdoor opening(s) size shall be calculated in accordance with the following:

(1) The ratio of the interior spaces shall be the available volume of all communicating spaces divided by the required volume.

(2) The outdoor size reduction factor shall be 1 minus the ratio of interior spaces.

(3) The minimum size of outdoor opening(s) shall be the full size of outdoor opening(s) calculated in accordance with Section 506.4, multiplied by the reduction factor. The minimum dimension of air openings shall not be less than 3 inches (76 mm). [NFPA 54:9.3.4(3)]

506.6 Engineered Installations. Engineered combustion air installations shall provide an adequate supply of combustion, ventilation, and dilution air and shall be approved by the Authority Having Jurisdiction. [NFPA 54:9.3.5]

506.7 Mechanical Combustion Air Supply. Where all combustion air is provided by a mechanical air supply system, the combustion air shall be supplied from outdoors at the minimum rate of 0.35 cubic feet per minute per 1000 Btu/h [0.034 (m³/min)/kW] for all appliances located within the space. [NFPA 54:9.3.6]

506.7.1 Exhaust Fans. Where exhaust fans are installed, additional air shall be provided to replace the exhausted air. [NFPA 54:9.3.6.1]

506.7.2 Interlock. Each of the appliances served shall be interlocked to the mechanical air supply system to prevent main burner operation where the mechanical air supply system is not in operation. [NFPA 54:9.3.6.2]

506.7.3 Specified Combustion Air. Where combustion air is provided by the building's mechanical ventilation system, the system shall provide the specified combustion air rate in addition to the required ventilation air. [NFPA 54:9.3.6.3]

506.8 Louvers, Grilles, and Screens. The required size of openings for combustion, ventilation, and dilution air shall be based on the net free area of each opening. Where the free area through a design of louver, grille, or screen is known, it shall be used in calculating the size opening required to provide the free area specified. Where the louver and grille design and free area are not known, it shall be assumed that wood louvers have 25 percent free area, and metal louvers and grilles have 75 percent free area. Nonmotorized louvers and grilles shall be fixed in the open position. [NFPA 54:9.3.7.1]

506.8.1 Minimum Screen Mesh Size. Screens shall not be smaller than ¼ of an inch (6.4 mm) mesh. [NFPA 54:9.3.7.2]

506.8.2 Motorized Louvers. Motorized louvers shall be interlocked with the appliance so they are proven in the full open position prior to main burner ignition and during main burner operation. Means shall be provided to prevent the main burner from igniting should the louver fail to open during burner startup and to shut down the main burner if the louvers close during burner operation. [NFPA 54:9.3.7.3]

506.9 Combustion Air Ducts. Combustion air ducts shall comply with the following [NFPA 54:9.3.8]:

(1) Ducts shall be constructed of galvanized steel or a material having equivalent corrosion resistance, strength, and rigidity.

Exception: Within dwellings units, unobstructed stud and joist spaces shall not be prohibited from conveying combustion air, provided that not more than one fireblock is removed. [NFPA 54:9.3.8.1]

(2) Ducts shall terminate in an unobstructed space, allowing free movement of combustion air to the appliances. [NFPA 54:9.3.8.2]

(3) Ducts shall serve a single space. [NFPA 54:9.3.8.3]

(4) Ducts shall not serve both upper and lower combustion air openings where both such openings are used. The separation between ducts serving upper and lower combustion air openings shall be maintained to the source of combustion air. [NFPA 54:9.3.8.4]

(5) Ducts shall not be screened where terminating in an attic space. [NFPA 54:9.3.8.5]

(6) Combustion air intake openings located on the exterior of the building shall have the lowest side of the combustion air intake openings located at least 12 inches (305 mm) vertically from the adjoining finished ground level. [NFPA 54:9.3.8.8]

(7) Horizontal upper combustion air ducts shall not slope downward toward the source of combustion air. [NFPA 54:9.3.8.6]

(8) The remaining space surrounding a chimney liner, gas vent, special gas vent, or plastic piping installed within a masonry, metal, or factory built chimney shall not be used to supply combustion air.

Exception: Direct-vent appliances designed for installation in a solid fuel-burning fireplace where installed in accordance with the manufacturer's installation instructions. [NFPA 54:9.3.8.7]

507.0 Appliance and Equipment Installation Requirements.

507.1 Dielectric Insulator. The Authority Having Jurisdiction shall have the authority to require the use of an approved dielectric insulator on the water piping connections of water heaters and related water heating appliances.

507.2 Seismic Provisions. Water heaters shall be anchored or strapped to resist horizontal displacement due to earthquake motion. Strapping shall be at points within the upper one-third and lower one-third of its vertical dimensions. At the lower point, a distance of not less than 4 inches (102 mm) shall be maintained from the controls with the strapping.

507.3 Appliance Support. Appliances and equipment shall be furnished either with load distributing bases or with a sufficient number of supports to prevent damage to either the building structure or the appliance and the equipment. [NFPA 54:9.1.8.1]

507.3.1 Structural Capacity. At the locations selected for installation of appliances and equipment, the dynamic and static load carrying capacities of the building structure shall be checked to determine whether they are adequate to carry the additional loads. The appliances and equipment shall be supported and shall be connected to the piping so as not to exert undue stress on the connections. [NFPA 54:9.1.8.2]

507.4 Ground Support. A water heater supported from the earth shall rest on level concrete or other approved base extending not less than 3 inches (76 mm) above the adjoining ground level.

507.5 Drainage Pan. Where a water heater is located in an attic, in or on an attic ceiling assembly, floor-ceiling assembly, or floor-subfloor assembly where damage results from a leaking water heater, a watertight pan of corrosion-resistant materials shall be installed beneath the water heater with not less than ¾ of an inch (20 mm) diameter drain to an approved location. Such pan shall be not less than 1½ inches (38 mm) in depth.

507.6 Added or Converted Equipment or Appliances. When additional or replacement appliances or equipment is installed or an appliance is converted to gas from another fuel, the location in which the appliances or equipment is to be operated shall be checked to verify the following:

(1) Air for combustion and ventilation is provided where required, in accordance with the provisions of Section 506.0. Where existing facilities are not adequate, they shall be upgraded to meet Section 506.0 specifications.

(2) The installation components and appliances meet the clearances to combustible material provisions of Section 507.27. It shall be determined that the installation and operation of the additional or replacement appliances do not render the remaining appliances unsafe for continued operation.

(3) The venting system is constructed and sized in accordance with the provisions of Section 509.0. Where the existing venting system is not adequate, it shall be upgraded to comply with Section 509.0. [NFPA 54:9.1.2]

507.7 Types of Gases. The appliance shall be connected to the fuel gas for which it was designed. No attempt shall be made to convert the appliance from the gas specified on the rating plate for use with a different gas without consulting the installation instructions, the serving gas supplier, or the appliance manufacturer for complete instructions. [NFPA 54:9.1.3]

507.8 Safety Shutoff Devices for Unlisted LP-Gas Appliance Used Indoors. Unlisted appliances for use with undiluted liquefied petroleum gases and installed indoors, except attended laboratory equipment, shall be equipped with safety shutoff devices of the complete shutoff type. [NFPA 54:9.1.4]

507.9 Use of Air or Oxygen Under Pressure. Where air or oxygen under pressure `is used in connection with the gas supply, effective means such as a backpressure regulator and relief valve shall be provided to prevent air or oxygen from passing back into the gas piping. Where oxygen is used, installation shall be in accordance with NFPA 51. [NFPA 54:9.1.5]

507.10 Protection of Gas Appliances from Fumes or Gases other than Products of Combustion. Non-direct-vent appliances installed in beauty shops, barber shops, or other facilities where chemicals that generate corrosive or flammable products such as aerosol sprays are routinely used shall be located in a mechanical room separate or partitioned off from other areas with provisions for combustion and dilution air from outdoors. Direct-vent appliances in such facilities shall be in accordance with the appliance manufacturer's installation instructions. [NFPA 54:9.1.6.2]

507.11 Process Air. In addition to air needed for combustion in commercial or industrial processes, process air shall be provided as required for cooling of appliances, equipment, or material; for controlling dew point, heating, drying, oxidation, dilution, safety exhaust, odor control, and air for compressors; and for comfort and proper working conditions for personnel. [NFPA 54:9.1.7]

507.12 Flammable Vapors. Appliances shall not be installed in areas where the open use, handling, or dispensing of flammable liquids occurs, unless the design, operation, or installation reduces the potential of ignition of the flammable vapors. Appliances installed in compliance with Section 507.13 through Section 507.15 shall be considered to comply with the intent of this provision. [NFPA 54:9.1.9]

507.13 Installation in Residential Garages. Appliances in residential garages and in adjacent spaces that open to the garage and are not part of the living space of a dwelling unit shall be installed so that all burners and burner-ignition devices are located not less than 18 inches (457 mm) above the floor unless listed as flammable vapor ignition resistant. [NFPA 54:9.1.10.1]

507.13.1 Physical Damage. Appliances installed in garages, warehouses, or other areas subject to mechanical damage shall be guarded against such damage by being installed behind protective barriers or by being elevated or located out of the normal path of vehicles.

507.13.2 Access from the Outside. Where appliances are installed in a separate, enclosed space having access only from outside of the garage, such appliances shall be permitted to be installed at floor level, providing the required combustion air is taken from the exterior of the garage. [NFPA 54:9.1.10.3]

507.14 Installation in Commercial Garages. Appliances installed in commercial garages shall comply with Section 507.14.1 and Section 507.14.2.

507.14.1 Parking Structures. Appliances installed in enclosed, basement, and underground parking structures shall be installed in accordance with NFPA 88A. [NFPA 54:9.1.11.1]

507.14.2 Repair Garages. Appliances installed in repair garages shall be installed in accordance with NFPA 30A. [NFPA 54:9.1.11.2]

507.15 Installation in Aircraft Hangars. Heaters in aircraft hangars shall be installed in accordance with NFPA 409. [NFPA 54:9.1.12]

507.16 Venting of Flue Gases. Appliances shall be vented in accordance with the provisions of Section 509.0. [NFPA 54:9.1.14]

507.17 Extra Device or Attachment. No device or attachment shall be installed on any appliance that could in any way impair the combustion of gas. [NFPA 54:9.1.15]

507.18 Adequate Capacity of Piping. When additional appliances are being connected to a gas piping system, the existing piping shall be checked to determine whether it has adequate capacity. Where the capacity is inadequate, the existing system shall be enlarged as necessary, or separate gas piping of adequate capacity shall be run from the point of delivery to the appliance. [NFPA 54:9.1.16]

507.19 Avoiding Strain on Gas Piping. Appliances shall be supported and connected to the piping so as not to exert undue strain on the connections. [NFPA 54:9.1.17]

507.20 Gas Appliance Pressure Regulators. Where the gas supply pressure is higher than that at which the appliance is designed to operate or varies beyond the design pressure limits of the appliance, a gas appliance pressure regulator shall be installed. [NFPA 54:9.1.18]

507.21 Venting of Gas Appliance Pressure Regulators. Venting of gas appliance pressure regulators shall comply with the following requirements:

(1) Appliance pressure regulators requiring access to the atmosphere for successful operation shall be equipped with vent piping leading outdoors or, if the regulator vent is an integral part of the appliance, into the combustion chamber adjacent to a continuous pilot, unless constructed or equipped with a vent limiting means to limit the escape of gas from the vent opening in the event of diaphragm failure.

(2) Vent limiting means shall be employed on listed appliance pressure regulators only.

(3) In the case of vents leading outdoors, means shall be employed to prevent water from entering this piping and also to prevent blockage of vents by insects and foreign matter.

(4) Under no circumstances shall a regulator be vented to the appliance flue or exhaust system.

(5) In the case of vents entering the combustion chamber, the vent shall be located so the escaping gas is readily ignited by the pilot and the heat liberated thereby does not adversely affect the normal operation of the safety shutoff system. The terminus of the vent shall be securely held in a fixed position relative to the pilot. For manufactured gas, the need for a flame arrester in the vent piping shall be determined.

(6) A vent line(s) from an appliance pressure regulator and a bleed line(s) from a diaphragm-type valve shall not be connected to a common manifold terminating in a combustion chamber. Vent lines shall not terminate in positive-pressure-type combustion chambers. [NFPA 54:9.1.19]

507.22 Bleed Lines for Diaphragm-Type Valves. Bleed lines shall comply with the following requirements:

(1) Diaphragm-type valves shall be equipped to convey bleed gas to the outdoors or into the combustion chamber adjacent to a continuous pilot.

(2) In the case of bleed lines leading outdoors, means shall be employed to prevent water from entering this piping and also to prevent blockage of vents by insects and foreign matter.

(3) Bleed lines shall not terminate in the appliance flue or exhaust system.

(4) In the case of bleed lines entering the combustion chamber, the bleed line shall be located so the bleed gas is readily ignited by the pilot and the heat liberated thereby does not adversely affect the normal operation of the safety shutoff system. The terminus of the bleed line shall be securely held in a fixed position relative to the pilot. For manufactured gas, the need for a flame arrester in the bleed line piping shall be determined.

(5) A bleed line(s) from a diaphragm-type valve and a vent line(s) from an appliance pressure regulator shall not be connected to a common manifold terminating in a combustion chamber. Bleed lines shall not terminate in positive-pressure-type combustion chambers. [NFPA 54:9.1.20]

507.23 Combination of Appliances and Equipment. Any combination of appliances, equipment, attachments, or devices used together in any manner shall comply with the standards that apply to the individual appliance and equipment. [NFPA 54:9.1.21]

507.24 Installation Instructions. The installing agency shall conform to the appliance and equipment manufacturer's recommendations in completing an installation. The installing agency shall leave the manufacturer's installation, operating, and maintenance instructions in a location on the premises where they are readily available for reference and guidance of the Authority Having Jurisdiction, service personnel, and the owner or operator. [NFPA 54:9.1.22]

507.25 Protection of Outdoor Appliances. Appliances not listed for outdoor installation but installed outdoors shall be provided with protection to the degree that the environment requires. Appliances listed for outdoor installation shall be permitted to be installed without protection in accordance with the provisions of its listing and the manufacturer's installation instructions.

507.26 Accessibility for Service. All appliances shall be located with respect to building construction and other equipment so as to permit access to the appliance. Sufficient clearance shall be maintained to permit cleaning of heating surfaces; the replacement of filters, blowers, motors, burners, controls, and vent connections; the lubrication of moving parts where necessary; the adjustment and cleaning of burners and pilots; and the proper functioning of explosion vents, if provided. For attic installation, the passageway and servicing area adjacent to the appliance shall be floored. [NFPA 54:9.2.1]

507.27 Clearance to Combustible Materials. Appliances and their vent connectors shall be installed with clearances from combustible material so their operation does not create a hazard to persons or property. Minimum clearances between combustible walls and the back and sides of various conventional types of appliances and their vent connectors are specified in Section 509.0. [NFPA 54:9.2.2]

508.0 Appliances on Roofs.

508.1 General. Appliances on roofs shall be designed or enclosed so as to withstand climatic conditions in the area in which they are installed. Where enclosures are provided, each enclosure shall permit easy entry and movement, shall be of reasonable height, and shall have at least a 30 inch (762 mm) clearance between the entire service access panel(s) of the appliance, and the wall of the enclosure. [NFPA 54:9.4.1.1]

508.1.1 Load Capacity. Roofs on which appliances are to be installed shall be capable of supporting the additional load or shall be reinforced to support the additional load. [NFPA 54:9.4.1.2]

508.1.2 Fasteners. All access locks, screws, and bolts shall be of corrosion-resistant material. [NFPA 54:9.4.1.3]

508.2 Installation of Appliances on Roofs. Appliances shall be installed in accordance with the manufacturer's installation instructions. [NFPA 54:9.4.2.1]

508.2.1 Edge of Roof Clearance. Appliances shall be installed on a well-drained surface of the roof. At least 6 feet (1829 mm) of clearance shall be available between any part of the appliance and the edge of a roof or similar hazard, or rigidly fixed rails, guards, parapets, or other building structures at least 42 inches (1067 mm) in height shall be provided on the exposed side. [NFPA 54:9.4.2.2]

508.2.1.1 Guards and Rails. Guards or rails shall be required where the following exist:

(1) The clearance between the appliance and a roof edge or open end of an equipment platform is less than 6 feet (1829 mm).

(2) The open end of the equipment platform is located more than 30 inches (762 mm) above the roof, floor, or grade below.

Where guards or rails are installed, they shall be constructed so as to prevent the passage of a 21 inch (533 mm) diameter ball, resist the imposed loading conditions, and shall extend not less than 30 inches (762 mm) beyond each side of the equipment or appliance.

Exception: Guards shall not be required where a permanent fall arrest anchorage connector system in accordance with ASSE Z359.1 is installed.

508.2.2 Electrical Power. All appliances requiring an external source of electrical power for its operation shall be provided with the following:

(1) A readily accessible electrical disconnecting means within sight of the appliance that completely de-energizes the appliance.

(2) A 120 V-ac grounding-type receptacle outlet on the roof adjacent to the appliance on the supply side of the disconnect switch. [NFPA 54:9.4.2.3]

508.2.3 Platform or Walkway. Where water stands on the roof at the appliance or in the passageways to the appliance, or where the roof is of a design having a water seal, a suitable platform, walkway, or both shall be provided above the water line. Such platform(s) or walkway(s) shall be located adjacent to the appliance and control panels so that the appliance can be safely serviced where water stands on the roof. [NFPA 54:9.4.2.4]

508.3 Appliances on Roofs. Appliances located on roofs or other elevated locations shall be accessible. [NFPA 54:9.4.3.1]

508.3.1 Access. Buildings of more than 15 feet (4572 mm) in height shall have an inside means of access to the roof, unless other means acceptable to the Authority Having Jurisdiction are used. [NFPA 54:9.4.3.2]

508.3.2 Access Type. The inside means of access shall be a permanent or foldaway inside stairway or ladder, terminating in an enclosure, scuttle, or trapdoor. Such scuttles or trapdoors shall be at least 22 inches by 24 inches (559 mm by 610 mm) in size, shall open easily and safely under all conditions, especially snow, and shall be constructed so as to permit access from the roof side unless deliberately locked on the inside.

At least 6 feet (1829 mm) of clearance shall be available between the access opening and the edge of the roof or similar hazard, or rigidly fixed rails or guards a minimum of 42 inches (1067 mm) in height shall be provided on the exposed side. Where parapets or other building structures are utilized in lieu of guards or rails, they shall be a minimum of 42 inches (1067 mm) in height. [NFPA 54:9.4.3.3]

508.3.3 Permanent Lighting. Permanent lighting shall be provided at the roof access. The switch for such lighting shall be located inside the building near the access means leading to the roof. [NFPA 54:9.4.3.4]

508.4 Appliances in Attics and Under-Floor Spaces. An attic or under-floor space in which an appliance is installed shall be accessible through an opening and passageway, not less than as large as the largest component of the appliance, and not less than 22 inches by 30 inches (559 mm by 762 mm).

508.4.1 Length of Passageway. Where the height of the passageway is less than 6 feet (1829 mm), the distance from the passageway access to the appliance shall not exceed 20 feet (6096 mm) measured along the centerline of the passageway. [NFPA 54:9.5.1.1]

508.4.2 Width of Passageway. The passageway shall be unobstructed and shall have solid flooring not less than 24 inches (610 mm) wide from the entrance opening to the appliance. [NFPA 54:9.5.1.2]

508.4.3 Work Platform. A level working platform not less than 30 inches by 30 inches (762 mm by 762 mm) shall be provided in front of the service side of the appliance. [NFPA 54:9.5.2]

508.4.4 Lighting and Convenience Outlet. A permanent 120 V receptacle outlet and a lighting fixture shall be installed near the appliance. The switch controlling the lighting fixture shall be located at the entrance to the passageway. [NFPA 54:9.5.3]

509.0 Venting of Appliances.

509.1 Listing. Type B and Type B-W gas vents shall comply with UL 441, and Type L gas vents shall comply with UL 641.

509.1.1 Installation. Listed vents shall be installed in accordance with Section 509.0 and the manufacturer's installation instructions. [NFPA 54:12.2.1]

509.1.2 Prohibited Discharge. Appliance vents shall not discharge into a space enclosed by screens having openings less than ¼ of an inch (6.4 mm) mesh.

509.2 Connection to Venting Systems. Except as permitted in Section 509.2.1 through Section 509.2.7, all appliances shall be connected to venting systems. [NFPA 54:12.3.1]

509.2.1 Appliances Not Required to be Vented. The following appliances shall not be required to be vented:

(1) A single listed booster-type (automatic instantaneous) water heater, when designed and used solely for the sanitizing rinse requirements of a dishwashing machine, provided that the appliance is installed with the draft hood in place and unaltered, if a draft hood is required, in a commercial kitchen having a mechanical exhaust system. [Where installed in this manner, the draft hood outlet shall not be less than 36 inches (914 mm) vertically and 6 inches (152 mm) horizontally from any surface other than the appliance.] [NFPA 54:12.3.2(5)]

(2) Other appliances listed for unvented use and not provided with flue collars. [NFPA 54:12.3.2(10)]

509.2.2 Maximum Input Rating. Where any or all of the appliances in Section 509.2.1(1) and Section 509.2.1(2) are installed so the aggregate input rating exceeds 20 (Btu/h)/ft^3 (207 W/m^3) of room or space in which it is installed, one or more shall be provided with venting systems or other approved means for conveying the vent gases to the outdoors so that the aggregate input rating of the remaining unvented appliances does not exceed 20 (Btu/h)/ft^3 (207 W/m^3). [NFPA 54:12.3.2.1]

509.2.3 Adjacent Room or Space. Where the calculation includes the volume of an adjacent room or space, the room or space in which the appliances are installed shall be directly connected to the adjacent room or space by a doorway, archway, or other opening of comparable size that cannot be closed. [NFPA 54:12.3.2.2]

509.2.4 Ventilating Hoods. The use of ventilating hoods and exhaust systems to vent appliances shall be limited to industrial appliances and appliances installed in commercial applications. [NFPA 54:12.3.3]

509.2.5 Well-Ventilated Spaces. The flue gases from industrial-type appliances shall not be required to be vented to the outdoors where such gases are discharged into a large and well-ventilated industrial space. [NFPA 54:12.3.4]

509.2.6 Direct-Vent Appliances. Listed direct-vent appliances shall be installed in accordance with the manufacturer's installation instructions and Section 509.8.2. [NFPA 54:12.3.5]

509.2.7 Appliances with Integral Vents. Appliances incorporating integral venting means shall be installed in accordance with the manufacturer's installation instructions, Section 509.8 and Section 509.8.1. [NFPA 54:12.3.6]

509.3 Minimum Safe Performance. Venting systems shall be designed and constructed to convey all flue and vent gases to the outdoors. [NFPA 54:12.1]

509.3.1 Appliance Draft Requirements. A venting system shall satisfy the draft requirements of the appliance in accordance with the manufacturer's instructions. [NFPA 54:12.4.1]

509.3.2 Appliance Venting Requirements. Appliances required to be vented shall be connected to a venting system designed and installed in accordance with the provisions of Section 509.4 through Section 509.15. [NFPA 54:12.4.2]

509.3.3 Mechanical Draft Systems. Mechanical draft systems shall be listed in accordance with UL 378 and installed in accordance with both the appliance and the mechanical draft system manufacturer's installation instructions. [NFPA 54:12.4.3.1]

509.3.3.1 Venting. Appliances requiring venting shall be permitted to be vented by means of mechanical draft systems of either forced or induced draft design. [NFPA 54:12.4.3.2]

509.3.3.2 Leakage. Forced draft systems and all portions of induced draft systems under positive pressure during operation shall be designed and installed so as to prevent leakage of flue or vent gases into a building. [NFPA 54:12.4.3.3]

509.3.3.3 Vent Connectors. Vent connectors serving appliances vented by natural draft shall not be connected into any portion of mechanical draft systems operating under positive pressure. [NFPA 54:12.4.3.4]

509.3.3.4 Operation. Where a mechanical draft system is employed, provision shall be made to prevent the flow of gas to the main burners when the draft system is not performing so as to satisfy the operating requirements of the appliance for safe performance. [NFPA 54:12.4.3.5]

509.3.3.5 Exit Terminals. The exit terminals of mechanical draft systems shall be not less than 7 feet (2134 mm) above finished ground level where located adjacent to public walkways and shall be located as specified in Section 509.8 and Section 509.8.1 of this code. [NFPA 54:12.4.3.6]

509.3.4 Ventilating Hoods and Exhaust Systems. Where automatically operated appliances, other than commercial cooking appliances, are vented through a ventilating hood or exhaust system equipped with a damper or with a power means of exhaust, provisions shall be made to allow the flow of gas to the main burners only when the damper is open to a position to properly vent the appliance and when the power means of exhaust is in operation. [NFPA 54:12.4.4.1]

509.3.5 Circulating Air Ducts, Above-Ceiling Air-Handling Spaces, and Furnace Plenums. Venting systems shall not extend into or pass through any fabricated air duct or furnace plenum. [NFPA 54:12.4.5.1]

509.3.6 Above-Ceiling or Nonducted Air Handling System. Where a venting system passes through an above-ceiling air space or other nonducted portion of an air-handling system, it shall conform to one of the following requirements:

(1) The venting system shall be a listed special gas vent, other system serving a Category III or Category IV appliance, or other positive pressure vent, with joints sealed in accordance with the appliance or vent manufacturer's instructions.

(2) The vent system shall be installed such that no fittings or joints between sections are installed in the above-ceiling space.

(3) The venting system shall be installed in a conduit or enclosure with joints between the interior of the enclosure and the ceiling space sealed. [NFPA 54:12.4.5.2]

509.4 Type of Venting System to be Used. The type of venting system to be used shall be in accordance with Table 509.4. [NFPA 54:12.5.1]

509.4.1 Plastic Piping. Where plastic piping is used to vent an appliance, the appliance shall be listed for use with such venting materials and the appliance manufacturer's installation instructions shall identify the specific plastic piping material. The plastic pipe venting materials shall be labeled in accordance with the product standards specified by the appliance manufacturer or shall be listed and labeled in accordance with UL 1738. [NFPA 54:12.5.2]

509.4.2 Plastic Vent Joints. Plastic pipe and fittings used to vent appliances shall be installed in accordance with the appliance manufacturer's installation instructions. Plastic pipe venting materials listed and labeled in accordance with UL 1738 shall be installed in accordance with the vent manufacturer's installation instructions. Where primer is required, it shall be of a contrasting color. [NFPA 54:12.5.3]

509.4.3 Special Gas Vents. Special gas vents shall be listed and labeled in accordance with UL 1738 and installed in accordance with the special gas vent manufacturer's installation instructions. [NFPA 54:12.5.4]

TABLE 509.4
TYPE OF VENTING SYSTEM TO BE USED
[NFPA 54: TABLE 12.5.1]

APPLIANCES	TYPE OF VENTING SYSTEM	LOCATION OF REQUIREMENTS
Listed Category I appliances Listed appliances equipped with draft hood Appliances listed for use with Type B gas vent	Type B gas vent	Section 509.6
	Chimney	Section 509.5
	Single-wall metal pipe	Section 509.7
	Listed chimney lining system for gas venting	Section 509.5.3
	Special gas vent listed for these appliances	Section 509.4.3
Listed vented wall furnaces	Type B-W gas vent	Section 509.6
Category II, Category III , and Category IV appliances	As specified or furnished by manufacturers of listed appliances	Section 509.4.1 and Section 509.4.3
Appliances that can be converted to use solid fuel Unlisted combination gas- and oil-burning appliances Combination gas- and solid fuel-burning appliances Appliances listed for use with chimneys only Unlisted appliances	Chimney	Section 509.5
Listed combination gas- and oil-burning appliances	Type L vent	Section 509.6
	Chimney	Section 509.5
Decorative appliance in vented fireplace	Chimney	UMC Section 911.2
Gas-fired toilets	Single-wall metal pipe	Section 509.7
Direct-vent appliances	—	Section 509.2.6
Appliances with integral vents	—	Section 509.2.7

509.5 Masonry, Metal, and Factory-Built Chimneys. Chimneys shall be installed in accordance with Section 509.5.1 through Section 509.5.3.

509.5.1 Factory-Built Chimneys. Factory-built chimneys shall be listed in accordance with UL 103, UL 959, or UL 2561. Factory-built chimneys shall be installed in accordance with the manufacturer's installation instructions. Factory-built chimneys used to vent appliances that operate at positive vent pressure shall be listed for such application. [NFPA 54:12.6.1.1]

509.5.1.1 Decorative Shrouds. Decorative shrouds addressed in Section 509.5.4.3 shall comply with UL 103 for factory -built residential chimneys.

509.5.2 Metal Chimneys. Metal chimneys shall be built and installed in accordance with NFPA 211. [NFPA 54:12.6.1.2]

509.5.3 Masonry Chimneys. Masonry chimneys shall be built and installed in accordance with NFPA 211 and lined with one of the following:

(1) Approved clay flue lining

(2) A chimney lining system listed and labeled in accordance with UL 1777

(3) Other approved material that resists corrosion, erosion, softening, or cracking from vent gases at temperatures up to 1800°F (982°C)

Exception: Masonry chimney flues lined with a chimney lining system specifically listed for use with listed appliances with draft hoods, Category I appliances, and other appliances listed for use with Type B vents shall be permitted. The liner shall be installed in accordance with the liner manufacturer's installation instructions. A permanent identifying label shall be attached at the point where the connection is to be made to the liner. The label shall read "This chimney liner is for appliances that burn gas only. Do not connect to solid or liquid fuel-burning appliances or incinerators." [NFPA 54:12.6.1.3]

509.5.4 Termination. A chimney for residential-type or low-heat appliances shall extend at least 3 feet (914 mm) above the highest point where it passes through a roof of a building and at least 2 feet (610 mm) higher than any portion of a building within a horizontal distance of 10 feet (3048 mm). [NFPA 54:12.6.2.1] (See Figure 509.5.4)

509.5.4.1 Medium-Heat Gas Appliances. A chimney for medium-heat appliances shall extend at least 10 feet (3048 mm) higher than any portion of any building within 25 feet (7620 mm). [NFPA 54:12.6.2.2]

509.5.4.2 Chimney Height. A chimney shall extend at least 5 feet (1524 mm) above the highest connected appliance draft hood outlet or flue collar. [NFPA 54:12.6.2.3]

509.5.4.3 Decorative Shrouds. Decorative shrouds shall not be installed at the termination of factory-built chimneys except where such shrouds are listed and labeled for use with the specific factory-built chimney system and are installed in accordance with the manufacturer's installation instructions. [NFPA 54:12.6.2.4]

509.5.5 Size of Chimneys. The effective area of a chimney venting system serving listed appliances with draft hoods, Category I appliances, and other appliances listed for use with Type B vents shall be in accordance with one of the following methods:

(1) Those listed in Section 510.0.

(2) For sizing an individual chimney venting system for a single appliance with a draft hood, the effective areas of the vent connector and chimney flue shall be not less than the area of the appliance flue collar or draft hood outlet or greater than seven times the draft hood outlet area.

(3) For sizing a chimney venting system connected to two appliances with draft hoods, the effective area of the chimney flue shall be not less than the area of the larger draft hood outlet plus 50 percent of the area of the smaller draft hood outlet or greater than seven times the smaller draft hood outlet area.

(4) Chimney venting systems using mechanical draft shall be sized in accordance with approved engineering methods.

(5) Other approved engineering methods. [NFPA 54:12.6.3.1]

509.5.6 Inspection of Chimneys. Before replacing an existing appliance or connecting a vent connector to a chimney, the chimney passageway shall be examined to ascertain that it is clear and free of obstructions and shall be cleaned if previously used for venting solid or liquid fuel-burning appliances or fireplaces. [NFPA 54:12.6.4.1]

509.5.6.1 Standard. Chimneys shall be lined in accordance with NFPA 211.

Exception: Existing chimneys shall be permitted to have their use continued when an appliance is replaced by an appliance of similar type, input rating, and efficiency, where the chimney complies with Section 509.5.6 through Section 509.5.6.3 and the sizing of the chimney is in accordance with Section 509.5.5. [NFPA 54:12.6.4.2]

509.5.6.2 Cleanouts. Cleanouts shall be examined and where they do not remain tightly closed when not in use they shall be repaired or replaced. [NFPA 54:12.6.4.3]

509.5.6.3 Existing Chimney. When inspection reveals that an existing chimney is not safe for the intended application, it shall be repaired, rebuilt, lined, relined, or replaced with a vent or chimney to conform to NFPA 211 and shall be suitable for the appliances to be attached. [NFPA 54:12.6.4.4]

509.5.7 Chimney Serving Appliances Burning Other Fuels. An appliance shall not be connected to a chimney flue serving a separate appliance designed to burn solid fuel. [NFPA 54:12.6.5.1]

509.5.7.1 Gas and Liquid Fuel-Burning Appliances. Where one chimney serves gas appliances and liquid fuel-burning appliances, the appliances shall be connected through separate openings or connected through a single opening where joined by a suitable fitting located as close as practical to the chimney. Where two or more openings are provided into one chimney flue, they shall be at different levels. Where the gas appliance is automatically controlled, it shall be equipped with a safety shutoff device. [NFPA 54:12.6.5.2]

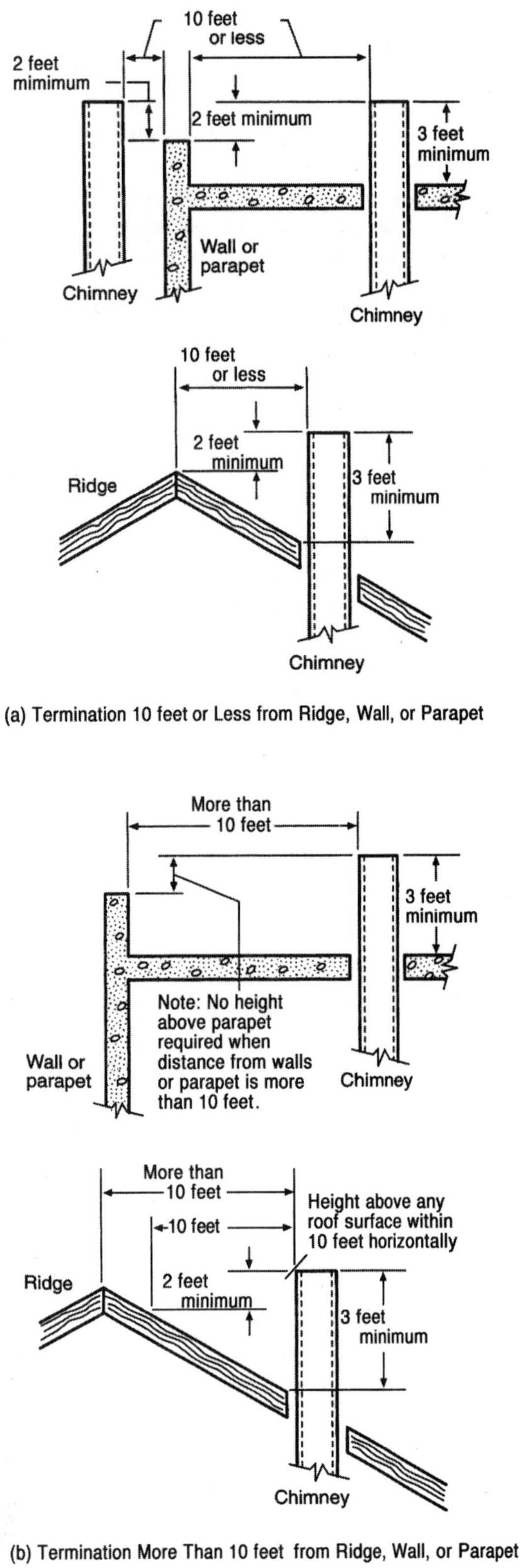

For SI units: 1 foot = 304.8 mm

FIGURE 509.5.4
TYPICAL TERMINATION LOCATIONS FOR CHIMNEYS AND SINGLE-WALL METAL PIPES SERVING RESIDENTIAL-TYPE AND LOW-HEAT APPLIANCE
[NFPA: 54: FIGURE A.12.6.2.1]

509.5.7.2 Gas and Solid Fuel-Burning Appliances. A listed combination gas- and solid-fuel-burning appliance connected to a single chimney flue shall be equipped with a manual reset device to shut off gas to the main burner in the event of sustained backdraft or flue gas spillage. The chimney flue shall be sized to properly vent the appliance. [NFPA 54:12.6.5.3]

509.5.7.3 Combination Gas- and Oil-Burning Appliances. A single chimney flue serving a listed combination gas- and oil-burning appliance shall be sized in accordance with the appliance manufacturer's instructions. [NFPA 54:12.6.5.4]

509.5.8 Support of Chimneys. All portions of chimneys shall be supported for the design and weight of the materials employed. Listed factory-built chimneys shall be supported and spaced in accordance with the manufacturer's installation instructions. [NFPA 54:12.6.6]

509.5.9 Cleanouts. Where a chimney that formerly carried flue products from liquid or solid fuel-burning appliances is used with an appliance using fuel gas, an accessible cleanout shall be provided. The cleanout shall have a tight-fitting cover and be installed so its upper edge is at least 6 inches (152 mm) below the lower edge of the lowest chimney inlet opening. [NFPA 54:12.6.7]

509.5.10 Space Surrounding Lining or Vent. The remaining space surrounding a chimney liner, gas vent, special gas vent, or plastic piping installed within a masonry chimney shall not be used to vent another appliance.

Exception: The insertion of another liner or vent within the chimney as provided in this code and the liner or vent manufacturer's instructions. [NFPA 54:12.6.8.1]

509.5.10.1 Combustion Air. The remaining space surrounding a chimney liner, gas vent, special gas vent, or plastic piping installed within a masonry, metal, or factory-built chimney flue shall not be used to supply combustion air.

Exception: Direct-vent appliances designed for installation in a solid fuel-burning fireplace where installed in accordance with the manufacturer's installation instructions. [NFPA 54:12.6.8.2]

509.6 Gas Vents. The installation of gas vents shall meet the following requirements:

(1) Gas vents shall be installed in accordance with the manufacturer's installation instructions.

(2) A Type B-W gas vent shall have a listed capacity not less than that of the listed vented wall furnace to which it is connected.

(3) Gas vents installed within masonry chimneys shall be installed in accordance with the manufacturer's installation instructions. Gas vents installed within masonry chimneys shall be identified with a permanent label installed at the point where the vent enters the chimney. The label shall contain the following language: "This gas vent is for appliances that burn gas. Do not connect to solid or liquid fuel–burning appliances or incinerators."

(4) Screws, rivets, and other fasteners shall not penetrate the inner wall of double-wall gas vents, except at the transition from the appliance draft hood outlet, flue collar, or single-wall metal connector to a double-wall vent. [NFPA 54:12.7.2]

509.6.1 Gas Vent Termination. The termination of gas vents shall comply with the following requirements:

(1) A gas vent shall terminate in accordance with one of the following:

(a) Gas vents that are 12 inches (300 mm) or less in size and located not less than 8 feet (2438 mm) from a vertical wall or similar obstruction shall terminate above the roof in accordance with Figure 509.6.1 and Table 509.6.1.

(b) Gas vents that are over 12 inches (300 mm) in size or are located less than 8 feet (2438 mm) from a vertical wall or similar obstruction shall terminate not less than 2 feet (610 mm) above the highest point where they pass through the roof and not less than 2 feet (610 mm) above any portion of a building within 10 feet (3048 mm) horizontally.

(c) Industrial appliances as provided in Section 509.2.5.

(d) Direct-vent systems as provided in Section 509.2.6.

(e) Appliances with integral vents as provided in Section 509.2.7.

(f) Mechanical draft systems as provided in Section 509.3.3 through Section 509.3.3.5.

(g) Ventilating hoods and exhaust systems as provided in Section 509.3.4.

(2) A Type B or a Type L gas vent shall terminate at least 5 feet (1524 mm) in vertical height above the highest connected appliance draft hood or flue collar.

(3) A Type B-W gas vent shall terminate at least 12 feet (3658 mm) in vertical height above the bottom of the wall furnace.

(4) A gas vent extending through an exterior wall shall not terminate adjacent to the wall or below eaves or parapets, except as provided in Section 509.2.6 and Section 509.3.3 through Section 509.3.3.5.

(5) Decorative shrouds shall not be installed at the termination of gas vents except where such shrouds are listed for use with the specific gas venting system and are installed in accordance with the manufacturer's installation instructions.

(6) All gas vents shall extend through the roof flashing, roof jack, or roof thimble and terminate with a listed cap or listed roof assembly.

(7) A gas vent shall terminate at least 3 feet (914 mm) above a forced air inlet located within 10 feet (3048 mm). [NFPA 54:12.7.3]

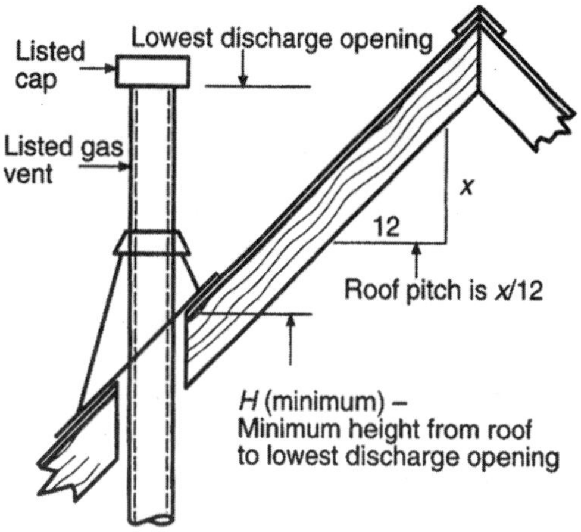

FIGURE 509.6.1
TERMINATION LOCATIONS FOR GAS VENTS WITH LISTED CAPS 12 INCHES OR LESS IN SIZE AT LEAST 8 FEET FROM A VERTICAL WALL [NFPA 54: FIGURE 12.7.3]

TABLE 509.6.1
ROOF SLOPE HEIGHTS
[NFPA 54: TABLE 12.7.3]

ROOF SLOPE	H (minimum) (feet)
Flat to 6/12	1.0
Over 6/12 to 7/12	1.25
Over 7/12 to 8/12	1.5
Over 8/12 to 9/12	2.0
Over 9/12 to 10/12	2.5
Over 10/12 to 11/12	3.25
Over 11/12 to 12/12	4.0
Over 12/12 to 14/12	5.0
Over 14/12 to 16/12	6.0
Over 16/12 to 18/12	7.0
Over 18/12 to 20/12	7.5
Over 20/12 to 21/12	8.0

For SI Units: 1 inch = 25.4 mm, 1 foot = 304.8 mm

509.6.1.1 Insulation Shield. Where a vent passes through an insulated assembly, an approved metal shield shall be installed between the vent and insulation. The shield shall extend not less than 2 inches (51 mm) above the insulation and be secured to the structure in accordance with the manufacturer's installation instructions.

509.6.2 Size of Gas Vents. Venting systems shall be sized and constructed in accordance with Section 509.6.2.1 through Section 509.6.2.3 and the appliance manufacturer's instructions. [NFPA 54:12.7.4]

509.6.2.1 Category I Appliances. The sizing of natural draft venting systems serving one or more listed appliances equipped with a draft hood or appliances listed for use with a Type B gas vent, installed in a single story of a building, shall be in accordance with one of the following:

(1) The provisions of Section 510.0.

(2) Vents serving fan-assisted combustion system appliances, or combinations of fan-assisted combustion system and draft hood-equipped appliances, shall be sized in accordance with Section 510.0 or other approved engineering methods.

(3) For sizing an individual gas vent for a single, draft hood-equipped appliance, the effective area of the vent connector and the gas vent shall be not less than the area of the appliance draft hood outlet or greater than seven times the draft hood outlet area.

(4) For sizing a gas vent connected to two appliances with draft hoods, the effective area of the vent shall be not less than the area of the larger draft hood outlet plus 50 percent of the area of the smaller draft hood outlet or greater than seven times the smaller draft hood outlet area.

(5) Other approved engineering practices. [NFPA 54:12.7.4.1]

509.6.2.2 Vent Offsets. Type B and Type L vents sized in accordance with Section 509.6.2.1(3) or Section 509.6.2.1(4) shall extend in a generally vertical direction with offsets not exceeding 45 degrees (0.79 rad), except that a vent system having not more than one 60 degree (1.05 rad) offset shall be permitted. Any angle greater than 45 degrees (0.79 rad) from the vertical is considered horizontal. The total horizontal distance of a vent plus the horizontal vent connector serving draft hood-equipped appliances shall not be greater than 75 percent of the vertical height of the vent. [NFPA 54:12.7.4.2]

509.6.2.3 Category II, Category III, and Category IV Appliances. The sizing of gas vents for Category II, Category III, and Category IV appliances shall be in accordance with the appliance manufacturer's instructions. The sizing of plastic pipe specified by the appliance manufacturer as a venting material for Category II, III, and IV appliances shall be in accordance with the appliance manufacturers' instructions. [NFPA 54:12.7.4.3]

509.6.2.4 Sizing. Chimney venting systems using mechanical draft shall be sized in accordance with approved engineering methods. [NFPA 54:12.7.4.4]

509.6.3 Gas Vents Serving Appliances on More than One Floor. A common vent shall be permitted in multistory installations to vent Category I appliances located on more than one floor level, provided the venting system is designed and installed in accordance with approved engineering methods.

For the purpose of this section, crawl spaces, basements, and attics shall be considered floor levels. [NFPA 54:12.7.5.1]

509.6.3.1 Occupiable Space. All appliances connected to the common vent shall be located in rooms separated from occupiable space. Each of these rooms shall have provisions for an adequate supply of combustion, ventilation, and dilution air that is not supplied from occupiable space. [NFPA 54:12.7.5.2] (See Figure 509.6.3.1)

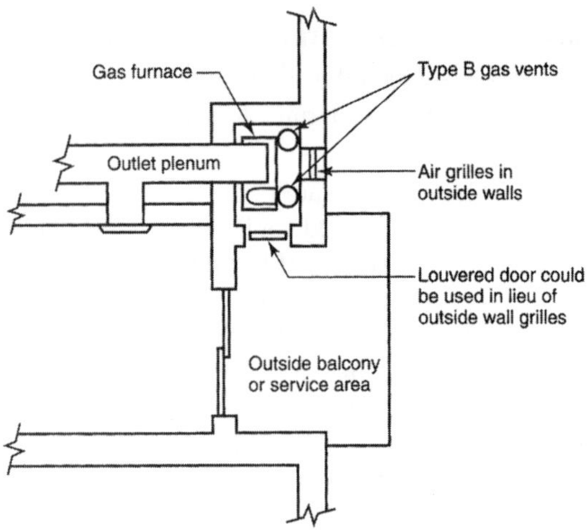

FIGURE 509.6.3.1
PLAN VIEW OF PRACTICAL SEPARATION METHOD FOR MULTISTORY GAS VENTING [NFPA 54: FIGURE A.12.7.5.2]

509.6.3.2 Multistory Venting System. The size of the connectors and common segments of multistory venting systems for appliances listed for use with a Type B double-wall gas vent shall be in accordance with Table 510.2(1), provided all of the following apply:

(1) The available total height (*H*) for each segment of a multistory venting system is the vertical distance between the level of the highest draft hood outlet or flue collar on that floor and the centerline of the next highest interconnection tee.

(2) The size of the connector for a segment is determined from the appliance's gas input rate and available connector rise and shall not be smaller than the draft hood outlet or flue collar size.

(3) The size of the common vertical vent segment, and of the interconnection tee at the base of that segment, is based on the total appliance's gas input rate entering that segment and its available total height. [NFPA 54:12.7.5.3]

509.6.4 Support of Gas Vents. Gas vents shall be supported and spaced in accordance with the manufacturer's installation instructions. [NFPA 54:12.7.6]

509.6.5 Marking. In those localities where solid and liquid fuels are used extensively, gas vents shall be permanently identified by a label attached to the wall or ceiling at a point where the vent connector enters the gas vent. The label shall read: "This gas vent is for appliances that burn gas. Do not connect to solid or liquid fuel-burning appliances or incinerators." The Authority Having Jurisdiction shall determine whether its area constitutes such a locality. [NFPA 54:12.7.7]

509.7 Single-Wall Metal Pipe. Single-wall metal pipe shall be constructed of galvanized sheet steel not less than 0.0304 of an inch (0.7722 mm) thick or of other approved, noncombustible, corrosion-resistant material. [NFPA 54:12.8.1]

509.7.1 Cold Climate. Uninsulated single-wall metal pipe shall not be used outdoors for venting appliances in regions where the 99 percent winter design temperature is below 32°F (0°C). [NFPA 54:12.8.2]

509.7.2 Termination. The termination of single-wall metal pipe shall meet the following requirements:

(1) Single-wall metal pipe shall terminate at least 5 feet (1524 mm) in vertical height above the highest connected appliance draft hood outlet or flue collar. [NFPA 54:12.8.3(1)]

(2) Single-wall metal pipe shall extend at least 2 feet (610 mm) above the highest point where it passes through a roof of a building and at least 2 feet (610 mm) higher than any portion of a building within a horizontal distance of 10 feet (3048 mm). [NFPA 54:12.8.3(2)] (See Figure 509.5.4)

(3) An approved cap or roof assembly shall be attached to the terminus of a single-wall metal pipe. [NFPA 54:12.8.3(3)]

509.7.3 Installation with Appliances Permitted by Section 509.4. Single-wall metal pipe shall not be used as a vent in dwellings and residential occupancies. [NFPA 54:12.8.4.1]

509.7.3.1 Limitations. Single-wall metal pipe shall be used only for runs directly from the space in which the appliance is located through the roof or exterior wall to the outer air. A pipe passing through a roof shall extend without interruption through the roof flashing, roof jacket, or roof thimble. [NFPA 54:12.8.4.2]

509.7.3.2 Attic or Concealed Space. Single-wall metal pipe shall not originate in any unoccupied attic or concealed space and shall not pass through any attic, inside wall, concealed space, or floor. [NFPA 54:12.8.4.3]

509.7.3.3 Incinerator. Single-wall metal pipe used for venting an incinerator shall be exposed and readily examinable for its full length and shall have required clearances maintained.

509.7.3.4 Clearances. Minimum clearances from single-wall metal pipe to combustible material shall be in accordance with Table 509.7.3.4(1). Reduced clearances from single-wall metal pipe to combustible material shall be as specified for vent connectors in Table 509.7.3.4(2). [NFPA 54:12.8.4.4]

509.7.3.5 Combustible Exterior Wall. Single-wall metal pipe shall not pass through a combustible exterior wall unless guarded at the point of passage by a ventilated metal thimble not smaller than the following:

(1) For listed appliances with draft hoods and appliances listed for use with Type B gas vents, the thimble shall be a minimum of 4 inches (100 mm) larger in diameter than the metal pipe. Where there is a run of not less than 6 feet (1829 mm) of metal pipe in the opening between the draft hood outlet and the thimble, the thimble shall be a minimum of 2 inches (50 mm) larger in diameter than the metal pipe.

(2) For unlisted appliances having draft hoods, the thimble shall be a minimum of 6 inches (150 mm) larger in diameter than the metal pipe.

(3) For residential and low-heat appliances, the thimble shall be a minimum of 12 inches (300 mm) larger in diameter than the metal pipe.

Exception: In lieu of thimble protection, all combustible material in the wall shall be removed a sufficient distance from the metal pipe to provide the specified clearance from such metal pipe to combustible material. Any material used to close up such opening shall be noncombustible. [NFPA 54:12.8.4.6]

509.7.3.6 Roof Thimble. Where a single-wall metal pipe passes through a roof constructed of combustible material, a noncombustible, nonventilating

TABLE 509.7.3.4(1)
CLEARANCES FOR CONNECTORS
[NFPA 54: TABLE 12.8.4.4]*

	MINIMUM DISTANCE FROM COMBUSTIBLE MATERIAL (inches)			
APPLIANCE	LISTED TYPE B GAS VENT MATERIAL	LISTED TYPE L VENT MATERIAL	SINGLE-WALL METAL PIPE	FACTORY-BUILT CHIMNEY SECTIONS
Listed appliances with draft hoods and appliances listed for use with Type B gas vents	As listed	As listed	6	As listed
Residential boilers and furnaces with listed gas conversion burner and with draft hood	6	6	9	As listed
Residential appliances listed for use with Type L vents	Not permitted	As listed	9	As listed
Unlisted residential appliances with draft hood	Not permitted	6	9	As listed
Residential and low-heat appliances other than those above	Not permitted	9	18	As listed
Medium-heat appliances	Not permitted	Not permitted	36	As listed

For SI units: 1 inch = 25.4 mm

* These clearances shall apply unless the installation instructions of a listed appliance or connector specify different clearances, in which case the listed clearances shall apply.

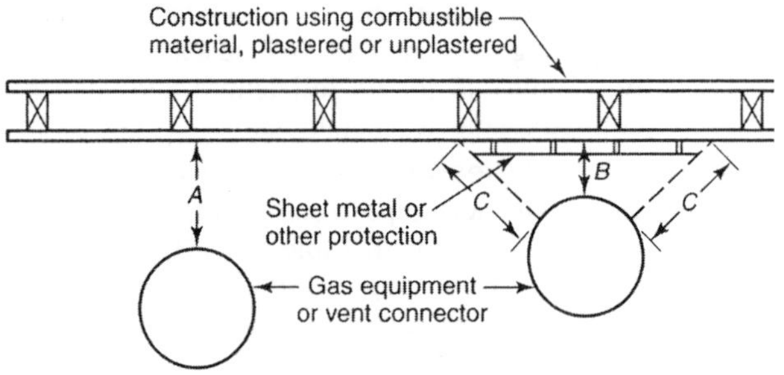

Notes:

[1] *A* – Equals the clearance with no protection specified in Table 509.7.3.4(1) and Table 509.7.3.4(2) and in the sections applying to various types of equipment.

[2] *B* – Equals the reduced clearance permitted in accordance with Table 509.7.3.4(2).

[3] The protection applied to the construction using combustible material shall extend far enough in each direction to make C equal to A.

FIGURE 509.7.3.4(1)[1, 2, 3]
EXTENT OF PROTECTION NECESSARY TO REDUCE CLEARANCES FROM GAS APPLIANCE OR VENT CONNECTORS
[NFPA 54: FIGURE 10.3.2.3(a)]

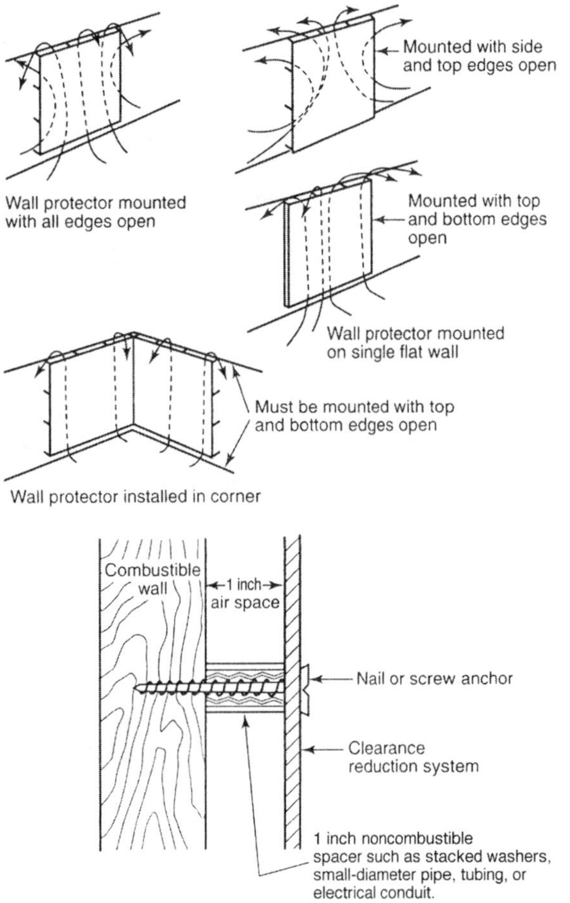

For SI units: 1 inch = 25.4 mm

Note: Masonry walls shall be attached to combustible walls using wall ties. Spacers shall not be used directly behind appliance or connector.

FIGURE 509.7.3.4(2)
WALL PROTECTOR CLEARANCE REDUCTION SYSTEM
[NFPA 54: FIGURE 10.3.2.3(b)]

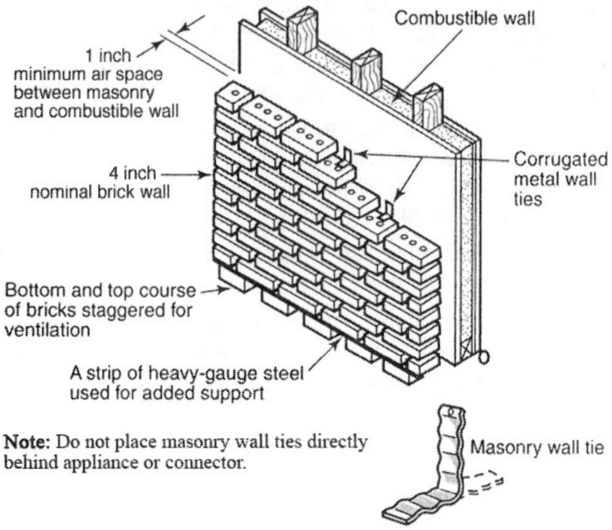

Note: Do not place masonry wall ties directly behind appliance or connector.

For SI units: 1 inch = 25.4 mm

FIGURE 509.7.3.4(3)
MASONRY CLEARANCE REDUCTION SYSTEM
[NFPA 54: FIGURE 10.3.2.3(c)]

thimble shall be used at the point of passage. The thimble shall extend at least 18 inches (457 mm) above and 6 inches (152 mm) below the roof with the annular space open at the bottom and closed only at the top. The thimble shall be sized in accordance with Section 509.7.3.5. [NFPA 54:12.8.4.5]

509.7.4 Size of Single-Wall Metal Pipe. Single-wall metal piping shall comply with the following requirements:

(1) A venting system of a single-wall metal pipe shall be sized in accordance with one of the following methods and the appliance manufacturer's instructions:

(a) For a draft hood-equipped appliance, in accordance with Section 510.0.

(b) For a venting system for a single appliance with a draft hood, the areas of the connector and the pipe each shall not be less than the area of the appliance flue collar or draft hood outlet, whichever is smaller. The vent area shall not be greater than seven times the draft hood outlet area.

(c) Other approved engineering methods.

(2) Where a single-wall metal pipe is used and has a shape other than round, it shall have an equivalent effective area equal to the effective area of the round pipe for which it is substituted and the minimum internal dimension of the pipe shall be 2 inches (50 mm).

(3) The vent cap or a roof assembly shall have a venting capacity not less than that of the pipe to which it is attached. [NFPA 54:12.8.5]

509.7.5 Support of Single-Wall Metal Pipe. All portions of single-wall metal pipe shall be supported for the design and weight of the material employed. [NFPA 54:12.8.6]

TABLE 509.7.3.4(2)
REDUCTION OF CLEARANCES WITH SPECIFIED FORMS OF PROTECTION[1, 2, 3, 4, 5, 6, 7, 8, 9, 10, 11]
[NFPA 54: TABLE 10.2.3]

TYPE OF PROTECTION APPLIED TO AND COVERING ALL SURFACES OF COMBUSTIBLE MATERIAL WITHIN THE DISTANCE SPECIFIED AS THE REQUIRED CLEARANCE WITH NO PROTECTION [SEE FIGURE 509.7.3.4(1) THROUGH FIGURE 509.7.3.4(3)]	WHERE THE REQUIRED CLEARANCE WITH NO PROTECTION FROM APPLIANCE, VENT CONNECTOR, OR SINGLE-WALL METAL PIPE IS:									
	36 (inches)		18 (inches)		12 (inches)		9 (inches)		6 (inches)	
	ALLOWABLE CLEARANCES WITH SPECIFIED PROTECTION (inches)									
	USE COLUMN 1 FOR CLEARANCES ABOVE APPLIANCE OR HORIZONTAL CONNECTOR. USE COLUMN 2 FOR CLEARANCES FROM APPLIANCES, VERTICAL CONNECTOR, AND SINGLE-WALL METAL PIPE.									
	ABOVE COLUMN 1	SIDES AND REAR COLUMN 2	ABOVE COLUMN 1	SIDES AND REAR COLUMN 2	ABOVE COLUMN 1	SIDES AND REAR COLUMN 2	ABOVE COLUMN 1	SIDES AND REAR COLUMN 2	ABOVE COLUMN 1	SIDES AND REAR COLUMN 2
(1) 3½ inch thick masonry wall without ventilated air space	—	24	—	12	—	9	—	6	—	5
(2) ½ of an inch insulation board over 1 inch glass fiber or mineral wool batts	24	18	12	9	9	6	6	5	4	3
(3) 0.024 inch (nominal 24 gauge) sheet metal over 1 inch glass fiber or mineral wool batts reinforced with wire on rear face with ventilated air space	18	12	9	6	6	4	5	3	3	3
(4) 3½ inch thick masonry wall with ventilated air space	—	12	—	6	—	6	—	6	—	6
(5) 0.024 inch (nominal 24 gauge) sheet metal with ventilated air space	18	12	9	6	6	4	5	3	3	2
(6) ½ of an inch thick insulation board with ventilated air space	18	12	9	6	6	4	5	3	3	3
(7) 0.024 inch (nominal 24 gauge) sheet metal with ventilated air space over 0.024 inch (nominal 24 gauge) sheet metal with ventilated air space	18	12	9	6	6	4	5	3	3	3
(8) 1 inch glass fiber or mineral wool batts sandwiched between two sheets 0.024 inch (nominal 24 gauge) sheet metal with ventilated air space	18	12	9	6	6	4	5	3	3	3

For SI units: 1 inch = 25.4 mm, °C = (°F-32)/1.8

Notes:

[1] Reduction of clearances from combustible materials shall not interfere with combustion air, draft hood clearance and relief, and accessibility of servicing.

[2] All clearances shall be measured from the outer surface of the combustible material to the nearest point on the surface of the appliance, disregarding any intervening protection applied to the combustible material.

[3] Spacers and ties shall be of noncombustible material. No spacer or tie shall be used directly opposite the appliance or connector.

[4] Where all clearance reduction systems use a ventilated air space, adequate provision for air circulation shall be provided as described. [See Figure 509.7.3.4(2) and Figure 509.7.3.4(3)]

[5] At least 1 inch (25.4 mm) shall be between clearance reduction systems and combustible walls and ceilings for reduction systems using a ventilated air space.

[6] Where a wall protector is mounted on a single flat wall away from corners, it shall have a minimum 1 inch (25.4 mm) air gap. To provide adequate air circulation, the bottom and top edges, or only the side and top edges, or edges shall be left open.

[7] Mineral wool batts (blanket or board) shall have a minimum density of 8 pounds per cubic foot (lb/ft^3) (128 kg/m^3) and a minimum melting point of 1500°F (816°C).

[8] Insulation material used as part of a clearance reduction system shall have a thermal conductivity of 1 British thermal unit inch per hour square foot degree Fahrenheit [Btu•in/(h•ft^2•°F)] [0.1 W/(m•K)] or less.

[9] At least 1 inch (25.4 mm) shall be between the appliance and the protector. In no case shall the clearance between the appliance and the combustible surface be reduced below that allowed in this table.

[10] All clearances and thicknesses are minimum; larger clearances and thicknesses are acceptable.

[11] Listed single-wall connectors shall be installed in accordance with the terms of their listing and the manufacturer's installation instructions.

509.7.6 Marking. Single-wall metal pipe shall comply with the marking provisions of Section 509.6.5. [NFPA 54:12.8.7]

509.8 Through-the-Wall Vent Termination. A mechanical draft venting system shall terminate at least 3 feet (914 mm) above any forced air inlet located within 10 feet (3048 mm) (See Figure 509.8).

Exceptions:

(1) This provision shall not apply to the combustion air intake of a direct-vent appliance.

(2) This provision shall not apply to the separation of the integral outdoor air inlet and flue gas discharge of listed outdoor appliances. [NFPA 54:12.9.1]

509.8.1 Mechanical Draft Venting System. A mechanical draft venting system of other than direct-vent type shall terminate at least 4 feet (1219 mm) below, 4 feet (1219 mm) horizontally from, or 1 foot (305 mm) above any door, operable window, or gravity air inlet into any building. The bottom of the vent terminal shall be located at least 12 inches (305 mm) above finished ground level. [NFPA 54:12.9.2]

509.8.2 Direct-Vent Appliance. The clearances for through-the-wall direct-vent terminals shall be in accordance with Table 509.8.2. The bottom of the vent terminal and the air intake shall be located not less than 12 inches (305 mm) above finished ground level. [NFPA 54:12.9.3]

TABLE 509.8.2
THROUGH-THE-WALL DIRECT-VENT TERMINATION CLEARANCES
[NFPA 54: TABLE 12.9.3]

DIRECT-VENT APPLIANCE INPUT RATING	THROUGH-THE-WALL VENT TERMINAL CLEARANCE FROM ANY AIR OPENING INTO A BUILDING (inches)
10 000 Btu/h and less	6
Greater than 10 000 Btu/h and not exceeding 50 000 Btu/h	9
Greater than 50 000 Btu/h and not exceeding 150 000 Btu/h	12
> 150 000 Btu/h	In accordance with the appliance manufacturer's instructions and in no case less than the clearances specified in Section 509.8.1.

For SI Units: 1 inch = 25.4 mm, 1000 British thermal units per hour = 0.293 kW

509.8.3 Category I through Category IV and Noncategorized Appliances. Through-the-wall vents for Category II and Category IV appliances and noncategorized condensing appliances shall not terminate over public walkways or over an area where condensate or vapor could create a nuisance or hazard or could be detrimental to the operation of regulators, relief valves, or other equipment. Where local experience indicates that condensate is a problem with Category I and Category III appliances, this provision shall also apply.

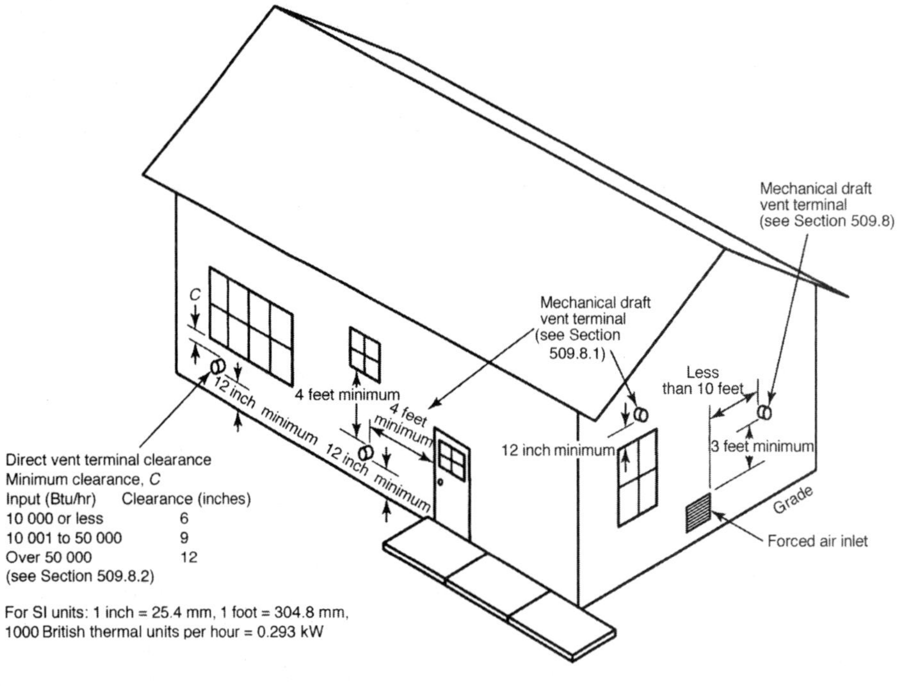

FIGURE 509.8
EXIT TERMINALS OF MECHANICAL DRAFT AND DIRECT-VENT VENTING SYSTEMS
[NFPA 54: FIGURE A.12.9]

Drains for condensate shall be installed in accordance with the appliance and the vent manufacturer's installation instructions. [NFPA 54:12.9.4]

509.8.4 Annular Spaces. Where vents, including those for direct-vent appliances or combustion air intake pipes, penetrate outside walls of buildings, the annular spaces around such penetrations shall be permanently sealed using approved materials to prevent entry of combustion products into the building. [NFPA 54:12.9.5]

509.8.5 Vent Terminals. Vent systems for Category IV appliances that terminate through an outside wall of a building and discharge flue gases perpendicular to the adjacent wall shall be located not less than 10 feet (3048 mm) horizontally from an operable opening in an adjacent building.

Exception: This shall not apply to vent terminals that are 2 feet (610 mm) or more above or 25 feet (7620 mm) or more below operable openings. [NFPA 54:12.9.6]

509.9 Condensation Drain. Provision shall be made to collect and dispose of condensate from venting systems serving Category II and Category IV appliances and noncategorized condensing appliances in accordance with Section 509.8.3. [NFPA 54:12.10.1]

509.9.1 Local Experience. Where local experience indicates that condensation is a problem, provision shall be made to drain off and dispose of condensate from venting systems serving Category I and Category III appliances in accordance with Section 509.8.3. [NFPA 54:12.10.2]

509.10 Vent Connectors for Category I Appliances. A vent connector shall be used to connect an appliance to a gas vent, chimney, or single-wall metal pipe, except where the gas vent, chimney, or single-wall metal pipe is directly connected to the appliance. [NFPA 54:12.11.1]

509.10.1 Materials. A vent connector shall be made of noncombustible, corrosion-resistant material capable of withstanding the vent gas temperature produced by the appliance and of sufficient thickness to withstand physical damage. [NFPA 54:12.11.2.1]

509.10.1.1 Unconditioned Area. Where the vent connector used for an appliance having a draft hood or a Category I appliance is located in or passes through an unconditioned area, attic, or crawl space, that portion of the vent connector shall be listed Type B, Type L, or listed vent material having equivalent insulation qualities.

Exception: Single-wall metal pipe located within the exterior walls of the building and located in an unconditioned area other than an attic or a crawl space having a local 99 percent winter design temperature of 5°F (-15°C) or higher. [NFPA 54:12.11.2.2]

509.10.1.2 Residential-Type Appliances. Vent connectors for residential-type appliances shall comply with the following:

(1) Vent connectors for listed appliances having draft hoods, appliances having draft hoods and equipped with listed conversion burners, and Category I appliances that are not installed in attics, crawl spaces, or other unconditioned areas shall be one of the following:
  - (a) Type B or Type L vent material.
  - (b) Galvanized sheet steel not less than 0.018 of an inch (0.457 mm) thick.
  - (c) Aluminum (1100 or 3003 alloy or equivalent) sheet not less than 0.027 of an inch (0.686 mm) thick.
  - (d) Stainless steel sheet not less than 0.012 of an inch (0.305 mm) thick.
  - (e) Smooth interior wall metal pipe having resistance to heat and corrosion equal to or greater than that of Section 509.10.1.2(1)(b), Section 509.10.1.2(1)(c), or Section 509.10.1.2(1)(d).
  - (f) A listed vent connector.

(2) Vent connectors shall not be covered with insulation.

Exception: Listed insulated vent connectors shall be installed in accordance with the manufacturer's installation instructions. [NFPA 54:12.11.2.3]

509.10.1.3 Nonresidential Low-Heat Appliances. A vent connector for a nonresidential low-heat appliance shall be a factory-built chimney section or steel pipe having resistance to heat and corrosion equivalent to that for the appropriate galvanized pipe as specified in Table 509.10.1.3. Factory-built chimney sections shall be joined together in accordance with the chimney manufacturer's instructions. [NFPA 54:12.11.2.4]

TABLE 509.10.1.3
MINIMUM THICKNESS FOR GALVANIZED STEEL VENT CONNECTORS FOR LOW-HEAT APPLIANCES
[NFPA 54: TABLE 12.11.2.4]

DIAMETER OF CONNECTOR (inches)	MINIMUM THICKNESS (inches)
Less than 6	0.019
6 to less than 10	0.023
10 to 12 inclusive	0.029
14 to 16 inclusive	0.034
Over 16	0.056

For SI units: 1 inch = 25.4 mm, 1 square inch = 0.000645 m^2

509.10.1.4 Medium-Heat Appliances. Vent connectors for medium-heat appliances shall be constructed of factory-built, medium-heat chimney sections or steel of a thickness not less than that specified in Table 509.10.1.4 and shall comply with the following:

(1) A steel vent connector for an appliance with a vent gas temperature in excess of 1000°F (538°C) measured at the entrance to the connector shall be lined with medium-duty fire brick or the equivalent.

(2) The lining shall be at least 2½ inches (64 mm) thick for a vent connector having a diameter or greatest cross-sectional dimension of 18 inches (457 mm) or less.

(3) The lining shall be at least 4½ inches (114 mm) thick laid on the 4½ inches (114 mm) bed for a vent connector having a diameter or greatest cross-sectional dimension greater than 18 inches (457 mm).

(4) Factory-built chimney sections, if employed, shall be joined together in accordance with the chimney manufacturer's instructions. [NFPA 54:12.11.2.5]

TABLE 509.10.1.4
MINIMUM THICKNESS FOR STEEL VENT CONNECTORS FOR MEDIUM-HEAT APPLIANCES
[NFPA 54: TABLE 12.11.2.5]

VENT CONNECTOR SIZE		
DIAMETER (inches)	**AREA (square inches)**	**MINIMUM THICKNESS (inches)**
Up to 14	Up to 154	0.053
Over 14 to 16	154 to 201	0.067
Over 16 to 18	201 to 254	0.093
Over 18	Exceeding 254	0.123

For SI units: 1 inch = 25.4 mm, 1 square inch = 0.000645 m^2

509.10.2 Size of Vent Connector. A vent connector for an appliance with a single draft hood or for a Category I fan-assisted combustion system appliance shall be sized and installed in accordance with Section 510.0 or other approved engineering methods. [NFPA 54:12.11.3.1]

509.10.2.1 Manifold. For a single appliance having more than one draft hood outlet or flue collar, the manifold shall be constructed according to the instructions of the appliance manufacturer. Where there are no instructions, the manifold shall be designed and constructed in accordance with approved engineering practices. As an alternative method, the effective area of the manifold shall equal the combined area of the flue collars or draft hood outlets, and the vent connectors shall have a minimum 1 foot (305 mm) rise. [NFPA 54:12.11.3.2]

509.10.2.2 Size. Where two or more appliances are connected to a common vent or chimney, each vent connector shall be sized in accordance with Section 510.0 or other approved engineering methods. [NFPA 54:12.11.3.3]

As an alternative method applicable only where all of the appliances are draft hood-equipped, each vent connector shall have an effective area not less than the area of the draft hood outlet of the appliance to which it is connected. [NFPA 54:12.11.3.4]

509.10.2.3 Height. Where two or more appliances are vented through a common vent connector or vent manifold, the common vent connector or vent manifold shall be located at the highest level consistent with available headroom and clearance to combustible material and sized in accordance with Section 510.0 or other approved engineering methods. [NFPA 54:12.11.3.5]

As an alternative method applicable only where there are two draft hood-equipped appliances, the effective area of the common vent connector or vent manifold and all junction fittings shall be not less than the area of the larger vent connector plus 50 percent of the area of the smaller flue collar outlet. [NFPA 54:12.11.3.6]

509.10.2.4 Size Increase. Where the size of a vent connector is increased to overcome installation limitations and obtain connector capacity equal to the appliance input, the size increase shall be made at the appliance draft hood outlet. [NFPA 54:12.11.3.7]

509.10.3 Two or More Appliances Connected to a Single Vent. Where two or more openings are provided into one chimney flue or vent, either of the following shall apply:

(1) The openings shall be at different levels.

(2) The connectors shall be attached to the vertical portion of the chimney or vent at an angle of 45 degrees (0.79 rad) or less relative to the vertical. [NFPA 54:12.11.4.1]

509.10.3.1 Height of Connector. Where two or more vent connectors enter a common vent, chimney flue, or single-wall metal pipe, the smaller connector shall enter at the highest level consistent with the available headroom or clearance to combustible material. [NFPA 54:12.11.4.2]

509.10.3.2 Pressure. Vent connectors serving Category I appliances shall not be connected to any portion of a mechanical draft system operating under positive static pressure, such as those serving Category III or Category IV appliances. [NFPA 54:12.11.4.3]

509.10.4 Clearance. Minimum clearances from vent connectors to combustible material shall be in accordance with Table 509.7.3.4(1).

Exception: The clearance between a vent connector and combustible material shall be permitted to be reduced where the combustible material is protected as specified for vent connectors in Table 509.7.3.4(2). [NFPA 54:12.11.5]

509.10.5 Joints. Joints between sections of connector piping and connections to flue collars or draft hood outlets shall be fastened in accordance with one of the following methods:

(1) Sheet metal screws.

(2) Vent connectors of listed vent material assembled and connected to flue collars or draft hood outlets in accordance with the manufacturer's instructions.

(3) Other approved means. [NFPA 54:12.11.6]

509.10.6 Slope. A vent connector shall be installed without any dips or sags and shall slope upward toward the vent or chimney at least ¼ inch per foot (20.8 mm/m).

Exception: Vent connectors attached to a mechanical draft system installed in accordance with appliance and the draft system manufacturer's instructions. [NFPA 54:12.11.7]

509.10.7 Length of Vent Connector. The length of vent connectors shall comply with Section 509.10.7.1 or Section 509.10.7.2.

509.10.7.1 Single Wall Connector. The maximum horizontal length of a single-wall connector shall be 75 percent of the height of the chimney or vent, except for engineered systems. [NFPA 54:12.11.8.1]

509.10.7.2 Type B Double Wall Connector. The maximum horizontal length of a Type B double-wall connector shall be 100 percent of the height of the chimney or vent, except for engineered systems. The maximum length of an individual connector for a chimney or vent system serving multiple appliances, from the appliance outlet to the junction with the common vent or another connector, shall be 100 percent of the height of the chimney or vent. [NFPA 54:12.11.8.2]

509.10.8 Support. A vent connector shall be supported for the design and weight of the material employed to maintain clearances and prevent physical damage and separation of joints. [NFPA 54:12.11.9]

509.10.9 Chimney Connection. Where entering a flue in a masonry or metal chimney, the vent connector shall be installed above the extreme bottom to avoid stoppage. Where a thimble or slip joint is used to facilitate removal of the connector, the connector shall be firmly attached to or inserted into the thimble or slip joint to prevent the connector from falling out. Means shall be employed to prevent the connector from entering so far as to restrict the space between its end and the opposite wall of the chimney flue. [NFPA 54:12.11.10]

509.10.10 Inspection. The entire length of a vent connector shall be readily accessible for inspection, cleaning, and replacement. [NFPA 54:12.11.11]

509.10.11 Fireplaces. A vent connector shall not be connected to a chimney flue serving a fireplace unless the fireplace flue opening is permanently sealed. [NFPA 54:12.11.12]

509.10.12 Passage through Ceilings, Floors, or Walls. A vent connector shall not pass through a ceiling, floor, or fire-resistance-rated wall. A single-wall metal pipe connector shall not pass through an interior wall.

Exception: Vent connectors made of listed Type B or Type L vent material and serving listed appliances with draft hoods and other appliances listed for use with Type B gas vents that pass through walls or partitions constructed of combustible material shall be installed with not less than the listed clearance to combustible material.

509.10.12.1 Medium-Heat Appliances. Vent connectors for medium-heat appliances shall not pass through walls or partitions constructed of combustible material. [NFPA 54:12.11.13.2]

509.11 Vent Connectors for Category II, Category III, and Category IV Appliances. The vent connectors for Category II, Category III, and Category IV appliances shall be in accordance with Section 509.4 through Section 509.4.3. [NFPA 54:12.12]

509.12 Draft Hoods and Draft Controls. Vented appliances shall be installed with draft hoods.

Exception: Dual oven-type combination ranges; incinerators; direct-vent appliances; fan-assisted combustion system appliances; appliances requiring chimney draft for operation; single firebox boilers equipped with conversion burners with inputs exceeding 400 000 Btu/h (117 kW); appliances equipped with blast, power, or pressure burners that are not listed for use with draft hoods; and appliances designed for forced venting. [NFPA 54:12.13.1]

509.12.1 Installation. A draft hood supplied with or forming a part of a listed vented appliance shall be installed without alteration, exactly as furnished and specified by the appliance manufacturer. [NFPA 54:12.13.2]

If a draft hood is not supplied by the appliance manufacturer where one is required, a draft hood shall be installed, be of a listed or approved type, and, in the absence of other instructions, be of the same size as the appliance flue collar. Where a draft hood is required with a conversion burner, it shall be of a listed or approved type. [NFPA 54:12.13.2.1]

Where a draft hood of special design is needed or preferable, the installation shall be approved and in accordance with the recommendations of the appliance manufacturer. [NFPA 54:12.13.2.2]

509.12.2 Draft Control Devices. Where a draft control device is part of the appliance or is supplied by the appliance manufacturer, it shall be installed in accordance with the manufacturer's instructions. In the absence of manufacturer's instructions, the device shall be attached to the flue collar of the appliance or as near to the appliance as practical. [NFPA 54:12.13.3]

509.12.3 Additional Devices. Appliances requiring controlled chimney draft shall be permitted to be equipped with listed double-acting barometric draft regulators installed and adjusted in accordance with the manufacturer's instructions. [NFPA 54:12.13.4]

509.12.4 Location. Draft hoods and barometric draft regulators shall be installed in the same room or enclosure as the appliance in such a manner as to prevent any difference in pressure between the hood or regulator and the combustion air supply. [NFPA 54:12.13.5]

509.12.5 Positioning. Draft hoods and draft regulators shall be installed in the position for which they were designed with reference to the horizontal and vertical planes and shall be located so that the relief opening is not obstructed by any part of the appliance or adjacent construction. The appliance and its draft hood shall be located so that the relief opening is accessible for checking vent operation. [NFPA 54:12.13.6]

509.12.6 Clearance. A draft hood shall be located so that its relief opening is not less than 6 inches (152 mm) from any surface except that of the appliance it serves and the venting system to which the draft hood is connected. Where a greater or lesser clearance is indicated on the appliance label, the clearance shall not be less than that specified on the label. Such clearances shall not be reduced. [NFPA 54:12.13.7]

509.13 Manually Operated Dampers. A manually operated damper shall not be placed in any appliance vent connector. Fixed baffles shall not be classified as manually operated dampers. [NFPA 54:12.14]

509.14 Automatically Operated Vent Dampers. An automatically operated vent damper shall be of a listed type. [NFPA 54:12.15]

509.14.1 Listing. Automatically operated vent dampers for oil fired appliances shall comply with UL 17. The automatic damper control shall comply with UL 378.

509.15 Obstructions. Devices that retard the flow of vent gases shall not be installed in a vent connector, chimney, or vent. The following shall not be considered as obstructions:

(1) Draft regulators and safety controls specifically listed for installation in venting systems and installed in accordance with the manufacturer's installation instructions.

(2) Approved draft regulators and safety controls designed and installed in accordance with approved engineering methods.

(3) Listed heat reclaimers and automatically operated vent dampers installed in accordance with the manufacturer's installation instructions.

(4) Vent dampers serving listed appliances installed in accordance with Section 510.1 or Section 510.2 or other approved engineering methods.

(5) Approved economizers, heat reclaimers, and recuperators installed in venting systems of appliances not required to be equipped with draft hoods, provided the appliance manufacturer's instructions cover the installation of such a device in the venting system and performance in accordance with Section 509.3 and Section 509.3.1 is obtained. [NFPA 54:12.16]

510.0 Sizing of Category I Venting Systems.

510.1 Single Appliance Vent Table 510.1.2(1) through Table 510.1.2(6). Venting Table 510.1.2(1) through Table 510.1.2(6) shall not be used where obstructions are installed in the venting system. The installation of vents serving listed appliances with vent dampers shall be in accordance with the appliance manufacturer's instructions or in accordance with the following:

(1) The maximum capacity of the vent system shall be determined using the "NAT Max" column.

(2) The minimum capacity shall be determined as though the appliance were a fan-assisted appliance, using the "FAN Min" column to determine the minimum capacity of the vent system. Where the corresponding "FAN Min" is "NA," the vent configuration shall not be permitted and an alternative venting configuration shall be utilized. [NFPA 54:13.1.1]

510.1.1 Vent Downsizing. Where the vent size determined from the tables is smaller than the appliance draft hood outlet or flue collar, the use of the smaller size shall be permitted, provided that the installation complies with all of the following requirements:

(1) The total vent height (*H*) is at least 10 feet (3048 mm).

(2) Vents for appliance draft hood outlets or flue collars 12 inches (300 mm) in diameter or smaller are not reduced more than one table size.

(3) Vents for appliance draft hood outlets or flue collars larger than 12 inches (300 mm) in diameter are not reduced more than two table sizes.

(4) The maximum capacity listed in the tables for a fan-assisted appliance is reduced by 10 percent (0.90 x maximum table capacity).

(5) The draft hood outlet is greater than 4 inches (100 mm) in diameter. A 3 inch (80 mm) diameter vent shall not be connected to a 4 inch (100 mm) diameter draft hood outlet. This provision shall not apply to fan-assisted appliances. [NFPA 54:13.1.2]

510.1.2 Elbows. Single-appliance venting configurations with zero (0) lateral lengths in Table 510.1.2(1), Table 510.1.2(2), and Table 510.1.2(5) shall not have elbows in the venting system. Single-appliance venting with lateral lengths include two 90 degree (1.57 rad) elbows. For each additional elbow up to and including 45 degrees (0.79 rad), the maximum capacity listed in the venting tables shall be reduced by 5 percent. For each additional elbow greater than 45 degrees (0.79 rad) up to and including 90 degrees (1.57 rad), the maximum capacity listed in the venting tables shall be reduced by 10 percent. Where multiple offsets occur in a vent, the total lateral length of all offsets combined shall not exceed that specified in Table 510.1.2(1) through Table 510.1.2(5). [NFPA 54:13.1.3]

510.1.3 Zero Lateral. Zero (0) lateral (*L*) shall apply only to a straight vertical vent attached to a top outlet draft hood or flue collar. [NFPA 54:13.1.4]

510.1.4 High-Altitude Installations. Sea level input ratings shall be used when determining maximum capacity for high-altitude installation. Actual input (derated for altitude) shall be used for determining minimum capacity for high-altitude installation. [NFPA 54:13.1.5]

510.1.5 Multiple Input Ratings. For appliances with more than one input rate, the minimum vent capacity (FAN Min) determined from Table 510.1.2(1) through Table 510.2(9) shall be less than the lowest appliance input rating, and the maximum vent capacity (FAN Max/NAT Max) determined from the tables shall be greater than the highest appliance rating input. [NFPA 54:13.1.6]

510.1.6 Corrugated Chimney Liner Reduction. Listed corrugated metallic chimney liner systems in masonry chimneys shall be sized by using Table 510.1.2(1) or Table 510.1.2(2) for Type B vents with the maximum capacity reduced by 20 percent (0.80 x maximum capacity) and the minimum capacity as shown in Table 510.1.2(1) or Table 510.1.2(2).

Corrugated metallic liner systems installed with bends or offsets shall have their maximum capacity further reduced in accordance with Section 510.1.2. The 20 percent reduction for corrugated metallic chimney liner systems includes an allowance for one long radius 90 degree (1.57 rad) turn at the bottom of the liner. [NFPA 54:13.1.7]

510.1.7 Connection to Chimney Liners. Connections between chimney liners and listed double-wall connectors shall be made with listed adapters designed for such purpose. [NFPA 54:13.1.8]

510.1.8 Vertical Vent Upsizing Using 7 x Rule. Where the vertical vent has a larger diameter than the vent connector, the vertical vent diameter shall be used to determine the minimum vent capacity, and the connector diameter shall be used to determine the maximum vent capacity. The flow area of the vertical vent shall not exceed seven times the flow area of the listed appliance categorized vent area, flue collar area, or draft hood outlet area unless designed in accordance with approved engineering methods. [NFPA 54:13.1.9]

510.1.9 Draft Hood Conversion Accessories. Draft hood conversion accessories for use with masonry chimneys venting listed Category I fan-assisted appliances shall be listed and installed in accordance with the listed accessory manufacturer's installation instructions. [NFPA 54:13.1.10]

510.1.10 Chimney and Vent Locations. Table 510.1.2(1) through Table 510.1.2(5) shall be used only for chimneys and vents not exposed to the outdoors below the roof line. A Type B vent or listed chimney lining system passing through an unused masonry chimney flue shall not be considered to be exposed to the outdoors. Where vents extend outdoors above the roof more than 5 feet (1524 mm) higher than required by Table 509.6.1, and where vents terminate in accordance with Section 509.6.1(1)(b), the outdoor portion of the vent shall be enclosed as required by this paragraph for vents not considered to be exposed to the outdoors, or such venting system shall be engineered. A Type B vent passing through an unventilated enclosure or chase insulated to a value of not less than R-8 shall not be considered to be exposed to the outdoors. Table 510.1.2(3) in combination with Table 510.1.2(6) shall be used for claytile-lined exterior masonry chimneys, provided all of the following requirements are met:

(1) The vent connector is Type B double wall.

(2) The vent connector length is limited to 18 inches per inch (18 mm/mm) of vent connector diameter.

(3) The appliance is draft hood equipped.

(4) The input rating is less than the maximum capacity given in Table 510.1.2(3).

(5) For a water heater, the outdoor design temperature shall not be less than 5°F (-15°C).

(6) For a space-heating appliance, the input rating is greater than the minimum capacity given by Table 510.1.2(6). [NFPA 54:13.1.11]

510.1.11 Corrugated Vent Connector Size. Corrugated vent connectors shall not be smaller than the listed appliance categorized vent diameter, flue collar diameter, or draft hood outlet diameter. [NFPA 54:13.1.12]

510.1.12 Upsizing. Vent connectors shall not be upsized more than two sizes greater than the listed appliance categorized vent diameter, flue collar diameter, or draft hood outlet diameter. [NFPA 54:13.1.13]

510.1.13 Single Run of Vent. In a single run of vent or vent connector, more than one diameter and type shall be permitted to be used, provided that all the sizes and types are permitted by the tables. [NFPA 54:13.1.14]

510.1.14 Interpolation. Interpolation shall be permitted in calculating capacities for vent dimensions that fall between table entries. [NFPA 54:13.1.15]

510.1.15 Extrapolation. Extrapolation beyond the table entries shall not be permitted. [NFPA 54:13.1.16]

510.1.16 Engineering Methods. For vent heights lower than 6 feet (1829 mm) and higher than shown in Table 510.1.2(1) through Table 510.2(9), engineering methods shall be used to calculate vent capacities. [NFPA 54:13.1.17]

510.1.17 Height Entries. Where the actual height of a vent falls between entries in the height column of the applicable table in Table 510.1.2(1) through Table 510.1.2(6), either of the following shall be used:

(1) Interpolation.

(2) The lower appliance input rating shown in the table entries for FAN Max and NAT Max column values; and the higher appliance input rating for the FAN Min column values. [NFPA 54:13.1.18]

510.2 Multiple Appliance Vent Table 510.2(1) through Table 510.2(9). Venting Table 510.2(1) through Table 510.2(9) shall not be used where obstructions are installed in the venting system. The installation of vents serving listed appliances with vent dampers shall be in accordance with the appliance manufacturer's instructions, or in accordance with the following:

(1) The maximum capacity of the vent connector shall be determined using the NAT Max column.

(2) The maximum capacity of the vertical vent or chimney shall be determined using the FAN + NAT column when the second appliance is a fan-assisted appliance, or the NAT + NAT column when the second appliance is equipped with a draft hood.

(3) The minimum capacity shall be determined as if the appliance were a fan-assisted appliance, as follows:

  (a) The minimum capacity of the vent connector shall be determined using the FAN Min column.

  (b) The FAN + FAN column shall be used when the second appliance is a fan-assisted appliance, and the

FAN + NAT column shall be used when the second appliance is equipped with a draft hood, to determine whether the vertical vent or chimney configuration is not permitted (NA). Where the vent configuration is NA, the vent configuration shall not be permitted and an alternative venting configuration shall be utilized. [NFPA 54:13.2.1]

510.2.1 Vent Connector Maximum Length. The maximum vent connector horizontal length shall be 18 inches per inch (18 mm/mm) of connector diameter as shown in Table 510.2.1, or as permitted by Section 510.2.2. [NFPA 54:13.2.2]

TABLE 510.2.1
VENT CONNECTOR MAXIMUM LENGTH
[NFPA 54: TABLE 13.2.2]

CONNECTOR DIAMETER (inches)	MAXIMUM CONNECTOR HORIZONTAL LENGTH (feet)
3	4½
4	6
5	7½
6	9
7	10½
8	12
9	13½
10	15
12	18
14	21
16	24
18	27
20	30
22	33
24	36

For SI units: 1 inch = 25.4 mm, 1 foot = 304.8 mm

510.2.2 Vent Connector Exceeding Maximum Length. The vent connector shall be routed to the vent utilizing the shortest possible route. Connectors with longer horizontal lengths than those listed in Table 510.2.1 are permitted under the following conditions:

(1) The maximum capacity (FAN Max or NAT Max) of the vent connector shall be reduced 10 percent for each additional multiple of the length listed in Table 510.2.1. For example, the maximum length listed for a 4 inch (100 mm) connector is 6 feet (1829 mm). With a connector length greater than 6 feet (1829 mm) but not exceeding 12 feet (3658 mm), the maximum capacity must be reduced by 10 percent (0.90 x maximum vent connector capacity). With a connector length greater than 12 feet (3658 mm) but not exceeding 18 feet (5486 mm), the maximum capacity shall be reduced by 20 percent (0.80 x maximum vent capacity).

(2) For a connector serving a fan-assisted appliance, the minimum capacity (FAN Min) of the connector shall be determined by referring to the corresponding single appliance table. For Type B double-wall connectors, Table 510.1.2(1) shall be used. For single-wall connectors, Table 510.1.2(2) shall be used. The height (*H*) and lateral (*L*) shall be measured according to the procedures for a single appliance vent, as if the other appliances were not present. [NFPA 54:13.2.3]

510.2.3 Vent Connector Manifolds. Where the vent connectors are combined prior to entering the vertical portion of the common vent to form a common vent manifold, the size of the common vent manifold and the common vent shall be determined by applying a 10 percent reduction (0.90 x maximum common vent capacity) to the common vent capacity part of the common vent tables. The length of the common vent manifold (*LM*) shall not exceed 18 inches per inch (18 mm/mm) of common vent diameter (*D*). [NFPA 54:13.2.4]

510.2.4 Vent Offsets. Where the common vertical vent is offset, the maximum capacity of the common vent shall be reduced in accordance with Section 510.2.5, and the horizontal length of the common vent offset shall not exceed 18 inches per inch (18 mm/mm) of common vent diameter (*D*). Where multiple offsets occur in a common vent, the total horizontal length of all offsets combined shall not exceed 18 inches per inch (18 mm/mm) of the common vent diameter. [NFPA 54:13.2.5]

510.2.5 Elbows in Vents. For each elbow up to and including 45 degrees (0.79 rad) in the common vent, the maximum common vent capacity listed in the venting tables shall be reduced by 5 percent. For each elbow greater than 45 degrees (0.79 rad) up to and including 90 degrees (1.57 rad), the maximum common vent capacity listed in the venting tables shall be reduced by 10 percent. [NFPA 54:13.2.6]

510.2.6 Elbows in Connectors. The vent connector capacities listed in the common vent sizing tables include allowance for two 90 degree (1.57 rad) elbows. For each additional elbow up to and including 45 degrees (0.79 rad), the maximum vent connector capacity listed in the venting tables shall be reduced by 5 percent. For each elbow greater than 45 degrees (0.79 rad) up to and including 90 degrees (1.57 rad), the maximum vent connector capacity listed in the venting tables shall be reduced by 10 percent. [NFPA 54:13.2.7]

510.2.7 Common Vent Minimum Size. The cross-sectional area of the common vent shall be equal to or greater than the cross-sectional area of the largest connector. [NFPA 54:13.2.8]

510.2.8 Tee and Wye Fittings. Tee and wye fittings connected to a common gas vent shall be considered as part of the common gas vent and constructed of materials consistent with that of the common gas vent. [NFPA 54:13.2.9]

510.2.9 Size of Fittings. At the point where tee or wye fittings connect to a common gas vent, the opening size of the fitting shall be equal to the size of the common vent. Such fittings shall not be prohibited from having reduced size openings at the point of connection of appliance gas vent connectors. [NFPA 54:13.2.10]

510.2.10 High-Altitude Installations. Sea level input ratings shall be used when determining maximum capacity for high-altitude installation. Actual input (derated for altitude) shall be used for determining minimum capacity for high-altitude installation. [NFPA 54:13.2.11]

510.2.11 Connector Rise. The connector rise (*R*) for each appliance connector shall be measured from the draft hood outlet or flue collar to the centerline where the vent gas streams come together. [NFPA 54:13.2.12]

510.2.12 Vent Height. For multiple appliances all located on one floor, available total height (*H*) shall be measured from the highest draft hood outlet or flue collar up to the level of the outlet of the common vent. [NFPA 54:13.2.13]

510.2.13 Multistory Installations. For multistory installations, available total height (*H*) for each segment of the system shall be the vertical distance between the highest draft hood outlet or flue collar entering that segment and the centerline of the next higher interconnection tee. [NFPA 54:13.2.14]

510.2.14 Size of Vents for Multistory Installations. The size of the lowest connector and of the vertical vent leading to the lowest interconnection of a multistory system shall be in accordance with Table 510.1.2(1) or Table 510.1.2(2) for available total height (*H*) up to the lowest interconnection. [NFPA 54:13.2.15]

510.2.15 Vent Type Multistory Installations. Where used in multistory systems, vertical common vents shall be Type B double wall and shall be installed with a listed vent cap. [NFPA 54:13.2.16]

510.2.16 Offsets in Multistory Installations. Offsets in multistory common vent systems shall be limited to a single offset in each system, and systems with an offset shall comply with all of the following:

(1) The offset angle shall not exceed 45 degrees (0.79 rad) from vertical.

(2) The horizontal length of the offset shall not exceed 18 inches per inch (18 mm/mm) of common vent diameter of the segment in which the offset is located.

(3) For the segment of the common vertical vent containing the offset, the common vent capacity listed in the common venting tables shall be reduced by 20 percent (0.80 x maximum common vent capacity).

(4) A multistory common vent shall not be reduced in size above the offset. [NFPA 54:13.2.17]

510.2.17 Vertical Vent Size Limitation. Where two or more appliances are connected to a vertical vent or chimney, the flow area of the largest section of vertical vent or chimney shall not exceed seven times the smallest listed appliance categorized vent areas, flue collar area, or draft hood outlet area unless designed in accordance with approved engineering methods. [NFPA 54:13.2.18]

510.2.18 Multiple Input Ratings. For appliances with more than one input rate, the minimum vent connector capacity (FAN Min) determined from the tables shall be less than the lowest appliance input rating, and the maximum vent connector capacity (FAN Max or NAT Max) determined from the tables shall be greater than the highest appliance input rating. [NFPA 54:13.2.19]

510.2.19 Corrugated Metallic Chimney Liner Reduction. Listed corrugated metallic chimney liner systems in masonry chimneys shall be sized by using Table 510.2(1) or Table 510.2(2) for Type B vents, with the maximum capacity reduced by 20 percent (0.80 x maximum capacity) and the minimum capacity as shown in Table 510.2(1) or Table 510.2(2). Corrugated metallic liner systems installed with bends or offsets shall have their maximum capacity further reduced in accordance with Section 510.2.4 and Section 510.2.5. The 20 percent reduction for corrugated metallic chimney liner systems includes an allowance for one long radius 90 degree (1.57 rad) turn at the bottom of the liner. [NFPA 54:13.2.20]

510.2.20 Chimneys and Vents. Table 510.2(1) through Table 510.2(5) shall be used only for chimneys and vents not exposed to the outdoors below the roof line. A Type B vent or listed chimney lining system passing through an unused masonry chimney flue shall not be considered to be exposed to the outdoors. A Type B vent passing through an unventilated enclosure or chase insulated to a value of not less than R8 shall not be considered to be exposed to the outdoors. Where vents extend outdoors above the roof more than 5 feet (1524 mm) higher than required by Table 509.6.1, and where vents terminate in accordance with Section 509.6.1(1)(b), the outdoor portion of the vent shall be enclosed as required by this section for vents not considered to be exposed to the outdoors, or such venting system shall be engineered. Table 510.2(6) through Table 510.2(9) shall be used for clay tile lined exterior masonry chimneys, provided all the following conditions are met:

(1) The vent connector is Type B double wall.

(2) At least one appliance is draft hood equipped.

(3) The combined appliance input rating is less than the maximum capacity given by Table 510.2(6) (for NAT + NAT) or Table 510.2(8) (for FAN + NAT).

(4) The input rating of each space-heating appliance is greater than the minimum input rating given by Table 510.2(7) (for NAT + NAT) or Table 510.2(9) (for FAN + NAT).

(5) The vent connector sizing is in accordance with Table 510.2(3). [NFPA 54:13.2.22]

510.2.21 Vent Connector Sizing. Vent connectors shall not be increased more than two sizes greater than the listed appliance categorized vent diameter, flue collar diameter, or draft hood outlet diameter. Vent connectors for draft hood-equipped appliances shall not be smaller than the draft hood outlet diameter. Where a vent connector size(s) determined from the tables for a fan-assisted appliance(s) is smaller than the flue collar diameter, the use of the smaller size(s) shall be permitted, provided that the installation complies with all of the following conditions:

(1) Vent connectors for fan-assisted appliance flue collars 12 inches (300 mm) in diameter or smaller are

not reduced by more than one table size [e.g., 12 inches to 10 inches (300 mm to 250 mm) is a one-size reduction], and those larger than 12 inches (300 mm) in diameter are not reduced more than two table sizes [e.g., 24 inches to 20 inches (600 mm to 500 mm) is a two-size reduction].

(2) The fan-assisted appliance(s) is common vented with a draft hood-equipped appliance(s).

(3) The vent connector has a smooth interior wall. [NFPA 54:13.2.24]

510.2.22 Combination of Pipe Types and Sizes. All combinations of pipe sizes, single-wall metal pipe, and double-wall metal pipe shall be allowed within any connector run(s) or within the common vent, provided ALL of the appropriate tables permit ALL of the desired sizes and types of pipe, as if they were used for the entire length of the subject connector or vent. Where single-wall and Type B double-wall metal pipes are used for vent connectors within the same venting system, the common vent shall be sized using Table 510.2(2) or Table 510.2(4) as appropriate. [NFPA 54:13.2.25]

510.2.23 Multiple Connector and Vent Sizes. Where Table 510.2(1) through Table 510.2(9) permits more than one diameter of pipe to be used for a connector or vent, all the permitted sizes shall be permitted to be used. [NFPA 54:13.2.26]

510.2.24 Interpolation. Interpolation shall be permitted in calculating capacities for vent dimensions that fall between table entries. [NFPA 54:13.2.27]

510.2.25 Extrapolation. Extrapolation beyond the table entries shall not be permitted. [NFPA 54:13.2.28]

510.2.26 Engineering Methods. For vent heights lower than 6 feet (1829 mm) and higher than shown in the tables, engineering methods shall be used to calculate vent capacities. [NFPA 54:13.2.29]

510.2.27 Height Entries. Where the actual height of a vent falls between entries in the height column of the applicable table in Table 510.2(1) through Table 510.2(9), either of the following shall be used:

(1) Interpolation.

(2) The lower appliance input rating shown in the table entries, for FAN Max and NAT Max column values; and the higher appliance input rating for the FAN Min column values. [NFPA 54:13.2.30]

99% Winter Design Temperatures for the Contiguous United States

This map is a necessarily generalized guide to temperatures in the contiguous United States. Temperatures shown for areas such as mountainous regions and large urban centers are not necessarily accurate. The climate data used to develop this map are from the ASHRAE Handbook — Fundamentals (Climate Conditions for the United States).

For 99% winter design temperatures in Alaska, consult the ASHRAE Handbook — Fundamentals.
99% winter design temperatures for Hawaii are greater than 37°F
For SI units: °C = (°F-32)/1.8

FIGURE 510.1.10
RANGE OF WINTER DESIGN TEMPERATURES USED IN ANALYZING EXTERIOR MASONRY CHIMNEYS IN THE UNITED STATES
[NFPA 54: FIGURE F.2.4]

TABLE 510.1.2(1)
TYPE B DOUBLE-WALL GAS VENT [NFPA 54: TABLE 13.1(a)]*

							NUMBER OF APPLIANCES:		SINGLE							
							APPLIANCE TYPE:		CATEGORY I							
							APPLIANCE VENT CONNECTION:		CONNECTED DIRECTLY TO VENT							
		VENT DIAMETER – *D* (inch)														
		3			4			5			6			7		
		APPLIANCE INPUT RATING IN THOUSANDS OF BTU PER HOUR														
HEIGHT *H* (feet)	LATERAL *L* (feet)	FAN		NAT	FAN		NAT	FAN		NAT	FAN		NAT	FAN		NAT
		Min	Max	Max	Min	Max	Max	Min	Max	Max	Min	Max	Max	Min	Max	Max
6	0	0	78	46	0	152	86	0	251	141	0	375	205	0	524	285
	2	13	51	36	18	97	67	27	157	105	32	232	157	44	321	217
	4	21	49	34	30	94	64	39	153	103	50	227	153	66	316	211
	6	25	46	32	36	91	61	47	149	100	59	223	149	78	310	205
8	0	0	84	50	0	165	94	0	276	155	0	415	235	0	583	320
	2	12	57	40	16	109	75	25	178	120	28	263	180	42	365	247
	5	23	53	38	32	103	71	42	171	115	53	255	173	70	356	237
	8	28	49	35	39	98	66	51	164	109	64	247	165	84	347	227
10	0	0	88	53	0	175	100	0	295	166	0	447	255	0	631	345
	2	12	61	42	17	118	81	23	194	129	26	289	195	40	402	273
	5	23	57	40	32	113	77	41	187	124	52	280	188	68	392	263
	10	30	51	36	41	104	70	54	176	115	67	267	175	88	376	245
15	0	0	94	58	0	191	112	0	327	187	0	502	285	0	716	390
	2	11	69	48	15	136	93	20	226	150	22	339	225	38	475	316
	5	22	65	45	30	130	87	39	219	142	49	330	217	64	463	300
	10	29	59	41	40	121	82	51	206	135	64	315	208	84	445	288
	15	35	53	37	48	112	76	61	195	128	76	301	198	98	429	275
20	0	0	97	61	0	202	119	0	349	202	0	540	307	0	776	430
	2	10	75	51	14	149	100	18	250	166	20	377	249	33	531	346
	5	21	71	48	29	143	96	38	242	160	47	367	241	62	519	337
	10	28	64	44	38	133	89	50	229	150	62	351	228	81	499	321
	15	34	58	40	46	124	84	59	217	142	73	337	217	94	481	308
	20	48	52	35	55	116	78	69	206	134	84	322	206	107	464	295
30	0	0	100	64	0	213	128	0	374	220	0	587	336	0	853	475
	2	9	81	56	13	166	112	14	283	185	18	432	280	27	613	394
	5	21	77	54	28	160	108	36	275	176	45	421	273	58	600	385
	10	27	70	50	37	150	102	48	262	171	59	405	261	77	580	371
	15	33	64	NA	44	141	96	57	249	163	70	389	249	90	560	357
	20	56	58	NA	53	132	90	66	237	154	80	374	237	102	542	343
	30	NA	NA	NA	73	113	NA	88	214	NA	104	346	219	131	507	321
50	0	0	101	67	0	216	134	0	397	232	0	633	363	0	932	518
	2	8	86	61	11	183	122	14	320	206	15	497	314	22	715	445
	5	20	82	NA	27	177	119	35	312	200	43	487	308	55	702	438
	10	26	76	NA	35	168	114	45	299	190	56	471	298	73	681	426
	15	59	70	NA	42	158	NA	54	287	180	66	455	288	85	662	413
	20	NA	NA	NA	50	149	NA	63	275	169	76	440	278	97	642	401
	30	NA	NA	NA	69	131	NA	84	250	NA	99	410	259	123	605	376
100	0	NA	NA	NA	0	218	NA	0	407	NA	0	665	400	0	997	560
	2	NA	NA	NA	10	194	NA	12	354	NA	13	566	375	18	831	510
	5	NA	NA	NA	26	189	NA	33	347	NA	40	557	369	52	820	504
	10	NA	NA	NA	33	182	NA	43	335	NA	53	542	361	68	801	493
	15	NA	NA	NA	40	174	NA	50	321	NA	62	528	353	80	782	482
	20	NA	NA	NA	47	166	NA	59	311	NA	71	513	344	90	763	471
	30	NA	NA	NA	NA	NA	NA	78	290	NA	92	483	NA	115	726	449
	50	NA	NA	NA	NA	NA	NA	NA	NA	NA	147	428	NA	180	651	405

For SI units: 1 inch = 25.4 mm, 1 foot = 304.8 mm, 1000 British thermal units per hour = 0.293 kW, 1 square inch = 0.000645 m^2
* NA: Not applicable.

TABLE 510.1.2(1)
TYPE B DOUBLE-WALL GAS VENT [NFPA 54: TABLE 13.1(a)] (continued)

						NUMBER OF APPLIANCES:				SINGLE						
						APPLIANCE TYPE:				CATEGORY I						
						APPLIANCE VENT CONNECTION:				CONNECTED DIRECTLY TO VENT						
		VENT DIAMETER – *D* (inch)														
		8			9			10			12			14		
		APPLIANCE INPUT RATING IN THOUSANDS OF BTU PER HOUR														
HEIGHT *H* (feet)	LATERAL *L* (feet)	FAN		NAT	FAN		NAT	FAN		NAT	FAN		NAT	FAN		NAT
		Min	Max	Max	Min	Max	Max	Min	Max	Max	Min	Max	Max	Min	Max	Max
6	0	0	698	370	0	897	470	0	1121	570	0	1645	850	0	2267	1170
	2	53	425	285	63	543	370	75	675	455	103	982	650	138	1346	890
	4	79	419	279	93	536	362	110	668	445	147	975	640	191	1338	880
	6	93	413	273	110	530	354	128	661	435	171	967	630	219	1330	870
8	0	0	780	415	0	1006	537	0	1261	660	0	1858	970	0	2571	1320
	2	50	483	322	60	619	418	71	770	515	98	1124	745	130	1543	1020
	5	83	473	313	99	607	407	115	758	503	154	1110	733	199	1528	1010
	8	99	463	303	117	596	396	137	746	490	180	1097	720	231	1514	1000
10	0	0	847	450	0	1096	585	0	1377	720	0	2036	1060	0	2825	1450
	2	48	533	355	57	684	457	68	852	560	93	1244	850	124	1713	1130
	5	81	522	346	95	671	446	112	839	547	149	1229	829	192	1696	1105
	10	104	504	330	122	651	427	142	817	525	187	1204	795	238	1669	1080
15	0	0	970	525	0	1263	682	0	1596	840	0	2380	1240	0	3323	1720
	2	45	633	414	53	815	544	63	1019	675	86	1495	985	114	2062	1350
	5	76	620	403	90	800	529	105	1003	660	140	1476	967	182	2041	1327
	10	99	600	386	116	777	507	135	977	635	177	1446	936	227	2009	1289
	15	115	580	373	134	755	491	155	953	610	202	1418	905	257	1976	1250
20	0	0	1057	575	0	1384	752	0	1756	930	0	2637	1350	0	3701	1900
	2	41	711	470	50	917	612	59	1150	755	81	1694	1100	107	2343	1520
	5	73	697	460	86	902	599	101	1133	738	135	1674	1079	174	2320	1498
	10	95	675	443	112	877	576	130	1105	710	172	1641	1045	220	2282	1460
	15	111	654	427	129	853	557	150	1078	688	195	1609	1018	248	2245	1425
	20	125	634	410	145	830	537	167	1052	665	217	1578	990	273	2210	1390
30	0	0	1173	650	0	1548	855	0	1977	1060	0	3004	1550	0	4252	2170
	2	33	826	535	42	1072	700	54	1351	865	74	2004	1310	98	2786	1800
	5	69	811	524	82	1055	688	96	1332	851	127	1981	1289	164	2759	1775
	10	91	788	507	107	1028	668	125	1301	829	164	1944	1254	209	2716	1733
	15	105	765	490	124	1002	648	143	1272	807	187	1908	1220	237	2674	1692
	20	119	743	473	139	977	628	160	1243	784	207	1873	1185	260	2633	1650
	30	149	702	444	171	929	594	195	1189	745	246	1807	1130	305	2555	1585
50	0	0	1297	708	0	1730	952	0	2231	1195	0	3441	1825	0	4934	2550
	2	26	975	615	33	1276	813	41	1620	1010	66	2431	1513	86	3409	2125
	5	65	960	605	77	1259	798	90	1600	996	118	2406	1495	151	3380	2102
	10	86	935	589	101	1230	773	118	1567	972	154	2366	1466	196	3332	2064
	15	100	911	572	117	1203	747	136	1536	948	177	2327	1437	222	3285	2026
	20	113	888	556	131	1176	722	151	1505	924	195	2288	1408	244	3239	1987
	30	141	844	522	161	1125	670	183	1446	876	232	2214	1349	287	3150	1910
100	0	0	1411	770	0	1908	1040	0	2491	1310	0	3925	2050	0	5729	2950
	2	21	1155	700	25	1536	935	30	1975	1170	44	3027	1820	72	4313	2550
	5	60	1141	692	71	1519	926	82	1955	1159	107	3002	1803	136	4282	2531
	10	80	1118	679	94	1492	910	108	1923	1142	142	2961	1775	180	4231	2500
	15	93	1095	666	109	1465	895	126	1892	1124	163	2920	1747	206	4182	2469
	20	105	1073	653	122	1438	880	141	1861	1107	181	2880	1719	226	4133	2438
	30	131	1029	627	149	1387	849	170	1802	1071	215	2803	1663	265	4037	2375
	50	197	944	575	217	1288	787	241	1688	1000	292	2657	1550	350	3856	2250

For SI units: 1 inch = 25.4 mm, 1 foot = 304.8 mm, 1000 British thermal units per hour = 0.293 kW, 1 square inch = 0.000645 m^2

TABLE 510.1.2(1)
TYPE B DOUBLE-WALL GAS VENT [NFPA 54: TABLE 13.1(a)] (continued)

								NUMBER OF APPLIANCES:		SINGLE						
								APPLIANCE TYPE:		CATEGORY I						
								APPLIANCE VENT CONNECTION:		CONNECTED DIRECTLY TO VENT						
		VENT DIAMETER – *D* (inch)														
		16			18			20			22			24		
		APPLIANCE INPUT RATING IN THOUSANDS OF BTU PER HOUR														
HEIGHT *H* (feet)	LATERAL *L* (feet)	FAN		NAT	FAN		NAT	FAN		NAT	FAN		NAT	FAN		NAT
		Min	Max	Max	Min	Max	Max	Min	Max	Max	Min	Max	Max	Min	Max	Max
6	0	0	2983	1530	0	3802	1960	0	4721	2430	0	5737	2950	0	6853	3520
	2	178	1769	1170	225	2250	1480	296	2782	1850	360	3377	2220	426	4030	2670
	4	242	1761	1160	300	2242	1475	390	2774	1835	469	3370	2215	555	4023	2660
	6	276	1753	1150	341	2235	1470	437	2767	1820	523	3363	2210	618	4017	2650
8	0	0	3399	1740	0	4333	2220	0	5387	2750	0	6555	3360	0	7838	4010
	2	168	2030	1340	212	2584	1700	278	3196	2110	336	3882	2560	401	4634	3050
	5	251	2013	1330	311	2563	1685	398	3180	2090	476	3863	2545	562	4612	3040
	8	289	2000	1320	354	2552	1670	450	3163	2070	537	3850	2530	630	4602	3030
10	0	0	3742	1925	0	4782	2450	0	5955	3050	0	7254	3710	0	8682	4450
	2	161	2256	1480	202	2868	1890	264	3556	2340	319	4322	2840	378	5153	3390
	5	243	2238	1461	300	2849	1871	382	3536	2318	458	4301	2818	540	5132	3371
	10	298	2209	1430	364	2818	1840	459	3504	2280	546	4268	2780	641	5099	3340
15	0	0	4423	2270	0	5678	2900	0	7099	3620	0	8665	4410	0	10 393	5300
	2	147	2719	1770	186	3467	2260	239	4304	2800	290	5232	3410	346	6251	4080
	5	229	2696	1748	283	3442	2235	355	4278	2777	426	5204	3385	501	6222	4057
	10	283	2659	1712	346	3402	2193	432	4234	2739	510	5159	3343	599	6175	4019
	15	318	2623	1675	385	3363	2150	479	4192	2700	564	5115	3300	665	6129	3980
20	0	0	4948	2520	0	6376	3250	0	7988	4060	0	9785	4980	0	11 753	6000
	2	139	3097	2000	175	3955	2570	220	4916	3200	269	5983	3910	321	7154	4700
	5	219	3071	1978	270	3926	2544	337	4885	3174	403	5950	3880	475	7119	4662
	10	273	3029	1940	334	3880	2500	413	4835	3130	489	5896	3830	573	7063	4600
	15	306	2988	1910	372	3835	2465	459	4786	3090	541	5844	3795	631	7007	4575
	20	335	2948	1880	404	3791	2430	495	4737	3050	585	5792	3760	689	6953	4550
30	0	0	5725	2920	0	7420	3770	0	9341	4750	0	11 483	5850	0	13 848	7060
	2	127	3696	2380	159	4734	3050	199	5900	3810	241	7194	4650	285	8617	5600
	5	206	3666	2350	252	4701	3020	312	5863	3783	373	7155	4622	439	8574	5552
	10	259	3617	2300	316	4647	2970	386	5803	3739	456	7090	4574	535	8505	5471
	15	292	3570	2250	354	4594	2920	431	5744	3695	507	7026	4527	590	8437	5391
	20	319	3523	2200	384	4542	2870	467	5686	3650	548	6964	4480	639	8370	5310
	30	369	3433	2130	440	4442	2785	540	5574	3565	635	6842	4375	739	8239	5225
50	0	0	6711	3440	0	8774	4460	0	11 129	5635	0	13 767	6940	0	16 694	8430
	2	113	4554	2840	141	5864	3670	171	7339	4630	209	8980	5695	251	10 788	6860
	5	191	4520	2813	234	5826	3639	283	7295	4597	336	8933	5654	394	10 737	6818
	10	243	4464	2767	295	5763	3585	355	7224	4542	419	8855	5585	491	10 652	6749
	15	274	4409	2721	330	5701	3534	396	7155	4511	465	8779	5546	542	10 570	6710
	20	300	4356	2675	361	5641	3481	433	7086	4479	506	8704	5506	586	10 488	6670
	30	347	4253	2631	412	5523	3431	494	6953	4421	577	8557	5444	672	10 328	6603
100	0	0	7914	4050	0	10 485	5300	0	13 454	6700	0	16 817	8600	0	20 578	10 300
	2	95	5834	3500	120	7591	4600	138	9577	5800	169	11 803	7200	204	14 264	8800
	5	172	5797	3475	208	7548	4566	245	9528	5769	293	11 748	7162	341	14 204	8756
	10	223	5737	3434	268	7478	4509	318	9447	5717	374	11 658	7100	436	14 105	8683
	15	252	5678	3392	304	7409	4451	358	9367	5665	418	11 569	7037	487	14 007	8610
	20	277	5619	3351	330	7341	4394	387	9289	5613	452	11 482	6975	523	13 910	8537
	30	319	5505	3267	378	7209	4279	446	9136	5509	514	11 310	6850	592	13 720	8391
	50	415	5289	3100	486	6956	4050	572	8841	5300	659	10 979	6600	752	13 354	8100

For SI units: 1 inch = 25.4 mm, 1 foot = 304.8 mm, 1000 British thermal units per hour = 0.293 kW, 1 square inch = 0.000645 m^2

TABLE 510.1.2(2)
TYPE B DOUBLE-WALL GAS VENT [NFPA 54: TABLE 13.1(b)]*

							NUMBER OF APPLIANCES:	SINGLE								
							APPLIANCE TYPE:	CATEGORY I								
							APPLIANCE VENT CONNECTION:	SINGLE-WALL METAL CONNECTOR								
		VENT DIAMETER – *D* (inch)														
		3			4			5			6			7		
		APPLIANCE INPUT RATING IN THOUSANDS OF BTU PER HOUR														
HEIGHT *H* (feet)	LATERAL *L* (feet)	FAN		NAT	FAN		NAT	FAN		NAT	FAN		NAT	FAN		NAT
		Min	Max	Max	Min	Max	Max	Min	Max	Max	Min	Max	Max	Min	Max	Max
6	0	38	77	45	59	151	85	85	249	140	126	373	204	165	522	284
	2	39	51	36	60	96	66	85	156	104	123	231	156	159	320	213
	4	NA	NA	33	74	92	63	102	152	102	146	225	152	187	313	208
	6	NA	NA	31	83	89	60	114	147	99	163	220	148	207	307	203
8	0	37	83	50	58	164	93	83	273	154	123	412	234	161	580	319
	2	39	56	39	59	108	75	83	176	119	121	261	179	155	363	246
	5	NA	NA	37	77	102	69	107	168	114	151	252	171	193	352	235
	8	NA	NA	33	90	95	64	122	161	107	175	243	163	223	342	225
10	0	37	87	53	57	174	99	82	293	165	120	444	254	158	628	344
	2	39	61	41	59	117	80	82	193	128	119	287	194	153	400	272
	5	52	56	39	76	111	76	105	185	122	148	277	186	190	388	261
	10	NA	NA	34	97	100	68	132	171	112	188	261	171	237	369	241
15	0	36	93	57	56	190	111	80	325	186	116	499	283	153	713	388
	2	38	69	47	57	136	93	80	225	149	115	337	224	148	473	314
	5	51	63	44	75	128	86	102	216	140	144	326	217	182	459	298
	10	NA	NA	39	95	116	79	128	201	131	182	308	203	228	438	284
	15	NA	NA	NA	NA	NA	72	158	186	124	220	290	192	272	418	269
20	0	35	96	60	54	200	118	78	346	201	114	537	306	149	772	428
	2	37	74	50	56	148	99	78	248	165	113	375	248	144	528	344
	5	50	68	47	73	140	94	100	239	158	141	363	239	178	514	334
	10	NA	NA	41	93	129	86	125	223	146	177	344	224	222	491	316
	15	NA	NA	NA	NA	NA	80	155	208	136	216	325	210	264	469	301
	20	NA	NA	NA	NA	NA	NA	186	192	126	254	306	196	309	448	285
30	0	34	99	63	53	211	127	76	372	219	110	584	334	144	849	472
	2	37	80	56	55	164	111	76	281	183	109	429	279	139	610	392
	5	49	74	52	72	157	106	98	271	173	136	417	271	171	595	382
	10	NA	NA	NA	91	144	98	122	255	168	171	397	257	213	570	367
	15	NA	NA	NA	115	131	NA	151	239	157	208	377	242	255	547	349
	20	NA	NA	NA	NA	NA	NA	181	223	NA	246	357	228	298	524	333
	30	NA	NA	NA	NA	NA	NA	NA	NA	NA	NA	NA	NA	389	477	305
50	0	33	99	66	51	213	133	73	394	230	105	629	361	138	928	515
	2	36	84	61	53	181	121	73	318	205	104	495	312	133	712	443
	5	48	80	NA	70	174	117	94	308	198	131	482	305	164	696	435
	10	NA	NA	NA	89	160	NA	118	292	186	162	461	292	203	671	420
	15	NA	NA	NA	112	148	NA	145	275	174	199	441	280	244	646	405
	20	NA	NA	NA	NA	NA	NA	176	257	NA	236	420	267	285	622	389
	30	NA	NA	NA	NA	NA	NA	NA	NA	NA	315	376	NA	373	573	NA
100	0	NA	NA	NA	49	214	NA	69	403	NA	100	659	395	131	991	555
	2	NA	NA	NA	51	192	NA	70	351	NA	98	563	373	125	828	508
	5	NA	NA	NA	67	186	NA	90	342	NA	125	551	366	156	813	501
	10	NA	NA	NA	85	175	NA	113	324	NA	153	532	354	191	789	486
	15	NA	NA	NA	132	162	NA	138	310	NA	188	511	343	230	764	473
	20	NA	NA	NA	NA	NA	NA	168	295	NA	224	487	NA	270	739	458
	30	NA	NA	NA	NA	NA	NA	231	264	NA	301	448	NA	355	685	NA
	50	NA	NA	NA	NA	NA	NA	NA	NA	NA	NA	NA	NA	540	584	NA

For SI units: 1 inch = 25.4 mm, 1 foot = 304.8 mm, 1000 British thermal units per hour = 0.293 kW, 1 square inch = 0.000645 m^2
* NA: Not applicable.

TABLE 510.1.2(2)
TYPE B DOUBLE-WALL GAS VENT [NFPA 54: TABLE 13.1(b)] (continued)*

					NUMBER OF APPLIANCES:	SINGLE							
					APPLIANCE TYPE:	CATEGORY I							
					APPLIANCE VENT CONNECTION:	SINGLE-WALL METAL CONNECTOR							
		VENT DIAMETER – *D* (inch)											
		8			9			10			12		
		APPLIANCE INPUT RATING IN THOUSANDS OF BTU PER HOUR											
HEIGHT *H* (feet)	LATERAL *L* (feet)	FAN		NAT	FAN		NAT	FAN		NAT	FAN		NAT
		Min	Max	Max	Min	Max	Max	Min	Max	Max	Min	Max	Max
6	0	211	695	369	267	894	469	371	1118	569	537	1639	849
	2	201	423	284	251	541	368	347	673	453	498	979	648
	4	237	416	277	295	533	360	409	664	443	584	971	638
	6	263	409	271	327	526	352	449	656	433	638	962	627
8	0	206	777	414	258	1002	536	360	1257	658	521	1852	967
	2	197	482	321	246	617	417	339	768	513	486	1120	743
	5	245	470	311	305	604	404	418	754	500	598	1104	730
	8	280	458	300	344	591	392	470	740	486	665	1089	715
10	0	202	844	449	253	1093	584	351	1373	718	507	2031	1057
	2	193	531	354	242	681	456	332	849	559	475	1242	848
	5	241	518	344	299	667	443	409	834	544	584	1224	825
	10	296	497	325	363	643	423	492	808	520	688	1194	788
15	0	195	966	523	244	1259	681	336	1591	838	488	2374	1237
	2	187	631	413	232	812	543	319	1015	673	457	1491	983
	5	231	616	400	287	795	526	392	997	657	562	1469	963
	10	284	592	381	349	768	501	470	966	628	664	1433	928
	15	334	568	367	404	742	484	540	937	601	750	1399	894
20	0	190	1053	573	238	1379	750	326	1751	927	473	2631	1346
	2	182	708	468	227	914	611	309	1146	754	443	1689	1098
	5	224	692	457	279	896	596	381	1126	734	547	1665	1074
	10	277	666	437	339	866	570	457	1092	702	646	1626	1037
	15	325	640	419	393	838	549	526	1060	677	730	1587	1005
	20	374	616	400	448	810	526	592	1028	651	808	1550	973
30	0	184	1168	647	229	1542	852	312	1971	1056	454	2996	1545
	2	175	823	533	219	1069	698	296	1346	863	424	1999	1308
	5	215	806	521	269	1049	684	366	1324	846	524	1971	1283
	10	265	777	501	327	1017	662	440	1287	821	620	1927	1243
	15	312	750	481	379	985	638	507	1251	794	702	1884	1205
	20	360	723	461	433	955	615	570	1216	768	780	1841	1166
	30	461	670	426	541	895	574	704	1147	720	937	1759	1101
50	0	176	1292	704	220	1724	948	295	2223	1189	428	3432	1818
	2	168	971	613	209	1273	811	280	1615	1007	401	2426	1509
	5	204	953	602	257	1252	795	347	1591	991	496	2396	1490
	10	253	923	583	313	1217	765	418	1551	963	589	2347	1455
	15	299	894	562	363	1183	736	481	1512	934	668	2299	1421
	20	345	866	543	415	1150	708	544	1473	906	741	2251	1387
	30	442	809	502	521	1086	649	674	1399	848	892	2159	1318
100	0	166	1404	765	207	1900	1033	273	2479	1300	395	3912	2042
	2	158	1152	698	196	1532	933	259	1970	1168	371	3021	1817
	5	194	1134	688	240	1511	921	322	1945	1153	460	2990	1796
	10	238	1104	672	293	1477	902	389	1905	1133	547	2938	1763
	15	281	1075	656	342	1443	884	447	1865	1110	618	2888	1730
	20	325	1046	639	391	1410	864	507	1825	1087	690	2838	1696
	30	418	988	NA	491	1343	824	631	1747	1041	834	2739	1627
	50	617	866	NA	711	1205	NA	895	1591	NA	1138	2547	1489

For SI units: 1 inch = 25.4 mm, 1 foot = 304.8 mm, 1000 British thermal units per hour = 0.293 kW, 1 square inch = 0.000645 m^2
* NA: Not applicable.

TABLE 510.1.2(3)
MASONRY CHIMNEY [NFPA 54: TABLE 13.1(c)]*

							NUMBER OF APPLIANCES:			SINGLE						
							APPLIANCE TYPE:			CATEGORY I						
							APPLIANCE VENT CONNECTION:			TYPE B DOUBLE-WALL CONNECTOR						
		TYPE B DOUBLE-WALL CONNECTOR DIAMETER – *D* (inch) TO BE USED WITH CHIMNEY AREAS WITHIN THE SIZE LIMITS AT BOTTOM														
		3			4			5			6			7		
		APPLIANCE INPUT RATING IN THOUSANDS OF BTU PER HOUR														
HEIGHT *H* (feet)	LATERAL *L* (feet)	FAN		NAT	FAN		NAT	FAN		NAT	FAN		NAT	FAN		NAT
		Min	Max	Max	Min	Max	Max	Min	Max	Max	Min	Max	Max	Min	Max	Max
6	2	NA	NA	28	NA	NA	52	NA	NA	86	NA	NA	130	NA	NA	180
	5	NA	NA	25	NA	NA	49	NA	NA	82	NA	NA	117	NA	NA	165
8	2	NA	NA	29	NA	NA	55	NA	NA	93	NA	NA	145	NA	NA	198
	5	NA	NA	26	NA	NA	52	NA	NA	88	NA	NA	134	NA	NA	183
	8	NA	NA	24	NA	NA	48	NA	NA	83	NA	NA	127	NA	NA	175
10	2	NA	NA	31	NA	NA	61	NA	NA	103	NA	NA	162	NA	NA	221
	5	NA	NA	28	NA	NA	57	NA	NA	96	NA	NA	148	NA	NA	204
	10	NA	NA	25	NA	NA	50	NA	NA	87	NA	NA	139	NA	NA	191
15	2	NA	NA	35	NA	NA	67	NA	NA	114	NA	NA	179	53	475	250
	5	NA	NA	35	NA	NA	62	NA	NA	107	NA	NA	164	NA	NA	231
	10	NA	NA	28	NA	NA	55	NA	NA	97	NA	NA	153	NA	NA	216
	15	NA	NA	NA	NA	NA	48	NA	NA	89	NA	NA	141	NA	NA	201
20	2	NA	NA	38	NA	NA	74	NA	NA	124	NA	NA	201	51	522	274
	5	NA	NA	36	NA	NA	68	NA	NA	116	NA	NA	184	80	503	254
	10	NA	NA	NA	NA	NA	60	NA	NA	107	NA	NA	172	NA	NA	237
	15	NA	NA	NA	NA	NA	NA	NA	NA	97	NA	NA	159	NA	NA	220
	20	NA	NA	NA	NA	NA	NA	NA	NA	83	NA	NA	148	NA	NA	206
30	2	NA	NA	41	NA	NA	82	NA	NA	137	NA	NA	216	47	581	303
	5	NA	NA	NA	NA	NA	76	NA	NA	128	NA	NA	198	75	561	281
	10	NA	NA	NA	NA	NA	67	NA	NA	115	NA	NA	184	NA	NA	263
	15	NA	NA	NA	NA	NA	NA	NA	NA	107	NA	NA	171	NA	NA	243
	20	NA	NA	NA	NA	NA	NA	NA	NA	91	NA	NA	159	NA	NA	227
	30	NA	NA	NA	NA	NA	NA	NA	NA	NA	NA	NA	NA	NA	NA	188
50	2	NA	NA	NA	NA	NA	92	NA	NA	161	NA	NA	251	NA	NA	351
	5	NA	NA	NA	NA	NA	NA	NA	NA	151	NA	NA	230	NA	NA	323
	10	NA	NA	NA	NA	NA	NA	NA	NA	138	NA	NA	215	NA	NA	304
	15	NA	NA	NA	NA	NA	NA	NA	NA	127	NA	NA	199	NA	NA	282
	20	NA	NA	NA	NA	NA	NA	NA	NA	NA	NA	NA	185	NA	NA	264
	30	NA	NA	NA	NA	NA	NA	NA	NA	NA	NA	NA	NA	NA	NA	NA
Minimum internal area of chimney (square inches)		12			19			28			38			50		
Maximum internal area of chimney (square inches)		Seven times the listed appliance categorized vent area, flue collar area, or draft hood outlet areas.														

For SI units: 1 inch = 25.4 mm, 1 foot = 304.8 mm, 1000 British thermal units per hour = 0.293 kW, 1 square inch = 0.000645 m^2
* NA: Not applicable.

TABLE 510.1.2(3)
MASONRY CHIMNEY [NFPA 54: TABLE 13.1(c)] (continued)*

		NUMBER OF APPLIANCES: SINGLE											
		APPLIANCE TYPE: CATEGORY I											
		APPLIANCE VENT CONNECTION: TYPE B DOUBLE-WALL CONNECTOR											
		TYPE B DOUBLE-WALL CONNECTOR DIAMETER – *D* (inch) TO BE USED WITH CHIMNEY AREAS WITHIN THE SIZE LIMITS AT BOTTOM											
		8			9			10			12		
		APPLIANCE INPUT RATING IN THOUSANDS OF BTU PER HOUR											
HEIGHT *H* (feet)	**LATERAL *L* (feet)**	**FAN**		**NAT**	**FAN**		**NAT**	**FAN**		**NAT**	**FAN**		**NAT**
		Min	**Max**	**Max**	**Min**	**Max**	**Max**	**Min**	**Max**	**Max**	**Min**	**Max**	**Max**
6	2	NA	NA	247	NA	NA	320	NA	NA	401	NA	NA	581
	5	NA	NA	231	NA	NA	298	NA	NA	376	NA	NA	561
8	2	NA	NA	266	84	590	350	100	728	446	139	1024	651
	5	NA	NA	247	NA	NA	328	149	711	423	201	1007	640
	8	NA	NA	239	NA	NA	318	173	695	410	231	990	623
10	2	68	519	298	82	655	388	98	810	491	136	1144	724
	5	NA	NA	277	124	638	365	146	791	466	196	1124	712
	10	NA	NA	263	155	610	347	182	762	444	240	1093	668
15	2	64	613	336	77	779	441	92	968	562	127	1376	841
	5	99	594	313	118	759	416	139	946	533	186	1352	828
	10	126	565	296	148	727	394	173	912	567	229	1315	777
	15	NA	NA	281	171	698	375	198	880	485	259	1280	742
20	2	61	678	375	73	867	491	87	1083	627	121	1548	953
	5	95	658	350	113	845	463	133	1059	597	179	1523	933
	10	122	627	332	143	811	440	167	1022	566	221	1482	879
	15	NA	NA	314	165	780	418	191	987	541	251	1443	840
	20	NA	NA	296	186	750	397	214	955	513	277	1406	807
30	2	57	762	421	68	985	558	81	1240	717	111	1793	1112
	5	90	741	393	106	962	526	125	1216	683	169	1766	1094
	10	115	709	373	135	927	500	158	1176	648	210	1721	1025
	15	NA	NA	353	156	893	476	181	1139	621	239	1679	981
	20	NA	NA	332	176	860	450	203	1103	592	264	1638	940
	30	NA	NA	288	NA	NA	416	249	1035	555	318	1560	877
50	2	51	840	477	61	1106	633	72	1413	812	99	2080	1243
	5	83	819	445	98	1083	596	116	1387	774	155	2052	1225
	10	NA	NA	424	126	1047	567	147	1347	733	195	2006	1147
	15	NA	NA	400	146	1010	539	170	1307	702	222	1961	1099
	20	NA	NA	376	165	977	511	190	1269	669	246	1916	1050
	30	NA	NA	327	NA	NA	468	233	1196	623	295	1832	984
Minimum internal area of chimney (square inches)		63			78			95			132		
Maximum internal area of chimney (square inches)		Seven times the listed appliance categorized vent area, flue collar area, or draft hood outlet areas.											

For SI units: 1 inch = 25.4 mm, 1 foot = 304.8 mm, 1000 British thermal units per hour = 0.293 kW, 1 square inch = 0.000645 m^2

* NA: Not applicable.

TABLE 510.1.2(4)
MASONRY CHIMNEY [NFPA 54: TABLE 13.1(d)]*

							NUMBER OF APPLIANCES:		SINGLE							
							APPLIANCE TYPE:		CATEGORY I							
							APPLIANCE VENT CONNECTION:		SINGLE-WALL METAL CONNECTOR							
		SINGLE-WALL METAL CONNECTOR DIAMETER – *D* (inch) TO BE USED WITH CHIMNEY AREAS WITHIN THE SIZE LIMITS AT BOTTOM														
		3			4			5			6			7		
		APPLIANCE INPUT RATING IN THOUSANDS OF BTU PER HOUR														
HEIGHT *H* (feet)	LATERAL *L* (feet)	FAN		NAT	FAN		NAT	FAN		NAT	FAN		NAT	FAN		NAT
		Min	Max	Max	Min	Max	Max	Min	Max	Max	Min	Max	Max	Min	Max	Max
6	2	NA	NA	28	NA	NA	52	NA	NA	86	NA	NA	130	NA	NA	180
	5	NA	NA	25	NA	NA	48	NA	NA	81	NA	NA	116	NA	NA	164
8	2	NA	NA	29	NA	NA	55	NA	NA	93	NA	NA	145	NA	NA	197
	5	NA	NA	26	NA	NA	51	NA	NA	87	NA	NA	133	NA	NA	182
	8	NA	NA	23	NA	NA	47	NA	NA	82	NA	NA	126	NA	NA	174
10	2	NA	NA	31	NA	NA	61	NA	NA	102	NA	NA	161	NA	NA	220
	5	NA	NA	28	NA	NA	56	NA	NA	95	NA	NA	147	NA	NA	203
	10	NA	NA	24	NA	NA	49	NA	NA	86	NA	NA	137	NA	NA	189
15	2	NA	NA	35	NA	NA	67	NA	NA	113	NA	NA	178	166	473	249
	5	NA	NA	32	NA	NA	61	NA	NA	106	NA	NA	163	NA	NA	230
	10	NA	NA	27	NA	NA	54	NA	NA	96	NA	NA	151	NA	NA	214
	15	NA	NA	NA	NA	NA	46	NA	NA	87	NA	NA	138	NA	NA	198
20	2	NA	NA	38	NA	NA	73	NA	NA	123	NA	NA	200	163	520	273
	5	NA	NA	35	NA	NA	67	NA	NA	115	NA	NA	183	NA	NA	252
	10	NA	NA	NA	NA	NA	59	NA	NA	105	NA	NA	170	NA	NA	235
	15	NA	NA	NA	NA	NA	NA	NA	NA	95	NA	NA	156	NA	NA	217
	20	NA	NA	NA	NA	NA	NA	NA	NA	80	NA	NA	144	NA	NA	202
30	2	NA	NA	41	NA	NA	81	NA	NA	136	NA	NA	215	158	578	302
	5	NA	NA	NA	NA	NA	75	NA	NA	127	NA	NA	196	NA	NA	279
	10	NA	NA	NA	NA	NA	66	NA	NA	113	NA	NA	182	NA	NA	260
	15	NA	NA	NA	NA	NA	NA	NA	NA	105	NA	NA	168	NA	NA	240
	20	NA	NA	NA	NA	NA	NA	NA	NA	88	NA	NA	155	NA	NA	223
	30	NA	NA	NA	NA	NA	NA	NA	NA	NA	NA	NA	NA	NA	NA	182
50	2	NA	NA	NA	NA	NA	91	NA	NA	160	NA	NA	250	NA	NA	350
	5	NA	NA	NA	NA	NA	NA	NA	NA	149	NA	NA	228	NA	NA	321
	10	NA	NA	NA	NA	NA	NA	NA	NA	136	NA	NA	212	NA	NA	301
	15	NA	NA	NA	NA	NA	NA	NA	NA	124	NA	NA	195	NA	NA	278
	20	NA	NA	NA	NA	NA	NA	NA	NA	NA	NA	NA	180	NA	NA	258
	30	NA	NA	NA	NA	NA	NA	NA	NA	NA	NA	NA	NA	NA	NA	NA
Minimum internal area of chimney (square inches)		12			19			28			38			50		
Maximum internal area of chimney (square inches)		Seven times the listed appliance categorized vent area, flue collar area, or draft hood outlet areas.														

For SI units: 1 inch = 25.4 mm, 1 foot = 304.8 mm, 1000 British thermal units per hour = 0.293 kW, 1 square inch = 0.000645 m^2

* NA: Not applicable.

TABLE 510.1.2(4)
MASONRY CHIMNEY [NFPA 54: TABLE 13.1(d)] (continued)*

				NUMBER OF APPLIANCES:				SINGLE					
				APPLIANCE TYPE:				CATEGORY I					
				APPLIANCE VENT CONNECTION:				SINGLE-WALL METAL CONNECTOR					
		SINGLE-WALL METAL CONNECTOR DIAMETER – *D* (inch) TO BE USED WITH CHIMNEY AREAS WITHIN THE SIZE LIMITS AT BOTTOM											
		8			9			10			12		
		APPLIANCE INPUT RATING IN THOUSANDS OF BTU PER HOUR											
HEIGHT *H* (feet)	LATERAL *L* (feet)	FAN		NAT	FAN		NAT	FAN		NAT	FAN		NAT
		Min	Max	Max	Min	Max	Max	Min	Max	Max	Min	Max	Max
6	2	NA	NA	247	NA	NA	319	NA	NA	400	NA	NA	580
	5	NA	NA	230	NA	NA	297	NA	NA	375	NA	NA	560
8	2	NA	NA	265	NA	NA	349	382	725	445	549	1021	650
	5	NA	NA	246	NA	NA	327	NA	NA	422	673	1003	638
	8	NA	NA	237	NA	NA	317	NA	NA	408	747	985	621
10	2	216	518	297	271	654	387	373	808	490	536	1142	722
	5	NA	NA	276	334	635	364	459	789	465	657	1121	710
	10	NA	NA	261	NA	NA	345	547	758	441	771	1088	665
15	2	211	611	335	264	776	440	362	965	560	520	1373	840
	5	261	591	312	325	755	414	444	942	531	637	1348	825
	10	NA	NA	294	392	722	392	531	907	504	749	1309	774
	15	NA	NA	278	452	692	372	606	873	481	841	1272	738
20	2	206	675	374	258	864	490	252	1079	625	508	1544	950
	5	255	655	348	317	842	461	433	1055	594	623	1518	930
	10	312	622	330	382	806	437	517	1016	562	733	1475	875
	15	NA	NA	311	442	773	414	591	979	539	823	1434	835
	20	NA	NA	292	NA	NA	392	663	944	510	911	1394	800
30	2	200	759	420	249	982	556	340	1237	715	489	1789	1110
	5	245	737	391	306	958	524	417	1210	680	600	1760	1090
	10	300	703	370	370	920	496	500	1168	644	708	1713	1020
	15	NA	NA	349	428	884	471	572	1128	615	798	1668	975
	20	NA	NA	327	NA	NA	445	643	1089	585	883	1624	932
	30	NA	NA	281	NA	NA	408	NA	NA	544	1055	1539	865
50	2	191	837	475	238	1103	631	323	1408	810	463	2076	1240
	5	NA	NA	442	293	1078	593	398	1381	770	571	2044	1220
	10	NA	NA	420	355	1038	562	447	1337	728	674	1994	1140
	15	NA	NA	395	NA	NA	533	546	1294	695	761	1945	1090
	20	NA	NA	370	NA	NA	504	616	1251	660	844	1898	1040
	30	NA	NA	318	NA	NA	458	NA	NA	610	1009	1805	970
Minimum internal area of chimney (square inches)		63			78			95			132		
Maximum internal area of chimney (square inches)		Seven times the listed appliance categorized vent area, flue collar area, or draft hood outlet areas.											

For SI units: 1 inch = 25.4 mm, 1 foot = 304.8 mm, 1000 British thermal units per hour = 0.293 kW, 1 square inch = 0.000645 m^2
* NA: Not applicable.

TABLE 510.1.2(5)
SINGLE-WALL METAL PIPE OR TYPE B ASBESTOS-CEMENT VENT [NFPA 54: TABLE 13.1(e)]*

					NUMBER OF APPLIANCES:	SINGLE			
					APPLIANCE TYPE:	DRAFT HOOD-EQUIPPED			
					APPLIANCE VENT CONNECTION:	CONNECTED DIRECTLY TO PIPE OR VENT			
		DIAMETER – *D* (inch) TO BE USED WITH CHIMNEY AREAS WITHIN THE SIZE LIMITS AT BOTTOM							
		3	4	5	6	7	8	10	12
HEIGHT *H* (feet)	LATERAL *L* (feet)	APPLIANCE INPUT RATING IN THOUSANDS OF BTU PER HOUR							
		MAXIMUM APPLIANCE INPUT RATING IN THOUSANDS OF BTU PER HOUR							
6	0	39	70	116	170	232	312	500	750
	2	31	55	94	141	194	260	415	620
	5	28	51	88	128	177	242	390	600
8	0	42	76	126	185	252	340	542	815
	2	32	61	102	154	210	284	451	680
	5	29	56	95	141	194	264	430	648
	10	24	49	86	131	180	250	406	625
10	0	45	84	138	202	279	372	606	912
	2	35	67	111	168	233	311	505	760
	5	32	61	104	153	215	289	480	724
	10	27	54	94	143	200	274	455	700
	15	NA	46	84	130	186	258	432	666
15	0	49	91	151	223	312	420	684	1040
	2	39	72	122	186	260	350	570	865
	5	35	67	110	170	240	325	540	825
	10	30	58	103	158	223	308	514	795
	15	NA	50	93	144	207	291	488	760
	20	NA	NA	82	132	195	273	466	726
20	0	53	101	163	252	342	470	770	1190
	2	42	80	136	210	286	392	641	990
	5	38	74	123	192	264	364	610	945
	10	32	65	115	178	246	345	571	910
	15	NA	55	104	163	228	326	550	870
	20	NA	NA	91	149	214	306	525	832
30	0	56	108	183	276	384	529	878	1370
	2	44	84	148	230	320	441	730	1140
	5	NA	78	137	210	296	410	694	1080
	10	NA	68	125	196	274	388	656	1050
	15	NA	NA	113	177	258	366	625	1000
	20	NA	NA	99	163	240	344	596	960
	30	NA	NA	NA	NA	192	295	540	890
50	0	NA	120	210	310	443	590	980	1550
	2	NA	95	171	260	370	492	820	1290
	5	NA	NA	159	234	342	474	780	1230
	10	NA	NA	146	221	318	456	730	1190
	15	NA	NA	NA	200	292	407	705	1130
	20	NA	NA	NA	185	276	384	670	1080
	30	NA	NA	NA	NA	222	330	605	1010

For SI units: 1 inch = 25.4 mm, 1 foot = 304.8 mm, 1000 British thermal units per hour = 0.293 kW, 1 square inch = 0.000645 m^2
* NA: Not applicable.

TABLE 510.1.2(6)
EXTERIOR MASONRY CHIMNEY [NFPA 54: TABLE 13.1(f)][1, 2]

			NUMBER OF APPLIANCES:	SINGLE			
			APPLIANCE TYPE:	NAT			
			APPLIANCE VENT CONNECTION:	TYPE B DOUBLE-WALL CONNECTOR			
MINIMUM ALLOWABLE INPUT RATING OF SPACE-HEATING APPLIANCE IN THOUSANDS OF BTU PER HOUR								
VENT HEIGHT *H* (feet)	**INTERNAL AREA OF CHIMNEY (square inches)**							
	12	**19**	**28**	**38**	**50**	**63**	**78**	**113**
	Local 99% winter design temperature: 37°F or greater							
6	0	0	0	0	0	0	0	0
8	0	0	0	0	0	0	0	0
10	0	0	0	0	0	0	0	0
15	NA	0	0	0	0	0	0	0
20	NA	NA	123	190	249	184	0	0
30	NA	NA	NA	NA	NA	393	334	0
50	NA	NA	NA	NA	NA	NA	NA	579
	Local 99% winter design temperature: 27°F to 36°F							
6	0	0	68	116	156	180	212	266
8	0	0	82	127	167	187	214	263
10	0	51	97	141	183	201	225	265
15	NA	NA	NA	NA	233	253	274	305
20	NA	NA	NA	NA	NA	307	330	362
30	NA	NA	NA	NA	NA	419	445	485
50	NA	NA	NA	NA	NA	NA	NA	763
	Local 99% winter design temperature: 17°F to 26°F							
6	NA	NA	NA	NA	NA	215	259	349
8	NA	NA	NA	NA	197	226	264	352
10	NA	NA	NA	NA	214	245	278	358
15	NA	NA	NA	NA	NA	296	331	398
20	NA	NA	NA	NA	NA	352	387	457
30	NA	NA	NA	NA	NA	NA	507	581
50	NA	NA	NA	NA	NA	NA	NA	NA
	Local 99% winter design temperature: 5°F to 16°F							
6	NA	NA	NA	NA	NA	NA	NA	416
8	NA	NA	NA	NA	NA	NA	312	423
10	NA	NA	NA	NA	NA	289	331	430
15	NA	NA	NA	NA	NA	NA	393	485
20	NA	NA	NA	NA	NA	NA	450	547
30	NA	NA	NA	NA	NA	NA	NA	682
50	NA	NA	NA	NA	NA	NA	NA	972
	Local 99% winter design temperature: -10°F to 4°F							
6	NA	NA	NA	NA	NA	NA	NA	484
8	NA	NA	NA	NA	NA	NA	NA	494
10	NA	NA	NA	NA	NA	NA	NA	513
15	NA	NA	NA	NA	NA	NA	NA	586
20	NA	NA	NA	NA	NA	NA	NA	650
30	NA	NA	NA	NA	NA	NA	NA	805
50	NA	NA	NA	NA	NA	NA	NA	1003
Local 99% winter design temperature: -11°F or lower Not recommended for any vent configurations								

For SI units: 1 inch = 25.4 mm, 1 foot = 304.8 mm, 1000 British thermal units per hour = 0.293 kW, 1 square inch = 0.000645 m^2, °C = (°F-32)/1.8

Notes:

[1] See Figure 510.1.10 for a map showing local 99 percent winter design temperatures in the United States.

[2] NA: Not applicable.

TABLE 510.2(1)
TYPE B DOUBLE-WALL VENT [NFPA 54: TABLE 13.2(a)]*

		NUMBER OF APPLIANCES: TWO OR MORE														
		APPLIANCE TYPE: CATEGORY I														
		APPLIANCE VENT CONNECTION: TYPE B DOUBLE-WALL CONNECTOR														
		VENT CONNECTOR CAPACITY														
		TYPE B DOUBLE-WALL VENT AND CONNECTOR DIAMETER – *D* (inch)														
		3			4			5			6			7		
		APPLIANCE INPUT RATING LIMITS IN THOUSANDS OF BTU PER HOUR														
VENT HEIGHT *H* (feet)	CONNECTOR RISE *R* (feet)	FAN		NAT	FAN		NAT	FAN		NAT	FAN		NAT	FAN		NAT
		Min	Max	Max	Min	Max	Max	Min	Max	Max	Min	Max	Max	Min	Max	Max
6	1	22	37	26	35	66	46	46	106	72	58	164	104	77	225	142
	2	23	41	31	37	75	55	48	121	86	60	183	124	79	253	168
	3	24	44	35	38	81	62	49	132	96	62	199	139	82	275	189
8	1	22	40	27	35	72	48	49	114	76	64	176	109	84	243	148
	2	23	44	32	36	80	57	51	128	90	66	195	129	86	269	175
	3	24	47	36	37	87	64	53	139	101	67	210	145	88	290	198
10	1	22	43	28	34	78	50	49	123	78	65	189	113	89	257	154
	2	23	47	33	36	86	59	51	136	93	67	206	134	91	282	182
	3	24	50	37	37	92	67	52	146	104	69	220	150	94	303	205
15	1	21	50	30	33	89	53	47	142	83	64	220	120	88	298	163
	2	22	53	35	35	96	63	49	153	99	66	235	142	91	320	193
	3	24	55	40	36	102	71	51	163	111	68	248	160	93	339	218
20	1	21	54	31	33	99	56	46	157	87	62	246	125	86	334	171
	2	22	57	37	34	105	66	48	167	104	64	259	149	89	354	202
	3	23	60	42	35	110	74	50	176	116	66	271	168	91	371	228
30	1	20	62	33	31	113	59	45	181	93	60	288	134	83	391	182
	2	21	64	39	33	118	70	47	190	110	62	299	158	85	408	215
	3	22	66	44	34	123	79	48	198	124	64	309	178	88	423	242
50	1	19	71	36	30	133	64	43	216	101	57	349	145	78	477	197
	2	21	73	43	32	137	76	45	223	119	59	358	172	81	490	234
	3	22	75	48	33	141	86	46	229	134	61	366	194	83	502	263
100	1	18	82	37	28	158	66	40	262	104	53	442	150	73	611	204
	2	19	83	44	30	161	79	42	267	123	55	447	178	75	619	242
	3	20	84	50	31	163	89	44	272	138	57	452	200	78	627	272

	COMMON VENT CAPACITY											
	TYPE B DOUBLE-WALL COMMON VENT DIAMETER – *D* (inch)											
	4			5			6			7		
	COMBINED APPLIANCE INPUT RATING IN THOUSANDS OF BTU PER HOUR											
VENT HEIGHT *H* (feet)	FAN +FAN	FAN +NAT	NAT +NAT	FAN +FAN	FAN +NAT	NAT +NAT	FAN +FAN	FAN +NAT	NAT +NAT	FAN +FAN	FAN +NAT	NAT +NAT
6	92	81	65	140	116	103	204	161	147	309	248	200
8	101	90	73	155	129	114	224	178	163	339	275	223
10	110	97	79	169	141	124	243	194	178	367	299	242
15	125	112	91	195	164	144	283	228	206	427	352	280
20	136	123	102	215	183	160	314	255	229	475	394	310
30	152	138	118	244	210	185	361	297	266	547	459	360
50	167	153	134	279	244	214	421	353	310	641	547	423
100	175	163	NA	311	277	NA	489	421	NA	751	658	479

For SI units: 1 inch = 25.4 mm, 1 foot = 304.8 mm, 1000 British thermal units per hour = 0.293 kW, 1 square inch = 0.000645 m^2
* NA: Not applicable.

TABLE 510.2(1)
TYPE B DOUBLE-WALL VENT [NFPA 54: TABLE 13.2(a)] (continued)

		NUMBER OF APPLIANCES: TWO OR MORE								
		APPLIANCE TYPE: CATEGORY I								
		APPLIANCE VENT CONNECTION: TYPE B DOUBLE-WALL CONNECTOR								
		VENT CONNECTOR CAPACITY								
		TYPE B DOUBLE-WALL VENT AND CONNECTOR DIAMETER – *D* (inch)								
		8			9			10		
		APPLIANCE INPUT RATING LIMITS IN THOUSANDS OF BTU PER HOUR								
VENT HEIGHT *H* (feet)	CONNECTOR RISE *R* (feet)	FAN		NAT	FAN		NAT	FAN		NAT
		Min	Max	Max	Min	Max	Max	Min	Max	Max
6	1	92	296	185	109	376	237	128	466	289
	2	95	333	220	112	424	282	131	526	345
	3	97	363	248	114	463	317	134	575	386
8	1	100	320	194	118	408	248	138	507	303
	2	103	356	230	121	454	294	141	564	358
	3	105	384	258	123	492	330	143	612	402
10	1	106	341	200	125	436	257	146	542	314
	2	109	374	238	128	479	305	149	596	372
	3	111	402	268	131	515	342	152	642	417
15	1	110	389	214	134	493	273	162	609	333
	2	112	419	253	137	532	323	165	658	394
	3	115	445	286	140	565	365	167	700	444
20	1	107	436	224	131	552	285	158	681	347
	2	110	463	265	134	587	339	161	725	414
	3	113	486	300	137	618	383	164	764	466
30	1	103	512	238	125	649	305	151	802	372
	2	105	535	282	129	679	360	155	840	439
	3	108	555	317	132	706	405	158	874	494
50	1	97	627	257	120	797	330	144	984	403
	2	100	645	306	123	820	392	148	1014	478
	3	103	661	343	126	842	441	151	1043	538
100	1	91	810	266	112	1038	341	135	1285	417
	2	94	822	316	115	1054	405	139	1306	494
	3	97	834	355	118	1069	455	142	1327	555

	COMMON VENT CAPACITY								
	TYPE B DOUBLE-WALL COMMON VENT DIAMETER – *D* (inch)								
	8			9			10		
	COMBINED APPLIANCE INPUT RATING IN THOUSANDS OF BTU PER HOUR								
VENT HEIGHT *H* (feet)	FAN +FAN	FAN +NAT	NAT +NAT	FAN +FAN	FAN +NAT	NAT +NAT	FAN +FAN	FAN +NAT	NAT +NAT
6	404	314	260	547	434	335	672	520	410
8	444	348	290	602	480	378	740	577	465
10	477	377	315	649	522	405	800	627	495
15	556	444	365	753	612	465	924	733	565
20	621	499	405	842	688	523	1035	826	640
30	720	585	470	979	808	605	1209	975	740
50	854	706	550	1164	977	705	1451	1188	860
100	1025	873	625	1408	1215	800	1784	1502	975

For SI units: 1 inch = 25.4 mm, 1 foot = 304.8 mm, 1000 British thermal units per hour = 0.293 kW, 1 square inch = 0.000645 m^2

TABLE 510.2(1)
TYPE B DOUBLE-WALL VENT [NFPA 54: TABLE 13.2(a)] (continued)*

		NUMBER OF APPLIANCES:	TWO OR MORE										
		APPLIANCE TYPE:	CATEGORY I										
		APPLIANCE VENT CONNECTION:	TYPE B DOUBLE-WALL CONNECTOR										
		VENT CONNECTOR CAPACITY											
		TYPE B DOUBLE-WALL VENT AND CONNECTOR DIAMETER – *D* (inch)											
		12			14			16			18		
		APPLIANCE INPUT RATING LIMITS IN THOUSANDS OF BTU PER HOUR											
VENT HEIGHT *H* (feet)	CONNECTOR RISE *R* (feet)	FAN		NAT	FAN		NAT	FAN		NAT	FAN		NAT
		Min	Max	Max	Min	Max	Max	Min	Max	Max	Min	Max	Max
6	2	174	764	496	223	1046	653	281	1371	853	346	1772	1080
	4	180	897	616	230	1231	827	287	1617	1081	352	2069	1370
	6	NA	NA	NA	NA	NA	NA	NA	NA	NA	NA	NA	NA
8	2	186	822	516	238	1126	696	298	1478	910	365	1920	1150
	4	192	952	644	244	1307	884	305	1719	1150	372	2211	1460
	6	198	1050	772	252	1445	1072	313	1902	1390	380	2434	1770
10	2	196	870	536	249	1195	730	311	1570	955	379	2049	1205
	4	201	997	664	256	1371	924	318	1804	1205	387	2332	1535
	6	207	1095	792	263	1509	1118	325	1989	1455	395	2556	1865
15	2	214	967	568	272	1334	790	336	1760	1030	408	2317	1305
	4	221	1085	712	279	1499	1006	344	1978	1320	416	2579	1665
	6	228	1181	856	286	1632	1222	351	2157	1610	424	2796	2025
20	2	223	1051	596	291	1443	840	357	1911	1095	430	2533	1385
	4	230	1162	748	298	1597	1064	365	2116	1395	438	2778	1765
	6	237	1253	900	307	1726	1288	373	2287	1695	450	2984	2145
30	2	216	1217	632	286	1664	910	367	2183	1190	461	2891	1540
	4	223	1316	792	294	1802	1160	376	2366	1510	474	3110	1920
	6	231	1400	952	303	1920	1410	384	2524	1830	485	3299	2340
50	2	206	1479	689	273	2023	1007	350	2659	1315	435	3548	1665
	4	213	1561	860	281	2139	1291	359	2814	1685	447	3730	2135
	6	221	1631	1031	290	2242	1575	369	2951	2055	461	3893	2605
100	2	192	1923	712	254	2644	1050	326	3490	1370	402	4707	1740
	4	200	1984	888	263	2731	1346	336	3606	1760	414	4842	2220
	6	208	2035	1064	272	2811	1642	346	3714	2150	426	4968	2700

	COMMON VENT CAPACITY											
	TYPE B DOUBLE-WALL COMMON VENT DIAMETER – *D* (inch)											
	12			14			16			18		
	COMBINED APPLIANCE INPUT RATING IN THOUSANDS OF BTU PER HOUR											
VENT HEIGHT *H* (feet)	FAN +FAN	FAN +NAT	NAT +NAT	FAN +FAN	FAN +NAT	NAT +NAT	FAN +FAN	FAN +NAT	NAT +NAT	FAN +FAN	FAN +NAT	NAT +NAT
6	900	696	588	1284	990	815	1735	1336	1065	2253	1732	1345
8	994	773	652	1423	1103	912	1927	1491	1190	2507	1936	1510
10	1076	841	712	1542	1200	995	2093	1625	1300	2727	2113	1645
15	1247	986	825	1794	1410	1158	2440	1910	1510	3184	2484	1910
20	1405	1116	916	2006	1588	1290	2722	2147	1690	3561	2798	2140
30	1658	1327	1025	2373	1892	1525	3220	2558	1990	4197	3326	2520
50	2024	1640	1280	2911	2347	1863	3964	3183	2430	5184	4149	3075
100	2569	2131	1670	3732	3076	2450	5125	4202	3200	6749	5509	4050

For SI units: 1 inch = 25.4 mm, 1 foot = 304.8 mm, 1000 British thermal units per hour = 0.293 kW, 1 square inch = 0.000645 m^2
* NA: Not applicable.

TABLE 510.2(1)
TYPE B DOUBLE-WALL VENT [NFPA 54: TABLE 13.2(a)] (continued)*

				NUMBER OF APPLIANCES:	TWO OR MORE					
				APPLIANCE TYPE:	CATEGORY I					
				APPLIANCE VENT CONNECTION:	TYPE B DOUBLE-WALL CONNECTOR					
		VENT CONNECTOR CAPACITY								
		TYPE B DOUBLE-WALL VENT AND CONNECTOR DIAMETER – *D* (inch)								
		20			22			24		
		APPLIANCE INPUT RATING LIMITS IN THOUSANDS OF BTU PER HOUR								
VENT HEIGHT *H* (feet)	CONNECTOR RISE *R* (feet)	FAN		NAT	FAN		NAT	FAN		NAT
		Min	Max	Max	Min	Max	Max	Min	Max	Max
6	2	NA	NA	NA	NA	NA	NA	NA	NA	NA
	4	NA	NA	NA	NA	NA	NA	NA	NA	NA
	6	NA	NA	NA	NA	NA	NA	NA	NA	NA
8	2	NA	NA	NA	NA	NA	NA	NA	NA	NA
	4	471	2737	1800	560	3319	2180	662	3957	2590
	6	478	3018	2180	568	3665	2640	669	4373	3130
10	2	NA	NA	NA	NA	NA	NA	NA	NA	NA
	4	486	2887	1890	581	3502	2280	686	4175	2710
	6	494	3169	2290	589	3849	2760	694	4593	3270
15	2	NA	NA	NA	NA	NA	NA	NA	NA	NA
	4	523	3197	2060	624	3881	2490	734	4631	2960
	6	533	3470	2510	634	4216	3030	743	5035	3600
20	2	NA	NA	NA	NA	NA	NA	NA	NA	NA
	4	554	3447	2180	661	4190	2630	772	5005	3130
	6	567	3708	2650	671	4511	3190	785	5392	3790
30	2	NA	NA	NA	NA	NA	NA	NA	NA	NA
	4	619	3840	2365	728	4861	2860	847	5606	3410
	6	632	4080	2875	741	4976	3480	860	5961	4150
50	2	NA	NA	NA	NA	NA	NA	NA	NA	NA
	4	580	4601	2633	709	5569	3185	851	6633	3790
	6	594	4808	3208	724	5826	3885	867	6943	4620
100	2	NA	NA	NA	NA	NA	NA	NA	NA	NA
	4	523	5982	2750	639	7254	3330	769	8650	3950
	6	539	6143	3350	654	7453	4070	786	8892	4810

	COMMON VENT CAPACITY								
	TYPE B DOUBLE-WALL COMMON VENT DIAMETER – *D* (inch)								
	20			22			24		
	COMBINED APPLIANCE INPUT RATING IN THOUSANDS OF BTU PER HOUR								
VENT HEIGHT *H* (feet)	FAN +FAN	FAN +NAT	NAT +NAT	FAN +FAN	FAN +NAT	NAT +NAT	FAN +FAN	FAN +NAT	NAT +NAT
6	2838	2180	1660	3488	2677	1970	4206	3226	2390
8	3162	2439	1860	3890	2998	2200	4695	3616	2680
10	3444	2665	2030	4241	3278	2400	5123	3957	2920
15	4026	3133	2360	4971	3862	2790	6016	4670	3400
20	4548	3552	2640	5573	4352	3120	6749	5261	3800
30	5303	4193	3110	6539	5157	3680	7940	6247	4480
50	6567	5240	3800	8116	6458	4500	9837	7813	5475
100	8597	6986	5000	10 681	8648	5920	13 004	10 499	7200

For SI units: 1 inch = 25.4 mm, 1 foot = 304.8 mm, 1000 British thermal units per hour = 0.293 kW, 1 square inch = 0.000645 m^2
* NA: Not applicable.

TABLE 510.2(2)
TYPE B DOUBLE-WALL VENT [NFPA 54: TABLE 13.2(b)]*

							NUMBER OF APPLIANCES:			TWO OR MORE						
							APPLIANCE TYPE:			CATEGORY I						
							APPLIANCE VENT CONNECTION:			SINGLE-WALL METAL CONNECTOR						
		VENT CONNECTOR CAPACITY														
		SINGLE-WALL METAL VENT CONNECTOR DIAMETER – *D* (inch)														
		3			4			5			6			7		
		APPLIANCE INPUT RATING LIMITS IN THOUSANDS OF BTU PER HOUR														
VENT HEIGHT *H* (feet)	CONNECTOR RISE *R* (feet)	FAN Min	FAN Max	NAT Max	FAN Min	FAN Max	NAT Max	FAN Min	FAN Max	NAT Max	FAN Min	FAN Max	NAT Max	FAN Min	FAN Max	NAT Max
6	1	NA	NA	26	NA	NA	46	NA	NA	71	NA	NA	102	207	223	140
	2	NA	NA	31	NA	NA	55	NA	NA	85	168	182	123	215	251	167
	3	NA	NA	34	NA	NA	62	121	131	95	175	198	138	222	273	188
8	1	NA	NA	27	NA	NA	48	NA	NA	75	NA	NA	106	226	240	145
	2	NA	NA	32	NA	NA	57	125	126	89	184	193	127	234	266	173
	3	NA	NA	35	NA	NA	64	130	138	100	191	208	144	241	287	197
10	1	NA	NA	28	NA	NA	50	119	121	77	182	186	110	240	253	150
	2	NA	NA	33	84	85	59	124	134	91	189	203	132	248	278	183
	3	NA	NA	36	89	91	67	129	144	102	197	217	148	257	299	203
15	1	NA	NA	29	79	87	52	116	138	81	177	214	116	238	291	158
	2	NA	NA	34	83	94	62	121	150	97	185	230	138	246	314	189
	3	NA	NA	39	87	100	70	127	160	109	193	243	157	255	333	215
20	1	49	56	30	78	97	54	115	152	84	175	238	120	233	325	165
	2	52	59	36	82	103	64	120	163	101	182	252	144	243	346	197
	3	55	62	40	87	107	72	125	172	113	190	264	164	252	363	223
30	1	47	60	31	77	110	57	112	175	89	169	278	129	226	380	175
	2	51	62	37	81	115	67	117	185	106	177	290	152	236	397	208
	3	54	64	42	85	119	76	122	193	120	185	300	172	244	412	235
50	1	46	69	34	75	128	60	109	207	96	162	336	137	217	460	188
	2	49	71	40	79	132	72	114	215	113	170	345	164	226	473	223
	3	52	72	45	83	136	82	119	221	123	178	353	186	235	486	252
100	1	45	79	34	71	150	61	104	249	98	153	424	140	205	585	192
	2	48	80	41	75	153	73	110	255	115	160	428	167	212	593	228
	3	51	81	46	79	157	85	114	260	129	168	433	190	222	603	256

	COMMON VENT CAPACITY											
	TYPE B DOUBLE-WALL COMMON VENT DIAMETER – *D* (inch)											
	4			5			6			7		
	COMBINED APPLIANCE INPUT RATING IN THOUSANDS OF BTU PER HOUR											
VENT HEIGHT *H* (feet)	FAN +FAN	FAN +NAT	NAT +NAT	FAN +FAN	FAN +NAT	NAT +NAT	FAN +FAN	FAN +NAT	NAT +NAT	FAN +FAN	FAN +NAT	NAT +NAT
6	NA	78	64	NA	113	99	200	158	144	304	244	196
8	NA	87	71	NA	126	111	218	173	159	331	269	218
10	NA	94	76	163	137	120	237	189	174	357	292	236
15	121	108	88	189	159	140	275	221	200	416	343	274
20	131	118	98	208	177	156	305	247	223	463	383	302
30	145	132	113	236	202	180	350	286	257	533	446	349
50	159	145	128	268	233	208	406	337	296	622	529	410
100	166	153	NA	297	263	NA	469	398	NA	726	633	464

For SI units: 1 inch = 25.4 mm, 1 foot = 304.8 mm, 1000 British thermal units per hour = 0.293 kW, 1 square inch = 0.000645 m^2

* NA: Not applicable.

TABLE 510.2(2)
TYPE B DOUBLE-WALL VENT [NFPA 54: TABLE 13.2(b)] (continued)

		NUMBER OF APPLIANCES: TWO OR MORE								
		APPLIANCE TYPE: CATEGORY I								
		APPLIANCE VENT CONNECTION: SINGLE-WALL METAL CONNECTOR								
		VENT CONNECTOR CAPACITY								
		SINGLE-WALL METAL VENT CONNECTOR DIAMETER – *D* (inch)								
		8			9			10		
		APPLIANCE INPUT RATING LIMITS IN THOUSANDS OF BTU PER HOUR								
VENT HEIGHT *H* (feet)	CONNECTOR RISE *R* (feet)	FAN		NAT	FAN		NAT	FAN		NAT
		Min	Max	Max	Min	Max	Max	Min	Max	Max
6	1	262	293	183	325	373	234	447	463	286
	2	271	331	219	334	422	281	458	524	344
	3	279	361	247	344	462	316	468	574	385
8	1	285	316	191	352	403	244	481	502	299
	2	293	353	228	360	450	292	492	560	355
	3	302	381	256	370	489	328	501	609	400
10	1	302	335	196	372	429	252	506	534	308
	2	311	369	235	381	473	302	517	589	368
	3	320	398	265	391	511	339	528	637	413
15	1	312	380	208	397	482	266	556	596	324
	2	321	411	248	407	522	317	568	646	387
	3	331	438	281	418	557	360	579	690	437
20	1	306	425	217	390	538	276	546	664	336
	2	317	453	259	400	574	331	558	709	403
	3	326	476	294	412	607	375	570	750	457
30	1	296	497	230	378	630	294	528	779	358
	2	307	521	274	389	662	349	541	819	425
	3	316	542	309	400	690	394	555	855	482
50	1	284	604	245	364	768	314	507	951	384
	2	294	623	293	376	793	375	520	983	458
	3	304	640	331	387	816	423	535	1013	518
100	1	269	774	249	345	993	321	476	1236	393
	2	279	788	299	358	1011	383	490	1259	469
	3	289	801	339	368	1027	431	506	1280	527

	COMMON VENT CAPACITY								
	TYPE B DOUBLE-WALL COMMON VENT DIAMETER – *D* (inch)								
	8			9			10		
	COMBINED APPLIANCE INPUT RATING IN THOUSANDS OF BTU PER HOUR								
VENT HEIGHT *H* (feet)	FAN +FAN	FAN +NAT	NAT +NAT	FAN +FAN	FAN +NAT	NAT +NAT	FAN +FAN	FAN +NAT	NAT +NAT
6	398	310	257	541	429	332	665	515	407
8	436	342	285	592	473	373	730	569	460
10	467	369	309	638	512	398	787	617	487
15	544	434	357	738	599	456	905	718	553
20	606	487	395	824	673	512	1013	808	626
30	703	570	459	958	790	593	1183	952	723
50	833	686	535	1139	954	689	1418	1157	838
100	999	846	606	1378	1185	780	1741	1459	948

For SI units: 1 inch = 25.4 mm, 1 foot = 304.8 mm, 1000 British thermal units per hour = 0.293 kW, 1 square inch = 0.000645 m^2

TABLE 510.2(3)
MASONRY CHIMNEY [NFPA 54: TABLE 13.2(c)]*

		NUMBER OF APPLIANCES: TWO OR MORE														
		APPLIANCE TYPE: CATEGORY I														
		APPLIANCE VENT CONNECTION: TYPE B DOUBLE-WALL CONNECTOR														
		VENT CONNECTOR CAPACITY														
		TYPE B DOUBLE-WALL VENT CONNECTOR DIAMETER – *D* (inch)														
		3			4			5			6			7		
		APPLIANCE INPUT RATING LIMITS IN THOUSANDS OF BTU PER HOUR														
VENT HEIGHT *H* (feet)	CONNECTOR RISE *R* (feet)	FAN		NAT	FAN		NAT	FAN		NAT	FAN		NAT	FAN		NAT
		Min	Max	Max	Min	Max	Max	Min	Max	Max	Min	Max	Max	Min	Max	Max
6	1	24	33	21	39	62	40	52	106	67	65	194	101	87	274	141
	2	26	43	28	41	79	52	53	133	85	67	230	124	89	324	173
	3	27	49	34	42	92	61	55	155	97	69	262	143	91	369	203
8	1	24	39	22	39	72	41	55	117	69	71	213	105	94	304	148
	2	26	47	29	40	87	53	57	140	86	73	246	127	97	350	179
	3	27	52	34	42	97	62	59	159	98	75	269	145	99	383	206
10	1	24	42	22	38	80	42	55	130	71	74	232	108	101	324	153
	2	26	50	29	40	93	54	57	153	87	76	261	129	103	366	184
	3	27	55	35	41	105	63	58	170	100	78	284	148	106	397	209
15	1	24	48	23	38	93	44	54	154	74	72	277	114	100	384	164
	2	25	55	31	39	105	55	56	174	89	74	299	134	103	419	192
	3	26	59	35	41	115	64	57	189	102	76	319	153	105	448	215
20	1	24	52	24	37	102	46	53	172	77	71	313	119	98	437	173
	2	25	58	31	39	114	56	55	190	91	73	335	138	101	467	199
	3	26	63	35	40	123	65	57	204	104	75	353	157	104	493	222
30	1	24	54	25	37	111	48	52	192	82	69	357	127	96	504	187
	2	25	60	32	38	122	58	54	208	95	72	376	145	99	531	209
	3	26	64	36	40	131	66	56	221	107	74	392	163	101	554	233
50	1	23	51	25	36	116	51	51	209	89	67	405	143	92	582	213
	2	24	59	32	37	127	61	53	225	102	70	421	161	95	604	235
	3	26	64	36	39	135	69	55	237	115	72	435	180	98	624	260
100	1	23	46	24	35	108	50	49	208	92	65	428	155	88	640	237
	2	24	53	31	37	120	60	51	224	105	67	444	174	92	660	260
	3	25	59	35	38	130	68	53	237	118	69	458	193	94	679	285

	COMMON VENT CAPACITY														
	MINIMUM INTERNAL AREA OF MASONRY CHIMNEY FLUE (square inches)														
	12			19			28			38			50		
	COMBINED APPLIANCE INPUT RATING IN THOUSANDS OF BTU PER HOUR														
VENT HEIGHT *H* (feet)	FAN +FAN	FAN +NAT	NAT +NAT	FAN +FAN	FAN +NAT	NAT +NAT	FAN +FAN	FAN +NAT	NAT +NAT	FAN +FAN	FAN +NAT	NAT +NAT	FAN +FAN	FAN +NAT	NAT +NAT
6	NA	74	25	NA	119	46	NA	178	71	NA	257	103	NA	351	143
8	NA	80	28	NA	130	53	NA	193	82	NA	279	119	NA	384	163
10	NA	84	31	NA	138	56	NA	207	90	NA	299	131	NA	409	177
15	NA	NA	36	NA	152	67	NA	233	106	NA	334	152	523	467	212
20	NA	NA	41	NA	NA	75	NA	250	122	NA	368	172	565	508	243
30	NA	NA	NA	NA	NA	NA	NA	270	137	NA	404	198	615	564	278
50	NA	NA	NA	NA	NA	NA	NA	NA	NA	NA	NA	NA	NA	620	328
100	NA	NA	NA	NA	NA	NA	NA	NA	NA	NA	NA	NA	NA	NA	348

For SI units: 1 inch = 25.4 mm, 1 foot = 304.8 mm, 1000 British thermal units per hour = 0.293 kW, 1 square inch = 0.000645 m^2
* NA: Not applicable.

TABLE 510.2(3)
MASONRY CHIMNEY [NFPA 54: TABLE 13.2(c)] (continued)*

		NUMBER OF APPLIANCES: TWO OR MORE								
		APPLIANCE TYPE: CATEGORY I								
		APPLIANCE VENT CONNECTION: TYPE B DOUBLE-WALL CONNECTOR								
		VENT CONNECTOR CAPACITY								
		TYPE B DOUBLE-WALL VENT CONNECTOR DIAMETER – *D* (inch)								
		8			9			10		
		APPLIANCE INPUT RATING LIMITS IN THOUSANDS OF BTU PER HOUR								
VENT HEIGHT *H* (feet)	CONNECTOR RISE *R* (feet)	FAN		NAT	FAN		NAT	FAN		NAT
		Min	Max	Max	Min	Max	Max	Min	Max	Max
6	1	104	370	201	124	479	253	145	599	319
	2	107	436	232	127	562	300	148	694	378
	3	109	491	270	129	633	349	151	795	439
8	1	113	414	210	134	539	267	156	682	335
	2	116	473	240	137	615	311	160	776	394
	3	119	517	276	139	672	358	163	848	452
10	1	120	444	216	142	582	277	165	739	348
	2	123	498	247	145	652	321	168	825	407
	3	126	540	281	147	705	366	171	893	463
15	1	125	511	229	153	658	297	184	824	375
	2	128	558	260	156	718	339	187	900	432
	3	131	597	292	159	760	382	190	960	486
20	1	123	584	239	150	752	312	180	943	397
	2	126	625	270	153	805	354	184	1011	452
	3	129	661	301	156	851	396	187	1067	505
30	1	119	680	255	145	883	337	175	1115	432
	2	122	715	287	149	928	378	179	1171	484
	3	125	746	317	152	968	418	182	1220	535
50	1	115	798	294	140	1049	392	168	1334	506
	2	118	827	326	143	1085	433	172	1379	558
	3	121	854	357	147	1118	474	176	1421	611
100	1	109	907	334	134	1222	454	161	1589	596
	2	113	933	368	138	1253	497	165	1626	651
	3	116	956	399	141	1282	540	169	1661	705

	COMMON VENT CAPACITY								
	MINIMUM INTERNAL AREA OF MASONRY CHIMNEY FLUE (square inches)								
	63			78			113		
	COMBINED APPLIANCE INPUT RATING IN THOUSANDS OF BTU PER HOUR								
VENT HEIGHT *H* (feet)	FAN +FAN	FAN +NAT	NAT +NAT	FAN +FAN	FAN +NAT	NAT +NAT	FAN +FAN	FAN +NAT	NAT +NAT
6	NA	458	188	NA	582	246	1041	853	NA
8	NA	501	218	724	636	278	1144	937	408
10	606	538	236	776	686	302	1226	1010	454
15	682	611	283	874	781	365	1374	1156	546
20	742	668	325	955	858	419	1513	1286	648
30	816	747	381	1062	969	496	1702	1473	749
50	879	831	461	1165	1089	606	1905	1692	922
100	NA	NA	499	NA	NA	669	2053	1921	1058

For SI units: 1 inch = 25.4 mm, 1 foot = 304.8 mm, 1000 British thermal units per hour = 0.293 kW, 1 square inch = 0.000645 m^2
* NA: Not applicable.

TABLE 510.2(4)
MASONRY CHIMNEY [NFPA 54: TABLE 13.2(d)]*

						NUMBER OF APPLIANCES:			TWO OR MORE							
						APPLIANCE TYPE:			CATEGORY I							
						APPLIANCE VENT CONNECTION:			SINGLE-WALL METAL CONNECTOR							
		VENT CONNECTOR CAPACITY														
		SINGLE-WALL METAL VENT CONNECTOR DIAMETER – *D* (inch)														
		3			4			5			6			7		
		APPLIANCE INPUT RATING LIMITS IN THOUSANDS OF BTU PER HOUR														
VENT HEIGHT *H* (feet)	CONNECTOR RISE *R* (feet)	FAN		NAT	FAN		NAT	FAN		NAT	FAN		NAT	FAN		NAT
		Min	Max	Max	Min	Max	Max	Min	Max	Max	Min	Max	Max	Min	Max	Max
6	1	NA	NA	21	NA	NA	39	NA	NA	66	179	191	100	231	271	140
	2	NA	NA	28	NA	NA	52	NA	NA	84	186	227	123	239	321	172
	3	NA	NA	34	NA	NA	61	134	153	97	193	258	142	247	365	202
8	1	NA	NA	21	NA	NA	40	NA	NA	68	195	208	103	250	298	146
	2	NA	NA	28	NA	NA	52	137	139	85	202	240	125	258	343	177
	3	NA	NA	34	NA	NA	62	143	156	98	210	264	145	266	376	205
10	1	NA	NA	22	NA	NA	41	130	151	70	202	225	106	267	316	151
	2	NA	NA	29	NA	NA	53	136	150	86	210	255	128	276	358	181
	3	NA	NA	34	97	102	62	143	166	99	217	277	147	284	389	207
15	1	NA	NA	23	NA	NA	43	129	151	73	199	271	112	268	376	161
	2	NA	NA	30	92	103	54	135	170	88	207	295	132	277	411	189
	3	NA	NA	34	96	112	63	141	185	101	215	315	151	286	439	213
20	1	NA	NA	23	87	99	45	128	167	76	197	303	117	265	425	169
	2	NA	NA	30	91	111	55	134	185	90	205	325	136	274	455	195
	3	NA	NA	35	96	119	64	140	199	103	213	343	154	282	481	219
30	1	NA	NA	24	86	108	47	126	187	80	193	347	124	259	492	183
	2	NA	NA	31	91	119	57	132	203	93	201	366	142	269	518	205
	3	NA	NA	35	95	127	65	138	216	105	209	381	160	277	540	229
50	1	NA	NA	24	85	113	50	124	204	87	188	392	139	252	567	208
	2	NA	NA	31	89	123	60	130	218	100	196	408	158	262	588	230
	3	NA	NA	35	94	131	68	136	231	112	205	422	176	271	607	255
100	1	NA	NA	23	84	104	49	122	200	89	182	410	151	243	617	232
	2	NA	NA	30	88	115	59	127	215	102	190	425	169	253	636	254
	3	NA	NA	34	93	124	67	133	228	115	199	438	188	262	654	279

	COMMON VENT CAPACITY														
	MINIMUM INTERNAL AREA OF MASONRY CHIMNEY FLUE (square inches)														
	12			19			28			38			50		
	COMBINED APPLIANCE INPUT RATING IN THOUSANDS OF BTU PER HOUR														
VENT HEIGHT *H* (feet)	FAN +FAN	FAN +NAT	NAT +NAT	FAN +FAN	FAN +NAT	NAT +NAT	FAN +FAN	FAN +NAT	NAT +NAT	FAN +FAN	FAN +NAT	NAT +NAT	FAN +FAN	FAN +NAT	NAT +NAT
6	NA	NA	25	NA	118	45	NA	176	71	NA	255	102	NA	348	142
8	NA	NA	28	NA	128	52	NA	190	81	NA	276	118	NA	380	162
10	NA	NA	31	NA	136	56	NA	205	89	NA	295	129	NA	405	175
15	NA	NA	36	NA	NA	66	NA	230	105	NA	335	150	NA	400	210
20	NA	NA	NA	NA	NA	74	NA	247	120	NA	362	170	NA	503	240
30	NA	NA	NA	NA	NA	NA	NA	NA	135	NA	398	195	NA	558	275
50	NA	NA	NA	NA	NA	NA	NA	NA	NA	NA	NA	NA	NA	612	325
100	NA	NA	NA	NA	NA	NA	NA	NA	NA	NA	NA	NA	NA	NA	NA

For SI units: 1 inch = 25.4 mm, 1 foot = 304.8 mm, 1000 British thermal units per hour = 0.293 kW, 1 square inch = 0.000645 m^2

* NA: Not applicable.

TABLE 510.2(4)
MASONRY CHIMNEY [NFPA 54: TABLE 13.2(d)] (continued)*

		NUMBER OF APPLIANCES: TWO OR MORE								
		APPLIANCE TYPE: CATEGORY I								
		APPLIANCE VENT CONNECTION: SINGLE-WALL METAL CONNECTOR								
		VENT CONNECTOR CAPACITY								
		SINGLE-WALL METAL VENT CONNECTOR DIAMETER – *D* (inch)								
		8			9			10		
		APPLIANCE INPUT RATING LIMITS IN THOUSANDS OF BTU PER HOUR								
VENT HEIGHT *H* (feet)	CONNECTOR RISE *R* (feet)	FAN		NAT	FAN		NAT	FAN		NAT
		Min	Max	Max	Min	Max	Max	Min	Max	Max
6	1	292	366	200	362	474	252	499	594	316
	2	301	432	231	373	557	299	509	696	376
	3	309	491	269	381	634	348	519	793	437
8	1	313	407	207	387	530	263	529	672	331
	2	323	465	238	397	607	309	540	766	391
	3	332	509	274	407	663	356	551	838	450
10	1	333	434	213	410	571	273	558	727	343
	2	343	489	244	420	640	317	569	813	403
	3	352	530	279	430	694	363	580	880	459
15	1	349	502	225	445	646	291	623	808	366
	2	359	548	256	456	706	334	634	884	424
	3	368	586	289	466	755	378	646	945	479
20	1	345	569	235	439	734	306	614	921	387
	2	355	610	266	450	787	348	627	986	443
	3	365	644	298	461	831	391	639	1042	496
30	1	338	665	250	430	864	330	600	1089	421
	2	348	699	282	442	908	372	613	1145	473
	3	358	729	312	452	946	412	626	1193	524
50	1	328	778	287	417	1022	383	582	1302	492
	2	339	806	320	429	1058	425	596	1346	545
	3	349	831	351	440	1090	466	610	1386	597
100	1	315	875	328	402	1181	444	560	1537	580
	2	326	899	361	415	1210	488	575	1570	634
	3	337	921	392	427	1238	529	589	1604	687

	COMMON VENT CAPACITY								
	MINIMUM INTERNAL AREA OF MASONRY CHIMNEY FLUE (square inches)								
	63			78			113		
	COMBINED APPLIANCE INPUT RATING IN THOUSANDS OF BTU PER HOUR								
VENT HEIGHT *H* (feet)	FAN +FAN	FAN +NAT	NAT +NAT	FAN +FAN	FAN +NAT	NAT +NAT	FAN +FAN	FAN +NAT	NAT +NAT
6	NA	455	187	NA	579	245	NA	846	NA
8	NA	497	217	NA	633	277	1136	928	405
10	NA	532	234	771	680	300	1216	1000	450
15	677	602	280	866	772	360	1359	1139	540
20	765	661	321	947	849	415	1495	1264	640
30	808	739	377	1052	957	490	1682	1447	740
50	NA	821	456	1152	1076	600	1879	1672	910
100	NA	NA	494	NA	NA	663	2006	1885	1046

For SI units: 1 inch = 25.4 mm, 1 foot = 304.8 mm, 1000 British thermal units per hour = 0.293 kW, 1 square inch = 0.000645 m^2

* NA: Not applicable.

TABLE 510.2(5)
SINGLE-WALL METAL PIPE OR TYPE B ASBESTOS-CEMENT VENT [NFPA 54: TABLE 13.2(e)]*

				NUMBER OF APPLIANCES:	TWO OR MORE		
				APPLIANCE TYPE:	DRAFT HOOD-EQUIPMENT		
				APPLIANCE VENT CONNECTION:	DIRECT TO PIPE OR VENT		
		VENT CONNECTOR CAPACITY					
		VENT CONNECTOR DIAMETER – *D* (inch)					
TOTAL VENT HEIGHT *H* (feet)	**CONNECTOR RISE *R* (feet)**	**3**	**4**	**5**	**6**	**7**	**8**
		MAXIMUM APPLIANCE INPUT RATING IN THOUSANDS OF BTU PER HOUR					
6-8	1	21	40	68	102	146	205
	2	28	53	86	124	178	235
	3	34	61	98	147	204	275
15	1	23	44	77	117	179	240
	2	30	56	92	134	194	265
	3	35	64	102	155	216	298
30 and up	1	25	49	84	129	190	270
	2	31	58	97	145	211	295
	3	36	68	107	164	232	321

	COMMON VENT CAPACITY						
	COMMON VENT DIAMETER – *D* (inch)						
TOTAL VENT HEIGHT *H* (feet)	**4**	**5**	**6**	**7**	**8**	**10**	**12**
	COMBINED APPLIANCE INPUT RATING IN THOUSANDS OF BTU PER HOUR						
6	48	78	111	155	205	320	NA
8	55	89	128	175	234	365	505
10	59	95	136	190	250	395	560
15	71	115	168	228	305	480	690
20	80	129	186	260	340	550	790
30	NA	147	215	300	400	650	940
50	NA	NA	NA	360	490	810	1190

For SI units: 1 inch = 25.4 mm, 1 foot = 304.8 mm, 1000 British thermal units per hour = 0.293 kW, 1 square inch = 0.000645 m^2
* NA: Not applicable.

TABLE 510.2(6)
EXTERIOR MASONRY CHIMNEY [NFPA 54: TABLE 13.2(f)]*

				NUMBER OF APPLIANCES:	TWO OR MORE			
				APPLIANCE TYPE:	NAT + NAT			
				APPLIANCE VENT CONNECTION:	TYPE B DOUBLE-WALL CONNECTOR			
COMBINED APPLIANCE MAXIMUM INPUT RATING IN THOUSANDS OF BTU PER HOUR								
VENT HEIGHT *H* (feet)	INTERNAL AREA OF CHIMNEY (square inches)							
	12	**19**	**28**	**38**	**50**	**63**	**78**	**113**
6	25	46	71	103	143	188	246	NA
8	28	53	82	119	163	218	278	408
10	31	56	90	131	177	236	302	454
15	NA	67	106	152	212	283	365	546
20	NA	NA	NA	NA	NA	325	419	648
30	NA	NA	NA	NA	NA	NA	496	749
50	NA	NA	NA	NA	NA	NA	NA	922
100	NA	NA	NA	NA	NA	NA	NA	NA

For SI units: 1 inch = 25.4 mm, 1 foot = 304.8 mm, 1000 British thermal units per hour = 0.293 kW, 1 square inch = 0.000645 m^2
* NA: Not applicable.

TABLE 510.2(7)
EXTERIOR MASONRY CHIMNEY [NFPA 54: TABLE 13.2(g)][1, 2]

				NUMBER OF APPLIANCES:	**TWO OR MORE**			
				APPLIANCE TYPE:	**NAT + NAT**			
				APPLIANCE VENT CONNECTION:	**TYPE B DOUBLE-WALL CONNECTOR**			
MINIMUM ALLOWABLE INPUT RATING OF SPACE-HEATING APPLIANCE IN THOUSANDS OF BTU PER HOUR								
VENT HEIGHT *H* (feet)	**INTERNAL AREA OF CHIMNEY (square inches)**							
	12	**19**	**28**	**38**	**50**	**63**	**78**	**113**
Local 99% winter design temperature: 37°F or greater								
6	0	0	0	0	0	0	0	NA
8	0	0	0	0	0	0	0	0
10	0	0	0	0	0	0	0	0
15	NA	0	0	0	0	0	0	0
20	NA	NA	NA	NA	NA	184	0	0
30	NA	NA	NA	NA	NA	393	334	0
50	NA	NA	NA	NA	NA	NA	NA	579
100	NA	NA	NA	NA	NA	NA	NA	NA
Local 99% winter design temperature: 27°F to 36°F								
6	0	0	68	NA	NA	180	212	NA
8	0	0	82	NA	NA	187	214	263
10	0	51	NA	NA	NA	201	225	265
15	NA	NA	NA	NA	NA	253	274	305
20	NA	NA	NA	NA	NA	307	330	362
30	NA	NA	NA	NA	NA	NA	445	485
50	NA	NA	NA	NA	NA	NA	NA	763
100	NA	NA	NA	NA	NA	NA	NA	NA
Local 99% winter design temperature: 17°F to 26°F								
6	NA	NA	NA	NA	NA	NA	NA	NA
8	NA	NA	NA	NA	NA	NA	264	352
10	NA	NA	NA	NA	NA	NA	278	358
15	NA	NA	NA	NA	NA	NA	331	398
20	NA	NA	NA	NA	NA	NA	387	457
30	NA	NA	NA	NA	NA	NA	NA	581
50	NA	NA	NA	NA	NA	NA	NA	862
100	NA	NA	NA	NA	NA	NA	NA	NA
Local 99% winter design temperature: 5°F to 16°F								
6	NA	NA	NA	NA	NA	NA	NA	NA
8	NA	NA	NA	NA	NA	NA	NA	NA
10	NA	NA	NA	NA	NA	NA	NA	430
15	NA	NA	NA	NA	NA	NA	NA	485
20	NA	NA	NA	NA	NA	NA	NA	547
30	NA	NA	NA	NA	NA	NA	NA	682
50	NA	NA	NA	NA	NA	NA	NA	NA
100	NA	NA	NA	NA	NA	NA	NA	NA
Local 99% winter design temperature: 4°F or lower Not recommended for any vent configurations								

For SI units: 1 inch = 25.4 mm, 1 foot = 304.8 mm, 1000 British thermal units per hour = 0.293 kW, 1 square inch = 0.000645 m^2, °C = (°F-32)/1.8

Notes:

1 See Figure 510.1.10 for a map showing local 99 percent winter design temperatures in the United States.

2 NA: Not applicable.

TABLE 510.2(8)
EXTERIOR MASONRY CHIMNEY [NFPA 54: TABLE 13.2(h)]*

				NUMBER OF APPLIANCES:	TWO OR MORE			
				APPLIANCE TYPE:	FAN + NAT			
				APPLIANCE VENT CONNECTION:	TYPE B DOUBLE-WALL CONNECTOR			
COMBINED APPLIANCE MAXIMUM INPUT RATING IN THOUSANDS OF BTU PER HOUR								
VENT HEIGHT *H* (feet)	INTERNAL AREA OF CHIMNEY (square inches)							
	12	19	28	38	50	63	78	113
6	74	119	178	257	351	458	582	853
8	80	130	193	279	384	501	636	937
10	84	138	207	299	409	538	686	1010
15	NA	152	233	334	467	611	781	1156
20	NA	NA	250	368	508	668	858	1286
30	NA	NA	NA	404	564	747	969	1473
50	NA	NA	NA	NA	NA	831	1089	1692
100	NA	NA	NA	NA	NA	NA	NA	1921

For SI units: 1 inch = 25.4 mm, 1 foot = 304.8 mm, 1000 British thermal units per hour = 0.293 kW, 1 square inch = 0.000645 m^2
* NA: Not applicable.

TABLE 510.2(9)
EXTERIOR MASONRY CHIMNEY [NFPA 54: TABLE 13.2(i)][1, 2]

				NUMBER OF APPLIANCES:	TWO OR MORE			
				APPLIANCE TYPE:	FAN + NAT			
				APPLIANCE VENT CONNECTION:	TYPE B DOUBLE-WALL CONNECTOR			
MINIMUM ALLOWABLE INPUT RATING OF SPACE-HEATING APPLIANCE IN THOUSANDS OF BTU PER HOUR								
VENT HEIGHT *H* (feet)	INTERNAL AREA OF CHIMNEY (square inches)							
	12	19	28	38	50	63	78	113
Local 99% winter design temperature: 37°F or greater								
6	0	0	0	0	0	0	0	0
8	0	0	0	0	0	0	0	0
10	0	0	0	0	0	0	0	0
15	NA	0	0	0	0	0	0	0
20	NA	NA	123	190	249	184	0	0
30	NA	NA	NA	334	398	393	334	0
50	NA	NA	NA	NA	NA	714	707	579
100	NA	NA	NA	NA	NA	NA	NA	1600
Local 99% winter design temperature: 27°F to 36°F								
6	0	0	68	116	156	180	212	266
8	0	0	82	127	167	187	214	263
10	0	51	97	141	183	201	225	265
15	NA	111	142	183	233	253	274	305
20	NA	NA	187	230	284	307	330	362
30	NA	NA	NA	330	319	419	445	485
50	NA	NA	NA	NA	NA	672	705	763
100	NA	NA	NA	NA	NA	NA	NA	1554
Local 99% winter design temperature: 17°F to 26°F								
6	0	55	99	141	182	215	259	349
8	52	74	111	154	197	226	264	352
10	NA	90	125	169	214	245	278	358
15	NA	NA	167	212	263	296	331	398
20	NA	NA	212	258	316	352	387	457
30	NA	NA	NA	362	429	470	507	581
50	NA	NA	NA	NA	NA	723	766	862
100	NA	NA	NA	NA	NA	NA	NA	1669
Local 99% winter design temperature: 5°F to 16°F								
6	NA	78	121	166	214	252	301	416
8	NA	94	135	182	230	269	312	423
10	NA	111	149	198	250	289	331	430
15	NA	NA	193	247	305	346	393	485
20	NA	NA	NA	293	360	408	450	547
30	NA	NA	NA	377	450	531	580	682
50	NA	NA	NA	NA	NA	797	853	972
100	NA	NA	NA	NA	NA	NA	NA	1833
Local 99% winter design temperature: -10°F to 4°F								
6	NA	NA	145	196	249	296	349	484
8	NA	NA	159	213	269	320	371	494
10	NA	NA	175	231	292	339	397	513
15	NA	NA	NA	283	351	404	457	586
20	NA	NA	NA	333	408	468	528	650
30	NA	NA	NA	NA	NA	603	667	805
50	NA	NA	NA	NA	NA	NA	955	1003
100	NA	NA	NA	NA	NA	NA	NA	NA
Local 99% winter design temperature: -11°F or lower Not recommended for any vent configurations								

For SI units: 1 inch = 25.4 mm, 1 foot = 304.8 mm, 1000 British thermal units per hour = 0.293 kW, 1 square inch = 0.000645 m^2, °C = (°F-32)/1.8

Notes:

[1] See Figure 510.1.10 for a map showing local 99 percent winter design temperatures in the United States.

[2] NA: Not applicable.

CHAPTER 6

WATER SUPPLY AND DISTRIBUTION

601.0 General.

601.1 Applicability. This chapter shall govern the materials, design, and installation of water supply systems, including methods and devices used for backflow prevention.

601.2 Water Supply and Flushing. Each plumbing fixture shall be provided with an adequate supply of potable running water piped thereto in an approved manner, so arranged as to flush and keep it in a clean and sanitary condition without danger of backflow or cross-connection. Water closets and urinals shall be flushed using an approved flush tank or flushometer valve.

Exceptions:

(1) Listed fixtures that do not require water for their operation and are not connected to the water supply.

(2) Where not deemed necessary for safety and sanitation by the Authority Having Jurisdiction.

601.2.1 Hot and Cold Water Required. In occupancies where plumbing fixtures are installed for private use, hot water shall be required for bathing, washing, laundry, cooking purposes, dishwashing or maintenance. In occupancies where plumbing fixtures are installed for public use, hot water shall be required for bathing and washing purposes. This requirement shall not supersede the requirements for individual temperature control limitations for public lavatories and public and private bidets, bathtubs, whirlpool bathtubs, and shower control valves.

601.3 Identification of a Potable and Nonpotable Water System. In buildings where potable water and nonpotable water systems are installed, each system shall be clearly identified in accordance with Section 601.3.1 through Section 601.3.5.

601.3.1 Potable Water. Green background with white lettering.

601.3.2 Color and Information. Each system shall be identified with a colored pipe or band and coded with paints, wraps, and materials compatible with the piping.

Except as required by Section 601.3.3, nonpotable water systems shall have a yellow background with black uppercase lettering, with the words "CAUTION: NONPOTABLE WATER, DO NOT DRINK." Each nonpotable system shall be identified to designate the liquid being conveyed, and the direction of normal flow shall be clearly shown. The minimum size of the letters and length of the color field shall comply with Table 601.3.2.

The background color and required information shall be indicated every 20 feet (6096 mm) but not less than once per room, and shall be visible from the floor level.

601.3.3 Alternate Water Sources. Alternate water source systems shall have a purple (Pantone color No. 512, 522C, or equivalent) background with uppercase lettering and shall be field or factory marked as follows:

(1) Gray water systems shall be marked in accordance with this section with the words "CAUTION: NONPOTABLE GRAY WATER, DO NOT DRINK" in black letters.

(2) Reclaimed (recycled) water systems shall be marked in accordance with this section with the words: "CAUTION: NONPOTABLE RECLAIMED (RECYCLED) WATER, DO NOT DRINK" in black letters.

(3) On-site treated water systems shall be marked in accordance with this section with the words: "CAUTION: ON-SITE TREATED NONPOTABLE WATER, DO NOT DRINK" in black letters.

(4) Rainwater catchment systems shall be marked in accordance with this section with the words: "CAUTION: NONPOTABLE RAINWATER, DO NOT DRINK" in black letters.

TABLE 601.3.2
MINIMUM LENGTH OF COLOR FIELD AND SIZE OF LETTERS

OUTSIDE DIAMETER OF PIPE OR COVERING (inches)	MINIMUM LENGTH OF COLOR FIELD (inches)	MINIMUM SIZE OF LETTERS (inches)
½ to 1¼	8	½
1½ to 2	8	¾
2½ to 6	12	1¼
8 to 10	24	2½
Over 10	32	3½

For SI units: 1 inch = 25.4 mm

601.3.4 Fixtures. Where vacuum breakers or backflow preventers are installed with fixtures listed in Chapter 17, identification of the discharge side shall be permitted to be omitted.

601.3.5 Outlets. Each outlet on the nonpotable water line that is used for special purposes shall be posted with black uppercase lettering as follows: "CAUTION: NONPOTABLE WATER, DO NOT DRINK."

602.0 Unlawful Connections.

602.1 Prohibited Installation. No installation of potable water supply piping, or part thereof, shall be made in such a manner that it will be possible for used, unclean, polluted, or contaminated water, mixtures, or substances to enter a portion of such piping from a tank, receptor, equipment, or plumbing fixture by reason of backsiphonage, suction, or other cause, either during normal use and operation thereof, or where such tank, receptor, equipment, or plumbing fixture is flooded or subject to pressure exceeding the operating pressure in the hot or cold water piping.

602.2 Cross-Contamination. No person shall make a connection or allow one to exist between pipes or conduits carrying domestic water supplied by a public or private building supply system, and pipes, conduits, or fixtures containing or carrying water from any other source or containing or carrying water that has been used for any purpose whatsoever, or piping carrying chemicals, liquids, gases, or substances whatsoever, unless there is provided a backflow prevention device approved for the potential hazard and maintained in accordance with this code. Each point of use shall be separately protected where potential cross-contamination of individual units exists.

602.3 Backflow Prevention. No plumbing fixture, device, or construction shall be installed or maintained, or shall be connected to a domestic water supply, where such installation or connection provides a possibility of polluting such water supply or cross-connection between a distributing system of water for drinking and domestic purposes and water that becomes contaminated by such plumbing fixture, device, or construction unless there is provided a backflow prevention device approved for the potential hazard.

602.4 Approval by Authority. No water piping supplied by a private water supply system shall be connected to any other source of supply without the approval of the Authority Having Jurisdiction, Health Department, or other department having jurisdiction.

603.0 Cross-Connection Control.

603.1 General. Cross-connection control shall be provided in accordance with the provisions of this chapter.

No person shall install a water-operated equipment or mechanism, or use a water-treating chemical or substance, where it is found that such equipment, mechanism, chemical, or substance causes pollution or contamination of the domestic water supply. Such equipment or mechanism shall be permitted where equipped with an approved backflow prevention device or assembly.

603.2 Approval of Devices or Assemblies. Before a device or an assembly is installed for the prevention of backflow, it shall have first been approved by the Authority Having Jurisdiction. Devices or assemblies shall be tested in accordance with recognized standards or other standards acceptable to the Authority Having Jurisdiction. Backflow prevention devices and assemblies shall comply with Table 603.2, except for specific applications and provisions as stated in Section 603.5.1 through Section 603.5.21.

Devices or assemblies installed in a potable water supply system for protection against backflow shall be maintained in good working condition by the person or persons having control of such devices or assemblies. Such devices or assemblies shall be tested at the time of installation, repair, or relocation and not less than on an annual schedule thereafter, or more often where required by the Authority Having Jurisdiction. Where found to be defective or inoperative, the device or assembly shall be repaired or replaced. No device or assembly shall be removed from use or relocated or other device or assembly substituted, without the approval of the Authority Having Jurisdiction.

Testing or maintenance shall be performed by a certified backflow assembly tester or repairer in accordance with ASSE Series 5000 or otherwise approved by the Authority Having Jurisdiction.

603.3 Backflow Prevention Devices, Assemblies, and Methods. Backflow prevention devices, assemblies, and methods shall comply with Section 603.3.1 through Section 603.3.12.

603.3.1 Air Gap. The minimum air gap to afford backflow protection shall be in accordance with Table 603.3.1.

603.3.2 Atmospheric Vacuum Breaker (AVB). An atmospheric vacuum breaker consists of a body, a checking member, and an atmospheric port.

603.3.3 Hose Connection Backflow Preventer. A hose connection backflow preventer consists of two independent check valves with an independent atmospheric vent between and a means of field testing and draining.

603.3.4 Double Check Valve Backflow Prevention Assembly (DC). A double check valve backflow prevention assembly consists of two independently acting internally loaded check valves, four properly located test cocks, and two isolation valves.

603.3.5 Pressure Vacuum Breaker Backflow Prevention Assembly (PVB). A pressure vacuum breaker backflow prevention assembly consists of a loaded air inlet valve, an internally loaded check valve, two properly located test cocks, and two isolation valves. This device shall be permitted to be installed indoors where provisions for spillage are provided.

603.3.6 Spill-Resistant Pressure Vacuum Breaker (SVB). A pressure-type vacuum breaker backflow prevention assembly consists of one check valve force loaded closed and an air inlet vent valve force loaded open to atmosphere, positioned downstream of the check valve and located between and including two tightly closing shutoff valves and test cocks.

603.3.7 Reduced-Pressure Principle Backflow Prevention Assembly (RP). A reduced-pressure principle backflow prevention assembly consists of two independently acting internally loaded check valves, a differential pressure relief valve, four properly located test cocks, and two isolation valves.

603.3.8 Double Check Detector Fire Protection Backflow Prevention Assembly. A double check valve backflow prevention assembly with a parallel detector assembly consisting of a water meter and a double check valve backflow prevention assembly (DC).

603.3.9 Reduced Pressure Detector Fire Protection Backflow Prevention Assembly. A reduced-pressure principle backflow prevention assembly with a parallel detector assembly consisting of a water meter and a reduced-pressure principle backflow prevention assembly (RP).

603.3.10 Dual Check Backflow Preventer. A dual check backflow preventer consists of two independently acting check valves, force loaded to a normally closed position.

TABLE 603.2
BACKFLOW PREVENTION DEVICES, ASSEMBLIES, AND METHODS

DEVICE, ASSEMBLY, OR METHOD[1]	APPLICABLE STANDARDS	DEGREE OF HAZARD: POLLUTION (LOW HAZARD) BACK-SIPHONAGE	POLLUTION (LOW HAZARD) BACK-PRESSURE	CONTAMINATION (HIGH HAZARD) BACK-SIPHONAGE	CONTAMINATION (HIGH HAZARD) BACK-PRESSURE	INSTALLATION[2,3]
Air gap	ASME A112.1.2	X	—	X	—	See Table 603.3.1 in this chapter.
Air gap fittings for use with plumbing fixtures, appliances, and appurtenances	ASME A112.1.3	X	—	X	—	Air gap fitting is a device with an internal air gap, and typical installation includes plumbing fixtures, appliances, and appurtenances. The critical level shall not be installed below the flood level rim.
Atmospheric vacuum breaker (consists of a body, checking member and atmospheric port)	ASSE 1001 or CSA B64.1.1	X	—	X	—	Upright position. No valve downstream. Minimum of 6 inches or listed distance above all downstream piping and flood level rim of receptor.[4,5]
Antisiphon fill valve (ballcocks) for gravity water closet flush tanks and urinal tanks	ASSE 1002/ ASME A112.1002/ CSA B125.12	X	—	X	—	Installation of gravity water closet flush tank and urinal tanks with the fill valve installed with the critical level not less than 1 inch above the opening of the overflow pipe.[4,5]
Vacuum breaker wall hydrants, hose bibbs, freeze resistant, automatic draining type	ASSE 1019 or CSA B64.2.1.1	X	—	X	—	Installation includes wall hydrants and hose bibbs. Such devices are not for use under continuous pressure conditions (means of shutoff downstream of device is prohibited).[4,5]
Hose connection vacuum breakers	ASSE 1011	X	—	X	—	Such devices are not for use under continuous pressure conditions. No valve downstream.[4,6]
Hose connection backflow preventers	ASSE 1052	X	—	X	—	Such devices are not for use under continuous pressure conditions.[4,6]
Dual check backflow preventer	ASSE 1024	X	X	—	—	Installation does not include carbonated drink dispensers.
Dual check backflow preventer wall hydrants, freeze resistant	ASSE 1053	X	—	X	—	Such devices are not for use under continuous pressure conditions.[4]
Freeze resistant sanitary yard hydrants	ASSE 1057	X	—	X	—	Such devices are not for use under continuous pressure conditions.[4]
Backflow preventer with intermediate atmospheric vent	ASSE 1012	X	X	—	—	Installation of potable water connections to water boilers. No high-hazard chemicals shall be introduced into the system using such devices. Designed to operate under continuous pressure conditions. May discharge water.
Backflow preventer with intermediate atmospheric vent and pressure reducing valve	ASSE 1081	X	X	—	—	Installation of potable water connections to water boilers. No high-hazard chemicals shall be introduced into the system using such devices. Designed to operate under continuous pressure conditions. May discharge water.
Backflow preventer for Carbonated Beverage Dispensers (two independent check valves with a vent to the atmosphere)	ASSE 1022	X	—	—	—	Installation includes carbonated beverage machines or dispensers. These devices operate under intermittent or continuous pressure conditions.
Spill-Resistant Pressure Vacuum Breaker (single check valve with air inlet vent and means of field testing)	ASSE 1056	X	—	X	—	Upright position. Minimum of 12 inches or listed distance above all downstream piping and flood-level rim of receptor.[5]

TABLE 603.2
BACKFLOW PREVENTION DEVICES, ASSEMBLIES, AND METHODS (continued)

DEVICE, ASSEMBLY, OR METHOD[1]	APPLICABLE STANDARDS	DEGREE OF HAZARD				INSTALLATION[2,3]
		POLLUTION (LOW HAZARD)		CONTAMINATION (HIGH HAZARD)		
		BACK-SIPHONAGE	BACK-PRESSURE	BACK-SIPHONAGE	BACK-PRESSURE	
Double Check Valve Backflow Prevention Assembly (two independent check valves and means of field testing)	ASSE 1015; AWWA C510; CSA B64.5 or CSA B64.5.1	X	X	—	—	Horizontal unless otherwise listed. Access and clearance shall be in accordance with the manufacturer's instructions, and not less than a 12 inch clearance at the bottom for maintenance. May need platform/ladder for test and repair. Does not discharge water.
Double Check Detector Fire Protection Backflow Prevention Assembly (two independent check valves with a parallel detector assembly consisting of a water meter and a double check valve backflow prevention assembly and means for field testing)	ASSE 1048	X	X	—	—	Horizontal unless otherwise listed. Access and clearance shall be in accordance with the manufacturer's instructions, and not less than a 12 inch clearance at the bottom for maintenance. May need platform/ladder for test and repair. Does not discharge water. Installation includes a fire protection system and is designed to operate under continuous pressure conditions.
Pressure Vacuum Breaker Backflow Prevention Assembly (loaded air inlet valve, internally loaded check valve and means for field testing)	ASSE 1020 or CSA B64.1.2	X	—	X	—	Upright position. May have valves downstream. Minimum of 12 inches above all downstream piping and flood-level rim of the receptor. May discharge water.
Reduced Pressure Principle Backflow Prevention Assembly (two independently acting loaded check valves, a differential pressure relief valve and means for field testing)	ASSE 1013; AWWA C511; CSA B64.4 or CSA B64.4.1	X	X	X	X	Horizontal unless otherwise listed. Access and clearance shall be in accordance with the manufacturer's instructions, and not less than a 12 inch clearance at the bottom for maintenance. May need platform/ladder for test and repair. May discharge water.
Reduced Pressure Detector Fire Protection Backflow Prevention Assembly (two independently acting loaded check valves, a differential pressure relief valve, with a parallel detector assembly consisting of a water meter and a reduced-pressure principle backflow prevention assembly, and means for field testing)	ASSE 1047	X	X	X	X	Horizontal unless otherwise listed. Access and clearance shall be in accordance with the manufacturer's instructions, and not less than a 12 inch clearance at the bottom for maintenance. May need platform/ladder for test and repair. May discharge water. Installation includes a fire protection system and is designed to operate under continuous pressure conditions.
Laboratory faucet backflow preventer	ASSE 1035	—	—	X	X	Installation includes laboratory faucets. Such devices are not for use under continuous pressure conditions. No valve downstream.[4]

For SI units: 1 inch = 25.4 mm

Notes:

[1] See the description of devices and assemblies in this chapter.

[2] Installation in pit or vault requires previous approval by the Authority Having Jurisdiction.

[3] Refer to the general and specific requirement for installation.

[4] Not to be subjected to operating pressure for more than 12 hours in a 24 hour period.

[5] For deck-mounted and equipment-mounted vacuum breaker, see Section 603.5.13.

[6] Shall be installed in accordance with Section 603.5.7.

TABLE 603.3.1
MINIMUM AIR GAPS FOR WATER DISTRIBUTION[4]

FIXTURES	WHERE NOT AFFECTED BY SIDEWALLS[1] (inches)	WHERE AFFECTED BY SIDEWALLS[2] (inches)
Effective openings[3] not greater than ½ of an inch in diameter	1	1½
Effective openings[3] not greater than ¾ of an inch in diameter	1½	2¼
Effective openings[3] not greater than 1 inch in diameter	2	3
Effective openings[3] greater than 1 inch in diameter	Two times the diameter of effective opening	Three times the diameter of effective opening

For SI units: 1 inch = 25.4 mm

Notes:

[1] Sidewalls, ribs, or similar obstructions do not affect air gaps where spaced from the inside edge of the spout opening a distance exceeding three times the diameter of the effective opening for a single wall, or a distance exceeding four times the effective opening for two intersecting walls.

[2] Vertical walls, ribs, or similar obstructions extending from the water surface to or above the horizontal plane of the spout opening other than specified in Footnote 1 above. The effect of three or more such vertical walls or ribs has not been determined. In such cases, the air gap shall be measured from the top of the wall.

[3] The effective opening shall be the minimum cross-sectional area at the seat of the control valve or the supply pipe or tubing that feeds the device or outlet. Where two or more lines supply one outlet, the effective opening shall be the sum of the cross-sectional areas of the individual supply lines or the area of the single outlet, whichever is smaller.

[4] Air gaps less than 1 inch (25.4 mm) shall be approved as a permanent part of a listed assembly that has been tested under actual backflow conditions with vacuums of 0 to 25 inches of mercury (85 kPa).

603.3.11 Laboratory Faucet Backflow Preventers. Laboratory faucet backflow preventers shall comply with ASSE 1035.

603.3.12 Backflow Preventer with Intermediate Atmospheric Vent. A backflow preventer with intermediate atmospheric vent consists of two independently acting check valves, force loaded to a normally closed position, and an intermediate chamber with a means for automatically venting to atmosphere, force loaded to a normally open position.

603.4 General Requirements. Assemblies shall comply with listed standards and be acceptable to the Authority Having Jurisdiction, with jurisdiction over the selection and installation of backflow prevention assemblies.

603.4.1 Backflow Prevention Valve. Where more than one backflow prevention valve is installed on a single premise, and the valves are installed in one location, each separate valve shall be permanently identified by the permittee in a manner satisfactory to the Authority Having Jurisdiction.

603.4.2 Testing. The premise owner or responsible person shall have the backflow prevention assembly tested by a certified backflow assembly tester at the time of installation, repair, or relocation and not less than on an annual schedule thereafter, or more often where required by the Authority Having Jurisdiction. The periodic testing shall be performed in accordance with the procedures referenced in ASSE Series 5000 by a tester qualified in accordance with those standards. The field test kit used shall comply with ASSE 1064.

603.4.3 Access and Clearance. Access and clearance shall be provided for the required testing, maintenance, and repair. Access and clearance shall be in accordance with the manufacturer's instructions, and not less than 12 inches (305 mm) between the lowest portion of the assembly and grade, floor, or platform. Installations elevated that exceed 5 feet (1524 mm) above the floor or grade shall be provided with a platform capable of supporting a tester or maintenance person.

603.4.4 Connections. Direct connections between potable water piping and sewer-connected wastes shall not be permitted to exist under any condition with or without backflow protection. Where potable water is discharged to the drainage system, it shall be by means of an approved air gap of two pipe diameters of the supply inlet, but in no case shall the gap be less than 1 inch (25.4 mm). Connection shall be permitted to be made to the inlet side of a trap provided that an approved vacuum breaker is installed not less than 6 inches (152 mm), or the distance according to the device's listing, above the flood-level rim of such trapped fixture, so that at no time will such device be subjected to backpressure.

603.4.5 Hot Water Backflow Preventers. Backflow preventers for hot water exceeding 110°F (43°C) shall be a type designed to operate at temperatures exceeding 110°F (43°C) without rendering a portion of the assembly inoperative.

603.4.6 Integral Backflow Preventers. Fixtures, appliances, or appurtenances with integral backflow preventers or integral air gaps manufactured as a unit shall be installed in accordance with their listing requirements and the manufacturer's installation instructions.

603.4.7 Freeze Protection. In cold climate areas, backflow assemblies and devices shall be protected from freezing with an outdoor enclosure that complies with ASSE 1060 or by a method acceptable to the Authority Having Jurisdiction.

603.4.8 Drain Lines. Drain lines serving backflow devices or assemblies shall be sized in accordance with the discharge rates of the manufacturer's flow charts of such devices or assemblies.

603.4.9 Prohibited Locations. Backflow prevention devices with atmospheric vents or ports shall not be installed in pits, underground, or submerged locations. Backflow preventers shall not be located in an area containing fumes that are toxic, poisonous, or corrosive.

603.5 Specific Requirements. Specific requirements for backflow prevention shall comply with Section 603.5.1 through Section 603.5.21.

603.5.1 Atmospheric Vacuum Breaker. Water closet and urinal flushometer valves shall be protected against backflow by an approved backflow prevention assembly, device, or method. Where the valves are equipped with an atmospheric vacuum breaker, the vacuum breaker shall be installed on the discharge side of the flushometer valve with the critical level not less than 6 inches (152 mm), or the distance according to its listing, above the overflow rim of a water closet bowl or the highest part of a urinal.

603.5.2 Ballcock. Water closet and urinal tanks shall be equipped with a ballcock. The ballcock shall be installed with the critical level not less than 1 inch (25.4 mm) above the full opening of the overflow pipe. In cases where the ballcock has no hush tube, the bottom of the water supply inlet shall be installed 1 inch (25.4 mm) above the full opening of the overflow pipe.

603.5.3 Backflow Prevention. Water closet flushometer tanks shall be protected against backflow by an approved backflow prevention assembly, device, or method.

603.5.4 Heat Exchangers. Heat exchangers used for heat transfer, heat recovery, or solar heating shall protect the potable water system from being contaminated by the heat-transfer medium. Single-wall heat exchangers used in indirect-fired water heaters shall meet the requirements of Section 505.4.1. Double-wall heat exchangers shall separate the potable water from the heat-transfer medium by providing a space between the two walls that are vented to the atmosphere.

603.5.5 Water Supply Inlets. Water supply inlets to tanks, vats, sumps, swimming pools, and other receptors shall be protected by one of the following means:

(1) An approved air gap.

(2) A listed vacuum breaker installed on the discharge side of the last valve with the critical level not less than 6 inches (152 mm) or in accordance with its listing.

(3) A backflow preventer suitable for the degree of hazard, installed in accordance with the requirements for that type of device or assembly as set forth in this chapter.

603.5.6 Protection from Lawn Sprinklers and Irrigation Systems. Potable water supplies to systems having no pumps or connections for pumping equipment, and no chemical injection or provisions for chemical injection, shall be protected from backflow by one of the following devices:

(1) Atmospheric vacuum breaker (AVB)

(2) Pressure vacuum breaker backflow prevention assembly (PVB)

(3) Spill-resistant pressure vacuum breaker (SVB)

(4) Reduced-pressure principle backflow prevention assembly (RP)

603.5.6.1 Systems with Pumps. Where sprinkler and irrigation systems have pumps, connections for pumping equipment, or auxiliary air tanks, or are otherwise capable of creating backpressure, the potable water supply shall be protected by the following type of device where the backflow device is located upstream from the source of backpressure:

(1) Reduced-pressure principle backflow prevention assembly (RP)

603.5.6.2 Systems with Backflow Devices. Where systems have a backflow device installed downstream from a potable water supply pump or a potable water supply pump connection, the device shall be one of the following:

(1) Atmospheric vacuum breaker (AVB)

(2) Pressure vacuum breaker backflow prevention assembly (PVB)

(3) Spill-resistant pressure vacuum breaker (SVB)

(4) Reduced-pressure principle backflow prevention assembly (RP)

603.5.6.3 Systems with Chemical Injectors. Where systems include a chemical injector or provisions for chemical injection, the potable water supply shall be protected by a reduced-pressure principle backflow prevention assembly (RP).

603.5.7 Outlets with Hose Attachments. Potable water outlets with hose attachments, other than water heater drains, boiler drains, and clothes washer connections, shall be protected by a nonremovable hose bibb-type backflow preventer, a nonremovable hose bibb-type vacuum breaker, or by an atmospheric vacuum breaker installed not less than 6 inches (152 mm) above the highest point of usage located on the discharge side of the last valve. In climates where freezing temperatures occur, a listed self-draining frost-proof hose bibb with an integral backflow preventer or vacuum breaker shall be used.

603.5.8 Water-Cooled Equipment. Water-cooled compressors, degreasers, or other water-cooled equipment shall be protected by a backflow preventer installed in accordance with the requirements of this chapter. Water-cooled equipment that produces backpressure shall be equipped with the appropriate protection.

603.5.9 Aspirators. Water inlets to water-supplied aspirators shall be equipped with a vacuum breaker installed in accordance with its listing requirements and this chapter. The discharge shall drain through an air gap. Where the tailpiece of a fixture to receive the discharge of an aspirator is used, the air gap shall be located above the flood-level rim of the fixture.

603.5.10 Steam or Hot Water Boilers. Potable water connections to steam or hot water boilers shall be protected from backflow by a double check valve backflow prevention assembly, backflow preventer with intermediate atmospheric vent and pressure reducing valve, or reduced pressure principle backflow prevention assembly in accordance with Table 603.2. Where chemicals are introduced into the system a reduced pressure principle backflow prevention assembly shall be provided in accordance with Table 603.2.

603.5.11 Nonpotable Water Piping. In cases where it is impractical to correct individual cross-connections on the domestic waterline, the line supplying such outlets shall be considered a nonpotable water line. No drinking or domestic water outlets shall be connected to the nonpotable waterline. Where possible, portions of the nonpotable waterline shall be exposed, and exposed portions shall be properly identified in a manner satisfactory to the Authority Having Jurisdiction. Each outlet on the nonpotable waterline that is permitted to be used for drinking or domestic purposes shall be posted: "CAUTION: NONPOTABLE WATER, DO NOT DRINK."

603.5.12 Beverage Dispensers. Potable water supply to beverage dispensers carbonated beverage dispensers, or coffee machines shall be protected by an air gap or a vented backflow preventer that complies with ASSE 1022. For carbonated beverage dispensers, piping material installed downstream of the backflow preventer shall not be affected by carbon dioxide gas.

603.5.13 Deck-Mounted and Equipment-Mounted Vacuum Breakers. Deck-mounted or equipment-mounted vacuum breakers shall be installed in accordance with their listing and the manufacturer's installation instructions, with the critical level not less than 1 inch (25.4 mm) above the flood-level rim.

603.5.14 Protection from Fire Systems. Except as provided in Section 603.5.14.1 and Section 603.5.14.2, potable water supplies to fire protection systems that are normally under pressure, including but not limited to standpipes and automatic sprinkler systems, except in one- or two-family or townhouse residential sprinkler systems, piped in materials approved for potable water distribution systems shall be protected from backpressure and backsiphonage by one of the following testable devices:

(1) Double check valve backflow prevention assembly (DC)

(2) Double check detector fire protection backflow prevention assembly

(3) Reduced pressure principle backflow prevention assembly (RP)

(4) Reduced pressure detector fire protection backflow prevention assembly

Potable water supplies to fire protection systems that are not normally under pressure shall be protected from backflow and shall be in accordance with the requirements of the appropriate standards referenced in Chapter 17.

603.5.14.1 Fire Department Connection. Where fire protection systems supplied from a potable water system include a fire department (siamese) connection that is located less than 1700 feet (518.2 m) from a nonpotable water source that is capable of being used by the fire department as a secondary water supply, the potable water supply shall be protected by one of the following:

(1) Reduced pressure principle backflow prevention assembly (RP)

(2) Reduced pressure detector fire protection backflow prevention assembly

Nonpotable water sources include fire department vehicles carrying water of questionable quality or water that is treated with antifreeze, corrosion inhibitors, or extinguishing agents.

603.5.14.2 Chemicals. Where antifreeze, corrosion inhibitors, or other chemicals are added to a fire protection system supplied from a potable water supply, the potable water system shall be protected by one of the following:

(1) Reduced pressure principle backflow prevention assembly (RP)

(2) Reduced pressure detector fire protection backflow prevention assembly

603.5.14.3 Hydraulic Design. Where a backflow device is installed in the potable water supply to a fire protection system, the hydraulic design of the system shall account for the pressure drop through the backflow device. Where such devices are retrofitted for an existing fire protection system, the hydraulics of the sprinkler system design shall be checked to verify that there will be sufficient water pressure available for satisfactory operation of the fire sprinklers.

603.5.15 Health Care or Laboratory Areas. Vacuum breakers for washer-hose bedpans shall be located not less than 5 feet (1524 mm) above the floor. Hose connections in health care or laboratory areas shall be not less than 6 feet (1829 mm) above the floor.

603.5.16 Special Equipment. Portable cleaning equipment and dental vacuum pumps shall be protected from backflow by an air gap, an atmospheric vacuum breaker, a spill-resistant vacuum breaker, or a reduced pressure principle backflow preventer.

603.5.17 Potable Water Outlets and Valves. Potable water outlets, freeze-proof yard hydrants, combination stop-and-waste valves, or other fixtures that incorporate a stop and waste feature that drains into the ground shall not be installed underground.

603.5.18 Pure Water Process Systems. The water supply to a pure water process system, such as dialysis water systems, semiconductor washing systems, and similar process piping systems, shall be protected from backpressure and backsiphonage by a reduced-pressure principle backflow preventer.

603.5.18.1 Dialysis Water Systems. The individual connections of the dialysis related equipment to the dialysis pure water system shall not require additional backflow protection.

603.5.19 Plumbing Fixture Fittings. Plumbing fixture fittings with integral backflow protection shall comply with ASME A112.18.1/CSA B125.1.

603.5.20 Swimming Pools, Spas, and Hot Tubs. Potable water supply to swimming pools, spas, and hot tubs shall be protected by an air gap or a reduced pressure principle backflow preventer in accordance with the following:

(1) The unit is equipped with a submerged fill line.

(2) The potable water supply is directly connected to the unit circulation system.

603.5.21 Chemical Dispensers. The water supply to chemical dispensers shall be protected against backflow. TIA The chemical dispenser shall comply with ASSE/IAPMO TIA 1055 or the water supply shall be protected by one of the following methods:

(1) Air gap

(2) Atmospheric vacuum breaker (AVB)

(3) Pressure vacuum breaker backflow prevention assembly (PVB)

(4) Spill-resistant pressure vacuum breaker (SVB)

(5) Reduced-pressure principle backflow prevention assembly (RP)

604.0 Materials.

604.1 Pipe, Tube, and Fittings. Pipe, tube, fittings, solvent cement, thread sealants, solders, and flux used in potable water systems intended to supply drinking water shall comply with NSF 61. Where pipe fittings and valves are made from copper alloys containing more than 15 percent zinc by weight and are used in plastic piping systems, they shall be resistant to dezincification and stress corrosion cracking in compliance with NSF 14.

Materials used in the water supply system, except valves and similar devices, shall be of a like material, except where otherwise approved by the Authority Having Jurisdiction.

Materials for building water piping and building supply piping shall comply with the applicable standards referenced in Table 604.1.

604.2 Lead Content. The maximum allowable lead content in pipes, pipe fittings, plumbing fittings, and fixtures intended to convey or dispense water for human consumption shall be not more than a weighted average of 0.25 percent with respect to the wetted surfaces of pipes, pipe fittings, plumbing fittings, and fixtures. For solder and flux, the lead content shall be not more than 0.2 percent where used in piping systems that convey or dispense water for human consumption.

Exceptions:

(1) Pipes, pipe fittings, plumbing fittings, fixtures, or backflow preventers used for nonpotable services such as manufacturing, industrial processing, irrigation, outdoor watering, or any other uses where the water is not used for human consumption.

(2) Flush valves, fill valves, flushometer valves, tub fillers, shower valves, service saddles, or water distribution main gate valves that are 2 inches (50 mm) in diameter or larger.

604.2.1 Lead Content of Water Supply Pipe and Fittings. Pipes, pipe fittings, valves, and faucets utilized in the water supply system for non-drinking water applications shall have a maximum of 8 percent lead content.

604.3 Copper or Copper Alloy Tube. Copper or copper alloy tube for water piping shall have a weight of not less than Type L.

Exception: Type M copper or copper alloy tubing shall be permitted to be used for water piping where piping is aboveground in, or on, a building or underground outside of structures.

604.4 Hard-Drawn Copper or Copper Alloy Tubing. Hard-drawn copper or copper alloy tubing for water supply and distribution in addition to the required incised marking shall be marked in accordance with ASTM B88. The colors shall be: Type K, green; Type L, blue; and Type M, red.

604.5 Flexible Connectors. Flexible water connectors shall be installed in readily accessible locations, and where under continuous pressure shall comply with ASME A112.18.6/CSA B125.6. Flexible water connectors with an excess flow shutoff device shall comply with CSA B125.5/IAPMO Z600.

604.6 Cast-Iron Fittings. Cast-iron fittings up to and including 2 inches (50 mm) in size, where used in connection with potable water piping, shall be galvanized.

604.7 Malleable Iron Fittings. Malleable iron water fittings shall be galvanized.

604.8 Previously Used Piping and Tubing. Piping and tubing that has previously been used for a purpose other than for potable water systems shall not be used.

604.9 Epoxy Coating. The epoxy coating used on existing, underground steel building supply piping shall comply with NSF 61 and AWWA C210.

604.10 Plastic Materials. Approved plastic materials shall be permitted to be used in building supply piping, provided that where metal building supply piping is used for electrical grounding purposes, replacement piping, therefore, shall be of like materials.

Exception: Where a grounding system acceptable to the Authority Having Jurisdiction is installed, inspected, and approved, the metallic pipe shall be permitted to be replaced with nonmetallic pipe.

604.10.1 Tracer Wire. Plastic materials for building supply piping outside underground shall have an electrically continuous corrosion-resistant blue insulated copper tracer wire, or other approved conductor installed

TABLE 604.1
MATERIALS FOR BUILDING SUPPLY AND WATER DISTRIBUTION PIPING AND FITTINGS

MATERIAL	BUILDING SUPPLY PIPE AND FITTINGS	WATER DISTRIBUTION PIPE AND FITTINGS	REFERENCED STANDARD(S) PIPE	REFERENCED STANDARD(S) FITTINGS
Copper and Copper Alloys	X	X	ASTM B42, ASTM B43, ASTM B75, ASTM B88, ASTM B135, ASTM B251, ASTM B302, ASTM B447	ASME B16.15, ASME B16.18, ASME B16.22, ASME B16.26, ASME B16.50[2], ASME B16.51, ASSE 1061, ASTM F3226, IAPMO PS 117
CPVC	X	X	ASTM D2846, ASTM F441, ASTM F442, CSA B137.6	ASSE 1061, ASTM D2846, ASTM F437, ASTM F438, ASTM F439, ASTM F1970, CSA B137.6
CPVC-AL-CPVC	X	X	ASTM F2855	ASTM D2846
Ductile-Iron	X	X	AWWA C151	ASME B16.4, AWWA C110, AWWA C153
Galvanized Steel	X	X	ASTM A53	—
Malleable Iron	X	X	—	ASME B16.3
PE	X[1]	—	ASTM D2239, ASTM D2737, ASTM D3035, AWWA C901, CSA B137.1	ASTM D2609, ASTM D2683, ASTM D3261, ASTM F1055, CSA B137.1
PE-AL-PE	X	X	ASTM F1282, CSA B137.9	ASTM F1282, ASTM F1974, CSA B137.9
PE-AL-PEX	X	X	ASTM F1986	ASTM F1986
PE-RT	X	X	ASTM F2769, CSA B137.18	ASSE 1061, ASTM D3261, ASTM F1055, ASTM F1807, ASTM F2098, ASTM F2159, ASTM F2735, ASTM F2769, CSA B137.18
PEX	X	X	ASTM F876, CSA B137.5, AWWA C904[1]	ASSE 1061, ASTM F877, ASTM F1807, ASTM F1960, ASTM F2080, ASTM F2159, ASTM F2735, CSA B137.5
PEX-AL-PEX	X	X	ASTM F1281, CSA B137.10	ASTM F1281, ASTM F1974, ASTM F2434, CSA B137.10
PP	X	X	ASTM F2389, CSA B137.11	ASTM F2389, CSA B137.11
PVC	X[1]	—	ASTM D1785, ASTM D2241, AWWA C900	ASTM D2464, ASTM D2466, ASTM D2467, ASTM F1970, AWWA C907
Stainless Steel	X	X	ASTM A269, ASTM A312, ASTM A554, ASTM A778	ASTM F3226, IAPMO PS 117

Notes:
[1] For building supply or exterior cold-water applications, not for water distribution piping.
[2] For brazed fittings only.

adjacent to the piping. Access shall be provided to the tracer wire, or the tracer wire shall terminate aboveground at each end of the nonmetallic piping. The tracer wire size shall be not less than 14 AWG, and the insulation type shall be suitable for direct burial.

604.11 Solder. Solder shall comply with the requirements of Section 604.2.

604.12 Flexible Corrugated Connectors. Flexible corrugated connectors of copper, copper alloy, or stainless steel shall be limited to the following connector lengths:

(1) Fixture Connectors – 30 inches (762 mm)
(2) Washing Machine Connectors – 72 inches (1829 mm)
(3) Dishwasher and Icemaker Connectors – 120 inches (3048 mm)

604.13 Water Heater Connectors. Flexible metallic (copper and stainless steel), reinforced flexible, braided stainless steel, or polymer braided with EPDM core connectors that connect a water heater to the piping system shall comply with ASME A112.18.6/CSA B125.6. Copper, copper alloy, or stainless steel flexible connectors shall not exceed 24 inches

(610 mm). PEX, PEX-AL-PEX, PE-AL-PE, or PE-RT tubing shall not be installed within the first 18 inches (457 mm) of piping connected to a water heater.

605.0 Joints and Connections.

605.1 Copper or Copper Alloy Pipe, Tubing, and Joints. Joining methods for copper or copper alloy pipe, tubing, and fittings shall be installed in accordance with the manufacturer's installation instructions and shall comply with Section 605.1.1 through Section 605.1.5.

605.1.1 Brazed Joints. Brazed joints between copper or copper alloy pipe or tubing and fittings shall be made with brazing alloys having a liquid temperature above 1000°F (538°C). The joint surfaces to be brazed shall be cleaned bright by either manual or mechanical means. Tubing shall be cut square and reamed to full inside diameter. Brazing flux shall be applied to the joint surfaces where required by manufacturer's recommendation. Brazing filler metal shall conform to AWS A5.8 and shall be applied at the point where the pipe or tubing enters the socket of the fitting.

605.1.2 Flared Joints. Flared joints for soft copper or copper alloy water tubing shall be made with fittings that comply with the applicable standards referenced in Table 604.1. Pipe or tubing shall be cut square using an appropriate tubing cutter. The tubing shall be reamed to full inside diameter, resized to round, and expanded with a proper flaring tool.

605.1.3 Mechanical Joints. Mechanical joints shall include, but are not limited to, compression, flanged, grooved, pressed, and push fit fittings.

605.1.3.1 Mechanically Formed Tee Fittings. Mechanically formed tee fittings shall have extracted collars that shall be formed in a continuous operation consisting of drilling a pilot hole and drawing out the pipe or tube surface to form a collar having a height not less than three times the thickness of the branch tube wall. The branch pipe or tube shall be notched to conform to the inner curve of the run pipe or tube and shall have two dimple depth stops to ensure that penetration of the branch pipe or tube into the collar is of a depth for brazing and that the branch pipe or tube does not obstruct the flow in the main line pipe or tube. Dimple depth stops shall be in line with the run of the pipe or tube. The second dimple shall be ¼ of an inch (6.4 mm) above the first and shall serve as a visual point of inspection. Fittings and joints shall be made by brazing. Soldered joints shall not be permitted.

605.1.3.2 Press-Connect Fittings. Press-connect fittings for copper or copper alloy pipe or tubing shall have an elastomeric o-ring that forms the joint. The pipe or tubing shall be fully inserted into the fitting, and the pipe or tubing marked at the shoulder of the fitting. Pipe or tubing shall be cut square, chamfered, and reamed to full inside diameter. The fitting alignment shall be checked against the mark on the pipe or tubing to ensure the pipe or tubing is inserted into the fitting. The joint shall be pressed using the tool recommended by the manufacturer.

605.1.3.3 Push Fit Fittings. Removable and nonremovable push fit fittings for copper or copper alloy tubing or pipe that employ quick assembly push fit connectors shall comply with ASSE 1061. Push fit fittings for copper or copper alloy pipe or tubing shall have an approved elastomeric o-ring that forms the joint. Pipe or tubing shall be cut square, chamfered, and reamed to full inside diameter. The tubing shall be fully inserted into the fitting, and the tubing marked at the shoulder of the fitting. The fitting alignment shall be checked against the mark on the tubing to ensure the tubing is inserted into the fitting and gripping mechanism has engaged on the pipe.

605.1.4 Soldered Joints. Soldered joints between copper or copper alloy pipe or tubing and fittings shall be made in accordance with ASTM B828 with the following sequence of joint preparation and operation as follows: measuring and cutting, reaming, cleaning, fluxing, assembly and support, heating, applying the solder, cooling and cleaning. Pipe or tubing shall be cut square and reamed to the full inside diameter including the removal of burrs on the outside of the pipe or tubing. Surfaces to be joined shall be cleaned bright by manual or mechanical means. Flux shall be applied to pipe or tubing and fittings and shall conform to ASTM B813, and shall become noncorrosive and nontoxic after soldering. Insert pipe or tubing into the base of the fitting and remove excess flux. Pipe or tubing and fitting shall be supported to ensure a uniform capillary space around the joint. Heat shall be applied using an air or fuel torch with the flame perpendicular to the pipe or tubing using acetylene or an LP gas. Preheating shall depend on the size of the joint. The flame shall be moved to the fitting cup and alternate between the pipe or tubing and fitting. Solder conforming to ASTM B32 shall be applied to the joint surfaces until capillary action draws the molten solder into the cup. Solder and fluxes with a lead content that exceeds 0.2 percent shall be prohibited in piping systems conveying potable water. Joint surfaces shall not be disturbed until cool and any remaining flux residue shall be cleaned.

605.1.5 Threaded Joints. Threaded joints for copper or copper alloy pipe shall be made with pipe threads that comply with ASME B1.20.1. Thread sealant tape or compound shall be applied only on male threads, and such material shall be of approved types, insoluble in water, and nontoxic.

605.2 CPVC Plastic Pipe and Joints. CPVC plastic pipe and fitting joining methods shall be installed in accordance with the manufacturer's installation instructions and shall comply with Section 605.2.1 through Section 605.2.3.

605.2.1 Mechanical Joints. Mechanical joints shall include compression, flanged, grooved and push fit fittings.

605.2.1.1 Push Fit Fittings. Removable and nonremovable push fit fittings that employ a quick assembly push fit connector shall comply with ASSE 1061.

605.2.2 Solvent Cement Joints. Solvent cement joints for CPVC pipe and fittings shall be clean from dirt and moisture. Solvent cements shall comply with ASTM F493, requiring the use of a primer shall be orange in color. The primer shall be colored and shall comply with ASTM F656. Listed solvent cement that complies with ASTM F493 and that does not require the use of primers, yellow or red in color, shall be permitted for pipe and fittings that comply with ASTM D2846, ½ of an inch (15 mm) through 2 inches (50 mm) in diameter or ASTM F442, ½ of an inch (15 mm) through 3 inches (80 mm) in diameter. Apply primer where required inside the fitting and to the depth of the fitting on pipe. Apply liberal coat of cement to the outside surface of pipe to depth of fitting and inside of fitting. Place pipe inside fitting to forcefully bottom the pipe in the socket and hold together until joint is set.

605.2.3 Threaded Joints. Threads shall comply with ASME B1.20.1. A minimum of Schedule 80 shall be permitted to be threaded; however, the pressure rating shall be reduced by 50 percent. The use of molded fittings shall not result in a 50 percent reduction in the pressure rating of the pipe provided that the molded fittings shall be fabricated so that the wall thickness of the material is maintained at the threads. Thread sealant compound that is compatible with the pipe and fitting, insoluble in water, and nontoxic shall be applied to male threads. Caution shall be used during assembly to prevent over tightening of the CPVC components once the thread sealant has been applied. Female CPVC threaded fittings shall be used with plastic male threads only.

605.3 CPVC/AL/CPVC Plastic Pipe and Joints. Chlorinated polyvinyl chloride/aluminum/chlorinated polyvinyl chloride (CPVC/AL/CPVC) plastic pipe and fitting joining methods shall be installed in accordance with the manufacturer's installation instructions and shall comply with Section 605.3.1 or Section 605.3.2.

605.3.1 Solvent Cement Joints. Solvent cement joints for CPVC/AL/CPVC pipe and fittings shall be clean from dirt and moisture. Solvent cements that comply with ASTM F493, requiring the use of a primer shall be orange in color. The primer shall be colored and shall comply with ASTM F656. Listed solvent cement that complies with ASTM F493 and that does not require the use of primers, yellow in color, shall be permitted to join pipe that comply with ASTM F2855 and fittings that comply with ASTM D2846, ½ of an inch (15 mm) through 2 inches (50 mm) in diameter. Apply primer where required inside the fitting and to the depth of the fitting on pipe. Apply liberal coat of cement to the outside surface of pipe to depth of fitting and inside of fitting. Place pipe inside fitting to forcefully bottom the pipe in the socket and hold together until joint is set.

605.3.2 Mechanical Joints. Mechanical joints shall include flanged, grooved, and push fit fittings.

605.3.2.1 Push Fit Fittings. Removable and nonremovable push fit fittings that employ a quick assembly push fit connector shall comply with ASSE 1061.

605.4 Ductile Iron Pipe and Joints. Ductile iron pipe and fitting joining methods shall be installed in accordance with the manufacturer's installation instructions and shall comply with Section 605.4.1 or Section 605.4.2.

605.4.1 Mechanical Joints. Mechanical joints for ductile iron pipe and fittings shall consist of a bell that is cast integrally with the pipe or fitting and provided with an exterior flange having bolt holes and a socket with annular recesses for the sealing gasket and the plain end of the pipe or fitting. The elastomeric gasket shall comply with AWWA C111. Lubricant recommended for potable water application by the pipe manufacturer shall be applied to the gasket and plain end of the pipe.

605.4.2 Push-On Joints. Push-on joints for ductile iron pipe and fittings shall consist of a single elastomeric gasket that shall be assembled by positioning the elastomeric gasket in an annular recess in the pipe or fitting socket and forcing the plain end of the pipe or fitting into the socket. The plain end shall compress the elastomeric gasket to form a positive seal and shall be designed so that the elastomeric gasket shall be locked in place against displacement. The elastomeric gasket shall comply with AWWA C111. Lubricant recommended for potable water application by the pipe manufacturer shall be applied to the gasket and plain end of the pipe.

605.5 Galvanized Steel Pipe and Joints. Galvanized steel pipe and fitting joining methods shall be installed in accordance with the manufacturer's installation instructions and shall comply with Section 605.5.1 or Section 605.5.2.

605.5.1 Mechanical Joints. Mechanical joints shall be made with an approved and listed elastomeric gasket.

605.5.2 Threaded Joints. Threaded joints shall be made with pipe threads that comply with ASME B1.20.1. Thread sealant tape or compound shall be applied only on male threads, and such material shall be of approved types, insoluble in water, and nontoxic.

605.6 PE Plastic Pipe/Tubing and Joints. PE plastic pipe or tubing and fitting joining methods shall be installed in accordance with the manufacturer's installation instructions and shall comply with Section 605.6.1 or Section 605.6.2.

605.6.1 Heat-Fusion Joints. Heat-fusion joints between PE pipe or tubing and fittings shall be assembled in accordance with Section 605.6.1.1 through Section 605.6.1.3 using butt, socket, or electro-fusion heat methods.

605.6.1.1 Butt-Fusion Joints. Butt-fusion joints shall be made in accordance with ASTM F2620. Joints shall be made by heating the squared ends of two pipes, pipe and fitting, or two fittings by holding ends against a heated element. The heated element shall be removed where the proper melt is obtained and joined ends shall be placed together with applied force.

605.6.1.2 Electro-Fusion Joints. Electro-fusion joints shall be heated internally by a conductor at the interface of the joint. Align and restrain fitting to pipe to prevent movement and apply electric current to the fitting. Turn off the current when the proper time has elapsed to heat the joint. The joint shall fuse together and remain undisturbed until cool.

605.6.1.3 Socket-Fusion Joints. Socket-fusion joints shall be made in accordance with ASTM F2620. Joints shall be made by simultaneously heating the outside surface of a pipe end and the inside of a fitting socket. Where the proper melt is obtained, the pipe and fitting shall be joined by inserting one into the other with applied force. The joint shall fuse together and remain undisturbed until cool.

605.6.2 Mechanical Joints. Mechanical joints between PE pipe or tubing and fittings shall include insert and mechanical compression fittings that provide a pressure seal resistance to pullout. Joints for insert fittings shall be made by cutting the pipe square, using a cutter designed for plastic piping, and removal of sharp edges. Two stainless steel clamps shall be placed over the end of the pipe. Fittings shall be checked for proper size based on the diameter of the pipe. The end of pipe shall be placed over the barbed insert fitting, making contact with the fitting shoulder. Clamps shall be positioned equal to 180 degrees (3.14 rad) apart and shall be tightened to provide a leak tight joint. Compression type couplings and fittings shall be permitted for use in joining PE piping and tubing. Stiffeners that extend beyond the clamp or nut shall be prohibited. Bends shall be not less than 30 pipe diameters, or the coil radius where bending with the coil. Bends shall not be permitted closer than 10 pipe diameters of a fitting or valve. Mechanical joints shall be designed for their intended use.

605.7 PE-AL-PE Plastic Pipe/Tubing and Joints. PE-AL-PE plastic pipe or tubing and fitting joining methods shall be installed in accordance with the manufacturer's installation instructions and shall comply with Section 605.7.1 and Section 605.7.1.1.

605.7.1 Mechanical Joints. Mechanical joints for PE-AL-PE pipe or tubing and fittings shall be either of the metal insert fittings with a split ring and compression nut or metal insert fittings with copper crimp rings. Metal insert fittings shall comply with ASTM F1974. Crimp insert fittings shall be joined to the pipe by placing the copper crimp ring around the outer circumference of the pipe, forcing the pipe material into the space formed by the ribs on the fitting until the pipe contacts the shoulder of the fitting. The crimp ring shall then be positioned on the pipe so the edge of the crimp ring is ⅛ of an inch (3.2 mm) to ¼ of an inch (6.4 mm) from the end of the pipe. The jaws of the crimping tool shall be centered over the crimp ring and tool perpendicular to the barb. The jaws shall be closed around the crimp ring and shall not be crimped more than once.

605.7.1.1 Compression Joints. Compression joints for PE-AL-PE pipe or tubing and fittings shall be joined through the compression of a split ring, by a compression nut around the circumference of the pipe. The compression nut and split ring shall be placed around the pipe. The ribbed end of the fitting shall be inserted into the pipe until the pipe contacts the shoulder of the fitting. Position and compress the split ring by tightening the compression nut onto the insert fitting.

605.8 PE-RT. Polyethylene of raised temperature (PE-RT) tubing and fitting joining methods and shall comply with Section 605.8.1.

605.8.1 Mechanical Joints. Fittings for PE-RT tubing shall comply with the applicable standards listed in Table 604.1. Mechanical joints for PE-RT tubing shall be installed in accordance with the manufacturer's installation instructions.

605.9 PEX Plastic Tubing and Joints. PEX plastic tubing and fitting joining methods shall be installed in accordance with the manufacturer's installation instructions and shall comply with Section 605.9.1 through Section 605.9.3.

605.9.1 Fittings. Fittings for PEX tubing shall comply with the applicable standards referenced in Table 604.1. PEX tubing that complies with ASTM F876 shall be marked with the applicable standard designation for the fittings, specified by the tubing manufacturer for use with the tubing.

605.9.2 Mechanical Joints. Mechanical joints shall be installed in accordance with the manufacturer's installation instructions.

605.9.3 Push Fit Fittings. Removable and nonremovable push fit fittings that employ a quick assembly push fit connector shall comply with ASSE 1061.

605.10 PEX-AL-PEX Plastic Tubing and Joints. PEX-AL-PEX plastic pipe or tubing and fitting joining methods shall be installed in accordance with the manufacturer's installation instructions and shall comply with Section 605.10.1 and Section 605.10.1.1.

605.10.1 Mechanical Joints. Mechanical joints between PEX-AL-PEX tubing and fittings shall include mechanical and compression type fittings and insert fittings with a crimping ring. Insert fittings utilizing a crimping ring shall comply with ASTM F1974 or ASTM F2434. Crimp joints for crimp insert fittings shall be joined to PEX-AL-PEX pipe by the compression of a crimp ring around the outer circumference of the pipe, forcing the pipe material into annular spaces formed by ribs on the fitting.

605.10.1.1 Compression Joints. Compression joints shall include compression insert fittings and shall be joined to PEX-AL-PEX pipe through the compression of a split ring or compression nut around the outer circumference of the pipe, forcing the pipe material into the annular space formed by the ribs on the fitting.

605.11 Polypropylene (PP) Piping and Joints. PP pipe and fittings shall be installed in accordance with the manufacturer's installation instructions and shall comply with Section 605.11.1 through Section 605.11.3.

605.11.1 Heat-Fusion Joints. Heat-fusion joints for polypropylene (PP) pipe and fitting joints shall be installed with socket-type heat-fused polypropylene fittings, fusion outlets, butt-fusion polypropylene fittings or pipe, or electro-fusion polypropylene fittings. Joint surfaces shall be clean and free from moisture. The joint shall be undisturbed until cool. Joints shall be made in accordance with ASTM F2389 or CSA B137.11.

605.11.2 Mechanical and Compression Sleeve Joints. Mechanical and compression sleeve joints shall be installed in accordance with the manufacturer's installation instructions.

605.11.3 Threaded Joints. PP pipe shall not be threaded. PP transition fittings for connection to other piping materials shall only be threaded by use of copper alloy or stainless steel inserts molded in the fitting.

605.12 PVC Plastic Pipe and Joints. PVC plastic pipe and fitting joining methods shall be installed in accordance with the manufacturer's installation instructions and shall comply with Section 605.12.1 through Section 605.12.3.

PVC piping shall not be exposed to direct sunlight unless the piping does not exceed 24 inches (610 mm) and is wrapped with not less than 0.04 of an inch (1.02 mm) thick tape or otherwise protected from UV degradation.

605.12.1 Mechanical Joints. Mechanical joints shall be designed to provide a permanent seal and shall be of the mechanical or push-on joint. The mechanical joint shall include a pipe spigot that has a wall thickness to withstand without deformation or collapse; the compressive force exerted where the fitting is tightened. The push-on joint shall have a minimum wall thickness of the bell at any point between the ring and the pipe barrel. The elastomeric gasket shall comply with ASTM D3139, and be of such size and shape as to provide a compressive force against the spigot and socket after assembly to provide a positive seal.

605.12.2 Solvent Cement Joints. Solvent cement joints for PVC pipe and fittings shall be clean from dirt and moisture. Pipe shall be cut square and pipe shall be deburred. Where surfaces to be joined are cleaned and free of dirt, moisture, oil, and other foreign material, apply primer purple in color that complies with ASTM F656. Primer shall be applied to the surface of the pipe and fitting is softened. Solvent cement that complies with ASTM D2564 shall be applied to all joint surfaces. Joints shall be made while both the inside socket surface and outside surface of pipe are wet with solvent cement. Hold joint in place and undisturbed for 1 minute after assembly.

605.12.3 Threaded Joints. Threads shall comply with ASME B1.20.1. A minimum of Schedule 80 shall be permitted to be threaded; however, the pressure rating shall be reduced by 50 percent. The use of molded fittings shall not result in a 50 percent reduction in the pressure rating of the pipe provided that the molded fittings shall be fabricated so that the wall thickness of the material is maintained at the threads. Thread sealant compound that is compatible with the pipe and fitting, insoluble in water and nontoxic shall be applied to male threads. Caution shall be used during assembly to prevent over tightening of the PVC components once the thread sealant has been applied. Female PVC threaded fittings shall be used with plastic male threads only.

605.13 Stainless Steel Pipe and Joints. Joining methods for stainless steel pipe and fittings shall be installed in accordance with the manufacturer's installation instructions and shall comply with Section 605.13.1 or Section 605.13.2.

605.13.1 Mechanical Joints. Mechanical joints shall be designed for their intended use. Such joints shall include compression, flanged, grooved, press-connect, and threaded.

605.13.2 Welded Joints. Welded joints shall be either fusion or resistance welded based on the selection of the base metal. The chemical composition of the filler metal shall comply with AWS A5.9 based on the alloy content of the piping material.

605.14 Slip Joints. In water piping, slip joints shall be permitted to be used only on the exposed fixture supply.

605.15 Dielectric Unions. Dielectric unions where installed at points of connection where there is a dissimilarity of metals shall be in accordance with ASSE 1079.

605.16 Joints Between Various Materials. Joints between various materials shall be installed in accordance with the manufacturer's installation instructions and shall comply with Section 605.16.1 through Section 605.16.3.

605.16.1 Copper or Copper Alloy Pipe or Tubing to Threaded Pipe Joints. Joints from copper or copper alloy pipe or tubing to threaded pipe shall be made using copper alloy adapter, copper alloy nipple [minimum 6 inches (152 mm)], dielectric fitting, or dielectric union in accordance with ASSE 1079. The joint between the copper or copper alloy pipe or tubing and the fitting shall be a soldered, brazed, flared, or press-connect joint and the connection between the threaded pipe and the fitting shall be made with a standard pipe size threaded joint.

605.16.2 Plastic Pipe to Other Materials. Where connecting plastic pipe to other types of piping, approved types of adapter or transition fittings designed for the specific transition intended shall be used.

605.16.3 Stainless Steel to Other Materials. Where connecting stainless steel pipe to other types of piping, mechanical joints of the compression type, dielectric fitting, or dielectric union in accordance with ASSE 1079 and designed for the specific transition intended shall be used.

606.0 Valves.

606.1 General. Valves up to and including 2 inches (50 mm) in size shall be copper alloy or other approved material. Sizes exceeding 2 inches (50 mm) shall be permitted to have cast iron or copper alloy bodies. Each gate or ball valve shall be a fullway or full-port type with working parts of the non-corrosive material. Valves carrying water used in potable water systems intended to supply drinking water shall comply with the requirements of NSF 61 and ASME A112.4.14, ASME B16.34, ASTM F1970, ASTM F2389, AWWA C500, AWWA C504, AWWA C507, IAPMO Z1157, MSS SP-67, MSS SP-70, MSS SP-71, MSS SP-72, MSS SP-78, MSS SP-80, MSS SP-110, MSS SP-122, or NSF 359.

606.2 Fullway Valve. A fullway valve controlling outlets shall be installed on the discharge side of each water meter and each unmetered water supply. Water piping supplying more than one building on one premise shall be equipped with a separate fullway valve to each building, so arranged that the water supply can be turned on or off to an individual or separate building provided; however, that supply piping to a single-family residence and building accessory thereto shall be permitted to be controlled by one valve. Such shutoff valves shall be accessible. A fullway valve shall be installed on the discharge piping from water supply tanks at or near the tank. A fullway valve shall be installed on the cold water supply pipe to each water heater at or near the water heater.

606.3 Multidwelling Units. In multidwelling units, one or more shutoff valves shall be provided in each dwelling unit so that the water supply to a plumbing fixture or group of fixtures in that dwelling unit can be shut off without stopping water supply to fixtures in other dwelling units. These valves shall be accessible in the dwelling unit that they control.

606.4 Multiple Openings. Valves used to control two or more openings shall be fullway gate valves, ball valves, or other approved valves designed and approved for the service intended.

606.5 Control Valve. A control valve shall be installed immediately ahead of each water-supplied appliance and immediately ahead of each slip joint or appliance supply.

Parallel water distribution systems shall provide a control valve either immediately ahead of each fixture being supplied or installed at the manifold, and shall be identified with the fixture being supplied. Where parallel water distribution system manifolds are located in attics, crawl spaces, or other locations not readily accessible, a separate shutoff valve shall be required immediately ahead of each individual fixture or appliance served.

606.6 Accessible. Required shutoff or control valves shall be accessible.

606.7 Multiple Fixtures. A single control valve shall be installed on a water supply line ahead of an automatic metering valve that supplies a battery of fixtures.

606.8 Check Valve Required. All systems that circulate water by means of a pump or other mechanical device or method shall have a check valve(s) or equal device(s) installed so as to ensure the direction of flow.

606.9 Leak Detection Devices. Where leak detection devices for water supply and distribution are installed, they shall comply with IAPMO IGC 115 or IAPMO IGC 349.

607.0 Potable Water Supply Tanks.

607.1 General. Potable water supply tanks shall be installed in accordance with the manufacturer's installation instructions and supported in accordance with the building code.

607.2 Potable Water Tanks. Potable water supply tanks, interior tank coatings, or tank liners intended to supply drinking water shall comply with NSF 61.

607.3 Venting. Tanks used for potable water shall be tightly covered and vented in accordance with the manufacturer's installation instructions. Such vent shall be screened with a corrosion-resistant material of not less than number 24 mesh.

607.4 Overflow. Tanks shall have not less than a 16 square inch (0.01 m^2) overflow that is screened with a corrosion-resistant material of not less than number 24 mesh.

607.5 Valves. Pressurized tanks shall be provided with a listed pressure-relief valve installed in accordance with the manufacturer's installation instructions. The relief valve shall be discharged in accordance with Section 608.5. Where a potable water supply tank is located above the fixtures, appliances, or system components it serves, it shall be equipped with a vacuum relief valve that complies with CSA Z21.22.

608.0 Water Pressure, Pressure Regulators, Pressure Relief Valves, and Vacuum Relief Valves.

608.1 Inadequate Water Pressure. Where the water pressure in the main or other source of supply will not provide a residual water pressure of not less than 15 pounds force per square inch (psi) (103 kPa), after allowing for friction and other pressure losses, a tank and a pump or other means that will provide said 15 psi (103 kPa) pressure shall be installed. Where fixtures, fixture fittings, or both are installed that, require residual pressure exceeding 15 psi (103 kPa), that minimum residual pressure shall be provided.

608.2 Excessive Water Pressure. Where static water pressure in the water supply piping is exceeding 80 psi (552 kPa), an approved-type pressure regulator preceded by an adequate strainer shall be installed and the static pressure reduced to 80 psi (552 kPa) or less. Pressure regulators for potable water distribution systems shall comply with ASSE 1003. Pressure regulator(s) equal to or exceeding 1½ inches (40 mm) shall not require a strainer. Such regulator(s) shall control the pressure to water outlets in the building unless otherwise approved by the Authority Having Jurisdiction. Each such regulator and strainer shall be accessibly located aboveground or in a vault equipped with a properly sized and sloped bore-sighted drain to daylight, shall be protected from freezing, and shall have the strainer readily accessible for cleaning without removing the regulator or strainer body or disconnecting the supply piping.

Pipe size determinations shall be based on 80 percent of the reduced pressure where using Table 610.4.

An approved expansion tank shall be installed in the cold water distribution piping downstream of each such regulator to prevent excessive pressure from developing due to thermal expansion and to maintain the pressure setting of the regulator. Expansion tanks used in potable water systems intended to supply drinking water shall comply with NSF 61. The expansion tank shall be properly sized and installed in accordance with the manufacturer's installation instructions and listing. Systems designed by registered design professionals shall be permitted to use approved pressure relief valves in lieu of expansion tanks provided such relief valves have a maximum pressure relief setting of 100 psi (689 kPa) or less.

608.3 Expansion Tanks, and Combination Temperature and Pressure-Relief Valves. A water system provided with a check valve, backflow preventer, or other normally closed device that prevents dissipation of building pressure back into the water main, independent of the type of water heater used, shall be provided with an approved, listed, and adequately sized expansion tank or other approved device having a similar function to control thermal expansion. Prepressurized water expansion tanks shall comply with IAPMO Z1088. Such expansion tank or other approved device shall be installed on the building side of the check valve, backflow preventer, or other device and shall be sized and installed in accordance with the manufacturer's installation instructions.

A water system containing storage water heating equipment shall be provided with an approved, listed, adequately sized combination temperature and pressure-relief valve, except for listed nonstorage instantaneous heaters having an inside diameter of not more than 3 inches (80 mm). Each such approved combination temperature and pressure-relief valve shall be installed on the water-heating device in an approved location based on its listing requirements and the manufacturer's installation instructions. Each such combination temperature and pressure-relief valve shall be provided with a drain in accordance with Section 608.5.

608.4 Pressure Relief Valves. Each pressure relief valve shall be an approved automatic type with drain, and each such relief valve shall be set at a pressure of not more than 150 psi (1034 kPa). No shutoff valve shall be installed between the relief valve and the system.

608.5 Discharge Piping. The discharge piping serving a temperature relief valve, pressure relief valve, or combination of both shall have no valves, obstructions, or means of isolation and be provided with the following:

(1) Not less than the size of the valve outlet and shall discharge full size to the flood level of the area receiving the discharge and pointing down.

(2) Materials shall be rated at not less than the operating temperature of the system and approved for such use or shall comply with ASME A112.4.1.

(3) Discharge pipe shall discharge independently by gravity through an air gap into the drainage system or outside of the building with the end of the pipe not exceeding 2 feet (610 mm) and not less than 6 inches (152 mm) above the ground and pointing downwards.

(4) Discharge in such a manner that does not cause personal injury or structural damage.

(5) No part of such discharge pipe shall be trapped or subject to freezing.

(6) The terminal end of the pipe shall not be threaded.

(7) Discharge from a relief valve into a water heater pan shall be prohibited.

(8) The discharge termination point shall be readily observable.

608.6 Water-Heating Devices. A water-heating device connected to a separate storage tank and having valves between said heater and tank shall be provided with an approved water pressure relief valve.

608.7 Vacuum Relief Valves. Where a hot-water storage tank or an indirect water heater is located at an elevation above the fixture outlets in the hot-water system, a vacuum relief valve that complies with CSA Z21.22 shall be installed on the storage tank or heater.

609.0 Installation, Testing, Unions, and Location.

609.1 Installation. Water piping shall be adequately supported in accordance with Table 313.3. Burred ends shall be reamed to the full bore of the pipe or tube. Changes in direction shall be made by the appropriate use of fittings, except that changes in direction in copper or copper alloy tubing shall be permitted to be made with bends, provided that such bends are made with bending equipment that does not deform or create a loss in the cross-sectional area of the tubing. Changes in direction are allowed with flexible pipe and tubing without fittings in accordance with the manufacturer's instructions. Provisions shall be made for expansion in hot-water piping. Piping, equipment, appurtenances, and devices shall be installed in a workmanlike manner in accordance with the provisions and intent of the code. Building supply yard piping shall be not less than 12 inches (305 mm) below the average local frost depth. The cover shall be not less than 12 inches (305 mm) below finish grade.

609.2 Trenches. Water pipes shall not be run or laid in the same trench as building sewer or drainage piping constructed of clay or materials that are not approved for use within a building unless both of the following conditions are met:

(1) The bottom of the water pipe shall be not less than 12 inches (305 mm) above the top of the sewer or drain line.

(2) The water pipe shall be placed on a solid shelf excavated at one side of the common trench with a clear horizontal distance of not less than 12 inches (305 mm) from the sewer or drain line.

Water pipes crossing sewer or drainage piping constructed of clay or materials that are not approved for use within a building shall be laid not less than 12 inches (305 mm) above the sewer or drain pipe.

609.3 Under Concrete Slab. Water piping installed within a building and in or under a concrete floor slab resting on the ground shall be installed in accordance with the following requirements:

(1) Ferrous piping shall have a protective coating of an approved type; machine applied and in accordance with recognized standards. Field wrapping shall provide equivalent protection and shall be restricted to those short sections and fittings necessarily stripped for threading.

Zinc coating (galvanizing) shall not be deemed adequate protection for piping or fittings. Approved nonferrous piping shall not be required to be wrapped.

(2) Copper or copper alloy tubing shall be installed without joints where possible. Where joints are permitted, they shall be brazed, and fittings shall be wrought copper.

For the purpose of this section, "within a building" shall mean within the fixed limits of the building foundation.

609.4 Testing. Upon completion of a section or of the entire hot and cold water supply system, the system shall be tested with water or air. The potable water test pressure shall be greater than or equal to the working pressure under which the system is to be used. The air pressure shall be a minimum of 50 psi (345 kPa). Plastic pipe shall not be tested with air. The piping system shall withstand the test pressure without showing evidence of leakage for a period of not less than 15 minutes.

Exception: PEX, PP or PE-RT tube shall be permitted to be tested with air where permitted by the manufacturer's instructions.

609.5 Unions. Unions shall be installed in the water supply piping not more than 12 inches (305 mm) of regulating equipment, water heating, conditioning tanks, and similar equipment that requires service by removal or replacement in a manner that will facilitate its ready removal.

609.6 Location. Except as provided in Section 609.7, no building supply shall be located in a lot other than the lot that is the site of the building or structure served by such building supply.

609.7 Abutting Lot. Nothing contained in this code shall be construed to prohibit the use of an abutting lot to:

(1) Provide access to connect a building supply to an available public water service where proper cause and legal easement not in violation of other requirements have been first established to the satisfaction of the Authority Having Jurisdiction.

(2) Provide additional space for a building supply where the proper cause, transfer of ownership, or change of boundary not in violation of other requirements have been first established to the satisfaction of the Authority Having Jurisdiction. The instrument recording such action shall constitute an agreement with the Authority Having Jurisdiction, which shall clearly state and show that the areas so joined or used shall be maintained as a unit during the time they are so used. Such an agreement shall be recorded in the office of the County Recorder as a part of the conditions of ownership of said properties, and shall be binding on heirs, successors, and assigns to such properties. A copy of the instrument recording such proceedings shall be filed with the Authority Having Jurisdiction.

609.8 Pumps. Pumps shall be installed in accordance with the manufacturer's installation instructions.

609.8.1 Access. Pumps shall be accessible for repairs.

609.8.2 Potable Water Pumps. Pumps intended to supply drinking water shall be in accordance with NSF 61.

609.9 Low-Pressure Cutoff Required on Booster Pumps for Water Distribution Systems. Where a booster pump (excluding a fire pump) is connected to a building supply or underground water pipe, a low-pressure cutoff switch on the inlet side of the pump shall be installed not more than 5 feet (1524 mm) of the inlet. The cutoff switch shall be set for not less than 10 psi (69 kPa). A pressure gauge shall be installed between the shutoff valve and the pump.

609.10 Disinfection of Potable Water System. New or repaired potable water systems shall be disinfected prior to use where required by the Authority Having Jurisdiction. The method to be followed shall be that prescribed by the Health Authority or, in case no method is prescribed by it, the following:

(1) The pipe system shall be flushed with clean, potable water until potable water appears at the points of the outlet.

(2) The system or parts thereof shall be filled with a water-chlorine solution containing not less than 50 parts per million of chlorine, and the system or part thereof shall be valved-off and allowed to stand for 24 hours; or, the system or part thereof shall be filled with a water-chlorine solution containing not less than 200 parts per million of chlorine and allowed to stand for 3 hours.

(3) Following the allowed standing time, the system shall be flushed with clean, potable water until the chlorine residual in the water coming from the system does not exceed the chlorine residual in the flushing water.

(4) The procedure shall be repeated where it is shown by a bacteriological examination made by an approved agency that contamination persists in the system.

609.11 Water Hammer. Building water supply systems where quick-acting valves are installed shall be provided with water hammer arrester(s) to absorb high pressures resulting from the quick closing of these valves. Water hammer arresters shall be approved mechanical devices that comply with ASSE 1010 or PDI-WH 201 and shall be installed as close as possible to quick-acting valves.

609.11.1 Mechanical Devices. Where listed mechanical devices are used, the manufacturer's specifications as to location and method of installation shall be followed.

609.12 Pipe Insulation. Insulation of domestic hot water piping shall be in accordance with Section 609.12.1 and Section 609.12.2.

609.12.1 Insulation Requirements. Domestic hot water piping shall be insulated.

609.12.2 Pipe Insulation Wall Thickness. Hot water pipe insulation shall have a minimum wall thickness of not less than the diameter of the pipe for a pipe up to 2 inches (50 mm) in diameter. Insulation wall thickness shall be not less than 2 inches (51 mm) for a pipe of 2 inches (50 mm) or more in diameter.

Exceptions:

(1) Piping that penetrates framing members shall not be required to have pipe insulation for the distance of the framing penetration.

(2) Hot water piping between the fixture control valve or supply stop and the fixture or appliance shall not be required to be insulated.

610.0 Size of Potable Water Piping.

610.1 Size. The size of each water meter and each potable water supply pipe from the meter or other source of supply to the fixture supply branches, risers, fixtures, connections, outlets, or other uses shall be based on the total demand and shall be determined according to the methods and procedures outlined in this section. Water piping systems shall be designed to ensure that the maximum velocities allowed by the code and the applicable standard are not exceeded.

610.2 Pressure Loss. Where a water filter, water softener, backflow prevention device, tankless water heater, or similar device is installed in a water supply line, the pressure loss through such devices shall be included in the pressure loss calculations of the system, and the water supply pipe and meter shall be adequately sized to provide for such a pressure loss.

No water filter, water softener, backflow prevention device, or similar device regulated by this code shall be installed in a potable water supply piping where the installation of such device produces an excessive pressure drop in such water supply piping. In the absence of specific pressure drop information, the diameter of the inlet or outlet of such device or its connecting piping shall be not less than the diameter of such water distribution piping to the fixtures served by the device.

Such devices shall be of a type approved by the Authority Having Jurisdiction and shall be tested for flow rating and pressure loss by an approved laboratory or recognized testing agency to standards consistent with the intent of this chapter.

610.3 Quantity of Water. The quantity of water required to be supplied to every plumbing fixture shall be represented by fixture units, as shown in Table 610.3. Equivalent fixture values shown in Table 610.3 include both hot and cold water demand.

610.4 Sizing Water Supply and Distribution Systems. Systems within the range of Table 610.4 shall be permitted to be sized from that table or by the method in accordance with Section 610.5.

Listed parallel water distribution systems shall be installed in accordance with their listing, but at no time shall a portion of the system exceed the maximum velocities allowed by the code.

610.5 Sizing per Appendices A and C. Except as provided in Section 610.4, the size of each water piping system shall be determined in accordance with the procedure set forth in Appendix A. For alternate methods of sizing water supply systems, see Appendix C.

610.6 Friction and Pressure Loss. Except where the type of pipe used and the water characteristics are such that no decrease in capacity due to the length of service (age of system) is expected, friction-loss data shall be obtained from the "Fairly Rough" or "Rough" charts in Appendix A of this code. Friction or pressure losses in a water meter, valve, and fittings shall be obtained from the same sources. Pressure losses through water-treating equipment, backflow prevention devices, or other flow-restricting devices shall be computed in accordance with Section 610.2.

610.7 Conditions for Using Table 610.4. On a proposed water piping installation sized using Table 610.4, the following conditions shall be determined:

(1) Total number of fixture units as determined from Table 610.3, Equivalent Fixture Units, for the fixtures to be installed.

(2) Developed length of supply pipe from meter to the most remote outlet.

(3) Difference in elevation between the meter or other source of supply and the highest fixture or outlet.

(4) Pressure in the street main or another source of supply at the locality where the installation is to be made.

(5) In localities where there is a fluctuation of pressure in the main throughout the day, the water piping system shall be designed on the basis of the minimum pressure available.

610.8 Size of Meter and Building Supply Pipe Using Table 610.4. The size of the meter and the building supply pipe shall be determined as follows:

(1) Determine the available pressure at the water meter or other source of supply.

(2) Add or subtract depending on positive or negative elevation change, ½ psi (3.4 kPa) for each foot (305 mm) of difference in elevation between such source of supply and the highest water supply outlet in the building or on the premises.

(3) Use the "pressure range" group within which this pressure will fall using Table 610.4.

(4) Select the "length" column that is equal to or longer than the required length.

(5) Follow down the column to a fixture unit value equal to or exceeding the total number of fixture units required by the installation.

(6) Having located the proper fixture unit value for the required length, sizes of meter and building supply pipe as found in the two left-hand columns shall be applied.

No building supply pipe shall be less than ¾ of an inch (20 mm) in diameter.

610.9 Size of Branches. Where Table 610.4 is used, the minimum size of each branch shall be determined by the total fixture units served by that branch and then following the steps in Section 610.8. No branch piping shall exceed the total demand in fixture units for the system computed from Table 610.3.

610.10 Sizing for Flushometer Valves. Where using Table 610.4 to size water supply systems serving flushometer valves, the number of flushometer fixture units assigned to every section of pipe, whether branch or main, shall be determined by the number and category of flushometer valves served by that section of pipe, in accordance with Table 610.10. Piping supplying a flushometer valve shall be not less in size than the valve inlet.

Where using Table 610.10 to size water piping, care shall be exercised to assign flushometer fixture units based on the number and category of fixtures served.

TABLE 610.10
FLUSHOMETER FIXTURE UNITS FOR WATER SIZING USING TABLE 610.3

FIXTURE CATEGORY: WATER CLOSET WITH FLUSHOMETER VALVES		
NUMBER OF FLUSHOMETER VALVES	INDIVIDUAL FIXTURE UNITS ASSIGNED IN DECREASING VALUE	FIXTURE UNITS ASSIGNED FOR WATER CLOSETS AND SIMILAR 10-UNIT FIXTURES IN ACCUMULATIVE VALUES
1	40	40
2	30	70
3	20	90
4	15	105
5 or more	10 each	115 plus 10 for each additional fixture in excess of 5
FIXTURE CATEGORY: URINALS WITH FLUSHOMETER VALVES		
NUMBER OF FLUSHOMETER VALVES	INDIVIDUAL FIXTURE UNITS ASSIGNED IN DECREASING VALUE	FIXTURE UNITS ASSIGNED FOR URINALS AND SIMILAR 5-UNIT FIXTURES IN ACCUMULATIVE VALUES
1	20	20
2	15	35
3	10	45
4	8	53
5 or more	5 each	58 plus 5 for each additional fixture in excess of 5

In the example below, fixture units assigned to each section of pipe are computed. Each capital letter refers to the section of pipe above it unless otherwise shown.

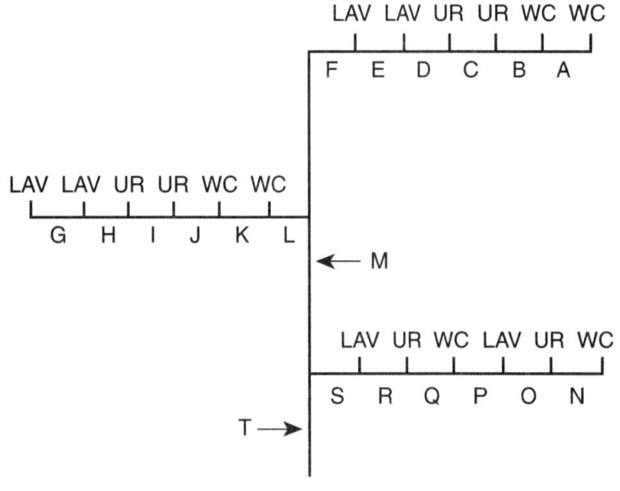

A: 1 WC = 40 F.U.
B: 2 WC = 70 F.U.
C: 2 WC (70) + 1 UR (20) = 90 F.U.
D: 2 WC (70) + 2 UR (35) = 105 F.U.
E: 2 WC (70) + 2 UR (35) + 1 LAV (1) = 106 F.U.
F: 2 WC (70) + 2 UR (35) + 2 LAV (2) = 107 F.U.
G: 1 LAV = 1 F.U.
H: 2 LAV = 2 F.U.
I: 2 LAV (2) + 1 UR (20) = 22 F.U.
J: 2 LAV (2)+ 2 UR (35) = 37 F.U.
K: 2 LAV (2) + 2 UR (35) + 1 WC (40) = 77 F.U.
L: 2 LAV (2) + 2 UR (35) + 2 WC (70) = 107 F.U.
M: 4 WC (105) + 4 UR (53) + 4 LAV (4) = 162 F.U.
N: 1 WC = 40 F.U.
O: 1 WC (40) + 1 UR (20) = 60 F.U.
P: 1 WC (40) + 1 UR (20) + 1 LAV (1) = 61 F.U.
Q: 2 WC (70) + 1 UR (20) + 1 LAV (1) = 91 F.U.
R: 2 WC (70) + 2 UR (35) + 1 LAV (1) = 106 F.U.
S: 2 WC (70) + 2 UR (35) + 2 LAV (2) = 107 F.U.
T: 6 WC (125) + 6 UR (63) + 6 LAV (6) = 194 F.U.

EXAMPLE 610.10
SIZING METHOD FOR PUBLIC USE FIXTURES USING TABLE 610.10

610.11 Sizing Systems for Flushometer Tanks. The size of branches and mains serving flushometer tanks shall be consistent with the sizing procedures for flush tank water closets.

610.12 Sizing for Velocity. Water piping systems shall not exceed the maximum velocities listed in this section or Appendix A.

610.12.1 Copper Tube Systems. Maximum velocities in copper and copper alloy tube and fitting systems shall not exceed 8 feet per second (ft/s) (2.4 m/s) in cold water and 5 ft/s (1.5 m/s) in hot water.

610.12.2 Tubing Systems Using Copper Fittings. Maximum velocities through copper fittings in tubing other than copper shall not exceed 8 ft/s (2.4 m/s) in cold water and 5 ft/s (1.5 m/s) in hot water.

610.13 Exceptions. The provisions of this section relative to the size of water piping shall not apply to the following:

(1) Water supply piping systems designed in accordance with recognized engineering procedures acceptable to the Authority Having Jurisdiction.
(2) Alteration of or minor additions to existing installations provided the Authority Having Jurisdiction finds that there will be an adequate supply of water to operate fixtures.
(3) Replacement of existing fixtures or appliances.
(4) Piping that is part of fixture equipment.
(5) Unusual conditions where, in the judgment of the Authority Having Jurisdiction, an adequate supply of water is provided to operate fixtures and equipment.

TABLE 610.3
WATER SUPPLY FIXTURE UNITS (WSFU) AND MINIMUM FIXTURE BRANCH PIPE SIZES[3]

APPLIANCES, APPURTENANCES OR FIXTURES[2]	MINIMUM FIXTURE BRANCH PIPE SIZE[1,4] (inches)	PRIVATE	PUBLIC	ASSEMBLY[6]
Bathtub or Combination Bath/Shower (fill)	½	4.0	4.0	—
¾ inch Bathtub Fill Valve	¾	10.0	10.0	—
Bidet	½	1.0	—	—
Clothes Washer	½	4.0	4.0	—
Dental Unit, cuspidor	½	—	1.0	—
Dishwasher, domestic	½	1.5	1.5	—
Drinking Fountain or Water Cooler	½	0.5	0.5	0.75
Hose Bibb	½	2.5	2.5	—
Hose Bibb, each additional[8]	½	1.0	1.0	—
Lavatory	½	1.0	1.0	1.0
Lawn Sprinkler, each head[5]	—	1.0	1.0	—
Mobile Home, each (minimum)	—	12.0	—	—
Sinks	—	—	—	—
Bar	½	1.0	2.0	—
Clinical Faucet	½	—	3.0	—
Clinical Flushometer Valve with or without faucet	1	—	8.0	—
Kitchen, domestic with or without dishwasher	½	1.5	1.5	—
Laundry	½	1.5	1.5	—
Service or Mop Basin	½	1.5	3.0	—
Washup, each set of faucets	½	—	2.0	—
Shower, per head	½	2.0	2.0	—
Urinal, 1.0 GPF Flushometer Valve	¾	See Footnote[7]		—
Urinal, greater than 1.0 GPF Flushometer Valve	¾	See Footnote[7]		—
Urinal, flush tank	½	2.0	2.0	3.0
Urinal with Drain Cleansing Action	½	1.0	1.0	1.0
Wash Fountain, circular spray	¾	—	4.0	—
Water Closet, 1.6 GPF Gravity Tank	½	2.5	2.5	3.5
Water Closet, 1.6 GPF Flushometer Tank	½	2.5	2.5	3.5
Water Closet, 1.6 GPF Flushometer Valve	1	See Footnote[7]		—
Water Closet, greater than 1.6 GPF Gravity Tank	½	3.0	5.5	7.0
Water Closet, greater than 1.6 GPF Flushometer Valve	1	See Footnote[7]		—

For SI units: 1 inch = 25 mm

Notes:

1 Size of the cold branch pipe, or both the hot and cold branch pipes.

2 Appliances, appurtenances, or fixtures not referenced in this table shall be permitted to be sized by reference to fixtures having a similar flow rate and frequency of use.

3 The listed fixture unit values represent their load on the cold water building supply. The separate cold water and hot water fixture unit value for fixtures having both hot and cold water connections shall be permitted to be each taken as three-quarter of the listed total value of the fixture.

4 The listed minimum supply branch pipe sizes for individual fixtures are the nominal (I.D.) pipe size.

5 For fixtures or supply connections likely to impose continuous flow demands, determine the required flow in gallons per minute (gpm) (L/s), and add it separately to the demand in gpm (L/s) for the distribution system or portions thereof.

6 Assembly [Public Use (See Table 422.1)].

7 Where sizing flushometer systems, see Section 610.10.

8 Reduced fixture unit loading for additional hose bibbs is to be used where sizing total building demand and for pipe sizing where more than one hose bibb is supplied by a segment of water distribution pipe. The fixture branch to each hose bibb shall be sized on the basis of 2.5 fixture units.

TABLE 610.4
FIXTURE UNIT TABLE FOR DETERMINING WATER PIPE AND METER SIZES

METER AND STREET SERVICE (inches)	BUILDING SUPPLY AND BRANCHES (inches)	MAXIMUM ALLOWABLE LENGTH (feet)														
		40	60	80	100	150	200	250	300	400	500	600	700	800	900	1000
PRESSURE RANGE – 30 to 45 psi[1]																
3/4	1/2[2]	6	5	4	3	2	1	1	1	0	0	0	0	0	0	0
3/4	3/4	16	16	14	12	9	6	5	5	4	4	3	2	2	2	1
3/4	1	29	25	23	21	17	15	13	12	10	8	6	6	6	6	6
1	1	36	31	27	25	20	17	15	13	12	10	8	6	6	6	6
3/4	1 1/4	36	33	31	28	24	23	21	19	17	16	13	12	12	11	11
1	1 1/4	54	47	42	38	32	28	25	23	19	17	14	12	12	11	11
1 1/2	1 1/4	78	68	57	48	38	32	28	25	21	18	15	12	12	11	11
1	1 1/2	85	84	79	65	56	48	43	38	32	28	26	22	21	20	20
1 1/2	1 1/2	150	124	105	91	70	57	49	45	36	31	26	23	21	20	20
2	1 1/2	151	129	129	110	80	64	53	46	38	32	27	23	21	20	20
1	2	85	85	85	85	85	85	82	80	66	61	57	52	49	46	43
1 1/2	2	220	205	190	176	155	138	127	120	104	85	70	61	57	54	51
2	2	370	327	292	265	217	185	164	147	124	96	70	61	57	54	51
2	2 1/2	445	418	390	370	330	300	280	265	240	220	198	175	158	143	133
PRESSURE RANGE – 46 to 60 psi[1]																
3/4	1/2[2]	7	7	6	5	4	3	2	2	1	1	1	0	0	0	0
3/4	3/4	20	20	19	17	14	11	9	8	6	5	4	4	3	3	3
3/4	1	39	39	36	33	28	23	21	19	17	14	12	10	9	8	8
1	1	39	39	39	36	30	25	23	20	18	15	12	10	9	8	8
3/4	1 1/4	39	39	39	39	39	39	34	32	27	25	22	19	19	17	16
1	1 1/4	78	78	76	67	52	44	39	36	30	27	24	20	19	17	16
1 1/2	1 1/4	78	78	78	78	66	52	44	39	33	29	24	20	19	17	16
1	1 1/2	85	85	85	85	85	85	80	67	55	49	41	37	34	32	30
1 1/2	1 1/2	151	151	151	151	128	105	90	78	62	52	42	38	35	32	30
2	1 1/2	151	151	151	151	150	117	98	84	67	55	42	38	35	32	30
1	2	85	85	85	85	85	85	85	85	85	85	85	85	85	83	80
1 1/2	2	370	370	340	318	272	240	220	198	170	150	135	123	110	102	94
2	2	370	370	370	370	368	318	280	250	205	165	142	123	110	102	94
2	2 1/2	654	640	610	580	535	500	470	440	400	365	335	315	285	267	250
PRESSURE RANGE – Over 60 psi[1]																
3/4	1/2[2]	7	7	7	6	5	4	3	3	2	1	1	1	1	1	0
3/4	3/4	20	20	20	20	17	13	11	10	8	7	6	6	5	4	4
3/4	1	39	39	39	39	35	30	27	24	21	17	14	13	12	12	11
1	1	39	39	39	39	38	32	29	26	22	18	14	13	12	12	11
3/4	1 1/4	39	39	39	39	39	39	39	39	34	28	26	25	23	22	21
1	1 1/4	78	78	78	78	74	62	53	47	39	31	26	25	23	22	21
1 1/2	1 1/4	78	78	78	78	78	74	65	54	43	34	26	25	23	22	21
1	1 1/2	85	85	85	85	85	85	85	85	81	64	51	48	46	43	40
1 1/2	1 1/2	151	151	151	151	151	151	130	113	88	73	51	51	46	43	40
2	1 1/2	151	151	151	151	151	151	142	122	98	82	64	51	46	43	40
1	2	85	85	85	85	85	85	85	85	85	85	85	85	85	85	85
1 1/2	2	370	370	370	370	360	335	305	282	244	212	187	172	153	141	129
2	2	370	370	370	370	370	370	370	340	288	245	204	172	153	141	129
2	2 1/2	654	654	654	654	654	650	610	570	510	460	430	404	380	356	329

For SI units: 1 inch = 25 mm, 1 foot = 304.8 mm, 1 pound-force per square inch = 6.8947 kPa

Notes:

[1] Available static pressure after head loss.

[2] Building supply, not less than 3/4 of an inch (20 mm) nominal size.

(6) The size and material of irrigation water piping installed outside of a building or structure and separated from the potable water supply by means of an approved air gap or backflow prevention device are not regulated by this code. The potable water piping system supplying each such irrigation system shall be adequately sized as required elsewhere in this chapter to deliver the full connected demand of both the domestic use and the irrigation systems.

611.0 Drinking Water Treatment Units.

611.1 Application. Drinking water treatment units shall comply with the applicable referenced standards in Table 611.1.

611.1.1 Alkaline Water Treatment. Alkaline water treatment devices shall comply with IAPMO IGC 322.

611.1.2 Scale Reduction Devices. Scale reduction devices shall comply with IAPMO Z601.

611.2 Air Gap Discharge. Discharge from drinking water treatment units shall enter the drainage system through an air gap in accordance with Table 603.3.1 or an air gap device that complies with Table 603.2, NSF 58, or IAPMO PS 65.

611.3 Connection Tubing. The tubing to and from drinking water treatment units shall be of a size and material as recommended by the manufacturer.

611.4 Sizing of Residential Softeners. Residential-use water softeners shall be sized in accordance with Table 611.4.

TABLE 611.4
SIZING OF RESIDENTIAL WATER SOFTENERS[4]

REQUIRED SIZE OF SOFTENER CONNECTION (inches)	NUMBER OF BATHROOM GROUPS SERVED[1]
¾	up to 2[2]
1	up to 4[3]

For SI units: 1 inch = 25 mm

Notes:

[1] Installation of a kitchen sink and dishwasher, laundry tray, and automatic clothes washer permitted without additional size increase.

[2] An additional water closet and lavatory permitted.

[3] Over four bathroom groups, the softener size shall be engineered for the specific installation.

[4] See also Appendix A, Recommended Rules for Sizing the Water Supply System, and Appendix C, Alternate Plumbing Systems, for alternate methods of sizing water supply systems.

612.0 Residential Fire Sprinkler Systems.

612.1 Where Required. Where residential sprinkler systems are required in one and two-family dwellings or townhouses, the systems shall be installed by personnel, installer, or both, certified in accordance with ASSE Series 7000 in accordance with this section or NFPA 13D. This section shall be considered equivalent to NFPA 13D. Partial residential sprinkler systems shall be permitted to be installed in buildings not required to be equipped with a residential sprinkler system.

612.2 Types of Systems. This section shall apply to stand-alone and multipurpose wet-pipe sprinkler systems that do not include the use of antifreeze. A multipurpose fire sprinkler system shall provide potable water to both fire sprinklers and plumbing fixtures. A stand-alone sprinkler system shall be separate and independent from the potable water distribution system. A backflow preventer shall not be required to separate a stand-alone sprinkler system from the water distribution system where the sprinkler system material is in accordance with the requirements of Section 604.0.

612.3 Sprinklers. Sprinklers shall be installed in accordance with Section 612.3.1 through Section 612.3.7.

612.3.1 Required Sprinkler Locations. Sprinklers shall be installed to protect all floor areas of a dwelling unit in one and two-family dwellings or townhouses.

Exceptions:

(1) Attics, crawl spaces, and normally unoccupied concealed spaces that do not contain fuel-fired appliances do not require sprinklers. In attics, crawl spaces, and normally unoccupied concealed spaces that contain fuel-fired equipment, a sprinkler shall be provided to protect the equipment; however, sprinklers shall not be required in the remainder of the space.

(2) Clothes closets, linen closets, and pantries that do not exceed 24 square feet (2.2 m^2) in area, with the smallest dimension not exceeding 3 feet (914 mm) and having wall and ceiling surfaces of gypsum board.

(3) Bathrooms and toilet rooms that do not exceed 55 square feet (5.1 m^2) in area.

TABLE 611.1
DRINKING WATER TREATMENT UNITS

APPLICATION	RESIDENTIAL		COMMERCIAL
	POINT OF USE	POINT OF ENTRY	
Aesthetic Contaminant Reduction (filters)	NSF 42	NSF 42	ASSE 1087 and NSF 42*
Health Related Contaminant Reduction (filters)	NSF 53	NSF 53	ASSE 1087 and NSF 53*
Water Softener	–	NSF 44	ASSE 1087
Ultraviolet Water Treatment	NSF 55	NSF 55	ASSE 1087
Reverse Osmosis	NSF 58	NSF 61	ASSE 1087
Distillation	NSF 62	NSF 62	ASSE 1087

* Required for commercial modular systems only.

(4) Garages; carports; exterior porches; unheated entry areas, such as mud rooms, that are adjacent to an exterior door; and similar areas.

(5) Covered unheated projections of the building at entrances/exits provided it is not the only means of egress from the dwelling unit.

(6) Ceiling pockets that meet the following requirements:

(a) The total volume of an unprotected ceiling pocket does not exceed 100 cubic feet (2.83 m^3).

(b) The entire floor under the unprotected ceiling pocket is protected by the sprinklers at the lower ceiling elevation.

(c) Each unprotected ceiling pocket is separated from an adjacent unprotected ceiling pocket by not less than a 10 feet (3048 mm) horizontal distance.

(d) The interior finish of the unprotected ceiling pocket is noncombustible material.

(e) Skylights not exceeding 32 square feet (2.97 m^2).

612.3.2 Sprinkler Installation. Sprinklers shall be listed residential sprinklers and shall be installed in accordance with the sprinkler manufacturer's installation instructions.

612.3.3 Temperature Rating and Separation from Heat Sources. Sprinklers shall have a temperature rating of not less than 135°F (57°C) and not more than 170°F (77°C). Sprinklers shall be separated from heat sources in accordance with the sprinkler manufacturer's installation instructions.

Exception: Sprinklers located close to a heat source in accordance with Section 612.3.3.1 shall be intermediate temperature sprinklers.

612.3.3.1 Intermediate Temperature Sprinklers. Sprinklers shall have an intermediate temperature rating of not less than 175°F (79°C) and not more than 225°F (107°C) where installed in the following locations:

(1) Directly under skylights, where the sprinkler is exposed to direct sunlight.

(2) In attics and concealed spaces located directly beneath a roof.

(3) Within the distance to a heat source in accordance with Table 612.3.3.1.

612.3.4 Freezing Areas. The piping system shall be protected in accordance with the requirements of Chapter 3. Where sprinklers are required in areas that are subject to freezing, dry-sidewall or dry-pendent sprinklers extending from a non-freezing area into a freezing area shall be installed.

612.3.5 Coverage Area Limit. The area of coverage of a single sprinkler shall be based on the sprinkler listing and the sprinkler manufacturer's installation instructions. The area of coverage of a single sprinkler shall not exceed 400 square feet (37.16 m^2).

612.3.6 Obstructions to Sprinkler Coverage. The water discharge from a sprinkler shall not be blocked by obstructions unless additional sprinklers are installed to protect the obstructed area. Additional sprinklers shall not be required where sprinkler separation from obstructions is in accordance with the requirements of Table 612.3.6, or the minimum distances specified in the sprinkler manufacturer's installation instructions.

TABLE 612.3.3.1
LOCATIONS WHERE INTERMEDIATE TEMPERATURE SPRINKLERS ARE REQUIRED

HEAT SOURCE	DISTANCE FROM HEAT SOURCE[1]	
	MINIMUM DISTANCE[2] (inches)	MAXIMUM DISTANCE (inches)
Fireplace, Side of Open or Recessed Fireplace	12	36
Fireplace, Front of Recessed Fireplace	36	60
Coal and Wood Burning Stove	12	42
Kitchen Range Top	9	18
Oven	9	18
Vent Connector or Chimney Connector	9	18
Heating Duct, Not Insulated	9	18
Hot Water Pipe, Not Insulated	6	12
Side of Ceiling or Wall Warm Air Register	12	24
Front of Wall Mounted Warm Air Register	18	36
Water Heater, Furnace, or Boiler	3	6
Luminaire up to 250 Watts	3	6
Luminaire 250 Watts up to 499 Watts	6	12

For SI units: 1 inch = 25.4 mm

Notes:

1 Distances shall be measured in a straight line from the nearest edge of the heat source to the nearest edge of the sprinkler.

2 Sprinklers shall not be located at distances less than the minimum table distance unless the sprinkler listing allows a lesser distance.

612.3.6.1 Additional Requirements for Pendent Sprinklers. Pendent sprinklers located within 3 feet (914 mm) of the center of a ceiling fan, surface-mounted ceiling luminaire, or similar object shall be considered to be obstructed, and additional sprinklers shall be provided.

612.3.6.2 Additional Requirements for Sidewall Sprinklers. Sidewall sprinklers located within 5 feet (1524 mm) of the center of a ceiling fan, surface-mounted ceiling luminaire, or similar object shall be considered to be obstructed, and additional sprinklers shall be provided.

612.3.7 Sprinkler Modifications Prohibited. Sprinklers shall not be painted, caulked, or modified. A sprinkler that has been painted, caulked, modified, or damaged shall be replaced with a new sprinkler.

TABLE 612.3.6
MINIMUM SEPARATION FROM OBSTRUCTION

PENDENT SPRINKLERS	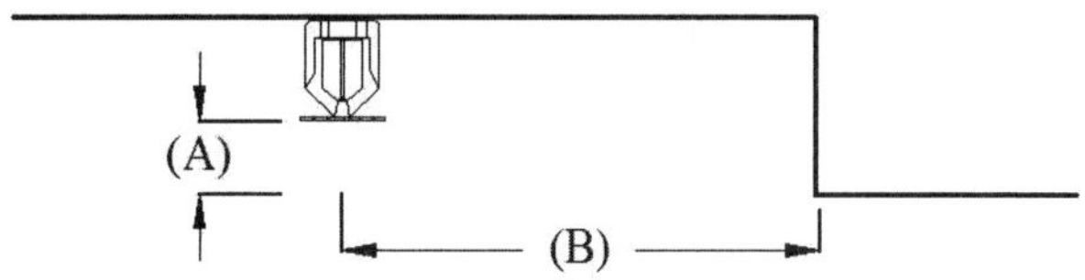
DISTANCE FROM DEFLECTOR TO PLANE AT BOTTOM OF OBSTRUCTION (A) (inches)	**MINIMUM DISTANCE TO OBSTRUCTION (B) (feet)**
1	1½
3	3
5	4
7	4½
9	6
11	6½
14	7

SIDEWALL SPRINKLER SIDE OBSTRUCTION	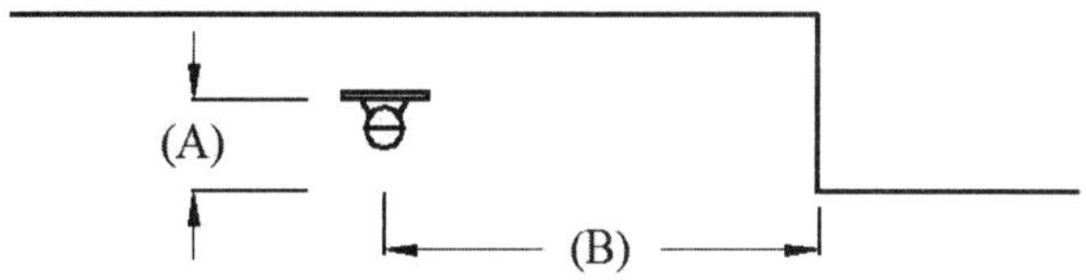
DISTANCE FROM DEFLECTOR TO PLANE AT BOTTOM OF OBSTRUCTION (A) (inches)	**MINIMUM DISTANCE TO OBSTRUCTION (B) (feet)**
1	1½
3	3
5	4
7	4½
9	6
11	6½
14	7

SIDEWALL SPRINKLER FORWARD OBSTRUCTION	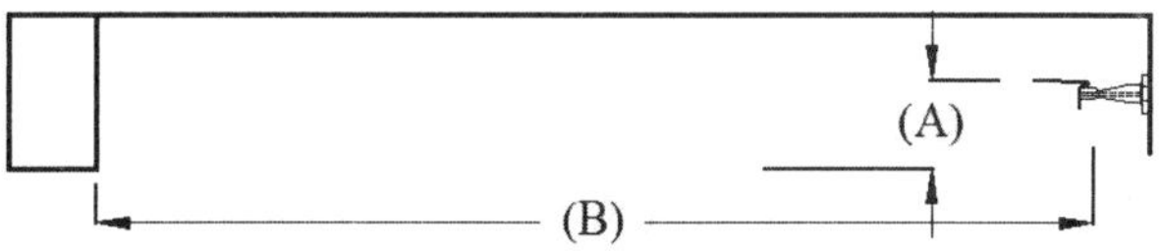
DISTANCE FROM DEFLECTOR TO PLANE AT BOTTOM OF OBSTRUCTION (A) (inches)	**MINIMUM DISTANCE TO OBSTRUCTION (B) (feet)**
1	8
2	10
3	11
4	12
6	13
7	14
9	15
11	16
14	17

For SI units: 1 inch = 25.4 mm, 1 foot = 304.8 mm

612.4 Sprinkler Piping System. Sprinkler piping systems shall be installed in accordance with Section 612.4.1 through Section 612.4.5.

612.4.1 General. Sprinkler piping shall be installed in accordance with the requirements for water distribution piping. Sprinkler piping shall comply with the material requirements for cold water distribution piping. For multipurpose piping systems, the sprinkler piping shall connect to and be a part of the cold water distribution piping system.

612.4.2 Nonmetallic Pipe and Tubing. Nonmetallic pipe and tubing, such as CPVC, PEX-AL-PEX, PE-RT, and PEX, shall be certified for residential sprinkler installations and shall have a pressure rating of not less than 130 psi (896 kPa) at 120°F (49°C).

612.4.2.1 Nonmetallic Pipe Protection. Nonmetallic pipe and tubing systems shall be protected from exposure to the occupied space by a layer of not less than ⅜ of an inch (9.5 mm) thick gypsum wallboard, ½ of an inch (12.7 mm) thick plywood, or other material having a 15-minute fire rating.

Exceptions:

(1) Pipe protection shall not be required in areas that are not required to be protected with sprinklers in accordance with Section 612.3.1.

(2) Pipe protection shall not be required where exposed piping is permitted by the pipe third party listing.

612.4.2.2 Sprinkler Installation on Systems Assembled with Solvent Cement. The solvent cementing of fittings shall be completed, and threaded adapters for sprinklers shall be verified as being clear of excess cement before the installation of sprinklers on systems assembled with solvent cement.

612.4.3 Shutoff Valves Prohibited. Shutoff valves shall not be installed in a location where the valve would isolate piping serving one or more sprinklers. Shutoff valves shall only be permitted for the entire water distribution system.

612.4.4 Single Dwelling Limit. The sprinkler piping beyond the service valve located at the beginning of the water distribution system shall serve only one dwelling unit.

612.4.5 Drain. A ½ inch (15 mm) drain for the sprinkler system shall be provided on the system side of the water distribution shutoff valve.

612.5 Sprinkler Piping Design. Sprinkler piping systems shall be sized in accordance with Section 612.5.1 through Section 612.5.3.2.2.

612.5.1 Determining System Design Flow. The sizing of the sprinkler piping system shall be based on the flow rate and pressure of each sprinkler in accordance with Section 612.5.1.1 and the number of sprinklers in accordance with Section 612.5.1.2.

612.5.1.1 Determining Required Flow Rate for Each Sprinkler. The minimum flow rate and pressure for each residential sprinkler shall be in accordance with the manufacturer's published data for the specific sprinkler model based on the following:

(1) The area of coverage.

(2) The ceiling configuration.

(3) The temperature rating.

(4) Additional conditions specified by the sprinkler manufacturer.

612.5.1.2 System Flow Rate. The flow rate used for sizing the sprinkler piping system shall be based on the following:

(1) The flow rate for a room having only one sprinkler shall be the flow rate required for the sprinkler in accordance with Section 612.5.1.1.

(2) The flow rate for a room having two or more sprinklers shall be determined by identifying the sprinkler in the room with the highest required flow rate in accordance with Section 612.5.1.1 and multiplying that flow rate by 2.

(3) Where the sprinkler manufacturer specifies different criteria for ceiling configurations that are not smooth, flat, and horizontal the required flow rate for that room shall be in accordance with the sprinkler manufacturer's instructions.

(4) The flow rate used for sizing the sprinkler system shall be the flow required by the room with the largest flow rate in accordance with Section 612.5.1.2(1), Section 612.5.1.2(2), and Section 612.5.1.2(3).

(5) For the purpose of this section, it shall be permissible to reduce the flow rate for a room by subdividing the space into two or more rooms, where each room is evaluated separately on the required design flow rate. Each room shall be bounded by walls and a ceiling. Openings in walls shall have a lintel not less than 8 inches (203 mm) in depth, and each lintel shall form a solid barrier between the ceiling and the top of the opening.

612.5.2 Sprinkler Pipe Water Supply. The water supply for a multipurpose or stand-alone sprinkler system shall be provided by the public water main, private water main, private well system, or storage tank. The water supply required shall be determined in accordance with Section 612.5.1.2 at a pressure not less than that used in accordance with Section 612.5.3.

612.5.2.1 Water Pressure from Individual Sources. Where a dwelling unit water supply is from a tank system, a private well system, or a combination of these, the available water pressure shall be based on the minimum pressure control setting of the pump.

612.5.2.2 Required Capacity. The water supply shall have the capacity to provide the required flow rate to the sprinklers for a period of time as follows:

(1) Seven minutes for one story dwelling units less than 2000 square feet (185.8 m^2) in area.

(2) Ten minutes for multi-level dwelling units and one story dwelling units not less than 2000 square feet (185.8 m^2) in the area.

Where a well system, a water supply tank system, or a combination thereof is used a combination of well capacity and tank storage shall be permitted to meet the capacity requirement.

612.5.3 Sprinkler Pipe Sizing. The sprinkler piping shall be sized for the flow rate in accordance with Section 612.5.1. The flow rate required to supply the plumbing fixtures shall not be required to be added to the sprinkler design flow for multipurpose or stand alone piping systems. The sizing of the water supply to the plumbing fixtures shall be determined in accordance with this chapter. For multipurpose piping systems, the largest pipe size required based on either the sprinkler piping calculations or the water distribution piping calculations shall be installed.

612.5.3.1 Sprinkler Pipe Sizing Method. The sprinkler pipe shall be sized using the prescriptive method in Section 612.5.3.2 or by hydraulic calculation in accordance with NFPA 13D. The sprinkler pipe size from the water supply source to a sprinkler shall be not less than ¾ of an inch (20 mm) in diameter. Threaded adapter fittings at the point where sprinklers are attached to the piping shall be not less than ½ of an inch (15 mm) in diameter.

612.5.3.2 Prescriptive Pipe Sizing Method. The sprinkler pipe shall be sized by determining the available pressure to offset friction loss in piping and based on the piping material, diameter and length using the equation in Section 612.5.3.2.1 and the procedure in Section 612.5.3.2.2.

612.5.3.2.1 Available Pressure Equation. The available system pressure (P_t) for sizing the sprinkler piping shall be determined in accordance with the Equation 612.5.3.2.1.

(Equation 612.5.3.2.1)

$$P_t = P_{sup} - PL_{ws} - PL_m - PL_d - PL_e - P_{sp}$$

Where:

P_t = Pressure used for sizing the system in Table 612.5.3.2(4) through Table 612.5.3.2(9)

P_{sup} = Pressure available from the water supply source

PL_{ws} = Pressure loss in the water service pipe

PL_m = Pressure loss through the water meter

PL_d = Pressure loss from devices other than the water meter

PL_e = Pressure loss associated with changes in elevation

P_{sp} = Maximum pressure required by a sprinkler

TABLE 612.5.3.2(1)
WATER SERVICE PRESSURE LOSS (PL_{ws})[1, 2, 3]

FLOW RATE (gpm)	¾ INCH WATER SERVICE PRESSURE LOSS (psi)				1 INCH WATER SERVICE PRESSURE LOSS (psi)				1¼ INCH WATER SERVICE PRESSURE LOSS (psi)			
	40 FEET OR LESS	41 FEET TO 75 FEET	76 FEET TO 100 FEET	101 FEET TO 150 FEET	40 FEET OR LESS	41 FEET TO 75 FEET	76 FEET TO 100 FEET	101 FEET TO 150 FEET	40 FEET OR LESS	41 FEET TO 75 FEET	76 FEET TO 100 FEET	101 FEET TO 150 FEET
8	5.1	8.7	11.8	17.4	1.5	2.5	3.4	5.1	0.6	1.0	1.3	1.9
10	7.7	13.1	17.8	26.3	2.3	3.8	5.2	7.7	0.8	1.4	2.0	2.9
12	10.8	18.4	24.9	NP	3.2	5.4	7.3	10.7	1.2	2.0	2.7	4.0
14	14.4	24.5	NP	NP	4.2	7.1	9.6	14.3	1.6	2.7	3.6	5.4
16	18.4	NP	NP	NP	5.4	9.1	12.4	18.3	2.0	3.4	4.7	6.9
18	22.9	NP	NP	NP	6.7	11.4	15.4	22.7	2.5	4.3	5.8	8.6
20	27.8	NP	NP	NP	8.1	13.8	18.7	27.6	3.1	5.2	7.0	10.4
22	NP	NP	NP	NP	9.7	16.5	22.3	NP	3.7	6.2	8.4	12.4
24	NP	NP	NP	NP	11.4	19.3	26.2	NP	4.3	7.3	9.9	14.6
26	NP	NP	NP	NP	13.2	22.4	NP	NP	5.0	8.5	11.4	16.9
28	NP	NP	NP	NP	15.1	25.7	NP	NP	5.7	9.7	13.1	19.4
30	NP	NP	NP	NP	17.2	NP	NP	NP	6.5	11.0	14.9	22.0
32	NP	NP	NP	NP	19.4	NP	NP	NP	7.3	12.4	16.8	24.8
34	NP	NP	NP	NP	21.7	NP	NP	NP	8.2	13.9	18.8	NP
36	NP	NP	NP	NP	24.1	NP	NP	NP	9.1	15.4	20.9	NP

For SI units: 1 gallon per minute = 0.06 L/s, 1 pound-force per square inch = 6.8947 kPa, 1 inch = 25 mm, 1 foot = 304.8 mm

Notes:

[1] Values are applicable for underground piping materials and are based on polyethylene pipe having an SDR of 11 and a Hazen Williams C Factor of 150.

[2] Values include the following length allowances for fittings: 25 percent length increase for actual lengths up to 100 feet (30 480 mm) and 15 percent length increase for actual lengths over 100 feet (30 480 mm).

[3] NP – Means not permitted.

TABLE 612.5.3.2(2)
MINIMUM WATER METER PRESSURE LOSS (PL_m)[1, 2]

FLOW RATE (gpm)	⅝ INCH METER PRESSURE LOSS (psi)	¾ INCH METER PRESSURE LOSS (psi)	1 INCH METER PRESSURE LOSS (psi)
8	2	1	1
10	3	1	1
12	4	1	1
14	5	2	1
16	7	3	1
18	9	4	1
20	11	4	2
22	NP	5	2
24	NP	5	2
26	NP	6	2
28	NP	6	2
30	NP	7	2
32	NP	7	3
34	NP	8	3
36	NP	8	3

For SI units: 1 gallon per minute = 0.06 L/s, 1 pound-force per square inch = 6.8947 kPa, 1 inch = 25 mm

Notes:

1 Table 612.5.3.2(2) establishes conservative values for water meter pressure loss for installations where the water meter loss is unknown. Where the actual water meter pressure loss is known, PL_m shall be the pressure loss as specified by the meter manufacturer.

2 NP – Means not permitted.

612.5.3.2.2 Calculation Procedure. The following procedure shall be used to determine the minimum size of the residential sprinkler piping:

Step 1 - Determine P_{sup}

Obtain the supply pressure available from the water main from the water purveyor, or for an individual source; the available supply pressure shall be in accordance with Section 612.5.2.1. The pressure shall be the flowing pressure available at the flow rate used when applying Table 612.5.3.2(1).

Step 2 – Determine PL_{ws}

Use Table 612.5.3.2(1) to determine the pressure loss in the water service pipe based on the size of the water service. Where the water service supplies more than one dwelling unit, 5 gpm (0.3 L/s) shall be added to the sprinkler flow rate.

Step 3 – Determine PL_m

Use Table 612.5.3.2(2) to determine the pressure loss from the water meter based on the water meter size.

Step 4 – Determine PL_d

Determine the pressure loss from devices, other than the water meter, installed in the piping system supplying sprinklers such as pressure-reducing valves, backflow preventers, water softeners, or water filters. Device pressure losses shall be based on the device manufacturer's specifications. The flow rate used to determine pressure loss shall be the sprinkler flow rate from Section 612.5.1.2. As an alternative to deducting pressure loss for a device, an automatic bypass valve shall be installed to divert flow around the device when a sprinkler activates.

TABLE 612.5.3.2(3)
ELEVATION LOSS (PL_e)

ELEVATION (feet)	PRESSURE LOSS (psi)
5	2.2
10	4.4
15	6.5
20	8.7
25	10.9
30	13.0
35	15.2
40	17.4

For SI units: 1 foot = 304.8 mm, 1 pound-force per square inch = 6.8947 kPa

Step 5 – Determine PL_e

Use Table 612.5.3.2(3) to determine the pressure loss associated with changes in elevation. The elevation used in applying the table shall be the difference between the elevation where the water source pressure was measured and the elevation of the highest sprinkler.

Step 6 – Determine P_{sp}

Determine the maximum pressure required by an individual sprinkler based on the flow rate from Section 612.5.1.1. The minimum pressure required is specified in the sprinkler manufacturer's published data for the specific sprinkler model based on the selected flow rate.

Step 7 – Calculate P_t

Using Equation 612.5.3.2.1, calculate the available system pressure for sizing the sprinkler piping.

Step 8 – Determine the maximum allowable pipe length

Use Table 612.5.3.2(4) through Table 612.5.3.2(9) to select a material and size for the residential sprinkler piping. The piping material and size shall be acceptable where the developed length of pipe between the inside water service valve and the most remote sprinkler does not exceed the maximum allowable length specified by the applicable table. Interpolation of P_t between the tabular values shall be permitted.

The maximum allowable length of piping in Table 612.5.3.2(4) through Table 612.5.3.2(9) incorporates an adjustment for pipe fittings, and no additional consideration of friction losses associated with pipe fittings shall be required.

612.6 Instructions and Signs. An owner's manual for the fire sprinkler system shall be provided to the owner. A sign or valve tag shall be installed at the main shutoff valve to the water distribution system stating the following: *"Warning, the water system for this home supplies fire sprinklers that require certain flow and pressure to fight a fire. Devices that restrict the flow decrease the pressure, or automatically shut-off the water to the fire sprinkler system, such as water softeners, filtration systems, and automatic shutoff valves shall not be added to this system without a review of the fire sprinkler system by a fire protection specialist. Do not remove this sign."*

612.7 Inspection and Testing. The inspection and testing of sprinkler systems shall be in accordance with Section 612.7.1 and Section 612.7.2.

612.7.1 Pre-Concealment Inspection. The following shall be verified prior to the concealment of any sprinkler system piping:

(1) Sprinklers are installed in all areas in accordance with Section 612.3.1.

(2) Where sprinkler water spray patterns are obstructed by construction features, luminaires or ceiling fans, additional sprinklers are installed in accordance with Section 612.3.6.

(3) Sprinklers are the correct temperature rating and are installed at or beyond the required separation distances from heat sources in accordance with Section 612.3.3 and Section 612.3.3.1.

(4) The minimum pipe size in accordance with the requirements of Table 612.5.3.2(4) through Table 612.5.3.2(9) or, where the piping system was hydraulically calculated in accordance with Section 612.5.3.1, the size used in the hydraulic calculation.

(5) The pipe length does not exceed the length permitted by Table 612.5.3.2(4) through Table 612.5.3.2(9) or, where the piping system was hydraulically calculated in accordance with Section 612.5.3.1, pipe lengths and fittings shall not exceed those used in the hydraulic calculation.

(6) Nonmetallic piping that conveys water to sprinklers is certified as having a pressure rating of not less than 130 psi (896 kPa) at 120°F (49°C).

(7) Piping is properly supported.

(8) The piping system is tested in accordance with Section 609.4.

612.7.2 Final Inspection. Upon completion of the residential sprinkler system, the system shall be inspected. The following shall be verified during the final inspection:

(1) Sprinklers are not painted, damaged, or otherwise hindered from the operation.

(2) Where a pump is required to provide water to the system, the pump starts automatically upon system water demand.

(3) Pressure reducing valves, water softeners, water filters, or other impairments to water flow that were not part of the original design has not been installed.

(4) The sign or valve tag in accordance with Section 612.6 is installed, and the owner's manual for the system is present.

TABLE 612.5.3.2(4)
ALLOWABLE PIPE LENGTH FOR ¾ INCH TYPE M COPPER WATER TUBING*

SPRINKLER FLOW RATE (gpm)	WATER DISTRIBUTION SIZE (inch)	AVAILABLE PRESSURE - P_t (psi)									
		15	20	25	30	35	40	45	50	55	60
		ALLOWABLE LENGTH OF PIPE FROM SERVICE VALVE TO FARTHEST SPRINKLER (feet)									
8	¾	217	289	361	434	506	578	650	723	795	867
9	¾	174	232	291	349	407	465	523	581	639	697
10	¾	143	191	239	287	335	383	430	478	526	574
11	¾	120	160	200	241	281	321	361	401	441	481
12	¾	102	137	171	205	239	273	307	341	375	410
13	¾	88	118	147	177	206	235	265	294	324	353
14	¾	77	103	128	154	180	205	231	257	282	308
15	¾	68	90	113	136	158	181	203	226	248	271
16	¾	60	80	100	120	140	160	180	200	220	241
17	¾	54	72	90	108	125	143	161	179	197	215
18	¾	48	64	81	97	113	129	145	161	177	193
19	¾	44	58	73	88	102	117	131	146	160	175
20	¾	40	53	66	80	93	106	119	133	146	159
21	¾	36	48	61	73	85	97	109	121	133	145
22	¾	33	44	56	67	78	89	100	111	122	133
23	¾	31	41	51	61	72	82	92	102	113	123
24	¾	28	38	47	57	66	76	85	95	104	114
25	¾	26	35	44	53	61	70	79	88	97	105
26	¾	24	33	41	49	57	65	73	82	90	98
27	¾	23	30	38	46	53	61	69	76	84	91
28	¾	21	28	36	43	50	57	64	71	78	85
29	¾	20	27	33	40	47	53	60	67	73	80
30	¾	19	25	31	38	44	50	56	63	69	75
31	¾	18	24	29	35	41	47	53	59	65	71
32	¾	17	22	28	33	39	44	50	56	61	67
33	¾	16	21	26	32	37	42	47	53	58	63
34	¾	NP	20	25	30	35	40	45	50	55	60
35	¾	NP	19	24	28	33	38	42	47	52	57
36	¾	NP	18	22	27	31	36	40	45	49	54
37	¾	NP	17	21	26	30	34	38	43	47	51
38	¾	NP	16	20	24	28	32	36	40	45	49
39	¾	NP	15	19	23	27	31	35	39	42	46
40	¾	NP	NP	18	22	26	29	33	37	40	44

For SI units: 1 pound-force per square inch = 6.8947 kPa, 1 gallon per minute = 0.06 L/s, 1 inch = 25 mm, 1 foot = 304.8 mm

* NP – Means not permitted.

TABLE 612.5.3.2(5)
ALLOWABLE PIPE LENGTH FOR 1 INCH TYPE M COPPER WATER TUBING

SPRINKLER FLOW RATE (gpm)	WATER DISTRIBUTION SIZE (inch)	AVAILABLE PRESSURE - P_t (psi)									
		15	20	25	30	35	40	45	50	55	60
		ALLOWABLE LENGTH OF PIPE FROM SERVICE VALVE TO FARTHEST SPRINKLER (feet)									
8	1	806	1075	1343	1612	1881	2149	2418	2687	2955	3224
9	1	648	864	1080	1296	1512	1728	1945	2161	2377	2593
10	1	533	711	889	1067	1245	1422	1600	1778	1956	2134
11	1	447	596	745	894	1043	1192	1341	1491	1640	1789
12	1	381	508	634	761	888	1015	1142	1269	1396	1523
13	1	328	438	547	657	766	875	985	1094	1204	1313
14	1	286	382	477	572	668	763	859	954	1049	1145
15	1	252	336	420	504	588	672	756	840	924	1008
16	1	224	298	373	447	522	596	671	745	820	894
17	1	200	266	333	400	466	533	600	666	733	799
18	1	180	240	300	360	420	479	539	599	659	719
19	1	163	217	271	325	380	434	488	542	597	651
20	1	148	197	247	296	345	395	444	493	543	592
21	1	135	180	225	270	315	360	406	451	496	541
22	1	124	165	207	248	289	331	372	413	455	496
23	1	114	152	190	228	267	305	343	381	419	457
24	1	106	141	176	211	246	282	317	352	387	422
25	1	98	131	163	196	228	261	294	326	359	392
26	1	91	121	152	182	212	243	273	304	334	364
27	1	85	113	142	170	198	226	255	283	311	340
28	1	79	106	132	159	185	212	238	265	291	318
29	1	74	99	124	149	174	198	223	248	273	298
30	1	70	93	116	140	163	186	210	233	256	280
31	1	66	88	110	132	153	175	197	219	241	263
32	1	62	83	103	124	145	165	186	207	227	248
33	1	59	78	98	117	137	156	176	195	215	234
34	1	55	74	92	111	129	148	166	185	203	222
35	1	53	70	88	105	123	140	158	175	193	210
36	1	50	66	83	100	116	133	150	166	183	199
37	1	47	63	79	95	111	126	142	158	174	190
38	1	45	60	75	90	105	120	135	150	165	181
39	1	43	57	72	86	100	115	129	143	158	172
40	1	41	55	68	82	96	109	123	137	150	164

For SI units: 1 pound-force per square inch = 6.8947 kPa, 1 gallon per minute = 0.06 L/s, 1 inch = 25 mm, 1 foot = 304.8 mm

TABLE 612.5.3.2(6)
ALLOWABLE PIPE LENGTH FOR ¾ INCH IPS CPVC PIPE

SPRINKLER FLOW RATE (gpm)	WATER DISTRIBUTION SIZE (inch)	AVAILABLE PRESSURE - P_t (psi)									
		15	20	25	30	35	40	45	50	55	60
		ALLOWABLE LENGTH OF PIPE FROM SERVICE VALVE TO FARTHEST SPRINKLER (feet)									
8	¾	348	465	581	697	813	929	1045	1161	1278	1394
9	¾	280	374	467	560	654	747	841	934	1027	1121
10	¾	231	307	384	461	538	615	692	769	845	922
11	¾	193	258	322	387	451	515	580	644	709	773
12	¾	165	219	274	329	384	439	494	549	603	658
13	¾	142	189	237	284	331	378	426	473	520	568
14	¾	124	165	206	247	289	330	371	412	454	495
15	¾	109	145	182	218	254	290	327	363	399	436
16	¾	97	129	161	193	226	258	290	322	354	387
17	¾	86	115	144	173	202	230	259	288	317	346
18	¾	78	104	130	155	181	207	233	259	285	311
19	¾	70	94	117	141	164	188	211	234	258	281
20	¾	64	85	107	128	149	171	192	213	235	256
21	¾	58	78	97	117	136	156	175	195	214	234
22	¾	54	71	89	107	125	143	161	179	197	214
23	¾	49	66	82	99	115	132	148	165	181	198
24	¾	46	61	76	91	107	122	137	152	167	183
25	¾	42	56	71	85	99	113	127	141	155	169
26	¾	39	52	66	79	92	105	118	131	144	157
27	¾	37	49	61	73	86	98	110	122	135	147
28	¾	34	46	57	69	80	92	103	114	126	137
29	¾	32	43	54	64	75	86	96	107	118	129
30	¾	30	40	50	60	70	81	91	101	111	121
31	¾	28	38	47	57	66	76	85	95	104	114
32	¾	27	36	45	54	63	71	80	89	98	107
33	¾	25	34	42	51	59	68	76	84	93	101
34	¾	24	32	40	48	56	64	72	80	88	96
35	¾	23	30	38	45	53	61	68	76	83	91
36	¾	22	29	36	43	50	57	65	72	79	86
37	¾	20	27	34	41	48	55	61	68	75	82
38	¾	20	26	33	39	46	52	59	65	72	78
39	¾	19	25	31	37	43	50	56	62	68	74
40	¾	18	24	30	35	41	47	53	59	65	71

For SI units: 1 pound-force per square inch = 6.8947 kPa, 1 gallon per minute = 0.06 L/s, 1 inch = 25 mm, 1 foot = 304.8 mm

**TABLE 612.5.3.2(7)
ALLOWABLE PIPE LENGTH FOR 1 INCH IPS CPVC PIPE**

SPRINKLER FLOW RATE (gpm)	WATER DISTRIBUTION SIZE (inch)	AVAILABLE PRESSURE - P_t (psi)									
		15	20	25	30	35	40	45	50	55	60
		ALLOWABLE LENGTH OF PIPE FROM SERVICE VALVE TO FARTHEST SPRINKLER (feet)									
8	1	1049	1398	1748	2098	2447	2797	3146	3496	3845	4195
9	1	843	1125	1406	1687	1968	2249	2530	2811	3093	3374
10	1	694	925	1157	1388	1619	1851	2082	2314	2545	2776
11	1	582	776	970	1164	1358	1552	1746	1940	2133	2327
12	1	495	660	826	991	1156	1321	1486	1651	1816	1981
13	1	427	570	712	854	997	1139	1281	1424	1566	1709
14	1	372	497	621	745	869	993	1117	1241	1366	1490
15	1	328	437	546	656	765	874	983	1093	1202	1311
16	1	291	388	485	582	679	776	873	970	1067	1164
17	1	260	347	433	520	607	693	780	867	954	1040
18	1	234	312	390	468	546	624	702	780	858	936
19	1	212	282	353	423	494	565	635	706	776	847
20	1	193	257	321	385	449	513	578	642	706	770
21	1	176	235	293	352	410	469	528	586	645	704
22	1	161	215	269	323	377	430	484	538	592	646
23	1	149	198	248	297	347	396	446	496	545	595
24	1	137	183	229	275	321	366	412	458	504	550
25	1	127	170	212	255	297	340	382	425	467	510
26	1	118	158	197	237	276	316	355	395	434	474
27	1	111	147	184	221	258	295	332	368	405	442
28	1	103	138	172	207	241	275	310	344	379	413
29	1	97	129	161	194	226	258	290	323	355	387
30	1	91	121	152	182	212	242	273	303	333	364
31	1	86	114	143	171	200	228	257	285	314	342
32	1	81	108	134	161	188	215	242	269	296	323
33	1	76	102	127	152	178	203	229	254	280	305
34	1	72	96	120	144	168	192	216	240	265	289
35	1	68	91	114	137	160	182	205	228	251	273
36	1	65	87	108	130	151	173	195	216	238	260
37	1	62	82	103	123	144	165	185	206	226	247
38	1	59	78	98	117	137	157	176	196	215	235
39	1	56	75	93	112	131	149	168	187	205	224
40	1	53	71	89	107	125	142	160	178	196	214

For SI units: 1 pound-force per square inch = 6.8947 kPa, 1 gallon per minute = 0.06 L/s, 1 inch = 25 mm, 1 foot = 304.8 mm

TABLE 612.5.3.2(8)
ALLOWABLE PIPE LENGTH FOR ¾ INCH PEX TUBING*

SPRINKLER FLOW RATE (gpm)	WATER DISTRIBUTION SIZE (inch)	AVAILABLE PRESSURE - P_t (psi)									
		15	20	25	30	35	40	45	50	55	60
		ALLOWABLE LENGTH OF PIPE FROM SERVICE VALVE TO FARTHEST SPRINKLER (feet)									
8	¾	93	123	154	185	216	247	278	309	339	370
9	¾	74	99	124	149	174	199	223	248	273	298
10	¾	61	82	102	123	143	163	184	204	225	245
11	¾	51	68	86	103	120	137	154	171	188	205
12	¾	44	58	73	87	102	117	131	146	160	175
13	¾	38	50	63	75	88	101	113	126	138	151
14	¾	33	44	55	66	77	88	99	110	121	132
15	¾	29	39	48	58	68	77	87	96	106	116
16	¾	26	34	43	51	60	68	77	86	94	103
17	¾	23	31	38	46	54	61	69	77	84	92
18	¾	21	28	34	41	48	55	62	69	76	83
19	¾	19	25	31	37	44	50	56	62	69	75
20	¾	17	23	28	34	40	45	51	57	62	68
21	¾	16	21	26	31	36	41	47	52	57	62
22	¾	NP	19	24	28	33	38	43	47	52	57
23	¾	NP	17	22	26	31	35	39	44	48	52
24	¾	NP	16	20	24	28	32	36	40	44	49
25	¾	NP	NP	19	22	26	30	34	37	41	45
26	¾	NP	NP	17	21	24	28	31	35	38	42
27	¾	NP	NP	16	20	23	26	29	33	36	39
28	¾	NP	NP	15	18	21	24	27	30	33	36
29	¾	NP	NP	NP	17	20	23	26	28	31	34
30	¾	NP	NP	NP	16	19	21	24	27	29	32
31	¾	NP	NP	NP	15	18	20	23	25	28	30
32	¾	NP	NP	NP	NP	17	19	21	24	26	28
33	¾	NP	NP	NP	NP	16	18	20	22	25	27
34	¾	NP	NP	NP	NP	NP	17	19	21	23	25
35	¾	NP	NP	NP	NP	NP	16	18	20	22	24
36	¾	NP	NP	NP	NP	NP	15	17	19	21	23
37	¾	NP	NP	NP	NP	NP	NP	16	18	20	22
38	¾	NP	NP	NP	NP	NP	NP	16	17	19	21
39	¾	NP	NP	NP	NP	NP	NP	NP	16	18	20
40	¾	NP	NP	NP	NP	NP	NP	NP	16	17	19

For SI units: 1 pound-force per square inch = 6.8947 kPa, 1 gallon per minute = 0.06 L/s, 1 inch = 25 mm, 1 foot = 304.8 mm

* NP – Means not permitted.

TABLE 612.5.3.2(9)
ALLOWABLE PIPE LENGTH FOR 1 INCH PEX TUBING

SPRINKLER FLOW RATE (gpm)	WATER DISTRIBUTION SIZE (inch)	AVAILABLE PRESSURE - P_t (psi)									
		15	20	25	30	35	40	45	50	55	60
		ALLOWABLE LENGTH OF PIPE FROM SERVICE VALVE TO FARTHEST SPRINKLER (feet)									
8	1	314	418	523	628	732	837	941	1046	1151	1255
9	1	252	336	421	505	589	673	757	841	925	1009
10	1	208	277	346	415	485	554	623	692	761	831
11	1	174	232	290	348	406	464	522	580	638	696
12	1	148	198	247	296	346	395	445	494	543	593
13	1	128	170	213	256	298	341	383	426	469	511
14	1	111	149	186	223	260	297	334	371	409	446
15	1	98	131	163	196	229	262	294	327	360	392
16	1	87	116	145	174	203	232	261	290	319	348
17	1	78	104	130	156	182	208	233	259	285	311
18	1	70	93	117	140	163	187	210	233	257	280
19	1	63	84	106	127	148	169	190	211	232	253
20	1	58	77	96	115	134	154	173	192	211	230
21	1	53	70	88	105	123	140	158	175	193	211
22	1	48	64	80	97	113	129	145	161	177	193
23	1	44	59	74	89	104	119	133	148	163	178
24	1	41	55	69	82	96	110	123	137	151	164
25	1	38	51	64	76	89	102	114	127	140	152
26	1	35	47	59	71	83	95	106	118	130	142
27	1	33	44	55	66	77	88	99	110	121	132
28	1	31	41	52	62	72	82	93	103	113	124
29	1	29	39	48	58	68	77	87	97	106	116
30	1	27	36	45	54	63	73	82	91	100	109
31	1	26	34	43	51	60	68	77	85	94	102
32	1	24	32	40	48	56	64	72	80	89	97
33	1	23	30	38	46	53	61	68	76	84	91
34	1	22	29	36	43	50	58	65	72	79	86
35	1	20	27	34	41	48	55	61	68	75	82
36	1	19	26	32	39	45	52	58	65	71	78
37	1	18	25	31	37	43	49	55	62	68	74
38	1	18	23	29	35	41	47	53	59	64	70
39	1	17	22	28	33	39	45	50	56	61	67
40	1	16	21	27	32	37	43	48	53	59	64

For SI units: 1 pound-force per square inch = 6.8947 kPa, 1 gallon per minute = 0.06 L/s, 1 inch = 25 mm, 1 foot = 304.8 mm

CHAPTER 7
SANITARY DRAINAGE

Part I – Drainage Systems.

701.0 General.

701.1 Applicability. This chapter shall govern the materials, design, and installation of sanitary drainage systems and building sewers.

701.2 Drainage Piping. Materials for drainage piping shall be in accordance with one of the referenced standards in Table 701.2 except that:

(1) No galvanized wrought-iron or galvanized steel pipe shall be used underground and shall be kept not less than 6 inches (152 mm) aboveground.

(2) ABS and PVC DWV piping installations shall be installed in accordance with applicable standards referenced in Table 701.2 and Chapter 14 "Firestop Protection." Except for individual single-family dwelling units, materials exposed within ducts or plenums shall have a flame-spread index of not more than 25 and a smoke-developed index of not more than 50, where tested in accordance with ASTM E84 or UL 723. Plastic piping installed in plenums shall be tested in accordance with all requirements of ASTM E84 or UL 723. Mounting methods, supports and sample sizes of materials for testing that are not specified in ASTM E84 or UL 723 shall be prohibited.

(3) No vitrified clay pipe or fittings shall be used aboveground or where pressurized by a pump or ejector. They shall be kept not less than 12 inches (305 mm) belowground.

(4) Copper or copper alloy tube for drainage and vent piping shall have a weight of not less than that of copper or copper alloy drainage tube type DWV.

(5) Stainless steel 304 pipe and fittings shall not be installed underground and shall be kept not less than 6 inches (152 mm) aboveground.

(6) Cast-iron soil pipe and fittings and the stainless steel couplings used to join these products shall be listed and tested in accordance with standards referenced in Table 701.2. Such pipe and fittings shall be marked with the country of origin, manufacturer's name or registered trademark as defined in the product standards, the third party certifier's mark, and the class of the pipe or fitting.

701.3 Drainage Fittings. Materials for drainage fittings shall comply with the applicable standards referenced in Table 701.2 of the same diameter as the piping served, and such fittings shall be compatible with the type of pipe used.

701.3.1 Screwed Pipe. Fittings on screwed pipe shall be of the recessed drainage type. Burred ends shall be reamed to the full bore of the pipe.

701.3.2 Threads. The threads of drainage fittings shall be tapped to allow ¼ inch per foot (20.8 mm/m) grade.

701.3.3 Type. Fittings used for drainage shall be of the drainage type, have a smooth interior water-way, and be constructed to allow ¼ inch per foot (20.8 mm/m) grade.

701.4 Continuous Wastes. Continuous wastes and fixture tailpieces shall be constructed from the materials specified in Section 701.2 for drainage piping, provided, however, that such connections where exposed or accessible shall be permitted to be of seamless drawn brass not less than No. 20 B & S Gauge (0.032 inches) (0.8 mm).

701.5 Lead. (See Chapter 17) Sheet lead shall comply with the following:

(1) For safe pans – not less than 4 pounds per square foot (lb/ft^2) (19 kg/m^2) or $1/16$ of an inch (1.6 mm) thick.

(2) For flashings or vent terminals – not less than 3 lb/ft^2 (15 kg/m^2) or 0.0472 of an inch (1.2 mm) thick.

(3) Lead bends and lead traps shall be not less than ⅛ of an inch (3.2 mm) in wall thickness.

701.6 Caulking Ferrules. Caulking ferrules shall be manufactured from copper or copper alloy and shall be in accordance with Table 701.6.

701.7 Soldering Bushings. Soldering bushings shall be of copper or copper alloy and shall be in accordance with Table 701.7.

TABLE 701.6
CAULKING FERRULES

PIPE SIZE (inches)	INSIDE DIAMETER (inches)	LENGTH (inches)	MINIMUM WEIGHT EACH	
			pounds	ounces
2	2¼	4½	1	0
3	3¼	4½	1	12
4	4¼	4½	2	8

For SI units: 1 inch = 25 mm, 1 pound = 0.453 kg, 1 ounce = 0.02834 kg

TABLE 701.7
SOLDERING BUSHINGS

PIPE SIZE (inches)	MINIMUM WEIGHT EACH		PIPE SIZE (inches)	MINIMUM WEIGHT EACH	
	pounds	ounces		pounds	ounces
1¼	0	6	2½	1	6
1½	0	8	3	2	0
2	0	14	4	3	8

For SI units: 1 inch = 25 mm, 1 pound = 0.453 kg, 1 ounce = 0.02834 kg

702.0 Fixture Unit Equivalents.

702.1 Trap Size. The unit equivalent of plumbing fixtures shown in Table 702.1 shall be based on the size of the trap required, and the unit equivalent of fixtures and devices not shown in Table 702.1 shall be based on the size of trap or trap arm.

Maximum drainage fixture units for a fixture trap and trap arm loadings for sizes up to 4 inches (100 mm) shall be in accordance with Table 702.1(1).

TABLE 701.2
MATERIALS FOR DRAIN, WASTE, VENT PIPE AND FITTINGS

MATERIAL	UNDERGROUND DRAIN, WASTE, VENT PIPE AND FITTINGS	ABOVEGROUND DRAIN, WASTE, VENT PIPE AND FITTINGS	BUILDING SEWER PIPE AND FITTINGS	REFERENCED STANDARD(S) PIPE	REFERENCED STANDARD(S) FITTINGS
ABS (Schedule 40)	X	X	X	ASTM D2661, ASTM D2680*	ASME A112.4.4, ASTM D2661, ASTM D2680*
Cast-Iron	X	X	X	ASTM A74, ASTM A888, CISPI 301	ASME B16.12, ASTM A74, ASTM A888, CISPI 301
Co-Extruded ABS (Schedule 40)	X	X	X	ASTM F628	ASME A112.4.4, ASTM D2661, ASTM D2680*
Co-Extruded Composite (Schedule 40)	X	X	X	ASTM F1488	ASME A112.4.4, ASTM D2661, ASTM D2665, ASTM F794*, ASTM F1866
Co-Extruded PVC (Schedule 40)	X	X	X	ASTM F891, ASTM F1760	ASME A112.4.4, ASTM D2665, ASTM F794*, ASTM F1336*, ASTM F1866
Copper and Copper Alloys (Type DWV)	X	X	X	ASTM B43, ASTM B75, ASTM B251, ASTM B302, ASTM B306	ASME B16.23, ASME B16.29
Galvanized Malleable Iron	—	X	—	—	ASME B16.3
Galvanized Steel	—	X	—	ASTM A53	—
Polyethylene	—	—	X	ASTM F714, ASTM F894	—
PVC (Schedule 40)	X	X	X	ASTM D1785, ASTM D2665, ASTM F794*	ASME A112.4.4, ASTM D2665, ASTM F794*, ASTM F1866
PVC (Sewer and Drain)	—	—	X	ASTM D2729	ASTM D2729
PVC PSM	—	—	X	ASTM D3034	ASTM D3034
Stainless Steel 304	—	X	—	ASME A112.3.1	ASME A112.3.1
Stainless Steel 316L	X	X	X	ASME A112.3.1	ASME A112.3.1
Vitrified Clay (Extra strength)	—	—	X	ASTM C700	ASTM C700

* For building sewer applications.

TABLE 702.1
DRAINAGE FIXTURE UNIT VALUES (DFU)

PLUMBING APPLIANCES, APPURTENANCES, OR FIXTURES	MINIMUM SIZE TRAP AND TRAP ARM[7] (inches)	PRIVATE	PUBLIC	ASSEMBLY[8]
Bathtub or Combination Bath/Shower	1½	2.0	2.0	—
Bidet	1¼	1.0	—	—
Bidet	1½	2.0	—	—
Clothes Washer, domestic, standpipe[5]	2	3.0	3.0	3.0
Dental Unit, cuspidor	1¼	—	1.0	1.0
Dishwasher, domestic, with independent drain[2]	1½	2.0	2.0	2.0
Drinking Fountain or Water Cooler	1¼	0.5	0.5	1.0
Food Waste Disposer, commercial	2	—	3.0	3.0
Floor Drain, emergency	2	—	0.0	0.0
Floor Drain (for additional sizes see Section 702.0)	2	2.0	2.0	2.0
Shower, single-head trap	2[9]	2.0	2.0	2.0
Multi-head, each additional	2	1.0	1.0	1.0
Lavatory	1¼	1.0	1.0	1.0
Lavatories in sets	1½	2.0	2.0	2.0
Washfountain	1½	—	2.0	2.0
Washfountain	2	—	3.0	3.0
Mobile Home, trap	3	12.0	—	—
Receptor, indirect waste[1,3]	1½	See footnote[1,3]		
Receptor, indirect waste[1,4]	2	See footnote[1,4]		
Receptor, indirect waste[1]	3	See footnote[1]		
Sinks	—	—	—	—
Bar	1½	1.0	—	—
Bar[2]	1½	—	2.0	2.0
Clinical	3	—	6.0	6.0
Commercial with food waste[2]	1½	—	3.0	3.0
Exam Room	1½	—	1.0	—
Special Purpose[2]	1½	2.0	3.0	3.0
Special Purpose	2	3.0	4.0	4.0
Special Purpose	3	—	6.0	6.0
Kitchen, domestic[2] (with or without food waste disposer, dishwasher, or both)	1½	2.0	2.0	—
Laundry[2] (with or without discharge from a clothes washer)	1½	2.0	2.0	2.0
Service or Mop Basin	2	—	3.0	3.0
Service or Mop Basin	3	—	3.0	3.0
Service, flushing rim	3	—	6.0	6.0
Wash, each set of faucets	—	—	2.0	2.0
Nonwater Urinal with Drain Cleansing Action	2	1.0	1.0	1.0
Urinal, integral trap 1.0 GPF[2]	2	2.0	2.0	5.0
Urinal, integral trap greater than 1.0 GPF	2	2.0	2.0	6.0
Urinal, exposed trap[2]	1½	2.0	2.0	5.0
Water Closet, 1.6 GPF Gravity Tank[6]	3	3.0	4.0	6.0
Water Closet, 1.6 GPF Flushometer Tank[6]	3	3.0	4.0	6.0
Water Closet, 1.6 GPF Flushometer Valve[6]	3	3.0	4.0	6.0
Water Closet, greater than 1.6 GPF Gravity Tank[6]	3	4.0	6.0	8.0
Water Closet, greater than 1.6 GPF Flushometer Valve[6]	3	4.0	6.0	8.0

For SI units: 1 inch = 25 mm

Notes:

1 Indirect waste receptors shall be sized based on the total drainage capacity of the fixtures that drain thereinto, in accordance with Table 702.2.
2 Provide a 2 inch (50 mm) minimum drain.
3 For refrigerators, coffee urns, water stations, and similar low demands.
4 For commercial sinks, dishwashers, and similar moderate or heavy demands.
5 Buildings having a clothes-washing area with clothes washers in a battery of three or more clothes washers shall be rated at 6 fixture units each for purposes of sizing common horizontal and vertical drainage piping.
6 Water closets shall be computed as 6 fixture units where determining septic tank sizes based on Appendix H of this code.
7 Trap sizes shall not be increased to the point where the fixture discharge is capable of being inadequate to maintain their self-scouring properties.
8 Assembly [Public Use (see Table 422.1)].
9 For a bathtub to shower retrofit, a 1½ inch (40 mm) trap and trap arm shall be permitted with a maximum shower size of 36 inches (914 mm) in width and 60 inches (1524 mm) in length. See Section 408.5 and Section 408.6.

TABLE 702.1(1)
MAXIMUM DRAINAGE FIXTURE UNITS FOR A TRAP AND TRAP ARM*

SIZE OF TRAP AND TRAP ARM (inches)	DRAINAGE FIXTURE UNIT VALUES (DFU)
1¼	1 unit
1½	3 units
2	4 units
3	6 units
4	8 units

For SI Units: 1 inch = 25 mm
* **Exception:** On self-service laundries.

702.2 Intermittent Flow. Drainage fixture units for intermittent flow into the drainage system shall be computed on the rated discharge capacity in gallons per minute (gpm) (L/s) in accordance with Table 702.2.

TABLE 702.2
DISCHARGE CAPACITY IN GALLONS PER MINUTE FOR INTERMITTENT FLOW ONLY*

GPM	FIXTURE UNITS
Up to 7½	Equals 1 Fixture Unit
Greater than 7½ to 15	Equals 2 Fixture Units
Greater than 15 to 30	Equals 4 Fixture Units
Greater than 30 to 50	Equals 6 Fixture Units

For SI units: 1 gallon per minute = 0.06 L/s
* Discharge capacity exceeding 50 gallons per minute (3.15 L/s) shall be determined by the Authority Having Jurisdiction.

702.3 Continuous Flow. For a continuous flow into a drainage system, such as from a pump, sump ejector, air conditioning equipment, or similar device, 2 fixture units shall be equal to each gallon per minute (gpm) (L/s) of flow.

703.0 Size of Drainage Piping.

703.1 Minimum Size. The minimum sizes of vertical, horizontal, or both drainage piping shall be determined from the total of fixture units connected thereto, and additionally, in the case of vertical drainage pipes, in accordance with their length.

703.2 Maximum Number of Fixture Units. Table 703.2 shows the maximum number of fixture units allowed on a vertical or horizontal drainage pipe, building drain, or building sewer of a given size; the maximum number of fixture units allowed on a branch interval of a given size; and the maximum length (in feet and meters) of a vertical drainage pipe of a given size.

703.3 Sizing per Appendix C. For alternate method of sizing drainage piping, see Appendix C.

704.0 Fixture Connections (Drainage).

704.1 Inlet Fittings. Drainage piping shall be provided with approved inlet fittings for fixture connections, correctly located according to the size and type of fixture proposed to be connected.

704.2 Single Vertical Drainage Pipe. Two fixtures set back-to-back, or side-by-side, within the distance allowed between a trap and its vent, shall be permitted to be served by a single vertical drainage pipe provided that each fixture wastes separately into an approved double-fixture fitting having inlet openings at the same level.

704.3 Commercial Sinks. Pot sinks, scullery sinks, dishwashing sinks, silverware sinks, and other similar fixtures shall be connected directly to the drainage system. A floor drain shall be provided adjacent to the fixture and shall be connected on the sewer side of the sink. No other drainage line shall be connected between the floor drain waste connection and the fixture drain. The fixture and floor drain shall be trapped and vented in accordance with this code.

705.0 Joints and Connections.

705.1 ABS and ABS Co-Extruded Plastic Pipe and Joints. Joining methods for ABS plastic pipe and fittings shall be installed in accordance with the manufacturer's installation instructions and shall comply with Section 705.1.1 through Section 705.1.3.

705.1.1 Mechanical Joints. Mechanical joints shall be designed to provide a permanent seal and shall be of the mechanical or push-on joint. The push-on joint shall include an elastomeric gasket that complies with ASTM D3212 and shall provide a compressive force against the spigot and socket after assembly to provide a permanent seal.

705.1.2 Solvent Cement Joints. Solvent cement joints for ABS pipe and fittings shall be clean from dirt and moisture. Pipe shall be cut square and shall be deburred. Where surfaces to be joined are cleaned, and free of dirt, moisture, oil, and other foreign material, the solvent cement that complies with ASTM D2235 shall be applied to all joint surfaces. Joints shall be made while both the inside socket surface and outside surface of pipe are wet with solvent cement. Hold joint in place and undisturbed for 1 minute after assembly.

705.1.3 Threaded Joints. Threads shall comply with ASME B1.20.1. A minimum of Schedule 80 shall be permitted to be threaded. Molded threads on adapter fittings for the transition to threaded joints shall be permitted. Thread sealant compound shall be applied to male threads, insoluble in water, and nontoxic. The joint between the pipe and transition fitting shall be of the solvent cement type. Caution shall be used during assembly to prevent over tightening of the ABS components once the thread sealant compound has been applied.

705.2 Cast-Iron Pipe and Joints. Joining methods for cast-iron pipe and fittings shall be installed in accordance with the manufacturer's installation instructions and shall comply with Section 705.2.1 or Section 705.2.2.

705.2.1 Caulked Joints. Caulked joints shall be firmly packed with oakum or hemp and filled with molten lead to a depth of not less than 1 inch (25.4 mm) in one continuous pour. The lead shall be caulked thoroughly at the inside and outside edges of the joint. After

TABLE 703.2
MAXIMUM UNIT LOADING AND MAXIMUM LENGTH OF DRAINAGE AND VENT PIPING

SIZE OF PIPE (inches)	1¼	1½	2	3	4	5	6	8	10	12
Maximum Units										
Drainage Piping[1]										
Vertical	1	2[2, 7]	16[3]	48[4]	256	600	1380	3600	5600	8400
Horizontal	1	1[7]	8[3]	35[4]	216[5]	428[5]	720[5]	2640[5]	4680[5]	8200[5]
Maximum Length										
Drainage Piping										
Vertical, (feet)	45	65	85	212	300	390	510	750	–	–
Horizontal (unlimited)										
Vent Piping										
Horizontal and Vertical[6]										
Maximum Units	1	8[3]	24	84	256	600	1380	3600	–	–
Maximum Lengths, (feet)	45	60	120	212	300	390	510	750		

For SI units: 1 inch = 25 mm, 1 foot = 304.8 mm

Notes:

[1] Excluding trap arm.

[2] Except for sinks, urinals, and dishwashers – exceeding 1 fixture unit.

[3] Except for six-unit traps or water closets.

[4] Not to exceed five water closets or five six-unit traps.

[5] Based on ¼ inch per foot (20.8 mm/m) slope. For ⅛ of an inch per foot (10.4 mm/m) slope, multiply horizontal fixture units by a factor of 0.8.

[6] The diameter of an individual vent shall be not less than 1¼ inches (32 mm) nor less than one-half the diameter of the drain to which it is connected. Fixture unit load values for drainage and vent piping shall be computed from Table 702.1 and Table 702.2. Not to exceed one-third of the total permitted length of a vent shall be permitted to be installed in a horizontal position. Where vents are increased one pipe size for their entire length, the maximum length limitations specified in this table do not apply. This table is in accordance with the requirements of Section 901.3.

[7] Up to 8 public lavatories are permitted to be installed on a 1½ inch (40 mm) vertical branch or horizontal sanitary branch sloped at ¼ inch per foot (20.8 mm/m).

caulking, the finished joint shall not exceed ⅛ of an inch (3.2 mm) below the rim of the hub. No paint, varnish, or other coatings shall be permitted on the joining material until after the joint has been tested and approved.

705.2.2 Mechanical Joints and Compression Joints. Mechanical joints for cast-iron pipe and fittings shall be of the elastomeric compression type or mechanical joint couplings. Compression type joints with an elastomeric gasket for cast-iron hub and spigot pipe shall comply with ASTM C564 and be tested in accordance with ASTM C1563. Hub and spigot shall be clean and free of dirt, mud, sand, and foreign materials. Cut pipe shall be free from sharp edges. Fold and insert gasket into the hub. Lubricate the joint following manufacturer's instructions. Insert spigot into hub until the spigot end of the pipe bottom out in the hub. Use the same procedure for the installation of fittings.

A mechanical joint shielded coupling type for hubless cast-iron pipe and fittings shall have a metallic shield that complies with ASTM A1056, ASTM C1277, ASTM C1540, or CISPI 310. The elastomeric gasket shall comply with ASTM C564. Hubless cast-iron pipe and fittings shall be clean and free of dirt, mud, sand, and foreign materials. Cut pipe shall be free from sharp edges. Gasket shall be placed on the end of the pipe or fitting and the stainless steel shield and clamp assembly on the end of the other pipe or fitting. Pipe or fittings shall be seated against the center stop inside the elastomeric sleeve. Slide the stainless steel shield and clamp assembly into a position centered over the gasket and tighten. Bands shall be tightened using an approved calibrated torque wrench specifically set by the manufacturer of the couplings.

705.3 Copper or Copper Alloy Pipe (DWV) and Joints. Joining methods for copper or copper alloy pipe and fittings shall be installed in accordance with the manufacturer's installation instructions and shall comply with Section 705.3.1 through Section 705.3.4.

705.3.1 Brazed Joints. Brazed joints between copper or copper alloy pipe and fittings shall be made with brazing alloys having a liquid temperature above 1000°F (538°C). The joint surfaces to be brazed shall be cleaned bright by either manual or mechanical means. Piping shall be cut square and reamed to full inside diameter. Brazing flux shall be applied to the joint surfaces where required by manufacturer's recommendation. Brazing filler metal shall conform to AWS A5.8 and shall be applied at the point where the pipe or tubing enters the socket of the fitting.

705.3.2 Mechanical Joints. Mechanical joints in copper or copper alloy piping shall be made with a mechanical coupling with grooved end piping or approved joint designed for the specific application.

705.3.3 Soldered Joints. Soldered joints between copper or copper alloy pipe and fittings shall be made in accordance with ASTM B828 with the following sequence of joint preparation and operation as follows: measuring and cutting, reaming, cleaning, fluxing, assembly and support, heating, applying the solder, cooling, and cleaning. Pipe shall be cut square and reamed to the full inside diameter including the removal of burrs on the outside of the pipe. Surfaces to be joined shall be cleaned bright by manual or mechanical means. Flux shall be applied to pipe and fittings and shall conform to ASTM B813, and shall become noncorrosive and nontoxic after soldering. Insert pipe into the base of the fitting and remove excess flux. Pipe and fitting shall be supported to ensure a uniform capillary space around the joint. Heat shall be applied using air or fuel torch with the flame perpendicular to the pipe using acetylene or an LP gas. Preheating shall depend on the size of the joint. The flame shall be moved to the fitting cup and alternate between the pipe and fitting. Solder conforming to ASTM B32 shall be applied to the joint surfaces until capillary action draws the molten solder into the cup. Joint surfaces shall not be disturbed until cool, and any remaining flux residue shall be cleaned.

705.3.4 Threaded Joints. Threaded joints for copper or copper alloy pipe shall be made with pipe threads that comply with ASME B1.20.1. Thread sealant tape or compound shall be applied only to male threads, and such material shall be approved types, insoluble in water, and nontoxic.

705.4 Galvanized Steel Pipe and Joints. Joining methods for galvanized steel pipe and fittings shall be installed in accordance with the manufacturer's installation instructions and shall comply with Section 705.4.1 or Section 705.4.2.

705.4.1 Mechanical Joints. Mechanical joints shall be made with an elastomeric gasket.

705.4.2 Threaded Joints. Threaded joints shall be made with pipe threads that comply with ASME B1.20.1. Thread sealant tape or compound shall be applied only to male threads, and such material shall be of approved types, insoluble in water, and nontoxic.

705.5 Polyethylene (PE) Sewer Pipe. Polyethylene (PE) sewer pipe or tubing and fitting joining methods shall be installed in accordance with the manufacturer's installation instructions and shall comply with Section 705.5.1 through Section 705.5.1.3.

705.5.1 Heat-Fusion Joints. Heat-fusion joints between PE sewer pipe or tubing and fittings shall be assembled in accordance with Section 705.5.1.1 through Section 705.5.1.3 using butt-fusion, electro-fusion, or socket-fusion heat methods. Do not disturb the joint until cooled to ambient temperature.

705.5.1.1 Butt-Fusion Joints. Butt-fusion joints for PE pipe shall be installed in accordance with ASTM F2620 and shall be made by heating the prepared ends of two pipes, pipe and fitting, or two fittings by holding ends against a heated element. The heated element shall be removed when the required melt or times are obtained and heated ends shall be placed together with applied force. Do not disturb the joint until cooled to ambient temperature.

705.5.1.2 Electro-Fusion Joints. Electro-fusion joints shall be heated internally by a conductor at the interface of the joint. Fittings shall comply with ASTM F1055 for the performance requirements of polyethylene electro-fusion fittings. The specified electro-fusion cycle used to form the joint requires consideration of the properties of the materials being joined, the design of the fitting being used, and the environmental conditions. Align and restrain fitting to pipe to prevent movement and apply electric current to the fitting. Turn off the current when the required time has elapsed to heat the joint. Do not disturb the joint until cooled to ambient temperature.

705.5.1.3 Socket-Fusion Joints. Socket fusion joints shall be installed in accordance with ASTM F2620 and shall be made by simultaneously heating the outside surface of a pipe end and the inside of a fitting socket. Where the required melt is obtained, the pipe and fitting shall be joined by inserting one into the other with applied force. Do not disturb the joint until cooled to ambient temperature.

705.6 PVC and PVC Co-Extruded Plastic Pipe and Joining Methods. Joining methods for PVC plastic pipe and fittings shall be installed in accordance with the manufacturer's installation instructions and shall comply with Section 705.6.1 through Section 705.6.3.

705.6.1 Mechanical Joints. Mechanical joints shall be designed to provide a permanent seal and shall be of the mechanical or push-on joint type. The push-on joint shall include an elastomeric gasket that complies with ASTM D3212 and shall provide a compressive force against the spigot and socket after assembly to provide a permanent seal.

705.6.2 Solvent Cement Joints. Solvent cement joints for PVC pipe and fittings shall be clean from dirt and moisture. Pipe shall be cut square, and pipe shall be deburred. Where surfaces to be joined are cleaned and free of dirt, moisture, oil, and other foreign material, apply primer purple in color that complies with ASTM F656. Primer shall be applied to the surface of the pipe and fitting is softened. Solvent cement that comply with ASTM D2564 shall be applied to all joint surfaces. Joints shall be made while both the inside socket surface and outside surface of pipe are wet with solvent cement. Hold joint in place and undisturbed for 1 minute after assembly.

705.6.3 Threaded Joints. Threads shall comply with ASME B1.20.1. A minimum of Schedule 80 shall be permitted to be threaded. Molded threads on adapter fittings for the transition to threaded joints shall be permitted. Thread sealant compound that is compatible with the pipe and fitting, insoluble in water and nontoxic shall be applied to male threads. The joint between the pipe and transition fitting shall be of the solvent cement type. Caution shall be used during assembly to prevent over tightening of the PVC components once the thread sealant has been applied. Female PVC threaded fittings shall be used with plastic male threads only.

705.7 Stainless Steel Pipe and Joints. Joining methods for stainless steel pipe and fittings shall be installed in accordance with the manufacturer's installation instructions and shall comply with Section 705.7.1 or Section 705.7.2.

705.7.1 Mechanical Joints. Mechanical joints between stainless steel pipe and fittings shall be of the compression, grooved coupling, hydraulic press-connect fittings, or flanged.

705.7.2 Welded Joints. Welded joints between stainless steel pipe and fittings shall be made in accordance with ASME A112.3.1 and shall be welded autogenously. Pipe shall be cleaned, free of scale and contaminating particles. Pipe shall be cut with a combination cutting and beveling tool that provides a square cut, and free of burrs. Mineral oil lubricant shall be used during the cutting and beveling process.

705.8 Vitrified Clay Pipe and Joints. Joining methods for vitrified clay pipe and fittings shall be installed in accordance with the manufacturer's installation instructions and shall comply with Section 705.8.1.

705.8.1 Mechanical Joints. Mechanical joints shall be designed to provide a permanent seal and shall be of the mechanical or push-on joint type. The push-on joint shall include an elastomeric gasket that complies with ASTM C425 and shall provide a compressive force against the spigot and socket after assembly to provide a permanent seal.

705.9 Special Joints. Special joints shall comply with Section 705.9.1 through Section 705.9.4.

705.9.1 Slip Joints. In fixture drains and traps, slip joints of approved materials shall be permitted to be used in accordance with their approvals.

705.9.2 Expansion Joints. Expansion joints shall be accessible, except where in vent piping or drainage stacks, and shall be permitted to be used where necessary to provide for expansion and contraction of the pipes.

705.9.3 Ground Joint, Flared, or Ferrule Connections. Copper or copper alloy ground joint flared, or ferrule-type connections that allow adjustment of tubing, but provide a rigid joint where made up, shall not be considered as slip joints.

705.9.4 Transition Joint. A solvent cement transition joint between ABS and PVC building drain and building sewer shall be made using listed transition solvent cement in accordance with ASTM D3138.

705.10 Joints Between Various Materials. Joints between various materials shall be installed in accordance with the manufacturer's installation instructions and with Section 705.10.1 through Section 705.10.4. Mechanical couplings used to join different materials shall comply with ASTM C1173 for belowground use, ASTM C1460 for aboveground use, or ASTM C1461 for aboveground and belowground use.

705.10.1 Copper or Copper Alloy Pipe to Cast-Iron Pipe. Joints from copper or copper alloy pipe or tubing to cast-iron pipe shall be made with a listed compression-type joint or copper alloy ferrule. The copper or copper alloy pipe or tubing shall be soldered or brazed to the ferrule, and the ferrule shall be joined to the cast-iron hub by a compression or caulked joint.

705.10.2 Copper or Copper Alloy Pipe to Threaded Pipe Joints. Joints from copper or copper alloy pipe or tubing to threaded pipe shall be made by the use of a listed copper alloy adapter or dielectric fitting. The joint between the copper or copper alloy pipe and the fitting shall be a soldered or brazed, and the connection between the threaded and the fittings shall be made with a standard pipe size threaded joint.

705.10.3 Plastic Pipe to Other Materials. Where connecting plastic pipe to other types of plastic or other types of piping material; approved listed adapter or transition fittings and listed for the specific transition intended shall be used.

705.10.4 Stainless Steel Pipe to Other Materials. Where connecting stainless steel pipe to other types of piping, listed mechanical joints of the compression type and listed for the specific transition intended shall be used.

706.0 Changes in Direction of Drainage Flow.

706.1 Approved Fittings. Changes in the direction of drainage piping shall be made by the appropriate use of approved fittings and shall be of the angles presented by a one-sixteenth bend, one-eighth bend, or one-sixth bend, or other approved fittings of equivalent sweep.

706.2 Horizontal to Vertical. Horizontal drainage lines, connecting with a vertical stack, shall enter through 45 degree (0.79 rad) wye branches, 60 degree (1.05 rad) wye branches, combination wye and one-eighth bend branches, sanitary tee or sanitary tapped tee branches, or other approved fittings of equivalent sweep. No fitting having more than one inlet at the same level shall be used unless such fitting is constructed so that the discharge from one inlet cannot readily enter any other inlet. Double sanitary tees shall be permitted to be used where the barrel of the fitting is not less than two pipe sizes larger than the largest inlet, (pipe sizes recognized for this purpose are 2 inches, 2½ inches, 3 inches, 3½ inches, 4 inches, 4½ inches, 5 inches, 6 inches, etc.) (50 mm, 65 mm, 80 mm, 90 mm, 100 mm, 115 mm, 125 mm, 150 mm, etc.).

706.3 Horizontal to Horizontal. Horizontal drainage lines connecting with other horizontal drainage lines shall enter through 45 degree (0.79 rad) wye branches, combination wye and one-eighth bend branches, or other approved fittings of equivalent sweep.

706.4 Vertical to Horizontal. Vertical drainage lines connecting with horizontal drainage lines shall enter through 45 degree (0.79 rad) wye branches, combination wye and one-eighth bend branches, or other approved fittings of equivalent sweep. Branches or offsets of 60 degrees (1.05 rad) shall be permitted to be used where installed in a true vertical position.

707.0 Cleanouts.

707.1 Plug. Each cleanout fitting for cast-iron pipe shall consist of a cast-iron or copper alloy body and an approved plug.

Each cleanout for galvanized wrought iron, galvanized steel, copper, or copper alloy pipe shall consist of a plug as specified in Table 707.1, or a standard weight copper alloy cap, or an approved ABS or PVC plastic plug, or an approved stainless steel cleanout or plug. Plugs shall have raised square heads or approved countersunk rectangular slots.

TABLE 707.1
CLEANOUTS

SIZE OF PIPE (inches)	SIZE OF CLEANOUT (inches)	THREADS (per inches)
1½	1½	11½
2	1½	11½
2½	2½	8
3	2½	8
4 & larger	3½	8

For SI units: 1 inch = 25 mm

707.2 Approved. Each cleanout fitting and each cleanout plug or cap shall be of an approved type. A list of approved standards for cleanouts are referenced in Table 707.2.

TABLE 707.2
CLEANOUT MATERIALS FOR DRAIN, WASTE, AND VENT

MATERIAL	STANDARD
ABS	ASTM D2661, CSA B79, IAPMO IGC 78, IAPMO IGC 224
Cast Iron	ASME A112.36.2, ASTM A888, CISPI 301, CSA B79, IAPMO IGC 224
Copper or Copper Alloy	ASME A112.36.2, CSA B79
Ductile Iron	CSA B79
Elastomers	CSA B79, IAPMO PS 90
Polyethylene (PE)	CSA B79
Polypropylene (PP)	CSA B79
PVC	ASTM D2665, CSA B79, IAPMO IGC 78, IAPMO IGC 224
Polyvinylidene Fluoride (PVDF)	CSA B79
Stainless Steel	CSA B79

707.3 Watertight and Gastight. Cleanouts shall be designed to be watertight and gastight.

707.4 Location. Each horizontal drainage pipe shall be provided with a cleanout at its upper terminal, and each run of piping, that is more than 100 feet (30 480 mm) in total developed length, shall be provided with a cleanout for each 100 feet (30 480 mm), or fraction thereof, in length of such piping. An additional cleanout shall be provided in a drainage line for each aggregate horizontal change in direction exceeding 135 degrees (2.36 rad). A cleanout shall be installed above the fixture connection fitting, serving each urinal, regardless of the location of the urinal in the building.

Exceptions:

(1) Cleanouts shall be permitted to be omitted on a horizontal drain line less than 5 feet (1524 mm) in length unless such line is serving sinks or urinals.

(2) Cleanouts shall be permitted to be omitted on a horizontal drainage pipe installed on a slope of 72 degrees (1.26 rad) or less from the vertical angle (one-fifth bend).

(3) Excepting the building drain, its horizontal branches, kitchen sinks, and urinals, a cleanout shall not be required on a pipe or piping that is above the floor level of the lowest floor of the building.

(4) An approved type of two-way cleanout fitting, installed inside the building wall near the connection between the building drain and the building sewer or installed outside of a building at the lower end of a building drain and extended to grade, shall be permitted to be substituted for an upper terminal cleanout.

707.5 Cleaning. Each cleanout shall be installed so that it opens to allow cleaning in the direction of flow of the soil or waste or at right angles thereto and, except in the case of wye branch and end-of-line cleanouts, shall be installed vertically above the flow line of the pipe.

707.6 Extension. Each cleanout extension shall be considered as drainage piping and each 90 degree (1.57 rad) cleanout extension shall be extended from a wye-type fitting or other approved fitting of equivalent sweep.

707.7 Interceptor. Each cleanout for an interceptor shall be outside of such interceptor.

707.8 Access. Each cleanout, unless installed under an approved cover plate, shall be above grade, readily accessible, and so located as to serve the purpose for which it is intended. Cleanouts located under cover plates shall be so installed as to provide the clearances and accessibility required by this section.

707.9 Clearance. Each cleanout in piping 2 inches (50 mm) or less in size shall be so installed that there is a clearance of not less than 18 inches (457 mm) by 18 inches (457 mm) in front of the cleanout. Cleanouts in piping exceeding 2 inches (50 mm) shall have a clearance of not less than 24 inches (610 mm) by 24 inches (610 mm) in front of the cleanout. Cleanouts in under-floor piping shall be extended to or above the finished floor or shall be extended outside the building where there is less than 18 inches (457 mm) vertical overall, allowing for obstructions such as ducts, beams, and piping, and 30 inches of (762 mm) horizontal clearance from the means of access to such cleanout. No under-floor cleanout shall be located exceeding 5 feet (1524 mm) from an access door, trap door, or crawl hole.

707.10 Fittings. Cleanout fittings shall be not less in size than those given in Table 707.1.

707.11 Pressure Drainage Systems. Cleanouts shall be provided for pressure drainage systems as classified under Section 710.7.

707.12 Countersunk Cleanout Plugs. Countersunk cleanout plugs shall be installed where raised heads cause a hazard.

707.13 Hubless Blind Plugs. Where a hubless blind plug is used for a required cleanout, the complete coupling and plug shall be accessible for removal or replacement.

707.14 Trap Arms. Cleanouts for trap arms shall be installed in accordance with Section 1002.3.

708.0 Grade of Horizontal Drainage Piping.

708.1 General. Horizontal drainage piping shall be run in practical alignment and a uniform slope of not less than ¼ inch per foot (20.8 mm/m) or 2 percent toward the point of disposal provided that, where it is impractical due to the depth of the street sewer, to the structural features, or to the arrangement of a building or structure to obtain a slope of ¼ inch per foot (20.8 mm/m) or 2 percent, such pipe or piping 4 inches (100 mm) or larger in diameter shall be permitted to have a slope of not less than ⅛ inch per foot (10.4 mm/m) or 1 percent, where first approved by the Authority Having Jurisdiction.

709.0 Gravity Drainage Required.

709.1 General. Where practicable, plumbing fixtures shall be drained to the public sewer or private sewage disposal system by gravity.

710.0 Drainage of Fixtures Located Below the Next Upstream Manhole or Below the Main Sewer Level.

710.1 Backflow Protection. Fixtures installed on a floor level that is lower than the next upstream manhole cover of the public, or private sewer shall be protected from backflow of sewage by installing an approved type of backwater valve. Fixtures on such floor level that are not below the next upstream manhole cover shall not be required to be protected by a backwater valve. Fixtures on floor levels above such elevation shall not discharge through the backwater valve. Cleanouts for drains that pass through a backwater valve shall be clearly identified with a permanent label stating "backwater valve downstream."

710.2 Sewage Discharge. Drainage piping serving fixtures that are located below the crown level of the main sewer shall discharge into an approved watertight sump or receiving tank, so located as to receive the sewage or wastes by gravity. From such sump or receiving tank, the sewage or other liquid wastes shall be lifted and discharged into the building drain or building sewer by approved ejectors, pumps, or other equally efficient approved mechanical devices.

710.3 Sewage Ejector and Pumps. A sewage ejector or sewage pump receiving the discharge of water closets or urinals:

(1) Shall have a discharge capacity of not less than 20 gpm (1.26 L/s).

(2) In single dwelling units, the ejector or pump shall be capable of passing an 1½ inch (38 mm) diameter solid ball, and the discharge piping of each ejector or pump shall have a backwater valve and gate valve, and be not less than 2 inches (50 mm) in diameter.

(3) In other than single-dwelling units, the ejector or pump shall be capable of passing a 2 inch (51 mm) diameter solid ball, and the discharge piping of each ejector or pump shall have a backwater valve and gate valve, and be not less than 3 inches (80 mm) in diameter.

710.4 Discharge Line. The discharge line from such ejector, pump, or another mechanical device shall be of approved pressure rated material and be provided with an accessible backwater or swing check valve and gate or ball valve. Where the gravity drainage line to which such discharge line connects is horizontal, the method of connection shall be from the top through a wye branch fitting. The gate or ball valve shall be located on the discharge side of the backwater or check valve.

Gate or ball valves, where installed in drainage piping, shall be fullway type with working parts of corrosion-resistant metal. Sizes 4 inches (100 mm) or more in diameter shall have cast-iron bodies and sizes less than 4 inches (100 mm), cast-iron or copper alloy bodies.

710.5 Size of Building Drains and Sewers. Building drains or building sewers receiving a discharge from a pump or ejector shall be adequately sized to prevent overloading. Two fixture units shall be allowed for each gallon per minute (L/s) of flow.

710.6 Backwater Valves. Backwater valves, gate valves, fullway ball valves, unions, motors, compressors, air tanks, and other mechanical devices required by this section shall be located where they will be accessible for inspection and repair and, unless continuously exposed, shall be enclosed in a masonry pit fitted with an adequately sized removable cover.

Backwater valves shall comply with ASME A112.14.1, and have bodies of cast-iron, plastic, copper alloy, or other approved materials; shall have noncorrosive bearings, seats, and self-aligning discs; and shall be constructed to ensure a positive mechanical seal. Such backwater valves shall remain open during periods of low flows to avoid screening of solids and shall not restrict capacities or cause excessive turbulence during peak loads. Unless otherwise listed, valve access covers shall be bolted type with gasket, and each valve shall bear the manufacturer's name cast into the body and the cover.

710.7 Drainage and Venting Systems. The drainage and venting systems, in connection with fixtures, sumps, receiving tanks, and mechanical waste-lifting devices shall be installed under the same requirements as provided for in this code for gravity systems.

710.8 Sump and Receiving Tank Construction. Sumps and receiving tanks shall be watertight and shall be constructed of concrete, metal, or other approved materials. Where constructed of poured concrete, the walls and bottom shall be adequately reinforced and designed to recognized acceptable standards. Metal sumps or tanks shall be of such thickness as to serve their intended purpose and shall be treated internally and externally to resist corrosion.

710.9 Alarm. Such sumps and receiving tanks shall be automatically discharged and, wherein a "public use" occupancy, shall be provided with dual pumps or ejectors arranged to function alternately in normal use and independently in case of overload or mechanical failure. The pumps shall have an audio and visual alarm, readily accessible, that signals pump failure or an overload condition. The lowest inlet shall have a clearance of not less than 2 inches (51 mm) from the high-water or "starting" level of the sump.

710.10 Sump and Receiving Tank Covers and Vents. Sumps and receiving tanks shall be provided with substantial covers having a bolt-and-gasket-type manhole or equivalent opening to permit access for inspection, repairs, and cleaning. The top shall be provided with a vent pipe that shall extend separately through the roof or, where permitted, be combined with other vent pipes. Such vent shall be large enough to maintain atmospheric pressure within the sump under normal operating conditions and, in no case, shall be less in size than that required by Table 703.2 for the number and type of fixtures discharging into the sump, nor less than 1½ inches (40 mm) in diameter. Where the preceding requirements are met and the vent, after leaving the sump, is combined with vents from fixtures discharging into the sump, the size of the combined vent need not exceed that required for the total number of fixtures discharging into the sump. No vent from an air-operating sewage ejector shall combine with other vents.

710.11 Air Tanks. Air tanks shall be so proportioned as to be of equal cubical capacity to the ejectors connected in addition to that in which there shall be maintained an air pressure of not less than 2 pounds per foot (lb/ft) (3 kg/m) of height the sewage is to be raised. No water-operated ejectors shall be permitted.

710.12 Grinder Pump Ejector. Grinder pumps shall be permitted to be used.

710.12.1 Discharge Piping. The discharge piping shall be sized in accordance with the manufacturer's installation instructions and shall be not less than 1¼ inches (32 mm) in diameter. A check valve and fullway-type shutoff valve shall be located on the discharge line.

710.13 Macerating Toilet Systems and Pumped Waste Systems. Fixtures shall be permitted to discharge to a macerating toilet system, or pumped waste system shall be permitted as an alternate to a sewage pump system where approved by the Authority Having Jurisdiction. Such systems shall comply with ASME A112.3.4/CSA B45.9 and shall be installed in accordance with the manufacturer's installation instructions.

710.13.1 Sumps. The sump shall be watertight and gastight.

710.13.2 Discharge Piping. The discharge piping shall be sized in accordance with manufacturer's instructions and shall be not less than ¾ of an inch (20 mm) in diameter. The developed length of the discharge piping shall not exceed the manufacturer's instructions. A check valve and fullway-type shutoff valve shall be located within the discharge line or internally within the device.

710.13.3 Venting. The plumbing fixtures that discharge into the macerating device shall be vented in accordance with this code. The sump shall be vented in accordance with the manufacturer's instructions, and such vent shall be permitted to connect to the fixture venting.

711.0 Suds Relief.

711.1 General. Drainage connections shall not be made into a drainage piping system within 8 feet (2438 mm) of a vertical to horizontal change of direction of a stack containing suds-producing fixtures. Bathtubs, laundries, washing machine standpipes, kitchen sinks, and dishwashers shall be considered suds-producing fixtures. Where parallel vent stacks are required, they shall connect to the drainage stack at a point 8 feet (2438 mm) above the lowest point of the drainage stack.

Exceptions:

(1) Single-family residences.

(2) Stacks receiving the discharge from less than three stories of plumbing fixtures.

712.0 Testing.

712.1 Media. The piping of the plumbing, drainage, and venting systems shall be tested with water or air except that plastic pipe shall not be tested with air. The Authority Having Jurisdiction shall be permitted to require the removal of cleanouts, etc., to ascertain whether the pressure has reached all parts of the system. After the plumbing fixtures have been set and their traps filled with water, they shall be submitted to a final test.

712.2 Water Test. The water test shall be applied to the drainage and vent systems either in its entirety or in sections. Where the test is applied to the entire system, openings in the piping shall be tightly closed, except the highest opening, and the system filled with water to the point of overflow. Where the system is tested in sections, each opening shall be tightly plugged, except the highest opening of the section under test, and each section shall be filled with water, but no section shall be tested with less than a 10 foot head of water (30 kPa). In testing successive sections, not less than the upper 10 feet (3048 mm) of the next preceding section shall be tested, so that no joint or pipe in the building (except the uppermost 10 feet (3048 mm) of the system) shall have been submitted to a test of less than a 10 foot head of water (30 kPa). The water shall be kept in the system, or in the portion under test, for not less than 15 minutes before inspection starts. The system shall then be tight at all points.

712.3 Air Test. The air test shall be made by attaching an air compressor testing apparatus to a suitable opening and, after closing all other inlets and outlets to the system, forcing air into the system until there is a uniform gauge pressure of 5 pounds-force per square inch (psi) (34 kPa) or sufficient to balance a column of mercury 10 inches (34 kPa) in height. The pressure shall be held without the introduction of additional air for a period of not less than 15 minutes.

Part II – Building Sewers.

713.0 Sewer Required.

713.1 Where Required. A building in which plumbing fixtures are installed and premises having drainage piping thereon shall have a connection to a public or private sewer, except as provided in Section 713.2, and Section 713.4.

713.2 Private Sewage Disposal System. Where no public sewer intended to serve a lot or premises is available in a thoroughfare or right of way abutting such lot or premises, drainage piping from a building or works shall be connected to an approved private sewage disposal system.

713.3 Public Sewer. Within the limits prescribed by Section 713.4 hereof, the rearrangement or subdivision into smaller parcels of a lot that abuts and is served by a public sewer shall not be deemed cause to permit the construction of a private sewage disposal system, and plumbing or drainage systems on a smaller parcel or parcels shall connect to the public sewer.

713.4 Public Sewer Availability. The public sewer shall be permitted to be considered as not being available where such public sewer or a building or an exterior drainage facility connected thereto is located more than 200 feet (60 960 mm) from a proposed building or exterior drainage facility on a lot or premises that abut and is served by such public sewer.

713.5 Permit. No permit shall be issued for the installation, alteration, or repair of a private sewage disposal system, or part thereof, on a lot for which a connection with a public sewer is available.

713.6 Lot. On every lot or premises hereafter connected to a public sewer, plumbing, and drainage systems or parts thereof on such lot or premises shall be connected with such public sewer.

713.7 Installation. In cities, counties, or both where the installation of building sewers is under the jurisdiction of a department other than the Authority Having Jurisdiction, the provisions of this code relating to building sewers need not apply.

Exception: Single-family dwellings and buildings or structures accessory thereto, existing and connected to an approved private sewage disposal system prior to the time of connecting the premises to the public sewer shall be permitted, where no hazard, nuisance, or insanitary condition is evidenced, and written permission has been obtained from the Authority Having Jurisdiction, remain connected to such properly maintained private sewage disposal system where there is insufficient grade or fall to permit drainage to the sewer by gravity.

714.0 Damage to Public Sewer or Private Sewage Disposal System.

714.1 Unlawful Practices. It shall be unlawful for a person to deposit, by means whatsoever, into a plumbing fixture, floor drain, interceptor, sump, receptor, or device which is connected to a drainage system, public sewer, private sewer, septic tank, or cesspool, ashes; cinders; solids; rags; flammable, poisonous, or explosive liquids or gases; oils; grease; and whatsoever that is capable of causing damage to the public sewer, private sewer, or private sewage disposal system.

714.2 Prohibited Water Discharge. No rain, surface, or subsurface water shall be connected to or discharged into a drainage system unless first approved by the Authority Having Jurisdiction.

714.3 Prohibited Sewer Connection. No cesspool, septic tank, seepage pit, or drain field shall be connected to a public sewer or to a building sewer leading to such public sewer.

714.4 Commercial Food Waste Disposer. The Authority Having Jurisdiction shall review before approval, the installation of a commercial food waste disposer connecting to a private sewage disposal system.

714.5 Tanks. An approved type, watertight sewage or wastewater holding tank, the contents of which, due to their character, shall be periodically removed and disposed of at some approved off-site location, shall be installed where required by the Authority Having Jurisdiction or the Health Officer to prevent anticipated surface or subsurface contamination or pollution, damage to the public sewer, or other hazardous or nuisance conditions.

715.0 Building Sewer Materials.

715.1 Materials. The building sewer, beginning 2 feet (610 mm) from a building or structure, shall be of such materials as prescribed in this code.

715.2 Joining Methods and Materials. Joining methods and materials shall be as prescribed in this code.

715.3 Existing Sewers. Replacement of existing building sewer and building storm sewers using trenchless methodology and materials shall be installed in accordance with ASTM F1216, ASTM F2561, ASTM F2599, or ASTM F3240.

716.0 Markings.

716.1 General. Pipe, brick, block, prefabricated septic tanks, prefabricated septic tank or seepage pit covers, or other parts or appurtenances incidental to the installation of building sewers or private sewage disposal systems shall be in accordance with the approval requirements of Chapter 3 of this code.

717.0 Size of Building Sewers.

717.1 General. The minimum size of a building sewer shall be determined on the basis of the total number of fixture units drained by such sewer, in accordance with Table 717.1. No building sewer shall be smaller than the building drain.

For alternate methods of sizing building sewers, see Appendix C.

TABLE 717.1
MAXIMUM/MINIMUM FIXTURE UNIT LOADING ON BUILDING SEWER PIPING*

SIZE OF PIPE (inches)	SLOPE (inches per foot)		
	1/16	1/8	1/4
6 and smaller	(As specified in Table 703.2/ No minimum loading)		
8	1950/1500	2800/625	3900/275
10	3400/1600	4900/675	6800/300
12	5600/1700	8000/725	11 200/325

For SI units: 1 inch = 25 mm, 1 inch per foot = 83.3 mm/m

* See also Appendix H, Private Sewage Disposal Systems. For alternate methods of sizing drainage piping, see Appendix C.

718.0 Grade, Support, and Protection of Building Sewers.

718.1 Slope. Building sewers shall be run in practical alignment and at a uniform slope of not less than ¼ inch per foot (20.8 mm/m) toward the point of disposal.

Exception: Where approved by the Authority Having Jurisdiction and where it is impractical, due to the depth of the street sewer or to the structural features or the arrangement of a building or structure, to obtain a slope of ¼ inch per foot (20.8 mm/m), such pipe or piping 4 inches (100 mm) through 6 inches (150 mm) shall be permitted to have a slope of not less than ⅛ inch per foot (10.4 mm/m) and such piping 8 inches (200 mm) and larger shall be permitted to have a slope of not less than 1/16 inch per foot (5.2 mm/m).

718.2 Support. Building sewer piping shall be laid on a firm bed throughout its entire length, and such piping laid in made or filled-in ground shall be laid on a bed of approved materials and shall be properly supported as required by the Authority Having Jurisdiction.

718.3 Protection from Damage. No building sewer or other drainage piping or part thereof, which is constructed of materials other than those approved for use under or within a building, shall be installed under or within 2 feet (610 mm) of a building or structure, or part thereof, nor less than 1 foot (305 mm) below the surface of the ground. The provisions of this subsection include structures such as porches and steps, whether covered or uncovered; breezeways; roofed porte cochere; roofed patios; carports; covered walks; covered driveways; and similar structures or appurtenances.

719.0 Cleanouts.

719.1 Locations. Cleanouts shall be placed inside the building near the connection between the building drain and the building sewer or installed outside the building at the lower end of the building drain and extended to grade.

Additional building sewer cleanouts shall be installed at intervals not to exceed 100 feet (30 480 mm) in straight runs and for each aggregate horizontal change in direction exceeding 135 degrees (2.36 rad).

719.2 No Additional Cleanouts. Where a building sewer or a branch thereof does not exceed 10 feet (3048 mm) in length and is a straight-line projection from a building drain that is provided with a cleanout, no cleanout will be required at its point of connection to the building drain.

719.3 Building Sewer Cleanouts. Required building sewer cleanouts shall be extended to grade and shall be in accordance with the appropriate sections of cleanouts, Section 707.0, for sizing, construction, and materials. Where building sewers are located under buildings, the cleanout requirements of Section 707.0 shall apply.

719.4 Cleaning. Each cleanout shall be installed so that it opens to allow cleaning in the direction of flow of the soil or waste or at right angles thereto and, except in the case of wye branch and end-of-line cleanouts, shall be installed vertically above the flow line of the pipe.

719.5 Access. Cleanouts installed under concrete or asphalt paving shall be made accessible by yard boxes or by extending flush with paving with approved materials and shall be adequately protected.

719.6 Manholes. Approved manholes shall be permitted to be installed in lieu of cleanouts, where first approved by the Authority Having Jurisdiction. The maximum distance between manholes shall not exceed 300 feet (91 440 mm).

The inlet and outlet connections shall be made by the use of a flexible compression joint not less than 12 inches (305 mm) and not exceeding 3 feet (914 mm) from the manhole. No flexible compression joints shall be embedded in the manhole base.

720.0 Sewer and Water Pipes.

720.1 General. Building sewers or drainage piping of clay or materials that are not approved for use within a building shall not be run or laid in the same trench as the water pipes unless the following requirements are met:

(1) The bottom of the water pipe, at points, shall be not less than 12 inches (305 mm) above the top of the sewer or drain line.

(2) The water pipe shall be placed on a solid shelf excavated at one side of the common trench with a clear horizontal distance of not less than 12 inches (305 mm) from the sewer or drain line.

(3) Water pipes crossing sewer or drainage piping constructed of clay or materials that are not approved for use within a building shall be laid not less than 12 inches (305 mm) above the sewer or drain pipe.

For the purpose of this section, "within a building" shall mean within the fixed limits of the building foundation.

721.0 Location.

721.1 Building Sewer. Except as provided in Section 721.2, no building sewer shall be located in a lot other than the lot that is the site of the building or structure served by such sewer nor shall a building sewer be located at a point having less than the minimum distances referenced in Table 721.1.

TABLE 721.1
MINIMUM HORIZONTAL DISTANCE REQUIRED FROM BUILDING SEWER (feet)

Buildings or structures[1]	2
Property line adjoining private property	Clear[2]
Water supply wells	50[3]
Streams	50
On-site domestic water service line	1[4]
Public water main	10[5, 6]

For SI units: 1 foot = 304.8 mm

Notes:

[1] Including porches and steps, whether covered or uncovered; breezeways; roofed porte-cochere; roofed patios; carports; covered walks; covered driveways; and similar structures or appurtenances.

[2] See also Section 312.3.

[3] Drainage piping shall clear domestic water supply wells by not less than 50 feet (15 240 mm). This distance shall be permitted to be reduced to not less than 25 feet (7620 mm) where the drainage piping is constructed of materials approved for use within a building.

[4] See Section 720.0.

[5] For parallel construction.

[6] For crossings, approval by the Health Department or the Authority Having Jurisdiction shall be required.

721.2 Abutting Lot. Nothing contained in this code shall be construed to prohibit the use of all or part of an abutting lot to:

(1) Provide access to connect a building sewer to an available public sewer where proper cause and legal easement, not in violation of other requirements, has been first established to the satisfaction of the Authority Having Jurisdiction.

(2) Provide additional space for a building sewer where the proper cause, transfer of ownership, or change of boundary, not in violation of other requirements, has been first established to the satisfaction of the Authority Having Jurisdiction. The instrument recording such action shall constitute an agreement with the Authority Having Jurisdiction and shall clearly state and show that the areas so joined or used shall be maintained as a unit during the time they are so used. Such an agreement shall be recorded in the office of the County Recorder as part of the conditions of ownership of said properties, and shall be binding on heirs, successors, and assigns to such properties. A copy of the instrument recording such proceedings shall be filed with the Authority Having Jurisdiction.

722.0 Abandoned Sewers and Sewage Disposal Facilities.

722.1 Building (House) Sewer. An abandoned building (house) sewer, or part thereof, shall be plugged or capped in an approved manner within 5 feet (1524 mm) of the property line.

722.2 Cesspools, Septic Tanks, and Seepage Pits. A cesspool, septic tank, and seepage pit that has been abandoned or has been discontinued otherwise from further use, or to which no waste or soil pipe from a plumbing fixture is connected, shall have the sewage removed therefrom and be completely filled with earth, sand, gravel, concrete, or other approved material.

722.3 Filling. The top cover or arch over the cesspool, septic tank, or seepage pit shall be removed before filling, and the filling shall not extend above the top of the vertical portions of the sidewalls or above the level of the outlet pipe until inspection has been called and the cesspool, septic tank, or seepage pit has been inspected. After such inspection, the cesspool, septic tank, or seepage pit shall be filled to the level of the top of the ground.

722.4 Ownership. No person owning or controlling a cesspool, septic tank, or seepage pit on the premises of such person or in that portion of a public street, alley, or other public property abutting such premises, shall fail, refuse, or neglect to comply with the provisions of this section or upon receipt of notice so to comply from the Authority Having Jurisdiction.

722.5 Disposal Facilities. Where disposal facilities are abandoned consequent to connecting premises with the public sewer, the permittee making the connection shall fill abandoned facilities in accordance with the Authority Having Jurisdiction within 30 days from the time of connecting to the public sewer.

723.0 Building Sewer Test.

723.1 General. Building sewers shall be tested by plugging the end of the building sewer at its points of connection to the public sewer or private sewage disposal system and completely filling the building sewer with water from the lowest to the highest point thereof, or by approved equivalent low-pressure air test. Plastic DWV piping systems shall not be tested by the air test method. The building sewer shall be watertight.

CHAPTER 8
INDIRECT WASTES

801.0 General.

801.1 Applicability. This chapter shall govern the materials, design, and installation of indirect waste piping, receptors, and connections; and provisions for discharge and disposal of condensate wastes, chemical wastes, industrial wastes, and clear water wastes.

801.2 Air Gap or Air Break Required. Indirect waste piping shall discharge into the building drainage system through an air gap or air break as set forth in this code. Where a drainage air gap is required by this code, the minimum vertical distance as measured from the lowest point of the indirect waste pipe or the fixture outlet to the flood-level rim of the receptor shall be not less than 1 inch (25.4 mm).

801.3 Food and Beverage Handling Establishments. Establishments engaged in the storage, preparation, selling, serving, processing, or other handling of food and beverage involving the following equipment that requires drainage shall provide indirect waste piping for refrigerators, refrigeration coils, freezers, walk-in coolers, iceboxes, ice-making machines, steam tables, egg boilers, coffee urns and brewers, hot-and-cold drink dispensers, and similar equipment.

801.3.1 Size of Indirect Waste Pipes. Except for refrigeration coils and ice-making machines, the size of the indirect waste pipe shall be not smaller than the drain on the unit, but shall be not smaller than 1 inch (25 mm), and the maximum developed length shall not exceed 15 feet (4572 mm). Indirect waste pipe for ice-making machines shall be not less than the drain on the unit and in no case less than ¾ of an inch (20 mm).

801.3.2 Walk-In Coolers. For walk-in coolers, floor drains shall be permitted to be connected to a separate drainage line discharging into an outside receptor. The flood-level rim of the receptor shall be not less than 6 inches (152 mm) lower than the lowest floor drain. Such floor drains shall be trapped and individually vented. Cleanouts shall be provided at 90 degree (1.57 rad) turns and shall be accessibly located. Such waste shall discharge through an air gap or air break into a trapped and vented receptor, except that a full-size air gap is required where the indirect waste pipe is under vacuum.

801.3.3 Food-Handling Fixtures. Food-preparation sinks, steam kettles, potato peelers, ice cream dipper wells, and similar equipment shall be indirectly connected to the drainage system by means of an air gap. Bins, sinks, and other equipment having drainage connections and used for the storage of unpackaged ice used for human ingestion, or used in direct contact with ready-to-eat food, shall be indirectly connected to the drainage system by means of an air gap. Each indirect waste pipe from food-handling fixtures or equipment shall be separately piped to the indirect waste receptor and shall not combine with other indirect waste pipes. The piping from the equipment to the receptor shall be not less than the drain on the unit and in no case less than ½ of an inch (15 mm).

801.4 Bar and Fountain Sink Traps. Where the sink in a bar, soda fountain, or counter is so located that the trap serving the sink cannot be vented, the sink drain shall discharge through an air gap or air break (see Section 801.3.3) into an approved receptor that is vented. The developed length from the fixture outlet to the receptor shall not exceed 5 feet (1524 mm).

801.5 Connections from Water Distribution System. Indirect waste connections shall be provided for drains, overflows, or relief pipes from potable water pressure tanks, water heaters, boilers, and similar equipment that is connected to the potable water distribution system. Such indirect waste connections shall be made using a water-distribution air gap constructed in accordance with Table 603.3.1.

801.6 Sterilizers. Lines, devices, or apparatus such as stills, sterilizers, and similar equipment requiring waste connections and used for sterile materials shall be indirectly connected using an air gap. Each such indirect waste pipe shall be separately piped to the receptor and shall not exceed 15 feet (4572 mm). Such receptors shall be located in the same room.

801.7 Drip or Drainage Outlets. Appliances, devices, or apparatus not regularly classified as plumbing fixtures, but which have a drip or drainage outlets, shall be permitted to be drained by indirect waste pipes discharging into an open receptor through either an air gap or air break (see Section 801.3.1).

802.0 Approvals.

802.1 General. No plumbing fixtures served by indirect waste pipes or receiving discharge therefrom shall be installed until first approved by the Authority Having Jurisdiction.

803.0 Indirect Waste Piping.

803.1 Materials. Pipe, tube, and fittings conveying indirect waste shall be of such materials and design as to perform their intended function to the satisfaction of the Authority Having Jurisdiction.

803.2 Copper and Copper Alloys. Joints and connections in copper and copper alloy pipe and tube shall be installed in accordance with Section 705.3.

803.3 Pipe Size and Length. Except as hereinafter provided, the size of indirect waste piping shall be in accordance with other sections of this code applicable to drainage and vent piping. No vent from indirect waste piping shall combine with a sewer-connected vent, but shall extend separately to the outside air. Indirect waste pipes exceeding 5 feet (1524 mm), but less than 15 feet (4572 mm) in length shall be directly trapped, but such traps need not be vented.

Indirect waste pipes less than 15 feet (4572 mm) in length shall be not less than the diameter of the drain outlet or tailpiece of the fixture, appliance, or equipment served, and in no case less than ½ of an inch (15 mm). Angles and changes of direction in such indirect waste pipes shall be provided with cleanouts to permit flushing and cleaning.

804.0 Indirect Waste Receptors.

804.1 Standpipe Receptors. Plumbing fixtures or other receptors receiving the discharge of indirect waste pipes shall be approved for the use proposed and shall be of such shape and capacity as to prevent splashing or flooding and shall be located where they are readily accessible for inspection and cleaning. No standpipe receptor for a clothes washer shall extend more than 30 inches (762 mm), or not less than 18 inches (457 mm) above its trap weir. No trap for a clothes washer standpipe receptor shall be installed below the floor, but shall be roughed in not less than 6 inches (152 mm) and not more than 18 inches (457 mm) above the floor. No indirect waste receptor shall be installed in a toilet room, closet, cupboard, or storeroom, or in a portion of a building not in general use by the occupants thereof; except standpipes for clothes washers shall be permitted to be installed in toilet and bathroom areas where the clothes washer is installed in the same room.

805.0 Pressure Drainage Connections.

805.1 General. Indirect waste connections shall be provided for drains, overflows, or relief vents from the water supply system, and no piping or equipment carrying wastes or producing wastes or other discharges under pressure shall be directly connected to a part of the drainage system.

The preceding shall not apply to an approved sump pump or to an approved pressure-wasting plumbing fixture or device where the Authority Having Jurisdiction has been satisfied that the drainage system is adequately sized to accommodate the anticipated discharge thereof.

806.0 Sterile Equipment.

806.1 General. Appliances, devices, or apparatus such as stills, sterilizers, and similar equipment requiring water and waste and used for sterile materials shall be drained through an air gap.

807.0 Appliances.

807.1 Non-Classed Apparatus. Commercial dishwashing machines, silverware washing machines, and other appliances, devices, equipment, or other apparatus not regularly classed as plumbing fixtures, which are equipped with pumps, drips, or drainage outlets, shall be permitted to be drained by indirect waste pipes discharging into an approved type of open receptor.

807.2 Undiluted Condensate Waste. Where undiluted condensate waste from a fuel-burning condensing appliance is discharged into the drainage system, the material in the drainage system shall be cast-iron, galvanized iron, plastic, or other materials approved for this use.

Exceptions:

(1) Where the above condensate is discharged to an exposed fixture tailpiece and trap, such tailpiece and trap shall be permitted to be a copper alloy.

(2) Materials approved in Section 701.0 shall be permitted to be used where data is provided that the condensate waste is adequately diluted.

807.3 Domestic Dishwashing Machine. No domestic dishwashing machine shall be directly connected to a drainage system or food waste disposer without the use of an approved dishwasher air gap fitting on the discharge side of the dishwashing machine. Listed air gaps shall be installed with the flood-level (FL) marking at or above the flood level of the sink or drainboard, whichever is higher.

808.0 Cooling Water.

808.1 General. Where permitted by the Authority Having Jurisdiction, clean running water used exclusively as a cooling medium in an appliance, device, or apparatus shall be permitted to discharge into the drainage system through the inlet side of a fixture trap in the event that a suitable fixture is not available to receive such discharge. Such trap connection shall be by means of a pipe connected to the inlet side of an approved fixture trap, the upper end terminating in a funnel-shaped receptacle set adjacent, and not less than 6 inches (152 mm) above the overflow rim of the fixture.

809.0 Drinking Fountains.

809.1 General. Drinking fountains shall be permitted to be installed with indirect wastes.

810.0 Steam and Hot Water Drainage Condensers and Sumps.

810.1 High-Temperature Discharge. No steam pipe shall be directly connected to plumbing or drainage system, nor shall water having a temperature above 140°F (60°C) be discharged under pressure directly into a drainage system. Pipes from boilers shall discharge by means of indirect waste piping as determined by the Authority Having Jurisdiction or the boiler manufacturer's recommendations. Such pipes shall be permitted to be indirectly connected by discharging into an open or closed condenser or an intercepting sump of an approved type that will prevent the entrance of steam or such water under pressure into the drainage system. Closed condensers or sumps shall be provided with a vent that shall be taken off the top and extended separately, full size above the roof. Condensers and sumps shall be properly trapped at the outlet with a deep seal trap extending to within 6 inches (152 mm) of the bottom of the tank. The top of the deep seal trap shall have a ¾ of an inch (19.1 mm) opening located at the highest point of the trap to serve as a siphon breaker. Outlets shall be taken off from the side in such a manner as to allow a waterline to be maintained that will permanently occupy not less than one-half the capacity of the condenser or sump. Inlets shall enter above the waterline. Wearing plates or baffles shall be installed in the tank to protect the shell. The sizes of the blowoff line inlet, the water outlets, and the vent shall be as shown in Table 810.1. The contents of condensers receiving steam or hot water under pressure shall pass through an open sump before entering the drainage system.

TABLE 810.1
PIPE CONNECTIONS IN BLOWOFF CONDENSERS AND SUMPS
(inches)

BOILER BLOWOFF	WATER OUTLET	VENT
¾*	¾*	2
1	1	2½
1¼	1¼	3
1½	1½	4
2	2	5
2½	2½	6

For SI units: 1 inch = 25 mm

* To be used only with boilers of 100 square feet (9.29 m^2) of heating surface or less.

810.2 Sumps, Condensers, and Intercepting Tanks. Sumps, condensers, or intercepting tanks that are constructed of concrete shall have walls and bottom, not less than 4 inches (102 mm) in thickness, and the inside shall be cement plastered not less than ½ of an inch (12.7 mm) in thickness. Condensers constructed of metal shall be not less than No. 12 U.S. standard gauge (0.109 inch) (2.77 mm), and such metal condensers shall be protected from external corrosion by an approved bituminous coating.

810.3 Cleaning. Sumps and condensers shall be provided with suitable means of access for cleaning and shall contain a volume of not less than twice the volume of water removed from the boiler or boilers connected to it where the normal water level of such boiler or boilers is reduced not less than 4 inches (102 mm).

810.4 Strainers. An indirect waste interceptor is receiving discharge-containing particles that would clog the receptor drain shall have a readily removable beehive strainer.

811.0 Chemical Wastes.

811.1 Pretreatment. Chemical or liquid industrial wastes that are likely to damage or increase maintenance costs on the sanitary sewer system, detrimentally affect sewage treatment or contaminate surface or subsurface waters shall be pretreated to render them innocuous before discharge into a drainage system. Detailed construction documents of the pretreatment facilities shall be required by the Authority Having Jurisdiction.

Piping conveying industrial, chemical, or process wastes from their point of origin to sewer-connected pretreatment facilities shall be of such material and design as to adequately perform its intended function to the satisfaction of the Authority Having Jurisdiction. Drainage discharge piping from pretreatment facilities or interceptors shall be in accordance with standard drainage installation procedures.

Copper or copper alloy tube shall not be used for chemical or industrial wastes as defined in this section.

811.2 Waste and Vent Pipes. Each waste pipe receiving or intended to receive the discharge of a fixture into which acid or corrosive chemical is placed, and each vent pipe connected thereto, shall be constructed of chlorinated polyvinyl chloride (CPVC), polypropylene (PP), polyvinylidene fluoride (PVDF), chemical-resistant glass, high-silicon iron pipe, or lead pipe with a wall thickness of not less than ⅛ of an inch (3.2 mm); an approved type of ceramic glazed or unglazed vitrified clay; or other approved corrosion-resistant materials. CPVC pipe and fittings shall comply with ASTM F2618. PP pipe and fittings shall comply with ASTM F1412 or CSA B181.3. PVDF pipe and fittings shall comply with ASTM F1673 or CSA B181.3. Chemical-resistant glass pipe and fittings shall comply with ASTM C1053. High-silicon iron pipe and fittings shall comply with ASTM A861.

811.3 Joining Materials. Joining materials shall be of approved type and quality.

811.4 Access. Where practicable, the piping shall be readily accessible and installed with the maximum of clearance from other services.

811.5 Permanent Record. The owner shall make and keep a permanent record of the location of piping and venting carrying chemical waste.

811.6 Chemical Vent. No chemical vent shall intersect vents for other services.

811.7 Discharge. Chemical wastes shall be discharged in a manner approved by the Authority Having Jurisdiction.

811.8 Diluted Chemicals. The provisions of this section about materials and methods of construction shall not apply to installations such as photographic or x-ray darkrooms or research or control laboratories where minor amounts of adequately diluted chemicals are discharged.

812.0 Clear Water Wastes.

812.1 General. Water lifts, expansion tanks, cooling jackets, sprinkler systems, drip or overflow pans, or similar devices that discharge clear wastewater into the building drainage system shall discharge through an indirect waste.

813.0 Swimming Pools.

813.1 General. Pipes carrying wastewater from swimming or wading pools, including pool drainage and backwash from filters, shall be installed as an indirect waste. Where a pump is used to discharge pool waste water to the drainage system, the pump discharge shall be installed as an indirect waste.

814.0 Condensate Waste and Control.

814.1 Condensate Disposal. Condensate from air washers, air-cooling coils, condensing appliances, and the overflow from evaporative coolers and similar water-supplied equipment or similar air-conditioning equipment shall be collected and discharged to an approved plumbing fixture or disposal area. Where discharged into the drainage system, equipment shall drain using an indirect waste pipe. The waste pipe shall have a slope of not less than ⅛ inch per foot (10.4 mm/m) or 1 percent slope and shall be of an approved corrosion-resistant material not smaller than the outlet size in accordance with Section 814.3 or Section 814.4 for air-cooling coils or condensing appliances, respectively. Condensate or wastewater shall not drain over a public way.

814.1.1 Condensate Pumps. Where approved by the Authority Having Jurisdiction, condensate pumps shall be installed in accordance with the manufacturer's installation instructions. Pump discharge shall rise vertically to a point where it is possible to connect to a gravity condensate drain and discharged to an approved disposal point. Each condensing unit shall be provided with a separate sump and interlocked with the equipment to prevent the equipment from operating during a failure. Separate pumps shall be permitted to connect to a single gravity indirect waste where equipped with check valves and approved by the Authority Having Jurisdiction.

814.2 Condensate Control. Where any equipment or appliance is installed in a space where damage is capable of resulting from condensate overflow, a drain line shall be provided and shall be drained in accordance with Section 814.1. An additional protection method for condensate overflow shall be provided in accordance with one of the following:

(1) A water level detecting device that will shut off the equipment or appliance in the event the primary drain is blocked.

(2) An additional watertight pan of corrosion-resistant material, with a separate drain line, installed beneath the cooling coil, unit, or the appliance to catch the overflow condensate due to a clogged primary condensate drain.

(3) An additional drain line at a level that is higher than the primary drain line connection of the drain pan.

(4) An additional watertight pan of corrosion-resistant material with a water level detection device installed beneath the cooling coil, unit, or the appliance to catch the overflow condensate due to a clogged primary condensate drain and to shut off the equipment.

The additional pan or the additional drain line connection shall be provided with a drain pipe of not less than ¾ of an inch (20 mm) nominal pipe size, discharging at a point that is readily observed.

814.2.1 Protection of Appurtenances. Where insulation or appurtenances are installed where damage is capable of resulting from a condensate drain pan overfill, such installations shall occur above the rim of the drain pan with supports. Where the supports are in contact with the condensate waste, the supports shall be of approved corrosion-resistant material.

814.3 Condensate Waste Pipe Material and Sizing. Condensate waste pipes from air-cooling coils shall be sized in accordance with the equipment capacity as specified in Table 814.3. The material of the piping shall comply with the pressure and temperature rating of the appliance or equipment and shall be approved for use with the liquid being discharged.

TABLE 814.3
MINIMUM CONDENSATE PIPE SIZE

EQUIPMENT CAPACITY IN TONS OF REFRIGERATION	MINIMUM CONDENSATE PIPE DIAMETER (inches)
Up to 20	¾
21 – 40	1
41 – 90	1¼
91 – 125	1½
126 – 250	2

For SI units: 1 ton of refrigerant = 3.52 kW, 1 inch = 25 mm

The size of condensate waste pipes is for one unit or a combination of units, or as recommended by the manufacturer. The capacity of waste pipes assumes a ⅛ inch per foot (10.4 mm/m) or 1 percent slope, with the pipe running three-quarters full at the following pipe conditions:

Outside Air – 20%		Room Air – 80%	
DB	WB	DB	WB
90°F	73°F	75°F	62.5°F

For SI units: °C = (°F-32)/1.8

Condensate drain sizing for other slopes or other conditions shall be approved by the Authority Having Jurisdiction.

Air-conditioning waste pipes shall be constructed of materials specified in Chapter 7.

814.3.1 Cleanouts. Condensate drain lines shall be configured or provided with a cleanout to permit the clearing of blockages and for maintenance without requiring the drain line to be cut.

814.4 Appliance Condensate Drains. Condensate drain lines from individual condensing appliances shall be sized as required by the manufacturer's instructions. Condensate drain lines serving more than one appliance shall be approved by the Authority Having Jurisdiction prior to installation.

814.5 Point of Discharge. Air-conditioning condensate waste pipes shall connect indirectly, except where permitted in Section 814.6, to the drainage system through an air gap or air break to trapped and vented receptors, dry wells, leach pits, or the tailpiece of plumbing fixtures. A condensate drain shall be trapped in accordance with the appliance manufacturer's instructions or as approved.

814.6 Condensate Waste from Air-Conditioning Coils. Where the condensate waste from air-conditioning coils discharges by direct connection to a lavatory tailpiece or to an approved accessible inlet on a bathtub overflow, the connection shall be located in the area controlled by the same person controlling the air-conditioned space.

814.7 Plastic Fittings. Female plastic screwed fittings shall be used with male plastic fittings and plastic threads.

CHAPTER 9
VENTS

901.0 General.

901.1 Applicability. This chapter shall govern the materials, design, and installation of plumbing vent systems.

901.2 Vents Required. Each plumbing fixture trap, except as otherwise provided in this code, shall be protected against siphonage and backpressure, and air circulation shall be ensured throughout all parts of the drainage system by means of vent pipes installed in accordance with the requirements of this chapter and as otherwise required by this code.

901.3 Trap Seal Protection. The vent system shall be designed to prevent a trap seal from being exposed to a pressure differential that exceeds 1 inch water column (0.24 kPa) on the outlet side of the trap.

902.0 Vents Not Required.

902.1 Interceptor. Vent piping shall be permitted to be omitted on an interceptor where such interceptor acts as a primary settling tank and discharges through a horizontal indirect waste pipe into a secondary interceptor. The second interceptor shall be properly trapped and vented.

902.2 Bars, Soda Fountains, and Counter. Traps serving sinks that are part of the equipment of bars, soda fountains, and counters need not be vented where the location and construction of such bars, soda fountains, and counters are such as to make it impossible to do so. Where such conditions exist, said sinks shall discharge using approved indirect waste pipes into a floor sink or other approved type of receptor.

903.0 Materials.

903.1 Applicable Standards. Vent pipe and fittings shall comply with the applicable standards referenced in Table 701.2, except that:

(1) No galvanized steel or 304 stainless steel pipe shall be installed underground and shall be not less than 6 inches (152 mm) aboveground.

(2) ABS and PVC DWV piping installations shall be in accordance with Chapter 14 "Firestop Protection." Except for individual single-family dwelling units, materials exposed within ducts or plenums shall have a flame-spread index of not more than 25 and a smoke-developed index of not more than 50 where tested in accordance with ASTM E84 or UL 723. Plastic piping installed in plenums shall be tested in accordance with all requirements of ASTM E84 or UL 723. Mounting methods, supports and sample sizes of materials for testing that are not specified in ASTM E84 or UL 723 shall be prohibited.

903.2 Use of Copper or Copper Alloy Tubing. Copper or copper alloy tube for underground drainage and vent piping shall have a weight of not less than that of copper or copper alloy drainage tube type DWV.

903.2.1 Aboveground. Copper or copper alloy tube for aboveground drainage and vent piping shall have a weight of not less than that of copper or copper alloy drainage tube type DWV.

903.2.2 Prohibited Use. Copper or copper alloy tube shall not be used for chemical or industrial wastes as defined in Section 811.0.

903.2.3 Marking. Copper or copper alloy tubing, in addition to the required incised marking, shall be marked in accordance with either ASTM B306 or ASTM B88. The colors shall be Type K, green; Type L, blue; Type M, red; and Type DWV, yellow.

903.3 Changes in Direction. Changes in the direction of vent piping shall be made by the appropriate use of approved fittings, and no such pipe shall be strained or bent. Burred ends shall be reamed to the full bore of the pipe.

904.0 Size of Vents.

904.1 Size. The size of vent piping shall be determined from its length and the total number of fixture units connected thereto, in accordance with Table 703.2. The diameter of an individual vent shall be not less than 1¼ inches (32 mm) nor less than one-half the diameter of the drain to which it is connected. In addition, the drainage piping of each building and each connection to a public sewer or a private sewage disposal system shall be vented by means of one or more vent pipes, the aggregate cross-sectional area of which shall be not less than that of the largest required building sewer as determined from Table 703.2. Vent pipes from fixtures located upstream from pumps, ejectors, backwater valves, or other devices that obstruct the free flow of air and other gases between the building sewer and the outside atmosphere shall not be used for meeting the cross-sectional area venting requirements of this section.

Exception: Where connected to a common building sewer, the drainage piping of two or more buildings located on the same lot and under one ownership shall be permitted to be vented by means of piping sized in accordance with Table 703.2, provided the aggregate cross-sectional area of vents is not less than that of the largest required common building sewer.

904.2 Length. Not more than one-third of the total permitted length, in accordance with Table 703.2, of a minimum-sized vent shall be installed in a horizontal position. Where a minimum-sized vent is increased one pipe size for its entire length, the maximum length limitation shall not apply.

905.0 Vent Pipe Grades and Connections.

905.1 Grade. Vent and branch vent pipes shall be free from drops or sags, and each such vent shall be level or shall be so graded and connected as to drip back by gravity to the drainage pipe it serves.

905.2 Horizontal Drainage Pipe. Where vents connect to a horizontal drainage pipe, each vent pipe shall have its invert taken off above the drainage centerline of such pipe downstream of the trap being served.

905.3 Vent Pipe Rise. Unless prohibited by structural conditions, each vent shall rise vertically to a point not less than 6 inches (152 mm) above the flood-level rim of the fixture served before offsetting horizontally, and where two or more vent pipes converge, each such vent pipe shall rise to a point not less than 6 inches (152 mm) in height above the flood-level rim of the plumbing fixture it serves before being connected to any other vent. Vents less than 6 inches (152 mm) above the flood-level rim of the fixture shall be installed with approved drainage fittings, material, and grade to the drain.

905.4 Roof Termination. Vent pipes shall extend undiminished in size above the roof, or shall be reconnected with soil or waste vent of the proper size.

905.5 Location of Opening. The vent pipe opening from soil or waste pipe, except for water closets and similar fixtures, shall not be below the weir of the trap.

905.6 Common Vertical Pipe. Two fixtures shall be permitted to be served by a common vertical pipe where each such fixture wastes separately into an approved double fitting having inlet openings at the same level.

906.0 Vent Termination.

906.1 Roof Termination. Each vent pipe or stack shall extend through its flashing and shall terminate vertically not less than 6 inches (152 mm) above the roof nor less than 1 foot (305 mm) from a vertical surface. ABS and PVC piping exposed to sunlight shall be protected by water based synthetic latex paints.

906.2 Clearance. Each vent shall terminate not less than 10 feet (3048 mm) from, or not less than 3 feet (914 mm) above, an openable window, door, opening, air intake, or vent shaft, or not less than 3 feet (914 mm) in every direction from a lot line, alley and street excepted.

906.3 Use of Roof. Vent pipes shall be extended separately or combined, of full required size, not less than 6 inches (152 mm) above the roof or firewall. Flagpoling of vents shall be prohibited except where the roof is used for assembly purposes or parking. Vents within 10 feet (3048 mm) of a part of the roof that is used for assembly purposes or parking shall extend not less than 7 feet (2134 mm) above such roof and shall securely stay.

906.4 Outdoor Installations. Vent pipes for outdoor installations shall extend not less than 10 feet (3048 mm) above the surrounding ground and shall be securely supported.

906.5 Joints. Joints at the roof around vent pipes shall be made watertight by the use of approved flashings or flashing material.

906.6 Lead. (See Chapter 17) Sheet lead shall comply with the following:

(1) For safe pans – not less than 4 pounds per square foot (lb/ft^2) (19 kg/m^2) or 1⁄16 of an inch (1.6 mm) thick.

(2) For flashings or vent terminals – not less than 3 lb/ft^2 (15 kg/m^2) or 0.0472 of an inch (1.2 mm) thick.

(3) Lead bends and lead traps shall be not less than 1⁄8 of an inch (3.2 mm) in wall thickness.

906.7 Frost or Snow Closure. Where frost or snow closure is likely to occur in locations having minimum design temperature below 0°F (-17.8°C), vent terminals shall be not less than 2 inches (50 mm) in diameter, but in no event smaller than the required vent pipe. The change in diameter shall be made inside the building not less than 1 foot (305 mm) below the roof in an insulated space and terminate not less than 10 inches (254 mm) above the roof, or in accordance with the Authority Having Jurisdiction.

907.0 Vent Stacks and Relief Vents.

907.1 Drainage Stack. Each drainage stack that extends 10 or more stories shall be served by a parallel vent stack, which shall extend undiminished in size from its upper terminal and connect to the drainage stack at or immediately below the lowest fixture drain. Each such vent stack shall also be connected to the drainage stack at each fifth floor, counting down from the uppermost fixture drain, using a yoke vent, the size of which shall be not less in diameter than either the drainage or the vent stack, whichever is smaller.

907.2 Yoke Vent. The yoke vent connection to the vent stack shall be placed not less than 42 inches (1067 mm) above the floor level, and the yoke vent connection to the drainage stack shall be using a wye-branch fitting placed below the lowest drainage branch connection serving that floor.

908.0 Wet Venting.

908.1 Vertical Wet Venting. Wet venting is limited to vertical drainage piping receiving the discharge from the trap arm of one and two fixture unit fixtures that also serves as a vent not exceeding four fixtures. Wet-vented fixtures shall be within the same story; provided, further, that fixtures with a continuous vent discharging into a wet vent shall be within the same story as the wet-vented fixtures. No wet vent shall exceed 6 feet (1829 mm) in developed length.

908.1.1 Size. The vertical piping between two consecutive inlet levels shall be considered a wet-vented section. Each wet-vented section shall be not less than one pipe size exceeding the required minimum waste pipe size of the upper fixture or shall be one pipe size exceeding the required minimum pipe size for the sum of the fixture units served by such wet-vented section, whichever is larger, but in no case less than 2 inches (50 mm) in diameter.

908.1.2 Vent Connection. Common vent sizing shall be the sum of the fixture units served but, in no case, smaller than the minimum vent pipe size required for a fixture served, or by Section 904.0.

908.2 Horizontal Wet Venting for a Bathroom Group. A bathroom group located on the same floor level shall be permitted to be vented by a horizontal wet vent where all of the conditions of Section 908.2.1 through Section 908.2.5 are met.

908.2.1 Vent Connection. The dry vent connection to the wet vent shall be an individual vent for the bidet, shower, or bathtub. One or two vented lavatory(s) shall be permitted to serve as a wet vent for a bathroom group. Only one wet-vented fixture drain or trap arm shall discharge upstream of the dry-vented fixture drain connection. Dry vent connections to the horizontal wet vent shall be in accordance with Section 905.2 and Section 905.3.

908.2.2 Size. The wet vent shall be sized based on the fixture unit discharge into the wet vent. The wet vent shall be not less than 2 inches (50 mm) in diameter for 4 drainage fixture units (dfu) or less, and not less than 3 inches (80 mm) in diameter for 5 dfu or more. The dry vent shall be sized in accordance with Table 702.1 and Table 703.2 based on the total fixture units discharging into the wet vent.

908.2.3 Trap Arm. The length of the trap arm shall not exceed the limits in Table 1002.2. The trap size shall be in accordance with Section 1003.3. The vent pipe opening from the horizontal wet vent, except for water closets and similar fixtures, shall not be below the weir of the trap.

908.2.4 Water Closet. The water closet fixture drain or trap arm connection to the wet vent shall be downstream of fixture drain or trap arm connections to the horizontal wet vent.

908.2.5 Additional Fixtures. Additional fixtures shall discharge downstream of the wet vent system and be conventionally vented. Only the fixtures within the bathroom group shall connect to the wet-vented horizontal branch.

909.0 Special Venting for Island Fixtures.

909.1 General. Traps for island sinks and similar equipment shall be roughed in above the floor and shall be permitted to be vented by extending the vent as high as possible, but not less than the drainboard height and then returning it downward and connecting it to the horizontal sink drain immediately downstream from the vertical fixture drain. The return vent shall be connected to the horizontal drain through a wye-branch fitting and shall, in addition, be provided with a foot vent taken off the vertical fixture vent by means of a wye branch immediately below the floor and extending to the nearest partition and then through the roof to the open air, or shall be permitted to be connected to other vents at a point not less than 6 inches (152 mm) above the flood-level rim of the fixtures served. Drainage fittings shall be used on the vent below the floor level, and a slope of not less than ¼ inch per foot (20.8 mm/m) back to the drain shall be maintained. The return bend used under the drainboard shall be a one-piece fitting or an assembly of a 45 degree (0.79 rad), a 90 degree (1.57 rad), and a 45 degree (0.79 rad) elbow in the order named. Pipe sizing shall be as elsewhere required in this code. The island sink drain, upstream of the returned vent, shall serve no other fixtures. An accessible cleanout shall be installed in the vertical portion of the foot vent.

910.0 Combination Waste and Vent Systems.

910.1 Where Permitted. Combination waste and vent systems shall be permitted where structural conditions preclude the installation of conventional systems as otherwise prescribed by this code.

910.2 Approval. Construction documents for each combination waste and vent system shall first be approved by the Authority Having Jurisdiction before a portion of such system is installed.

910.3 Vents. Each combination waste and vent system, as defined in Chapter 2, shall be provided with a vent or vents adequate to ensure free circulation of air. A branch exceeding 15 feet (4572 mm) in length shall be separately vented in an approved manner. The area of a vent installed in a combination waste and vent system shall be not less than one-half the inside cross-sectional area of the drain pipe served. The vent connection shall be downstream of the uppermost fixture.

910.4 Size. Each waste pipe and each trap in such a system shall be not less than two pipe sizes exceeding the sizes required by Chapter 7 of this code, and not less than two pipe sizes exceeding a fixture tailpiece or connection.

910.5 Vertical Waste Pipe. No vertical waste pipe shall be used in such a system, except the tailpiece or connection between the outlet of a plumbing fixture and the trap. Such tailpieces or connections shall be as short as possible, and in no case shall exceed 2 feet (610 mm).

Exception: Branch lines shall be permitted to have 45 degree (0.79 rad) vertical offsets.

910.6 Cleanouts. An accessible cleanout shall be installed in each vent for the combination waste and vent system. Cleanouts shall not be required on a wet-vented branch serving a single trap where the fixture tailpiece or connection is not less than 2 inches (50 mm) in diameter and provides ready access for cleaning through the trap.

910.7 Fixtures. No water closet or urinal shall be installed on such a system. Other one, two, or three unit fixtures remotely located from the sanitary system and adjacent to a combination waste and vent system shall be permitted to be connected to such system in the conventional manner by means of waste and vent pipes of regular sizes, providing that the two pipe size increase required in Section 910.4 is based on the total fixture unit load connected to the system.

See Appendix B of this code for explanatory notes on the design of combination waste and vent systems.

911.0 Circuit Venting.

911.1 Circuit Vent Permitted. A maximum of eight floor-outlet water closets, showers, bathtubs, or floor drains connected to a horizontal branch shall be permitted to be circuit vented. Each trap arm shall connect horizontally to the horizontal branch being circuit vented in accordance with Table 1002.2. The horizontal branch shall be classified as a drain and a vent from the most downstream trap arm connection to the most upstream trap arm connection to the horizontal branch.

Exception: Back-outlet and wall-hung water closets shall be permitted to be circuit vented provided that no floor-outlet fixtures are connected to the same horizontal branch.

911.2 Circuit Vent Size and Connection. The circuit vent size shall be in accordance with Table 703.2 according to the number of circuit vented fixtures connected to the horizontal branch but shall be not less than 2 inches (50 mm) in diameter. The vent shall connect to the horizontal branch on the vertical between the two most upstream trap arms. The circuit vent pipe shall not receive the discharge of soil or waste.

911.2.1 Multiple Circuit Vents. When multiple circuit vents are interconnected according to Section 911.4.1, each individual circuit vent shall be sized according to Section 911.2. The vent pipe connecting each circuit vent shall be sized according to Table 703.2.

911.3 Relief Vent. A 2 inch (50 mm) relief vent shall be provided for circuit-vented horizontal branches receiving the discharge of four or more water closets when connecting to a drainage stack that receives the discharge of soil or waste from upper horizontal branches.

911.3.1 Connection and Installation. The relief vent shall connect to the horizontal branch between the stack and the most downstream trap arm of the circuit vent. The relief vent shall be installed on the vertical to the horizontal branch.

911.3.2 Fixture Drain. The relief vent is permitted to serve as a fixture drain. Fixtures discharging to a relief vent shall be one or two fixture unit fixtures but shall not exceed a total of 4 fixture units.

911.4 Slope and Size of Horizontal Branch. The vented section of the horizontal branch shall be uniformly sloped and not more than 1 inch per foot (83.3 mm/m). The entire length of the vented section of the horizontal branch shall be sized for the total drainage discharge to the branch according to Table 703.2.

911.4.1 Multiple Circuit-Vented Branches. Circuit-vented horizontal branches are permitted to be connected together. Each group of a maximum of eight fixtures shall be considered a separate circuit vent and shall be in accordance with Section 911.4.1.1 and Section 911.4.1.2.

911.4.1.1 Size of Parallel Horizontal Branches. Parallel horizontal circuit vented branches shall be permitted to connect on the same floor level. Each separate circuit-vented horizontal branch that is interconnected shall be sized independently in accordance with Section 911.4.

911.4.1.2 Size of Continuous Horizontal Branches. Two or more circuit vented systems continuous on the same horizontal branch shall be uniformly sized for the total discharge into the branch.

911.5 Additional Fixtures. Fixtures, other than the circuit-vented fixtures, are permitted to discharge to the horizontal branch drain. Such fixtures shall be located on the same floor as the circuit-vented fixtures and shall be either individually or common vented.

912.0 Engineered Vent System.

912.1 General. The design and sizing of a vent system shall be permitted to be determined by accepted engineering practices. The system shall be designed by a registered design professional and approved in accordance with Section 301.5.

912.2 Minimum Requirements. An engineered vent system shall provide protection of the trap seal in accordance with Section 901.3.

CHAPTER 10

TRAPS AND INTERCEPTORS

1001.0 General.

1001.1 Applicability. This chapter shall govern the materials, design, and installation of traps and interceptors.

1001.2 Where Required. Each plumbing fixture shall be separately trapped by an approved type of liquid seal trap. This section shall not apply to fixtures with integral traps. Not more than one trap shall be permitted on a trap arm. Food waste disposers installed with a set of restaurant, commercial, or industrial sinks shall be connected to a separate trap. Each domestic clothes washer and each laundry tub shall be connected to a separate and independent trap, except that a trap serving a laundry tub shall also be permitted to receive the waste from a clothes washer set adjacent to it. The vertical distance between a fixture outlet and the trap weir shall be as short as practicable, but in no case shall the tailpiece exceed 24 inches (610 mm) in length. One trap shall be permitted to serve a set of not more than three single compartment sinks or laundry tubs of the same depth or three lavatories immediately adjacent to each other and in the same room where the waste outlets are not more than 30 inches (762 mm) apart, and the trap is centrally located where three compartments are installed.

1002.0 Traps Protected by Vent Pipes.

1002.1 Vent Pipes. Each plumbing fixture trap, except as otherwise provided in this code, shall be protected against siphonage, backpressure, and air circulation shall be assured throughout the drainage system using a vent pipe installed in accordance with the requirements of this code.

1002.2 Fixture Traps. Each fixture trap shall have a protecting vent so located that the developed length of the trap arm from the trap weir to the inner edge of the vent shall be within the distance given in Table 1002.2 but in no case less than two times the diameter of the trap arm.

TABLE 1002.2
HORIZONTAL LENGTHS OF TRAP ARMS
(EXCEPT FOR WATER CLOSETS AND SIMILAR FIXTURES)[1, 2]

TRAP ARM PIPE DIAMETER (inches)	DISTANCE TRAP TO VENT MINIMUM (inches)	LENGTH MAXIMUM (inches)
1¼	2½	30
1½	3	42
2	4	60
3	6	72
4	8	120
Exceeding 4	2 x Diameter	120

For SI units: 1 inch = 25.4 mm

Notes:

[1] Maintain ¼ inch per foot slope (20.8 mm/m).

[2] The developed length between the trap of a water closet or similar fixture (measured from the top of the closet flange to the inner edge of the vent) and its vent shall not exceed 6 feet (1829 mm).

1002.3 Change of Direction. A trap arm shall be permitted to change direction without the use of a cleanout where such change of direction does not exceed 90 degrees (1.57 rad). Horizontal changes in the direction of trap arms shall be in accordance with Section 706.3.

Exception: For trap arms, 3 inches (80 mm) in diameter and larger, the change of direction shall not exceed 135 degrees (2.36 rad) without the use of a cleanout.

1002.4 Vent Pipe Opening. The vent pipe opening from soil or waste pipe, except for water closets and similar fixtures, shall not be below the weir of the trap.

1003.0 Traps – Described.

1003.1 General Requirements. Each trap, except for traps within an interceptor or similar device shall be self-cleaning. Traps for bathtubs, showers, lavatories, sinks, laundry tubs, floor drains, urinals, drinking fountains, dental units, and similar fixtures shall be of standard design, weight and shall be of ABS, cast-brass, cast-iron, lead, PP, PVC, or other approved material. An exposed and readily accessible drawn-copper alloy tubing trap, not less than 17 B & S Gauge (0.045 inch) (1.143 mm), shall be permitted to be used on fixtures discharging domestic sewage.

Exception: Drawn-copper alloy tubing traps shall not be used for urinals. Each trap shall have the manufacturer's name stamped legibly in the metal of the trap, and each tubing trap shall have the gauge of the tubing in addition to the manufacturer's name. A trap shall have a smooth and uniform interior waterway.

1003.2 Slip Joint Fittings. A maximum of one approved slip joint fitting shall be permitted to be used on the outlet side of a trap, and no tubing trap shall be installed without a listed tubing trap adapter. Listed plastic trap adapters shall be permitted to be used to connect listed metal tubing traps.

1003.3 Size. The size (nominal diameter) of a trap for a given fixture shall be sufficient to drain the fixture rapidly but in no case less than nor more than one pipe size larger than given in Table 702.1. The trap shall be the same size as the trap arm to which it is connected.

1004.0 Traps.

1004.1 Prohibited. No form of trap that depends for its seal upon the action of movable parts shall be used. No trap that has concealed interior partitions, except those of plastic, glass, or similar corrosion-resisting material, shall be used. "S" traps, bell traps, and crown-vented traps shall be prohibited. No fixture shall be double trapped. Drum and bottle traps shall be installed for special conditions. No trap shall be installed without a vent, except as otherwise provided in this code.

1004.2 Movable Parts. Bladders, check valves or another type of devices with moveable parts shall be prohibited to serve as a trap.

1005.0 Trap Seals.

1005.1 General. Each fixture trap shall have a liquid seal of not less than 2 inches (51 mm) and not more than 4 inches (102 mm), except where a deeper seal is found necessary by the Authority Having Jurisdiction. Traps shall be set true with respect to their liquid seals and, where necessary, they shall be protected from freezing.

1006.0 Floor Drain Traps.

1006.1 General. Floor drains shall connect into a trap so constructed that it can be readily cleaned and of a size to serve efficiently the purpose for which it is intended. The drain inlet shall be so located that it is in full view. Where subject to the reverse flow of sewage or liquid waste, such drains shall be equipped with an approved backwater valve.

1007.0 Trap Seal Protection.

1007.1 General. Floor drain or similar traps directly connected to the drainage system and subject to infrequent use shall be protected with a trap seal primer, except where not deemed necessary for safety or sanitation by the Authority Having Jurisdiction. Trap seal primers shall be accessible for maintenance.

1007.2 Trap Seal Primers. Potable water supply trap seal primer valves shall comply with ASSE 1018. Drainage and electronic design type trap seal primer devices shall comply with ASSE 1044.

1008.0 Building Traps.

1008.1 General. Building traps shall not be installed except where required by the Authority Having Jurisdiction. Each building trap where installed shall be provided with a cleanout and with a relieving vent or fresh-air intake on the inlet side of the trap, which needs not be larger than one-half the diameter of the drain to which it connects. Such relieving vent or fresh-air intake shall be carried above grade and terminate in a screened outlet located outside the building.

1009.0 Interceptors (Clarifiers) and Separators.

1009.1 Where Required. Interceptors (clarifiers) (including grease, oil, sand, solid interceptors, etc.) shall be required by the Authority Having Jurisdiction where they are necessary for the proper handling of liquid wastes containing grease, flammable wastes, sand, solids, acid or alkaline substances, or other ingredients harmful to the building drainage system, the public or private sewer, or to public or private sewage disposal.

1009.2 Approval. The size, type, and location of each interceptor (clarifier) or separator shall be approved by the Authority Having Jurisdiction. Except where otherwise specifically permitted, no wastes other than those requiring treatment or separation shall be discharged into an interceptor (clarifier).

1009.3 Design. Interceptors (clarifiers) for sand and similar heavy solids shall be so designed and located as to be readily accessible for cleaning and shall have a water seal of not less than 6 inches (152 mm).

1009.4 Relief Vent. Interceptors (clarifiers) shall be so designed that they will not become air-bound where closed covers are used. Each interceptor (clarifier) shall be properly vented.

1009.5 Location. Each interceptor (clarifier) cover shall be readily accessible for servicing and maintaining the interceptor (clarifier) in working and operating condition. The use of ladders or the removal of bulky equipment to service interceptors (clarifiers) shall constitute a violation of accessibility. Location of interceptors (clarifiers) shall be shown on the approved building plan.

1009.6 Maintenance of Interceptors. Interceptors shall be maintained in efficient operating condition by periodic removal of accumulated grease, scum, oil, or other floating substances and solids deposited in the interceptor.

1009.7 Discharge. The waste pipe from oil and sand interceptors shall discharge as approved by the Authority Having Jurisdiction.

1010.0 Slaughterhouses, Packing Establishments, etc.

1010.1 General. A fish, fowl, and animal slaughterhouse or establishment; a fish, fowl, and meat packing or curing establishment; a soap factory, tallow-rendering, fat-rendering, and a hide-curing establishment shall be connected to and shall drain or discharge into an approved grease interceptor (clarifier).

1011.0 Minimum Requirements for Auto Wash Racks.

1011.1 General. A private or public wash rack or floor or slab used for cleaning machinery or machine parts shall be adequately protected against storm or surface water and shall drain or discharge into an approved interceptor (clarifier).

1012.0 Commercial and Industrial Laundries.

1012.1 General. Laundry equipment in commercial and industrial buildings that do not have integral strainers shall discharge into an interceptor having a wire basket or similar device that is removable for cleaning and that will prevent passage into the drainage system of solids ½ of an inch (12.7 mm) or larger in maximum dimensions, such as string, rags, buttons, or other solid materials detrimental to the public sewerage system.

1013.0 Bottling Establishments.

1013.1 General. Bottling plants shall discharge their process wastes into an interceptor that will provide for the separation of broken glass or other solids, before discharging liquid wastes into the drainage system.

1014.0 Grease Interceptors.

1014.1 General. Where it is determined by the Authority Having Jurisdiction that waste pretreatment is required, an approved type of grease interceptor(s) shall comply with ASME A112.14.3, ASME A112.14.4, CSA B481, PDI G-101, or PDI G-102, and sized in accordance with Section 1014.2.1 or Section 1014.3.6, shall be installed in accordance with the manufacturer's installation instructions to receive the drainage from fixtures or equipment that produce grease-laden waste. Grease-laden waste fixtures shall include, but not be limited to, sinks and drains, such as floor drains, floor sinks, and other fixtures or equipment in serving establishments, such as restaurants, cafes, lunch counters, cafeterias, bars and clubs, hotels, hospitals, sanitariums, factory or school kitchens, or other establishments where grease is introduced into the drainage or sewage system in quantities that can effect line stoppage or hinder sewage treatment or private sewage disposal systems. A combination of hydromechanical, gravity grease interceptors and engineered systems shall be allowed to meet this code and other applicable requirements of the Authority Having Jurisdiction where space or existing physical constraints of existing buildings necessitate such installations. A grease interceptor shall not be required for individual dwelling units or private living quarters. Water closets, urinals, and other plumbing fixtures conveying human waste shall not drain into or through the grease interceptor.

1014.1.1 Trapped and Vented. Each fixture discharging into a grease interceptor shall be individually trapped and vented in an approved manner.

1014.1.2 Maintenance. Grease interceptors shall be maintained in efficient operating condition by periodic removal of the accumulated grease and latent material. No such collected grease shall be introduced into drainage piping or a public or private sewer. Where the Authority Having Jurisdiction determines that a grease interceptor is not being properly cleaned or maintained, the Authority Having Jurisdiction shall have the authority to mandate the installation of additional equipment or devices and to mandate a maintenance program.

1014.1.3 Food Waste Disposers and Dishwashers. No food waste disposer or dishwasher shall be connected to or discharge into a grease interceptor. Commercial food waste disposers shall be permitted to discharge directly into the building's drainage system.

Exception: Food waste disposers shall be permitted to discharge to grease interceptors that are designed to receive the discharge of food waste.

1014.2 Hydromechanical Grease Interceptors. Plumbing fixtures or equipment connected to a Type A and B hydromechanical grease interceptor shall discharge through an approved type of vented flow control installed in a readily accessible and visible location. Flow control devices shall be designed and installed so that the total flow through such device or devices shall at no time be greater than the rated flow of the connected grease interceptor. No flow control device having adjustable or removable parts shall be approved. The vented flow control device shall be located such that no system vent shall be between the flow control and the grease interceptor inlet. The vent or air inlet of the flow control device shall connect with the sanitary drainage vent system, as elsewhere required by this code, or shall terminate through the roof of the building, and shall not terminate to the free atmosphere inside the building.

Exception: Listed grease interceptors with integral flow controls or restricting devices shall be installed in an accessible location in accordance with the manufacturer's installation instructions.

1014.2.1 Capacity. The total capacity in gallons (gal) (L) of fixtures discharging into a hydromechanical grease interceptor shall not exceed two and one-half times the certified gallon per minute (gpm) (L/s) flow rate of the interceptor in accordance with Table 1014.2.1.

For this section, the term "fixture" shall mean and include each plumbing fixture, appliance, apparatus, or other equipment required to be connected to or discharged into a grease interceptor by a provision of this section.

1014.2.2 Vent. A vent shall be installed downstream of hydromechanical grease interceptors in accordance with the requirements of this code.

TABLE 1014.2.1
HYDROMECHANICAL GREASE INTERCEPTOR SIZING USING GRAVITY FLOW RATES[1]

DIAMETER OF GREASE WASTE PIPE (inches)	MAXIMUM FULL PIPE FLOW (gpm)[2]	SIZE OF GREASE INTERCEPTOR	
		ONE-MINUTE DRAINAGE PERIOD (gpm)	TWO-MINUTE DRAINAGE PERIOD (gpm)
2	20	20	10
3	60	75	35
4	125	150	75
5	230	250	125
6	375	400	200

For SI units: 1 inch = 25 mm, 1 gallon per minute = 0.06 L/s

Notes:

1 For interceptor sizing by the fixture capacity see the example below.

2 ¼ inch slope per foot (20.8 mm/m) based on Manning's formula with friction factor N = 0.012.

EXAMPLE 1014.2.1
SIZING HYDROMECHANICAL GREASE INTERCEPTOR(S) USING FIXTURE CAPACITY

Step 1: Determine the flow rate from each fixture.

[Length] X [Width] X [Depth] / [231] = Gallons X [0.75 fill factor] / [Drain Period (1 minute or 2 minutes)]

Step 2: Calculate the total load from fixtures that discharge into the interceptor.

FIXTURES	COMPARTMENTS	LOAD (gallons)	SIZE OF GREASE INTERCEPTOR ONE-MINUTE DRAINAGE PERIOD (gpm)	TWO-MINUTE DRAINAGE PERIOD (gpm)
Compartment size	—	—	—	—
24 inches x 24 inches x 12 inches	2	44.9	—	—
Hydrant	—	3	—	—
Rated Appliance	—	2	—	—
—	—	49.9	50	25

For SI units: 1 inch = 25.4 mm, 1 gallon per minute = 0.06 L/s, 1 gallon = 3.785 L

1014.3 Gravity Grease Interceptors. Required gravity grease interceptors shall comply with the provisions of Section 1014.3.1 through Section 1014.3.7.

1014.3.1 General. The provisions of this section shall apply to the design, construction, installation, and testing of commercial kitchen gravity grease interceptors.

1014.3.2 Waste Discharge Requirements. Waste discharge in establishments from fixtures and equipment which contain grease, including but not limited to, scullery sinks, pot and pan sinks, dishwashers, soup kettles, and floor drains located in areas where grease-containing materials exist, shall be permitted to be drained into the sanitary waste through the interceptor where approved by the Authority Having Jurisdiction.

1014.3.2.1 Toilets and Urinals. Toilets, urinals, and other similar fixtures shall not drain through the interceptor.

1014.3.2.2 Inlet Pipe. Waste shall enter the interceptor through the inlet pipe.

1014.3.3 Design. Gravity interceptors shall be constructed in accordance with the applicable standard in Chapter 17 or the design approved by the Authority Having Jurisdiction.

1014.3.4 Location. Each grease interceptor shall be so installed and connected that it shall be easily accessible for inspection, cleaning, and removal of the intercepted grease. A gravity grease interceptor that complies with IAPMO Z1001 shall not be installed in a building where food is handled. Location of the grease interceptor shall meet the approval of the Authority Having Jurisdiction.

1014.3.4.1 Interceptors. Interceptors shall be placed as close as practical to the fixtures they serve.

1014.3.4.2 Business Establishment. Each business establishment for which a gravity grease interceptor is required shall have an interceptor which shall serve that establishment unless otherwise approved by the Authority Having Jurisdiction.

1014.3.4.3 Access. Each gravity grease interceptor shall be located to be readily accessible to the equipment required for maintenance.

1014.3.5 Construction Requirements. Gravity grease interceptors shall be designed to remove grease from effluent and shall be sized in accordance with this section. Gravity grease interceptors shall also be designed to retain grease until accumulations can be removed by pumping the interceptor. It is recommended that a sample box is located at the outlet end of gravity grease interceptors so that the Authority Having Jurisdiction can periodically sample effluent quality.

1014.3.6 Sizing Criteria. The volume of the interceptor shall be determined by using Table 1014.3.6. Where drainage fixture units (DFUs) are not known, the interceptor shall be sized based on the maximum DFUs allowed for the pipe size connected to the inlet of the interceptor. Refer to Table 703.2, Drainage Piping, Horizontal.

TABLE 1014.3.6
GRAVITY GREASE INTERCEPTOR SIZING

DRAINAGE FIXTURE UNITS[1, 3] (DFUs)	INTERCEPTOR VOLUME[2] (gallons)
8	500
21	750
35	1000
90	1250
172	1500
216	2000
307	2500
342	3000
428	4000
576	5000
720	7500
2112	10 000
2640	15 000

For SI units: 1 gallon = 3.785 L

Notes:

[1] The maximum allowable DFUs plumbed to the kitchen drain lines that will be connected to the grease interceptor.

[2] This size is based on DFUs, the pipe size from this code; Table 703.2; Useful Tables for flow in half-full pipes (ref: Mohinder Nayyar Piping Handbook, 3rd Edition, 1992). Based on 30-minute retention time (ref.: George Tchobanoglous and Metcalf & Eddy. Wastewater Engineering Treatment, Disposal, and Reuse, 3rd Ed. 1991 & Ronald Crites and George Tchobanoglous. Small and Decentralized Wastewater Management Systems, 1998). Rounded up to nominal interceptor volume.

[3] Where the flow rate of directly connected fixture(s) or appliance(s) have no assigned DFU values, the additional grease interceptor volume shall be based on the known flow rate (gpm) (L/s) multiplied by 30 minutes.

EXAMPLE 1014.3.6
GRAVITY GREASE INTERCEPTOR SIZING EXAMPLE

Given: A restaurant with the following fixtures and equipment.

One food preparation sink; three-floor drains - one in the food prep area, one in the grill area, and one receiving the indirect waste from the ice machine and a mop sink.

Kitchen Drain Line DFU Count (from Table 702.1):

3 floor drains at 2 DFUs each	=	6 DFUs
Mop sink at 3 DFUs each	=	3 DFUs
Food prep sink at 3 DFUs each	=	3 DFUs
Total	=	12 DFUs

Using Table 1014.3.6, the grease interceptor will be sized at 750 gallons (2389 L).

1014.3.7 Abandoned Gravity Grease Interceptors. Abandoned grease interceptors shall be pumped and filled as required for abandoned sewers and sewage disposal facilities in Section 722.0.

1015.0 FOG (Fats, Oils, and Greases) Disposal System.

1015.1 Purpose. The purpose of this section is to provide the necessary criteria for the sizing, application, and installation of FOG disposal systems designated as a pretreatment or discharge water quality compliance strategy.

1015.2 Components, Materials, and Equipment. FOG disposal systems, including components, materials, and equipment necessary for the proper function of the system, shall comply with ASME A112.14.6.

1015.3 Sizing and Installation. FOG disposal systems shall be sized and installed in accordance with the manufacturer's installation instructions.

1015.4 Performance. FOG disposal systems shall produce an effluent quality not to exceed 5.84 grains per gallon (gr/gal) (100 mg/L) FOG.

1016.0 Sand Interceptors.

1016.1 Discharge. Where the discharge of a fixture or drain contains solids or semi-solids heavier than water that would be harmful to a drainage system or cause a stoppage within the system, the discharge shall be through a sand interceptor. Multiple floor drains shall be permitted to discharge into one sand interceptor.

1016.2 Authority Having Jurisdiction. Sand interceptors are required where the Authority Having Jurisdiction deems it advisable to have a sand interceptor to protect the drainage system.

1016.3 Construction and Size. Sand interceptors shall be built of brick or concrete, prefabricated coated steel, or other watertight material. The interceptor shall have an interior baffle for full separation of the interceptor into two sections. The outlet pipe shall be the same size as the inlet pipe of the sand interceptor, the minimum being 3 inches (80 mm), and the baffle shall have two openings of the same diameter as the outlet pipe and at the same invert as the outlet pipe. These openings shall be staggered so that there cannot be a straight line flow between the inlet pipe and the outlet pipe. The invert of the inlet pipe shall be no lower than the invert of the outlet pipe.

The sand interceptor shall have a minimum dimension of 2 square feet (0.2 m^2) for the net free opening of the inlet section and a minimum depth under the invert of the outlet pipe of 2 feet (610 mm).

For each 5 gpm (0.3 L/s) flow or fraction thereof over 20 gpm (1.26 L/s), the area of the sand interceptor inlet section is to be increased by 1 square foot (0.09 m^2). The outlet section shall at all times have a minimum area of 50 percent of the inlet section.

The outlet section shall be covered by a solid removable cover, set flush with the finished floor, and the inlet section shall have an open grating, set flush with the finished floor and suitable for the traffic in the area in which it is located.

1016.4 Separate Use. Sand and similar interceptors for every solid shall be so designed and located as to be readily accessible for cleaning, shall have a water seal of not less than 6 inches (152 mm), and shall be vented.

1017.0 Oil and Flammable Liquid Interceptors.

1017.1 Interceptors Required. Repair garages and gasoline stations with grease racks or grease pits, and factories that have oily, flammable, or both types of wastes as a result of manufacturing, storage, maintenance, repair, or testing processes, shall be provided with an oil or flammable liquid interceptor that shall be connected to necessary floor drains. The separation or vapor compartment shall be independently vented to the outer air. Where two or more separation or vapor compartments are used, each shall be vented to the outer air or shall be permitted to connect to a header that is installed at a minimum of 6 inches (152 mm) above the spill line of the lowest floor drain and vented independently to the outer air. The minimum size of a flammable vapor vent shall be not less than 2 inches (50 mm), and, where vented through a sidewall, the vent shall be not less than 10 feet (3048 mm) above the adjacent level at an approved location. The interceptor shall be vented on the sewer side and shall not connect to a flammable vapor vent. Oil and flammable interceptors shall be provided with gastight cleanout covers that shall be readily accessible. The waste line shall be not less than 3 inches (80 mm) in diameter with a full-size cleanout to grade. Where an interceptor is provided with an overflow, it shall be provided with an overflow line [not less than 2 inches (50 mm) in diameter] to an approved waste oil tank having a minimum capacity of 550 gallons (2082 L) and meeting the requirements of the Authority Having Jurisdiction. The waste oil from the separator shall flow by gravity or shall be pumped to a higher elevation by an automatic pump. Pumps shall be adequately sized and accessible. Waste oil tanks shall have a 2 inch (50 mm) minimum pump-out connection at grade and an 1½ inch (40 mm) minimum vent to atmosphere at an approved location not less than 10 feet (3048 mm) above grade.

1017.2 Design of Interceptors. Each manufactured interceptor that is rated shall be stamped or labeled by the manufacturer with an indication of its full discharge rate in gpm (L/s). The full discharge rate to such an interceptor shall be determined at full flow. Each interceptor shall be rated equal to or greater than the incoming flow and shall be provided with an overflow line to an underground tank.

Interceptors not rated by the manufacturer shall have a depth of not less than 2 feet (610 mm) below the invert of the discharge drain. The outlet opening shall have not less than an 18 inch (457 mm) water seal and shall have a minimum capacity as follows: Where not more than three motor vehicles are serviced, stored, or both, interceptors shall have a minimum capacity of 6 cubic feet (0.2 m^3), and 1 cubic foot (0.03 m^3) of capacity shall be added for each vehicle up to 10 vehicles. Above 10 vehicles, the Authority Having Jurisdiction shall determine the size of the interceptor required. Where vehicles are serviced and not stored, interceptor capacity shall be based on a net capacity of 1 cubic foot (0.03 m^3) for each 100 square feet (9.29 m^2) of the surface to be drained into the interceptor, with a minimum of 6 cubic feet (0.2 m^3).

CHAPTER 11

STORM DRAINAGE

1101.0 General.

1101.1 Applicability. This chapter shall govern the materials, design, and installation of storm water drainage systems.

1101.2 Where Required. Roofs, paved areas, yards, courts, courtyards, vent shafts, light wells, or similar areas having rainwater, shall be drained into a separate storm sewer system, or into a combined sewer system where a separate storm sewer system is not available, or to some other place of disposal satisfactory to the Authority Having Jurisdiction. In the case of one- and two-family dwellings, storm water shall be permitted to be discharged on flat areas, such as streets or lawns, so long as the storm water shall flow away from the building and away from adjoining property, and shall not create a nuisance.

1101.3 Storm Water Drainage to Sanitary Sewer Prohibited. Storm water shall not be drained into sewers intended for sanitary drainage.

1101.4 Material Uses. Pipe, tube, and fittings conveying rainwater shall be of such materials and design as to perform their intended function to the satisfaction of the Authority Having Jurisdiction. Conductors within a vent or shaft shall be of cast-iron, galvanized steel, wrought iron, copper, copper alloy, lead, Schedule 40 ABS DWV, Schedule 40 PVC DWV, stainless steel 304 or 316L [stainless steel 304 pipe and fittings shall not be installed underground and shall be kept not less than 6 inches (152 mm) aboveground], or other approved materials, and changes in direction shall be in accordance with the requirements of Section 706.0. ABS and PVC DWV piping installations shall be installed in accordance with applicable standards referenced in Chapter 17 and Chapter 14 "Firestop Protection." Except for individual single-family dwelling units, materials exposed within ducts or plenums shall have a flame-spread index of not more than 25 and a smoke-developed index of not more than 50, where tested in accordance with ASTM E84 or UL 723. Plastic piping installed in plenums shall be tested in accordance with all requirements of ASTM E84 or UL 723. Mounting methods, supports and sample sizes of materials for testing that are not specified in ASTM E84 or UL 723 shall be prohibited.

1101.4.1 Copper and Copper Alloys. Joints and connections in copper and copper alloy pipe and tube shall be installed in accordance with Section 705.3.

1101.4.2 Conductors. Conductors installed aboveground in buildings shall comply with the applicable standards referenced in Table 701.2 for aboveground drain, waste, and vent pipe. Conductors installed aboveground level shall be of seamless copper water tube, Type K, L, or M; Schedule 40 copper pipe or Schedule 40 copper alloy pipe; Type DWV copper drainage tube; service weight cast-iron soil pipe or hubless cast-iron soil pipe; standard weight galvanized steel pipe; stainless steel 304 or 316L [stainless steel 304 pipe and fittings shall not be installed underground and shall be kept not less than 6 inches (152 mm) aboveground], or Schedule 40 ABS or Schedule 40 PVC plastic pipe.

1101.4.3 Leaders. Leaders installed outside shall comply with the applicable standards referenced in Table 701.2 for aboveground drain, waste, and vent pipe; aluminum sheet metal; galvanized steel sheet metal; or copper sheet metal.

1101.4.4 Underground Building Storm Drains. Underground building storm drains shall comply with the applicable standards referenced in Table 701.2 for underground drain, waste, and vent pipe.

1101.4.5 Building Storm Sewers. Building storm sewers shall comply with the applicable standards referenced in Table 701.2 for building sewer pipe.

1101.4.6 Subsoil Drains. Subsoil drains shall be open jointed, perforated, or both and constructed of materials in conformance with Table 1101.4.6.

TABLE 1101.4.6
MATERIALS FOR SUBSOIL DRAIN PIPE AND FITTINGS

MATERIAL	REFERENCED STANDARD(S)
PE	ASTM F667
PVC	ASTM D2729
Vitrified Clay (Extra strength)	ASTM C4, ASTM C700

1101.5 Expansion Joints Required. Expansion joints or sleeves shall be provided where warranted by temperature variations or physical conditions.

1101.6 Subsoil Drains. Subsoil drains shall be provided around the perimeter of buildings having basements, cellars, crawl spaces, or floors below grade. Such subsoil drains shall be permitted to be positioned inside or outside of the footing, shall be of perforated or open-jointed approved drain tile or pipe, not less than 3 inches (80 mm) in diameter, and shall be laid in gravel, slag, crushed rock, approved ¾ of an inch (19.1 mm) crushed, recycled glass aggregate, or other approved porous material with not less than 4 inches (102 mm) surrounding the pipe. Filter media shall be provided for exterior subsoil piping.

1101.6.1 Discharge. Subsoil drains shall be piped to a storm drain, to an approved water course, to the front street curb or gutter, to an alley, or the discharge from the subsoil drains shall be conveyed to the alley by a concrete gutter. Where a continuously flowing spring or groundwater is encountered, subsoil drains shall be piped to a storm drain or an approved water course.

1101.6.2 Sump. Where it is not possible to convey the drainage by gravity, subsoil drains shall discharge to an accessible sump provided with an approved automatic electric pump. The sump shall be not less than 15 inches (381 mm) in diameter, 18 inches (457 mm) in depth, and provided with a fitted cover. The sump pump shall have an adequate capacity to discharge water coming into the sump as it accumulates to the required discharge point, and the capacity of the pump shall be not less than 15 gallons per minute (gpm) (0.95 L/s). The discharge piping from the sump pump shall be not less than 1½ inches (40 mm) in diameter and have a union or other approved quick-disconnect assembly to make the pump accessible for servicing.

1101.6.3 Splash Blocks. For separate dwellings not serving continuously flowing springs or groundwater, the sump discharge pipe shall be permitted to discharge onto a concrete splash block with a minimum length of 24 inches (610 mm). This pipe shall be within 4 inches (102 mm) of the splash block and positioned to direct the flow parallel to the recessed line of the splash block.

1101.6.4 Backwater Valve. Subsoil drains subject to backflow where discharging into a storm drain shall be provided with a backwater valve in the drain line so located as to be accessible for inspection and maintenance.

1101.6.5 Open Area. Nothing in Section 1101.6 shall prevent drains that serve either subsoil drains or areaways of a detached building from discharging to a properly graded open area, provided that:

(1) They do not serve continuously flowing springs or groundwater.

(2) The point of discharge is not less than 10 feet (3048 mm) from a property line.

(3) It is impracticable to discharge such drains to a storm drain, to an approved water course, to the front street curb or gutter, or to an alley.

1101.7 Building Subdrains. Building subdrains located below the public sewer level shall discharge into a sump or receiving tank, the contents of which shall be automatically lifted and discharged into the drainage system as required for building sumps.

1101.8 Areaway Drains. Open subsurface space adjacent to a building, serving as an entrance to the basement or cellar of a building, shall be provided with a drain or drains. The areaway drains shall be not less than 2 inches (50 mm) in diameter for areaways at a maximum of 100 square feet (9.29 m^2) in area, and shall be discharged in the manner provided for subsoil drains not serving continuously flowing springs or groundwater (see Section 1101.6.1). Areaways exceeding 100 square feet (9.29 m^2) shall not drain into subsoil drains. The drains for areaways exceeding 100 square feet (9.29 m^2) shall be sized in accordance with Table 1103.2.

1101.9 Window Areaway Drains. Window areaways at a maximum of 10 square feet (0.93 m^2) in area shall be permitted to discharge to the subsoil drains through a 2 inch (50 mm) diameter pipe. However, window areaways exceeding 10 square feet (0.93 m^2) in area shall be handled in the manner provided for entrance areaways (see Section 1101.8).

1101.10 Filling Stations and Motor Vehicle Washing Establishments. Public filling stations and motor vehicle washing establishments shall have the paved area sloped toward sumps or gratings within the property lines. Curbs not less than 6 inches (152 mm) high shall be placed where required to direct water to gratings or sumps.

1101.11 Paved Areas. Where the occupant creates surface water drainage, the sumps, gratings, or floor drains shall be piped to a storm drain or an approved water course.

1101.12 Roof Drainage. Roof drainage shall comply with Section 1101.12.1 and Section 1101.12.2.

1101.12.1 Primary Roof Drainage. Roof areas of a building shall be drained by roof drains or gutters. The location and sizing of drains and gutters shall be coordinated with the structural design and pitch of the roof. Unless otherwise required by the Authority Having Jurisdiction, roof drains, gutters, vertical conductors or leaders, and horizontal storm drains for primary drainage shall be sized based on a storm of 60 minutes duration and 100 year return period. Refer to Table D 101.1 (in Appendix D) for 100 years, 60-minute storms at various locations.

1101.12.2 Secondary Drainage. Secondary (emergency) roof drainage shall be provided by one of the methods specified in Section 1101.12.2.1 or Section 1101.12.2.2.

1101.12.2.1 Roof Scuppers or Open Side. Secondary roof drainage shall be provided by an open-sided roof or scuppers where the roof perimeter construction extends above the roof in such a manner that water will be entrapped. An open-sided roof or scuppers shall be sized to prevent the depth of ponding water from exceeding that for which the roof was designed as determined by Section 1101.12.1. Scupper openings shall be not less than 4 inches (102 mm) high and have a width equal to the circumference of the roof drain required for the area served, sized in accordance with Table 1103.1.

1101.12.2.2 Secondary Roof Drain. Secondary roof drains shall be provided. The secondary roof drains shall be located not less than 2 inches (51 mm) above the roof surface. The maximum height of the roof drains shall be a height to prevent the depth of ponding water from exceeding that for which the roof was designed as determined by Section 1101.12.1. The secondary roof drains shall connect to a piping system in accordance with Section 1101.12.2.2.1 or Section 1101.12.2.2.2.

1101.12.2.2.1 Separate Piping System. The secondary roof drainage system shall be a separate system of piping, independent of the primary roof drainage system. The discharge shall be above grade, in a location observable by the building occupants or maintenance personnel. Secondary roof drain systems shall be sized in accordance with Section 1101.12.1 based on the rainfall rate for which the primary system is sized.

1101.12.2.2.2 Combined System. The secondary roof drains shall connect to the vertical piping of the primary storm drainage conductor downstream of the last horizontal offset located below the roof. The primary storm drainage system shall connect to the building storm water that connects to an underground public storm sewer. The combined secondary and primary roof drain systems shall be sized in accordance with Section 1103.0 based on double the rainfall rate for the local area.

1101.13 Cleanouts. Cleanouts for building storm drains shall comply with the requirements of Section 719.0 of this code.

1101.13.1 Rain Leaders and Conductors. Rain leaders and conductors connected to a building storm sewer shall have a cleanout installed at the base of the outside leader or outside conductor before it connects to the horizontal drain.

1101.14 Rainwater Sumps. Rainwater sumps serving "public use" occupancy buildings shall be provided with dual pumps arranged to function alternately in the case of overload or mechanical failure. Pumps rated 600 V or less shall comply with UL 778 and shall be installed in accordance with the manufacturer's installation instructions.

1101.15 Traps on Storm Drains and Leaders. Leaders and storm drains, where connected to a combined sewer, shall be trapped. Floor and area drains connected to a storm drain shall be trapped.

Exception: Traps shall not be required where roof drains, rain leaders, and other inlets are at locations permitted under Section 906.0, Vent Termination.

1101.15.1 Where Not Required. No trap shall be required for leaders or conductors that are connected to a sewer carrying storm water exclusively.

1101.15.2 Trap Size. Traps, where installed for individual conductors, shall be the same size as the horizontal drain to which they are connected.

1101.15.3 Method of Installation of Combined Sewer. Individual storm-water traps shall be installed on the stormwater drain branch serving each storm-water inlet, or a single trap shall be installed in the main storm drain just before its connection with the combined building sewer. Such traps shall be provided with an accessible cleanout on the outlet side of the trap.

1101.16 Leaders, Conductors, and Connections. Leaders or conductors shall not be used as soil, waste, or vent pipes nor shall soil, waste, or vent pipes be used as leaders or conductors.

1101.16.1 Protection of Leaders. Leaders installed along alleyways, driveways, or other locations where exposed to damage shall be protected by metal guards, recessed into the wall, or constructed from the ferrous pipe.

1101.16.2 Combining Storm with Sanitary Drainage. The sanitary and storm drainage system of a building shall be entirely separate, except where a combined sewer is used, in which case the building storm drain shall be connected in the same horizontal plane through a single wye fitting to the combined building sewer not less than 10 feet (3048 mm) downstream from a soil stack.

1102.0 Roof Drains.

1102.1 Applications. Roof drains shall be constructed of aluminum, cast-iron, copper alloy of not more than 15 percent zinc, leaded nickel bronze, stainless steel, ABS, PVC, polypropylene, polyethylene, or nylon and shall comply with ASME A112.3.1 or ASME A112.6.4.

1102.2 Dome Strainers Required. Roof drains shall have domed strainers.

Exception: Roof drain strainers for use on sun decks, parking decks, and similar areas that are normally serviced and maintained, shall be permitted to be of the flat surface type. Such roof drain strainers shall be level with the deck.

1102.3 Roof Drain Flashings. The connection between the roof and roof drains that pass through the roof and into the interior of the building shall be made watertight by the use of proper flashing material.

1102.3.1 Lead Flashing. Where lead flashing material is used, it shall be not less than 4 pounds per square foot (lb/ft^2) (19 kg/m^2).

1102.3.2 Copper Flashing. Where copper flashing material is used, it shall be not less than 12 ounces per square foot (oz/ft^2) (3.7 kg/m^2).

1103.0 Size of Leaders, Conductors, and Storm Drains.

1103.1 Vertical Conductors and Leaders. Vertical conductors and leaders shall be sized by the maximum projected roof area and Table 1103.1.

1103.2 Size of Horizontal Storm Drains and Sewers. The size of building storm drains, or building storm sewers or their horizontal branches shall be based on the maximum projected roof or paved area to be handled and Table 1103.2.

1103.3 Size of Roof Gutters. The size of semi-circular gutters shall be based on the maximum projected roof area and Table 1103.3.

1103.4 Side Walls Draining onto a Roof. Where vertical walls project above a roof to permit storm water to drain into the roof area below, the adjacent roof area shall be permitted to be computed from Table 1103.1 as follows:

(1) For one wall – add 50 percent of the wall area to the roof area figures.

(2) For two adjacent walls of equal height – add 35 percent of the total wall areas.

(3) For two adjacent walls of unequal height – add 35 percent of the total common height and add 50 percent of the remaining height of the highest wall.

(4) Two opposite walls of same height – add no additional area.

(5) Two opposite walls of differing heights – add 50 percent of the wall area above the top of the lower wall.

(6) Walls on three sides – add 50 percent of the area of the inner wall below the top of the lowest wall, plus an allowance for the area of the wall above the top of the lowest wall, in accordance with Section 1103.4(3) and Section 1103.4(5) above.

TABLE 1103.1
SIZING ROOF DRAINS, LEADERS, AND VERTICAL RAINWATER PIPING[2, 3]

SIZE OF DRAIN, LEADER, OR PIPE	FLOW	MAXIMUM ALLOWABLE HORIZONTAL PROJECTED ROOF AREAS AT VARIOUS RAINFALL RATES (square feet)											
inches	gpm[1]	1 (in/h)	2 (in/h)	3 (in/h)	4 (in/h)	5 (in/h)	6 (in/h)	7 (in/h)	8 (in/h)	9 (in/h)	10 (in/h)	11 (in/h)	12 (in/h)
2	30	2880	1440	960	720	575	480	410	360	320	290	260	240
3	92	8800	4400	2930	2200	1760	1470	1260	1100	980	880	800	730
4	192	18 400	9200	6130	4600	3680	3070	2630	2300	2045	1840	1675	1530
5	360	34 600	17 300	11 530	8650	6920	5765	4945	4325	3845	3460	3145	2880
6	563	54 000	27 000	17 995	13 500	10 800	9000	7715	6750	6000	5400	4910	4500
8	1208	116 000	58 000	38 660	29 000	23 200	19 315	16 570	14 500	12 890	11 600	10 545	9600

For SI units: 1 inch = 25 mm, 1 gallon per minute = 0.06 L/s, 1 inch per hour = 25.4 mm/h, 1 square foot = 0.0929 m^2

Notes:

1 Maximum discharge capacity, gpm (L/s) with approximately 1¾ inch (44 mm) head of water at the drain.

2 For rainfall rates other than those listed, determine the allowable roof area by dividing the area given in the 1 inch per hour (25.4 mm/h) column by the desired rainfall rate.

3 Vertical piping shall be round, square, or rectangular. Square pipe shall be sized to enclose its equivalent round pipe. Rectangular pipe shall have not less than the same cross-sectional area as its equivalent round pipe, except that the ratio of its side dimensions shall not exceed 3 to 1.

(7) Walls on four sides – no allowance for wall areas below the top of the lowest wall – add for areas above the top of the lowest wall in accordance with Section 1103.4(1), Section 1103.4(3), Section 1103.4(5), and Section 1103.4(6) above.

1104.0 Values for Continuous Flow.

1104.1 General. Where there is a continuous or semi-continuous discharge into the building storm drain or building storm sewer, as from a pump, ejector, air-conditioning plant, or similar device, 1 gpm (0.06 L/s) of such discharge shall be computed as being equivalent to 24 square feet (2.2 m^2) of roof area, based upon a rate of rainfall of 4 inches per hour (in/h) (102 mm/h).

1105.0 Controlled-Flow Roof Drainage.

1105.1 Application. Instead of sizing the storm drainage system in accordance with Section 1103.0, the roof drainage shall be permitted to be sized by controlled flow and storage of the storm water on the roof, provided the following conditions are met:

(1) The water from a 25-year frequency storm shall not be stored on the roof exceeding 24 hours.

(2) During the storm, the water depth on the roof shall not exceed the depths specified in Table 1105.1(1).

(3) Not less than two drains shall be installed in roof areas of 10 000 square feet (929 m^2) or less, and not less than one additional drain shall be installed for each 10 000 square feet (929 m^2) of roof area exceeding 10 000 square feet (929 m^2).

(4) Each roof drain shall have a precalibrated, fixed (non-adjustable), and proportional weir (notched) in a standing water collar inside the strainer. No mechanical devices or valves shall be permitted.

(5) Pipe sizing shall be based on the pre-calibrated rate of flow (gpm) (L/s) of the pre-calibrated weir for the maximum allowable water depth, and Table 1103.1 and Table 1103.2.

(6) The height of stones or other granular material above the waterproofed surface shall not be considered in water depth measurement, and the roof surface in the vicinity of the drain shall not be recessed to create a reservoir.

(7) Roof design, where controlled-flow roof drainage is used, shall be such that the design roof live load is not less than 30 lb/ft^2 (146 kg/m^2) to provide a safety factor exceeding the 15 lb/ft^2 (73 kg/m^2) represented by the depth of water stored on the roof in accordance with Table 1105.1(1).

(8) Scuppers shall be provided in parapet walls. The distance of scupper bottoms above the roof level at the drains shall not exceed the maximum distances specified in Table 1105.1(2).

(9) Scupper openings shall be not less than 4 inches (102 mm) high and have a width equal to the circumference of the roof drain required for the area served, sized in accordance with Table 1103.1.

(10) Flashings shall extend above the top of the scuppers.

(11) At a wall or parapet, 45 degree (0.79 rad) cants shall be installed.

(12) Separate storm and sanitary drainage systems shall be provided within the building.

(13) Calculations for the roof drainage system shall be submitted along with the plans to the Authority Having Jurisdiction for approval.

TABLE 1105.1(1)
CONTROLLED-FLOW MAXIMUM ROOF WATER DEPTH

ROOF RISE* (inches)	MAXIMUM WATER DEPTH AT DRAIN (inches)
Flat	3
2	4
4	5
6	6

For SI units: 1 inch = 25.4 mm

* Vertical measurement from the roof surface at the drain to the highest point of the roof surface served by the drain, ignoring a local depression immediately adjacent to the drain.

TABLE 1105.1(2)
DISTANCE OF SCUPPER BOTTOMS ABOVE ROOF

ROOF RISE* (inches)	ABOVE ROOF LEVEL AT DRAIN (inches)
Flat	3
2	4
4	5
6	6

For SI units: 1 inch = 25.4 mm

* Vertical measurement from the roof surface at the drain to the highest point of the roof surface served by the drain, ignoring a local depression immediately adjacent to the drain.

1105.2 Setback Roofs. Drains on setback roofs shall be permitted to be connected to the controlled-flow drainage systems provided:

(1) The setback is designed for storing water, or

(2) The square footage of the setback drainage area is converted as outlined in Section 1105.0 to gpm, and the storm-water pipe sizes in the controlled-flow system are based on the sum of the loads.

(3) The branch from each of the roof drains that are not provided with controlled flow shall be sized in accordance with Table 1103.1.

1106.0 Engineered Storm Drainage System.

1106.1 General. The design and sizing of a storm drainage system shall be permitted to be determined by accepted engineering practices. The system shall be designed by a registered design professional and approved in accordance with Section 301.5.

1106.2 Siphonic Roof Drainage Systems. The design of a siphonic roof drainage system shall comply with ASPE 45.

1106.3 Siphonic Roof Drains. Siphonic roof drains shall comply with ASME A112.6.9.

1107.0 Testing.

1107.1 Testing Required. New building storm drainage systems and parts of existing systems that have been altered, extended, or repaired shall be tested in accordance with Section 1107.2.1 or Section 1107.2.2 to disclose leaks and defects.

1107.2 Methods of Testing Storm Drainage Systems. Except for outside leaders and perforated or open-jointed drain tile, the piping of storm drain systems shall be tested upon completion of the rough piping installation by water or air, except that plastic pipe shall not be tested with air, and proved tight. The Authority Having Jurisdiction shall be permitted to require the removal of cleanout plugs to ascertain whether the pressure has reached parts of the system. One of the following test methods shall be used in accordance with Section 1107.2.1 through Section 1107.2.3.

1107.2.1 Water Test. After piping has been installed, the water test shall be applied to the drainage system, either to the entire system or sections. Where the test is applied to the entire system, all openings in the piping shall be tightly closed except for the highest opening, and the system shall be filled with water to the point of overflow. Where the system is tested in sections, each opening shall be tightly plugged except for the highest opening of the section under test, and each section shall be filled with water, but no section shall be tested with less than a 10 foot (3048 mm) head of water. In testing successive sections, not less than the upper 10 feet (3048 mm) of the next preceding section shall be tested so that no joint of pipe in the building except the uppermost 10 feet (3048 mm) of a roof drainage system, which shall be filled with water to the flood level of the uppermost roof drain, shall have been submitted to a test of less than 10 foot (3048 mm) head of water. The water shall be kept in the system or the portion of the test for not less than 15 minutes before inspection starts; the system shall then be tight.

1107.2.2 Air Test. The air test shall be made by attaching an air compressor testing apparatus to a suitable opening after closing other inlets and outlets to the system, forcing air into the system until there is a uniform gauge pressure of 5 pounds-force per square inch (psi) (34 kPa) or sufficient pressure to balance a column of mercury 10 inches (34 kPa) in height. This pressure shall be held without the introduction of additional air for not less than 15 minutes.

1107.2.3 Exceptions. Where circumstances exist that make air and water tests described in Section 1107.2.1 and Section 1107.2.2 impractical, see Section 105.3.

TABLE 1103.2
SIZING OF HORIZONTAL RAINWATER PIPING[1, 2]

SIZE OF PIPE	FLOW (1/8 inch per foot slope)	MAXIMUM ALLOWABLE HORIZONTAL PROJECTED ROOF AREAS AT VARIOUS RAINFALL RATES (square feet)					
inches	gpm	1 (in/h)	2 (in/h)	3 (in/h)	4 (in/h)	5 (in/h)	6 (in/h)
3	34	3288	1644	1096	822	657	548
4	78	7520	3760	2506	1880	1504	1253
5	139	13 360	6680	4453	3340	2672	2227
6	222	21 400	10 700	7133	5350	4280	3566
8	478	46 000	23 000	15 330	11 500	9200	7670
10	860	82 800	41 400	27 600	20 700	16 580	13 800
12	1384	133 200	66 600	44 400	33 300	26 650	22 200
15	2473	238 000	119 000	79 333	59 500	47 600	39 650

SIZE OF PIPE	FLOW (1/4 inch per foot slope)	MAXIMUM ALLOWABLE HORIZONTAL PROJECTED ROOF AREAS AT VARIOUS RAINFALL RATES (square feet)					
inches	gpm	1 (in/h)	2 (in/h)	3 (in/h)	4 (in/h)	5 (in/h)	6 (in/h)
3	48	4640	2320	1546	1160	928	773
4	110	10 600	5300	3533	2650	2120	1766
5	196	18 880	9440	6293	4720	3776	3146
6	314	30 200	15 100	10 066	7550	6040	5033
8	677	65 200	32 600	21 733	16 300	13 040	10 866
10	1214	116 800	58 400	38 950	29 200	23 350	19 450
12	1953	188 000	94 000	62 600	47 000	37 600	31 350
15	3491	336 000	168 000	112 000	84 000	67 250	56 000

SIZE OF PIPE	FLOW (1/2 inch per foot slope)	MAXIMUM ALLOWABLE HORIZONTAL PROJECTED ROOF AREAS AT VARIOUS RAINFALL RATES (square feet)					
inches	gpm	1 (in/h)	2 (in/h)	3 (in/h)	4 (in/h)	5 (in/h)	6 (in/h)
3	68	6576	3288	2192	1644	1310	1096
4	156	15 040	7520	5010	3760	3010	2500
5	278	26 720	13 360	8900	6680	5320	4450
6	445	42 800	21 400	14 267	10 700	8580	7140
8	956	92 000	46 000	30 650	23 000	18 400	15 320
10	1721	165 600	82 800	55 200	41 400	33 150	27 600
12	2768	266 400	133 200	88 800	66 600	53 200	44 400
15	4946	476 000	238 000	158 700	119 000	95 200	79 300

For SI units: 1 inch = 25 mm, 1 gallon per minute = 0.06 L/s, 1/8 inch per foot = 10.4 mm/m, 1 inch per hour = 25.4 mm/h, 1 square foot = 0.0929 m^2

Notes:

1 The sizing data for horizontal piping are based on the pipes flowing full.

2 For rainfall rates other than those listed, determine the allowable roof area by dividing the area given in the 1 inch per hour (25.4 mm/h) column by the desired rainfall rate.

TABLE 1103.3
SIZE OF GUTTERS

DIAMETER OF GUTTER (1/16 inch per foot slope)	MAXIMUM RAINFALL RATES BASED ON ROOF AREA (square feet)				
inches	2 (in/h)	3 (in/h)	4 (in/h)	5 (in/h)	6 (in/h)
3	340	226	170	136	113
4	720	480	360	288	240
5	1250	834	625	500	416
6	1920	1280	960	768	640
7	2760	1840	1380	1100	918
8	3980	2655	1990	1590	1325
10	7200	4800	3600	2880	2400

DIAMETER OF GUTTER (1/8 inch per foot slope)	MAXIMUM RAINFALL RATES BASED ON ROOF AREA (square feet)				
inches	2 (in/h)	3 (in/h)	4 (in/h)	5 (in/h)	6 (in/h)
3	480	320	240	192	160
4	1020	681	510	408	340
5	1760	1172	880	704	587
6	2720	1815	1360	1085	905
7	3900	2600	1950	1560	1300
8	5600	3740	2800	2240	1870
10	10 200	6800	5100	4080	3400

DIAMETER OF GUTTER (1/4 inch per foot slope)	MAXIMUM RAINFALL RATES BASED ON ROOF AREA (square feet)				
inches	2 (in/h)	3 (in/h)	4 (in/h)	5 (in/h)	6 (in/h)
3	680	454	340	272	226
4	1440	960	720	576	480
5	2500	1668	1250	1000	834
6	3840	2560	1920	1536	1280
7	5520	3680	2760	2205	1840
8	7960	5310	3980	3180	2655
10	14 400	9600	7200	5750	4800

DIAMETER OF GUTTER (1/2 inch per foot slope)	MAXIMUM RAINFALL RATES BASED ON ROOF AREA (square feet)				
inches	2 (in/h)	3 (in/h)	4 (in/h)	5 (in/h)	6 (in/h)
3	960	640	480	384	320
4	2040	1360	1020	816	680
5	3540	2360	1770	1415	1180
6	5540	3695	2770	2220	1850
7	7800	5200	3900	3120	2600
8	11 200	7460	5600	4480	3730
10	20 000	13 330	10 000	8000	6660

For SI units: 1 inch = 25 mm, 1/16 inch per foot = 5.2 mm/m, 1 inch per hour = 25.4 mm/h, 1 square foot = 0.0929 m^2

CHAPTER 12
FUEL GAS PIPING

1201.0 General.

1201.1 Applicability. The regulations of this chapter shall govern the installation of fuel gas piping in or in connection with a building, structure or within the property lines of premises up to 5 pounds-force per square inch (psi) (34 kPa) for natural gas and 10 psi (69 kPa) for undiluted propane, other than service pipe. Fuel oil piping systems shall be installed in accordance with NFPA 31.

1202.0 Coverage of Piping System.

1202.1 General. Coverage of piping systems shall extend from the point of delivery to the appliance connections. For other than undiluted liquefied petroleum gas (LP-Gas) systems, the point of delivery shall be the outlet of the service meter assembly or the outlet of the service regulator or service shutoff valve where no meter is provided. For undiluted LP-Gas systems, the point of delivery shall be considered to be the outlet of the final pressure regulator, exclusive of line gas regulators where no meter is installed. Where a meter is installed, the point of delivery shall be the outlet of the meter. [NFPA 54:1.1.1.1(A)]

1202.2 Piping System Requirements. Requirements for piping systems shall include design, materials, components, fabrication, assembly, installation, testing, inspection, operation, and maintenance. [NFPA 54:1.1.1.1(E)]

1202.3 Applications. This code shall not apply to the following items:

(1) Portable LP-Gas appliances and equipment of all types that are not connected to a fixed fuel piping system.

(2) Installation of appliances such as brooders, dehydrators, dryers, and irrigation equipment used for agricultural purposes.

(3) Raw material (feedstock) applications except for piping to special atmosphere generators.

(4) Oxygen-fuel gas cutting and welding systems.

(5) Industrial gas applications using such gases as acetylene and acetylenic compounds, hydrogen, ammonia, carbon monoxide, oxygen, and nitrogen.

(6) Petroleum refineries, pipeline compressor or pumping stations, loading terminals, compounding plants, refinery tank farms, and natural gas processing plants.

(7) Large integrated chemical plants or portions of such plants where flammable or combustible liquids or gases are produced by chemical reactions or used in chemical reactions.

(8) LP-Gas installations at utility gas plants.

(9) Liquefied natural gas (LNG) installations.

(10) Fuel gas piping in electric utility power plants.

(11) Proprietary items of equipment, apparatus, or instruments such as gas-generating sets, compressors, and calorimeters.

(12) LP-Gas equipment for vaporization, gas mixing, and gas manufacturing.

(13) LP-Gas piping for buildings under construction or renovations that is not to become part of the permanent building piping system—that is, temporary fixed piping for building heat.

(14) Installation of LP-Gas systems for railroad switch heating.

(15) Installation of LP-Gas and compressed natural gas (CNG) systems on vehicles.

(16) Gas piping, meters, gas pressure regulators, and other appurtenances used by the serving gas supplier in distribution of gas, other than undiluted LP-Gas.

(17) Building design and construction, except as specified herein.

(18) Fuel gas systems on recreational vehicles manufactured in accordance with NFPA 1192.

(19) Fuel gas systems using hydrogen as a fuel.

(20) Construction of appliances. [NFPA 54:1.1.1.2]

1203.0 Inspection.

1203.1 Inspection Notification. Upon completion of the installation, alteration, or repair of gas piping, and prior to the use thereof, the Authority Having Jurisdiction shall be notified that such gas piping is ready for inspection.

1203.2 Excavation. Excavations required for the installation of underground piping shall be kept open until the piping has been inspected and approved. Where such piping is covered or concealed before such approval, it shall be exposed upon the direction of the Authority Having Jurisdiction.

1203.3 Type of Inspections. The Authority Having Jurisdiction shall make the following inspections and either shall approve that portion of the work as completed or shall notify the permit holder wherein the same fails to be in accordance with this code.

1203.3.1 Rough Piping Inspection. This inspection shall be made after gas piping authorized by the permit has been installed and before such piping has been covered or concealed or fixture or appliance has been attached thereto. This inspection shall include a determination that the gas piping size, material, and installation meet the requirements of this code.

1203.3.2 Final Piping Inspection. This inspection shall be made after piping authorized by the permit has been installed, and after portions, thereof that are to be covered or concealed are so concealed, and before fixture, appliance, or shutoff valve has been attached thereto.

This inspection shall comply with Section 1213.1. Test gauges used in conducting tests shall be in accordance with Section 318.0.

1203.4 Inspection Waived. In cases where the work authorized by the permit consists of a minor installation of additional piping to piping already connected to a gas meter, the preceding inspections shall be permitted to be waived at the discretion of the Authority Having Jurisdiction. In this event, the Authority Having Jurisdiction shall make such inspection as deemed advisable to be assured that the work has been performed in accordance with the intent of this code.

1204.0 Certificate of Inspection.

1204.1 Issuance. Whereupon final piping inspection, the installation is found to be in accordance with the provisions of this code, a certificate of inspection shall be permitted to be issued by the Authority Having Jurisdiction.

1204.2 Gas Supplier. A copy of the certificate of such final piping inspection shall be issued to the serving gas supplier supplying gas to the premises.

1204.3 Unlawful. It shall be unlawful for a serving gas supplier, or person is furnishing gas, to turn on or cause to be turned on, a fuel gas or a gas meter or meters, until such certificate of final inspection, as herein provided, has been issued.

1205.0 Authority to Render Gas Service.

1205.1 Authorized Personnel. It shall be unlawful for a person, firm, or corporation, excepting an authorized agent or employee of a person, firm, or corporation engaged in the business of furnishing or supplying gas and whose service pipes supply or connect with the particular premises, to turn on or reconnect gas service in or on a premises where and when gas service is, at the time, not being rendered.

1205.2 Outlets. It shall be unlawful to turn on or connect gas in or on the premises unless outlets are securely connected to gas appliances or capped or plugged with screw joint fittings.

1206.0 Authority to Disconnect.

1206.1 Disconnection. The Authority Having Jurisdiction or the serving gas supplier is hereby authorized to disconnect gas piping or appliance or both that shall be found not to be in accordance with the requirements of this code or that are found defective and in such condition as to endanger life or property.

1206.2 Notice. Where such disconnection has been made, a notice shall be attached to such gas piping or appliance or both that shall state the same has been disconnected, together with the reasons thereof.

1206.3 Capped Outlets. It shall be unlawful to remove or disconnect gas piping or gas appliance without capping or plugging with a screw joint fitting, the outlet from which said pipe or appliance was removed. Outlets to which gas appliances are not connected shall be left capped and gastight on a piping system that has been installed, altered, or repaired.

Exception: Where an approved listed quick-disconnect device is used.

1207.0 Temporary Use of Gas.

1207.1 General. Where temporary use of gas is desired, and the Authority Having Jurisdiction deems the use necessary, a permit shall be permitted to be issued for such use for a period not to exceed that designated by the Authority Having Jurisdiction, provided that such gas piping system otherwise is in accordance with the requirements of this code regarding material, sizing, and safety.

1208.0 Gas Piping System Design, Materials, and Components.

1208.1 Installation of Piping System. Where required by the Authority Having Jurisdiction, a piping sketch or plan shall be prepared before proceeding with the installation. The plan shall show the proposed location of piping, the size of different branches, the various load demands, and the location of the point of delivery. [NFPA 54:5.1.1]

1208.1.1 Addition to Existing System. When additional appliances are being connected to a gas piping system, the existing piping shall be checked to determine whether it has adequate capacity. If the capacity of the system is determined to be inadequate for the additional appliances, the existing system shall be enlarged as required, or separate gas piping of adequate capacity shall be provided. [NFPA 54:5.1.2]

1208.2 Provision for Location of Point of Delivery. The location of the point of delivery shall be acceptable to the serving gas supplier. [NFPA 54:5.2]

1208.3 Interconnections Between Gas Piping Systems. Where two or more meters, or two or more service regulators where meters are not provided, are located on the same premises and supply separate users, the gas piping systems shall not be interconnected on the outlet side of the meters or service regulators. [NFPA 54:5.3.1]

1208.3.1 Interconnections for Standby Fuels. Where a supplementary gas for standby use is connected downstream from a meter or a service regulator where a meter is not provided, equipment to prevent backflow shall be installed. A three-way valve installed to admit the standby supply and at the same time shut off the regular supply shall be permitted to be used for this purpose. [NFPA 54:5.3.2.1 – 5.3.2.2]

1208.4 Sizing of Gas Piping Systems. Gas piping systems shall be of such size and so installed as to provide a supply of gas sufficient to meet the maximum demand and supply gas to each appliance inlet at not less than the minimum supply pressure required by the appliance. [NFPA 54:5.4.1]

1208.4.1 Maximum Gas Demand. The volumetric flow rate of gas to be provided shall be the sum of the maximum input of the appliances served. The volumetric flow rate of gas to be provided shall be adjusted for altitude where the installation is above 2000 feet (610 m). [NFPA 54:5.4.2.1 – 5.4.2.2] Where the input rating is not indicated, the gas supplier, appliance manufacturer, or a qualified agency shall be contacted, or the rating from Table 1208.4.1 shall be used for estimating the volumetric flow rate of gas to be supplied.

The total connected hourly load shall be used as the basis for piping sizing, assuming all appliances are operating at full capacity simultaneously.

Exception: Sizing shall be permitted to be based upon established load diversity factors. [NFPA 54:5.4.2.3]

TABLE 1208.4.1
APPROXIMATE GAS INPUT FOR TYPICAL APPLIANCES
[NFPA 54: TABLE A.5.4.2.1]

APPLIANCE	INPUT (Btu/h approx.)
Space Heating Units	
Warm air furnace	
Single family	100 000
Multifamily, per unit	60 000
Hydronic boiler	
Single family	100 000
Multifamily, per unit	60 000
Space and Water Heating Units	
Hydronic boiler	
Single-family	120 000
Multifamily, per unit	75 000
Water Heating Appliances	
Water heater, automatic storage	
30 to 40 gallon tank	35 000
Water heater, automatic storage	
50 gallon tank	50 000
Water heater, automatic instantaneous	
Capacity at 2 gallons per minute	142 800
Capacity at 4 gallons per minute	285 000
Capacity at 6 gallons per minute	428 400
Water heater, domestic, circulating or side-arm	35 000
Cooking Appliances	
Range, freestanding, domestic	65 000
Built-in oven or broiler unit, domestic	25 000
Built-in top unit, domestic	40 000
Other Appliances	
Refrigerator	3000
Clothes dryer, Type 1 (domestic)	35 000
Gas fireplace direct-vent	40 000
Gas log	80 000
Barbecue	40 000
Gaslight	2500

For SI units: 1000 British thermal units per hour = 0.293 kW

1208.4.2 Sizing Methods. Gas piping shall be sized in accordance with one of the following:

(1) Pipe sizing tables or sizing equations in this chapter.

(2) Other approved engineering methods.

(3) Sizing tables included in a listed piping system manufacturer's installation instructions. [NFPA 54:5.4.3]

1208.4.3 Allowable Pressure Drop. The design pressure loss in any piping system under maximum probable flow conditions, from the point of delivery to the inlet connection of the appliance, shall be such that the supply pressure at the appliance is greater than or equal to the minimum pressure required by the appliance. [NFPA 54:5.4.4]

1208.5 Maximum Operating Pressure in Buildings. The maximum operating pressure for any piping systems located inside buildings shall not exceed 5 psi (34 kPa) unless one or more of the following conditions are met:

(1) The piping joints are welded or brazed.

(2) The piping joints are flanged and all pipe-to-flange connections are made by welding or brazing.

(3) The piping is located in a ventilated chase or otherwise enclosed for protection against accidental gas accumulation.

(4) The piping is located inside buildings or separate areas of buildings used exclusively for one of the following:

(a) Industrial processing or heating

(b) Research

(c) Warehousing

(d) Boiler or mechanical rooms

(5) The piping is a temporary installation for buildings under construction.

(6) The piping serves appliances or equipment used for agricultural purposes.

(7) The piping system is an LP-Gas piping system with an operating pressure greater than 20 psi (138 kPa) and complies with NFPA 58. [NFPA 54:5.5.4]

1208.5.1 LP-Gas Systems. LP-Gas systems designed to operate below -5°F (-21°C) or with butane or a propane-butane mix shall be designed to either accommodate liquid LP-Gas or to prevent LP-Gas vapor from condensing back into a liquid. [NFPA 54:5.5.5]

1208.6 Acceptable Piping Materials and Joining Methods. Materials used for piping systems shall either comply with the requirements of this chapter or be acceptable to the Authority Having Jurisdiction. [NFPA 54:5.6.1.1]

1208.6.1 Used Materials. Pipe, fittings, valves, or other materials shall not be used again unless they are free of foreign materials and have been ascertained to be adequate for the service intended. [NFPA 54:5.6.1.2]

1208.6.2 Other Materials. Material not covered by the standards specifications listed herein shall meet the following criteria:

(1) Be investigated and tested to determine that it is safe and suitable for the proposed service.

(2) Be recommended for that service by the manufacturer.

(3) Be acceptable to the Authority Having Jurisdiction. [NFPA 54:5.6.1.3]

1208.6.3 Metallic Pipe. Cast-iron pipe shall not be used. [NFPA 54:5.6.2.1]

1208.6.3.1 Steel, Stainless Steel, and Wrought-Iron Pipe. Steel, stainless steel, and wrought-iron pipe shall be at least Schedule 40 and shall comply with the dimensional standards of ASME B36.10M and one of the following:

(1) ASTM A53

(2) ASTM A106

(3) ASTM A312 {NFPA 54:5.6.2.2}

1208.6.3.2 Copper and Copper Alloy Pipe. Copper and copper alloy pipe shall not be used if the gas contains more than an average of 0.3 grains of hydrogen sulfide per 100 standard cubic feet (scf) of gas (0.7 mg/100 L).

Threaded copper, copper alloy, or aluminum alloy pipe shall not be used with gases corrosive to such material. [NFPA 54:5.6.2.3 – 5.6.2.4]

1208.6.3.3 Aluminum Alloy Pipe. Aluminum alloy pipe shall comply with ASTM B241 (except that the use of alloy 5456 is prohibited), and shall be marked at each end of each length indicating compliance. Aluminum alloy pipe shall be coated to protect against external corrosion where it is in contact with masonry, plaster, or insulation or is subject to repeated wettings by such liquids as water, detergents, or sewage. [NFPA 54:5.6.2.5]

Aluminum alloy pipe shall not be used in exterior locations or underground. [NFPA 54:5.6.2.6]

1208.6.4 Metallic Tubing. Tubing shall not be used with gases corrosive to the tubing material. [NFPA 54:5.6.3.1]

1208.6.4.1 Stainless Steel. Stainless steel tubing shall comply with one of the following:

(1) ASTM A268

(2) ASTM A269 [NFPA 54:5.6.3.3]

1208.6.4.2 Steel Tubing. Steel tubing shall comply with ASTM A254. [NFPA 54:5.6.3.2]

1208.6.4.3 Copper and Copper Alloy Tubing. Copper and copper alloy tubing shall not be used if the gas contains more than an average of 0.3 grains of hydrogen sulfide per 100 scf of gas (0.7 mg/100 L). Copper tubing shall comply with standard Type K or L of ASTM B88 or ASTM B280. [NFPA 54:5.6.3.4]

1208.6.4.4 Aluminum Alloy Tubing. Aluminum alloy tubing shall comply with ASTM B210 or ASTM B241. Aluminum alloy tubing shall be coated to protect against external corrosion where it is in contact with masonry, plaster, or insulation or is subject to repeated wettings by such liquids as water, detergent, or sewage. Aluminum alloy tubing shall not be used in exterior locations or underground. [NFPA 54:5.6.3.5]

1208.6.4.5 Corrugated Stainless Steel Tubing. Corrugated stainless steel tubing shall be listed in accordance with CSA LC 1. [NFPA 54:5.6.3.6]

1208.6.5 Plastic Pipe, Tubing, and Fittings. Polyethylene plastic pipe, tubing, and fittings used to supply fuel gas shall conform to ASTM D2513. Pipe to be used shall be marked "gas" and "ASTM D2513." Polyamide pipe, tubing, and fittings shall be identified in and conform to ASTM F2945. Pipe to be used shall be marked "gas" and "ASTM F2945." Polyvinyl chloride (PVC) and chlorinated polyvinyl chloride (CPVC) plastic pipe, tubing, and fittings shall not be used to supply fuel gas. [NFPA 54:5.6.4.1.1 – 5.6.4.1.3]

1208.6.6 Regulator Vent Piping. Plastic pipe and fittings used to connect regulator vents to remote vent terminations shall be PVC conforming to UL 651 (Schedule 40 and 80). PVC vent piping shall not be installed indoors. [NFPA 54:5.6.4.2]

1208.6.7 Anodeless Risers. Anodeless risers shall comply with Section 1208.6.7.1 through Section 1208.6.7.3.

1208.6.7.1 Factory-Assembled Anodeless Risers. Factory-assembled anodeless risers shall be recommended by the manufacturer for the gas used and shall be leak tested by the manufacturer in accordance with written procedures. [NFPA 54:5.6.4.3(1)]

1208.6.7.2 Service Head Adapters and Field-Assembled Anodeless Risers. Service head adapters and field-assembled anodeless risers incorporating service head adapters shall be recommended by the manufacturer for the gas used and shall be design-certified to meet the requirements of Category I of ASTM D2513 and 49 CFR 192.281(e). The manufacturer shall provide the user qualified installation instructions as prescribed by 49 CFR 192.283(b). [NFPA 54:5.6.4.3(2)]

1208.6.7.3 Undiluted Liquified Petroleum Gas Piping. The use of plastic pipe, tubing, and fittings in undiluted LP-Gas piping systems shall be in accordance with NFPA 58. [NFPA 54:5.6.4.3(3)]

1208.6.8 Workmanship and Defects. Gas pipe, tubing, and fittings shall be clear and free from cutting burrs and defects in structure or threading, and shall be thoroughly brushed and chip and scale blown. Defects in pipe, tubing, and fittings shall not be repaired. Defective pipe, tubing, and fittings shall be replaced. [NFPA 54:5.6.5]

1208.6.9 Metallic Pipe Threads. Metallic pipe and fitting threads shall be taper pipe threads and shall comply with ASME B1.20.1. [NFPA 54:5.6.6.1]

1208.6.9.1 Damaged Threads. Pipe with threads that are stripped, chipped, corroded, or otherwise damaged shall not be used. Where a weld opens during the operation of cutting or threading, that portion of the pipe shall not be used. [NFPA 54:5.6.6.2]

1208.6.9.2 Number of Threads. Field threading of metallic pipe shall be in accordance with Table 1208.6.9.2. [NFPA 54:5.6.6.3]

TABLE 1208.6.9.2
SPECIFICATIONS FOR THREADING METALLIC PIPE
[NFPA 54: TABLE 5.6.6.3]

IRON PIPE SIZE (inches)	APPROXIMATE LENGTH OF THREADED PORTION (inches)	APPROXIMATE NUMBER OF THREADS TO BE CUT
½	¾	10
¾	¾	10
1	⅞	10
1¼	1	11
1½	1	11
2	1	11
2½	1½	12
3	1½	12
4	1⅝	13

For SI units: 1 inch = 25.4 mm

1208.6.9.3 Thread Joint Compounds. Thread joint compounds shall be resistant to the action of LP-Gas or to any other chemical constituents of the gases to be conducted through the piping. [NFPA 54:5.6.6.4]

1208.6.10 Metallic Piping Joints and Fittings. The type of piping joint used shall be suitable for the pressure and temperature conditions and shall be selected giving consideration to joint tightness and mechanical strength under the service conditions. The joint shall be able to sustain the maximum end force due to the internal pressure and any additional forces due to temperature expansion or contraction, vibration, fatigue, or the weight of the pipe and its contents. [NFPA 54:5.6.7]

1208.6.10.1 Pipe Joints. Schedule 40 and heavier pipe joints shall be threaded, flanged, brazed, welded, or assembled with press-connect fittings listed to CSA LC 4.

(1) Where nonferrous pipe is brazed, the brazing materials shall have a melting point in excess of 1000°F (538°C).

(2) Brazing alloys shall not contain more than 0.05 percent phosphorus. {NFPA 54:5.6.7.1}

1208.6.10.2 Copper Tubing Joints. Copper tubing joints shall be assembled with approved gas tubing fittings, shall be brazed with a material having a melting point in excess of 1000°F (538°C), or shall be assembled with press-connect fittings listed to CSA LC 4, Press-Connect Metallic Fittings for Use in Fuel Gas Distribution Systems. Brazing alloys shall not contain more than 0.05 percent phosphorus. [NFPA 54:5.6.7.2]

1208.6.10.3 Stainless Steel Tubing Joints. Stainless steel joints shall be welded, assembled with approved tubing fittings, brazed with a material having a melting point in excess of 1000°F (538°C), or assembled with press-connect fittings listed to CSA LC 4, Press-Connect Metallic Fittings for Use in Fuel Gas Distribution Systems. Brazing alloys and fluxes shall be recommended by the manufacturer for use on stainless steel alloys. [NFPA 54:5.6.7.3]

1208.6.10.4 Flared Joints. Flared joints shall be used only in systems constructed from nonferrous pipe and tubing where experience or tests have demonstrated that the joint is suitable for the conditions and where provisions are made in the design to prevent separation of the joints. [NFPA 54:5.6.7.4]

1208.6.10.5 Metallic Pipe Fittings. Metallic fittings shall comply with the following:

(1) Threaded fittings in sizes larger than 4 inches (100 mm) shall not be used.

(2) Fittings used with steel, stainless steel, or wrought-iron pipe shall be steel, stainless steel, copper alloy, malleable iron, or cast iron.

(3) Fittings used with copper or copper alloy pipe shall be copper or copper alloy.

(4) Fittings used with aluminum alloy pipe shall be aluminum alloy.

(5) Cast-iron fittings shall comply with the following:
    (a) Flanges shall be permitted.
    (b) Bushings shall not be used.
    (c) Fittings shall not be used in systems containing flammable gas-air mixtures.
    (d) Fittings in sizes 4 inches (100 mm) and larger shall not be used indoors unless approved by the Authority Having Jurisdiction.
    (e) Fittings in sizes 6 inches (150 mm) and larger shall not be used unless approved by the Authority Having Jurisdiction.

(6) Aluminum alloy fitting threads shall not form the joint seal.

(7) Zinc-aluminum alloy fittings shall not be used in systems containing flammable gas-air mixtures.

(8) Special fittings such as couplings, proprietary-type joints, saddle tees, gland-type compression fittings, and flared, flareless, or compression-type tubing fittings shall be as follows:
    (a) Used within the fitting manufacturer's pressure-temperature recommendations.
    (b) Used within the service conditions anticipated with respect to vibration, fatigue, thermal expansion, or contraction.
    (c) Acceptable to the Authority Having Jurisdiction.

(9) When pipe fittings are drilled and tapped in the field, the operation shall be in accordance with the following:
   (a) The operation shall be performed on systems having operating pressures of 5 psi (34 kPa) or less.
   (b) The operation shall be performed by the gas supplier or their designated representative.
   (c) The drilling and tapping operation shall be performed in accordance with written procedures prepared by the gas supplier.
   (d) The fittings shall be located outdoors.
   (e) The tapped fitting assembly shall be inspected and proven to be free of leaks. [NFPA 54:5.6.7.5]

1208.6.11 Plastic Piping, Joints, and Fittings. Plastic pipe, tubing, and fittings shall be joined in accordance with the manufacturer's instructions. Section 1208.6.11.1 through Section 1208.6.11.4 shall be observed when making such joints. [NFPA 54:5.6.8]

1208.6.11.1 Joint Design. The joint shall be designed and installed so that the longitudinal pull-out resistance of the joint will be at least equal to the tensile strength of the plastic piping material. [NFPA 54:5.6.8(1)]

1208.6.11.2 Heat-Fusion Joint. Heat-fusion joints shall be made in accordance with qualified procedures that have been established and proven by test to produce gastight joints at least as strong as the pipe or tubing being joined. Joints shall be made with the joining method recommended by the pipe manufacturer. Heat fusion fittings shall be marked "ASTM D2513." [NFPA 54:5.6.8(2)]

1208.6.11.3 Compression-Type Mechanical Joints. Where compression-type mechanical joints are used, the gasket material in the fitting shall be compatible with the plastic piping and with the gas distributed by the system. An internal tubular rigid stiffener shall be used in conjunction with the fitting. The stiffener shall be flush with the end of the pipe or tubing and shall extend at least to the outside end of the compression fitting when installed. The stiffener shall be free of rough or sharp edges and shall not be a force fit in the plastic. Split tubular stiffeners shall not be used. [NFPA 54:5.6.8(3)]

1208.6.11.4 Liquefied Petroleum Gas Piping Systems. Plastic piping joints and fittings for use in LP-Gas piping systems shall be in accordance with NFPA 58. [NFPA 54:5.6.8(4)]

1208.6.12 Flange Specification. Flanges shall comply with Section 1208.6.12.1 through Section 1208.6.12.7.

1208.6.12.1 Cast Iron Flanges. Cast iron flanges shall be in accordance with ASME B16.1. [NFPA 54:5.6.9.1.1]

1208.6.12.2 Steel Flanges. Steel flanges shall be in accordance with the following:

(1) ASME B16.5 or
(2) ASME B16.47. [NFPA 54:5.6.9.1.2]

1208.6.12.3 Non-Ferrous Flanges. Non-ferrous flanges shall be in accordance with ASME B16.24. [NFPA 54:5.6.9.1.3]

1208.6.12.4 Ductile Iron Flanges. Ductile iron flanges shall be in accordance with ASME B16.42. [NFPA 54:5.6.9.1.4]

1208.6.12.5 Dissimilar Flange Connections. Raised-face flanges shall not be joined to flat-faced cast iron, ductile iron or nonferrous material flanges. [NFPA 54:5.6.9.2]

1208.6.12.6 Flange Facings. Standard facings shall be permitted for use under this code. Where 150 psi (1034 kPa) steel flanges are bolted to Class 125 cast-iron flanges, the raised face on the steel flange shall be removed. [NFPA 54:5.6.9.3]

1208.6.12.7 Lapped Flanges. Lapped flanges shall be used only aboveground or in exposed locations accessible for inspection. [NFPA 54:5.6.9.4]

1208.6.13 Flange Gaskets. The material for gaskets shall be capable of withstanding the design temperature and pressure of the piping system and the chemical constituents of the gas being conducted without change to its chemical and physical properties. The effects of fire exposure to the joint shall be considered in choosing the material. [NFPA 54:5.6.10]

1208.6.13.1 Flange Gasket Materials. Acceptable materials shall include the following:

(1) Metal (plain or corrugated)
(2) Composition
(3) Aluminum "O" rings
(4) Spiral-wound metal gaskets
(5) Rubber-faced phenolic
(6) Elastomeric [NFPA 54:5.6.10.1]

1208.6.13.2 Metallic Flange Gaskets. Metallic flange gaskets shall be in accordance with ASME B16.20. [NFPA 54:5.6.10.2.1]

1208.6.13.3 Non-Metallic Flange Gaskets. Non-metallic flange gaskets shall be in accordance with ASME B16.21. [NFPA 54:5.6.10.2.2]

1208.6.13.4 Full-Face Flange Gasket. Full-face flange gaskets shall be used with all non-steel flanges. [NFPA 54:5.6.10.3]

1208.6.13.5 Separated Flanges. When a flanged joint is separated, the gasket shall be replaced. [NFPA 54:5.6.10.4]

1208.7 Gas Meters. Gas meters shall be selected for the maximum expected pressure and permissible pressure drop. [NFPA 54:5.7.1]

1208.7.1 Location. Gas meters shall be located in ventilated spaces readily accessible for examination, reading, replacement, or necessary maintenance. [NFPA 54:5.7.2.1]

1208.7.1.1 Subject to Damage. Gas meters shall not be placed where they will be subjected to damage, such as adjacent to a driveway, under a fire escape, in public passages, halls, or where they will be subject to excessive corrosion or vibration. [NFPA 54:5.7.2.2]

1208.7.1.2 Extreme Temperatures. Gas meters shall not be located where they will be subjected to extreme temperatures or sudden extreme changes in temperature or in areas where they are subjected to temperatures beyond those recommended by the manufacturer. [NFPA 54:5.7.2.3]

1208.7.2 Supports. Gas meters shall be supported or connected to rigid piping so as not to exert a strain on the meters. Where flexible connectors are used to connect a gas meter to downstream piping at mobile homes in mobile home parks, the meter shall be supported by a post or bracket placed in a firm footing or by other means providing equivalent support. [NFPA 54:5.7.3]

1208.7.3 Meter Protection. Meters shall be protected against overpressure, backpressure, and vacuum. [NFPA 54:5.7.4]

1208.7.4 Identification. Gas piping at multiple meter installations shall be marked by a metal tag or other permanent means designating the building or the part of the building being supplied and attached by the installing agency. [NFPA 54:5.7.5]

1208.8 Gas Pressure Regulators. A line pressure regulator shall be installed where the gas supply pressure exceeds the maximum allowable inlet pressure of the appliance served. [NFPA 54:5.8.1]

1208.8.1 Listing. Line pressure regulators shall be listed in accordance with CSA Z21.80 where the outlet pressure is set to 2 psi (14 kPa) or less. [NFPA 54:5.8.2]

1208.8.2 Location. The gas pressure regulator shall be accessible for servicing. [NFPA 54:5.8.3]

1208.8.3 Regulator Protection. Pressure regulators shall be protected against physical damage. [NFPA 54:5.8.4]

1208.8.4 Venting of Line Pressure Regulators. Line pressure regulators shall comply with all of the following:

(1) An independent vent to the exterior of the building, sized in accordance with the regulator manufacturer's instructions, shall be provided where the location of a regulator is such that a ruptured diaphragm will cause a hazard.

 (a) Where more than one regulator is at a location, each regulator shall have a separate vent to the outdoors or, if approved by the Authority Having Jurisdiction, the vent lines shall be permitted to be manifolded in accordance with accepted engineering practices to minimize back pressure in the event of diaphragm failure.

 (b) Materials for vent piping shall be in accordance with Section 1208.6 through Section 1208.6.13.5.

 Exception: A regulator and vent limiting means combination listed as complying with CSA Z21.80 shall be permitted to be used without a vent to the outdoors.

(2) The vent shall be designed to prevent the entry of water, insects, or other foreign materials that could cause blockage.

(3) The regulator vent shall terminate at least 3 feet (914 mm) from a source of ignition.

(4) At locations where regulators might be submerged during floods, a special antiflood-type breather vent fitting shall be installed, or the vent line shall be extended above the height of the expected flood waters.

(5) A regulator shall not be vented to the appliance flue or exhaust system. [NFPA 54:5.8.5.1]

1208.8.5 Venting of Gas Appliance Pressure Regulators. For venting of gas appliance pressure regulators see Section 507.21. [NFPA 54:5.8.5.2]

1208.8.6 Bypass Piping. Valved and regulated bypasses shall be permitted to be placed around gas line pressure regulators where continuity of service is imperative. [NFPA 54:5.8.6]

1208.8.7 Identification. Line pressure regulators at multiple regulator installations shall be marked by a metal tag or other permanent means designating the building or the part of the building being supplied. [NFPA 54:5.8.7]

1208.9 Overpressure Protection Required. Where the serving gas supplier delivers gas at a pressure greater than 2 psi (14 kPa) for piping systems serving appliances designed to operate at a gas pressure of 14 inches water column (3.5 kPa) or less, overpressure protection devices shall be installed. Piping systems serving equipment designed to operate at inlet pressures greater than 14 inches water column (3.5 kPa) shall be equipped with overpressure protection devices as required by the appliance manufacturer's installation instructions. [NFPA 54:5.9.1]

1208.10 Overpressure Protection Devices. Overpressure protection devices shall be one of the following:

(1) Pressure relief valve.

(2) Monitor regulator.

(3) Series regulator installed upstream from the line regulator and set to continuously limit the pressure on the inlet of the line regulator to the maximum values specified by Section 1208.11 or less.

(4) Automatic shutoff device installed in series with the line pressure regulator and set to shut off when the pressure on the downstream piping system reaches the maximum values specified by Section 1208.11 or less. This device shall be designed so that it will remain closed until manually reset. [NFPA 54:5.9.3.1]

1208.10.1 Separate Devices. The devices in Section 1208.10 shall be installed either as an integral part of the service or line pressure regulator or as separate units. Where separate overpressure protection devices are installed, they shall comply with Section 1208.10.2 through Section 1208.10.7. [NFPA 54:5.9.3.2]

1208.10.2 Construction and Installation. All overpressure protection devices shall meet the following requirements:

(1) Be constructed of materials so that the operation of the device is not impaired by corrosion of external parts by the atmosphere or of internal parts by the gas.

(2) Be designed and installed so they can be operated to determine whether the valve is free. The devices shall also be designed and installed so they can be tested to determine the pressure at which they operate and be examined for leakage when in the closed position. [NFPA 54:5.9.4]

1208.10.3 External Control Piping. External control piping shall be designed and installed so that damage to the control piping of one device does not render both the regulator and the overpressure protective device inoperative. [NFPA 54:5.9.5]

1208.10.4 Setting. Each pressure limiting or pressure relieving device shall be set so that the gas pressure supplied to the connected appliance(s) does not exceed the limits specified in Section 1208.11 and Section 1208.11.1. [NFPA 54:5.9.6]

1208.10.5 Unauthorized Operation. Where unauthorized operation of any shutoff valve could render a pressure relieving valve or pressure limiting device inoperative, one of the following shall be accomplished:

(1) The valve shall be locked in the open position. Instruct authorized personnel in the importance of leaving the shutoff valve open and of being present while the shutoff valve is closed so that it can be locked in the open position before leaving the premises.

(2) Duplicate relief valves shall be installed, each having adequate capacity to protect the system, and arrange the isolating valves or three-way valve so that only one relief valve can be rendered inoperative at a time. [NFPA 54:5.9.7]

1208.10.6 Discharge of Vents. The discharge stacks, vents, or outlet parts of all pressure relieving and pressure limiting devices shall be located so that gas is safely discharged to the outdoors. Discharge stacks or vents shall be designed to prevent the entry of water, insects, or other foreign material that could cause blockage.

The discharge stack or vent line shall be at least the same size as the outlet of the pressure relieving device. [NFPA 54:5.9.8.1 – 5.9.8.2]

1208.10.7 Size of Fittings, Pipe, and Openings. The fittings, pipe, and openings located between the system to be protected and the pressure relieving device shall be sized to prevent hammering of the valve and to prevent impairment of relief capacity. [NFPA 54:5.9.9]

1208.11 Pressure Limitation Requirements. Where piping systems serving appliances designed to operate with a gas supply pressure of 14 inches water column (3.5 kPa) or less are required to be equipped with overpressure protection by Section 1208.9, each overpressure protection device shall be adjusted to limit the gas pressure to each connected appliance to 2 psi (14 kPa) or less upon a failure of the line pressure regulator. [NFPA 54:5.9.2.1]

1208.11.1 Overpressure Protection Required. Where piping systems serving appliances designed to operate with a gas supply pressure greater than 14 inches water column (3.5 kPa) are required to be equipped with overpressure protection by Section 1208.9, each overpressure protection device shall be adjusted to limit the gas pressure to each connected appliance as required by the appliance manufacturer's installation instructions. [NFPA 54:5.9.2.2]

1208.11.2 Overpressure Protection Devices. Each overpressure protection device installed to meet the requirements of this section shall be capable of limiting the pressure to its connected appliance(s) as required by this section independently of any other pressure control equipment in the piping system. [NFPA 54:5.9.2.3]

1208.11.3 Detection of Failure. Each gas piping system for which an overpressure protection device is required by this section shall be designed and installed so that a failure of the primary pressure control device(s) is detectable. [NFPA 54:5.9.2.4]

1208.11.4 Flow Capacity. If a pressure relief valve is used to meet the requirements of this section, it shall have a flow capacity such that the pressure in the protected system is maintained at or below the limits specified in Section 1208.11 under the following conditions:

(1) The line pressure regulator for which the relief valve is providing overpressure protection has failed wide open.

(2) The gas pressure at the inlet of the line pressure regulator for which the relief valve is providing overpressure protection is not less than the regulator's normal operating inlet pressure. [NFPA 54:5.9.2.5]

1208.12 Backpressure Protection. Protective devices shall be installed as close to the equipment as practical where the design of equipment connected is such that air, oxygen, or standby gases could be forced into the gas supply system.

Gas and air combustion mixers incorporating double diaphragm "zero" or "atmosphere" governors or regulators shall require no further protection unless connected directly to compressed air or oxygen at pressures of 5 psi (34 kPa) or more. [NFPA 54:5.10.1.1 – 5.10.1.2]

1208.12.1 Protective Devices. Protective devices shall include but not be limited to the following:

(1) Check valves.

(2) Three-way valves (of the type that completely closes one side before starting to open the other side).

(3) Reverse flow indicators controlling positive shutoff valves.

(4) Normally closed air-actuated positive shutoff pressure regulators. [NFPA 54:5.10.2]

1208.13 Low-Pressure Protection. A protective device shall be installed between the meter and the appliance or equipment if the operation of the appliance or equipment is such that it could produce a vacuum or a dangerous reduction in gas pressure at the meter. Such protective devices include, but are not limited to, mechanical, diaphragm-operated, or electrically operated low-pressure shutoff valves. [NFPA 54:5.11]

1208.14 Shutoff Valves. Shutoff valves shall be approved and shall be selected giving consideration to pressure drop, service involved, emergency use, and reliability of operation. Shutoff valves of size 1 inch (25 mm) National Pipe Thread and smaller shall be listed. [NFPA 54:5.12]

1208.15 Expansion and Flexibility. Piping systems shall be designed to prevent failure from thermal expansion or contraction. [NFPA 54:5.14.1]

1208.15.1 Special Local Conditions. Where local conditions include earthquake, tornado, unstable ground, or flood hazards, special consideration shall be given to increased strength and flexibility of piping supports and connections. [NFPA 54:5.14.2]

1209.0 Excess Flow Valve.

1209.1 General. Where automatic excess flow valves are installed, they shall be listed to CSA Z21.93 and shall be sized and installed in accordance with the manufacturer's instructions. [NFPA 54:5.13]

1210.0 Gas Piping Installation.

1210.1 Piping Underground. Underground gas piping shall be installed with sufficient clearance from any other underground structure to avoid contact therewith, to allow maintenance, and to protect against damage from proximity to other structures. In addition, underground plastic piping shall be installed with sufficient clearance or shall be insulated from any source of heat so as to prevent the heat from impairing the serviceability of the pipe. [NFPA 54:7.1.1]

1210.1.1 Cover Requirements. Underground piping systems shall be installed with a minimum of 12 inches (305 mm) of cover. The minimum cover shall be increased to 18 inches (457 mm) if external damage to the pipe or tubing from external forces is likely to result. Where a minimum of 12 inches (305 mm) of cover cannot be provided, the pipe shall be installed in conduit or bridged (shielded). [NFPA 54:7.1.2.1 – 7.1.2.1(B)]

1210.1.2 Trenches. The trench shall be graded so that the pipe has a firm, substantially continuous bearing on the bottom of the trench. [NFPA 54:7.1.2.2]

1210.1.2.1 Backfilling. Where flooding of the trench is done to consolidate the backfill, care shall be exercised to see that the pipe is not floated from its firm bearing on the trench bottom. [NFPA 54:7.1.2.3]

1210.1.3 Protection Against Corrosion. Steel pipe and steel tubing installed underground shall be installed in accordance with Section 1210.1.3.1 through Section 1210.1.3.9. [NFPA 54:7.1.3]

1210.1.3.1 Zinc Coating. Zinc coating (galvanizing) shall not be deemed adequate protection for underground gas piping. [NFPA 54:7.1.3.1]

1210.1.3.2 Underground Piping. Underground piping shall comply with one or more of the following unless approved technical justification is provided to demonstrate that protection is unnecessary:

(1) The piping shall be made of corrosion-resistant material that is suitable for the environment in which it will be installed.

(2) Pipe shall have a factory-applied, electrically insulating coating. Fittings and joints between sections of coated pipe shall be coated in accordance with the coating manufacturer's instructions.

(3) The piping shall have a cathodic protection system installed, and the system shall be maintained in accordance with Section 1210.1.3.3 or Section 1210.1.3.6. [NFPA 54:7.1.3.2]

1210.1.3.3 Cathodic Protection. Cathodic protection systems shall be monitored by testing and the results shall be documented. The test results shall demonstrate one of the following:

(1) A pipe-to-soil voltage of −0.85 volts or more negative is produced, with reference to a saturated copper-copper sulfate half cell.

(2) A pipe-to-soil voltage of −0.78 volts or more negative is produced, with reference to a saturated KCl calomel half cell.

(3) A pipe-to-soil voltage of −0.80 volts or more negative is produced, with reference to a silver-silver chloride half cell.

(4) Compliance with a method described in Appendix D of Title 49 of the code of Federal Regulations, Part 192. [NFPA 54:7.1.3.3]

1210.1.3.4 Sacrificial Anodes. Sacrificial anodes shall be tested in accordance with the following:

(1) Upon installation of the cathodic protection system, except where prohibited by climatic conditions, in which case the testing shall be performed not later than 180 days after the installation of the system.

(2) 12 to 18 months after the initial test.

(3) Upon successful verification testing in accordance with Section 1210.1.3.4(1) and Section 1210.1.3.4(2), periodic follow-up testing shall be performed at intervals not to exceed 36 months. [NFPA 54:7.1.3.4]

1210.1.3.5 System Failing Tests. Systems failing a test shall be repaired not more than 180 days after the date of the failed testing. The testing schedule shall be restarted as required in Section 1210.1.3.4(1) and Section 1210.1.3.4(2), and the results shall comply with Section 1210.1.3.3. [NFPA 54:7.1.3.5]

1210.1.3.6 Impressed Current Cathodic Protection. Impressed current cathodic protection systems shall be inspected and tested in accordance with the following schedule:

(1) The impressed current rectifier voltage output shall be checked at intervals not exceeding two months.

(2) The pipe-to-soil voltage shall be tested at least annually. [NFPA 54:7.1.3.6]

1210.1.3.7 Documentation. Documentation of the results of the two most recent tests shall be retained. [NFPA 54:7.1.3.7]

1210.1.3.8 Dissimilar Metals. Where dissimilar metals are joined underground, an insulating coupling or fitting shall be used. [NFPA 54:7.1.3.8]

1210.1.3.9 Steel Risers. Steel risers, other than anodeless risers, connected to plastic piping shall be cathodically protected by means of a welded anode. [NFPA 54:7.1.3.9]

1210.1.4 Protection Against Freezing. Where the formation of hydrates or ice is known to occur, piping shall be protected against freezing. [NFPA 54:7.1.4]

1210.1.5 Piping through Foundation Wall. Underground piping, where installed through the outer foundation or basement wall of a building, shall be encased in a protective sleeve or protected by an approved device or method. The space between the gas piping and the sleeve and between the sleeve and the wall shall be sealed to prevent entry of gas and water. [NFPA 54:7.1.5]

1210.1.6 Piping Underground Beneath Buildings. Where gas piping is installed underground beneath buildings, the piping shall be either of the following:

(1) Encased in an approved conduit designed to withstand the imposed loads and installed in accordance with Section 1210.1.6.1 or Section 1210.1.6.2.

(2) A piping/encasement system listed for installation beneath buildings. [NFPA 54:7.1.6]

1210.1.6.1 Conduit with One End Terminating Outdoors. The conduit shall extend into an accessible portion of the building and, at the point where the conduit terminates in the building, the space between the conduit and the gas piping shall be sealed to prevent the possible entrance of any gas leakage. Where the end sealing is of a type that retains the full pressure of the pipe, the conduit shall be designed for the same pressure as the pipe. The conduit shall extend at least 4 inches (102 mm) outside the building, be vented outdoors above finished ground level, and be installed so as to prevent the entrance of water and insects. [NFPA 54:7.1.6.1]

1210.1.6.2 Conduit with Both Ends Terminating Indoors. Where the conduit originates and terminates within the same building, the conduit shall originate and terminate in an accessible portion of the building and shall not be sealed. [NFPA 54:7.1.6.2]

1210.1.7 Plastic Piping. Plastic piping shall be installed outdoors, underground only.

Exceptions:

(1) Plastic piping shall be permitted to terminate aboveground where an anodeless riser is used.

(2) Plastic piping shall be permitted to terminate with a wall head adapter aboveground in buildings, including basements, where the plastic piping is inserted in a piping material permitted for use in buildings. [NFPA 54:7.1.7.1]

1210.1.7.1 Connections Between Metallic and Plastic Piping. Connections made between metallic and plastic piping shall be made with fittings conforming to one of the following:

(1) ASTM D2513, Category I transition fittings

(2) ASTM F1973

(3) ASTM F2509 [NFPA 54:7.1.7.2]

1210.1.7.2 Tracer Wire. An electrically continuous corrosion-resistant tracer shall be buried with the plastic pipe to facilitate locating. The tracer shall be one of the following:

(1) A product specifically designed for that purpose.

(2) Insulated copper conductor not less than 14 AWG.

Where tracer wire is used, access shall be provided from aboveground, or one end of the tracer wire or tape shall be brought aboveground at a building wall or riser. [NFPA 54:7.1.7.3 – 7.1.7.3.2]

1210.2 CSST Piping Systems. CSST piping systems shall be installed in accordance with this code and the manufacturer's installation instructions. [NFPA 54:7.1.8]

1210.3 Installation of Aboveground Piping. Piping installed aboveground shall be securely supported and located where it will be protected from physical damage. Where passing through an exterior wall, the piping shall also be protected from corrosion by coating or wrapping with an inert material approved for such applications. The piping shall be sealed around its circumference at the point of the exterior penetration to prevent the entry of water, insects, and rodents. Where piping is encased in a protective pipe sleeve, the annular spaces between the gas piping and the sleeve and between the sleeve and the wall opening shall be sealed. [NFPA 54:7.2.1]

1210.3.1 Protective Coating. Where piping is in contact with a material or an atmosphere corrosive to the piping system, the piping and fittings shall be coated with a corrosion-resistant material. Any such coating used on piping or components shall not be considered as adding strength to the system. [NFPA 54:7.2.2]

1210.3.2 Building Structure. The installation of gas piping shall not cause structural stresses within building components to exceed allowable design limits. Approval shall be obtained before any beams or joists are cut or notched. [NFPA 54:7.2.3.1 – 7.2.3.2]

1210.3.3 Gas Piping to be Sloped. Piping for other than dry gas conditions shall be sloped not less than ¼ inch in 15 feet (6.4 mm in 4.6 m) to prevent traps. [NFPA 54:7.2.4]

1210.3.3.1 Ceiling Locations. Gas piping shall be permitted to be installed in accessible spaces between a fixed ceiling and a dropped ceiling, whether or not such spaces are used as a plenum. Valves shall not be located in such spaces.

Exception: Appliance or equipment shutoff valves required by this code shall be permitted to be installed in accessible spaces containing vented appliances.

1210.3.4 Prohibited Locations. Gas piping inside any building shall not be installed in or through a clothes chute, chimney or gas vent, dumbwaiter, elevator shaft, or air duct, other than combustion air ducts. [NFPA 54:7.2.5]

Exception: Ducts used to provide ventilation air in accordance with Section 506.0 or to above-ceiling spaces in accordance with Section 1210.3.3.1.

1210.3.5 Hangers, Supports, and Anchors. Piping shall be supported with metal pipe hooks, metal pipe straps, metal bands, metal brackets, metal hangers, or building structural components, suitable for the size of piping, of adequate strength and quality, and located at intervals so as to prevent or damp out excessive vibration. Piping shall be anchored to prevent undue strains on connected appliances and equipment and shall not be supported by other piping. Pipe hangers and supports shall conform to the requirements of MSS SP-58. [NFPA 54:7.2.6.1]

1210.3.5.1 Spacing. Spacing of supports in gas piping installations shall not be greater than shown in Table 1210.3.5.1. Spacing of supports of CSST shall be in accordance with the CSST manufacturer's instructions. [NFPA 54:7.2.6.2]

TABLE 1210.3.5.1
SUPPORT OF PIPING
[NFPA 54: TABLE 7.2.6.2]

STEEL PIPE, NOMINAL SIZE OF PIPE (inches)	SPACING OF SUPPORTS (feet)	NOMINAL SIZE OF TUBING SMOOTH-WALL (inches O.D.)	SPACING OF SUPPORTS (feet)
½	6	½	4
¾ or 1	8	⅝ or ¾	6
1¼ or larger (horizontal)	10	⅞ or 1 (horizontal)	8
1¼ or larger (vertical)	Every floor level	1 or larger (vertical)	Every floor level

For SI units: 1 inch = 25 mm, 1 foot = 304.8 mm

1210.3.5.2 Expansion and Contraction. Supports, hangers, and anchors shall be installed so as not to interfere with the free expansion and contraction of the piping between anchors. All parts of the supporting system shall be designed and installed so they are not disengaged by movement of the supported piping. [NFPA 54:7.2.6.3]

1210.3.5.3 Piping on Roof Tops. Gas piping installed on the roof surfaces shall be elevated above the roof surface and shall be supported in accordance with Table 1210.3.5.1. [NFPA 54:7.2.6.4]

1210.4 Concealed Piping in Buildings. Gas piping in concealed locations shall be installed in accordance with this section. [NFPA 54:7.3.1]

1210.4.1 Connections. Where gas piping is to be concealed, connections shall be of the following type:

(1) Pipe fittings such as elbows, tees, couplings, and right/left nipple/couplings.

(2) Joining tubing by brazing (see Section 1208.6.10.1).

(3) Press-connect fittings listed to CSA LC 4.

(4) CSST fittings listed to CSA LC 1.

(5) Where necessary to insert fittings in the gas pipe that has been installed in a concealed location, the pipe shall be reconnected by welding, flanges, or the use of a right/left nipple/coupling.

1210.4.2 Piping in Partitions. Concealed gas piping shall not be located in solid partitions. [NFPA 54:7.3.3]

1210.4.3 Tubing in Partitions. This provision shall not apply to tubing that pierces walls, floors, or partitions. Tubing installed vertically and horizontally inside hollow walls or partitions without protection along its entire concealed length shall meet the following requirements:

(1) A steel striker barrier not less than 0.0508 of an inch (1.3 mm) thick, or equivalent, is installed between the tubing and the finished wall and extends at least 4 inches (102 mm) beyond concealed penetrations of plates, firestops, wall studs, and so on.

(2) The tubing is installed in single runs and is not rigidly secured. [NFPA 54:7.3.4]

1210.4.4 Piping in Floors. In industrial occupancies, gas piping in solid floors such as concrete shall be laid in channels in the floor and covered to permit access to the piping with a minimum of damage to the building. Where piping in floor channels could be exposed to excessive moisture or corrosive substances, the piping shall be protected in an approved manner.

In other than industrial occupancies and where approved by the Authority Having Jurisdiction, gas piping embedded in concrete floor slabs constructed with portland cement shall be surrounded with a minimum of 1½ inches (38 mm) of concrete and shall not be in physical contact with other metallic structures such as reinforcing rods or electrically neutral conductors. All piping, fittings, and risers shall be protected against corrosion in accordance with Section 1210.3.1. Piping shall not be embedded in concrete slabs containing quick set additives or cinder aggregate. [NFPA 54:7.3.5.1 – 7.3.5.2]

1210.5 Piping in Vertical Chases. Where gas piping exceeding 5 psi (34 kPa) is located within vertical chases in accordance with Section 1208.5(3), the requirements of Section 1210.5.1 through Section 1210.5.3 shall apply. [NFPA 54:7.4]

1210.5.1 Pressure Reduction. Where pressure reduction is required in branch connections for compliance with Section 1208.5, such reduction shall take place either inside the chase or immediately adjacent to the outside wall of the chase. Regulator venting and downstream overpressure protection shall comply with Section 1208.8.4, Section 1208.8.5 and Section 1208.9 through Section 1208.11.4. The regulator shall be accessible for service and repair and vented in accordance with one of the following:

(1) Where the fuel gas is lighter than air, regulators equipped with a vent limiting means shall be permitted to be vented into the chase. Regulators not equipped with a vent limiting means shall be permitted to be vented either directly to the outdoors or to a point within the top 1 foot (305 mm) of the chase.

(2) Where the fuel gas is heavier than air, the regulator vent shall be vented only directly to the outdoors. [NFPA 54:7.4.1]

1210.5.2 Chase Construction. Chase construction shall comply with local building codes with respect to fire resistance and protection of horizontal and vertical openings. [NFPA 54:7.4.2]

1210.5.3 Ventilation. A chase shall be ventilated to the outdoors and only at the top. The opening(s) shall have a minimum free area [in square inches (square meters)] equal to the product of one-half of the maximum pressure in the piping [in pounds per square inch (kilopascals)] times the largest nominal diameter of that piping [in inches (millimeters)], or the cross-sectional area of the chase, whichever is smaller. Where more than one fuel gas piping system is present, the free area for each system shall be calculated and the largest area used. [NFPA 54:7.4.3]

1210.6 Gas Pipe Turns. Changes in direction of gas pipe shall be made by the use of fittings, factory bends, or field bends. [NFPA 54:7.5]

1210.6.1 Metallic Pipe. Metallic pipe bends shall comply with the following:

(1) Bends shall be made only with bending tools and procedures intended for that purpose.

(2) All bends shall be smooth and free from buckling, cracks, or other evidence of mechanical damage.

(3) The longitudinal weld of the pipe shall be near the neutral axis of the bend.

(4) Pipe shall not be bent through an arc of more than 90 degrees.

(5) The inside radius of a bend shall be not less than 6 times the outside diameter of the pipe. [NFPA 54:7.5.1]

1210.6.2 Plastic Pipe. Plastic pipe bends shall comply with the following:

(1) The pipe shall not be damaged, and the internal diameter of the pipe shall not be effectively reduced.

(2) Joints shall not be located in pipe bends.

(3) The radius of the inner curve of such bends shall not be less than 25 times the inside diameter of the pipe.

(4) Where the piping manufacturer specifies the use of special bending tools or procedures, such tools or procedures shall be used. [NFPA 54:7.5.2]

1210.6.3 Elbows. Factory-made welding elbows or transverse segments cut therefrom shall have an arc length measured along the crotch of at least 1 inch (25.4 mm) for pipe sizes 2 inches (50 mm) and larger. [NFPA 54:7.5.3]

1210.7 Drips and Sediment Traps. For other than dry gas conditions, a drip shall be provided at any point in the line of pipe where condensate could collect. Where required by the Authority Having Jurisdiction or the serving gas supplier, a drip shall also be provided at the outlet of the meter. This drip shall be installed so as to constitute a trap wherein an accumulation of condensate shuts off the flow of gas before it runs back into the meter. [NFPA 54:7.6.1]

1210.7.1 Location of Drips. All drips shall be installed only in such locations that they are readily accessible to permit cleaning or emptying. A drip shall not be located where the condensate is likely to freeze. [NFPA 54:7.6.2]

1210.7.2 Sediment Traps. The installation of sediment traps shall be in accordance with Section 1212.9. [NFPA 54:7.6.3]

1210.8 Outlets. Outlets shall be located and installed in accordance with the following requirements:

(1) The outlet fittings or piping shall be securely fastened in place.

(2) Outlets shall not be located behind doors.

(3) Outlets shall be located far enough from floors, walls, patios, slabs, and ceilings to permit the use of wrenches without straining, bending, or damaging the piping.

(4) The unthreaded portion of gas piping outlets shall extend not less than 1 inch (25.4 mm) through finished ceilings or indoor or outdoor walls.

(5) The unthreaded portion of gas piping outlets shall extend not less than 2 inches (51 mm) above the surface of floors or outdoor patios or slabs.

(6) The provisions of Section 1210.8(4) and Section 1210.8(5) shall not apply to listed quick-disconnect devices of the flush-mounted type or listed gas convenience outlets. Such devices shall be installed in accordance with the manufacturer's installation instructions. [NFPA 54:7.7.1.1 – 7.7.1.6]

1210.8.1 Cap Outlets. Each outlet, including a valve, shall be closed gastight with a threaded plug or cap immediately after installation and shall be left closed until the appliance or equipment is connected thereto. When an appliance or equipment is disconnected from an outlet and the outlet is not to be used again immediately, it shall be capped or plugged gastight.

Exceptions:

(1) Laboratory appliances installed in accordance with Section 1212.3.1 shall be permitted.

(2) The use of a listed quick-disconnect device with integral shutoff or listed gas convenience outlet shall be permitted. [NFPA 54:7.7.2.1]

1210.8.2 Appliance Shutoff Valves. Appliance shutoff valves installed in fireplaces shall be removed, and the piping capped gastight where the fireplace is used for solid fuel burning. [NFPA 54:7.7.2.2]

1210.9 Manual Gas Shutoff Valves. An accessible gas shutoff valve shall be provided upstream of each gas pressure regulator. Where two gas pressure regulators are installed in series in a single gas line, a manual valve shall not be required at the second regulator. [NFPA 54:7.8.1]

1210.9.1 Valves Controlling Multiple Systems. Main gas shutoff valves controlling several gas piping systems shall be readily accessible for operation and installed so as to be protected from physical damage. They shall be marked with a metal tag or other permanent means attached by the installing agency so that the gas piping systems supplied through them can be readily identified. [NFPA 54:7.8.2.1]

1210.9.2 Shutoff Valves for Multiple House Lines. In multiple-tenant buildings supplied through a master meter, through one service regulator where a meter is not provided, or where meters or service regulators are not readily accessible from the appliance or equipment location, an individual shutoff valve for each apartment or tenant line shall be provided at a convenient point of general accessibility. In a common system serving a number of individual buildings, shutoff valves shall be installed at each building. [NFPA 54:7.8.2.2]

1210.9.3 Emergency Shutoff Valves. An exterior shutoff valve to permit turning off the gas supply to each building in an emergency shall be provided. The emergency shutoff valves shall be plainly marked as such and their locations posted as required by the Authority Having Jurisdiction. [NFPA 54:7.8.2.3]

1210.9.4 Shutoff Valve for Laboratories. Each laboratory space containing two or more gas outlets installed on tables, benches, or in hoods in educational, research, commercial and industrial occupancies shall have a single shutoff valve through which all such gas outlets are supplied. The shutoff valve shall be accessible, located within the laboratory or adjacent to the laboratory's egress door, and identified. [NFPA 54:7.8.2.4]

1210.10 Prohibited Devices. Devices shall not be placed within the interior of gas piping or fittings where such devices reduce the cross-sectional area or otherwise obstruct the free flow of gas, except where allowance in the piping system design has been made for such devices. [NFPA 54:7.9]

1210.11 Systems Containing Gas-Air Mixtures Outside the Flammable Range. Where gas-air mixing machines are employed to produce mixtures above or below the flammable range, they shall be provided with stops to prevent adjustment of the mixture to within or approaching the flammable range. [NFPA 54:7.10]

1210.12 Systems Containing Flammable Gas-Air Mixtures. Systems containing flammable gas-air mixtures shall be in accordance with Section 1210.12.1 through Section 1210.12.6.

1210.12.1 Required Components. A central premix system with a flammable mixture in the blower or compressor shall consist of the following components:

(1) Gas-mixing machine in the form of an automatic gas-air proportioning device combined with a downstream blower or compressor.
(2) Flammable mixture piping, minimum Schedule 40.
(3) Automatic firecheck(s).
(4) Safety blowout(s) or backfire preventers for systems utilizing flammable mixture lines above $2\frac{1}{2}$ inches (65 mm) nominal pipe size (NPS) or the equivalent. [NFPA 54:7.11.1]

1210.12.2 Optional Components. The following components shall also be permitted to be utilized in any type of central premix system:

(1) Flowmeter(s)
(2) Flame arrester(s) [NFPA 54:7.11.2]

1210.12.3 Additional Requirements. Gas-mixing machines shall have nonsparking blowers and shall be constructed so that a flashback does not rupture machine casings. [NFPA 54:7.11.3]

1210.12.4 Special Requirements for Mixing Blowers. A mixing blower system shall be limited to applications with minimum practical lengths of mixture piping, limited to a maximum mixture pressure of 10 inches water column (2.5 kPa) and limited to gases containing no more than 10 percent hydrogen. The blower shall be equipped with a gas control valve at its air entrance arranged so that gas is admitted to the airstream, entering the blower in proper proportions for correct combustion by the type of burners employed, the said gas control valve being of either the zero governor or mechanical ratio valve type that controls the gas and air adjustment simultaneously. No valves or other obstructions shall be installed between the blower discharge and the burner or burners. [NFPA 54:7.11.4]

1210.12.5 Installation of Gas-Mixing Machines. Installation of gas-mixing machines shall comply with the following:

(1) The gas-mixing machine shall be located in a well-ventilated area or in a detached building or cutoff room provided with room construction and explosion vents in accordance with sound engineering principles. Such rooms or below-grade installations shall have adequate positive ventilation.
(2) Where gas-mixing machines are installed in well-ventilated areas, the type of electrical equipment shall be in accordance with NFPA 70 for general service conditions unless other hazards in the area prevail. Where gas-mixing machines are installed in small detached buildings or cutoff rooms, the electrical equipment and wiring shall be installed in accordance with NFPA 70 for hazardous locations (Articles 500 and 501, Class I, Division 2).
(3) Air intakes for gas-mixing machines using compressors or blowers shall be taken from outdoors whenever practical.
(4) Controls for gas-mixing machines shall include interlocks and a safety shutoff valve of the manual reset type in the gas supply connection to each machine arranged to automatically shut off the gas supply in

the event of high or low gas pressure. Except for open burner installations only, the controls shall be interlocked so that the blower or compressor stops operating following a gas supply failure. Where a system employs pressurized air, means shall be provided to shut off the gas supply in the event of air failure.

(5) Centrifugal gas-mixing machines in parallel shall be reviewed by the user and equipment manufacturer before installation, and means or plans for minimizing the effects of downstream pulsation and equipment overload shall be prepared and utilized as needed. [NFPA 54:7.11.5.1 – 7.11.5.5]

1210.12.6 Use of Automatic Firechecks, Safety Blowouts, or Backfire Preventers. Automatic firechecks and safety blowouts or backfire preventers shall be provided in piping systems distributing flammable air-gas mixtures from gas-mixing machines to protect the piping and the machines in the event of flashback, in accordance with the following:

(1) Approved automatic firechecks shall be installed upstream as close as practical to the burner inlets following the firecheck manufacturer's instructions.

(2) A separate manually operated gas valve shall be provided at each automatic firecheck for shutting off the flow of the gas-air mixture through the firecheck after a flashback has occurred. The valve shall be located upstream as close as practical to the inlet of the automatic firecheck.

Caution: These valves shall not be reopened after a flashback has occurred until the firecheck has cooled sufficiently to prevent re-ignition of the flammable mixture and has been reset properly.

(3) A safety blowout or backfiring preventer shall be provided in the mixture line near the outlet of each gas-mixing machine where the size of the piping is larger than 2½ inches (65 mm) NPS, or equivalent, to protect the mixing equipment in the event of an explosion passing through an automatic firecheck. The manufacturer's instructions shall be followed when installing these devices, particularly after a disc has burst. The discharge from the safety blowout or backfire preventer shall be located or shielded so that particles from the ruptured disc cannot be directed toward personnel. Wherever there are interconnected installations of gas-mixing machines with safety blowouts or backfire preventers, provision shall be made to keep the mixture from other machines from reaching any ruptured disc opening. Check valves shall not be used for this purpose.

(4) Large-capacity premix systems provided with explosion heads (rupture discs) to relieve excessive pressure in pipelines shall be located at and vented to a safe outdoor location. Provisions shall be provided for automatically shutting off the supply of the gas-air mixture in the event of rupture. [NFPA 54:7.11.6]

1211.0 Electrical Bonding and Grounding.

1211.1 Pipe and Tubing Other than CSST. Each above-ground portion of a gas piping system, other than CSST, that is likely to become energized shall be electrically continuous and bonded to an effective ground-fault current path. Gas piping, other than CSST, shall be considered to be bonded when it is connected to appliances that are connected to the appliance grounding conductor of the circuit supplying that appliance. [NFPA 54:7.12.1]

1211.2 Bonding of CSST Gas Piping. CSST gas piping systems, and gas piping systems containing one or more segments of CSST, shall be electrically continuous and bonded to the electrical service grounding electrode system or, where provided, lightning protection grounding electrode system. [NFPA 54:7.12.2]

1211.2.1 Bonding Jumper Connection. The bonding jumper shall connect to a metallic pipe, pipe fitting, or CSST fitting. [NFPA 54:7.12.2.1]

1211.2.2 Bonding Jumper Size. The bonding jumper shall not be smaller than 6 AWG copper wire or equivalent. [NFPA 54:7.12.2.2]

1211.2.3 Bonding Jumper Length. The length of the jumper between the connection to the gas piping system and the grounding electrode system shall not exceed 75 feet (22 860 mm). Any additional grounding electrodes installed to meet this requirement shall be bonded to the electrical service grounding electrode system or, where provided, lightning protection grounding electrode system. [NFPA 54:7.12.2.3]

1211.2.4 Bonding Connections. Bonding connections shall be in accordance with NFPA 70. [NFPA 54:7.12.2.4]

1211.2.5 Devices Used for Bonding. Devices used for the bonding connection shall be listed for the application in accordance with UL 467. [NFPA 54:7.12.2.5]

1211.3 Arc-Resistant Jacketed CSST. CSST listed with an arc resistant jacket or coating system in accordance with CSA LC 1 shall be electrically continuous and bonded to an effective ground fault current path. Where any CSST component of a piping system does not have an arc-resistant jacket or coating system, the bonding requirements of Section 1211.2 shall apply. Arc-resistant jacketed CSST shall be considered to be bonded when it is connected to appliances that are connected to the appliance grounding conductor of the circuit supplying that appliance. [NFPA 54:7.12.3]

1211.4 Prohibited Use. Gas piping shall not be used as a grounding conductor or electrode. [NFPA 54:7.12.4]

1211.5 Lightning Protection System. Where a lightning protection system is installed, the bonding of the gas piping shall be in accordance with NFPA 780. [NFPA 54:7.12.5]

1211.6 Electrical Circuits. Electrical circuits shall not utilize gas piping or components as conductors.

Exception: Low-voltage (50 V or less) control circuits, ignition circuits, and electronic flame detection device circuits shall be permitted to make use of piping or components as a part of an electric circuit. [NFPA 54:7.13]

1211.7 Electrical Connections. All electrical connections between wiring and electrically operated control devices in a piping system shall conform to the requirements of NFPA 70. [NFPA 54:7.14.1]

1211.7.1 Safety Control. Any essential safety control depending on electric current as the operating medium shall be of a type that shuts off (fail safe) the flow of gas in the event of current failure. [NFPA 54:7.14.2]

1212.0 Appliance and Equipment Connections to Building Piping.

1212.1 Connecting Appliances and Equipment. Appliances and equipment shall be connected to the building piping in compliance with Section 1212.6 through Section 1212.8 by one of the following:

(1) Rigid metallic pipe and fittings.

(2) Semirigid metallic tubing and metallic fittings. Aluminum alloy tubing shall not be used in exterior locations.

(3) A listed connector in compliance with CSA Z21.24. The connector shall be used in accordance with the manufacturer's installation instructions and shall be in the same room as the appliance. Only one connector shall be used per appliance.

(4) A listed connector in compliance with CSA Z21.75. Only one connector shall be used per appliance.

(5) CSST where installed in accordance with the manufacturer's installation instructions. CSST shall connect only to appliances that are fixed in place.

(6) Listed nonmetallic gas hose connectors in accordance with Section 1212.3.

(7) Unlisted gas hose connectors for use in laboratories and educational facilities in accordance with Section 1212.4. [NFPA 54:9.6.1]

1212.1.1 Commercial Cooking Appliances. Connectors used with commercial cooking appliances that are moved for cleaning and sanitation purposes shall be installed in accordance with the connector manufacturer's installation instructions. Such connectors shall be listed in accordance with CSA Z21.69. [NFPA 54:9.6.1.3]

1212.1.2 Restraining Device. Movement of appliances with casters shall be limited by a restraining device installed in accordance with the connector and appliance manufacturer's installation instructions. [NFPA 54:9.6.1.4]

1212.2 Suspended Low-Intensity Infrared Tube Heaters. Suspended low-intensity infrared tube heaters shall be connected to the building piping system with a connector listed for the application in accordance with CSA Z21.24 as follows:

(1) The connector shall be installed in accordance with the tube heater installation instructions and shall be in the same room as the appliance.

(2) Only one connector shall be used per appliance. [NFPA 54:9.6.1.5]

1212.3 Use of Nonmetallic Gas Hose Connectors. Listed gas hose connectors shall be used in accordance with the manufacturer's installation instructions and in accordance with Section 1212.3.1 and Section 1212.3.2. [NFPA 54:9.6.2]

1212.3.1 Indoor. Indoor gas hose connectors shall be used only to connect laboratory, shop, and ironing appliances requiring mobility during operation and installed in accordance with the following:

(1) An appliance shutoff valve shall be installed where the connector is attached to the building piping.

(2) The connector shall be of minimum length and shall not exceed 6 feet (1829 mm).

(3) The connector shall not be concealed and shall not extend from one room to another or pass through wall partitions, ceilings, or floors. [NFPA 54:9.6.2(1)]

1212.3.2 Outdoor. Where outdoor gas hose connectors are used to connect portable outdoor appliances, the connector shall be listed in accordance with CSA Z21.54 and installed in accordance with the following:

(1) An appliance shutoff valve, a listed quick-disconnect device, or a listed gas convenience outlet shall be installed where the connector is attached to the supply piping and in such a manner so as to prevent the accumulation of water or foreign matter.

(2) This connection shall be made only in the outdoor area where the appliance is to be used. [NFPA 54:9.6.2(2)]

(3) The connector length shall not exceed 15 feet (4572 mm).

1212.4 Injection (Bunsen) Burners. Injection (Bunsen) burners used in laboratories and educational facilities shall be permitted to be connected to the gas supply by an unlisted hose. [NFPA 54:9.6.3]

1212.5 Connection of Portable and Mobile Industrial Gas Appliances. Where portable industrial appliances or appliances requiring mobility or subject to vibration are connected to the building gas piping system by the use of a flexible hose, the hose shall be suitable and safe for the conditions under which it can be used. [NFPA 54:9.6.4.1]

1212.5.1 Swivel Joints or Couplings. Where industrial appliances requiring mobility are connected to the rigid piping by the use of swivel joints or couplings, the swivel joints or couplings shall be suitable for the service required, and only the minimum number required shall be installed. [NFPA 54:9.6.4.2]

1212.5.2 Metal Flexible Connectors. Where industrial appliances subject to vibration are connected to the building piping system by the use of all metal flexible connectors, the connectors shall be suitable for the service required. [NFPA 54:9.6.4.3]

1212.5.3 Flexible Connectors. Where flexible connections are used, they shall be of the minimum practical length and shall not extend from one room to another or pass through any walls, partitions, ceilings, or floors. Flexible connections shall not be used in any concealed location. They shall be protected against physical or thermal damage and shall be provided with gas shutoff valves in readily accessible locations in rigid piping upstream from the flexible connections. [NFPA 54:9.6.4.4]

1212.6 Appliance Shutoff Valves and Connections. Each appliance connected to a piping system shall have an accessible, approved manual shutoff valve with a nondisplaceable valve member, or a listed gas convenience outlet. Appliance shutoff valves and convenience outlets shall serve a single appliance only. [NFPA 54:9.6.5] The shutoff valve shall be located within 6 feet (1829 mm) of the appliance it serves. [NFPA 54:9.6.5.1] Where a connector is used, the valve shall be installed upstream of the connector. A union or flanged connection shall be provided downstream from the valve to permit removal of appliance controls. [NFPA 54:9.6.5.1(A)]

Exceptions:

(1) Shutoff valves serving decorative appliances in a fireplace shall not be located within the fireplace firebox except where the valve is listed for such use. [NFPA 54:9.6.5.1(B)]

(2) Shutoff valves shall be permitted to be accessibly located inside wall heaters and wall furnaces listed for recessed installation where necessary maintenance is performed without removal of the shutoff valve.

1212.7 Quick-Disconnect Devices. Quick-disconnect devices used to connect appliances to the building piping shall be listed to CSA Z21.41. Where installed indoors, an approved manual shutoff valve with a non-displaceable valve member shall be installed upstream of the quick-disconnect device. [NFPA 54:9.6.6.1 – 9.6.6.2]

1212.8 Gas Convenience Outlets. Appliances shall be permitted to be connected to the building piping by means of a listed gas convenience outlet, in conjunction with a listed appliance connector, installed in accordance with the manufacturer's installation instructions.

Gas convenience outlets shall be listed in accordance with CSA Z21.90 and installed in accordance with the manufacturer's installation instructions. [NFPA 54:9.6.7]

1212.9 Sediment Trap. Where a sediment trap is not incorporated as a part of the appliance, a sediment trap shall be installed downstream of the appliance shutoff valve as close to the inlet of the appliance as practical at the time of appliance installation. The sediment trap shall be either a tee fitting with a capped nipple in the bottom outlet, as illustrated in Figure 1212.9, or another device recognized as an effective sediment trap. Illuminating appliances, gas ranges, clothes dryers, decorative appliances for installation in vented fireplaces, gas fireplaces, and outdoor cooking appliances shall not be required to be so equipped. [NFPA 54:9.6.8]

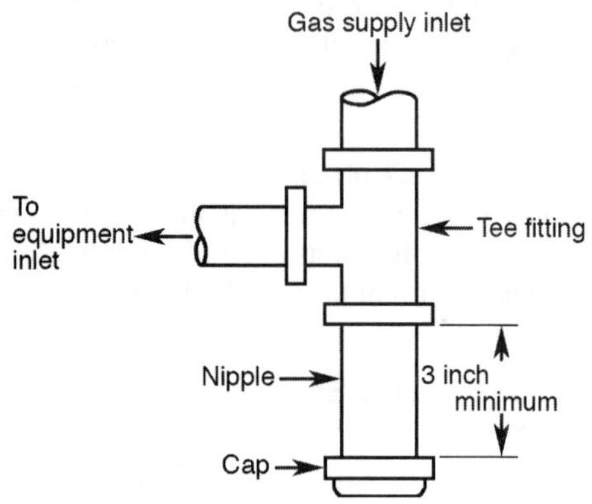

For SI units: 1 inch = 25.4 mm

FIGURE 1212.9
METHOD OF INSTALLING A TEE FITTING SEDIMENT TRAP [NFPA 54: FIGURE 9.6.8]

1212.10 Installation of Piping. Piping shall be installed in a manner not to interfere with inspection, maintenance, or servicing of the appliances. [NFPA 54:9.6.9]

1212.11 Liquefied Petroleum Gas Facilities and Piping. Liquefied petroleum gas facilities shall be in accordance with NFPA 58.

1213.0 Pressure Testing, Inspection, and Purging.

1213.1 Piping Installations. Prior to acceptance and initial operation, all piping installations shall be visually inspected and pressure tested to determine that the materials, design, fabrication, and installation practices comply with the requirements of this code. [NFPA 54:8.1.1.1]

1213.1.1 Inspection Requirements. Inspection shall consist of visual examination, during or after manufacture, fabrication, assembly, or pressure tests. [NFPA 54:8.1.1.2]

1213.1.2 Repairs and Additions. Where repairs or additions are made following the pressure test, the affected piping shall be tested. Minor repairs and additions are not required to be pressure tested, provided that the work is inspected and connections are tested with a noncorrosive leak-detecting fluid or other leak-detecting methods approved by the Authority Having Jurisdiction. [NFPA 54:8.1.1.3]

1213.1.3 New Branches. Where new branches are installed to new appliance(s), only the newly installed branch(es) shall be required to be pressure tested. Connections between the new piping and the existing piping shall be tested with a noncorrosive leak-detecting fluid or approved leak-detecting methods. [NFPA 54:8.1.1.4]

1213.1.4 Piping System. A piping system shall be tested as a complete unit or in sections. Under no circumstances shall a valve in a line be used as a bulkhead between gas in one section of the piping system and test medium in an adjacent section, unless a double block and bleed valve system is installed. A valve shall not be subjected to the test pressure unless it can be determined that the valve, including the valve closing mechanism, is designed to safely withstand the pressure. [NFPA 54:8.1.1.5]

1213.1.5 Regulators and Valves. Regulator and valve assemblies fabricated independently of the piping system in which they are to be installed shall be permitted to be tested with inert gas or air at the time of fabrication. [NFPA 54:8.1.1.6]

1213.1.6 Test Medium. The test medium shall be air, nitrogen, carbon dioxide, or an inert gas. Oxygen shall not be used as a test medium. [NFPA 54:8.1.2]

1213.2 Test Preparation. Test preparation shall comply with Section 1213.2.1 through Section 1213.2.6.

1213.2.1 Pipe Joints. Pipe joints, including welds, shall be left exposed for examination during the test.

Exception: Covered or concealed pipe end joints that have been previously tested in accordance with this code. [NFPA 54:8.1.3.1]

1213.2.2 Expansion Joints. Expansion joints shall be provided with temporary restraints, if required, for the additional thrust load under test. [NFPA 54:8.1.3.2]

1213.2.3 Appliances and Equipment. Appliances and equipment that are not to be included in the test shall be either disconnected from the piping or isolated by blanks, blind flanges, or caps. Flanged joints at which blinds are inserted to blank off other equipment during the test shall not be required to be tested. [NFPA 54:8.1.3.3]

1213.2.4 Designed for Operating Pressures Less Than Test Pressure. Where the piping system is connected to appliances or equipment designed for operating pressures of less than the test pressure, such appliances or equipment shall be isolated from the piping system by disconnecting them and capping the outlet(s). [NFPA 54:8.1.3.4]

1213.2.5 Designed for Operating Pressures Equal to or Greater Than Test Pressure. Where the piping system is connected to appliances or equipment designed for operating pressures equal to or greater than the test pressure, such appliances or equipment shall be isolated from the piping system by closing the individual appliance or equipment shutoff valve(s). [NFPA 54:8.1.3.5]

1213.2.6 Safety. All testing of piping systems shall be performed in a manner that protects the safety of employees and the public during the test. [NFPA 54:8.1.3.6]

1213.3 Test Pressure. This inspection shall include an air, CO_2, or nitrogen pressure test, at which time the gas piping shall stand a pressure of not less than 10 psi (69 kPa) gauge pressure. Test pressures shall be held for a length of time satisfactory to the Authority Having Jurisdiction but in no case less than 15 minutes with no perceptible drop in pressure. For welded piping, and for piping carrying gas at pressures in excess of 14 inches water column pressure (3.5 kPa), the test pressure shall be not less than 60 psi (414 kPa) and shall be continued for a length of time satisfactory to the Authority Having Jurisdiction, but in no case for less than 30 minutes. For CSST carrying gas at pressures in excess of 14 inches water column (3.5 kPa) pressure, the test pressure shall be not less than 30 psi (207 kPa) for 30 minutes. These tests shall be made using air, CO_2, or nitrogen pressure and shall be made in the presence of the Authority Having Jurisdiction. Necessary apparatus for conducting tests shall be furnished by the permit holder. Test gauges used in conducting tests shall be in accordance with Section 318.0.

1213.4 Detection of Leaks and Defects. The piping system shall withstand the test pressure specified without showing any evidence of leakage or other defects. Any reduction of test pressures as indicated by pressure gauges shall be deemed to indicate the presence of a leak unless such reduction can be readily attributed to some other cause. [NFPA 54:8.1.5.1]

1213.4.1 Detecting Leaks. The leakage shall be located by means of an approved gas detector, a noncorrosive leak detection fluid, or other approved leak detection methods. [NFPA 54:8.1.5.2]

1213.4.2 Repair or Replace. Where leakage or other defects are located, the affected portion of the piping system shall be repaired or replaced and retested. [NFPA 54:8.1.5.3]

1213.5 Piping System Leak Test. Leak checks using fuel gas shall be permitted in piping systems that have been pressure-tested in accordance with Section 1213.0 through Section 1213.4.2. [NFPA 54:8.2.1]

1213.5.1 Turning Gas On. During the process of turning gas on into a system of new gas piping, the entire system shall be inspected to determine that there are no open fittings or ends and that all valves at unused outlets are closed and plugged or capped. [NFPA 54:8.2.2]

1213.5.2 Leak Check. Immediately after the gas is turned on into a new system or into a system that has been initially restored after an interruption of service, the piping system shall be checked for leakage. Where leakage is indicated, the gas supply shall be shut off until the necessary repairs have been made. [NFPA 54:8.2.3]

1213.5.3 Placing Appliances and Equipment in Operation. Appliances and equipment shall not be placed in operation until after the piping system has been checked for leakage in accordance with Section 1213.5.2, the piping system is purged in accordance with Section 1213.6, and connections to the appliance are checked for leakage. [NFPA 54:8.2.4]

1213.6 Purging Requirements. The purging of piping shall be in accordance with Section 1213.6.1 through Section 1213.6.3. [NFPA 54:8.3]

1213.6.1 Piping Systems Required to be Purged Outdoors. The purging of piping systems shall be in accordance with Section 1213.6.1.1 through Section 1213.6.1.5 where the piping system meets either of the following:

(1) The design operating gas pressure is greater than 2 psig (14 kPa).

(2) The piping being purged contains one or more sections of pipe or tubing meeting the size and length criteria of Table 1213.6.1. [NFPA 54:8.3.1]

TABLE 1213.6.1
SIZE AND LENGTH OF PIPING
[NFPA 54: TABLE 8.3.1]*

NOMINAL PIPING SIZE (inches)	LENGTH OF PIPING (feet)
≥ 2½ < 3	> 50
≥ 3 < 4	> 30
≥ 4 < 6	> 15
≥ 6 < 8	> 10
≥ 8	Any length

For SI units: 1 inch = 25 mm, 1 foot = 304.8 mm
* CSST EHD size of 62 is equivalent to nominal 2 inches (50 mm) pipe or tubing size.

1213.6.1.1 Removal from Service. Where existing gas piping is opened, the section that is opened shall be isolated from the gas supply and the line pressure vented in accordance with Section 1213.6.1.4. Where gas piping meeting the criteria of Table 1213.6.1 is removed from service, the residual fuel gas in the piping shall be displaced with an inert gas. [NFPA 54:8.3.1.1]

1213.6.1.2 Removal of Piping. Where piping containing gas is to be removed, the line shall be first disconnected from sources of gas and then thoroughly purged with air, water, or inert gas before cutting, or welding is done.

1213.6.1.3 Placing in Operation. Where gas piping containing air and meeting the criteria of Table 1213.6.1 is placed in operation, the air in the piping shall first be displaced with an inert gas. The inert gas shall then be displaced with fuel gas in accordance with Section 1213.6.1.4. [NFPA 54:8.3.1.2]

1213.6.1.4 Outdoor Discharge of Purged Gases. The open end of a piping system being pressure vented or purged shall discharge directly to an outdoor location. Purging operations shall comply with all of the following requirements:

(1) The point of discharge shall be controlled with a shutoff valve.

(2) The point of discharge shall be located at least10 feet (3048 mm) from sources of ignition, at least 10 feet (3048 mm) from building openings and at least 25 feet (7620 mm) from mechanical air intake openings.

(3) During discharge, the open point of discharge shall be continuously attended and monitored with a combustible gas indicator that complies with Section 1213.6.1.5.

(4) Purging operations introducing fuel gas shall be stopped when 90 percent fuel gas by volume is detected within the pipe.

(5) Persons not involved in the purging operations shall be evacuated from all areas within 10 feet (3048 mm) of the point of discharge. [NFPA 54:8.3.1.3]

1213.6.1.5 Combustible Gas Indicator. Combustible gas indicators shall be listed and calibrated in accordance with the manufacturer's instructions. Combustible gas indicators shall numerically display a volume scale from 0 percent to 100 percent in 1 percent or smaller increments. [NFPA 54:8.3.1.4]

1213.6.2 Piping Systems Allowed to be Purged Indoors or Outdoors. The purging of piping systems shall be in accordance with the provisions of Section 1213.6.2.1 where the piping system meets both of the following:

(1) The design operating pressure is 2 psig (14 kPag) or less.

(2) The piping being purged is constructed entirely from pipe or tubing not meeting the size and length criteria of Table 1213.6.1. [NFPA 54:8.3.2]

1213.6.2.1 Purging Procedure. The piping system shall be purged in accordance with one or more of the following:

(1) The piping shall be purged with fuel gas and shall discharge to the outdoors.

(2) The piping shall be purged with fuel gas and shall discharge to the indoors or outdoors through an appliance burner not located in a combustion chamber. Such burner shall be provided with a continuous source of ignition.

(3) The piping shall be purged with fuel gas and shall discharge to the indoors or outdoors through a burner that has a continuous source of ignition and that is designed for such purpose.

(4) The piping shall be purged with fuel gas that is discharged to the indoors or outdoors, and the point of discharge shall be monitored with a listed combustible gas detector in accordance with Section 1213.6.2.2. Purging shall be stopped when fuel gas is detected.

(5) The piping shall be purged by the gas supplier in accordance with written procedures. [NFPA 54:8.3.2.1]

1213.6.2.2 Combustible Gas Detector. Combustible gas detectors shall be listed and calibrated or tested in accordance with the manufacturer's instructions. Combustible gas detectors shall be capable of indicating the presence of fuel gas. [NFPA 54:8.3.2.2]

1213.6.3 Purging Appliances and Equipment. After the piping system has been placed in operation, appliances and equipment shall be purged before being placed into operation. [NFPA 54:8.3.3]

1214.0 Required Gas Supply.

1214.1 General. The following regulations shall comply with this section and Section 1215.0, shall be the standard for the installation of gas piping. Natural gas regulations and tables are based on the use of a gas having a specific gravity of 0.60 and for undiluted liquefied petroleum gas, having a specific gravity of 1.50. Where gas of a different specific gravity is to be delivered, the specific gravity conversion factors provided by the serving gas supplier shall be used in sizing piping systems from the pipe sizing tables in this chapter.

1214.2 Volume. The hourly volume of gas required at each piping outlet shall be taken as not less than the maximum hourly rating as specified by the manufacturer of the appliance or appliances to be connected to each such outlet.

1214.3 Gas Appliances. Where the gas appliances to be installed have not been specified, Table 1208.4.1 shall be permitted to be used as a reference to estimate requirements of typical appliances.

To obtain the cubic feet per hour (m^3/h) of gas required, divide the input of the appliances by the average Btu (kW•h) heating value per cubic foot (m^3) of the gas. The average Btu (kW•h) per cubic foot (m^3) of the gas in the area of the installation shall be permitted to be obtained from the serving gas supplier.

1214.4 Size of Piping Outlets. The size of the supply piping outlet for a gas appliance shall be not less than ½ of an inch in diameter (15 mm).

The size of a piping outlet for a mobile home shall be not less than ¾ of an inch in diameter (20 mm).

1215.0 Required Gas Piping Size.

1215.1 Pipe Sizing Methods. Where the pipe size is to be determined using any of the methods in Section 1215.1.1 through Section 1215.1.3, the diameter of each pipe segment shall be obtained from the pipe sizing tables in Section 1215.2 or from the sizing equations in Section 1215.3. [NFPA 54:6.1]

1215.1.1 Longest Length Method. The pipe size of each section of gas piping shall be determined using the longest length of piping from the point of delivery to the most remote outlet and the load of the section. [NFPA 54:6.1.1] (see calculation example in Figure 1215.1.1)

1215.1.2 Branch Length Method. Pipe shall be sized as follows:

(1) Pipe size of each section of the longest pipe run from the point of delivery to the most remote outlet shall be determined using the longest run of piping and the load of the section.

(2) The pipe size of each section of branch piping not previously sized shall be determined using the length of piping from the point of delivery to the most remote outlet in each branch and the load of the section. [NFPA 54:6.1.2]

1215.1.3 Hybrid Pressure. The pipe size for each section of higher pressure gas piping shall be determined using the longest length of piping from the point of delivery to the most remote line pressure regulator. The pipe size from the line pressure regulator to each outlet shall be determined using the length of piping from the regulator to the most remote outlet served by the regulator. [NFPA 54:6.1.3]

1215.2 Sizing of Gas Piping Systems. Sizing of piping systems shall be in accordance with Section 1215.2.1 for natural gas piping systems and Section 1215.2.2 for propane piping systems.

1215.2.1 Natural Gas Piping Systems. Table 1215.2(1) through Table 1215.2(23) shall be used in conjunction with one of the methods described in Section 1215.1.1 through Section 1215.1.3 for piping materials other than non-corrugated stainless steel tubing. Section 1215.3 shall be used in conjunction with one of the methods described in Section 1215.1.1 through Section 1215.1.3 for non-corrugated stainless steel tubing. [NFPA 54:6.2.1, 6.2.2]

1215.2.2 Propane Piping Systems. Table 1215.2(24) through Table 1215.2(36) shall be used in conjunction with one of the methods described in Section 1215.1.1 through Section 1215.1.3 for piping materials other than non-corrugated stainless steel tubing. Section 1215.3 shall be used in conjunction with one of the methods described in Section 1215.1.1 through Section 1215.1.3 for non-corrugated stainless steel tubing. [NFPA 54:6.3.1, 6.3.2]

1215.3 Sizing Equations. The inside diameter of smooth wall pipe or tubing shall be determined by Equation 1215.3(1), Equation 1215.3(2) and Table 1215.3 using the equivalent pipe length determined by the methods in Section 1215.1.1 through Section 1215.1.3. [NFPA 54:6.4]

LOW-PRESSURE GAS FORMULA (LESS THAN 1.5 psi) [NFPA 54:6.4.1]

[Equation 1215.3(1)]

$$D = \frac{Q^{0.381}}{19.17\left(\frac{\Delta H}{Cr \times L}\right)^{0.206}}$$

Where:

D = inside diameter of pipe, inches

Q = input rate appliance(s), cubic feet per hour at 60°F and 30 inch mercury column

L = equivalent length of pipe, feet

ΔH = pressure drop, inches water column

Cr = in accordance with Table 1215.3

HIGH-PRESSURE GAS FORMULA (1.5 psi AND ABOVE) [NFPA 54:6.4.2]

[Equation 1215.3(2)]

$$D = \frac{Q^{0.381}}{18.93\left[\frac{(P_1^2 - P_2^2) \cdot Y}{Cr \times L}\right]^{0.206}}$$

Where:

D = inside diameter of pipe, inches

Q = input rate appliance(s), cubic feet per hour at 60°F and 30 inch mercury column

P_1 = upstream pressure, psia (P_1 + 14.7)

P_2 = downstream pressure, psia (P_2 + 14.7)

L = equivalent length of pipe, feet

Cr = in accordance with Table 1215.3

Y = in accordance with Table 1215.3

For SI units: 1 cubic foot = 0.0283 m^3, 1000 British thermal units per hour = 0.293 kW, 1 inch = 25 mm, 1 foot = 304.8 mm, 1 pound-force per square inch = 6.8947 kPa, °C = (°F-32)/1.8, 1 inch mercury column = 3.39 kPa, 1 inch water column = 0.249 kPa

TABLE 1215.3
***Cr* AND Y FOR NATURAL GAS AND UNDILUTED PROPANE AT STANDARD CONDITIONS**
[NFPA 54: TABLE 6.4.2]

GAS	FORMULA FACTORS	
	Cr	Y
Natural Gas	0.6094	0.9992
Undiluted Propane	1.2462	0.9910

1215.4 Sizing of Piping Sections. To determine the size of each section of pipe in a system within the range of Table 1215.2(1) through Table 1215.2(36), proceed as follows:

(1) Measure the length of the pipe from the gas meter location to the most remote outlet on the system.

(2) Select the length in feet column and row showing the distance, or the next longer distance where the table does not give the exact length.

(3) Starting at the most remote outlet, find in the row just selected the gas demand for that outlet. Where the exact figure of demand is not shown, choose the next larger figure in the row.

(4) At the top of this column will be found the correct size of pipe.

(5) Using this same row, proceed in a similar manner to each section of pipe serving this outlet. For each section of pipe, determine the total gas demand supplied by that section. Where gas piping sections serve both heating and cooling appliances and the installation prevents both units from operating simultaneously, the larger of the two demand loads needs to be used in sizing these sections.

(6) Size each section of branch piping not previously sized by measuring the distance from the gas meter location to the most remote outlet in that branch and follow the procedures of steps 2, 3, 4, and 5 above. Size branch piping in the order of their distance from the meter location, beginning with the most distant outlet not previously sized.

1215.5 Engineering Methods. For conditions other than those covered by Section 1215.1, such as longer runs or greater gas demands, the size of each gas piping system shall be determined by standard engineering methods acceptable to the Authority Having Jurisdiction, and each such system shall be so designed that the total pressure drop between the meter or another point of supply and an outlet where full demand is being supplied to all outlets, shall be in accordance with the requirements of Section 1208.4.

1215.6 Variable Gas Pressure. Where the supply gas pressure exceeds 5 psi (34.6 kPa) for natural gas and 10 psi (69 kPa) for undiluted propane or is less than 6 inches (1.5 kPa) of water column, or where diversity demand factors are used, the design, pipe, sizing, materials, location, and use of such systems first shall be approved by the Authority Having Jurisdiction. Piping systems designed for pressures exceeding the serving gas supplier's standard delivery pressure shall have prior verification from the gas supplier of the availability of the design pressure.

FIGURE 1215.1.1
EXAMPLE ILLUSTRATING USE OF TABLE 1208.4.1 AND TABLE 1215.2(1)

Problem: Determine the required pipe size of each section and outlet of the piping system shown in Figure 1215.1.1. Gas to be used has a specific gravity of 0.60 and 1100 British thermal units (Btu) per cubic foot (0.0114 kW•h/L), delivered at 8 inch water column (2.0 kPa) pressure.

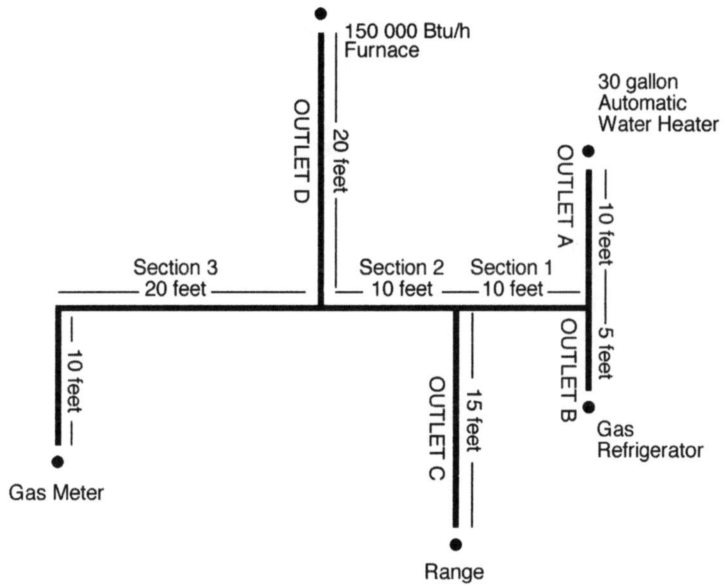

For SI units: 1 foot = 304.8 mm, 1 gallon = 3.785 L, 1000 British thermal units per hour = 0.293 kW, 1 cubic foot per hour = 0.0283 m³/h

Solution:

(1) Maximum gas demand of Outlet A —
32 cubic feet per hour (0.91 m³/h) (from Table 1208.4.1).
Maximum gas demand of Outlet B —
3 cubic feet per hour (0.08 m³/h) (from Table 1208.4.1).
Maximum gas demand of Outlet C —
59 cubic feet per hour (1.67 m³/h) (from Table 1208.4.1).
Maximum gas demand of Outlet D —
136 cubic feet per hour (3.85 m³/h) [150 000 Btu/hour (44 kW) divided by 1100 Btu per cubic foot (0.0114 kW•h/L)].

(2) The length of pipe from the gas meter to the most remote outlet (Outlet A) is 60 feet (18 288 mm).

(3) Using the length in feet column row marked 60 feet (18 288 mm) in Table 1215.2(1):
Outlet A, supplying 32 cubic feet per hour (0.91 m³/h), requires ½ of an inch (15 mm) pipe.
Section 1, supplying Outlets A and B, or 35 cubic feet per hour (0.99 m³/h) requires ½ of an inch (15 mm) pipe.
Section 2, supplying Outlets A, B, and C, or 94 cubic feet per hour (2.66 m³/h) requires ¾ of an inch (20 mm) pipe.
Section 3, supplying Outlets A, B, C, and D, or 230 cubic feet per hour (6.51 m³/h), requires 1 inch (25 mm) pipe.

(4) Using the column marked 60 feet (18 288 mm) in Table 1215.2(1):
Outlet B supplying 3 cubic feet per hour (0.08 m³/h), requires ½ of an inch (15 mm) pipe.
Outlet C, supplying 59 cubic feet per hour (1.67 m³/h), requires ½ of an inch (15 mm) pipe.

(5) Using the column marked 60 feet (18 288 mm) in Table 1215.2(1):
Outlet D, supplying 136 cubic feet per hour (3.85 m³/h), requires ¾ of an inch (20 mm) pipe.

TABLE 1215.2(1)
SCHEDULE 40 METALLIC PIPE [NFPA 54: TABLE 6.2.1(b)][1, 2]

									GAS:	NATURAL				
									INLET PRESSURE:	LESS THAN 2 psi				
									PRESSURE DROP:	0.5 in. w.c.				
									SPECIFIC GRAVITY:	0.60				
	PIPE SIZE (inch)													
NOMINAL:	½	¾	1	1¼	1½	2	2½	3	4	5	6	8	10	12
ACTUAL ID:	0.622	0.824	1.049	1.380	1.610	2.067	2.469	3.068	4.026	5.047	6.065	7.981	10.020	11.938
LENGTH (feet)	CAPACITY IN CUBIC FEET OF GAS PER HOUR													
10	172	360	678	1390	2090	4020	6400	11 300	23 100	41 800	67 600	139 000	252 000	399 000
20	118	247	466	957	1430	2760	4400	7780	15 900	28 700	46 500	95 500	173 000	275 000
30	95	199	374	768	1150	2220	3530	6250	12 700	23 000	37 300	76 700	139 000	220 000
40	81	170	320	657	985	1900	3020	5350	10 900	19 700	31 900	65 600	119 000	189 000
50	72	151	284	583	873	1680	2680	4740	9660	17 500	28 300	58 200	106 000	167 000
60	65	137	257	528	791	1520	2430	4290	8760	15 800	25 600	52 700	95 700	152 000
70	60	126	237	486	728	1400	2230	3950	8050	14 600	23 600	48 500	88 100	139 000
80	56	117	220	452	677	1300	2080	3670	7490	13 600	22 000	45 100	81 900	130 000
90	52	110	207	424	635	1220	1950	3450	7030	12 700	20 600	42 300	76 900	122 000
100	50	104	195	400	600	1160	1840	3260	6640	12 000	19 500	40 000	72 600	115 000
125	44	92	173	355	532	1020	1630	2890	5890	10 600	17 200	35 400	64 300	102 000
150	40	83	157	322	482	928	1480	2610	5330	9650	15 600	32 100	58 300	92 300
175	37	77	144	296	443	854	1360	2410	4910	8880	14 400	29 500	53 600	84 900
200	34	71	134	275	412	794	1270	2240	4560	8260	13 400	27 500	49 900	79 000
250	30	63	119	244	366	704	1120	1980	4050	7320	11 900	24 300	44 200	70 000
300	27	57	108	221	331	638	1020	1800	3670	6630	10 700	22 100	40 100	63 400
350	25	53	99	203	305	587	935	1650	3370	6100	9880	20 300	36 900	58 400
400	23	49	92	189	283	546	870	1540	3140	5680	9190	18 900	34 300	54 300
450	22	46	86	177	266	512	816	1440	2940	5330	8620	17 700	32 200	50 900
500	21	43	82	168	251	484	771	1360	2780	5030	8150	16 700	30 400	48 100
550	20	41	78	159	239	459	732	1290	2640	4780	7740	15 900	28 900	45 700
600	19	39	74	152	228	438	699	1240	2520	4560	7380	15 200	27 500	43 600
650	18	38	71	145	218	420	669	1180	2410	4360	7070	14 500	26 400	41 800
700	17	36	68	140	209	403	643	1140	2320	4190	6790	14 000	25 300	40 100
750	17	35	66	135	202	389	619	1090	2230	4040	6540	13 400	24 400	38 600
800	16	34	63	130	195	375	598	1060	2160	3900	6320	13 000	23 600	37 300
850	16	33	61	126	189	363	579	1020	2090	3780	6110	12 600	22 800	36 100
900	15	32	59	122	183	352	561	992	2020	3660	5930	12 200	22 100	35 000
950	15	31	58	118	178	342	545	963	1960	3550	5760	11 800	21 500	34 000
1000	14	30	56	115	173	333	530	937	1910	3460	5600	11 500	20 900	33 100
1100	14	28	53	109	164	316	503	890	1810	3280	5320	10 900	19 800	31 400
1200	13	27	51	104	156	301	480	849	1730	3130	5070	10 400	18 900	30 000
1300	12	26	49	100	150	289	460	813	1660	3000	4860	9980	18 100	28 700
1400	12	25	47	96	144	277	442	781	1590	2880	4670	9590	17 400	27 600
1500	11	24	45	93	139	267	426	752	1530	2780	4500	9240	16 800	26 600
1600	11	23	44	89	134	258	411	727	1480	2680	4340	8920	16 200	25 600
1700	11	22	42	86	130	250	398	703	1430	2590	4200	8630	15 700	24 800
1800	10	22	41	84	126	242	386	682	1390	2520	4070	8370	15 200	24 100
1900	10	21	40	81	122	235	375	662	1350	2440	3960	8130	14 800	23 400
2000	NA	20	39	79	119	229	364	644	1310	2380	3850	7910	14 400	22 700

For SI units: 1 inch = 25 mm, 1 foot = 304.8 mm, 1 cubic foot per hour = 0.0283 m^3/h, 1 pound-force per square inch = 6.8947 kPa, 1 inch water column = 0.249 kPa

Notes:

1 Table entries are rounded to 3 significant digits.

2 NA means a flow of less than 10 ft3/h (0.283 m^3/h).

TABLE 1215.2(2)
SCHEDULE 40 METALLIC PIPE [NFPA 54: TABLE 6.2.1(c)]*

							GAS:	NATURAL
							INLET PRESSURE:	LESS THAN 2 psi
							PRESSURE DROP:	3.0 in. w.c.
							SPECIFIC GRAVITY:	0.60
INTENDED USE: INITIAL SUPPLY PRESSURE OF 8.0 IN. W.C. OR GREATER								

	PIPE SIZE (inch)								
NOMINAL:	½	¾	1	1¼	1½	2	2½	3	4
ACTUAL ID:	0.622	0.824	1.049	1.380	1.610	2.067	2.469	3.068	4.026
LENGTH (feet)	CAPACITY IN CUBIC FEET OF GAS PER HOUR								
10	454	949	1790	3670	5500	10 600	16 900	29 800	60 800
20	312	652	1230	2520	3780	7280	11 600	20 500	41 800
30	250	524	986	2030	3030	5840	9310	16 500	33 600
40	214	448	844	1730	2600	5000	7970	14 100	28 700
50	190	397	748	1540	2300	4430	7060	12 500	25 500
60	172	360	678	1390	2090	4020	6400	11 300	23 100
70	158	331	624	1280	1920	3690	5890	10 400	21 200
80	147	308	580	1190	1790	3440	5480	9690	19 800
90	138	289	544	1120	1670	3230	5140	9090	18 500
100	131	273	514	1060	1580	3050	4860	8580	17 500
125	116	242	456	936	1400	2700	4300	7610	15 500
150	105	219	413	848	1270	2450	3900	6890	14 100
175	96	202	380	780	1170	2250	3590	6340	12 900
200	90	188	353	726	1090	2090	3340	5900	12 000
250	80	166	313	643	964	1860	2960	5230	10 700
300	72	151	284	583	873	1680	2680	4740	9660
350	66	139	261	536	803	1550	2470	4360	8890
400	62	129	243	499	747	1440	2290	4050	8270
450	58	121	228	468	701	1350	2150	3800	7760
500	55	114	215	442	662	1280	2030	3590	7330
550	52	109	204	420	629	1210	1930	3410	6960
600	50	104	195	400	600	1160	1840	3260	6640
650	47	99	187	384	575	1110	1760	3120	6360
700	46	95	179	368	552	1060	1690	3000	6110
750	44	92	173	355	532	1020	1630	2890	5890
800	42	89	167	343	514	989	1580	2790	5680
850	41	86	162	332	497	957	1530	2700	5500
900	40	83	157	322	482	928	1480	2610	5330
950	39	81	152	312	468	901	1440	2540	5180
1000	38	79	148	304	455	877	1400	2470	5040
1100	36	75	141	289	432	833	1330	2350	4780
1200	34	71	134	275	412	794	1270	2240	4560
1300	33	68	128	264	395	761	1210	2140	4370
1400	31	65	123	253	379	731	1160	2060	4200
1500	30	63	119	244	366	704	1120	1980	4050
1600	29	61	115	236	353	680	1080	1920	3910
1700	28	59	111	228	342	658	1050	1850	3780
1800	27	57	108	221	331	638	1020	1800	3670
1900	27	56	105	215	322	619	987	1750	3560
2000	26	54	102	209	313	602	960	1700	3460

For SI units: 1 inch = 25 mm, 1 foot = 304.8 mm, 1 cubic foot per hour = 0.0283 m^3/h, 1 pound-force per square inch = 6.8947 kPa, 1 inch water column = 0.249 kPa

* Table entries are rounded to 3 significant digits.

TABLE 1215.2(3)
SCHEDULE 40 METALLIC PIPE [NFPA 54: TABLE 6.2.1(d)]*

							GAS:	NATURAL	
							INLET PRESSURE:	LESS THAN 2 psi	
							PRESSURE DROP:	6.0 in. w.c.	
							SPECIFIC GRAVITY:	0.60	
INTENDED USE: INITIAL SUPPLY PRESSURE OF 11.0 IN. W.C. OR GREATER									
	PIPE SIZE (inch)								
NOMINAL:	½	¾	1	1¼	1½	2	2½	3	4
ACTUAL ID:	0.622	0.824	1.049	1.38	1.61	2.067	2.469	3.068	4.026
LENGTH (feet)	CAPACITY IN CUBIC FEET OF GAS PER HOUR								
10	660	1380	2600	5340	8000	15 400	24 600	43 400	88 500
20	454	949	1790	3670	5500	10 600	16 900	29 800	60 800
30	364	762	1440	2950	4410	8500	13 600	24 000	48 900
40	312	652	1230	2520	3780	7280	11 600	20 500	41 800
50	276	578	1090	2240	3350	6450	10 300	18 200	37 100
60	250	524	986	2030	3030	5840	9310	16 500	33 600
70	230	482	907	1860	2790	5380	8570	15 100	30 900
80	214	448	844	1730	2600	5000	7970	14 100	28 700
90	201	420	792	1630	2440	4690	7480	13 200	27 000
100	190	397	748	1540	2300	4430	7060	12 500	25 500
125	168	352	663	1360	2040	3930	6260	11 100	22 600
150	153	319	601	1230	1850	3560	5670	10 000	20 500
175	140	293	553	1140	1700	3270	5220	9230	18 800
200	131	273	514	1056	1580	3050	4860	8580	17 500
250	116	242	456	936	1400	2700	4300	7610	15 500
300	105	219	413	848	1270	2450	3900	6890	14 100
350	96	202	380	780	1170	2250	3590	6340	12 900
400	90	188	353	726	1090	2090	3340	5900	12 000
450	84	176	332	681	1020	1960	3130	5540	11 300
500	80	166	313	643	964	1860	2960	5230	10 700
550	76	158	297	611	915	1760	2810	4970	10 100
600	72	151	284	583	873	1680	2680	4740	9660
650	69	144	272	558	836	1610	2570	4540	9250
700	66	139	261	536	803	1550	2470	4360	8890
750	64	134	252	516	774	1490	2380	4200	8560
800	62	129	243	499	747	1440	2290	4050	8270
850	60	125	235	483	723	1390	2220	3920	8000
900	58	121	228	468	701	1350	2150	3800	7760
950	56	118	221	454	681	1310	2090	3690	7540
1000	55	114	215	442	662	1280	2030	3590	7330
1100	52	109	204	420	629	1210	1930	3410	6960
1200	50	104	195	400	600	1160	1840	3260	6640
1300	47	99	187	384	575	1110	1760	3120	6360
1400	46	95	179	368	552	1060	1690	3000	6110
1500	44	92	173	355	532	1020	1630	2890	5890
1600	42	89	167	343	514	989	1580	2790	5680
1700	41	86	162	332	497	957	1530	2700	5500
1800	40	83	157	322	482	928	1480	2610	5330
1900	39	81	152	312	468	901	1440	2540	5180
2000	38	79	148	304	455	877	1400	2470	5040

For SI units: 1 inch = 25 mm, 1 foot = 304.8 mm, 1 cubic foot per hour = 0.0283 m^3/h, 1 pound-force per square inch = 6.8947 kPa, 1 inch water column = 0.249 kPa

* Table entries are rounded to 3 significant digits.

TABLE 1215.2(4)
SCHEDULE 40 METALLIC PIPE [NFPA 54: TABLE 6.2.1(e)]*

							GAS:	NATURAL	
							INLET PRESSURE:	2.0 psi	
							PRESSURE DROP:	1.0 psi	
							SPECIFIC GRAVITY:	0.60	
	PIPE SIZE (inch)								
NOMINAL:	½	¾	1	1¼	1½	2	2½	3	4
ACTUAL ID:	0.622	0.824	1.049	1.380	1.610	2.067	2.469	3.068	4.026
LENGTH (feet)	CAPACITY IN CUBIC FEET OF GAS PER HOUR								
10	1510	3040	5560	11 400	17 100	32 900	52 500	92 800	189 000
20	1070	2150	3930	8070	12 100	23 300	37 100	65 600	134 000
30	869	1760	3210	6590	9880	19 000	30 300	53 600	109 000
40	753	1520	2780	5710	8550	16 500	26 300	46 400	94 700
50	673	1360	2490	5110	7650	14 700	23 500	41 500	84 700
60	615	1240	2270	4660	6980	13 500	21 400	37 900	77 300
70	569	1150	2100	4320	6470	12 500	19 900	35 100	71 600
80	532	1080	1970	4040	6050	11 700	18 600	32 800	67 000
90	502	1010	1850	3810	5700	11 000	17 500	30 900	63 100
100	462	934	1710	3510	5260	10 100	16 100	28 500	58 200
125	414	836	1530	3140	4700	9060	14 400	25 500	52 100
150	372	751	1370	2820	4220	8130	13 000	22 900	46 700
175	344	695	1270	2601	3910	7530	12 000	21 200	43 300
200	318	642	1170	2410	3610	6960	11 100	19 600	40 000
250	279	583	1040	2140	3210	6180	9850	17 400	35 500
300	253	528	945	1940	2910	5600	8920	15 800	32 200
350	232	486	869	1790	2670	5150	8210	14 500	29 600
400	216	452	809	1660	2490	4790	7640	13 500	27 500
450	203	424	759	1560	2330	4500	7170	12 700	25 800
500	192	401	717	1470	2210	4250	6770	12 000	24 400
550	182	381	681	1400	2090	4030	6430	11 400	23 200
600	174	363	650	1330	2000	3850	6130	10 800	22 100
650	166	348	622	1280	1910	3680	5870	10 400	21 200
700	160	334	598	1230	1840	3540	5640	9970	20 300
750	154	322	576	1180	1770	3410	5440	9610	19 600
800	149	311	556	1140	1710	3290	5250	9280	18 900
850	144	301	538	1100	1650	3190	5080	8980	18 300
900	139	292	522	1070	1600	3090	4930	8710	17 800
950	135	283	507	1040	1560	3000	4780	8460	17 200
1000	132	275	493	1010	1520	2920	4650	8220	16 800
1100	125	262	468	960	1440	2770	4420	7810	15 900
1200	119	250	446	917	1370	2640	4220	7450	15 200
1300	114	239	427	878	1320	2530	4040	7140	14 600
1400	110	230	411	843	1260	2430	3880	6860	14 000
1500	106	221	396	812	1220	2340	3740	6600	13 500
1600	102	214	382	784	1180	2260	3610	6380	13 000
1700	99	207	370	759	1140	2190	3490	6170	12 600
1800	96	200	358	736	1100	2120	3390	5980	12 200
1900	93	195	348	715	1070	2060	3290	5810	11 900
2000	91	189	339	695	1040	2010	3200	5650	11 500

For SI units: 1 inch = 25 mm, 1 foot = 304.8 mm, 1 cubic foot per hour = 0.0283 m^3/h, 1 pound-force per square inch = 6.8947 kPa

* Table entries are rounded to 3 significant digits.

TABLE 1215.2(5)
SCHEDULE 40 METALLIC PIPE [NFPA 54: TABLE 6.2.1(f)]*

					GAS:	NATURAL			
					INLET PRESSURE:	3.0 psi			
					PRESSURE DROP:	2.0 psi			
					SPECIFIC GRAVITY:	0.60			
	PIPE SIZE (inch)								
NOMINAL:	½	¾	1	1¼	1½	2	2½	3	4
ACTUAL ID:	0.622	0.824	1.049	1.380	1.610	2.067	2.469	3.068	4.026
LENGTH (feet)	CAPACITY IN CUBIC FEET OF GAS PER HOUR								
10	2350	4920	9270	19 000	28 500	54 900	87 500	155 000	316 000
20	1620	3380	6370	13 100	19 600	37 700	60 100	106 000	217 000
30	1300	2720	5110	10 500	15 700	30 300	48 300	85 400	174 000
40	1110	2320	4380	8990	13 500	25 900	41 300	73 100	149 000
50	985	2060	3880	7970	11 900	23 000	36 600	64 800	132 000
60	892	1870	3520	7220	10 800	20 800	33 200	58 700	120 000
70	821	1720	3230	6640	9950	19 200	30 500	54 000	110 000
80	764	1600	3010	6180	9260	17 800	28 400	50 200	102 000
90	717	1500	2820	5800	8680	16 700	26 700	47 100	96 100
100	677	1420	2670	5470	8200	15 800	25 200	44 500	90 800
125	600	1250	2360	4850	7270	14 000	22 300	39 500	80 500
150	544	1140	2140	4400	6590	12 700	20 200	35 700	72 900
175	500	1050	1970	4040	6060	11 700	18 600	32 900	67 100
200	465	973	1830	3760	5640	10 900	17 300	30 600	62 400
250	412	862	1620	3330	5000	9620	15 300	27 100	55 300
300	374	781	1470	3020	4530	8720	13 900	24 600	50 100
350	344	719	1350	2780	4170	8020	12 800	22 600	46 100
400	320	669	1260	2590	3870	7460	11 900	21 000	42 900
450	300	627	1180	2430	3640	7000	11 200	19 700	40 200
500	283	593	1120	2290	3430	6610	10 500	18 600	38 000
550	269	563	1060	2180	3260	6280	10 000	17 700	36 100
600	257	537	1010	2080	3110	5990	9550	16 900	34 400
650	246	514	969	1990	2980	5740	9150	16 200	33 000
700	236	494	931	1910	2860	5510	8790	15 500	31 700
750	228	476	897	1840	2760	5310	8470	15 000	30 500
800	220	460	866	1780	2660	5130	8180	14 500	29 500
850	213	445	838	1720	2580	4960	7910	14 000	28 500
900	206	431	812	1670	2500	4810	7670	13 600	27 700
950	200	419	789	1620	2430	4670	7450	13 200	26 900
1000	195	407	767	1580	2360	4550	7240	12 800	26 100
1100	185	387	729	1500	2240	4320	6890	12 200	24 800
1200	177	369	695	1430	2140	4120	6570	11 600	23 700
1300	169	353	666	1370	2050	3940	6290	11 100	22 700
1400	162	340	640	1310	1970	3790	6040	10 700	21 800
1500	156	327	616	1270	1900	3650	5820	10 300	21 000
1600	151	316	595	1220	1830	3530	5620	10 000	20 300
1700	146	306	576	1180	1770	3410	5440	9610	19 600
1800	142	296	558	1150	1720	3310	5270	9320	19 000
1900	138	288	542	1110	1670	3210	5120	9050	18 400
2000	134	280	527	1080	1620	3120	4980	8800	18 000

For SI units: 1 inch = 25 mm, 1 foot = 304.8 mm, 1 cubic foot per hour = 0.0283 m^3/h, 1 pound-force per square inch = 6.8947 kPa

* Table entries are rounded to 3 significant digits.

TABLE 1215.2(6)
SCHEDULE 40 METALLIC PIPE [NFPA 54: TABLE 6.2.1(g)]*

							GAS:	NATURAL	
							INLET PRESSURE:	5.0 psi	
							PRESSURE DROP:	3.5 psi	
							SPECIFIC GRAVITY:	0.60	
	PIPE SIZE (inch)								
NOMINAL:	½	¾	1	1¼	1½	2	2½	3	4
ACTUAL ID:	0.622	0.824	1.049	1.380	1.610	2.067	2.469	3.068	4.026
LENGTH (feet)	CAPACITY IN CUBIC FEET OF GAS PER HOUR								
10	3190	6430	11 800	24 200	36 200	69 700	111 000	196 000	401 000
20	2250	4550	8320	17 100	25 600	49 300	78 600	139 000	283 000
30	1840	3720	6790	14 000	20 900	40 300	64 200	113 000	231 000
40	1590	3220	5880	12 100	18 100	34 900	55 600	98 200	200 000
50	1430	2880	5260	10 800	16 200	31 200	49 700	87 900	179 000
60	1300	2630	4800	9860	14 800	28 500	45 400	80 200	164 000
70	1200	2430	4450	9130	13 700	26 400	42 000	74 300	151 000
80	1150	2330	4260	8540	12 800	24 700	39 300	69 500	142 000
90	1060	2150	3920	8050	12 100	23 200	37 000	65 500	134 000
100	979	1980	3620	7430	11 100	21 400	34 200	60 400	123 000
125	876	1770	3240	6640	9950	19 200	30 600	54 000	110 000
150	786	1590	2910	5960	8940	17 200	27 400	48 500	98 900
175	728	1470	2690	5520	8270	15 900	25 400	44 900	91 600
200	673	1360	2490	5100	7650	14 700	23 500	41 500	84 700
250	558	1170	2200	4510	6760	13 000	20 800	36 700	74 900
300	506	1060	1990	4090	6130	11 800	18 800	33 300	67 800
350	465	973	1830	3760	5640	10 900	17 300	30 600	62 400
400	433	905	1710	3500	5250	10 100	16 100	28 500	58 100
450	406	849	1600	3290	4920	9480	15 100	26 700	54 500
500	384	802	1510	3100	4650	8950	14 300	25 200	51 500
550	364	762	1440	2950	4420	8500	13 600	24 000	48 900
600	348	727	1370	2810	4210	8110	12 900	22 900	46 600
650	333	696	1310	2690	4030	7770	12 400	21 900	44 600
700	320	669	1260	2590	3880	7460	11 900	21 000	42 900
750	308	644	1210	2490	3730	7190	11 500	20 300	41 300
800	298	622	1170	2410	3610	6940	11 100	19 600	39 900
850	288	602	1130	2330	3490	6720	10 700	18 900	38 600
900	279	584	1100	2260	3380	6520	10 400	18 400	37 400
950	271	567	1070	2190	3290	6330	10 100	17 800	36 400
1000	264	551	1040	2130	3200	6150	9810	17 300	35 400
1100	250	524	987	2030	3030	5840	9320	16 500	33 600
1200	239	500	941	1930	2900	5580	8890	15 700	32 000
1300	229	478	901	1850	2770	5340	8510	15 000	30 700
1400	220	460	866	1780	2660	5130	8180	14 500	29 500
1500	212	443	834	1710	2570	4940	7880	13 900	28 400
1600	205	428	806	1650	2480	4770	7610	13 400	27 400
1700	198	414	780	1600	2400	4620	7360	13 000	26 500
1800	192	401	756	1550	2330	4480	7140	12 600	25 700
1900	186	390	734	1510	2260	4350	6930	12 300	25 000
2000	181	379	714	1470	2200	4230	6740	11 900	24 300

For SI units: 1 inch = 25 mm, 1 foot = 304.8 mm, 1 cubic foot per hour = 0.0283 m^3/h, 1 pound-force per square inch = 6.8947 kPa

* Table entries are rounded to 3 significant digits.

TABLE 1215.2(7)
SEMI-RIGID COPPER TUBING [NFPA 54: TABLE 6.2.1(h)][1, 2]

						GAS:	NATURAL		
						INLET PRESSURE:	LESS THAN 2 psi		
						PRESSURE DROP:	0.3 in. w.c.		
						SPECIFIC GRAVITY:	0.60		
	TUBE SIZE (inch)								
NOMINAL: K & L:	¼	⅜	½	⅝	¾	1	1¼	1½	2
NOMINAL: ACR:	⅜	½	⅝	¾	⅞	1⅛	1⅜	–	–
OUTSIDE:	0.375	0.500	0.625	0.750	0.875	1.125	1.375	1.625	2.125
INSIDE:[3]	0.305	0.402	0.527	0.652	0.745	0.995	1.245	1.481	1.959
LENGTH (feet)	CAPACITY IN CUBIC FEET OF GAS PER HOUR								
10	20	42	85	148	210	448	806	1270	2650
20	14	29	58	102	144	308	554	873	1820
30	11	23	47	82	116	247	445	701	1460
40	10	20	40	70	99	211	381	600	1250
50	NA	17	35	62	88	187	337	532	1110
60	NA	16	32	56	79	170	306	482	1000
70	NA	14	29	52	73	156	281	443	924
80	NA	13	27	48	68	145	262	413	859
90	NA	13	26	45	64	136	245	387	806
100	NA	12	24	43	60	129	232	366	761
125	NA	11	22	38	53	114	206	324	675
150	NA	10	20	34	48	103	186	294	612
175	NA	NA	18	31	45	95	171	270	563
200	NA	NA	17	29	41	89	159	251	523
250	NA	NA	15	26	37	78	141	223	464
300	NA	NA	13	23	33	71	128	202	420
350	NA	NA	12	22	31	65	118	186	387
400	NA	NA	11	20	28	61	110	173	360
450	NA	NA	11	19	27	57	103	162	338
500	NA	NA	10	18	25	54	97	153	319
550	NA	NA	NA	17	24	51	92	145	303
600	NA	NA	NA	16	23	49	88	139	289
650	NA	NA	NA	15	22	47	84	133	277
700	NA	NA	NA	15	21	45	81	128	266
750	NA	NA	NA	14	20	43	78	123	256
800	NA	NA	NA	14	20	42	75	119	247
850	NA	NA	NA	13	19	40	73	115	239
900	NA	NA	NA	13	18	39	71	111	232
950	NA	NA	NA	13	18	38	69	108	225
1000	NA	NA	NA	12	17	37	67	105	219
1100	NA	NA	NA	12	16	35	63	100	208
1200	NA	NA	NA	11	16	34	60	95	199
1300	NA	NA	NA	11	15	32	58	91	190
1400	NA	NA	NA	10	14	31	56	88	183
1500	NA	NA	NA	NA	14	30	54	84	176
1600	NA	NA	NA	NA	13	29	52	82	170
1700	NA	NA	NA	NA	13	28	50	79	164
1800	NA	NA	NA	NA	13	27	49	77	159
1900	NA	NA	NA	NA	12	26	47	74	155
2000	NA	NA	NA	NA	12	25	46	72	151

For SI units: 1 inch = 25 mm, 1 foot = 304.8 mm, 1 cubic foot per hour = 0.0283 m^3/h, 1 pound-force per square inch = 6.8947 kPa, 1 inch water column = 0.249 kPa

Notes:

[1] Table entries are rounded to 3 significant digits.

[2] NA means a flow of less than 10 ft^3/h (0.283 m^3/h).

[3] Table capacities are based on Type K copper tubing inside diameter (shown), which has the smallest inside diameter of the copper tubing products.

TABLE 1215.2(8)
SEMI-RIGID COPPER TUBING [NFPA 54: TABLE 6.2.1(i)][1, 2]

						GAS:	NATURAL		
						INLET PRESSURE:	LESS THAN 2 psi		
						PRESSURE DROP:	0.5 in. w.c.		
						SPECIFIC GRAVITY:	0.60		
	TUBE SIZE (inch)								
NOMINAL: K & L:	¼	⅜	½	⅝	¾	1	1¼	1½	2
NOMINAL: ACR:	⅜	½	⅝	¾	⅞	1⅛	1⅜	–	–
OUTSIDE:	0.375	0.500	0.625	0.750	0.875	1.125	1.375	1.625	2.125
INSIDE:[3]	0.305	0.402	0.527	0.652	0.745	0.995	1.245	1.481	1.959
LENGTH (feet)	CAPACITY IN CUBIC FEET OF GAS PER HOUR								
10	27	55	111	195	276	590	1060	1680	3490
20	18	38	77	134	190	406	730	1150	2400
30	15	30	61	107	152	326	586	925	1930
40	13	26	53	92	131	279	502	791	1650
50	11	23	47	82	116	247	445	701	1460
60	10	21	42	74	105	224	403	635	1320
70	NA	19	39	68	96	206	371	585	1220
80	NA	18	36	63	90	192	345	544	1130
90	NA	17	34	59	84	180	324	510	1060
100	NA	16	32	56	79	170	306	482	1000
125	NA	14	28	50	70	151	271	427	890
150	NA	13	26	45	64	136	245	387	806
175	NA	12	24	41	59	125	226	356	742
200	NA	11	22	39	55	117	210	331	690
250	NA	NA	20	34	48	103	186	294	612
300	NA	NA	18	31	44	94	169	266	554
350	NA	NA	16	28	40	86	155	245	510
400	NA	NA	15	26	38	80	144	228	474
450	NA	NA	14	25	35	75	135	214	445
500	NA	NA	13	23	33	71	128	202	420
550	NA	NA	13	22	32	68	122	192	399
600	NA	NA	12	21	30	64	116	183	381
650	NA	NA	12	20	29	62	111	175	365
700	NA	NA	11	20	28	59	107	168	350
750	NA	NA	11	19	27	57	103	162	338
800	NA	NA	10	18	26	55	99	156	326
850	NA	NA	10	18	25	53	96	151	315
900	NA	NA	NA	17	24	52	93	147	306
950	NA	NA	NA	17	24	50	90	143	297
1000	NA	NA	NA	16	23	49	88	139	289
1100	NA	NA	NA	15	22	46	84	132	274
1200	NA	NA	NA	15	21	44	80	126	262
1300	NA	NA	NA	14	20	42	76	120	251
1400	NA	NA	NA	13	19	41	73	116	241
1500	NA	NA	NA	13	18	39	71	111	232
1600	NA	NA	NA	13	18	38	68	108	224
1700	NA	NA	NA	12	17	37	66	104	217
1800	NA	NA	NA	12	17	36	64	101	210
1900	NA	NA	NA	11	16	35	62	98	204
2000	NA	NA	NA	11	16	34	60	95	199

For SI units: 1 inch = 25 mm, 1 foot = 304.8 mm, 1 cubic foot per hour = 0.0283 m^3/h, 1 pound-force per square inch = 6.8947 kPa, 1 inch water column = 0.249 kPa

Notes:

[1] Table entries are rounded to 3 significant digits.

[2] NA means a flow of less than 10 ft^3/h (0.283 m^3/h).

[3] Table capacities are based on Type K copper tubing inside diameter (shown), which has the smallest inside diameter of the copper tubing products.

TABLE 1215.2(9)
SEMI-RIGID COPPER TUBING [NFPA 54: TABLE 6.2.1(j)][1, 2]

						GAS:	NATURAL		
						INLET PRESSURE:	LESS THAN 2 psi		
						PRESSURE DROP:	1.0 in. w.c.		
						SPECIFIC GRAVITY:	0.60		
INTENDED USE: TUBE SIZING BETWEEN HOUSE LINE REGULATOR AND THE APPLIANCE									
	TUBE SIZE (inch)								
NOMINAL: K & L:	¼	⅜	½	⅝	¾	1	1¼	1½	2
NOMINAL: ACR:	⅜	½	⅝	¾	⅞	1⅛	1⅜	–	–
OUTSIDE:	0.375	0.500	0.625	0.750	0.875	1.125	1.375	1.625	2.125
INSIDE:[3]	0.305	0.402	0.527	0.652	0.745	0.995	1.245	1.481	1.959
LENGTH (feet)	CAPACITY IN CUBIC FEET OF GAS PER HOUR								
10	39	80	162	283	402	859	1550	2440	5080
20	27	55	111	195	276	590	1060	1680	3490
30	21	44	89	156	222	474	853	1350	2800
40	18	38	77	134	190	406	730	1150	2400
50	16	33	68	119	168	359	647	1020	2130
60	15	30	61	107	152	326	586	925	1930
70	13	28	57	99	140	300	539	851	1770
80	13	26	53	92	131	279	502	791	1650
90	12	24	49	86	122	262	471	742	1550
100	11	23	47	82	116	247	445	701	1460
125	NA	20	41	72	103	219	394	622	1290
150	NA	18	37	65	93	198	357	563	1170
175	NA	17	34	60	85	183	329	518	1080
200	NA	16	32	56	79	170	306	482	1000
250	NA	14	28	50	70	151	271	427	890
300	NA	13	26	45	64	136	245	387	806
350	NA	12	24	41	59	125	226	356	742
400	NA	11	22	39	55	117	210	331	690
450	NA	10	21	36	51	110	197	311	647
500	NA	NA	20	34	48	103	186	294	612
550	NA	NA	19	32	46	98	177	279	581
600	NA	NA	18	31	44	94	169	266	554
650	NA	NA	17	30	42	90	162	255	531
700	NA	NA	16	28	40	86	155	245	510
750	NA	NA	16	27	39	83	150	236	491
800	NA	NA	15	26	38	80	144	228	474
850	NA	NA	15	26	36	78	140	220	459
900	NA	NA	14	25	35	75	135	214	445
950	NA	NA	14	24	34	73	132	207	432
1000	NA	NA	13	23	33	71	128	202	420
1100	NA	NA	13	22	32	68	122	192	399
1200	NA	NA	12	21	30	64	116	183	381
1300	NA	NA	12	20	29	62	111	175	365
1400	NA	NA	11	20	28	59	107	168	350
1500	NA	NA	11	19	27	57	103	162	338
1600	NA	NA	10	18	26	55	99	156	326
1700	NA	NA	10	18	25	53	96	151	315
1800	NA	NA	NA	17	24	52	93	147	306
1900	NA	NA	NA	17	24	50	90	143	297
2000	NA	NA	NA	16	23	49	88	139	289

For SI units: 1 inch = 25 mm, 1 foot = 304.8 mm, 1 cubic foot per hour = 0.0283 m^3/h, 1 pound-force per square inch = 6.8947 kPa, 1 inch water column = 0.249 kPa

Notes:

1 Table entries are rounded to 3 significant digits.

2 NA means a flow of less than 10 ft^3/h (0.283 m^3/h).

3 Table capacities are based on Type K copper tubing inside diameter (shown), which has the smallest inside diameter of the copper tubing products.

TABLE 1215.2(10)
SEMI-RIGID COPPER TUBING [NFPA 54: TABLE 6.2.1(k)][2]

						GAS:	NATURAL		
						INLET PRESSURE:	LESS THAN 2 psi		
						PRESSURE DROP:	17.0 in. w.c.		
						SPECIFIC GRAVITY:	0.60		
	TUBE SIZE (inch)								
NOMINAL: K & L:	¼	⅜	½	⅝	¾	1	1¼	1½	2
NOMINAL: ACR:	⅜	½	⅝	¾	⅞	1⅛	1⅜	–	–
OUTSIDE:	0.375	0.500	0.625	0.750	0.875	1.125	1.375	1.625	2.125
INSIDE:[1]	0.305	0.402	0.527	0.652	0.745	0.995	1.245	1.481	1.959
LENGTH (feet)	CAPACITY IN CUBIC FEET OF GAS PER HOUR								
10	190	391	796	1390	1970	4220	7590	12 000	24 900
20	130	269	547	956	1360	2900	5220	8230	17 100
30	105	216	439	768	1090	2330	4190	6610	13 800
40	90	185	376	657	932	1990	3590	5650	11 800
50	79	164	333	582	826	1770	3180	5010	10 400
60	72	148	302	528	749	1600	2880	4540	9460
70	66	137	278	486	689	1470	2650	4180	8700
80	62	127	258	452	641	1370	2460	3890	8090
90	58	119	243	424	601	1280	2310	3650	7590
100	55	113	229	400	568	1210	2180	3440	7170
125	48	100	203	355	503	1080	1940	3050	6360
150	44	90	184	321	456	974	1750	2770	5760
175	40	83	169	296	420	896	1610	2540	5300
200	38	77	157	275	390	834	1500	2370	4930
250	33	69	140	244	346	739	1330	2100	4370
300	30	62	126	221	313	670	1210	1900	3960
350	28	57	116	203	288	616	1110	1750	3640
400	26	53	108	189	268	573	1030	1630	3390
450	24	50	102	177	252	538	968	1530	3180
500	23	47	96	168	238	508	914	1440	3000
550	22	45	91	159	226	482	868	1370	2850
600	21	43	87	152	215	460	829	1310	2720
650	20	41	83	145	206	441	793	1250	2610
700	19	39	80	140	198	423	762	1200	2500
750	18	38	77	135	191	408	734	1160	2410
800	18	37	74	130	184	394	709	1120	2330
850	17	35	72	126	178	381	686	1080	2250
900	17	34	70	122	173	370	665	1050	2180
950	16	33	68	118	168	359	646	1020	2120
1000	16	32	66	115	163	349	628	991	2060
1100	15	31	63	109	155	332	597	941	1960
1200	14	29	60	104	148	316	569	898	1870
1300	14	28	57	100	142	303	545	860	1790
1400	13	27	55	96	136	291	524	826	1720
1500	13	26	53	93	131	280	505	796	1660
1600	12	25	51	89	127	271	487	768	1600
1700	12	24	49	86	123	262	472	744	1550
1800	11	24	48	84	119	254	457	721	1500
1900	11	23	47	81	115	247	444	700	1460
2000	11	22	45	79	112	240	432	681	1420

For SI units: 1 inch = 25 mm, 1 foot = 304.8 mm, 1 cubic foot per hour = 0.0283 m^3/h, 1 pound-force per square inch = 6.8947 kPa, 1 inch water column = 0.249 kPa

Notes:

[1] Table capacities are based on Type K copper tubing inside diameter (shown), which has the smallest inside diameter of the copper tubing products.

[2] Table entries are rounded to 3 significant digits.

TABLE 1215.2(11)
SEMI-RIGID COPPER TUBING [NFPA 54: TABLE 6.2.1(I)][2]

						GAS:	NATURAL		
						INLET PRESSURE:	2.0 psi		
						PRESSURE DROP:	1.0 psi		
						SPECIFIC GRAVITY:	0.60		
	TUBE SIZE (inch)								
NOMINAL: K & L:	¼	⅜	½	⅝	¾	1	1¼	1½	2
NOMINAL: ACR:	⅜	½	⅝	¾	⅞	1⅛	1⅜	–	–
OUTSIDE:	0.375	0.500	0.625	0.750	0.875	1.125	1.375	1.625	2.125
INSIDE:[1]	0.305	0.402	0.527	0.652	0.745	0.995	1.245	1.481	1.959
LENGTH (feet)	CAPACITY IN CUBIC FEET OF GAS PER HOUR								
10	245	506	1030	1800	2550	5450	9820	15 500	32 200
20	169	348	708	1240	1760	3750	6750	10 600	22 200
30	135	279	568	993	1410	3010	5420	8550	17 800
40	116	239	486	850	1210	2580	4640	7310	15 200
50	103	212	431	754	1070	2280	4110	6480	13 500
60	93	192	391	683	969	2070	3730	5870	12 200
70	86	177	359	628	891	1900	3430	5400	11 300
80	80	164	334	584	829	1770	3190	5030	10 500
90	75	154	314	548	778	1660	2990	4720	9820
100	71	146	296	518	735	1570	2830	4450	9280
125	63	129	263	459	651	1390	2500	3950	8220
150	57	117	238	416	590	1260	2270	3580	7450
175	52	108	219	383	543	1160	2090	3290	6850
200	49	100	204	356	505	1080	1940	3060	6380
250	43	89	181	315	448	956	1720	2710	5650
300	39	80	164	286	406	866	1560	2460	5120
350	36	74	150	263	373	797	1430	2260	4710
400	33	69	140	245	347	741	1330	2100	4380
450	31	65	131	230	326	696	1250	1970	4110
500	30	61	124	217	308	657	1180	1870	3880
550	28	58	118	206	292	624	1120	1770	3690
600	27	55	112	196	279	595	1070	1690	3520
650	26	53	108	188	267	570	1030	1620	3370
700	25	51	103	181	256	548	986	1550	3240
750	24	49	100	174	247	528	950	1500	3120
800	23	47	96	168	239	510	917	1450	3010
850	22	46	93	163	231	493	888	1400	2920
900	22	44	90	158	224	478	861	1360	2830
950	21	43	88	153	217	464	836	1320	2740
1000	20	42	85	149	211	452	813	1280	2670
1100	19	40	81	142	201	429	772	1220	2540
1200	18	38	77	135	192	409	737	1160	2420
1300	18	36	74	129	183	392	705	1110	2320
1400	17	35	71	124	176	376	678	1070	2230
1500	16	34	68	120	170	363	653	1030	2140
1600	16	33	66	116	164	350	630	994	2070
1700	15	31	64	112	159	339	610	962	2000
1800	15	30	62	108	154	329	592	933	1940
1900	14	30	60	105	149	319	575	906	1890
2000	14	29	59	102	145	310	559	881	1830

For SI units: 1 inch = 25 mm, 1 foot = 304.8 mm, 1 cubic foot per hour = 0.0283 m^3/h, 1 pound-force per square inch = 6.8947 kPa

Notes:

[1] Table capacities are based on Type K copper tubing inside diameter (shown), which has the smallest inside diameter of the copper tubing products.

[2] Table entries are rounded to 3 significant digits.

TABLE 1215.2(12)
SEMI-RIGID COPPER TUBING [NFPA 54: TABLE 6.2.1(m)][3]

						GAS:	NATURAL		
						INLET PRESSURE:	2.0 psi		
						PRESSURE DROP:	1.5 psi		
						SPECIFIC GRAVITY:	0.60		
INTENDED USE: PIPE SIZING BETWEEN POINT OF DELIVERY AND THE HOUSE LINE REGULATOR. TOTAL LOAD SUPPLIED BY A SINGLE HOUSE LINE REGULATOR NOT EXCEEDING 150 CUBIC FEET PER HOUR.[2]									
	TUBE SIZE (inch)								
NOMINAL: K & L:	¼	⅜	½	⅝	¾	1	1¼	1½	2
NOMINAL: ACR:	⅜	½	⅝	¾	⅞	1⅛	1⅜	–	–
OUTSIDE:	0.375	0.500	0.625	0.750	0.875	1.125	1.375	1.625	2.125
INSIDE:[1]	0.305	0.402	0.527	0.652	0.745	0.995	1.245	1.481	1.959
LENGTH (feet)	CAPACITY IN CUBIC FEET OF GAS PER HOUR								
10	303	625	1270	2220	3150	6740	12 100	19 100	39 800
20	208	430	874	1530	2170	4630	8330	13 100	27 400
30	167	345	702	1230	1740	3720	6690	10 600	22 000
40	143	295	601	1050	1490	3180	5730	9030	18 800
50	127	262	532	931	1320	2820	5080	8000	16 700
60	115	237	482	843	1200	2560	4600	7250	15 100
70	106	218	444	776	1100	2350	4230	6670	13 900
80	98	203	413	722	1020	2190	3940	6210	12 900
90	92	190	387	677	961	2050	3690	5820	12 100
100	87	180	366	640	907	1940	3490	5500	11 500
125	77	159	324	567	804	1720	3090	4880	10 200
150	70	144	294	514	729	1560	2800	4420	9200
175	64	133	270	472	670	1430	2580	4060	8460
200	60	124	252	440	624	1330	2400	3780	7870
250	53	110	223	390	553	1180	2130	3350	6980
300	48	99	202	353	501	1070	1930	3040	6320
350	44	91	186	325	461	984	1770	2790	5820
400	41	85	173	302	429	916	1650	2600	5410
450	39	80	162	283	402	859	1550	2440	5080
500	36	75	153	268	380	811	1460	2300	4800
550	35	72	146	254	361	771	1390	2190	4560
600	33	68	139	243	344	735	1320	2090	4350
650	32	65	133	232	330	704	1270	2000	4160
700	30	63	128	223	317	676	1220	1920	4000
750	29	60	123	215	305	652	1170	1850	3850
800	28	58	119	208	295	629	1130	1790	3720
850	27	57	115	201	285	609	1100	1730	3600
900	27	55	111	195	276	590	1060	1680	3490
950	26	53	108	189	268	573	1030	1630	3390
1000	25	52	105	184	261	558	1000	1580	3300
1100	24	49	100	175	248	530	954	1500	3130
1200	23	47	95	167	237	505	910	1430	2990
1300	22	45	91	160	227	484	871	1370	2860
1400	21	43	88	153	218	465	837	1320	2750
1500	20	42	85	148	210	448	806	1270	2650
1600	19	40	82	143	202	432	779	1230	2560
1700	19	39	79	138	196	419	753	1190	2470
1800	18	38	77	134	190	406	731	1150	2400
1900	18	37	74	130	184	394	709	1120	2330
2000	17	36	72	126	179	383	690	1090	2270

For SI units: 1 inch = 25 mm, 1 foot = 304.8 mm, 1 cubic foot per hour = 0.0283 m^3/h, 1 pound-force per square inch = 6.8947 kPa

Notes:

1 Table capacities are based on Type K copper tubing inside diameter (shown), which has the smallest inside diameter of the copper tubing products.

2 Where this table is used to size the tubing upstream of a line pressure regulator, the pipe or tubing downstream of the line pressure regulator shall be sized using a pressure drop no greater than 1 inch water column (0.249 kPa).

3 Table entries are rounded to 3 significant digits.

TABLE 1215.2(13)
SEMI-RIGID COPPER TUBING [NFPA 54: TABLE 6.2.1(n)][2]

						GAS:	NATURAL		
						INLET PRESSURE:	5.0 psi		
						PRESSURE DROP:	3.5 psi		
						SPECIFIC GRAVITY:	0.60		
	TUBE SIZE (inch)								
NOMINAL: K & L:	¼	⅜	½	⅝	¾	1	1¼	1½	2
NOMINAL: ACR:	⅜	½	⅝	¾	⅞	1⅛	1⅜	–	–
OUTSIDE:	0.375	0.500	0.625	0.750	0.875	1.125	1.375	1.625	2.125
INSIDE:[1]	0.305	0.402	0.527	0.652	0.745	0.995	1.245	1.481	1.959
LENGTH (feet)	CAPACITY IN CUBIC FEET OF GAS PER HOUR								
10	511	1050	2140	3750	5320	11 400	20 400	32 200	67 100
20	351	724	1470	2580	3650	7800	14 000	22 200	46 100
30	282	582	1180	2070	2930	6270	11 300	17 800	37 000
40	241	498	1010	1770	2510	5360	9660	15 200	31 700
50	214	441	898	1570	2230	4750	8560	13 500	28 100
60	194	400	813	1420	2020	4310	7750	12 200	25 500
70	178	368	748	1310	1860	3960	7130	11 200	23 400
80	166	342	696	1220	1730	3690	6640	10 500	21 800
90	156	321	653	1140	1620	3460	6230	9820	20 400
100	147	303	617	1080	1530	3270	5880	9270	19 300
125	130	269	547	955	1360	2900	5210	8220	17 100
150	118	243	495	866	1230	2620	4720	7450	15 500
175	109	224	456	796	1130	2410	4350	6850	14 300
200	101	208	424	741	1050	2250	4040	6370	13 300
250	90	185	376	657	932	1990	3580	5650	11 800
300	81	167	340	595	844	1800	3250	5120	10 700
350	75	154	313	547	777	1660	2990	4710	9810
400	69	143	291	509	722	1540	2780	4380	9120
450	65	134	273	478	678	1450	2610	4110	8560
500	62	127	258	451	640	1370	2460	3880	8090
550	58	121	245	429	608	1300	2340	3690	7680
600	56	115	234	409	580	1240	2230	3520	7330
650	53	110	224	392	556	1190	2140	3370	7020
700	51	106	215	376	534	1140	2050	3240	6740
750	49	102	207	362	514	1100	1980	3120	6490
800	48	98	200	350	497	1060	1910	3010	6270
850	46	95	194	339	481	1030	1850	2910	6070
900	45	92	188	328	466	1000	1790	2820	5880
950	43	90	182	319	452	967	1740	2740	5710
1000	42	87	177	310	440	940	1690	2670	5560
1100	40	83	169	295	418	893	1610	2530	5280
1200	38	79	161	281	399	852	1530	2420	5040
1300	37	76	154	269	382	816	1470	2320	4820
1400	35	73	148	259	367	784	1410	2220	4630
1500	34	70	143	249	353	755	1360	2140	4460
1600	33	68	138	241	341	729	1310	2070	4310
1700	32	65	133	233	330	705	1270	2000	4170
1800	31	63	129	226	320	684	1230	1940	4040
1900	30	62	125	219	311	664	1200	1890	3930
2000	29	60	122	213	302	646	1160	1830	3820

For SI units: 1 inch = 25 mm, 1 foot = 304.8 mm, 1 cubic foot per hour = 0.0283 m^3/h, 1 pound-force per square inch = 6.8947 kPa

Notes:

[1] Table capacities are based on Type K copper tubing inside diameter (shown), which has the smallest inside diameter of the copper tubing products.

[2] Table entries are rounded to 3 significant digits.

TABLE 1215.2(14)
CORRUGATED STAINLESS STEEL TUBING (CSST) [NFPA 54: TABLE 6.2.1(o)][1, 2]

GAS:	NATURAL
INLET PRESSURE:	LESS THAN 2 psi
PRESSURE DROP:	0.5 in. w.c.
SPECIFIC GRAVITY:	0.60

	TUBE SIZE (EHD)[3]													
FLOW DESIGNATION:	13	15	18	19	23	25	30	31	37	39	46	48	60	62
LENGTH (feet)	CAPACITY IN CUBIC FEET OF GAS PER HOUR													
5	46	63	115	134	225	270	471	546	895	1037	1790	2070	3660	4140
10	32	44	82	95	161	192	330	383	639	746	1260	1470	2600	2930
15	25	35	66	77	132	157	267	310	524	615	1030	1200	2140	2400
20	22	31	58	67	116	137	231	269	456	536	888	1050	1850	2080
25	19	27	52	60	104	122	206	240	409	482	793	936	1660	1860
30	18	25	47	55	96	112	188	218	374	442	723	856	1520	1700
40	15	21	41	47	83	97	162	188	325	386	625	742	1320	1470
50	13	19	37	42	75	87	144	168	292	347	559	665	1180	1320
60	12	17	34	38	68	80	131	153	267	318	509	608	1080	1200
70	11	16	31	36	63	74	121	141	248	295	471	563	1000	1110
80	10	15	29	33	60	69	113	132	232	277	440	527	940	1040
90	10	14	28	32	57	65	107	125	219	262	415	498	887	983
100	9	13	26	30	54	62	101	118	208	249	393	472	843	933
150	7	10	20	23	42	48	78	91	171	205	320	387	691	762
200	6	9	18	21	38	44	71	82	148	179	277	336	600	661
250	5	8	16	19	34	39	63	74	133	161	247	301	538	591
300	5	7	15	17	32	36	57	67	95	148	226	275	492	540

For SI units: 1 inch = 25 mm, 1 foot = 304.8 mm, 1 cubic foot per hour = 0.0283 m^3/h, 1 pound-force per square inch = 6.8947 kPa, 1 inch water column = 0.249 kPa

Notes:

1 Table entries are rounded to 3 significant digits.

2 Table includes losses for four 90 degree (1.57 rad) bends and two end fittings. Tubing runs with larger numbers of bends, fittings, or both shall be increased by an equivalent length of tubing to the following equation: L = 1.3 n, where L is additional length (ft) of tubing and n is the number of additional fittings, bends, or both.

3 EHD = Equivalent Hydraulic Diameter, which is a measure of the relative hydraulic efficiency between different tubing sizes. The greater the value of EHD, the greater the gas capacity of the tubing.

TABLE 1215.2(15)
CORRUGATED STAINLESS STEEL TUBING (CSST) [NFPA 54: TABLE 6.2.1(p)][1, 2]

									GAS:	NATURAL				
									INLET PRESSURE:	LESS THAN 2 psi				
									PRESSURE DROP:	3.0 in. w.c.				
									SPECIFIC GRAVITY:	0.60				
INTENDED USE: INITIAL SUPPLY PRESSURE OF 8.0 INCH WATER COLUMN OR GREATER														
	TUBE SIZE (EHD)[3]													
FLOW DESIGNATION:	**13**	**15**	**18**	**19**	**23**	**25**	**30**	**31**	**37**	**39**	**46**	**48**	**60**	**62**
LENGTH (feet)	CAPACITY IN CUBIC FEET OF GAS PER HOUR													
5	120	160	277	327	529	649	1180	1370	2140	2423	4430	5010	8800	10 100
10	83	112	197	231	380	462	828	958	1530	1740	3200	3560	6270	7160
15	67	90	161	189	313	379	673	778	1250	1433	2540	2910	5140	5850
20	57	78	140	164	273	329	580	672	1090	1249	2200	2530	4460	5070
25	51	69	125	147	245	295	518	599	978	1123	1960	2270	4000	4540
30	46	63	115	134	225	270	471	546	895	1029	1790	2070	3660	4140
40	39	54	100	116	196	234	407	471	778	897	1550	1800	3180	3590
50	35	48	89	104	176	210	363	421	698	806	1380	1610	2850	3210
60	32	44	82	95	161	192	330	383	639	739	1260	1470	2600	2930
70	29	41	76	88	150	178	306	355	593	686	1170	1360	2420	2720
80	27	38	71	82	141	167	285	331	555	644	1090	1280	2260	2540
90	26	36	67	77	133	157	268	311	524	609	1030	1200	2140	2400
100	24	34	63	73	126	149	254	295	498	579	974	1140	2030	2280
150	19	27	52	60	104	122	206	240	409	477	793	936	1660	1860
200	17	23	45	52	91	106	178	207	355	415	686	812	1440	1610
250	15	21	40	46	82	95	159	184	319	373	613	728	1290	1440
300	13	19	37	42	75	87	144	168	234	342	559	665	1180	1320

For SI units: 1 foot = 304.8 mm, 1 cubic foot per hour = 0.0283 m^3/h, 1 pound-force per square inch = 6.8947 kPa, 1 inch water column = 0.249 kPa

Notes:

[1] Table entries are rounded to 3 significant digits.

[2] Table includes losses for four 90 degree (1.57 rad) bends and two end fittings. Tubing runs with larger numbers of bends, fittings, or both shall be increased by an equivalent length of tubing to the following equation: L = 1.3 n, where L is additional length (ft) of tubing and n is the number of additional fittings, bends, or both.

[3] EHD = Equivalent Hydraulic Diameter, which is a measure of the relative hydraulic efficiency between different tubing sizes. The greater the value of EHD, the greater the gas capacity of the tubing.

TABLE 1215.2(16)
CORRUGATED STAINLESS STEEL TUBING (CSST) [NFPA 54: TABLE 6.2.1(q)][1, 2]

											GAS:	NATURAL		
											INLET PRESSURE:	LESS THAN 2 psi		
											PRESSURE DROP:	6.0 in. w.c.		
											SPECIFIC GRAVITY:	0.60		
INTENDED USE: INITIAL SUPPLY PRESSURE OF 11.0 INCH WATER COLUMN OR GREATER														
	TUBE SIZE (EHD)[3]													
FLOW DESIGNATION:	13	15	18	19	23	25	30	31	37	39	46	48	60	62
LENGTH (feet)	CAPACITY IN CUBIC FEET OF GAS PER HOUR													
5	173	229	389	461	737	911	1690	1950	3000	3375	6280	7050	12 400	14 260
10	120	160	277	327	529	649	1180	1370	2140	2423	4430	5010	8800	10 100
15	96	130	227	267	436	532	960	1110	1760	1996	3610	4100	7210	8260
20	83	112	197	231	380	462	828	958	1530	1740	3120	3560	6270	7160
25	74	99	176	207	342	414	739	855	1370	1564	2790	3190	5620	6400
30	67	90	161	189	313	379	673	778	1250	1433	2540	2910	5140	5850
40	57	78	140	164	273	329	580	672	1090	1249	2200	2530	4460	5070
50	51	69	125	147	245	295	518	599	978	1123	1960	2270	4000	4540
60	46	63	115	134	225	270	471	546	895	1029	1790	2070	3660	4140
70	42	58	106	124	209	250	435	505	830	956	1660	1920	3390	3840
80	39	54	100	116	196	234	407	471	778	897	1550	1800	3180	3590
90	37	51	94	109	185	221	383	444	735	848	1460	1700	3000	3390
100	35	48	89	104	176	210	363	421	698	806	1380	1610	2850	3210
150	28	39	73	85	145	172	294	342	573	664	1130	1320	2340	2630
200	24	34	63	73	126	149	254	295	498	579	974	1140	2030	2280
250	21	30	57	66	114	134	226	263	447	520	870	1020	1820	2040
300	19	27	52	60	104	122	206	240	409	477	793	936	1660	1860

For SI units: 1 foot = 304.8 mm, 1 cubic foot per hour = 0.0283 m^3/h, 1 pound-force per square inch = 6.8947 kPa, 1 inch water column = 0.249 kPa

Notes:

[1] Table entries are rounded to 3 significant digits.

[2] Table includes losses for four 90 degree (1.57 rad) bends and two end fittings. Tubing runs with larger numbers of bends, fittings, or both shall be increased by an equivalent length of tubing to the following equation: L = 1.3 n, where L is additional length (ft) of tubing and n is the number of additional fittings, bends, or both.

[3] EHD = Equivalent Hydraulic Diameter, which is a measure of the relative hydraulic efficiency between different tubing sizes. The greater the value of EHD, the greater the gas capacity of the tubing.

TABLE 1215.2(17)
CORRUGATED STAINLESS STEEL TUBING (CSST) [NFPA 54: TABLE 6.2.1(r)][1, 2, 3, 4]

											GAS:	**NATURAL**		
											INLET PRESSURE:	**2.0 psi**		
											PRESSURE DROP:	**1.0 psi**		
											SPECIFIC GRAVITY:	**0.60**		
	TUBE SIZE (EHD)[5]													
FLOW DESIGNATION:	**13**	**15**	**18**	**19**	**23**	**25**	**30**	**31**	**37**	**39**	**46**	**48**	**60**	**62**
LENGTH (feet)	**CAPACITY IN CUBIC FEET OF GAS PER HOUR**													
10	270	353	587	700	1100	1370	2590	2990	4510	5037	9600	10 700	18 600	21 600
25	166	220	374	444	709	876	1620	1870	2890	3258	6040	6780	11 900	13 700
30	151	200	342	405	650	801	1480	1700	2640	2987	5510	6200	10 900	12 500
40	129	172	297	351	567	696	1270	1470	2300	2605	4760	5380	9440	10 900
50	115	154	266	314	510	624	1140	1310	2060	2343	4260	4820	8470	9720
75	93	124	218	257	420	512	922	1070	1690	1932	3470	3950	6940	7940
80	89	120	211	249	407	496	892	1030	1640	1874	3360	3820	6730	7690
100	79	107	189	222	366	445	795	920	1470	1685	3000	3420	6030	6880
150	64	87	155	182	302	364	646	748	1210	1389	2440	2800	4940	5620
200	55	75	135	157	263	317	557	645	1050	1212	2110	2430	4290	4870
250	49	67	121	141	236	284	497	576	941	1090	1890	2180	3850	4360
300	44	61	110	129	217	260	453	525	862	999	1720	1990	3520	3980
400	38	52	96	111	189	225	390	453	749	871	1490	1730	3060	3450
500	34	46	86	100	170	202	348	404	552	783	1330	1550	2740	3090

For SI units: 1 foot = 304.8 mm, 1 cubic foot per hour = 0.0283 m^3/h, 1 pound-force per square inch = 6.8947 kPa

Notes:

1 Table does not include effect of pressure drop across the line regulator. Where regulator loss exceeds 0.75 psi (5.17 kPa), DO NOT USE THIS TABLE. Consult with regulator manufacturer for pressure drops and capacity factors. Pressure drops across a regulator are capable of varying with flow rate.

2 CAUTION: Capacities shown in table are capable of exceeding maximum capacity for a selected regulator. Consult with regulator or tubing manufacturer for guidance.

3 Table includes losses for four 90 degree (1.57 rad) bends and two end fittings. Tubing runs with larger numbers of bends, fittings, or both shall be increased by an equivalent length of tubing according to the following equation: $L = 1.3\ n$, where L is additional length (ft) of tubing and n is the number of additional fittings, bends, or both.

4 Table entries are rounded to 3 significant digits.

5 EHD = Equivalent Hydraulic Diameter, which is a measure of the relative hydraulic efficiency between different tubing sizes. The greater the value of EHD, the greater the gas capacity of the tubing.

TABLE 1215.2(18)
CORRUGATED STAINLESS STEEL TUBING (CSST) [NFPA 54: TABLE 6.2.1(s)][1, 2, 3, 4]

										GAS:	NATURAL		
										INLET PRESSURE:	5.0 psi		
										PRESSURE DROP:	3.5 psi		
										SPECIFIC GRAVITY:	0.60		
	TUBE SIZE (EHD)[5]												
FLOW DESIGNATION:	13	15	18	19	23	25	30	31	37	39	46	48	60	62
LENGTH (feet)	CAPACITY IN CUBIC FEET OF GAS PER HOUR													
10	523	674	1080	1300	2000	2530	4920	5660	8300	9140	18 100	19 800	34 400	40 400
25	322	420	691	827	1290	1620	3080	3540	5310	5911	11 400	12 600	22 000	25 600
30	292	382	632	755	1180	1480	2800	3230	4860	5420	10 400	11 500	20 100	23 400
40	251	329	549	654	1030	1280	2420	2790	4230	4727	8970	10 000	17 400	20 200
50	223	293	492	586	926	1150	2160	2490	3790	4251	8020	8930	15 600	18 100
75	180	238	403	479	763	944	1750	2020	3110	3506	6530	7320	12 800	14 800
80	174	230	391	463	740	915	1690	1960	3020	3400	6320	7090	12 400	14 300
100	154	205	350	415	665	820	1510	1740	2710	3057	5650	6350	11 100	12 800
150	124	166	287	339	548	672	1230	1420	2220	2521	4600	5200	9130	10 500
200	107	143	249	294	478	584	1060	1220	1930	2199	3980	4510	7930	9090
250	95	128	223	263	430	524	945	1090	1730	1977	3550	4040	7110	8140
300	86	116	204	240	394	479	860	995	1590	1813	3240	3690	6500	7430
400	74	100	177	208	343	416	742	858	1380	1581	2800	3210	5650	6440
500	66	89	159	186	309	373	662	766	1040	1422	2500	2870	5060	5760

For SI units: 1 foot = 304.8 mm, 1 cubic foot per hour = 0.0283 m^3/h, 1 pound-force per square inch = 6.8947 kPa

Notes:

1 Table does not include effect of pressure drop across the line regulator. Where regulator loss exceeds 1 psi (7 kPa), DO NOT USE THIS TABLE. Consult with regulator manufacturer for pressure drops and capacity factors. Pressure drops across regulator are capable of varying with the flow rate.

2 CAUTION: Capacities shown in table are capable of exceeding the maximum capacity of selected regulator. Consult tubing manufacturer for guidance.

3 Table includes losses for four 90 degree (1.57 rad) bends and two end fittings. Tubing runs with larger numbers of bends, fittings, or both shall be increased by an equivalent length of tubing to the following equation: $L = 1.3\,n$, where L is additional length (feet) of tubing and n is the number of additional fittings, bends, or both.

4 Table entries are rounded to 3 significant digits.

5 EHD = Equivalent Hydraulic Diameter, which is a measure of the relative hydraulic efficiency between different tubing sizes. The greater the value of EHD, the greater the gas capacity of the tubing.

TABLE 1215.2(19)
POLYETHYLENE PLASTIC PIPE [NFPA 54: TABLE 6.2.1(t)]*

					GAS:	NATURAL		
					INLET PRESSURE:	LESS THAN 2 psi		
					PRESSURE DROP:	0.3 in. w.c.		
					SPECIFIC GRAVITY:	0.60		
	PIPE SIZE (inch)							
NOMINAL OD:	½	¾	1	1 ¼	1 ½	2	3	4
DESIGNATION:	SDR 9.3	SDR 11	SDR 11	SDR 10	SDR 11	SDR 11	SDR 11	SDR 11
ACTUAL ID:	0.660	0.860	1.077	1.328	1.554	1.943	2.864	3.682
LENGTH (feet)	CAPACITY IN CUBIC FEET OF GAS PER HOUR							
10	153	305	551	955	1440	2590	7170	13 900
20	105	210	379	656	991	1780	4920	9520
30	84	169	304	527	796	1430	3950	7640
40	72	144	260	451	681	1220	3380	6540
50	64	128	231	400	604	1080	3000	5800
60	58	116	209	362	547	983	2720	5250
70	53	107	192	333	503	904	2500	4830
80	50	99	179	310	468	841	2330	4500
90	46	93	168	291	439	789	2180	4220
100	44	88	159	275	415	745	2060	3990
125	39	78	141	243	368	661	1830	3530
150	35	71	127	221	333	598	1660	3200
175	32	65	117	203	306	551	1520	2940
200	30	60	109	189	285	512	1420	2740
250	27	54	97	167	253	454	1260	2430
300	24	48	88	152	229	411	1140	2200
350	22	45	81	139	211	378	1050	2020
400	21	42	75	130	196	352	974	1880
450	19	39	70	122	184	330	914	1770
500	18	37	66	115	174	312	863	1670

For SI units: 1 inch = 25 mm, 1 foot = 304.8 mm, 1 cubic foot per hour = 0.0283 m^3/h, 1 pound-force per square inch = 6.8947 kPa, 1 inch water column = 0.249 kPa
* Table entries are rounded to 3 significant digits.

TABLE 1215.2(20)
POLYETHYLENE PLASTIC PIPE [NFPA 54: TABLE 6.2.1(u)]*

					GAS:	NATURAL		
					INLET PRESSURE:	LESS THAN 2 psi		
					PRESSURE DROP:	0.5 in. w.c.		
					SPECIFIC GRAVITY:	0.60		
	PIPE SIZE (inch)							
NOMINAL OD:	½	¾	1	1 ¼	1 ½	2	3	4
DESIGNATION:	SDR 9.3	SDR 11	SDR 11	SDR 10	SDR 11	SDR 11	SDR 11	SDR 11
ACTUAL ID:	0.660	0.860	1.077	1.328	1.554	1.943	2.864	3.682
LENGTH (feet)	CAPACITY IN CUBIC FEET OF GAS PER HOUR							
10	201	403	726	1260	1900	3410	9450	18 260
20	138	277	499	865	1310	2350	6490	12 550
30	111	222	401	695	1050	1880	5210	10 080
40	95	190	343	594	898	1610	4460	8630
50	84	169	304	527	796	1430	3950	7640
60	76	153	276	477	721	1300	3580	6930
70	70	140	254	439	663	1190	3300	6370
80	65	131	236	409	617	1110	3070	5930
90	61	123	221	383	579	1040	2880	5560
100	58	116	209	362	547	983	2720	5250
125	51	103	185	321	485	871	2410	4660
150	46	93	168	291	439	789	2180	4220
175	43	86	154	268	404	726	2010	3880
200	40	80	144	249	376	675	1870	3610
250	35	71	127	221	333	598	1660	3200
300	32	64	115	200	302	542	1500	2900
350	29	59	106	184	278	499	1380	2670
400	27	55	99	171	258	464	1280	2480
450	26	51	93	160	242	435	1200	2330
500	24	48	88	152	229	411	1140	2200

For SI units: 1 inch = 25 mm, 1 foot = 304.8 mm, 1 cubic foot per hour = 0.0283 m^3/h, 1 pound-force per square inch = 6.8947 kPa, 1 inch water column = 0.249 kPa
* Table entries are rounded to 3 significant digits.

TABLE 1215.2(21)
POLYETHYLENE PLASTIC PIPE [NFPA 54: TABLE 6.2.1(v)]*

							GAS:	**NATURAL**
							INLET PRESSURE:	**2.0 psi**
							PRESSURE DROP:	**1.0 psi**
							SPECIFIC GRAVITY:	**0.60**
	PIPE SIZE (inch)							
NOMINAL OD:	½	¾	1	1¼	1½	2	3	4
DESIGNATION:	**SDR 9.3**	**SDR 11**	**SDR 11**	**SDR 10**	**SDR 11**	**SDR 11**	**SDR 11**	**SDR 11**
ACTUAL ID:	**0.660**	**0.860**	**1.077**	**1.328**	**1.554**	**1.943**	**2.864**	**3.682**
LENGTH (feet)	**CAPACITY IN CUBIC FEET OF GAS PER HOUR**							
10	1860	3720	6710	11 600	17 600	31 600	87 300	169 000
20	1280	2560	4610	7990	12 100	21 700	60 000	116 000
30	1030	2050	3710	6420	9690	17 400	48 200	93 200
40	878	1760	3170	5490	8300	14 900	41 200	79 700
50	778	1560	2810	4870	7350	13 200	36 600	70 700
60	705	1410	2550	4410	6660	12 000	33 100	64 000
70	649	1300	2340	4060	6130	11 000	30 500	58 900
80	603	1210	2180	3780	5700	10 200	28 300	54 800
90	566	1130	2050	3540	5350	9610	26 600	51 400
100	535	1070	1930	3350	5050	9080	25 100	48 600
125	474	949	1710	2970	4480	8050	22 300	43 000
150	429	860	1550	2690	4060	7290	20 200	39 000
175	395	791	1430	2470	3730	6710	18 600	35 900
200	368	736	1330	2300	3470	6240	17 300	33 400
250	326	652	1180	2040	3080	5530	15 300	29 600
300	295	591	1070	1850	2790	5010	13 900	26 800
350	272	544	981	1700	2570	4610	12 800	24 700
400	253	506	913	1580	2390	4290	11 900	22 900
450	237	475	856	1480	2240	4020	11 100	21 500
500	224	448	809	1400	2120	3800	10 500	20 300
550	213	426	768	1330	2010	3610	9990	19 300
600	203	406	733	1270	1920	3440	9530	18 400
650	194	389	702	1220	1840	3300	9130	17 600
700	187	374	674	1170	1760	3170	8770	16 900
750	180	360	649	1130	1700	3050	8450	16 300
800	174	348	627	1090	1640	2950	8160	15 800
850	168	336	607	1050	1590	2850	7890	15 300
900	163	326	588	1020	1540	2770	7650	14 800
950	158	317	572	990	1500	2690	7430	14 400
1000	154	308	556	963	1450	2610	7230	14 000
1100	146	293	528	915	1380	2480	6870	13 300
1200	139	279	504	873	1320	2370	6550	12 700
1300	134	267	482	836	1260	2270	6270	12 100
1400	128	257	463	803	1210	2180	6030	11 600
1500	124	247	446	773	1170	2100	5810	11 200
1600	119	239	431	747	1130	2030	5610	10 800
1700	115	231	417	723	1090	1960	5430	10 500
1800	112	224	404	701	1060	1900	5260	10 200
1900	109	218	393	680	1030	1850	5110	9900
2000	106	212	382	662	1000	1800	4970	9600

For SI units: 1 inch = 25 mm, 1 foot = 304.8 mm, 1 cubic foot per hour = 0.0283 m^3/h, 1 pound-force per square inch = 6.8947 kPa

* Table entries are rounded to 3 significant digits.

TABLE 1215.2(22)
POLYETHYLENE PLASTIC TUBING [NFPA 54: TABLE 6.2.1(w)][2, 3]

	GAS:	NATURAL
	INLET PRESSURE:	LESS THAN 2.0 psi
	PRESSURE DROP:	0.3 in. w.c.
	SPECIFIC GRAVITY:	0.60
	PLASTIC TUBING SIZE (CTS)[1] (inch)	
NOMINAL OD:	½	1
DESIGNATION:	SDR 7	SDR 11
ACTUAL ID:	0.445	0.927
LENGTH (feet)	CAPACITY IN CUBIC FEET OF GAS PER HOUR	
10	54	372
20	37	256
30	30	205
40	26	176
50	23	156
60	21	141
70	19	130
80	18	121
90	17	113
100	16	107
125	14	95
150	13	86
175	12	79
200	11	74
225	10	69
250	NA	65
275	NA	62
300	NA	59
350	NA	54
400	NA	51
450	NA	47
500	NA	45

For SI units: 1 inch = 25 mm, 1 foot = 304.8 mm, 1 cubic foot per hour = 0.0283m^3/h, 1 pound-force per square inch = 6.8947 kPa, 1 inch water column = 0.249 kPa

Notes:

[1] CTS = Copper tube size.

[2] Table entries are rounded to 3 significant digits.

[3] NA means a flow of less than 10 ft^3/h (0.283 m^3/h).

TABLE 1215.2(23)
POLYETHYLENE PLASTIC TUBING [NFPA 54: TABLE 6.2.1(x)][2, 3]

	GAS:	NATURAL
	INLET PRESSURE:	LESS THAN 2.0 psi
	PRESSURE DROP:	0.5 in. w.c.
	SPECIFIC GRAVITY:	0.60
	PLASTIC TUBING SIZE (CTS)[1] (inch)	
NOMINAL OD:	½	1
DESIGNATION:	SDR 7	SDR 11
ACTUAL ID:	0.445	0.927
LENGTH (feet)	CAPACITY IN CUBIC FEET OF GAS PER HOUR	
10	72	490
20	49	337
30	39	271
40	34	232
50	30	205
60	27	186
70	25	171
80	23	159
90	22	149
100	21	141
125	18	125
150	17	113
175	15	104
200	14	97
225	13	91
250	12	86
275	11	82
300	11	78
350	10	72
400	NA	67
450	NA	63
500	NA	59

For SI units: 1 inch = 25 mm, 1 foot = 304.8 mm, 1 cubic foot per hour = 0.0283m^3/h, 1 pound-force per square inch = 6.8947 kPa, 1 inch water column = 0.249 kPa

Notes:

[1] CTS = Copper tube size.

[2] Table entries are rounded to 3 significant digits.

[3] NA means a flow of less than 10 ft^3/h (0.283 m^3/h).

TABLE 1215.2(24)
SCHEDULE 40 METALLIC PIPE [NFPA 54: TABLE 6.3.1(a)]*

							GAS:	**UNDILUTED PROPANE**	
							INLET PRESSURE:	**10.0 psi**	
							PRESSURE DROP:	**1.0 psi**	
							SPECIFIC GRAVITY:	**1.50**	
INTENDED USE: PIPE SIZING BETWEEN FIRST STAGE (HIGH PRESSURE) REGULATOR AND SECOND STAGE (LOW PRESSURE) REGULATOR									
	PIPE SIZE (inch)								
NOMINAL INSIDE:	½	¾	1	1¼	1½	2	2½	3	4
ACTUAL:	0.622	0.824	1.049	1.380	1.610	2.067	2.469	3.068	4.026
LENGTH (feet)	**CAPACITY IN THOUSANDS OF BTU PER HOUR**								
10	3320	6950	13 100	26 900	40 300	77 600	124 000	219 000	446 000
20	2280	4780	9000	18 500	27 700	53 300	85 000	150 000	306 000
30	1830	3840	7220	14 800	22 200	42 800	68 200	121 000	246 000
40	1570	3280	6180	12 700	19 000	36 600	58 400	103 000	211 000
50	1390	2910	5480	11 300	16 900	32 500	51 700	91 500	187 000
60	1260	2640	4970	10 200	15 300	29 400	46 900	82 900	169 000
70	1160	2430	4570	9380	14 100	27 100	43 100	76 300	156 000
80	1080	2260	4250	8730	13 100	25 200	40 100	70 900	145 000
90	1010	2120	3990	8190	12 300	23 600	37 700	66 600	136 000
100	956	2000	3770	7730	11 600	22 300	35 600	62 900	128 000
125	848	1770	3340	6850	10 300	19 800	31 500	55 700	114 000
150	768	1610	3020	6210	9300	17 900	28 600	50 500	103 000
175	706	1480	2780	5710	8560	16 500	26 300	46 500	94 700
200	657	1370	2590	5320	7960	15 300	24 400	43 200	88 100
250	582	1220	2290	4710	7060	13 600	21 700	38 300	78 100
300	528	1100	2080	4270	6400	12 300	19 600	34 700	70 800
350	486	1020	1910	3930	5880	11 300	18 100	31 900	65 100
400	452	945	1780	3650	5470	10 500	16 800	29 700	60 600
450	424	886	1670	3430	5140	9890	15 800	27 900	56 800
500	400	837	1580	3240	4850	9340	14 900	26 300	53 700
550	380	795	1500	3070	4610	8870	14 100	25 000	51 000
600	363	759	1430	2930	4400	8460	13 500	23 900	48 600
650	347	726	1370	2810	4210	8110	12 900	22 800	46 600
700	334	698	1310	2700	4040	7790	12 400	21 900	44 800
750	321	672	1270	2600	3900	7500	12 000	21 100	43 100
800	310	649	1220	2510	3760	7240	11 500	20 400	41 600
850	300	628	1180	2430	3640	7010	11 200	19 800	40 300
900	291	609	1150	2360	3530	6800	10 800	19 200	39 100
950	283	592	1110	2290	3430	6600	10 500	18 600	37 900
1000	275	575	1080	2230	3330	6420	10 200	18 100	36 900
1100	261	546	1030	2110	3170	6100	9720	17 200	35 000
1200	249	521	982	2020	3020	5820	9270	16 400	33 400
1300	239	499	940	1930	2890	5570	8880	15 700	32 000
1400	229	480	903	1850	2780	5350	8530	15 100	30 800
1500	221	462	870	1790	2680	5160	8220	14 500	29 600
1600	213	446	840	1730	2590	4980	7940	14 000	28 600
1700	206	432	813	1670	2500	4820	7680	13 600	27 700
1800	200	419	789	1620	2430	4670	7450	13 200	26 900
1900	194	407	766	1570	2360	4540	7230	12 800	26 100
2000	189	395	745	1530	2290	4410	7030	12 400	25 400

For SI units: 1 inch = 25 mm, 1 foot = 304.8 mm, 1000 British thermal units per hour = 0.293 kW, 1 pound-force per square inch = 6.8947 kPa
* Table entries are rounded to 3 significant digits.

TABLE 1215.2(25)
SCHEDULE 40 METALLIC PIPE [NFPA 54: TABLE 6.3.1(b)]*

					GAS:	UNDILUTED PROPANE		
					INLET PRESSURE:	10.0 psi		
					PRESSURE DROP:	3.0 psi		
					SPECIFIC GRAVITY:	1.50		
INTENDED USE: PIPE SIZING BETWEEN FIRST STAGE (HIGH PRESSURE) REGULATOR AND SECOND STAGE (LOW PRESSURE) REGULATOR									
	PIPE SIZE (inch)								
NOMINAL INSIDE:	½	¾	1	1¼	1½	2	2½	3	4
ACTUAL:	0.622	0.824	1.049	1.380	1.610	2.067	2.469	3.068	4.026
LENGTH (feet)	CAPACITY IN THOUSANDS OF BTU PER HOUR								
10	5890	12 300	23 200	47 600	71 300	137 000	219 000	387 000	789 000
20	4050	8460	15 900	32 700	49 000	94 400	150 000	266 000	543 000
30	3250	6790	12 800	26 300	39 400	75 800	121 000	214 000	436 000
40	2780	5810	11 000	22 500	33 700	64 900	103 000	183 000	373 000
50	2460	5150	9710	19 900	29 900	57 500	91 600	162 000	330 000
60	2230	4670	8790	18 100	27 100	52 100	83 000	147 000	299 000
70	2050	4300	8090	16 600	24 900	47 900	76 400	135 000	275 000
80	1910	4000	7530	15 500	23 200	44 600	71 100	126 000	256 000
90	1790	3750	7060	14 500	21 700	41 800	66 700	118 000	240 000
100	1690	3540	6670	13 700	20 500	39 500	63 000	111 000	227 000
125	1500	3140	5910	12 100	18 200	35 000	55 800	98 700	201 000
150	1360	2840	5360	11 000	16 500	31 700	50 600	89 400	182 000
175	1250	2620	4930	10 100	15 200	29 200	46 500	82 300	167 800
200	1160	2430	4580	9410	14 100	27 200	43 300	76 500	156 100
250	1030	2160	4060	8340	12 500	24 100	38 400	67 800	138 400
300	935	1950	3680	7560	11 300	21 800	34 800	61 500	125 400
350	860	1800	3390	6950	10 400	20 100	32 000	56 500	115 300
400	800	1670	3150	6470	9690	18 700	29 800	52 600	107 300
450	751	1570	2960	6070	9090	17 500	27 900	49 400	100 700
500	709	1480	2790	5730	8590	16 500	26 400	46 600	95 100
550	673	1410	2650	5450	8160	15 700	25 000	44 300	90 300
600	642	1340	2530	5200	7780	15 000	23 900	42 200	86 200
650	615	1290	2420	4980	7450	14 400	22 900	40 500	82 500
700	591	1240	2330	4780	7160	13 800	22 000	38 900	79 300
750	569	1190	2240	4600	6900	13 300	21 200	37 400	76 400
800	550	1150	2170	4450	6660	12 800	20 500	36 200	73 700
850	532	1110	2100	4300	6450	12 400	19 800	35 000	71 400
900	516	1080	2030	4170	6250	12 000	19 200	33 900	69 200
950	501	1050	1970	4050	6070	11 700	18 600	32 900	67 200
1000	487	1020	1920	3940	5900	11 400	18 100	32 000	65 400
1100	463	968	1820	3740	5610	10 800	17 200	30 400	62 100
1200	442	923	1740	3570	5350	10 300	16 400	29 000	59 200
1300	423	884	1670	3420	5120	9870	15 700	27 800	56 700
1400	406	849	1600	3280	4920	9480	15 100	26 700	54 500
1500	391	818	1540	3160	4740	9130	14 600	25 700	52 500
1600	378	790	1490	3060	4580	8820	14 100	24 800	50 700
1700	366	765	1440	2960	4430	8530	13 600	24 000	49 000
1800	355	741	1400	2870	4300	8270	13 200	23 300	47 600
1900	344	720	1360	2780	4170	8040	12 800	22 600	46 200
2000	335	700	1320	2710	4060	7820	12 500	22 000	44 900

For SI units: 1 inch = 25 mm, 1 foot = 304.8 mm, 1000 British thermal units per hour = 0.293 kW, 1 pound-force per square inch = 6.8947 kPa
* Table entries are rounded to 3 significant digits.

TABLE 1215.2(26)
SCHEDULE 40 METALLIC PIPE [NFPA 54: TABLE 6.3.1(c)]*

						GAS:	UNDILUTED PROPANE		
						INLET PRESSURE:	2.0 psi		
						PRESSURE DROP:	1.0 psi		
						SPECIFIC GRAVITY:	1.50		
INTENDED USE: PIPE SIZING BETWEEN 2 PSIG SERVICE AND LINE PRESSURE REGULATOR									
	PIPE SIZE (inch)								
NOMINAL:	½	¾	1	1¼	1½	2	2½	3	4
ACTUAL ID:	0.622	0.824	1.049	1.380	1.610	2.067	2.469	3.068	4.026
LENGTH (feet)	CAPACITY IN THOUSANDS OF BTU PER HOUR								
10	2680	5590	10 500	21 600	32 400	62 400	99 500	176 000	359 000
20	1840	3850	7240	14 900	22 300	42 900	68 400	121 000	247 000
30	1480	3090	5820	11 900	17 900	34 500	54 900	97 100	198 000
40	1260	2640	4980	10 200	15 300	29 500	47 000	83 100	170 000
50	1120	2340	4410	9060	13 600	26 100	41 700	73 700	150 000
60	1010	2120	4000	8210	12 300	23 700	37 700	66 700	136 000
70	934	1950	3680	7550	11 300	21 800	34 700	61 400	125 000
80	869	1820	3420	7020	10 500	20 300	32 300	57 100	116 000
90	815	1700	3210	6590	9880	19 000	30 300	53 600	109 000
100	770	1610	3030	6230	9330	18 000	28 600	50 600	103 000
125	682	1430	2690	5520	8270	15 900	25 400	44 900	91 500
150	618	1290	2440	5000	7490	14 400	23 000	40 700	82 900
175	569	1190	2240	4600	6890	13 300	21 200	37 400	76 300
200	529	1110	2080	4280	6410	12 300	19 700	34 800	71 000
250	469	981	1850	3790	5680	10 900	17 400	30 800	62 900
300	425	889	1670	3440	5150	9920	15 800	27 900	57 000
350	391	817	1540	3160	4740	9120	14 500	25 700	52 400
400	364	760	1430	2940	4410	8490	13 500	23 900	48 800
450	341	714	1340	2760	4130	7960	12 700	22 400	45 800
500	322	674	1270	2610	3910	7520	12 000	21 200	43 200
550	306	640	1210	2480	3710	7140	11 400	20 100	41 100
600	292	611	1150	2360	3540	6820	10 900	19 200	39 200
650	280	585	1100	2260	3390	6530	10 400	18 400	37 500
700	269	562	1060	2170	3260	6270	9990	17 700	36 000
750	259	541	1020	2090	3140	6040	9630	17 000	34 700
800	250	523	985	2020	3030	5830	9300	16 400	33 500
850	242	506	953	1960	2930	5640	9000	15 900	32 400
900	235	490	924	1900	2840	5470	8720	15 400	31 500
950	228	476	897	1840	2760	5310	8470	15 000	30 500
1000	222	463	873	1790	2680	5170	8240	14 600	29 700
1100	210	440	829	1700	2550	4910	7830	13 800	28 200
1200	201	420	791	1620	2430	4680	7470	13 200	26 900
1300	192	402	757	1550	2330	4490	7150	12 600	25 800
1400	185	386	727	1490	2240	4310	6870	12 100	24 800
1500	178	372	701	1440	2160	4150	6620	11 700	23 900
1600	172	359	677	1390	2080	4010	6390	11 300	23 000
1700	166	348	655	1340	2010	3880	6180	10 900	22 300
1800	161	337	635	1300	1950	3760	6000	10 600	21 600
1900	157	327	617	1270	1900	3650	5820	10 300	21 000
2000	152	318	600	1230	1840	3550	5660	10 000	20 400

For SI units: 1 inch = 25 mm, 1 foot = 304.8 mm, 1000 British thermal units per hour = 0.293 kW, 1 pound-force per square inch = 6.8947 kPa
* Table entries are rounded to 3 significant digits.

TABLE 1215.2(27)
SCHEDULE 40 METALLIC PIPE [NFPA 54: TABLE 6.3.1(d)]*

					GAS:	UNDILUTED PROPANE			
					INLET PRESSURE:	11.0 in. w.c.			
					PRESSURE DROP:	0.5 in. w.c.			
					SPECIFIC GRAVITY:	1.50			
INTENDED USE: PIPE SIZING BETWEEN SINGLE OR SECOND STAGE (LOW PRESSURE) REGULATOR AND APPLIANCE									
	PIPE SIZE (inch)								
NOMINAL INSIDE:	½	¾	1	1¼	1½	2	2½	3	4
ACTUAL:	0.622	0.824	1.049	1.380	1.610	2.067	2.469	3.068	4.026
LENGTH (feet)	CAPACITY IN THOUSANDS OF BTU PER HOUR								
10	291	608	1150	2350	3520	6790	10 800	19 100	39 000
20	200	418	787	1620	2420	4660	7430	13 100	26 800
30	160	336	632	1300	1940	3750	5970	10 600	21 500
40	137	287	541	1110	1660	3210	5110	9030	18 400
50	122	255	480	985	1480	2840	4530	8000	16 300
60	110	231	434	892	1340	2570	4100	7250	14 800
80	101	212	400	821	1230	2370	3770	6670	13 600
100	94	197	372	763	1140	2200	3510	6210	12 700
125	89	185	349	716	1070	2070	3290	5820	11 900
150	84	175	330	677	1010	1950	3110	5500	11 200
175	74	155	292	600	899	1730	2760	4880	9950
200	67	140	265	543	814	1570	2500	4420	9010
250	62	129	243	500	749	1440	2300	4060	8290
300	58	120	227	465	697	1340	2140	3780	7710
350	51	107	201	412	618	1190	1900	3350	6840
400	46	97	182	373	560	1080	1720	3040	6190
450	42	89	167	344	515	991	1580	2790	5700
500	40	83	156	320	479	922	1470	2600	5300
550	37	78	146	300	449	865	1380	2440	4970
600	35	73	138	283	424	817	1300	2300	4700
650	33	70	131	269	403	776	1240	2190	4460
700	32	66	125	257	385	741	1180	2090	4260
750	30	64	120	246	368	709	1130	2000	4080
800	29	61	115	236	354	681	1090	1920	3920
850	28	59	111	227	341	656	1050	1850	3770
900	27	57	107	220	329	634	1010	1790	3640
950	26	55	104	213	319	613	978	1730	3530
1000	25	53	100	206	309	595	948	1680	3420
1100	25	52	97	200	300	578	921	1630	3320
1200	24	50	95	195	292	562	895	1580	3230
1300	23	48	90	185	277	534	850	1500	3070
1400	22	46	86	176	264	509	811	1430	2930
1500	21	44	82	169	253	487	777	1370	2800
1600	20	42	79	162	243	468	746	1320	2690
1700	19	40	76	156	234	451	719	1270	2590
1800	19	39	74	151	226	436	694	1230	2500
1900	18	38	71	146	219	422	672	1190	2420
2000	18	37	69	142	212	409	652	1150	2350

For SI units: 1 inch = 25 mm, 1 foot = 304.8 mm, 1000 British thermal units per hour = 0.293 kW, 1 inch water column = 0.249 kPa
* Table entries are rounded to 3 significant digits.

TABLE 1215.2(28)
SEMI-RIGID COPPER TUBING [NFPA 54: TABLE 6.3.1(e)][2]

							GAS:	**UNDILUTED PROPANE**	
							INLET PRESSURE:	**10.0 psi**	
							PRESSURE DROP:	**1.0 psi**	
							SPECIFIC GRAVITY:	**1.50**	
INTENDED USE: TUBE SIZING BETWEEN FIRST STAGE (HIGH PRESSURE) REGULATOR AND SECOND STAGE (LOW PRESSURE) REGULATOR									
	TUBE SIZE (inch)								
NOMINAL: K & L:	¼	⅜	½	⅝	¾	**1**	**1¼**	**1½**	**2**
NOMINAL: ACR:	⅜	½	⅝	¾	⅞	**1⅛**	**1⅜**	**–**	**–**
OUTSIDE:	**0.375**	**0.500**	**0.625**	**0.750**	**0.875**	**1.125**	**1.375**	**1.625**	**2.125**
INSIDE:[1]	**0.305**	**0.402**	**0.527**	**0.652**	**0.745**	**0.995**	**1.245**	**1.481**	**1.959**
LENGTH (feet)	**CAPACITY IN THOUSANDS OF BTU PER HOUR**								
10	513	1060	2150	3760	5330	11 400	20 500	32 300	67 400
20	352	727	1480	2580	3670	7830	14 100	22 200	46 300
30	283	584	1190	2080	2940	6290	11 300	17 900	37 200
40	242	500	1020	1780	2520	5380	9690	15 300	31 800
50	215	443	901	1570	2230	4770	8590	13 500	28 200
60	194	401	816	1430	2020	4320	7780	12 300	25 600
70	179	369	751	1310	1860	3980	7160	11 300	23 500
80	166	343	699	1220	1730	3700	6660	10 500	21 900
90	156	322	655	1150	1630	3470	6250	9850	20 500
100	147	304	619	1080	1540	3280	5900	9310	19 400
125	131	270	549	959	1360	2910	5230	8250	17 200
150	118	244	497	869	1230	2630	4740	7470	15 600
175	109	225	457	799	1130	2420	4360	6880	14 300
200	101	209	426	744	1060	2250	4060	6400	13 300
250	90	185	377	659	935	2000	3600	5670	11 800
300	81	168	342	597	847	1810	3260	5140	10 700
350	75	155	314	549	779	1660	3000	4730	9840
400	70	144	292	511	725	1550	2790	4400	9160
450	65	135	274	480	680	1450	2620	4130	8590
500	62	127	259	453	643	1370	2470	3900	8120
550	59	121	246	430	610	1300	2350	3700	7710
600	56	115	235	410	582	1240	2240	3530	7350
650	54	111	225	393	558	1190	2140	3380	7040
700	51	106	216	378	536	1140	2060	3250	6770
750	50	102	208	364	516	1100	1980	3130	6520
800	48	99	201	351	498	1060	1920	3020	6290
850	46	96	195	340	482	1030	1850	2920	6090
900	45	93	189	330	468	1000	1800	2840	5910
950	44	90	183	320	454	970	1750	2750	5730
1000	42	88	178	311	442	944	1700	2680	5580
1100	40	83	169	296	420	896	1610	2540	5300
1200	38	79	161	282	400	855	1540	2430	5050
1300	37	76	155	270	383	819	1470	2320	4840
1400	35	73	148	260	368	787	1420	2230	4650
1500	34	70	143	250	355	758	1360	2150	4480
1600	33	68	138	241	343	732	1320	2080	4330
1700	32	66	134	234	331	708	1270	2010	4190
1800	31	64	130	227	321	687	1240	1950	4060
1900	30	62	126	220	312	667	1200	1890	3940
2000	29	60	122	214	304	648	1170	1840	3830

For SI units: 1 inch = 25 mm, 1 foot = 304.8 mm, 1000 British thermal units per hour = 0.293 kW, 1 pound-force per square inch = 6.8947 kPa

Notes:

[1] Table capacities are based on Type K copper tubing inside diameter (shown), which has the smallest inside diameter of the copper tubing products.

[2] Table entries are rounded to 3 significant digits.

TABLE 1215.2(29)
SEMI-RIGID COPPER TUBING [NFPA 54: TABLE 6.3.1(f)][2, 3]

							GAS:	**UNDILUTED PROPANE**	
							INLET PRESSURE:	**11.0 in. w.c.**	
							PRESSURE DROP:	**0.5 in. w.c.**	
							SPECIFIC GRAVITY:	**1.50**	
INTENDED USE: TUBE SIZING BETWEEN SINGLE OR SECOND STAGE (LOW PRESSURE) REGULATOR AND APPLIANCE									
	TUBE SIZE (inch)								
NOMINAL: K & L:	**¼**	**⅜**	**½**	**⅝**	**¾**	**1**	**1¼**	**1½**	**2**
NOMINAL: ACR:	**⅜**	**½**	**⅝**	**¾**	**⅞**	**1⅛**	**1⅜**	**–**	**–**
OUTSIDE:	**0.375**	**0.500**	**0.625**	**0.750**	**0.875**	**1.125**	**1.375**	**1.625**	**2.125**
INSIDE:[1]	**0.305**	**0.402**	**0.527**	**0.652**	**0.745**	**0.995**	**1.245**	**1.481**	**1.959**
LENGTH (feet)	**CAPACITY IN THOUSANDS OF BTU PER HOUR**								
10	45	93	188	329	467	997	1800	2830	5890
20	31	64	129	226	321	685	1230	1950	4050
30	25	51	104	182	258	550	991	1560	3250
40	21	44	89	155	220	471	848	1340	2780
50	19	39	79	138	195	417	752	1180	2470
60	17	35	71	125	177	378	681	1070	2240
70	16	32	66	115	163	348	626	988	2060
80	15	30	61	107	152	324	583	919	1910
90	14	28	57	100	142	304	547	862	1800
100	13	27	54	95	134	287	517	814	1700
125	11	24	48	84	119	254	458	722	1500
150	10	21	44	76	108	230	415	654	1360
175	NA	20	40	70	99	212	382	602	1250
200	NA	18	37	65	92	197	355	560	1170
250	NA	16	33	58	82	175	315	496	1030
300	NA	15	30	52	74	158	285	449	936
350	NA	14	28	48	68	146	262	414	861
400	NA	13	26	45	63	136	244	385	801
450	NA	12	24	42	60	127	229	361	752
500	NA	11	23	40	56	120	216	341	710
550	NA	11	22	38	53	114	205	324	674
600	NA	10	21	36	51	109	196	309	643
650	NA	NA	20	34	49	104	188	296	616
700	NA	NA	19	33	47	100	180	284	592
750	NA	NA	18	32	45	96	174	274	570
800	NA	NA	18	31	44	93	168	264	551
850	NA	NA	17	30	42	90	162	256	533
900	NA	NA	17	29	41	87	157	248	517
950	NA	NA	16	28	40	85	153	241	502
1000	NA	NA	16	27	39	83	149	234	488
1100	NA	NA	15	26	37	78	141	223	464
1200	NA	NA	14	25	35	75	135	212	442
1300	NA	NA	14	24	34	72	129	203	423
1400	NA	NA	13	23	32	69	124	195	407
1500	NA	NA	13	22	31	66	119	188	392
1600	NA	NA	12	21	30	64	115	182	378
1700	NA	NA	12	20	29	62	112	176	366
1800	NA	NA	11	20	28	60	108	170	355
1900	NA	NA	11	19	27	58	105	166	345
2000	NA	NA	11	19	27	57	102	161	335

For SI units: 1 inch = 25 mm, 1 foot = 304.8 mm, 1000 British thermal units per hour = 0.293 kW, 1 inch water column = 0.249 kPa

Notes:

1 Table capacities are based on Type K copper tubing inside diameter (shown), which has the smallest inside diameter of the copper tubing products.

2 Table entries are rounded to 3 significant digits.

3 NA means a flow of less than 10 000 Btu/h (2.93 kW).

TABLE 1215.2(30)
SEMI-RIGID COPPER TUBING [NFPA 54: TABLE 6.3.1(g)][2]

					GAS:			**UNDILUTED PROPANE**	
					INLET PRESSURE:			**2.0 psi**	
					PRESSURE DROP:			**1.0 psi**	
					SPECIFIC GRAVITY:			**1.50**	
INTENDED USE: TUBE SIZING BETWEEN 2 PSIG SERVICE AND LINE PRESSURE REGULATOR									
	TUBE SIZE (inch)								
NOMINAL: K & L:	**¼**	**⅜**	**½**	**⅝**	**¾**	**1**	**1¼**	**1½**	**2**
NOMINAL: ACR:	**⅜**	**½**	**⅝**	**¾**	**⅞**	**1⅛**	**1⅜**	**–**	**–**
OUTSIDE:	**0.375**	**0.500**	**0.625**	**0.750**	**0.875**	**1.125**	**1.375**	**1.625**	**2.125**
INSIDE:[1]	**0.305**	**0.402**	**0.527**	**0.652**	**0.745**	**0.995**	**1.245**	**1.481**	**1.959**
LENGTH (feet)	**CAPACITY IN THOUSANDS OF BTU PER HOUR**								
10	413	852	1730	3030	4300	9170	16 500	26 000	54 200
20	284	585	1190	2080	2950	6310	11 400	17 900	37 300
30	228	470	956	1670	2370	5060	9120	14 400	29 900
40	195	402	818	1430	2030	4330	7800	12 300	25 600
50	173	356	725	1270	1800	3840	6920	10 900	22 700
60	157	323	657	1150	1630	3480	6270	9880	20 600
70	144	297	605	1060	1500	3200	5760	9090	18 900
80	134	276	562	983	1390	2980	5360	8450	17 600
90	126	259	528	922	1310	2790	5030	7930	16 500
100	119	245	498	871	1240	2640	4750	7490	15 600
125	105	217	442	772	1100	2340	4210	6640	13 800
150	95	197	400	700	992	2120	3820	6020	12 500
175	88	181	368	644	913	1950	3510	5540	11 500
200	82	168	343	599	849	1810	3270	5150	10 700
250	72	149	304	531	753	1610	2900	4560	9510
300	66	135	275	481	682	1460	2620	4140	8610
350	60	124	253	442	628	1340	2410	3800	7920
400	56	116	235	411	584	1250	2250	3540	7370
450	53	109	221	386	548	1170	2110	3320	6920
500	50	103	209	365	517	1110	1990	3140	6530
550	47	97	198	346	491	1050	1890	2980	6210
600	45	93	189	330	469	1000	1800	2840	5920
650	43	89	181	316	449	959	1730	2720	5670
700	41	86	174	304	431	921	1660	2620	5450
750	40	82	168	293	415	888	1600	2520	5250
800	39	80	162	283	401	857	1540	2430	5070
850	37	77	157	274	388	829	1490	2350	4900
900	36	75	152	265	376	804	1450	2280	4750
950	35	72	147	258	366	781	1410	2220	4620
1000	34	71	143	251	356	760	1370	2160	4490
1100	32	67	136	238	338	721	1300	2050	4270
1200	31	64	130	227	322	688	1240	1950	4070
1300	30	61	124	217	309	659	1190	1870	3900
1400	28	59	120	209	296	633	1140	1800	3740
1500	27	57	115	201	286	610	1100	1730	3610
1600	26	55	111	194	276	589	1060	1670	3480
1700	26	53	108	188	267	570	1030	1620	3370
1800	25	51	104	182	259	553	1000	1570	3270
1900	24	50	101	177	251	537	966	1520	3170
2000	23	48	99	172	244	522	940	1480	3090

For SI units: 1 inch = 25 mm, 1 foot = 304.8 mm, 1000 British thermal units per hour = 0.293 kW, 1 pound-force per square inch = 6.8947 kPa

Notes:

[1] Table capacities are based on Type K copper tubing inside diameter (shown), which has the smallest inside diameter of the copper tubing products.

[2] Table entries are rounded to 3 significant digits.

TABLE 1215.2(31)
CORRUGATED STAINLESS STEEL TUBING (CSST) [NFPA 54: TABLE 6.3.1(h)][1, 2]

										GAS:	UNDILUTED PROPANE			
										INLET PRESSURE:	11.0 in. w.c.			
										PRESSURE DROP:	0.5 in. w.c.			
										SPECIFIC GRAVITY:	1.50			
INTENDED USE: CSST SIZING BETWEEN SINGLE OR SECOND STAGE (LOW PRESSURE) REGULATOR AND APPLIANCE SHUTOFF VALVE														
	TUBE SIZE (EHD)[3]													
FLOW DESIGNATION:	13	15	18	19	23	25	30	31	37	39	46	48	60	62
LENGTH (feet)	CAPACITY IN THOUSANDS OF BTU PER HOUR													
5	72	99	181	211	355	426	744	863	1420	1638	2830	3270	5780	6550
10	50	69	129	150	254	303	521	605	971	1179	1990	2320	4110	4640
15	39	55	104	121	208	248	422	490	775	972	1620	1900	3370	3790
20	34	49	91	106	183	216	365	425	661	847	1400	1650	2930	3290
25	30	42	82	94	164	192	325	379	583	762	1250	1480	2630	2940
30	28	39	74	87	151	177	297	344	528	698	1140	1350	2400	2680
40	23	33	64	74	131	153	256	297	449	610	988	1170	2090	2330
50	20	30	58	66	118	137	227	265	397	548	884	1050	1870	2080
60	19	26	53	60	107	126	207	241	359	502	805	961	1710	1900
70	17	25	49	57	99	117	191	222	330	466	745	890	1590	1760
80	15	23	45	52	94	109	178	208	307	438	696	833	1490	1650
90	15	22	44	50	90	102	169	197	286	414	656	787	1400	1550
100	14	20	41	47	85	98	159	186	270	393	621	746	1330	1480
150	11	15	31	36	66	75	123	143	217	324	506	611	1090	1210
200	9	14	28	33	60	69	112	129	183	283	438	531	948	1050
250	8	12	25	30	53	61	99	117	163	254	390	476	850	934
300	8	11	23	26	50	57	90	107	147	234	357	434	777	854

For SI units: 1 foot = 304.8 mm, 1000 British thermal units per hour = 0.293 kW, 1 inch water column = 0.249 kPa

Notes:

1 Table includes losses for four 90 degree (1.57 rad) bends and two end fittings. Tubing runs with larger numbers of bends, fittings, or both shall be increased by an equivalent length of tubing to the following equation: L = 1.3 n, where L is additional length (ft) of tubing and n is the number of additional fittings, bends, or both.

2 Table entries are rounded to 3 significant digits.

3 EHD = Equivalent Hydraulic Diameter, which is a measure of the relative hydraulic efficiency between different tubing sizes. The greater the value of EHD, the greater the gas capacity of the tubing.

TABLE 1215.2(32)
CORRUGATED STAINLESS STEEL TUBING (CSST) [NFPA 54: TABLE 6.3.1(i)][1, 2, 3, 4]

								GAS:			**UNDILUTED PROPANE**		
								INLET PRESSURE:			**2.0 psi**		
								PRESSURE DROP:			**1.0 psi**		
								SPECIFIC GRAVITY:			**1.50**		
INTENDED USE: CSST SIZING BETWEEN 2 PSIG SERVICE AND LINE PRESSURE REGULATOR														
	TUBE SIZE (EHD)[5]													
FLOW DESIGNATION:	**13**	**15**	**18**	**19**	**23**	**25**	**30**	**31**	**37**	**39**	**46**	**48**	**60**	**62**
LENGTH (feet)	**CAPACITY IN THOUSANDS OF BTU PER HOUR**													
10	426	558	927	1110	1740	2170	4100	4720	7130	7958	15 200	16 800	29 400	34 200
25	262	347	591	701	1120	1380	2560	2950	4560	5147	9550	10 700	18 800	21 700
30	238	316	540	640	1030	1270	2330	2690	4180	4719	8710	9790	17 200	19 800
40	203	271	469	554	896	1100	2010	2320	3630	4116	7530	8500	14 900	17 200
50	181	243	420	496	806	986	1790	2070	3260	3702	6730	7610	13 400	15 400
75	147	196	344	406	663	809	1460	1690	2680	3053	5480	6230	11 000	12 600
80	140	189	333	393	643	768	1410	1630	2590	2961	5300	6040	10 600	12 200
100	124	169	298	350	578	703	1260	1450	2330	2662	4740	5410	9530	10 900
150	101	137	245	287	477	575	1020	1180	1910	2195	3860	4430	7810	8890
200	86	118	213	248	415	501	880	1020	1660	1915	3340	3840	6780	7710
250	77	105	191	222	373	448	785	910	1490	1722	2980	3440	6080	6900
300	69	96	173	203	343	411	716	829	1360	1578	2720	3150	5560	6300
400	60	82	151	175	298	355	616	716	1160	1376	2350	2730	4830	5460
500	53	72	135	158	268	319	550	638	1030	1237	2100	2450	4330	4880

For SI units: 1 foot = 304.8 mm, 1000 British thermal units per hour = 0.293 kW, 1 pound-force per square inch = 6.8947 kPa

Notes:

1 Table does not include effect of pressure drop across the line regulator. Where regulator loss exceeds 0.5 psi (3.4 kPa) [based on 13 inch water column (3.2 kPa) outlet pressure], DO NOT USE THIS TABLE. Consult with regulator manufacturer for pressure drops and capacity factors. Pressure drops across a regulator are capable of varying with flow rate.

2 CAUTION: Capacities shown in table are capable of exceeding the maximum capacity for a selected regulator. Consult with regulator or tubing manufacturer for guidance.

3 Table includes losses for four 90 degree (1.57 rad) bends and two end fittings. Tubing runs with larger numbers of bends, fittings, or both shall be increased by an equivalent length of tubing to the following equation: L = 1.3 n, where L is additional length (ft) of tubing and n is the number of additional fittings, bends, or both.

4 Table entries are rounded to 3 significant digits.

5 EHD = Equivalent Hydraulic Diameter, which is a measure of the relative hydraulic efficiency between different tubing sizes. The greater the value of EHD, the greater the gas capacity of the tubing.

TABLE 1215.2(33)
CORRUGATED STAINLESS STEEL TUBING (CSST) [NFPA 54: TABLE 6.3.1(j)][1, 2, 3, 4]

									GAS:	UNDILUTED PROPANE			
									INLET PRESSURE:	5.0 psi			
									PRESSURE DROP:	3.5 psi			
									SPECIFIC GRAVITY:	1.50			
	TUBE SIZE (EHD)[5]												
FLOW DESIGNATION:	**13**	**15**	**18**	**19**	**23**	**25**	**30**	**31**	**37**	**39**	**46**	**48**	**60**	**62**
LENGTH (feet)	**CAPACITY IN THOUSANDS OF BTU PER HOUR**													
10	826	1070	1710	2060	3150	4000	7830	8950	13 100	14 441	28 600	31 200	54 400	63 800
25	509	664	1090	1310	2040	2550	4860	5600	8400	9339	18 000	19 900	34 700	40 400
30	461	603	999	1190	1870	2340	4430	5100	7680	8564	16 400	18 200	31 700	36 900
40	396	520	867	1030	1630	2030	3820	4400	6680	7469	14 200	15 800	27 600	32 000
50	352	463	777	926	1460	1820	3410	3930	5990	6717	12 700	14 100	24 700	28 600
75	284	376	637	757	1210	1490	2770	3190	4920	5539	10 300	11 600	20 300	23 400
80	275	363	618	731	1170	1450	2680	3090	4770	5372	9990	11 200	19 600	22 700
100	243	324	553	656	1050	1300	2390	2760	4280	4830	8930	10 000	17 600	20 300
150	196	262	453	535	866	1060	1940	2240	3510	3983	7270	8210	14 400	16 600
200	169	226	393	464	755	923	1680	1930	3050	3474	6290	7130	12 500	14 400
250	150	202	352	415	679	828	1490	1730	2740	3124	5620	6390	11 200	12 900
300	136	183	322	379	622	757	1360	1570	2510	2865	5120	5840	10 300	11 700
400	117	158	279	328	542	657	1170	1360	2180	2498	4430	5070	8920	10 200
500	104	140	251	294	488	589	1050	1210	1950	2247	3960	4540	8000	9110

For SI units: 1 foot = 304.8 mm, 1000 British thermal units per hour = 0.293 kW, 1 pound-force per square inch = 6.8947 kPa

Notes:

1 Table does not include effect of pressure drop across the line regulator. Where regulator loss exceeds 0.5 psi (3.4 kPa) [based on 13 inch water column (3.2 kPa) outlet pressure], DO NOT USE THIS TABLE. Consult with regulator manufacturer for pressure drops and capacity factors. Pressure drops across a regulator are capable of varying with flow rate.

2 CAUTION: Capacities shown in table are capable of exceeding the maximum capacity for a selected regulator. Consult with regulator or tubing manufacturer for guidance.

3 Table includes losses for four 90 degree (1.57 rad) bends and two end fittings. Tubing runs with larger numbers of bends, fittings, or both shall be increased by an equivalent length of tubing to the following equation: $L = 1.3\ n$, where L is additional length (ft) of tubing and n is the number of additional fittings, bends, or both.

4 Table entries are rounded to 3 significant digits.

5 EHD = Equivalent Hydraulic Diameter, which is a measure of the relative hydraulic efficiency between different tubing sizes. The greater the value of EHD, the greater the gas capacity of the tubing.

TABLE 1215.2(34)
POLYETHYLENE PLASTIC PIPE [NFPA 54: TABLE 6.3.1(k)]*

					GAS:	UNDILUTED PROPANE		
					INLET PRESSURE:	11.0 in. w.c.		
					PRESSURE DROP:	0.5 in. w.c.		
					SPECIFIC GRAVITY:	1.50		
INTENDED USE: PE PIPE SIZING BETWEEN INTEGRAL SECOND-STAGE REGULATOR AT TANK OR SECOND-STAGE (LOW PRESSURE) REGULATOR AND BUILDING								
	PIPE SIZE (inch)							
NOMINAL OD:	½	¾	1	1¼	1½	2	3	4
DESIGNATION:	SDR 9.3	SDR 11	SDR 11	SDR 10	SDR 11	SDR 11	SDR 11	SDR 11
ACTUAL ID:	0.660	0.860	1.077	1.328	1.554	1.943	2.864	3.682
LENGTH (feet)	CAPACITY IN THOUSANDS OF BTU PER HOUR							
10	340	680	1230	2130	3210	5770	16 000	30 900
20	233	468	844	1460	2210	3970	11 000	21 200
30	187	375	677	1170	1770	3180	8810	17 000
40	160	321	580	1000	1520	2730	7540	14 600
50	142	285	514	890	1340	2420	6680	12 900
60	129	258	466	807	1220	2190	6050	11 700
70	119	237	428	742	1120	2010	5570	10 800
80	110	221	398	690	1040	1870	5180	10 000
90	103	207	374	648	978	1760	4860	9400
100	98	196	353	612	924	1660	4590	8900
125	87	173	313	542	819	1470	4070	7900
150	78	157	284	491	742	1330	3690	7130
175	72	145	261	452	683	1230	3390	6560
200	67	135	243	420	635	1140	3160	6100
250	60	119	215	373	563	1010	2800	5410
300	54	108	195	338	510	916	2530	4900
350	50	99	179	311	469	843	2330	4510
400	46	92	167	289	436	784	2170	4190
450	43	87	157	271	409	736	2040	3930
500	41	82	148	256	387	695	1920	3720

For SI units: 1 inch = 25 mm, 1 foot = 304.8 mm, 1000 British thermal units per hour = 0.293 kW, 1 inch water column = 0.249 kPa
* Table entries are rounded to 3 significant digits.

TABLE 1215.2(35)
POLYETHYLENE PLASTIC PIPE [NFPA 54: TABLE 6.3.1(l)]*

					GAS:	UNDILUTED PROPANE		
					INLET PRESSURE:	2.0 psi		
					PRESSURE DROP:	1.0 psi		
					SPECIFIC GRAVITY:	1.50		
INTENDED USE: PE PIPE SIZING BETWEEN 2 PSI SERVICE REGULATOR AND LINE PRESSURE REGULATOR								
	PIPE SIZE (inch)							
NOMINAL OD:	½	¾	1	1¼	1½	2	3	4
DESIGNATION:	SDR 9.3	SDR 11	SDR 11	SDR 10	SDR 11	SDR 11	SDR 11	SDR 11
ACTUAL ID:	0.660	0.860	1.077	1.328	1.554	1.943	2.864	3.682
LENGTH (feet)	CAPACITY IN THOUSANDS OF BTU PER HOUR							
10	3130	6260	11 300	19 600	29 500	53 100	147 000	284 000
20	2150	4300	7760	13 400	20 300	36 500	101 000	195 000
30	1730	3450	6230	10 800	16 300	29 300	81 100	157 000
40	1480	2960	5330	9240	14 000	25 100	69 400	134 100
50	1310	2620	4730	8190	12 400	22 200	61 500	119 000
60	1190	2370	4280	7420	11 200	20 100	55 700	108 000
70	1090	2180	3940	6830	10 300	18 500	51 300	99 100
80	1010	2030	3670	6350	9590	17 200	47 700	92 200
90	952	1910	3440	5960	9000	16 200	44 700	86 500
100	899	1800	3250	5630	8500	15 300	42 300	81 700
125	797	1600	2880	4990	7530	13 500	37 500	72 400
150	722	1450	2610	4520	6830	12 300	33 900	65 600
175	664	1330	2400	4160	6280	11 300	31 200	60 300
200	618	1240	2230	3870	5840	10 500	29 000	56 100
250	548	1100	1980	3430	5180	9300	25 700	49 800
300	496	994	1790	3110	4690	8430	23 300	45 100
350	457	914	1650	2860	4320	7760	21 500	41 500
400	425	851	1530	2660	4020	7220	12 000	38 600
450	399	798	1440	2500	3770	6770	18 700	36 200
500	377	754	1360	2360	3560	6390	17 700	34 200
550	358	716	1290	2240	3380	6070	16 800	32 500
600	341	683	1230	2140	3220	5790	16 000	31 000
650	327	654	1180	2040	3090	5550	15 400	29 700
700	314	628	1130	1960	2970	5330	14 700	28 500
750	302	605	1090	1890	2860	5140	14 200	27 500
800	292	585	1050	1830	2760	4960	13 700	26 500
850	283	566	1020	1770	2670	4800	13 300	25 700
900	274	549	990	1710	2590	4650	12 900	24 900
950	266	533	961	1670	2520	4520	12 500	24 200
1000	259	518	935	1620	2450	4400	12 200	23 500
1100	246	492	888	1540	2320	4170	11 500	22 300
1200	234	470	847	1470	2220	3980	11 000	21 300
1300	225	450	811	1410	2120	3810	10 600	20 400
1400	216	432	779	1350	2040	3660	10 100	19 600
1500	208	416	751	1300	1960	3530	9760	18 900
1600	201	402	725	1260	1900	3410	9430	18 200
1700	194	389	702	1220	1840	3300	9130	17 600
1800	188	377	680	1180	1780	3200	8850	17 100
1900	183	366	661	1140	1730	3110	8590	16 600
2000	178	356	643	1110	1680	3020	8360	16 200

For SI units: 1 inch = 25 mm, 1 foot = 304.8 mm, 1000 British thermal units per hour = 0.293 kW, 1 pound-force per square inch = 6.8947 kPa
* Table entries are rounded to 3 significant digits.

TABLE 1215.2(36)
POLYETHYLENE PLASTIC TUBING [NFPA 54: TABLE 6.3.1(m)][2]

	GAS:	UNDILUTED PROPANE
	INLET PRESSURE:	11.0 in. w.c.
	PRESSURE DROP:	0.5 in. w.c.
	SPECIFIC GRAVITY:	1.50
INTENDED USE: SIZING BETWEEN INTEGRAL 2-STAGE REGULATOR AT TANK OR SECOND-STAGE (LOW-PRESSURE REGULATOR) AND THE BUILDING		
	PLASTIC TUBING SIZE (CTS)[1] (inch)	
NOMINAL OD:	½	1
DESIGNATION:	SDR 7	SDR 11
ACTUAL ID:	0.445	0.927
LENGTH (feet)	CAPACITY IN THOUSANDS OF BTU PER HOUR	
10	121	828
20	83	569
30	67	457
40	57	391
50	51	347
60	46	314
70	42	289
80	39	269
90	37	252
100	35	238
125	31	211
150	28	191
175	26	176
200	24	164
225	22	154
250	21	145
275	20	138
300	19	132
350	18	121
400	16	113
450	15	106
500	15	100

For SI units: 1 inch = 25 mm, 1 foot = 304.8 mm, 1000 British thermal units per hour = 0.293 kW, 1 inch water column = 0.249 kPa

Notes:

[1] CTS = Copper tube size.

[2] Table entries are rounded to 3 significant digits.

CHAPTER 13

HEALTH CARE FACILITIES AND MEDICAL GAS AND MEDICAL VACUUM SYSTEMS

Part I – General Requirements.

1301.0 General Requirements.

1301.1 Applicability. This chapter applies to the special fixtures and systems in health care facilities; the special plumbing requirements for such facilities; and the installation, testing, and verification of Categories 1, 2, and 3 medical gas and medical vacuum piping systems, except as otherwise indicated in this chapter, from the central supply system to the station outlets or inlets in hospitals, clinics, and other health care facilities. Other plumbing in such facilities shall comply with other applicable sections of this code. For Category 3 medical gas systems, only oxygen and nitrous oxide shall be used.

1301.2 Where Not Applicable. This chapter does not apply to the following except as otherwise addressed in this chapter:

(1) Cylinder and container management, storage, and reserve requirements
(2) Bulk supply systems
(3) Electrical connections and requirements
(4) Motor requirements and controls
(5) Systems having nonstandard operating pressures
(6) Waste anesthetic gas disposal (WAGD) systems
(7) Surface-mounted medical gas rail systems
(8) Breathing air replenishment (BAR) systems
(9) Portable compressed gas systems
(10) Medical support gas systems
(11) Gas-powered device supply systems
(12) Scavenging systems

1301.3 Conflict of Requirements. The requirements of this chapter shall not be interpreted to conflict with the requirements of NFPA 99. For requirements of portions of medical gas and vacuum systems not addressed in this chapter or medical gas and vacuum systems beyond the scope of this chapter refer to NFPA 99.

1301.4 Where Required. Construction and equipment requirements shall be applied only to new construction and new equipment, except as modified in individual sections of this chapter. [NFPA 99:1.3.2]

1301.5 Existing Systems. Only the altered, renovated, or modernized portion of an existing system or individual component shall be required to meet the installation and equipment requirements stated in this code. If the alteration, renovation, or modernization adversely impacts the existing performance requirements of a system or component, additional upgrading shall be required. An existing system that is not in strict compliance with the provisions of this code shall be permitted to be continued in use, unless the Authority Having Jurisdiction has determined that such use constitutes a distinct hazard to life. [NFPA 99:1.3.2.1 – 1.3.2.3]

1302.0 Design Requirements.

1302.1 Risk Categories. Activities, systems, or equipment shall be designed to meet Category 1 through Category 4 requirements, as detailed in this chapter. [NFPA 99:4.1]

1302.1.1 Processes and Operations. The health care facility's governing body shall establish the processes and operations that are planned for the health care facility. [NFPA 99:4.2.1]

1302.1.1.1 Risk Categories. The governing body shall conduct risk assessments and shall determine risk categories based on the character of the processes and operations conducted in the health care facility. [NFPA 99:4.2.1.1]

1302.1.2 Risk Assessment. Risk categories shall be classified by the health care facility's governing body by following and documenting a defined risk assessment procedure. [NFPA 99:4.2.2]

1302.1.2.1 Documents to the Authority Having Jurisdiction. Where required by the Authority Having Jurisdiction (AHJ), the risk assessment shall be provided to the AHJ for review based on the character of the processes and operations conducted in the health care facility. [NFPA 99:4.2.2.1]

1302.1.3 Documented Risk Assessment. A documented risk assessment shall not be required where Category 1 is selected. [NFPA 99:4.2.3]

1302.2 Patient Care Spaces. The health care facility's governing body or its designee shall establish the following areas in accordance with the type of patient care anticipated (see definition of patient care space in Chapter 2):

(1) Category 1 spaces
(2) Category 2 spaces
(3) Category 3 spaces
(4) Category 4 spaces [NFPA 99:1.3.4.1]

1302.3 Anesthesia. It shall be the responsibility of the health care facility's governing body to designate anesthetizing locations. [NFPA 99:1.3.4.2]

1302.4 Wet Procedure Locations. It shall be the responsibility of the health care facility's governing body to designate wet procedure locations. [NFPA 99:1.3.4.3]

1303.0 Health Care Facilities.

1303.1 Drinking Fountain Control Valves. Drinking fountain control valves shall be flush-mounted or fully recessed where installed in corridors or other areas where patients are transported on a gurney, bed, or wheelchair.

1303.2 Psychiatric Patient Rooms. Piping and drain traps in psychiatric patient rooms shall be concealed. Fixtures and fittings shall be resistant to vandalism.

1303.3 Locations for Ice Storage. Ice makers or ice storage containers shall be located in nursing stations or similarly supervised areas to minimize potential contamination.

1303.4 Sterilizers and Bedpan Steamers. Sterilizers and bedpan steamers shall be installed in accordance with the manufacturer's installation instructions and comply with Section 1303.4.1 and Section 1303.4.2.

1303.4.1 Drainage Connections. Sterilizers and bedpan steamers shall be connected to the sanitary drainage system through an air gap in accordance with Section 801.2. The size of indirect waste piping shall be not less than the size of the drain connection on the fixture. Each such indirect waste pipe shall not exceed 15 feet (4572 mm) in length and shall be separately piped to a receptor. Such receptors shall be located in the same room as the equipment served. Except for bedpan steamers, such indirect waste pipes shall not require traps. A trap having a seal of not less than 3 inches (76 mm) shall be provided in the indirect waste pipe for a bedpan steamer.

1303.4.2 Vapor Vents and Stacks. Where a sterilizer or bedpan steamer has provision for a vapor vent and such a vent is required by the manufacturer, the vent shall be extended to the outdoors above the roof. Sterilizer and bedpan steamer vapor vents shall be installed in accordance with the manufacturer's installation instructions and shall not be connected to a drainage system vent.

1303.5 Aspirators. Provisions for aspirators or other water-supplied suction devices shall be installed with the specific approval of the Authority Having Jurisdiction. Where aspirators are used for removing body fluids, they shall include a collection container to collect liquids and solid particles. Aspirators shall indirectly discharge to the sanitary drainage system through an air gap in accordance with Section 806.1. The potable water supply to an aspirator shall be protected by a vacuum breaker or equivalent backflow protection device in accordance with Section 603.5.9.

1303.6 Drains. Drains shall be installed on dryers, aftercoolers, separators, and receivers.

1303.7 Clinical Sinks. Clinical sinks shall be installed in accordance with the manufacturer's installation instructions and shall comply with Section 1303.7.1.

1303.7.1 Drainage Connection. Clinical sinks shall be directly connected to the sanitary drainage system and shall be provided with approved flushing devices installed in accordance with Section 413.1.

1303.8 Water Supply for Hospitals. Hospitals shall be provided with not less than two approved potable water sources that are installed in such a manner as to prevent the interruption of water service.

1303.9 Work Performed in Occupied Healthcare Facilities. In existing, occupied, inpatient healthcare facilities, all plumbing systems installation and remodel work shall be performed by personnel certified in accordance with ASSE/IAPMO 12010, ASSE/IAPMO 12030 and ASSE/IAPMO 12040.

1304.0 Medical Gas and Medical Vacuum Piping Systems.

1304.1 General. The installation of medical gas and medical vacuum piping systems shall comply with the requirements of this chapter.

1304.2 Certification of Systems. Certification of medical gas and vacuum systems shall comply with the requirements of Section 1306.0.

1304.3 Construction Documents. Before a medical gas or medical vacuum system is installed or altered in a hospital, medical facility, or clinic, duplicate construction documents shall be filed with the Authority Having Jurisdiction. Approval of the plans shall be obtained before issuance of a permit by the Authority Having Jurisdiction.

1304.3.1 Requirements. Construction documents shall show the following:

(1) Plot plan of the site, drawn to scale, indicating the location of existing or new cylinder storage areas, property lines, driveways, and existing or proposed buildings.

(2) Piping layout of the proposed piping system or alteration, including alarms, valves, the origin of gases, user outlets, and user inlets. The demand and loading of piping, existing or future, shall also be indicated.

(3) Complete specification of materials.

1304.4 Extent of Work. Construction documents submitted to the Authority Having Jurisdiction shall clearly indicate the nature and extent of the work proposed and shall show in detail that such work will be in accordance with the provisions of this chapter.

1304.5 Record. A record of as-built plans and valve identification records shall remain on the site.

1305.0 System Performance.

1305.1 Required Operating Pressures. Medical gas and vacuum systems shall be capable of delivering service in the pressure ranges listed in Table 1305.1.

1305.2 Minimum Flow Rates. Medical gas and vacuum systems shall be capable of supplying the flow rates listed in Table 1305.2.

1305.3 Minimum Station Outlets and Inlets. Station outlets and inlets for medical gas and vacuum systems shall be provided as listed in Table 1305.3.

1306.0 System Certification.

1306.1 Certification. Prior to a medical gas or vacuum system being placed in service, such system shall be certified in accordance with Section 1306.2.

1306.2 Certification Tests. Certification tests, verified and attested to by the certification agency, shall include the following:

(1) Verifying in accordance with the installation requirements.

(2) Testing and checking for leakage, correct zoning, and identification of control valves.

TABLE 1305.1
STANDARD DESIGNATION COLORS AND OPERATING PRESSURES FOR GAS AND VACUUM SYSTEMS
[NFPA 99: TABLE 5.1.11]

GAS SERVICE	ABBREVIATED NAME	COLORS (BACKGROUND/TEXT)	STANDARD GAUGE PRESSURE
Medical air	Med Air	Yellow/black	50–55 psi
Carbon dioxide	CO_2	Gray/black or gray/white	50–55 psi
Helium	He	Brown/white	50–55 psi
Nitrogen	N_2	Black/white	160–185 psi
Nitrous oxide	N_2O	Blue/white	50–55 psi
Oxygen	O_2	Green/white or white/green	50–55 psi
Oxygen/carbon dioxide mixtures	O_2/CO_2 n% (n = % of CO_2)	Green/white	50–55 psi
Medical–surgical vacuum	Med Vac	White/black	15 inch to 30 inch HgV
Waste anesthetic gas disposal	WAGD	Violet/white	Varies with system type
Other mixtures	Gas A% / Gas B%	Colors as above; major gas for background/minor gas for text	None
Nonmedical air (Category 3 gas-powered device)	—	Yellow-and-white diagonal stripe/black	None
Nonmedical and Category 3 vacuum	—	White-and-black diagonal stripe/black boxed	None
Laboratory air	—	Yellow-and-white checker board/black	None
Laboratory vacuum	—	White-and-black checkerboard/black boxed	None
Instrument air	—	Red/white	160–185 psi

For SI units: 1 pound-force per square inch = 6.8947 kPa, 1 inch of mercury vacuum (HgV) = 3.386 kPa

(3) Checking for identification and labeling of pipelines, station outlets, and control valves.
(4) Testing for cross-connection, flow rate, system pressure drop, and system performance.
(5) Functional testing of pressure relief valves and safety valves.
(6) Functional testing of sources of supply.
(7) Functional testing of alarm systems, including accuracy of system components.
(8) Purge flushing of system and filling with specific source gases.
(9) Testing for purity and cleanliness of source gases.
(10) Testing for specific gas identity at each station outlet.

1306.3 Report Items. A report that includes the specific items addressed in Section 1306.2, and other information required by this chapter, shall be delivered to the Authority Having Jurisdiction prior to acceptance of the system.

1306.4 Components. Functioning of alarm components shall be verified in accordance with the testing and monitoring requirements of the manufacturer and the Authority Having Jurisdiction.

Part II – Category 1 Piped Gas and Vacuum Systems.

1307.0 Central Supply Systems.

1307.1 Terms. Where the terms medical gas or medical support gas occur, the provisions shall apply to all piped systems for oxygen, nitrous oxide, medical air, carbon dioxide, helium, nitrogen, instrument air, and mixtures thereof. Wherever the name of a specific gas service occurs, the provision shall apply only to that gas. [NFPA 99:5.1.1.3]

TABLE 1305.2
MINIMUM FLOW RATES
(cubic feet per minute)

MEDICAL SYSTEM	FLOW RATE
Oxygen	0.71 CFM per outlet[1]
Nitrous Oxide	0.71 CFM per outlet[1]
Medical Compressed Air	0.71 CFM per outlet[1]
Nitrogen	15 CFM free air per outlet
Vacuum	1 SCFM per inlet[2]
Carbon Dioxide	0.71 CFM per outlet[1]
Helium	0.71 CFM per outlet

For SI units: 1 cubic foot per minute (CFM) = 0.47 L/s

Notes:

1 A room designed for a permanently located respiratory ventilator or anesthesia machine shall have an outlet capable of a flow rate of 6.36 CFM (3.0 L/s) at the station outlet.

2 For testing and certification purposes, individual station inlets shall be capable of a flow rate of 3 SCFM (1.4 L/s), while maintaining a system pressure of not less than 12 inches of mercury (41 kPa) at the nearest adjacent vacuum inlet.

1307.2 Nature of Hazards of Gas and Vacuum Systems. Potential fire and explosion hazards associated with positive pressure gas central piping systems and medical–surgical vacuum systems shall be considered in the design, installation, testing, operation, and maintenance of these systems. [NFPA 99:5.1.2]

1307.3 Permitted Locations for Medical Gases. Central supply systems for oxygen, medical air, nitrous oxide,

TABLE 1305.3
MINIMUM OUTLETS AND INLETS PER STATION

LOCATION	OXYGEN	MEDICAL VACUUM	MEDICAL AIR	NITROUS OXIDE	NITROGEN	HELIUM	CARBON DIOXIDE
Patient rooms for medical/surgical, obstetrics, and pediatrics	1/bed	1/bed	1/bed	—	—	—	—
Examination/treatment for nursing units	1/bed	1/bed	—	—	—	—	—
Intensive care (all)	3/bed	3/bed	2/bed	—	—	—	—
Nursery[1]	2/bed	2/bed	1/bed	—	—	—	—
General operating rooms	2/room	3/room[4]	2/room	1/room	1/room	—	—
Cystoscopic and special invasive procedures	2/room	3/room[4]	2/room	—	—	—	—
Recovery delivery and labor/delivery/recovery rooms[2]	2/bed 2/room	2/bed 3/room[4]	1/bed 1/room	—	—	—	—
Labor rooms	1/bed	1/bed	1/bed	—	—	—	—
First aid and emergency treatment[3]	1/bed	1/bed[4]	1/bed	—	—	—	—
Autopsy	—	1/station	1/station	—	—	—	—
Anesthesia workroom	1/station	—	1/station	—	—	—	—

Notes:
[1] Includes pediatric nursery.
[2] Includes obstetric recovery.
[3] Emergency trauma rooms used for surgical procedures shall be classified as general operating rooms.
[4] Vacuum inlets required are in addition to inlets used as part of a scavenging system for removal of anesthetizing gases.

carbon dioxide, and all other patient medical gases shall be piped only to medical gas outlets complying with Section 1315.0, into areas where the gases will be used under the direction of licensed medical professionals for purposes congruent with the following:

(1) Direct respiration by patients.
(2) Clinical application of the gas to a patient, such as the use of an insufflator to inject carbon dioxide into patient body cavities during laparoscopic surgery and carbon dioxide used to purge heart-lung machine blood flow ways.
(3) Medical device applications directly related to respiration.
(4) Power for medical devices used directly on patients.
(5) Calibration of medical devices intended for Section 1307.3(1) through Section 1307.3(4).
(6) Simulation centers for the education, training, and assessment of health care professionals. [NFPA 99:5.1.3.5.2]

1307.4 Materials. Materials used in central supply systems shall meet the following requirements:

(1) In those portions of systems intended to handle oxygen at gauge pressures greater than 350 pounds-force per square inch (psi) (2413 kPa), interconnecting hose shall contain no polymeric materials.
(2) In those portions of systems intended to handle oxygen or nitrous oxide material, construction shall be compatible with oxygen under the temperatures and pressures to which the components can be exposed in the containment and use of oxygen, nitrous oxide, mixtures of these gases, or mixtures containing more than 23.5 percent oxygen.
(3) If potentially exposed to cryogenic temperatures, materials shall be designed for low temperature service.
(4) If intended for outdoor installation, materials shall be installed per the manufacturer's requirements. [NFPA 99:5.1.3.5.4]

1308.0 Pressure-Regulating Equipment.

1308.1 Where Required. Pressure-regulating equipment shall be installed in the supply main upstream of the final line-pressure valve. Where multiple piping systems for the same gas at different operating pressures are required, separate pressure-regulating equipment, relief valves, and source shutoff valves shall be provided for each pressure.

1308.2 Pressure Relief Valves. All pressure relief valves shall meet the following requirements:

(1) They shall be of brass, bronze, or stainless steel construction.
(2) They shall be designed for the specific gas service.
(3) They shall have a relief pressure setting not higher than the maximum allowable working pressure (MAWP) of the component with the lowest working pressure rating in the portion of the system being protected.
(4) They shall be vented to the outside of the building, except that relief valves for compressed air systems having less than 3000 cubic feet (84 950 L) at STP shall be permitted to be diffused locally by means that will not restrict the flow.
(5) They shall have a vent discharge line that is not smaller than the size of the relief valve outlet.
(6) Where two or more relief valves discharge into a common vent line, its internal cross-sectional area shall be not less than the aggregate cross-sectional area of all relief valve vent discharge lines served.

(7) They shall not discharge into locations creating potential hazards.

(8) They shall have the discharge terminal turned down and screened to prevent the entry of rain, snow, or vermin.

(9) They shall be designed in accordance with ASME B31.3. [NFPA 99:5.1.3.5.6.1]

1308.3 Pressure-Relief Valve Requirements. Central supply systems for positive pressure gases shall include one or more relief valves, all meeting the following requirements:

(1) They shall be located between each final line regulator and the source valve.

(2) They shall have a relief setting that is 50 percent above the normal system operating pressure, as indicated in Table 1305.1. [NFPA 99:5.1.3.5.6.3]

1309.0 Oxygen Concentrator Supply Units.

1309.1 Oxygen Requirements. Oxygen concentrator supply units for use with medical gas pipelines shall produce oxygen meeting the requirements of Oxygen 93 USP or Oxygen USP. [NFPA 99:5.1.3.5.11.1]

1309.2 Particulate Size. Output shall have less than or equal to 1.686 x 10^{-6} pounds per cubic yard (1 mg/m^3) of permanent particulates sized 1 micron or larger at normal atmospheric pressure. [NFPA 99:5.1.3.5.11.2]

1309.3 Suitability. Materials of construction on the air side of the oxygen concentrator unit shall be suitable for the service as determined by the manufacturer. [NFPA 99:5.1.3.5.11.3]

1309.4 Compatible Materials. Materials of construction on the oxygen side of the oxygen concentrator unit shall comply with Section 1307.4. [NFPA 99:5.1.3.5.11.4]

1309.5 Oxygen Concentrator Components. The components that make up the oxygen concentrator unit shall be as follows:

(1) The manufacturer of the concentrator unit shall be permitted to use such components and arrangement of such components as needed to produce oxygen complying with Section 1309.1 in the quantity as required by the facility, except where otherwise specifically defined in this code.

(2) Air receivers and oxygen accumulators, where used, shall comply with Section VIII, "Unfired Pressure Vessels," of the ASME Boiler and Pressure Vessels Code and be provided with overpressure relief valves. [NFPA 99:5.1.3.5.11.5]

1309.6 Supply Air Quality. The supply air to the concentrators shall be of a quality to ensure the oxygen concentrator unit can produce oxygen complying with Section 1309.1 and shall not be subject to normally anticipated contamination (e.g., vehicle or other exhausts, gas leakage, discharge from vents, flooding, and so forth). [NFPA 99:5.1.3.5.11.6]

1309.7 Electrical Components. The oxygen concentrator supply unit and any associated electrical equipment shall be provided, a minimum, with the following electrical components:

(1) Either a disconnect switch for each major electrical component or a single disconnect that deactivates all electrical components in the concentrator unit.

(2) Motor starting devices with overload protection for any component with an electrical motor over 2 hp (1.5 kW). [NFPA 99:5.1.3.5.11.7]

1309.8 Vent Valve. A vent valve shall be provided as follows:

(1) Located on the source side of the concentrator outlet isolation valve to permit the operation of the oxygen concentrator unit for validation, calibration, and testing while the unit is isolated from the pipeline system.

(2) Sized to allow for at least 25 percent of the oxygen concentrator unit flow.

(3) Vented to a location compliant with Section 1309.8.1. [NFPA 99:5.1.3.5.11.8]

1309.8.1 Venting of Relief Valves. Indoor supply systems shall have all relief valves vented per Section 1308.2(4) through Section 1308.2(9). [NFPA 99:5.1.3.3.3.2]

1309.9 Valved Sample Port. A DN8 (NPS 1/4) valved sample port shall be provided near the oxygen concentration monitor sensor connection for sampling of the gas from the oxygen concentrator unit. [NFPA 99:5.1.3.5.11.9]

1309.10 Suitable Filter. At least one 0.1 micron filter suitable for oxygen service shall be provided at the outlet of the oxygen concentrator supply unit. [NFPA 99:5.1.3.5.11.10]

1309.11 Check Valve. A check valve shall be provided at the outlet of the oxygen concentrator supply unit to prevent backflow into the oxygen concentrator supply unit and to allow service to the unit. [NFPA 99:5.1.3.5.11.11]

1309.12 Outlet Valve. An outlet valve shall be provided to isolate all components of the oxygen concentrator from the pipeline with the following characteristics:

(1) The valve shall have both manual and automatic actuation with visual indication of open or closed.

(2) The valve shall close automatically whenever the oxygen concentrator unit is not producing oxygen of a concentration equal to that in Section 1309.1.

(3) Continuing operation of the oxygen concentrator supply unit through the vent mode shall be permitted with the isolating valve closed.

(4) The isolating valve, when automatically closed due to low concentration, shall require manual reset to ensure the oxygen concentrator supply unit is examined prior to return to service.

(5) Closing the isolating valve, whether automatically or manually, shall activate an alarm signal at the master alarms (see Section 1317.1.1) indicating that the oxygen concentrator supply unit is disconnected. [NFPA 99:5.1.3.5.11.12]

1309.13 Oxygen Concentration Monitor. The oxygen concentrator supply unit shall be provided with an oxygen concentration monitor with the following characteristics:

(1) The monitor shall be capable of monitoring 99 percent oxygen concentration with 1 percent accuracy.

(2) The monitor shall continuously display the oxygen concentration and shall activate local alarm and master alarms per NFPA 99 when a concentration lower than 91 percent is observed.

(3) The monitor shall continuously display the oxygen concentration.

(4) It shall be permitted to insert the monitor into the pipeline without a demand check. [NFPA 99:5.1.3.5.11.13]

1310.0 Category 1 Medical Air Central Supply Systems.

1310.1 Quality of Medical Air. Medical air shall be required to have the following characteristics:

(1) It shall be supplied from cylinders, bulk containers, or medical air compressor sources, or it shall be reconstituted from oxygen USP and oil-free, dry nitrogen NF.

(2) It shall meet the requirements of medical air USP.

(3) It shall have no detectable liquid hydrocarbons.

(4) It shall have less than 25 ppm gaseous hydrocarbons.

(5) It shall have equal to or less than 1.686×10^{-6} pounds per cubic yard (1 mg/m^3) of permanent particulates sized 1 micron or larger in the air at normal atmospheric pressure. [NFPA 99:5.1.3.6.1]

1310.2 Uses of Medical Air. Medical air sources shall be connected to the medical air distribution system only and shall be used only for air in the application of human respiration and calibration of medical devices for respiratory application. [NFPA 99:5.1.3.6.2]

1310.3 Medical Air Compressors. Medical air compressors shall be installed in a well-lit, ventilated, and clean location and shall be accessible. The location shall be provided with drainage facilities in accordance with this code. The medical air compressor area shall be located separately from medical gas cylinder system sources, and shall be readily accessible for maintenance.

1310.3.1 Category 1 Medical Air Compressor. Medical air compressors shall be sufficient to serve the peak calculated demand with the largest single compressor out of service. In no case shall there be fewer than two compressors. [NFPA 99:5.1.3.6.3.9(B)]

1310.3.2 Required Components. Medical air compressor systems shall consist of the following:

(1) Components shall be arranged to allow service and a continuous supply of medical air in the event of a single fault failure.

Component arrangement shall be permitted to vary as required by the technology(ies) employed, provided that an equal level of operating redundancy and medical air quality is maintained. [NFPA 99:5.1.3.6.3.9(A)(1), 5.1.3.6.3.9(A)(2)]

(2) Automatic means to prevent backflow from all on-cycle compressors through all off-cycle compressors.

(3) Manual shutoff valve to isolate each compressor from the centrally piped system and from other compressors for maintenance or repair without loss of pressure in the system.

(4) Intake filter-muffler(s) of the dry type.

(5) Pressure relief valve(s) set at 50 percent above line pressure.

(6) Piping and components between the compressor and the source shutoff valve that do not contribute to contaminant levels.

(7) Except as defined in Section 1310.3.2(1) through Section 1310.3.2(6), materials and devices used between the medical air intake and the medical air source valve that are of any design or construction appropriate for the service as determined by the manufacturer. [NFPA 99:5.1.3.6.3.2 (2-7)]

1310.4 Medical Air Receivers. Receivers for medical air shall meet the following requirements:

(1) They shall be made of corrosion-resistant materials or otherwise be made corrosion resistant.

(2) They shall comply with Section VIII, "Unfired Pressure Vessels," of the ASME Boiler and Pressure Vessel Code.

(3) They shall be equipped with a pressure relief valve, automatic drain, manual drain, sight glass, and pressure indicator.

(4) They shall be of a capacity sufficient to prevent the compressors from short-cycling. [NFPA 99:5.1.3.6.3.6]

1310.5 Valves. A medical air receiver(s) shall be provided with proper valves to allow the flow of compressed air to enter and exit out of separate receiver ports during normal operation and allow the receiver to be bypassed during service without shutting down the supply of medical air. [NFPA 99:5.1.3.6.3.9(D)]

1311.0 Compressor Intake.

1311.1 Air Sources. Air sources for medical air compressors shall comply with Section 1311.2 through Section 1311.6.

1311.2 Medical Air Compressor Source. The medical air compressors shall draw their air from a source of clean air. [NFPA 99:5.1.3.6.3.11(A)]

If an air source equal to or better than outside air (e.g., air already filtered for use in operating room ventilating systems) is available, it shall be permitted to be used for the medical air compressors with the following provisions:

(1) This alternate source of supply air shall be available on a continuous 24 hours-per-day, 7 day-per-week basis.

(2) Ventilating systems having fans with motors or drive belts located in the airstream shall not be used as a source of medical air intake. [NFPA 99:5.1.3.6.3.11(E)]

1311.3 Air Intakes. Compressor intake piping shall be permitted to be made of materials and use a joining technique as permitted under Section 1319.0 and Section 1320.0. [NFPA 99:5.1.3.6.3.11(F)]

1311.4 Location. Medical air intakes shall be located as follows:

(1) The medical air intake shall be located a minimum of 25 feet (7620 mm) from ventilating system exhausts, fuel storage vents, combustion vents, plumbing vents, and

vacuum discharges, or areas that can collect vehicular exhausts or other noxious fumes.

(2) The medical air intake shall be located a minimum of 20 feet (6096 mm) above ground level.

(3) The medical air intake shall be located a minimum of 10 feet (3048 mm) from any door, window, or other opening in the building. [NFPA 99:5.1.3.6.3.11(B-D)]

1311.5 Separate Compressors. Air intakes for separate compressors shall be permitted to be joined together to one common intake where the following conditions are met:

(1) The common intake is sized to minimize backpressure in accordance with the manufacturer's recommendations.

(2) Each compressor can be isolated by manual or check valve, blind flange, or tube cap to prevent open inlet piping when the compressor(s) is removed for service from the consequent backflow of room air into the other compressor(s). [NFPA 99:5.1.3.6.3.11(G)]

1311.6 Screening. The end of the intake shall be turned down and screened or otherwise be protected against the entry of vermin, debris, or precipitation by screening fabricated or composed of a noncorroding material. [NFPA 99:5.1.3.6.3.11(H)]

1312.0 Medical Surgical Vacuum Central Supply Systems.

1312.1 General. The vacuum plant shall be installed in a well-lit, ventilated, and clean location with accessibility. The location shall be provided with drainage facilities in accordance with this code. The vacuum plant, where installed as a source, shall be located separately from other medical vacuum system sources and shall be readily accessible for maintenance.

1312.2 Medical-Surgical Vacuum Sources. Medical-surgical vacuum central supply systems shall consist of the following:

(1) Two or more vacuum pumps sufficient to serve the peak calculated demand with the largest single vacuum pump out of service.

(2) Automatic means to prevent backflow from any on-cycle vacuum pumps through any off-cycle vacuum pumps.

(3) Shutoff valve or other isolation means to isolate each vacuum pump from the centrally piped system, and other vacuum pumps for maintenance or repair without loss of vacuum in the system.

(4) Vacuum receiver.

(5) Piping between the vacuum pump(s), discharge(s), receiver(s), and vacuum source shutoff valve in accordance with Section 1319.0, except brass, galvanized, or black steel pipe, which is permitted to be used as recommended by the manufacturer.

(6) Except as defined in Section 1312.2(1) through Section 1312.2(5), materials and devices used between the medical vacuum exhaust and the medical vacuum source that are permitted to be of any design or construction appropriate for the service as determined by the manufacturer.

(7) Vacuum filtration per Section 1312.4. [NFPA 99:5.1.3.7.1.1]

1312.3 Vacuum Receivers. Receivers for vacuum shall meet the following requirements:

(1) They shall be made of materials deemed suitable by the manufacturer.

(2) They shall comply with Section VIII, "Unfired Pressure Vessels," of the ASME Boiler and Pressure Vessel Code.

(3) They shall be capable of withstanding a gauge pressure of 60 psi (414 kPa) and 30 inch (762 mm) gauge HgV.

(4) They shall be equipped with a manual drain.

(5) They shall be of a capacity based on the technology of the pumps. [NFPA 99:5.1.3.7.3]

1312.4 Vacuum Filtration. Central supply systems for vacuum shall be provided with inlet filtration with the following characteristics:

(1) Filtration shall be at least duplex to allow one filter to be exchanged without impairing vacuum system.

(2) Filtration shall be located on the patient side of the vacuum producer.

(3) Filters shall be efficient to 0.03 μ and 99.97 percent HEPA or better, per DOE-STD-3020.

(4) Filtration shall be sized for 100 percent of the peak calculated demand while one filter or filter bundle is isolated.

(5) It shall be permitted to group multiple filters into bundles to achieve the required capacities.

(6) The system shall be provided with isolation valves on the source side of each filter or filter bundle and isolation valves on the patient side of each filter or filter bundle, permitting the filters to be isolated without shutting off flow to the central supply system.

(7) A means shall be available to allow the user to observe any accumulations of liquids.

(8) A vacuum relief petcock shall be provided to allow vacuum to be relieved in the filter canister during filter replacement.

(9) Filter elements and canisters shall be permitted to be constructed of materials as deemed suitable by the manufacturer.

(10) In normal operation, one filter or filter bundle shall be isolated from the system to be available for service should a blockage in the operating filter occur or rotation of the filters be desired after filter element exchange. [NFPA 99:5.1.3.7.4]

1312.5 Piping Arrangement and Redundancies. Piping arrangement shall be as follows:

(1) Piping shall be arranged to allow service and a continuous supply of medical-surgical vacuum in the event of a single fault failure.

(2) Piping arrangement shall be permitted to vary based on the technology(ies) employed, provided that an equal level of operating redundancy is maintained.

(3) Where only one set of vacuum pumps is available for a combined medical-surgical vacuum system and an analysis, a research, or a teaching laboratory vacuum system,

such laboratories shall be connected separately from the medical-surgical system directly to the receiver tank through its own isolation valve and fluid trap located at the receiver, and between the isolation valve and fluid trap, a scrubber shall be permitted to be installed. [NFPA 99:5.1.3.7.5, 5.1.3.7.5.1]

1312.6 Piping Serviceability. The medical-surgical vacuum receiver(s) shall be serviceable without shutting down the medical-surgical vacuum system by any method to ensure continuation of service to the facility's medical-surgical pipeline distribution system. [NFPA 99:5.1.3.7.5.2]

1312.7 Shutoff Valve. Medical-surgical vacuum central supply systems shall be provided with a source shutoff valve per Section 1314.6. [NFPA 99:5.1.3.7.5.3]

1313.0 Medical-Surgical Vacuum Exhaust.

1313.1 Vacuum Source Exhausts. The medical-surgical vacuum pumps shall exhaust in a manner and location that minimizes the hazards of noise and contamination to the facility and its environment. [NFPA 99:5.1.3.7.7.1]

1313.2 Location. The exhaust shall be located as follows:

(1) Outdoors.

(2) At least 25 feet (7620 mm) from any door, window, air intake, or other openings in buildings or places of public assembly.

(3) At a level different from air intakes.

(4) Where prevailing winds, adjacent buildings, topography, or other influences will not divert the exhaust into occupied areas or prevent dispersion of the exhaust. [NFPA 99: 5.1.3.7.7.2]

1313.3 Screening. The end of the exhaust shall be turned down and screened or otherwise be protected against the entry of vermin, debris, or precipitation by screening fabricated or composed of a noncorroding material. [NFPA 99:5.1.3.7.7.3]

1313.4 Dips and Loops. The exhaust shall be free of dips and loops that might trap condensate or oil or provided with a drip leg and valved drain at the bottom of the low point. [NFPA 99:5.1.3.7.7.4]

1313.5 Multiple Pumps. Vacuum exhausts from multiple pumps shall be permitted to be joined together to one common exhaust where the following conditions are met:

(1) The common exhaust is sized to minimize backpressure in accordance with the pump manufacturer's recommendations.

(2) Each pump can be isolated by manual or check valve, blind flange, or tube cap to prevent open exhaust piping when the pump(s) is removed for service from consequent flow of exhaust air into the room. [NFPA 99:5.1.3.7.7.5]

1314.0 Valves.

1314.1 Gas and Vacuum Shutoff Valves. Shutoff valves shall be provided to isolate sections or portions of the piped distribution system for maintenance, repair, or planned future expansion need and to facilitate periodic testing. [NFPA 99:5.1.4.1.1]

1314.2 Security. All valves, except valves in zone valve box assemblies, shall be secured by any of the following means:

(1) Located in secured areas.

(2) Locked or latched in their operating position.

(3) Located above ceilings, but remaining accessible and not obstructed. [NFPA 99:5.1.4.1.2]

1314.3 Labeled. All valves shall be labeled as to gas supplied and the area(s) controlled, in accordance with Section 1323.14. [NFPA 99:5.1.4.1.3]

1314.4 Accessibility. Zone valves shall be installed in valve boxes with removable covers large enough to allow manual operation of valves.

Zone valves for use in certain areas, such as psychiatric or pediatric areas, shall be permitted to be secured with the approval of the Authority Having Jurisdiction to prevent inappropriate access. [NFPA 99:5.1.4.1.4]

1314.4.1 Flammable Gases. Valves for nonflammable medical gases shall not be installed with valves for flammable gases in the same zone valve box assembly with flammable gases. [NFPA 99:5.1.4.1.5]

1314.5 Valve Types. New or replacement valves shall be permitted to be of any type as long as they meet the following conditions:

(1) They have a minimum Cv factor in accordance with Table 1314.5.

(2) They use a quarter turn to off.

(3) They are constructed of materials suitable for the service.

(4) They are provided with copper tube extensions by the manufacturer for brazing or with corrugated medical tubing (CMT) fittings.

(5) They indicate to the operator if the valve is open or closed.

(6) They permit in-line serviceability.

(7) They are cleaned for oxygen service by the manufacturer if used for any positive-pressure service. [NFPA 99:5.1.4.1.6]

TABLE 1314.5
POSITIVE PRESSURE GASES
[NFPA 99:5.1.4.1.6(a)]

VALVE SIZE (inch)	MINIMUM Cv (full open)
1/2	17
3/4	31
1	60
1 1/4	110
1 1/2	169
2	357
2 1/2	390
3	912
4	1837

For SI units: 1 inch = 25.4 mm

1314.6 Source Valves. A shutoff valve shall be placed at the immediate connection of each central supply system to the piped distribution system to allow the entire central supply system, including all accessory devices (e.g., air dryers, final line regulators), to be isolated from the facility. [NFPA 99:5.1.4.2.1]

1314.6.1 Location. The source valve shall be located in the immediate vicinity of the central supply system. [NFPA 99:5.1.4.2.2]

1314.7 Main Line Valve. A shutoff valve shall be provided in the main supply line inside of the buildings being served, except where one or more of the following conditions exist:

(1) The source and source valve are located inside the building served.

(2) The source system is physically mounted to the wall of the building served, and the pipeline enters the building in the immediate vicinity of the source valve. [NFPA 99:5.1.4.3.1]

1314.7.1 Location. The main line valve shall be located on the facility side of the source valve and outside of the source room, the enclosure, or where the main line first enters the building. [NFPA 99:5.1.4.3.2]

1314.8 Riser Valves. Each riser supplied from the main line shall be provided with a shutoff valve in the riser adjacent to the main line. [NFPA 99:5.1.4.4]

1314.9 Service Valves. Service valves shall be installed to allow servicing or modification of lateral branch piping from a main or riser without shutting down the entire main, riser, or facility. [NFPA 99:5.1.4.5.1]

1314.9.1 Branch Piping. Only one service valve shall be required for each branch off of a riser, regardless of how many zone valve boxes are installed on that lateral.

Service valves shall be placed in the branch piping prior to any zone valve box assembly on that branch. [NFPA 99:5.1.4.5.2, 5.1.4.5.3]

1314.10 Zone Valves. All station outlets/inlets shall be supplied through a zone valve, which shall be placed as follows:

(1) It is installed so that a wall intervenes between the valve and the outlets/inlets that it controls.

(2) It is readily operable from a standing position.

(3) It is installed where it is visible and accessible at all times.

(4) It is not installed where it can be hidden from plain view, such as behind normally open or normally closed doors.

(5) It is not installed in a room with the station outlets/inlets that it controls.

(6) It is not installed in rooms, areas, or closets that can be closed or locked. [NFPA 99:5.1.4.6.1]

1314.10.1 Readily Accessible. A zone valve in each medical gas or vacuum line shall be provided for each Category 1 space and anesthetizing location for moderate sedation, deep sedation, or general anesthesia specific for the occupancy. These zone valves shall be located as follows:

(1) They are installed immediately outside the area controlled.

(2) They are readily accessible in an emergency. [NFPA 99:5.1.4.6.2]

1314.10.2 Arrangement. Piping on the patient side of zone valves shall be arranged to provide the following:

(1) Shutting off the supply of medical gas or vacuum to one zone will not affect the supply of medical gas or vacuum to another zone or the rest of the system.

(2) Service will only be to outlets/inlets located on that same story.

(3) All gas delivery columns, hose reels, ceiling tracks, control panels, pendants, booms, or other special installations are located on the patient side of the zone valve. [NFPA 99:5.1.4.6.3]

1314.10.3 Indicators. A pressure/vacuum indicator shall be provided on the station outlet/inlet side of each zone valve. [NFPA 99:5.1.4.6.4]

1314.11 In-Line Shutoff Valves. Optional in-line valves shall be permitted to be installed to isolate or shut off piping for servicing of individual rooms or areas. [NFPA 99:5.1.4.7]

1314.12 Valves for Future Connections. Future connection valves shall be labeled as to gas content. [NFPA 99:5.1.4.8.1]

1314.12.1 Downstream Piping. Downstream piping shall be closed with a brazed cap with tubing allowance for cutting and rebrazing. [NFPA 99:5.1.4.8.2]

1315.0 Station Outlets and Inlets.

1315.1 General. Each station outlet/inlet for medical gases or vacuums shall be gas-specific, whether the outlet/inlet is threaded or is a noninterchangeable quick coupler. [NFPA 99:5.1.5.1]

1315.2 Required Valves. Each station outlet shall consist of a primary and a secondary valve (or assembly).

Each station inlet shall consist of a primary valve (or assembly) and shall be permitted to include a secondary valve (or assembly). [NFPA 99:5.1.5.2, 5.1.5.3]

1315.3 Secondary Valve. The secondary valve (or assembly) shall close automatically to stop the flow of gas (or vacuum, if provided) when the primary valve (or assembly) is removed. [NFPA 99:5.1.5.4]

1315.4 Identification. Each outlet/inlet shall be legibly identified in accordance with Section 1323.15. [NFPA 99:5.1.5.5]

1315.5 Threaded Outlets/Fittings. Threaded outlets/inlets shall be noninterchangeable connections complying with the mandatory requirements of CGA V-5. [NFPA 99:5.1.5.6]

1315.6 Gas-Specific Station Outlet/Inlet. Each station outlet/inlet, including those mounted in columns, hose reels, ceiling tracks, or other special installations, shall be designed so that parts or components that are required to be gas-specific for compliance with Section 1315.1 and Section 1315.8 cannot be interchanged between the station outlet/inlet for different gases. [NFPA 99:5.1.5.7]

1315.7 Common Parts. The use of common parts in outlets/inlets, such as springs, O-rings, fasteners, seals, and shutoff poppets, shall be permitted. [NFPA 99:5.1.5.8]

1315.8 Marking of Components. Components of a vacuum station inlet necessary for the maintenance of vacuum specificity shall be legibly marked to identify them as components or parts of a vacuum or suction system. [NFPA 99:5.1.5.9]

1315.9 Components Not Specific to a Vacuum. Components of inlets not specific to a vacuum shall not be required to be marked. [NFPA 99:5.1.5.10]

1315.10 Factory-Installed Copper Inlet Tubes. Factory-installed copper inlet tubes on station outlets extending no further than 8 inches (203 mm) from the body of the terminal shall be not less than DN8 (NPS $^1/_4$) ($^3/_8$ inch O.D.) size, with 0.3 inch (7.6 mm) minimum inside diameter. [NFPA 99:5.1.5.11]

1315.11 Factory-Installed Copper Outlet Tubes. Factory-installed copper outlet tubes on station inlets extending no further than 8 inches (203 mm) from the body of the terminal shall be not less than DN10 (NPS $^3/_8$) ($^1/_2$ in. O.D.) size, with 0.4 inch (10.2 mm) minimum inside diameter. [NFPA 99:5.1.5.12]

1315.12 Protection from Damage. Station outlets/inlets shall be permitted to be recessed or otherwise protected from damage. [NFPA 99:5.1.5.13]

1315.13 Multiple Wall Outlets/Inlets. When multiple wall outlets/inlets are installed, they shall be spaced to allow the simultaneous use of adjacent outlets/inlets with any of the various types of therapy equipment. [NFPA 99:5.1.5.14]

1315.14 Nonstandard Operation Pressures. Station outlets in systems having nonstandard operating pressures shall meet the following additional requirements:

(1) They shall be gas-specific.

(2) They shall be pressure-specific where a single gas is piped at more than one operating pressure [e.g., a station outlet for oxygen at 80 psi (552 kPa) shall not accept an adapter for oxygen at 50 psi (345 kPa)].

(3) If operated at a pressure in excess of 80 psi (552 kPa), they shall be either D.I.S.S. connectors or comply with Section 1315.14(4).

(4) If operated at a gauge pressure between 200 psi and 300 psi (1379 kPa and 2068 kPa), the station outlet shall be designed so as to prevent the removal of the adapter until the pressure has been relieved to prevent the adapter injuring the user or others when removed from the outlet. [NFPA 99:5.1.5.15]

» **1315.15 Post Installation.** After installation of the piping, but before installation of the station outlets and inlets and other medical gas and medical gas system components (e.g., pressure-actuating switches for alarms, manifolds, pressure gauges, or pressure relief valves), the line shall be blown clear using oil-free, dry nitrogen NF.

1316.0 Pressure and Vacuum Indicator Locations.

» **1316.1 Isolation.** A pressure-relief valve shall not be isolated from its intended use by a valve.

» **1316.2 Pressure and Vacuum Indicator Locations.** Pressure/vacuum indicators shall be readable from a standing position. Pressure/vacuum indicators shall be provided at the following locations, as a minimum:

(1) Adjacent to the alarm-initiating device for source main line pressure and vacuum alarms in the master alarm system.

(2) At or in area alarm panels to indicate the pressure/vacuum at the alarm activating device for each system that is monitored by the panel.

(3) On the station outlet/inlet side of zone valves. [NFPA 99:5.1.8.2.1, 5.1.8.2.2]

1317.0 Warning Systems.

1317.1 Category 1. All master, area, and local alarm systems used for medical gas and vacuum systems shall include the following:

(1) Separate visual indicators for each condition monitored, except as permitted in Section 1317.1.2 for local alarms that are displayed on master alarm panels.

(2) Visual indicators that remain in alarm until the situation that has caused the alarm is resolved.

(3) Cancelable audible indication of each alarm condition that produces a sound with a minimum level of 80 decibels at 3 feet (914 mm).

(4) Means to indicate a lamp or LED failure and audible failure.

(5) Visual and audible indication that the communication with an alarm-initiating device is disconnected.

(6) Labeling of each indicator, indicating the condition monitored.

(7) Labeling of each alarm panel for its area of surveillance.

(8) Reinitiation of the audible signal if another alarm condition occurs while the audible alarm is silenced.

(9) Power for master, area alarms, sensors, and switches from the life safety branch of the essential electrical system as described in NFPA 99.

(10) Power for local alarms, dew point sensors, and carbon monoxide sensors permitted to be from the same essential electrical branch as is used to power the air compressor system.

(11) Where used for communications, wiring from switches or sensors that is supervised or protected as required by NFPA 70 for life safety and critical branches circuits in which protection is any of the following types:

   (a) Conduit
   (b) Free air
   (c) Wire
   (d) Cable tray
   (e) Raceways

(12) Communication devices that do not use electrical wiring for signal transmission will be supervised such that failure of communication shall initiate an alarm.

(13) Assurance by the responsible authority of the facility that the labeling of alarms, where room numbers or designations are used, is accurate and up-to-date.

(14) Provisions for automatic restart after a power loss of 10 seconds (e.g., during generator startup) without giving false signals or requiring manual reset.

(15) Alarm switches/sensors installed so as to be removable. [NFPA 99:5.1.9.1]

1317.1.1 Master Alarms. A master alarm system shall be provided to monitor the operation and condition of the source of supply, the reserve source (if any), and the pressure in the main lines of each medical gas and vacuum piping system. [NFPA 99:5.1.9.2]

1317.1.2 Master Alarm Signal. The master alarm shall include at least one signal from the source equipment to indicate a problem with the source equipment at this location. This master alarm signal shall activate when any of the required local alarm signals for this source equipment activates. [NFPA 99:5.1.9.5.2]

1318.0 Piping Materials for Field-Installed Positive Pressure Medical Gas Systems.

1318.1 General. The provisions of this section shall apply to field-installed piping for the distribution of medical gas systems.

1318.2 Cleaning. Tubes, valves, fittings, station outlets, and other piping components in medical gas systems shall have been cleaned for oxygen service by the manufacturer prior to installation in accordance with the mandatory requirements of CGA G-4.1, except that fittings shall be permitted to be cleaned by a supplier or agency other than the manufacturer. [NFPA 99:5.1.10.1.1]

Where tube ends, fittings or other components become contaminated before installation they shall be recleaned in accordance with Section 1321.8.7 and Section 1321.8.8.

1318.3 Delivery. Each length of tube shall be delivered plugged or capped by the manufacturer and kept sealed until prepared for installation. Fittings, valves, and other components shall be delivered sealed and labeled and kept sealed until prepared for installation. [NFPA 99:5.1.10.1.2, 5.1.10.1.3]

1318.4 Tubes for Medical Gas Systems. Tubes shall be hard-drawn seamless copper in accordance with ASTM B819, medical gas tube, Type L, except Type K shall be used where operating pressures are above a gauge pressure of 185 psi (1276 kPa) and the pipe sizes are larger than DN80 [(NPS 3) (3$^{1}/_{8}$ inches O.D.)]. {NFPA 99:5.1.10.1.4}

1318.5 Manufacturer Markings. ASTM B819, medical gas tube shall be identified by the manufacturer's markings "OXY," "MED," "OXY/MED," "OXY/ACR," or "ACR/MED" in blue (Type L) or green (Type K). [NFPA 99:5.1.10.1.7]

1318.6 Documentation. The installer shall furnish documentation certifying that all installed piping materials comply with the requirements of Section 1318.2. [NFPA 99:5.1.10.1.8]

1319.0 Piping Materials for Field-Installed Medical-Surgical Vacuum Systems.

1319.1 Tubes for Medical Vacuum Systems. Piping for vacuum systems shall be constructed of any of the following:

(1) Hard-drawn seamless copper tube in accordance with the following:
   (a) ASTM B88, copper tube (Type K, Type L, or Type M)
   (b) ASTM B280, copper ACR tube
   (c) ASTM B819, copper medical gas tubing (Type K or Type L)

(2) Stainless steel tube in accordance with the following:
   (a) ASTM A269 TP304L or 316L
   (b) ASTM A312 TP304L or 316L
   (c) ASTM A312 TP 304L/316L, Schedule 5S pipe, and ASTM A403 WP304L/316L, Schedule 5S fittings {NFPA 99:5.1.10.2.1}

1319.1.1 Where Not Required. If medical gas tube in accordance with ASTM B819, Standard Specification for Seamless Copper Tube for Medical Gas Systems, is used for vacuum piping, such special marking shall not be required. [NFPA 99:5.1.10.2.2.2]

1320.0 Joints and Connections.

1320.1 General. This section sets forth the requirements for pipe joint installations for a medical gas or vacuum system.

1320.2 Changes in Direction. Positive pressure patient gas systems, medical support gas systems, and vacuum systems constructed of hard-drawn seamless copper or stainless steel tubing shall have all turns, offsets, and other changes in direction made using fittings or techniques appropriate to any of the following acceptable joining methods:

(1) Brazing, as described in Section 1321.0.

(2) Welding, as described in Section 1322.1 through Section 1322.2.1.

(3) Memory metal fittings, as described in Section 1322.3.

(4) Axially swaged, elastic preload fittings, as described in Section 1322.4.

(5) Threaded, as described in Section 1322.5. [NFPA 99:5.1.10.3.1]

1320.2.1 Medical Vacuum Systems. Vacuum systems fabricated from copper tubing shall be permitted to have branch connections made using mechanically formed, drilled, and extruded tee-branch connections that are formed in accordance with the tool manufacturer's instructions. Such branch connections shall be joined by brazing, as described in Section 1321.0. [NFPA 99:5.1.10.3.3]

1321.0 Brazed Joints.

1321.1 Brazed Joints and Fittings. Fittings shall be wrought-copper capillary fittings complying with ASME B16.22, or brazed fittings complying with ASME B16.50. Cast copper alloy fittings shall not be permitted.

Brazed joints shall be made using a brazing alloy that exhibits a melting temperature in excess of 1000°F (538°C) to retain the integrity of the piping system in the event of fire exposure. [NFPA 99:5.1.10.4.1.1 – 5.1.10.4.1.3]

1321.2 Tube Joints. Brazed tube joints shall be the socket type. [NFPA 99:5.1.10.4.1.4]

1321.3 Filler Metals. Filler metals shall bond with and be metallurgically compatible with the base metals being joined.

Filler metals shall comply with AWS A5.8. [NFPA 99:5.1.10.4.1.5, 5.1.10.4.1.6]

1321.4 Copper-to-Copper Joints. Copper-to-copper joints shall be brazed using a copper–phosphorus or copper-phosphorus-silver brazing filler metal (BCuP series) without flux. [NFPA 99:5.1.10.4.1.7]

1321.5 Accessible. Joints to be brazed in place shall be accessible for necessary preparation, assembly, heating, filler application, cooling, cleaning, and inspection. [NFPA 99:5.1.10.4.1.9]

1321.6 Purging. Braze joints shall be continuously purged with nitrogen NF. [NFPA 99:5.1.10.4.1.10]

1321.7 Tube Ends. Tube ends shall be cut square using a sharp tubing cutter to avoid deforming the tube. [NFPA 99:5.1.10.4.2.1]

1321.7.1 Cutting Wheels. The cutting wheels on tubing cutters shall be free from grease, oil, or other lubricant not suitable for oxygen service. [NFPA 99:5.1.10.4.2.2]

1321.7.2 Cut Ends. The cut ends of the tube shall be permitted to be rolled smooth or deburred with a sharp, clean deburring tool, taking care to prevent chips from entering the tube. [NFPA 99:5.1.10.4.2.3]

1321.8 Cleaning Procedures. The interior surfaces of tubes, fittings, and other components that are cleaned for oxygen service shall be stored and handled to avoid contamination prior to assembly and brazing. [NFPA 99:5.1.10.4.3.1]

1321.8.1 Exterior Surfaces. The exterior surfaces of tube ends shall be cleaned prior to brazing to remove any surface oxides. When cleaning the exterior surfaces of tube ends, no matter shall be allowed to enter the tube. [NFPA 99:5.1.10.4.3.2, 5.1.10.4.3.3]

1321.8.2 Interior Surfaces. If the interior surfaces of fitting sockets become contaminated prior to brazing, they shall be recleaned for oxygen in accordance with Section 1321.8.7 and be cleaned for brazing with a clean, oil-free, stainless steel or brass wire brush. [NFPA 99:5.1.10.4.3.4]

1321.8.3 Abrasive Pads. Clean, nonshedding, abrasive pads shall be used to clean the exterior surfaces of the tube ends. [NFPA 99:5.1.10.4.3.5]

1321.8.4 Prohibited. The use of steel wool or sand cloth shall be prohibited. The cleaning process shall not result in grooving of the surfaces to be joined. [NFPA 99:5.1.10.4.3.6, 5.1.10.4.3.7]

1321.8.5 Wiped. After being abraded, the surfaces shall be wiped using a clean, lint-free white cloth. [NFPA 99:5.1.10.4.3.8]

1321.8.6 Examination. Tubes, fittings, valves, and other components shall be visually examined internally before being joined to verify that they have not become contaminated for oxygen service and that they are free of obstructions or debris. [NFPA 99:5.1.10.4.3.9]

1321.8.7 On-Site Recleaning. The interior surfaces of tube ends, fittings, and other components that were cleaned for oxygen service by the manufacturer, but that became contaminated prior to being installed, shall be permitted to be recleaned on-site by the installer by thoroughly scrubbing the interior surfaces with a clean, hot water–alkaline solution, such as sodium carbonate or trisodium phosphate, using a solution of 1 pound (0.5 kg) of sodium carbonate or trisodium phosphate to 3 gallons (11 L) of potable water, and thoroughly rinsing them with clean, hot, potable water.

Other aqueous cleaning solutions shall be permitted to be used for on-site recleaning provided that they are as recommended in the mandatory requirements of CGA G-4.1. [NFPA 99:5.1.10.4.3.10, 5.1.10.4.3.11]

1321.8.8 Contaminated Materials. Material that has become contaminated internally and is not clean for oxygen service shall not be installed. [NFPA 99:5.1.10.4.3.12]

1321.8.9 Timeframe for Brazing. Joints shall be brazed within 8 hours after the surfaces are cleaned for brazing. [NFPA 99:5.1.10.4.3.13]

1321.9 Brazing Dissimilar Metals. Flux shall only be used when brazing dissimilar metals, such as copper and bronze or brass, using a silver (BAg series) brazing filler metal. [NFPA 99:5.1.10.4.4.1]

1321.9.1 Surface Cleaning. Surfaces shall be cleaned for brazing in accordance with Section 1321.8. [NFPA 99:5.1.10.4.4.2]

1321.9.2 Flux. Flux shall be applied sparingly to minimize contamination of the inside of the tube with flux. The flux shall be applied and worked over the cleaned surfaces to be brazed using a stiff bristle brush to ensure complete coverage and wetting of the surfaces with flux. [NFPA 99:5.1.10.4.4.3, 5.1.10.4.4.4]

1321.9.3 Short Sections of Copper. Where possible, short sections of copper tube shall be brazed onto the non-copper component, and the interior of the sub-assembly shall be cleaned of flux prior to installation in the piping system. [NFPA 99:5.1.10.4.4.5]

1321.9.4 Flux-Coated Brazing Rods. On joints DN20 (NPS ¾) (⅞ inch O.D.) size and smaller, flux-coated brazing rods shall be permitted to be used in lieu of applying flux to the surfaces being joined. [NFPA 99:5.1.10.4.4.6]

1321.10 Nitrogen Purge. When brazing, joints shall be continuously purged with oil-free, dry nitrogen NF to prevent the formation of copper oxide on the inside surfaces of the joint. [NFPA 99:5.1.10.4.5.1]

1321.10.1 Source. The source of the purge gas shall be monitored, and the installer shall be audibly alerted when the source content is low. [NFPA 99:5.1.10.4.5.2]

1321.10.2 Flow Rate Control. The purge gas flow rate shall be controlled by the use of a pressure regulator and flowmeter, or combination thereof.

Pressure regulators alone shall not be used to control purge gas flow rates. [NFPA 99:5.1.10.4.5.3, 5.1.10.4.5.4]

1321.10.3 Oxygen Analyzer. In order to ensure that all ambient air has been removed from the pipeline prior to brazing, an oxygen analyzer shall be used to verify the effectiveness of the purge. The oxygen analyzer shall read below 1 percent oxygen concentration before brazing begins. [NFPA 99:5.1.10.4.5.5]

1321.10.4 During Installation. During and after installation, openings in the piping system shall be kept sealed to maintain a nitrogen atmosphere within the piping to prevent debris or other contaminants from entering the system. [NFPA 99:5.1.10.4.5.6]

1321.10.5 Discharge Opening. While a joint is being brazed, a discharge opening shall be provided on the opposite side of the joint from where the purge gas is being introduced. [NFPA 99:5.1.10.4.5.7]

1321.10.6 Temperature of Joint. The flow of purge gas shall be maintained until the joint is cool to the touch. [NFPA 99:5.1.10.4.5.8]

1321.10.7 Opening to Be Sealed. After the joint has cooled, the purge discharge opening shall be sealed to prevent contamination of the inside of the tube and maintain the nitrogen atmosphere within the piping system. [NFPA 99:5.1.10.4.5.9]

1321.10.8 Final Brazed Connection. The final brazed connection of new piping to an existing pipeline containing the system gas shall be permitted to be made without the use of a nitrogen purge. [NFPA 99:5.1.10.4.5.10]

1321.10.9 Final Tie-In Test. After a final brazed connection in a positive pressure medical gas pipeline is made without a nitrogen purge, an outlet in the immediate downstream zone of the affected portion(s) of both the new and existing piping shall be tested in accordance with the final tie-in test in Section 1324.5.9 through Section 1324.5.9.4. [NFPA 99:5.1.10.4.5.11]

1321.10.10 Autogenous Orbital Welding Process. When using the autogenous orbital welding process, joints shall be continuously purged inside and outside with inert gas(es) in accordance with the qualified welding procedure. [NFPA 99:5.1.10.4.5.12]

1321.11 Assembling and Heating Brazed Joints. Tube ends shall be inserted into the socket, either fully or to a mechanically limited depth that is not less than the minimum cup depth (overlap) specified by ASME B16.50. [NFPA 99:5.1.10.4.6.1]

1321.11.1 Heating of Joint. Where flux is permitted, the joint shall be heated slowly until the flux has liquefied. After flux is liquefied, or where flux is not permitted to be used, the joint shall be heated quickly to the brazing temperature, taking care not to overheat the joint. [NFPA 99:5.1.10.4.6.2, 5.1.10.4.6.3]

1321.12 Inspection of Brazed Joints. After brazing, the outside of all joints shall be cleaned by washing with water and a wire brush to remove any residue and allow clear visual inspection of the joint. [NFPA 99:5.1.10.4.7.1]

1321.12.1 Where Flux Is Used. Where flux has been used, the wash water shall be hot. [NFPA 99:5.1.10.4.7.2]

1321.12.2 Visually Inspected. Each brazed joint shall be visually inspected after cleaning the outside surfaces. [NFPA 99:5.1.10.4.7.3]

1321.12.3 Prohibited Brazed Joints. Joints exhibiting the following conditions shall not be permitted:

(1) Flux or flux residue (when flux or flux-coated BAg series rods are used with dissimilar metals).

(2) Base metal melting or erosion.

(3) Unmelted filler metal.

(4) Failure of the filler metal to be clearly visible all the way around the joint at the interface between the socket and the tube.

(5) Cracks in the tube or component.

(6) Cracks in the braze filler metal.

(7) Failure of the joint to hold the test pressure under the installer-performed initial pressure test (see Section 1324.5 through Section 1324.5.1.2) and standing pressure test (see Section 1324.5.4 or Section 1324.5.5). [NFPA 99:5.1.10.4.7.4]

1321.12.4 Defective Brazed Joints. Brazed joints that are identified as defective under the conditions of Section 1321.12.3(2) or Section 1321.12.3(5) shall be replaced. [NFPA 99:5.1.10.4.7.5]

Brazed joints that are identified as defective under the conditions of Section 1321.12.3(1), Section 1321.12.3(3), Section 1321.12.3(4), Section 1321.12.3(6) or Section 1321.12.3(7) shall be permitted to be repaired, except that no joint shall be reheated more than once before being replaced. [NFPA 99:5.1.10.4.7.6]

1322.0 Welded Joints.

1322.1 Welded Joints Procedure. Welded joints for medical gas and medical-surgical vacuum systems shall be permitted to be made using a gas tungsten arc welding (GTAW) autogenous orbital procedure. [NFPA 99:5.1.10.5.1.1]

1322.1.1 Welder Qualification Procedure. The GTAW autogenous orbital procedure and the welder qualification procedure shall be qualified in accordance with Section IX of the ASME Boiler and Pressure Vessel Code. Welder qualification procedures shall include a bend test and a tensile test in accordance with Section IX of the ASME Boiler and Pressure Vessel Code on each tube size diameter. [NFPA 99:5.1.10.5.1.2, 5.1.10.5.1.3]

1322.1.2 Welding Procedure Specification. Each welder shall qualify to a welding procedure specification (WPS) for each tube diameter. [NFPA 99:5.1.10.5.1.4]

1322.1.3 Purging of Joints. GTAW autogenous orbital welded joints shall be purged during welding with a commercially available mixture of 75 percent helium (+/- 5 percent) and 25 percent argon (+/- 5 percent). [NFPA 99:5.1.10.5.1.5]

1322.1.4 Shield Gas. The shield gas shall be as required in Section 1322.1.3. [NFPA 99:5.1.10.5.1.6]

1322.1.5 Test Coupons. Test coupons shall be welded and inspected, as a minimum, at the start of work and every 4 hours thereafter, or when the machine is idle for more than 30 minutes, and at the end of the work period. Test coupons shall be inspected on the I.D. and O.D. by a qualified quality control inspector. Test coupons shall also be welded at change of operator, weld head, welding power supply, or gas source. [NFPA 99:5.1.10.5.1.7 – 5.1.10.5.1.9]

1322.2 Welding for Stainless Tube. Stainless tube shall be welded using metal inert gas (MIG) welding, tungsten inert gas (TIG) welding, or other welding techniques suited to joining stainless tube. [NFPA 99:5.1.10.5.2.1]

1322.2.1 Qualifications. Welders shall be qualified to Section IX of the ASME Boiler and Pressure Vessel Code. [NFPA 99:5.1.10.5.2.2]

1322.3 Memory Metal Fittings. Memory metal fittings having a temperature rating not less than 1000°F (538°C) and a pressure rating not less than 300 psi (2068 kPa) shall be permitted to be used to join copper or stainless steel tube. Memory metal fittings shall be installed by qualified technicians in accordance with the manufacturer's instructions. [NFPA 99:5.1.10.6.1, 5.1.10.6.2]

1322.4 Axially Swaged Fittings. Axially swaged, elastic strain preload fittings providing metal-to-metal seals, having a temperature rating not less than 1000°F (538°C) and a pressure rating not less than 300 psi (2068 kPa), and that, when complete, are permanent and nonseparable shall be permitted to be used to join copper or stainless steel tube. Axially swaged, elastic strain preload fittings shall be installed by qualified technicians in accordance with the manufacturer's instructions. [NFPA 99:5.1.10.7.1, 5.1.10.7.2]

1322.5 Threaded Fittings. Threaded fittings shall meet the following criteria:

(1) They shall be limited to connections for pressure and vacuum indicators, alarm devices, gas-specific demand check fittings, and source equipment on the source side of the source valve.

(2) They shall be tapered pipe threads complying with ASME B1.20.1.

(3) They shall be made up with polytetrafluoroethylene (PTFE) tape or other thread sealant recommended for oxygen service, with sealant applied to the male threads only and care taken to ensure sealant does not enter the pipe. [NFPA 99:5.1.10.8]

1322.6 Other Types of Fittings. Listed or approved metallic gas tube fittings that, when made up, provide a permanent joint having the mechanical, thermal, and sealing integrity of a brazed joint shall be permitted to be used. [NFPA 99:5.1.10.9.1]

1322.6.1 Dielectric Fittings. Dielectric fittings that comply with the following shall be permitted only where required by the manufacturer of special medical equipment to electrically isolate the equipment from the system distribution piping:

(1) They shall be of brass or copper construction with an approved dielectric.

(2) They shall be permitted to be a union.

(3) They shall be clean for oxygen where used for medical gases and medical support gases. [NFPA 99:5.1.10.9.2]

1322.7 Prohibited Joints. The following joints shall be prohibited throughout medical gas and vacuum distribution pipeline systems:

(1) Flared and compression-type connections, including connections to station outlets and inlets, alarm devices, and other components.

(2) Other straight-threaded connections, including unions.

(3) Pipe-crimping tools used to permanently stop the flow of medical gas and vacuum piping.

(4) Removable and nonremovable push-fit fittings that employ a quick assembly push fit connector. [NFPA 99:5.1.10.10]

1323.0 Installation of Piping and Equipment.

1323.1 Required Pipe Sizing. Piping systems shall be designed and sized to deliver the required flow rates at the utilization pressures. [NFPA 99:5.1.10.11.1.1]

1323.1.1 Mains and Branches. Mains and branches in medical gas piping systems shall be not less than DN15 (NPS ½) (⅝ inch O.D.) size. Mains and branches in medical-surgical vacuum systems shall be not less than DN20 (NPS ¾) (⅞ inch O.D.) size. [NFPA 99:5.1.10.11.1.2, 5.1.10.11.1.3]

1323.1.2 Drops to Individual Stations. Drops to individual station outlets and inlets shall be not less than DN15 (NPS ½) (⅝ inch O.D.) size. [NFPA 99:5.1.10.11.1.4]

1323.1.3 Runouts and Connecting Tubing. Runouts to alarm panels and connecting tubing for gauges and alarm devices shall be permitted to be DN8 (NPS ¼) (⅜ inch O.D.) size. [NFPA 99:5.1.10.11.1.5]

1323.1.4 Maximum Demand. Where the maximum demand for each medical gas or vacuum system does not exceed the values in Table 1323.1.4(1) through Table 1323.1.4(6), the size of pipe of each section of the system shall be determined in accordance with Section 1323.1.5. The size for systems beyond the range of Table 1323.1.4(1) through Table 1323.1.4(6) shall be determined in accordance with Section 1323.1.6.

1323.1.5 Sizing Procedures. The size of each section of pipe in a system within the range of Table 1323.1.4(1) through Table 1324.1.4(6) shall be determined in accordance with the following:

(1) Determine the total flow rate and number of outlets or inlets for each section of pipe in accordance with Table 1305.2 and Table 1305.3.

(2) Measure the length of the section of pipe to each station outlet or inlet on the system. Multiply the measured pipe length by 1.5 (150 percent), to account for the number of fittings in the system, to determine the pipe equivalent length.

(3) Beginning with the most remote outlet or inlet, multiply the total flow rate by the diversity factor specified in Table 1323.1.5(1) for each section of pipe to determine the sizing flow rate for the piping.

(4) Select Table 1323.1.4(1) through Table 1324.1.4(6) based on the medical gas or vacuum being transported through the piping.

(5) Select an estimated pipe size for determining the system pressure loss. Multiply the pipe equivalent length, for a given section of pipe, by the pressure loss for the sizing flow rate in the applicable table. Divide that number by 100 to determine the system pressure loss for the section of pipe.

(6) Add the pressure loss for each section of piping, from the source equipment location to the outlet or inlet, to determine the total system pressure loss to each outlet or inlet. The total system pressure loss in the piping to each outlet or inlet shall not exceed the values specified in Table 1323.1.5(2).

»

TABLE 1323.1.5(1)
SYSTEM SIZING – FLOW REQUIREMENTS FOR STATION OUTLETS AND INLETS[1]

NUMBER OF OUTLETS AND INLETS TERMINAL UNITS PER FACILITY	DIVERSITY PERCENTAGE OF AVERAGE FLOW PER OUTLETS AND INLETS TERMINAL UNITS	MINIMUM PERMISSIBLE SYSTEM FLOW OF ALL PRESSURIZED MEDICAL GAS SYSTEMS[2] (standard cubic feet per minute)
1–10	100%	Actual Demand
11–25	75%	7.0
26–50	50%	13.1
51–100	50%	17.5

Notes:
[1] Flow rates of station outlets and inlets in accordance with Table 1305.2.
[2] The minimum system flow is the average outlets and inlets flow times the number of station outlets and inlets times the diversity percentage.

»

TABLE 1323.1.5(2)
MAXIMUM PERMITTED PRESSURE LOSS IN MEDICAL GAS AND MEDICAL VACUUM SYSTEMS

TYPE OF SYSTEM	MAXIMUM ALLOWABLE SYSTEM PRESSURE LOSS (psi)
Medical Air	5
Nitrogen	15
Nitrous Oxide	5
Carbon Dioxide	5
Oxygen	5
Medical Vacuum	4 inches of mercury

For SI units: 1 pound-force per square inch = 6.8947 kPa, 1 inch of mercury = 3.386 kPa

» **1323.1.6 Engineering Methods.** For conditions other than those covered by Section 1323.1.4, such as longer runs of greater gas or vacuum demands, the size of each medical gas or vacuum piping system shall be determined by standard engineering methods acceptable to the Authority Having Jurisdiction, and each system shall be so designed that the total pressure drop or gain between the source equipment and an outlet or inlet shall not exceed the allowable pressures shown in Table 1305.1.

1323.2 Pipe Protection. Piping shall be protected against freezing, corrosion, and physical damage. [NFPA 99:5.1.10.11.2] «

1323.2.1 Exposed Piping. Piping exposed in corridors and other areas where subject to physical damage from the movement of carts, stretchers, portable equipment, or vehicles shall be protected. [NFPA 99:5.1.10.11.2.1] «

1323.2.2 Underground Piping. Piping underground within buildings or embedded in concrete floors or walls shall be installed in a continuous conduit. [NFPA 99:5.1.10.11.2.2] «

1323.3 Location of Piping. Piping risers shall be permitted to be installed in pipe shafts if protected from physical damage, effects of excessive heat, corrosion, or contact with oil. [NFPA 99:5.1.10.11.3.1] «

1323.3.1 Prohibited Locations. Piping shall not be installed in kitchens, stairwells, elevator shafts, elevator machine rooms, areas with open flames, electrical service equipment over 600 volts, and areas prohibited under NFPA 70 except for the following locations: «

(1) Room locations for medical air compressor supply systems and medical-surgical vacuum pump supply systems

(2) Room locations for secondary distribution circuit panels and breakers having a maximum voltage rating of 600 volts [NFPA 99:5.1.10.11.3.2]

1323.3.2 Approved Locations. Medical gas piping shall be permitted to be installed in the same service trench or tunnel with fuel gas lines, fuel oil lines, electrical lines, steam lines, and similar utilities, provided that the space is ventilated (naturally or mechanically) and the ambient temperature around the medical gas piping is limited to 130°F (54°C) maximum. [NFPA 99:5.1.10.11.3.3] «

1323.3.3 Prohibited Contact with Oil. Medical gas piping shall not be located where subject to contact with oil, including a possible flooding area in the case of a major oil leak. [NFPA 99:5.1.10.11.3.4] «

1323.4 Pipe Support. Piping shall be supported from the building structure. [NFPA 99:5.1.10.11.4.1] «

1323.4.1 Hangers and Supports. Hangers and supports shall comply with and be installed in accordance with MSS SP-58. [NFPA 99:5.1.10.11.4.2] «

1323.4.2 Copper Tube. Supports for copper tube shall be sized for copper tube. [NFPA 99:5.1.10.11.4.3] «

1323.4.3 Damp Locations. In potentially damp locations, copper tube hangers or supports that are in contact with the tube shall be plastic-coated or otherwise be electrically insulated from the tube by a material that will not absorb moisture. [NFPA 99:5.1.10.11.4.5] «

1323.4.4 Maximum Spacing. Maximum support spacing shall be in accordance with Table 1323.4.4. [NFPA 99:5.1.10.11.4.6] «

»

TABLE 1323.1.4(1)
PRESSURE LOSS FOR MEDICAL AIR

FLOW RATE (SCFM)[1]	PRESSURE DROP (psi) PER 100 FEET[2]		
	½ INCH PIPE	¾ INCH PIPE	1 INCH PIPE
0.35	0.004	0.001	–
0.71	0.012	0.003	–
1.06	0.023	0.005	–
1.41	0.037	0.007	–
1.77	0.055	0.011	–
2.12	0.075	0.015	–
2.47	0.097	0.019	–
2.82	0.123	0.024	–
3.18	0.151	0.029	–
3.53	0.181	0.035	–
4.24	0.249	0.048	–
4.94	0.326	0.063	–
5.65	0.413	0.080	–
6.36	0.507	0.098	–
7.06	0.611	0.118	0.030
7.77	0.723	0.139	0.035
8.47	0.843	0.162	0.041
9.18	0.969	0.187	0.047
9.89	1.108	0.212	0.053
10.59	1.252	0.240	0.060
12.36	1.647	0.315	0.079
14.12	2.090	0.398	0.100
15.89	2.580	0.490	0.123
17.66	3.116	0.591	0.148
19.42	–	0.701	0.176
21.19	–	0.818	0.205
22.95	–	0.944	0.236
24.72	–	1.078	0.268
28.25	–	1.369	0.341
31.78	–	1.690	0.421
35.31	–	2.043	0.509
38.84	–	2.425	0.603
42.37	–	2.838	0.705
45.90	–	3.280	0.814
49.43	–	3.751	0.929
52.97	–	4.249	1.052
56.50	–	–	1.181
60.03	–	–	1.318
63.56	–	–	1.461
67.09	–	–	1.611
70.62	–	–	1.768
81.21	–	–	2.276
88.28	–	–	2.647
95.34	–	–	3.044

For SI units: 1 standard cubic foot per minute = 28.32 SLPM, 1 inch = 25 mm, 1 foot = 304.8 mm, 1 pound-force per square inch = 6.8947 kPa

Notes:

[1] Based on the pressure of 14.7 psig (101 kPa) at 68°F (20°C).

[2] Based on the pressure of 55 psig (379 kPa) at 68°F (20°C).

TABLE 1323.1.4(2)
PRESSURE LOSS FOR NITROGEN

«

FLOW RATE (SCFM)[1]	PRESSURE DROP (psi) PER 100 FEET[2]		
	½ INCH PIPE	¾ INCH PIPE	1 INCH PIPE
5.30	0.126	0.024	–
10.59	0.430	0.082	–
15.89	0.886	0.168	–
21.19	1.485	0.281	–
26.48	2.220	0.419	–
31.78	3.089	0.581	–
37.08	4.087	0.766	–
42.37	–	0.975	–
47.67	–	1.206	–
52.97	–	1.460	0.361
58.26	–	1.736	0.429
63.56	–	2.033	0.502
68.85	–	2.352	0.580
74.15	–	2.692	0.663
79.45	–	3.054	0.752
84.74	–	3.436	0.845
90.04	–	3.840	0.943
95.34	–	4.264	1.046
100.63	–	4.709	1.154
105.93	–	–	1.267
116.52	–	–	1.508
127.12	–	–	1.768
137.71	–	–	2.046
148.30	–	–	2.344
158.90	–	–	2.660
169.49	–	–	2.994
180.08	–	–	3.347
190.67	–	–	3.719
201.27	–	–	4.108
211.86	–	–	4.516
222.45	–	–	4.942
233.05	–	–	–
243.64	–	–	–
254.23	–	–	–
264.83	–	–	–
275.42	–	–	–
286.01	–	–	–
296.60	–	–	–
307.20	–	–	–
317.79	–	–	–

For SI units: 1 standard cubic foot per minute = 28.32 SLPM, 1 inch = 25 mm, 1 foot = 304.8 mm, 1 pound-force per square inch = 6.8947 kPa

Notes:

[1] Based on the pressure of 14.7 psig (101 kPa) at 68°F (20°C).

[2] Based on the pressure of 55 psig (379 kPa) at 68°F (20°C).

»

TABLE 1323.1.4(3)
PRESSURE LOSS FOR NITROUS OXIDE AND CARBON DIOXIDE

FLOW RATE (SCFM)[1]	PRESSURE DROP (psi) PER 100 FEET[2]		
	½ INCH PIPE	¾ INCH PIPE	1 INCH PIPE
0.35	0.004	–	–
0.71	0.014	–	–
1.06	0.029	–	–
1.41	0.047	–	–
1.77	0.070	–	–
2.12	0.096	–	–
2.47	0.125	–	–
2.82	0.159	–	–
3.18	0.195	–	–
3.53	0.235	0.045	–
4.24	0.324	0.062	–
4.94	0.425	0.081	–
5.65	0.539	0.103	–
6.36	0.664	0.127	–
7.06	0.802	0.153	0.038
7.77	0.950	0.181	0.045
8.47	1.110	0.211	0.053
9.18	1.281	0.243	0.061
9.89	1.463	0.278	0.070
10.59	1.656	0.314	0.079
12.36	2.186	0.413	0.103
14.12	2.752	0.525	0.131
15.89	3.442	0.648	0.162
17.66	4.166	0.783	0.195
19.42	–	0.929	0.231
21.19	–	0.744	0.270
22.95	–	0.858	0.312
24.72	–	0.980	0.356
28.25	–	1.244	0.453
31.78	–	1.537	0.560
35.31	–	1.858	0.677
38.84	–	2.205	0.804
42.37	–	2.581	0.941
45.90	–	2.982	1.088
49.43	–	3.411	1.245
52.97	–	4.249	1.411
56.50	–	–	1.587
60.03	–	–	1.772
63.56	–	–	1.967
67.09	–	–	2.174
70.62	–	–	2.385
79.45	–	–	2.959
88.28	–	–	3.589

For SI units: 1 standard cubic foot per minute = 28.32 SLPM, 1 inch = 25 mm, 1 foot = 304.8 mm, 1 pound-force per square inch = 6.8947 kPa

Notes:

1 Based on the pressure of 14.7 psig (101 kPa) at 68°F (20°C).

2 Based on the pressure of 55 psig (379 kPa) at 68°F (20°C).

«

TABLE 1323.1.4(4)
PRESSURE LOSS FOR OXYGEN

FLOW RATE (SCFM)[1]	PRESSURE DROP (psi) PER 100 FEET[2]		
	½ INCH PIPE	¾ INCH PIPE	1 INCH PIPE
0.35	0.004	–	–
0.71	0.013	0.003	–
1.06	0.025	0.005	–
1.41	0.041	0.008	–
1.77	0.060	0.012	–
2.12	0.082	0.016	–
2.47	0.107	0.021	–
2.82	0.135	0.026	–
3.18	0.166	0.032	–
3.53	0.199	0.038	–
4.24	0.274	0.053	–
4.94	0.359	0.069	–
5.65	0.454	0.087	–
6.36	0.558	0.107	–
7.06	0.672	0.129	0.033
7.77	0.795	0.153	0.039
8.47	0.927	0.179	0.045
9.18	1.066	0.205	0.052
9.89	1.218	0.233	0.059
10.59	1.377	0.263	0.066
12.36	1.811	0.346	0.087
14.12	2.298	0.438	0.110
15.89	2.837	0.539	0.135
17.66	3.456	0.650	0.163
19.42	–	0.771	0.193
21.19	–	0.900	0.225
22.95	–	1.038	0.260
24.72	–	1.185	0.295
28.25	–	1.505	0.375
31.78	–	1.859	0.463
35.31	–	2.247	0.559
38.84	–	2.667	0.663
42.37	–	3.121	0.775
45.90	–	3.607	0.895
49.43	–	4.125	1.022
52.97	–	–	1.157
56.50	–	–	1.299
60.03	–	–	1.449
63.56	–	–	1.607
67.09	–	–	1.772
70.62	–	–	1.944
81.21	–	–	2.503
91.81	–	–	3.127
102.40	–	–	3.813

For SI units: 1 standard cubic foot per minute = 28.32 SLPM, 1 inch = 25 mm, 1 foot = 304.8 mm, 1 pound-force per square inch = 6.8947 kPa

Notes:

1 Based on pressure of 14.7 psig (101 kPa) at 68°F (20°C).

2 Based on pressure of 55 psig (379 kPa) at 68°F (20 °C).

TABLE 1323.1.4(5)
PRESSURE LOSS FOR VACUUM

FLOW RATE (SCFM)[1]	VACUUM LOSS (inch of mercury) PER 100 FEET FOR COPPER TUBE[2]				
	¾ INCH TUBE	1 INCH TUBE	1¼ INCH TUBE	1½ INCH TUBE	2 INCH TUBE
0.35	0.019	–	–	–	–
0.71	0.061	–	–	–	–
1.06	0.120	–	–	–	–
1.41	0.194	–	–	–	–
1.77	0.284	–	–	–	–
2.12	0.387	–	–	–	–
2.47	0.504 –	–	–	–	
2.82	0.634	–	–	–	–
3.18	0.777	–	–	–	–
3.53	0.932	0.238	–	–	–
4.24	1.277	0.325	–	–	–
4.94	1.669	0.424	–	–	–
5.65	2.106	0.534	–	–	–
6.36	2.586	0.655	–	–	–
7.06	3.110	0.787	0.272	–	–
7.77	3.674	0.929	0.321	–	–
8.47	4.280	1.081	0.373	–	–
9.18	4.927	1.243	0.429	–	–
9.89	–	1.416	0.488	–	–
10.59	–	1.597	0.551	0.242	–
11.30	–	1.789	0.616	0.270	–
12.01	–	1.990	0.685	0.300	–
12.71	–	2.200	0.757	0.332	–
13.42	–	2.419	0.832	0.365	–
14.12	–	2.648	0.911	0.399	–
14.83	–	2.886	0.992	0.435	–
15.54	–	3.132	1.077	0.471	–
16.24	–	3.388	1.164	0.510	–
16.95	–	3.652	1.254	0.549	–
17.66	–	3.925	1.348	0.590	–
18.36	–	4.207	1.444	0.632	0.167
19.07	–	4.498	1.543	0.675	0.179
19.77	–	4.797	1.646	0.720	0.190
20.48	–	–	1.751	0.766	0.202
21.19	–	–	1.859	0.813	0.214
24.72	–	–	2.441	1.066	0.281
28.25	–	–	3.092	1.350	0.356
31.78	–	–	3.811	1.662	0.438
35.31	–	–	4.596	2.004	0.527
38.84	–	–	–	2.373	0.624
42.37	–	–	–	2.770	0.728
45.90	–	–	–	3.194	0.838

TABLE 1323.1.4(5)
PRESSURE LOSS FOR VACUUM (continued)

FLOW RATE (SCFM)[1]	VACUUM LOSS (inch of mercury) PER 100 FEET FOR COPPER TUBE[2]				
	¾ INCH TUBE	1 INCH TUBE	1¼ INCH TUBE	1½ INCH TUBE	2 INCH TUBE
49.43	–	–	–	3.645	0.956
52.97	–	–	–	4.122	1.081
56.50	–	–	–	4.626	1.212
63.56	–	–	–	–	1.495
70.62	–	–	–	–	1.803
77.68	–	–	–	–	2.138
84.74	–	–	–	–	2.497
91.81	–	–	–	–	2.882
98.87	–	–	–	–	3.291
105.93	–	–	–	–	3.724
112.99	–	–	–	–	4.181

For SI units: 1 standard cubic foot per minute = 28.32 SLPM, 1 inch = 25 mm, 1 foot = 304.8 mm, 1 inch of mercury = 3.386 kPa

Notes:

1 Based on the pressure of 14.7 psig (101 kPa) at 68°F (20°C).

2 Based on the pressure of 19 inches of mercury gauge vacuum (64 kPa) at 68°F (20°C).

TABLE 1323.1.4(6)
PRESSURE LOSS FOR VACUUM (CATEGORY 3)

FLOW RATE (SCFM)[1]	VACUUM LOSS (inch of mercury) PER 100 FEET FOR PLASTIC TUBE[2]				
	¾ INCH TUBE	1 INCH TUBE	1¼ INCH TUBE	1½ INCH TUBE	2 INCH TUBE
0.35	0.005	–	–	–	–
0.71	0.010	–	–	–	–
1.06	0.015	–	–	–	–
1.41	0.021	–	–	–	–
1.77	0.026	–	–	–	–
2.12	0.060	0.010	–	–	–
2.47	0.077	0.020	–	–	–
2.82	0.096	0.025	–	–	–
3.18	0.118	0.031	0.011	–	–
3.53	0.141	0.036	0.013	–	–
4.24	0.192	0.050	0.017	–	–
4.94	0.249	0.064	0.023	0.010	–
5.65	0.313	0.081	0.028	0.012	–
6.36	0.383	0.099	0.035	0.015	–
7.06	0.459	0.118	0.041	0.018	–
7.77	0.541	0.139	0.049	0.021	–
8.47	0.628	0.161	0.056	0.024	–
9.18	0.722	0.185	0.065	0.027	–
9.89	0.821	0.210	0.073	0.031	–
10.59	0.925	0.237	0.083	0.035	–
11.30	1.035	0.265	0.092	0.039	0.010
12.01	1.151	0.294	0.102	0.043	0.011
12.71	1.270	0.324	0.113	0.048	0.012

»

TABLE 1323.1.4(6)
PRESSURE LOSS FOR VACUUM (CATEGORY 3) (continued)

FLOW RATE (SCFM)[1]	VACUUM LOSS (inch of mercury) PER 100 FEET FOR PLASTIC TUBE[2]				
	¾ INCH TUBE	1 INCH TUBE	1¼ INCH TUBE	1½ INCH TUBE	2 INCH TUBE
13.42	1.396	0.356	0.124	0.052	0.014
14.12	1.525	0.389	0.135	0.057	0.015
14.83	1.662	0.424	0.147	0.062	0.016
15.54	1.803	0.460	0.160	0.068	0.017
16.24	1.948	0.496	0.172	0.073	0.019
16.95	2.099	0.535	0.186	0.078	0.020
17.66	2.256	0.574	0.199	0.084	0.022
18.36	2.415	0.615	0.213	0.090	0.023
19.07	2.581	0.657	0.228	0.096	0.025
19.77	2.750	0.699	0.243	0.102	0.026
20.48	2.925	0.744	0.258	0.109	0.028
21.19	3.106	0.790	0.274	0.115	0.030
24.72	4.074	1.034	0.358	0.151	0.039
28.25	–	1.307	0.452	0.190	0.049
31.78	–	1.608	0.556	0.234	0.060
35.31	–	1.936	0.669	0.281	0.072
38.84	–	2.291	0.791	0.332	0.085
42.37	–	2.672	0.922	0.387	0.099
45.90	–	3.078	1.062	0.446	0.113
49.43	–	3.510	1.211	0.508	0.129
52.97	–	3.969	1.368	0.574	0.146
56.50	–	4.450	1.534	0.643	0.163
63.56	–	–	1.890	0.792	0.201
70.62	–	–	2.278	0.954	0.242
77.68	–	–	2.699	1.130	0.286
84.74	–	–	3.151	1.318	0.334
91.81	–	–	3.634	1.520	0.385
98.87	–	–	4.148	1.734	0.439
105.93	–	–	4.691	1.961	0.496
112.99	–	–	–	2.200	0.556

For SI units: 1 standard cubic foot per minute = 28.32 SLPM, 1 inch = 25 mm, 1 foot = 304.8 mm, 1 inch of mercury = 3.386 kPa

Notes:

[1] Based on the pressure of 14.7 psig (101 kPa) at 68°F (20°C).

[2] Based on the pressure of 19 inches of mercury gauge vacuum (64 kPa) at 68°F (20°C).

» **1323.4.5 Seismic Provisions.** Where required, medical gas and vacuum piping shall be seismically restrained against earthquakes in accordance with the applicable building code. [NFPA 99:5.1.10.11.4.7]

» **1323.5 Frost Protection.** Buried piping outside of buildings shall be installed below the local level of frost penetration. [NFPA 99:5.1.10.11.5.1]

TABLE 1323.4.4
MAXIMUM PIPE SUPPORT SPACING
[NFPA 99: TABLE 5.1.10.11.4.6] «

PIPE SIZE			HANGER SPACING (feet)
DN8	(NPS ¼)	(⅜ of an inch O.D.)	5
DN10	(NPS ⅜)	(½ of an inch O.D.)	6
DN15	(NPS ½)	(⅝ of an inch O.D.)	6
DN20	(NPS ¾)	(⅞ of an inch O.D.)	7
DN25	(NPS 1)	(1⅛ of an inch O.D.)	8
DN32	(NPS 1¼)	(1⅜ of an inch O.D.)	9
DN40 and larger	(NPS 1½)	(1⅝ of an inch O.D.)	10
Vertical risers, all sizes, every floor, but not to exceed:			15

For SI units: 1 inch = 25 mm, 1 foot = 304.8 mm

1323.5.1 Backfilling and Trenching. The installation procedure for underground piping shall protect the piping from physical damage while being backfilled. [NFPA 99:5.1.10.11.5.2] «

1323.5.2 Conduit, Cover, or Enclosure. If underground piping is protected by a conduit, cover, or other enclosure, the following requirements shall be met: «

(1) Access shall be provided at the joints for visual inspection and leak testing.

(2) The conduit, cover, or enclosure shall be self-draining and not retain groundwater in prolonged contact with the pipe. [NFPA 99:5.1.10.11.5.3]

1323.5.3 Excessive Stresses. Buried piping that will be subject to surface loads shall be buried at a depth that will protect the piping or its enclosure from excessive stresses. [NFPA 99:5.1.10.11.5.4] «

1323.5.4 Minimum Backfill. The minimum backfilled cover above the top of the pipe or its enclosure for buried piping outside of buildings shall be 36 inches (914 mm), except that the minimum cover shall be permitted to be reduced to 18 inches (457 mm) where there is no potential for damage from surface loads or surface conditions. [NFPA 99:5.1.10.11.5.5] «

1323.5.5 Trenches. Trenches shall be excavated so that the pipe or its enclosure has firm, substantially continuous bearing on the bottom of the trench. [NFPA 99:5.1.10.11.5.6] «

1323.5.6 Composition of Backfill. Backfill shall be clean, free from material that can damage the pipe, and compacted. [NFPA 99:5.1.10.11.5.7] «

1323.5.7 Marker. A continuous tape or marker placed immediately above the pipe or its enclosure shall clearly identify the pipeline by specific name. [NFPA 99:5.1.10.11.5.8] «

1323.5.8 Warning. A continuous warning means shall also be provided above the pipeline at approximately one-half the depth of burial. [NFPA 99:5.1.10.11.5.9] «

1323.5.9 Wall Sleeve. Where underground piping is installed through a wall sleeve, the outdoor end of the sleeve shall be sealed to prevent the entrance of groundwater into the building. [NFPA 99:5.1.10.11.5.10]

1323.6 Connectors. Hose and flexible connectors, both metallic and nonmetallic, shall be no longer than necessary and shall not penetrate or be concealed in walls, floors, ceilings, or partitions. [NFPA 99:5.1.10.11.6.1]

1323.6.1 Flexible Connectors. Flexible connectors, metallic or nonmetallic, shall have a minimum burst pressure, with a gauge pressure of 1000 psi (6895 kPa). [NFPA 99:5.1.10.11.6.2]

1323.6.2 Metallic Flexible Joints. Metallic flexible joints shall be permitted in the pipeline where required for expansion joints, seismic protection, thermal expansion, or vibration control and shall be as follows:

(1) For all wetted surfaces, made of bronze, copper, or stainless steel.

(2) Cleaned at the factory for oxygen service and received on the job site with certification of cleanliness.

(3) Suitable for service at 300 psig (2068 kPa) or above and able to withstand temperatures of 1000°F (538°C).

(4) Provided with brazing extensions to allow brazing into the pipeline per Section 1321.0.

(5) Supported with pipe hangers and supports as required for their additional weight. [NFPA 99:5.1.10.11.6.3]

1323.7 Prohibited System Interconnections. Two or more medical gas or vacuum piping systems shall not be interconnected for installation, testing, or any other reason except as permitted by Section 1323.7.1. [NFPA 99:5.1.10.11.7.1]

1323.7.1 Medical Gas and Medical Vacuum. Medical gas and vacuum systems with the same contents shall be permitted to be interconnected with an inline valve installed between the systems. [NFPA 99:5.1.10.11.7.2]

1323.7.2 Leak Testing. Leak testing shall be accomplished by separately charging and testing each individual piping system. [NFPA 99:5.1.10.11.7.3]

1323.8 Manufacturer's Instructions. The installation of individual components shall be made in accordance with the instructions of the manufacturer. Manufacturer's instructions shall include directions and information deemed by the manufacturer to be adequate for attaining proper operation, testing, and maintenance of the medical gas and vacuum systems. Copies of the manufacturer's instructions shall be left with the system owner. [NFPA 99:5.1.10.11.8.1 – 5.1.10.11.8.3]

1323.9 Changes in System Use. Where a positive-pressure medical gas piping distribution system originally used or constructed for use at one pressure and for one gas is converted for operation at another pressure or for another gas, all provisions of Section 1318.0 through Section 1323.12 shall apply as if the system were new. [NFPA 99:5.1.10.11.9.1]

1323.9.1 Medical Vacuum System. A vacuum system shall not be permitted to be converted for use as a gas system. [NFPA 99:5.1.10.11.9.2]

1323.10 Qualifications of Installers. The installation of medical gas and vacuum systems shall be made by qualified, competent technicians who are experienced in performing such installations, including all personnel who actually install the piping system. Installers of medical gas and vacuum piped distribution systems, all appurtenant piping supporting pump and compressor source systems, and appurtenant piping supporting source gas manifold systems not including permanently installed bulk source systems, shall be certified in accordance with ASSE 6010. [NFPA 99:5.1.10.11.10.1, 5.1.10.11.10.2]

1323.10.1 Brazing. Brazing shall be performed by individuals who are qualified in accordance with Section 1323.11. [NFPA 99:5.1.10.11.10.5]

1323.10.2 Documentation. Prior to any installation work, the installer of medical gas and vacuum piping shall provide and maintain documentation on the job site for the qualification of brazing procedures and individual brazers that is required under Section 1323.11. [NFPA 99:5.1.10.11.10.6]

1323.10.3 Health Care Organization Personnel. Health care organization personnel shall be permitted to install piping systems if all of the requirements of Section 1323.10 are met during the installation. [NFPA 99:5.1.10.11.10.7]

1323.11 Qualification of Brazing Procedures and Brazing. Brazing procedures and brazer performance for the installation of medical gas and vacuum piping shall be qualified in accordance with either Section IX, "Welding and Brazing Qualifications," of the ASME Boiler and Pressure Vessel Code, or AWS B2.2, both as modified by Section 1323.11.1 through Section 1323.11.4. [NFPA 99:5.1.10.11.11.1]

1323.11.1 Examination. Brazers shall be qualified by visual examination of the test coupon followed by sectioning. [NFPA 99:5.1.10.11.11.2]

1323.11.2 Brazing Procedure Specification. The brazing procedure specification shall address cleaning, joint clearance, overlap, internal purge gas, purge gas flow rate, and filler metal. [NFPA 99:5.1.10.11.11.3]

1323.11.3 Documentation. The brazing procedure qualification record and the record of brazer performance qualification shall document filler metal used, base metals, cleaning, joint clearance, overlap, internal purge gas and flow rate during brazing of coupon, and absence of internal oxidation in the completed coupon. [NFPA 99:5.1.10.11.11.4]

1323.11.4 Procedures. Brazing procedures qualified by a technically competent group or agency shall be permitted under the following conditions:

(1) The brazing procedure specification and the procedure qualification records meet the requirements of this code.

(2) The employer obtains a copy of both the brazing procedure specification and the supporting qualification records from the group or agency and signs and dates these records, thereby accepting responsibility for the qualifications that were performed by the group or agency.

(3) The employer qualifies at least one brazer following each brazing procedure specification used. [NFPA 99:5.1.10.11.11.5]

1323.11.5 Conditions of Acceptance. An employer shall be permitted to accept brazer qualification records of a previous employer under the following conditions:

(1) The brazer has been qualified following the same or an equivalent procedure that the new employer uses.

(2) The new employer obtains a copy of the record of brazer performance qualification tests from the previous employer and signs and dates these records, thereby accepting responsibility for the qualifications performed by the previous employer. [NFPA 99:5.1.10.11.11.6]

1323.11.6 Qualifications. Performance qualifications of brazers shall remain in effect indefinitely, unless the brazer does not braze with the qualified procedure for a period exceeding 6 months or there is a specific reason to question the ability of the brazer. [NFPA 99:5.1.10.11.11.7]

1323.12 Breaching or Penetrating Medical Gas Piping. Positive pressure patient medical gas piping and medical support gas piping shall not be breached or penetrated by any means or process that will result in residual copper particles or other debris remaining in the piping or affect the oxygen-clean interior of the piping. The breaching or penetrating process shall ensure that any debris created by the process remains contained within the work area. [NFPA 99:5.1.10.11.12.1, 5.1.10.11.12.2]

1323.13 Labeling, Identification and Operating Pressure. Color and pressure requirements shall be in accordance with Table 1305.1. [NFPA 99:5.1.11]

1323.13.1 Pipe Labeling. Piping shall be labeled by stenciling or adhesive markers that identify the patient medical gas, the support gas, or the vacuum system and include the following:

(1) Name of the gas or vacuum system or the chemical symbol per Table 1305.1.

(2) Gas or vacuum system color code per Table 1305.1.

(3) Where positive pressure gas piping systems operate at pressures other than the standard gauge pressure in Table 1305.1, the operating pressure in addition to the name of the gas. [NFPA 99:5.1.11.1.1]

1323.13.2 Location of Pipe Labeling. Pipe labels shall be located as follows:

(1) At intervals of not more than 20 feet (6096 mm).

(2) At least once in or above every room.

(3) On both sides of walls or partitions penetrated by the piping.

(4) At least once in every story height traversed by risers. [NFPA 99:5.1.11.1.2]

1323.13.3 Paint. Medical gas piping shall not be painted. [NFPA 99:5.1.11.1.3]

1323.14 Identification of Shutoff Valves. Shutoff valves shall be identified with the following:

(1) Name or chemical symbol for the specific medical gas or vacuum system.

(2) Room or areas served.

(3) Caution to not close or open valve except in emergency. [NFPA 99:5.1.11.2.1]

1323.14.1 Nonstandard Operating Pressures. Where positive pressure gas piping systems operate at pressures other than the standard gauge pressure of 50 psi (345 kPa) to 55 psi (379 kPa) or a gauge pressure of 160 psi (1103 kPa) to 185 psi (1276 kPa) for nitrogen or instrument air, the valve identification shall also include the nonstandard operating pressure. [NFPA 99:5.1.11.2.2]

1323.14.2 Source Valves. Source valves shall be labeled in substance as follows:

SOURCE VALVE
FOR THE (SOURCE NAME)
[NFPA 99:5.1.11.2.3]

1323.14.3 Main Line Valves. Main line valves shall be labeled in substance as follows:

MAIN LINE VALVE FOR THE
(GAS/VACUUM NAME)
SERVING (NAME OF BUILDING)
[NFPA 99:5.1.11.2.4]

1323.14.4 Riser Valves. The riser valve(s) shall be labeled in substance as follows:

RISER FOR THE (GAS/VACUUM NAME)
SERVING (NAME OF THE AREA/BUILDING SERVED BY THE PARTICULAR RISER)
[NFPA 99:5.1.11.2.5]

1323.14.5 Service Valves. The service valve(s) shall be labeled in substance as follows:

SERVICE VALVE FOR THE
(GAS/VACUUM NAME) SERVING
(NAME OF THE AREA/BUILDING
SERVED BY THE PARTICULAR VALVE)
[NFPA 99:5.1.11.2.6]

1323.14.6 Zone Valve Box. Zone valve box assemblies shall be labeled with the rooms, areas, or spaces that they control as follows:

ZONE VALVES FOR THE
(GAS/VACUUM NAME)
SERVING (NAME OF ROOMS OR SPACES SERVED BY THE PARTICULAR VALVE)

Labeling shall either be visible from outside the zone valve box assembly through the cover or be replicated on the outside, but not affixed to the removable cover. [NFPA 99:5.1.11.2.7]

1323.15 Identification. Station outlets and inlets shall be identified as to the name or chemical symbol for the specific medical gas or vacuum provided.

In sleep labs, where the outlet is downstream of a flow control device, the station outlet identification shall include a warning not to use the outlet for ventilating patients.

Where medical gas systems operate at pressures other than the standard gauge pressure of 50 psi to 55 psi (345 kPa to 380 kPa) or a gauge pressure of 160 psi to 185 psi (1103 kPa to 1275 kPa) for nitrogen, the station outlet identification shall include the nonstandard operating pressure in addition to the name of the gas. [NFPA 99:5.1.11.3.1 – 5.1.11.3.2]

1324.0 Performance Criteria and Testing Category 1 (Gases, Medical Surgical Vacuum).

1324.1 Where Required. Inspection and testing shall be performed on components, or portions thereof, of new, piped medical gas or vacuum systems, additions, renovations, temporary installations, or repaired systems in accordance with Section 1324.2 through Section 1324.5.11, and certified in accordance with Section 1306.0.

1324.2 Breached Systems. All systems that are breached and components that are subject to additions, renovations, or replacement (e.g., new gas sources: bulk, manifolds, compressors, dryers, alarms) shall be inspected and tested. Systems shall be deemed breached at the point of pipeline intrusion by physical separation or by system component removal, replacement, or addition. Breached portions of the systems subject to inspection and testing shall be confined to only the specific altered zone and components in the immediate zone or area that is located upstream for vacuum systems and downstream for pressure gases at the point or area of intrusion. [NFPA 99:5.1.12.1.3 – 5.1.12.1.5]

1324.2.1 Reports. The inspection and testing reports shall be submitted directly to the party that contracted for the testing, who shall submit the report through channels to the responsible facility authority and any others that are required. Reports shall contain detailed listings of all findings and results. [NFPA 99:5.1.12.1.6, 5.1.12.1.7]

1324.3 Test Gas. The test gas shall be oil-free, dry nitrogen NF. [NFPA 99:5.1.12.2.1.2]

1324.4 Initial Piping Blowdown. Piping in medical gas and vacuum distribution systems shall be blown clear by means of oil-free, dry nitrogen NF after installation of the distribution piping but before installation of station outlet/inlet rough-in assemblies and other system components (e.g., pressure/vacuum alarm devices, pressure/vacuum indicators, pressure relief valves, manifolds, source equipment). [NFPA 99:5.1.12.2.2]

1324.5 Initial Pressure Tests – Medical Gas and Vacuum Systems. Each section of the piping in medical gas and vacuum systems shall be pressure tested. Initial pressure tests shall be conducted as follows:

(1) After blowdown of the distribution piping.

(2) After installation of station outlet/inlet rough-in assemblies.

(3) Prior to the installation of components of the distribution piping system that would be damaged by the test pressure (e.g., pressure/vacuum alarm devices, pressure/vacuum indicators, line pressure relief valves). [NFPA 99:5.1.12.2.3.1, 5.1.12.2.3.2]

1324.5.1 Shutoff Valve. The source shutoff valve shall remain closed during tests specified in Section 1324.5 through Section 1324.5.1.2. [NFPA 99:5.1.12.2.3.3]

1324.5.1.1 Required Test Pressure. The test pressure for pressure gases and vacuum systems shall be 1.5 times the system operating pressure but not less than a gauge pressure of 150 psi (1034 kPa). The test pressure shall be maintained until each joint has been examined for leakage by means of a leak detectant that is safe for use with oxygen and does not contain ammonia. [NFPA 99:5.1.12.2.3.4, 5.1.12.2.3.5]

1324.5.1.2 Leaks. Leaks, if any, shall be located, repaired (if permitted), replaced (if required), and retested. [NFPA 99:5.1.12.2.3.6]

1324.5.2 Initial Cross-Connection Test. It shall be determined that no cross-connections exist between the various medical gas and vacuum piping systems. [NFPA 99:5.1.12.2.4]

1324.5.2.1 Atmospheric Pressure. All piping systems shall be reduced to atmospheric pressure. [NFPA 99:5.1.12.2.4.1]

1324.5.2.2 Sources of Test Gas. Sources of test gas shall be disconnected from all piping systems, except for the one system being tested. [NFPA 99:5.1.12.2.4.2]

1324.5.2.3 System to Be Charged. The system under test shall be charged with oil-free, dry nitrogen NF to a gauge pressure of 50 psi (345 kPa). [NFPA 99:5.1.12.2.4.3]

1324.5.2.4 Check Outlets and Inlets. After the installation of the individual faceplates with appropriate adapters matching outlet/inlet labels, each individual outlet/inlet in each installed medical gas and vacuum piping system shall be checked to determine that the test gas is being dispensed only from the piping system being tested. [NFPA 99:5.1.12.2.4.4]

1324.5.2.5 Repeat Test. The cross-connection test referenced in Section 1324.5.2 shall be repeated for each installed medical gas and vacuum piping system. [NFPA 99:5.1.12.2.4.5]

1324.5.2.6 Identification of System. The proper labeling and identification of system outlets/inlets shall be confirmed during these tests. [NFPA 99:5.1.12.2.4.6]

1324.5.3 Initial Piping Purge Tests. The outlets in each medical gas piping system shall be purged to remove any particulate matter from the distribution piping. [NFPA 99:5.1.12.2.5]

1324.5.3.1 Procedure. Using appropriate adapters, each outlet shall be purged with an intermittent high-volume flow of test gas until the purge produces no discoloration in a clean white cloth. [NFPA 99:5.1.12.2.5.1]

1324.5.3.2 Location. The purging required in Section 1324.5.3.1 shall be started at the closest outlet/inlet to the zone valve and continue to the furthest outlet/inlet within the zone. [NFPA 99:5.1.12.2.5.2]

1324.5.4 Standing Pressure Tests – for Positive Pressure Medical Gas Piping Systems. After successful completion of the initial pressure tests under Section 1324.5 through Section 1324.5.1.2, medical gas distribution piping shall be subjected to a standing pressure test. [NFPA 99:5.1.12.2.6]

1324.5.4.1 Time Frame for Testing. Tests shall be conducted after the final installation of station outlet valve bodies, faceplates, and other distribution system components (e.g. pressure alarm devices, pressure indicators, line pressure relief valves, manufactured assemblies, hose). [NFPA 99:5.1.12.2.6.1]

1324.5.4.2 Source Valve. The source valve shall be closed during this test. [NFPA 99:5.1.12.2.6.2]

1324.5.4.3 Length of Testing. The piping systems shall be subjected to a 24 hour standing pressure test using oil-free, dry nitrogen NF. [NFPA 99:5.1.12.2.6.3]

1324.5.4.4 Test Pressure. Test pressures shall be 20 percent above the normal system operating line pressure. [NFPA 99:5.1.12.2.6.4]

1324.5.4.5 Conclusion of Test. The leakage over the 24 hour test shall not exceed 0.5 percent of the starting pressure [e.g., 0.3 psi (2 kPa) starting at 60 psig (414 kPa), 0.125 inch (3.2 mm) HgV starting at 25 inches (635 mm) HgV] except that attributed to specific changes of ambient temperature. [NFPA 99:5.1.12.2.6.5]

1324.5.4.6 Leaks. Leaks, if any, shall be located, repaired (if permitted) or replaced (if required), and retested. [NFPA 99:5.1.12.2.6.6]

1324.5.4.7 Proof of Testing. The 24 hour standing pressure test of the positive pressure system shall be witnessed by an ASSE 6020 inspector, an ASSE 6030 verifier, or the Authority Having Jurisdiction or its designee. A form indicating that this test has been performed and witnessed shall be provided to the verifier at the start of the tests required in Section 1324.5.7 through Section 1324.5.11. [NFPA 99:5.1.12.2.6.7]

1324.5.5 Standing Pressure Tests – Medical Vacuum Piping Systems. After successful completion of the initial pressure tests under Section 1324.5 through Section 1324.5.1.2, vacuum distribution piping shall be subjected to a standing vacuum test. [NFPA 99:5.1.12.2.7]

1324.5.5.1 Timeframe for Testing. Tests shall be conducted after installation of all components of the vacuum system. [NFPA 99:5.1.12.2.7.1]

1324.5.5.2 Length of Testing. The piping systems shall be subjected to a 24 hour standing vacuum test. [NFPA 99:5.1.12.2.7.2]

1324.5.5.3 Test Pressure. Test pressure shall be between 12 inches (305 mm) HgV and full vacuum. [NFPA 99:5.1.12.2.7.3]

1324.5.5.4 Disconnection of Testing Source. During the test, the source of test vacuum shall be disconnected from the piping system. [NFPA 99:5.1.12.2.7.4]

1324.5.5.5 Conclusion of Test. At the conclusion of the test, there shall be no change in the vacuum other than that attributed to changes of ambient temperature. [NFPA 99:5.1.12.2.7.5]

1324.5.5.6 Proof of Testing. The 24 hour standing pressure test of the vacuum system shall be witnessed by the Authority Having Jurisdiction or its designee. A form indicating that this test has been performed and witnessed shall be provided to the verifier at the start of the tests required in Section 1324.5.7 through Section 1324.5.11. [NFPA 99:5.1.12.2.7.6]

1324.5.5.7 Leaks. Leaks, if any, shall be located, repaired (if permitted) or replaced (if required), and retested. [NFPA 99:5.1.12.2.7.7]

1324.5.6 System Inspection. System inspections shall be performed prior to concealing piping distribution systems in walls, ceilings, chases, trenches, underground, or otherwise hidden from view. [NFPA 99:5.1.12.3.1.1]

1324.5.6.1 Test Gas. The test gas shall be nitrogen NF. [NFPA 99:5.1.12.3.1.2]

1324.5.6.2 Inspection Qualification. Inspections shall be conducted by a party technically competent and experienced in the field of medical gas and vacuum pipeline inspections and testing and meeting the requirements of ASSE 6020, or ASSE 6030. [NFPA 99:5.1.12.3.1.3]

1324.5.6.3 Inspection Personnel. Inspections shall be performed by a party other than the installing contractor. [NFPA 99:5.1.12.3.1.4]

1324.5.6.4 Qualifications. Where systems have not been installed by inhouse personnel, inspections shall be permitted by personnel of the organization who meet the requirements of Section 1324.5.6.2. [NFPA 99:5.1.12.3.1.5]

1324.5.6.5 Inspections. The initial pressure tests performed by the installing contractor shall be witnessed by an ASSE 6020 inspector, an ASSE 6030 verifier, or the Authority Having Jurisdiction or its designee. A form indicating that this test has been performed and witnessed shall be provided to the verifier at the start of the tests required in Section 1324.5.7 through Section 1324.5.11. The presence and correctness of labeling and valve tagging required by this code for all concealed components and piping distribution systems shall be inspected. [NFPA 99:5.1.12.3.2 – 5.1.12.3.2.2]

1324.5.7 System Verification. Verification tests shall be performed only after all tests required in Section 1324.3 through Section 1324.5.5.7, Installer Performed Tests, have been completed. [NFPA 99:5.1.12.4.1.1]

1324.5.7.1 Test Gas. The test gas shall be oil-free, dry nitrogen NF or the system gas where permitted. [NFPA 5.1.12.4.1.2]

1324.5.7.2 Approved Tester. Testing shall be conducted by a party technically competent and experienced in the field of medical gas and vacuum pipeline testing and meeting the requirements of ASSE 6030, except as required by Section 1324.5.7.3. [NFPA 99:5.1.12.4.1.3]

Testing shall be performed by a party other than the installing contractor. [NFPA 99:5.1.12.4.1.5]

Where systems have not been installed by in-house personnel, testing shall be permitted by personnel of that organization who meet the requirements of Section 1324.5.7.2. [NFPA 99:5.1.12.4.1.6]

1324.5.7.3 Cryogenic Fluid Testing. Testing of the cryogenic fluid central supply system shall be conducted by a party technically competent and experienced in the field of cryogenic fluid systems and meeting the requirements of ASSE 6035, in accordance with the mandatory requirements in CGA M-1. [NFPA 99:5.1.12.4.1.4]

1324.5.8 Particulate Matter. In order to remove any traces of particulate matter deposited in the pipelines as a result of construction, a heavy, intermittent purging of the pipeline shall be done. [NFPA 99:5.1.12.4.6]

1324.5.9 Final Tie-In Test. Each joint in the final connection between the new work and the existing system shall be leak-tested with the gas of system designation at the normal operating pressure by means of a leak detectant that is safe for use with oxygen and does not contain ammonia. [NFPA 99:5.1.12.4.9.2]

1324.5.9.1 Vacuum Joints. Vacuum joints shall be tested using an ultrasonic leak detector or other means that will allow detection of leaks in an active vacuum system. [NFPA 99:5.1.12.4.9.3]

1324.5.9.2 Pressure Gases. For pressure gases, immediately after the final brazed connection is made and leak-tested, an outlet in the new piping and an outlet in the existing piping that are immediately downstream from the point or area of intrusion shall be purged in accordance with the applicable requirements of Section 1324.5.8. [NFPA 99:5.1.12.4.9.4]

1324.5.9.3 Positive Pressure Gases. Before the new work is used for patient care, positive pressure gases shall be tested for operational pressure and gas concentration in accordance with Section 1324.5.10 and Section 1324.5.11. [NFPA 99:5.1.12.4.9.5]

1324.5.9.4 Permanent Records. Permanent records of these tests shall be maintained in accordance with NFPA 99. [NFPA 99:5.1.12.4.9.6]

1324.5.10 Operational Flow Pressure Drop Test. Operational flow pressure drop tests shall be performed at each station outlet/inlet or terminal where the user makes connections and disconnections. [NFPA 99: 5.1.12.4.10]

1324.5.10.1 Medical-Surgical Vacuum Inlets. Medical-surgical vacuum inlets shall draw 3 SCFM (85 Nl/min) without reducing the vacuum pressure below 12 inch (305 mm) gauge HgV at any adjacent station inlet. [NFPA 99:5.1.12.4.10.4]

1324.5.10.2 Oxygen and Medical Air Outlets. Oxygen and medical air outlets serving Category 1 space shall allow a transient flow rate of 6 SCFM (170 SLPM) for 3 seconds. [NFPA 99:5.1.12.4.10.5]

1324.5.11 Medical Gas Concentration Test. After purging each system with the gas of system designation, the following shall be performed:

(1) Each pressure gas source and outlet shall be analyzed for concentration of gas, by volume.

(2) Analysis shall be conducted with instruments designed to measure the specific gas dispensed.

(3) Allowable concentrations shall be as indicated in Table 1324.5.11. [NFPA 99:5.1.12.4.11]

TABLE 1324.5.11
GAS CONCENTRATIONS
[NFPA 99:5.1.12.4.11]

MEDICAL GAS	CONCENTRATION
Oxygen USP	≥99% oxygen
Oxygen 93 USP	≥90% oxygen ≤96%
Nitrous oxide USP	≥99% nitrous oxide
Nitrogen NF	≤1% oxygen or ≥99% nitrogen
Medical air USP	19.5% - 23.5% oxygen
Other gases	Named gases by ±1%, or per specification

Part III – Category 2 Piped Gas and Vacuum Systems.

1325.0 Category 2 Piped Gas and Vacuum Systems.

1325.1 General. Category 2 piped gas or piped vacuum system requirements shall be permitted when all of the following criteria are met:

(1) Only moderate sedation; minimal sedation, as defined in Chapter 2; or no sedation is performed. Deep sedation and general anesthesia shall not be permitted.

(2) The loss of the piped gas or piped vacuum systems is likely to cause minor injury to patients, staff, or visitors.

(3) The facility piped gas or piped vacuum systems are intended for Category 2 patient care space as defined in Chapter 2. [NFPA 99:5.2.1.2]

1325.2 Nature of Hazards of Gas and Vacuum Systems. The requirement of Section 1307.2 shall apply to the nature of hazards of gas and vacuum systems. [NFPA 99:5.2.2]

1325.3 Central Supply Systems. Category 2 systems shall comply with Section 1307.3 through Section 1309.13. [NFPA 99:5.2.3.4]

1325.4 Category 2 Medical Air Supply Systems. Category 2 systems shall comply with Section 1310.0 through Section 1311.6, except as follows:

(1) Medical air compressors, dryers, aftercoolers, filters, and regulators shall be permitted to be simplex.

(2) The facility staff shall develop their emergency plan to deal with the loss of medical air. [NFPA 99:5.2.3.5]

1325.5 Oxygen Concentrators. Oxygen supply systems using concentrators shall be permitted to consist of two sources, one of which shall be a cylinder header with sufficient cylinder connections for one average day's supply. [NFPA 99:5.2.3.6]

1325.6 Category 2 Medical-Surgical Vacuum. Category 2 systems shall comply with Section 1312.0 through Section 1313.5, except as follows:

(1) Medical-surgical vacuum systems shall be permitted to be simplex.

(2) The facility staff shall develop their emergency plan to deal with the loss of medical-surgical vacuum. [NFPA 99:5.2.3.7]

1325.7 Valves. Category 2 systems shall comply with Section 1314.0 through Section 1314.12.1. [NFPA 99:5.2.4]

1325.8 Station Outlets and Inlets. Category 2 systems shall comply with Section 1315.0. [NFPA 99:5.2.5]

1325.9 Pressure and Vacuum Indicators. Category 2 systems shall comply with Section 1316.2. [NFPA 99:5.2.8]

1325.10 Warning Systems (Category 2). Warning systems associated with Category 2 systems shall provide the master, area, and local alarm functions of a Category 1 system as required in Section 1317.0, except as follows:

(1) Warning systems shall be permitted to be a single alarm panel.

(2) The alarm panel shall be located in an area of continuous surveillance while the facility is in operation.

(3) Pressure and vacuum switches/sensors shall be mounted at the source equipment with a pressure indicator at the master alarm panel. [NFPA 99:5.2.9]

1325.11 Category 2 Distribution. Level 2 systems shall comply with Section 1318.0 through Section 1323.12. [NFPA 99:5.2.10]

1325.12 Labeling and Identification. Category 2 systems shall comply with Section 1323.13 through Section 1323.15. [NFPA 99:5.2.11]

1325.13 Performance Criteria and Testing — Category 2 (Gas, Medical–Surgical and Vacuum). Category 2 systems shall comply with Section 1324.0. [NFPA 99:5.2.12]

Part IV – Category 3 Piped Gas and Vacuum Systems.

1326.0 Category 3 Piped Gas and Vacuum Systems.

1326.1 General. Category 3 piped gas and vacuum systems shall be permitted when all of the following criteria are met:

(1) Only moderate sedation; minimal sedation, as defined in Chapter 2; or no sedation is performed. Deep sedation and general anesthesia are not performed.

(2) The loss of the piped gas and vacuum systems is not likely to cause injury to patients, staff, or visitors, but can cause discomfort.

(3) The facility piped gas and vacuum systems are intended for Category 3 or Category 4 patient care rooms as defined in Chapter 2. [NFPA 99:5.3.1.2]

1326.2 Nature of Hazards of Gas and Vacuum Systems. The requirement of Section 1307.2 shall apply to the nature of hazards of gas and vacuum systems. [NFPA 99:5.3.2]

1326.3 Medical Air Supply Systems. Category 3 systems shall comply with Section 1310.0 through Section 1311.6, except as follows:

(1) Medical air compressors, dryers, after coolers, filters, and regulators shall be permitted to be simplex.

(2) The facility staff shall develop their emergency plan to deal with the loss of medical air. [NFPA 99:5.3.3.5]

1326.4 Oxygen Central Supply Systems Using Concentrators. Category 3 oxygen supply systems using concentrators shall be permitted to consist of two sources, one of which shall be a cylinder header with sufficient cylinder connections for one average day's supply. [NFPA 99:5.3.3.6]

1326.5 Medical–Surgical Vacuum. Category 3 systems shall comply with Section 1312.0 through Section 1313.5, except as follows:

(1) Medical–surgical vacuum systems shall be permitted to be simplex.

(2) The facility staff shall develop their emergency plan to deal with the loss of medical–surgical vacuum. [NFPA 99:5.3.3.7]

1326.6 Valves. Category 3 systems shall comply with Section 1314.0. [NFPA 99:5.3.4]

1326.7 Station Outlets and Inlets. Category 3 systems shall comply with Section 1315.0. [NFPA 99:5.3.5]

1326.8 Pressure and Vacuum Indicators. Category 3 systems shall comply with Section 1316.2. [NFPA 99:5.3.8]

1326.9 Warning Systems. Warning systems associated with Category 3 systems shall provide the master, area, and local alarm functions of a Category 1 system as required in Section 1317.0, except as follows:

(1) Warning systems shall be permitted to be a single alarm panel.

(2) The alarm panel shall be located in an area of continuous surveillance while the facility is in operation.

(3) Pressure and vacuum switches/sensors shall be mounted at the source equipment with a pressure indicator at the master alarm panel. [NFPA 99:5.3.9]

1326.10 Distribution. Category 3 systems shall comply with Section 1318.0 through Section 1323.12. [NFPA 99:5.3.10]

1326.11 Labeling and Identification. Category 3 systems shall comply with Section 1323.13 through Section 1323.15. [NFPA 99:5.3.11]

Part V – Dental Gas and Vacuum Systems.

1327.0 Dental Gas and Vacuum Systems.

1327.1 General. Dental gas and vacuum systems shall comply with this code and NFPA 99.

1327.2 Emergency Shutoff Valves (Oxygen and Nitrous Oxide). Emergency shutoff valves shall be provided in accordance with the following:

(1) Where a central medical gas supply is remote from a single treatment facility, the main supply line shall be provided with an emergency shutoff valve located in the single treatment facility so as to be accessible from all use-point locations in an emergency.

(2) Where a central medical gas supply system supplies two treatment facilities, each facility shall be provided with an emergency shutoff valve located in that treatment facility so as to be accessible from all use-point locations in an emergency.

(3) Emergency shutoff valves shall be labeled to indicate the gas controlled by the shutoff valve and shall shut off only the gas to the treatment facility that they serve.

(4) A remotely activated shutoff valve at a gas supply manifold shall not be used for emergency shutoff. For clinical purposes, such a remote valve actuator shall not fail-close in the event of loss of electric power. Where remote actuators are the type that fail-open, it shall be mandatory that cylinder shutoff valves be closed whenever the system is not in use. [NFPA 99:15.4.2.6.1 – 15.4.2.6.4.2]

1327.3 Warning Systems (Oxygen and Nitrous Oxide). Category 2 warning systems shall comply with Section 1325.10 except as follows:

(1) Warning systems shall be permitted to be a single alarm panel.

(2) The alarm panel shall be located in an area of continuous surveillance while the facility is in operation.

(3) Pressure and vacuum switches/sensors shall be mounted at the source equipment with a pressure indicator at the master alarm panel.

(4) Warning systems for medical gas systems shall provide the following alarms:

(a) Oxygen main line pressure low.

(b) Oxygen main line pressure high.

(c) Oxygen changeover to secondary bank or about to changeover (if automatic).

(d) Nitrous oxide main line pressure low.

(e) Nitrous oxide main line pressure high.

(f) Nitrous oxide changeover to secondary bank or about to changeover (if automatic).

(5) Audible and noncancelable alarm visual signals shall indicate if the pressure in the main line increases or decreases 20 percent from the normal operating pressure.

(6) Visual indications shall remain until the situation that caused the alarm is resolved.

(7) Pressure switches/sensors shall be installed downstream of any emergency shutoff valves and any other shutoff valves in the system and shall cause an alarm for the medical gas if the pressure decreases or increases 20 percent from the normal operating pressure.

(8) A cancelable audible indication of each alarm condition that produces a sound at the alarm panel shall reinitiate the audible signal if another alarm condition occurs while the audible signal is silenced. [NFPA 99:15.4.2.10]

1327.4 Initial Pressure Test. Each section of the piping in positive-pressure gas systems and copper vacuum systems shall be pressure tested. Plastic vacuum and plastic scavenging piping shall not be pressure tested. [NFPA 99:15.4.7.4.4.1]

1327.4.1 Pressure Test. Initial pressure tests shall be conducted as follows:

(1) After blowdown of the distribution piping

(2) After installation of station outlet/inlet rough-in assemblies

(3) Prior to the installation of components of the distribution piping system that would be damaged by the test pressure (e.g., pressure/vacuum alarm devices, pressure/vacuum indicators, and line pressure relief valves) [NFPA 99:15.4.7.4.4.2]

1327.4.2 Source Shutoff Valve. The source shutoff valve shall remain closed during the pressure tests. [NFPA 99:15.4.7.4.4.3]

1327.4.3 Test Pressure. The test pressure for oxygen and nitrous oxide piping shall be 1.5 times the system operating pressure but not less than a gauge pressure of 150 psi (1035 kPa). [NFPA 99:15.4.7.4.4.4]

1327.4.4 Examine for Leaks. The test pressure shall be maintained until each joint has been examined for leakage by means of a leak detectant that is safe for use with oxygen and does not contain ammonia. [NFPA 99:15.4.7.4.4.5]

1327.4.5 Leaks Located. Any leaks shall be located, repaired (if permitted), or replaced (if required) by the installer, and retested. [NFPA 99:15.4.7.4.4.6]

1327.5 Maximum Copper Tube Support Spacing. The maximum support spacing for copper tube shall be in accordance with Table 1327.5. [NFPA 99:15.4.5.6.5]

TABLE 1327.5
MAXIMUM COPPER TUBE SUPPORT SPACING
[NFPA 99: TABLE 15.4.5.6.5]

PIPE SIZE			HANGER SPACING (feet)
DN8	(NPS 1/4)	(3/8 of an inch O.D.)	5
DN10	(NPS 3/8)	(1/2 of an inch O.D.)	6
DN15	(NPS 1/2)	(5/8 of an inch O.D.)	6
DN20	(NPS 3/4)	(7/8 of an inch O.D.)	7
DN25	(NPS 1)	(1 1/8 of an inch O.D.)	8
DN32	(NPS 1 1/4)	(1 3/8 of an inch O.D.)	9
DN40 and larger	(NPS 1 1/2)	(1 5/8 of an inch O.D.)	10
Vertical risers, all sizes, every floor, but not to exceed:			15

For SI units: 1 inch = 25 mm, 1 foot = 304.8 mm

1327.6 Maximum Plastic Pipe Support Spacing. The maximum support spacing for plastic pipe shall be in accordance with Table 1327.6. [NFPA 99:15.4.5.6.6]

TABLE 1327.6
MAXIMUM PLASTIC PIPE SUPPORT SPACING
[NFPA 99: TABLE 15.4.5.6.6]

PIPE SIZE			HANGER SPACING (feet)
DN15	(NPS 1/2)	(5/8 of an inch O.D.)	4
DN20	(NPS 3/4)	(7/8 of an inch O.D.)	4
DN25	(NPS 1)	(1 1/8 of an inch O.D.)	4.33
DN32	(NPS 1 1/4)	(1 3/8 of an inch O.D.)	4.33
DN40	(NPS 1 1/8)	(1 5/8 of an inch O.D.)	4.66
DN50	(NPS 2)	(2 3/8 of an inch O.D.)	4.66
DN65 and larger	(NPS 2 1/2)	(2 7/8 of an inch O.D.)	5
Vertical risers, all sizes, every floor, but not to exceed:			10

For SI units: 1 inch = 25 mm, 1 foot = 304.8 mm

1327.7 Standing Pressure Tests for Oxygen and Nitrous Oxide Piping. After successful completion of the initial pressure tests in Section 1327.4, the gas distribution piping shall be subject to a standing pressure test. [NFPA 99:15.4.7.4.6.1]

1327.7.1 Tests Required. Tests shall be conducted after the final installation of station outlet valve bodies, faceplates, and other distribution system components (e.g., pressure alarm devices, pressure indicators, line pressure relief valves, manufactured assemblies, and hoses). [NFPA 99:15.4.7.4.6.2]

1327.7.2 Source Valve. The source valve shall be closed during this test. [NFPA 99:15.4.7.4.6.3]

1327.7.3 Piping Systems. The piping systems shall be subjected to 24-hour standing pressure tests using oil-free, dry nitrogen NF. [NFPA 99:15.4.7.4.6.4]

1327.7.4 Test Pressure. Test pressures shall be 20 percent above the normal system operating line pressure. [NFPA 99:15.4.7.4.6.5]

1327.7.5 Change in Test Pressure. At the conclusion of the tests, there shall be no change in the test pressure except that attributed to specific changes in ambient temperature. [NFPA 99:15.4.7.4.6.6]

1327.7.6 Leaks. Any leaks shall be located, repaired (if permitted), or replaced (if required) by the installer, and retested. [NFPA 99:15.4.7.4.6.7]

1327.8 Verifier Operational Pressure Test. Operational pressure tests shall be performed at each station outlet or terminal where the user makes connections and disconnections. [NFPA 99:15.4.7.5.8.1]

1327.8.1 Test Gas. Tests shall be performed with the gas of system designation. [NFPA 99:15.4.7.5.8.2]

1327.8.2 Medical Gas Outlets. All medical gas outlets with a gauge pressure of 50 psi (345 kPa), including oxygen and nitrous oxide, shall deliver 1.8 standard cubic feet per minute (SCFM) (50 SLPM) with a pressure drop of not more than 5 psi (34 kPa) and static pressure of 50 psi (345 kPa) to 55 psi (379 kPa). [NFPA 99:15.4.7.5.8.3]

CHAPTER 14
FIRESTOP PROTECTION

1401.0 General.

1401.1 Applicability. Piping penetrations of required fire-resistance-rated walls, partitions, floors, floor/ceiling assemblies, roof/ceiling assemblies, or shaft enclosures shall be protected in accordance with the requirements of the building code, and this chapter.

1402.0 Construction Documents.

1402.1 Penetrations. Construction documents shall indicate with sufficient detail how penetrations of fire-resistance-rated assemblies shall be firestopped prior to obtaining design approval.

1403.0 Installation.

1403.1 Materials. Firestop systems shall be installed in accordance with this chapter, the building code, and the manufacturer's installation instructions.

1404.0 Combustible Piping Installations.

1404.1 General Requirements. Combustible piping installations shall be protected in accordance with the appropriate fire resistance rating requirements in the building code that list the acceptable area, height, and type of construction for use in specific occupancies to assure compliance and integrity of the fire resistance rating prescribed.

1404.2 Fire-Resistance Rating. Where penetrating a fire-resistance-rated wall, partition, floor, floor-ceiling assembly, roof-ceiling assembly, or shaft enclosure, the fire-resistance rating of the assembly shall be restored to its original rating.

1404.3 Firestop Systems. Penetrations shall be protected by an approved penetration firestop system installed as tested in accordance with ASTM E119, ASTM E814, UL 263, or UL 1479 with a positive pressure differential of not less than 0.01 of an inch of water (0.002 kPa). Systems shall have an F rating of not less than 1 hour but not less than the required fire-resistance rating of the assembly being penetrated. Systems protecting floor penetrations shall have a T rating of not less than 1 hour but not less than the required fire-resistance rating of the floor being penetrated. Floor penetrations contained within the cavity of a wall at the location of the floor penetration do not require a T rating. No T rating shall be required for floor penetrations by piping that is not in direct contact with combustible material.

1404.4 Connections. Where piping penetrates a rated assembly, the combustible piping shall not connect to noncombustible piping unless it is capable of being demonstrated that the transition is in accordance with Section 1404.3.

1404.5 Insulation and Coverings. Insulation and coverings on or in the penetrating item shall not be permitted unless the specific insulating or covering material has been tested as part of the penetrating firestop system.

1404.6 Sleeves. Where sleeves are used, the sleeves shall be securely fastened to the fire-resistance-rated assembly. The (inside) annular space between the sleeve and the penetrating item and the (outside) annular space between the sleeve and the fire-resistance-rated assembly shall be firestopped in accordance with this chapter.

1405.0 Noncombustible Piping Installations.

1405.1 General Requirements. Noncombustible piping installations shall be protected in accordance with the appropriate fire resistance rating requirements in the building code that list the acceptable area, height, and type of construction for use in specific occupancies to ensure compliance and integrity of the fire-resistance rating prescribed.

1405.2 Fire-Resistance Rating. Where penetrating a fire-resistance-rated wall, partition, floor, floor-ceiling assembly, roof-ceiling assembly, or shaft enclosure, the fire-resistance rating of the assembly shall be restored to its original rating.

Exceptions:

(1) Concrete, mortar, or grout shall be permitted to be used to fill the annular spaces around cast-iron, copper, copper alloy, or steel piping that penetrates concrete or masonry fire-resistant-rated assemblies. The nominal diameter of the penetrating item shall not exceed 6 inches (150 mm), and the opening size shall not exceed 144 square inches (0.093 m^2).

The thickness of concrete, mortar, or grout shall be the full thickness of the assembly or the thickness necessary to provide a fire-resistance rating not less than the required fire-resistance rating of the assembly penetrated.

(2) The material used to fill the annular space shall prevent the passage of flame and hot gases capable of igniting cotton waste for the time period equivalent to the fire-resistance rating of the assembly, where tested to standard(s) referenced in Section 1405.3.

1405.3 Firestop Systems. Penetrations shall be protected by an approved penetration firestop system installed as tested in accordance with ASTM E119, ASTM E814, UL 263, or UL 1479 with a positive pressure differential of not less than 0.01 of an inch of water (0.002 kPa). Systems shall have an F rating of not less than 1 hour but not less than the required fire-resistance rating of the assembly being penetrated. Systems protecting floor penetrations shall have a T rating of not less than 1 hour but not less than the required fire-resistance rating of the floor being penetrated. Floor penetrations contained within the cavity of a wall at the location of the floor penetration do not require a T rating. No T rating shall be required for floor penetrations by piping that is not in direct contact with combustible material.

1405.4 Connections. Where piping penetrates a rated assembly, the combustible piping shall not connect to noncombustible piping unless it is capable of being demonstrated that the transition is in accordance with the requirements of Section 1405.3.

1405.5 Unshielded Couplings. Unshielded couplings shall not be used to connect noncombustible piping unless it is capable of being demonstrated that the fire-resistive integrity of the penetration is maintained.

1405.6 Sleeves. Where sleeves are used, the sleeves shall be securely fastened to the fire-resistance-rated assembly. The (inside) annular space between the sleeve and the penetrating item and the (outside) annular space between the sleeve and the fire-resistance-rated assembly shall be firestopped in accordance with this chapter.

1405.7 Insulation and Coverings. Insulation and coverings on or in the penetrating item shall not be permitted unless the specific insulating or covering material has been tested as part of the penetrating firestop system.

1406.0 Required Inspection.

1406.1 General. Prior to being concealed, piping penetrations shall be inspected by the Authority Having Jurisdiction to verify compliance with the fire-resistance rating prescribed in the building code.

1406.2 Examination. The Authority Having Jurisdiction shall conduct a thorough examination of sufficient representative installations, including destructive inspection, to provide verification of satisfactory compliance with this chapter, the appropriate manufacturer's installation instructions applied by the installer, construction documents, specifications, and applicable manufacturer's product information.

1406.3 Penetrations. The Authority Having Jurisdiction shall determine the type, size, and quantity of penetrations to be inspected.

1406.4 Field Installations. The Authority Having Jurisdiction shall compare the field installations with the documentation supplied by the installer to determine the following:

(1) The required F ratings (1 hour, 2 hour, 3 hour, or 4 hour) and T ratings (0 hour, 1 hour, 2 hour, 3 hour, or 4 hour) of the penetration firestop systems are at least the same as the hourly rating of the assembly being penetrated.

(2) The penetrating firestop system includes the penetrating item as documented through testing of the systems conducted by an independent testing agency.

(3) The penetrating firestop system is installed as tested.

CHAPTER 15

ALTERNATE WATER SOURCES FOR NONPOTABLE APPLICATIONS

1501.0 General.

1501.1 Applicability. The provisions of this chapter shall apply to the construction, alteration, and repair of alternate water source systems for nonpotable applications.

1501.1.1 Allowable Use of Alternate Water. Where approved or required by the Authority Having Jurisdiction, alternate water sources [reclaimed (recycled) water, gray water, and on-site treated nonpotable water] shall be permitted to be used instead of potable water for the applications identified in this chapter.

1501.2 System Design. Alternate water source systems shall be designed in accordance with this chapter by a licensed plumbing contractor or a registered design professional. Components, piping, and fittings used in any alternate water source system shall be listed.

Exceptions:

(1) A registered design professional is not required to design gray water systems having a maximum discharge capacity of 250 gallons per day (gal/d) (0.011 L/s) for single family and multi-family dwellings.

(2) A registered design professional is not required to design an on-site treated nonpotable water system for single-family dwellings having a maximum discharge capacity of 250 gal/d (0.011 L/s).

1501.3 Permit. It shall be unlawful for a person to construct, install, alter, or cause to be constructed, installed, or altered an alternate water source system in a building or on a premise without first obtaining a permit to do such work from the Authority Having Jurisdiction.

1501.4 Component Identification. System components shall be properly identified as to the manufacturer.

1501.5 Maintenance and Inspection. Alternate water source systems and components shall be inspected and maintained in accordance with Section 1501.5.1 through Section 1501.5.3.

1501.5.1 Frequency. Alternate water source systems and components shall be inspected and maintained in accordance with Table 1501.5 unless more frequent inspection and maintenance are required by the manufacturer.

1501.5.2 Maintenance Log. A maintenance log for gray water and on-site treated nonpotable water systems is required to have a permit in accordance with Section 1501.3 and shall be maintained by the property owner and be available for inspection. The property owner or designated appointee shall ensure that a record of testing, inspection, and maintenance in accordance with Table 1501.5 is maintained in the log. The log will indicate the frequency of inspection and maintenance for each system.

1501.5.3 Maintenance Responsibility. The required maintenance and inspection of alternate water source systems shall be the responsibility of the property owner unless otherwise required by the Authority Having Jurisdiction.

TABLE 1501.5
MINIMUM ALTERNATE WATER SOURCE TESTING, INSPECTION, AND MAINTENANCE FREQUENCY

DESCRIPTION	MINIMUM FREQUENCY
Inspect and clean filters and screens, and replace (where necessary).	Every 3 months
Inspect and verify that disinfection, filters, and water quality treatment devices and systems are operational and maintaining minimum water quality requirements as determined by the Authority Having Jurisdiction.	In accordance with manufacturer's instructions, and the Authority Having Jurisdiction.
Inspect pumps and verify operation.	After initial installation and every 12 months thereafter
Inspect valves and verify operation.	After initial installation and every 12 months thereafter
Inspect pressure tanks and verify operation.	After initial installation and every 12 months thereafter
Clear debris from and inspect storage tanks, locking devices, and verify operation.	After initial installation and every 12 months thereafter
Inspect caution labels and marking.	After initial installation and every 12 months thereafter
Inspect and maintain mulch basins for gray water irrigation systems.	As needed to maintain mulch depth and prevent ponding and runoff.
Cross-connection inspection and test*	After initial installation and every 12 months thereafter

* The cross-connection test shall be performed in the presence of the Authority Having Jurisdiction in accordance with the requirements of this chapter.

1501.6 Operation and Maintenance Manual. An operation and maintenance manual for gray water and on-site treated water systems required to have a permit in accordance with Section 1501.3 shall be supplied to the building owner by the system designer. The operation and maintenance manual shall include the following:

(1) Detailed diagram of the entire system and the location of system components.

(2) Instructions for operating and maintaining the system.

(3) Details on maintaining the required water quality for on-site nonpotable water systems.

(4) Details on deactivating the system for maintenance, repair, or other purposes.

(5) Applicable testing, inspection, and maintenance frequencies in accordance with Table 1501.5.

(6) A method of contacting the manufacturer(s).

1501.7 Minimum Water Quality Requirements. The minimum water quality for alternate water source systems shall meet the applicable water quality requirements for the intended application as determined by the Authority Having Jurisdiction. In the absence of water quality requirements, for on-site treated nonpotable systems, the water quality requirements of NSF 350 shall apply. The EPA/625/R-04/108 contains recommended water reuse guidelines to assist regulatory agencies to develop, revise, or expand alternate water source water quality standards.

Exception: Water treatment is not required for gray water used for subsurface irrigation.

1501.8 Material Compatibility. Alternate water source systems shall be constructed of materials that are compatible with the type of pipe and fitting materials, water treatment, and water conditions in the system.

1501.9 Commercial, Industrial, and Institutional Restroom Signs. A sign shall be installed in restrooms in commercial, industrial, and institutional occupancies using reclaimed (recycled) water and on-site treated water, for water closets, urinals, or both. Each sign shall contain ½ of an inch (12.7 mm) letters of a highly visible color on a contrasting background. The location of the sign(s) shall be such that the sign(s) are visible to users. The location of the sign(s) shall be approved by the Authority Having Jurisdiction and shall contain the following text:

TO CONSERVE WATER, THIS BUILDING USES *___________* TO FLUSH TOILETS AND URINALS.

1501.9.1 Equipment Room Signs. Each room containing reclaimed (recycled) water and on-site treated water equipment shall have a sign posted in a location that is visible to anyone working on or near nonpotable water equipment with the following wording in 1 inch (25.4 mm) letters:

CAUTION: NONPOTABLE *___________*, DO NOT DRINK. DO NOT CONNECT TO DRINKING WATER SYSTEM. NOTICE: CONTACT BUILDING MANAGEMENT BEFORE PERFORMING ANY WORK ON THIS WATER SYSTEM.

*___________*Shall indicate RECLAIMED (RECYCLED) WATER or ON-SITE TREATED WATER, accordingly.

1501.10 System Controls. Controls for pumps, valves, and other devices that contain mercury that come in contact with alternate water source water supply shall not be permitted.

1502.0 Inspection and Testing.

1502.1 General. Alternate water source systems shall be inspected and tested in accordance with Section 1502.2 through Section 1502.3.4.

1502.2 Supply System Inspection and Test. Alternate water source systems shall be inspected and tested in accordance with this code for testing of potable water piping.

1502.3 Annual Cross-Connection Inspection and Testing. An initial and subsequent annual inspection and test shall be performed on both the potable and alternate water source systems. The potable and alternate water source system shall be isolated from each other and independently inspected and tested to ensure there is no cross-connection in accordance with Section 1502.3.1 through Section 1502.3.4.

1502.3.1 Visual System Inspection. Before commencing the cross-connection testing, a dual system inspection shall be conducted by the Authority Having Jurisdiction and other authorities having jurisdiction as follows:

(1) Meter locations of the alternate water source and potable water lines shall be checked to verify that no modifications were made and that no cross-connections are visible.

(2) Pumps and equipment, equipment room signs and exposed piping in equipment room shall be checked.

(3) Valves shall be checked to ensure that the valve lock seals are still in place and intact. Valve control door signs shall be checked to verify that no signs have been removed.

1502.3.2 Cross-Connection Test. The procedure for determining cross-connection shall be followed by the applicant in the presence of the Authority Having Jurisdiction and other authorities having jurisdiction to determine whether a cross-connection has occurred as follows:

(1) The potable water system shall be activated and pressurized. The alternate water source system shall be shut down, depressurized, and drained.

(2) The potable water system shall remain pressurized for a minimum period specified by the Authority Having Jurisdiction while the alternate water source system is empty. The minimum period the alternate water source system is to remain depressurized shall be determined on a case-by-case basis, taking into account the size and complexity of the potable and the alternate water source distribution systems, but in no case shall that period be less than 1 hour.

(3) The drain on the alternate water source system shall be checked for flow during the test and fixtures, potable and alternate water source, shall be tested and inspected for flow. Flow from an alternate water source system outlet indicates a cross-connection. No flow from a potable water outlet shall indicate that it is connected to the alternate water source system.

(4) The potable water system shall then be depressurized and drained.

(5) The alternate water source system shall then be activated and pressurized.

(6) The alternate water source system shall remain pressurized for a minimum period specified by the Authority Having Jurisdiction while the potable water system is empty. The minimum period the potable water system is to remain depressurized shall be determined on a case-by-case basis, but in no case shall that period be less than 1 hour.

(7) Fixtures, potable, and alternate water source shall be tested and inspected for flow. Flow from a potable water system outlet indicates a cross-connection. No flow from an alternate water source outlet will indicate that it is connected to the potable water system.

(8) The drain on the potable water system shall be checked for flow during the test and at the end of the test.

(9) Where there is no flow detected in the fixtures which would indicate a cross-connection, the potable water system shall be repressurized.

1502.3.3 Discovery of Cross-Connection. If a cross-connection is discovered, the following procedure, in the presence of the Authority Having Jurisdiction, shall be activated immediately:

(1) The alternate water source piping to the building shall be shutdown at the meter, and the alternate water source riser shall be drained.

(2) Potable water piping to the building shall be shutdown at the meter.

(3) The cross-connection shall be uncovered and disconnected.

(4) The building shall be retested in accordance with Section 1502.3.1 and Section 1502.3.2.

(5) The potable water system shall be chlorinated with 50 parts-per-million (ppm) chlorine for 24 hours.

(6) The potable water system shall be flushed after 24 hours, and a standard bacteriological test shall be performed. Where test results are acceptable, the potable water system shall be permitted to be recharged.

1502.3.4 Annual Inspection. An annual inspection of the alternate water source system, following the procedures listed in Section 1502.3.1 shall be required. Annual cross-connection testing, following the procedures listed in Section 1502.3.2 shall be required by the Authority Having Jurisdiction, unless site conditions do not require it. In no event shall the test occur less than once in 4 years. Alternate testing requirements shall be permitted by the Authority Having Jurisdiction.

1502.4 Separation Requirements. Underground alternate water source service piping other than gray water shall be separated from the building sewer in accordance with this code. Treated nonpotable water pipes shall be permitted to be run or laid in the same trench as potable water pipes with a 12 inch (305 mm) minimum vertical and horizontal separation where both pipe materials are approved for use within a building. Where horizontal piping materials do not comply with this requirement, the minimum separation shall be increased to 60 inches (1524 mm). The potable water piping shall be installed at an elevation above the treated nonpotable water piping.

1502.5 Abandonment. Alternate water source systems that are no longer in use or fail to be maintained in accordance with Section 1501.5 shall be abandoned. Abandonment shall comply with Section 1502.5.1 and Section 1502.5.2.

1502.5.1 General. An abandoned system or part thereof covered under the scope of this chapter shall be disconnected from remaining systems, drained, plugged, and capped in an approved manner.

1502.5.2 Underground Tank. An underground water storage tank that has been abandoned or otherwise discontinued from use in a system covered under the scope of this chapter shall be completely drained and filled with earth, sand, gravel, concrete, or other approved material or removed in a manner satisfactory to the Authority Having Jurisdiction.

1502.6 Sizing. Unless otherwise provided for in this chapter, alternate water source piping shall be sized in accordance with Chapter 6 for sizing potable water piping.

1503.0 Gray Water Systems.

1503.1 General. The provisions of this section shall apply to the construction, alteration, and repair of gray water systems.

1503.2 System Requirements. Gray water shall be permitted to be diverted away from a sewer or private sewage disposal system, and discharge to a subsurface irrigation or subsoil irrigation system. The gray water shall be permitted to discharge to a mulch basin for single-family and multi-family dwellings. Gray water shall not be used to irrigate root crops or food crops intended for human consumption that comes in contact with soil.

1503.2.1 Surge Capacity. Gray water systems shall be designed to have the capacity to accommodate peak flow rates and distribute the total amount of estimated gray water on a daily basis to a subsurface irrigation field, subsoil irrigation field, or mulch basin without surfacing, ponding, or runoff. A surge tank is required for systems that are unable to accommodate peak flow rates and distribute the total amount of gray water by gravity drainage. The water discharge for gray water systems shall be determined in accordance with Section 1503.8.1 or Section 1503.8.2.

1503.2.2 Diversion. The gray water system shall connect to the sanitary drainage system downstream of fixture traps and vent connections through a gray water diverter valve. The gray water diverter valve shall comply with IAPMO PS 59 and be installed in an accessible location and clearly indicate the direction of flow.

1503.2.3 Backwater Valves. Gray water drains subject to backflow shall be provided with a backwater valve so located as to be accessible for inspection and maintenance.

1503.2.4 Rainwater Diversion Valves. Rainwater diversion valves ranging from 6 inches (150 mm) to 12 inches (300 mm) in diameter shall comply with IAPMO IGC 352. Valves shall be accessible and include a filter located upstream of the valve when required.

1503.3 Connections to Potable and Reclaimed (Recycled) Water Systems. Gray water systems shall have no direct connection to a potable water supply, on-site treated nonpotable water supply, or reclaimed (recycled) water systems. Potable, on-site treated nonpotable, or reclaimed (recycled) water is permitted to be used as makeup water for a nonpressurized storage tank provided the connection is protected by an air gap in accordance with this code.

1503.4 Location. No gray water system or part thereof shall be located on a lot other than the lot that is the site of the building or structure that discharges the gray water, nor shall a gray water system or part thereof be located at a point having less than the minimum distances indicated in Table 1503.4.

TABLE 1503.4
LOCATION OF GRAY WATER SYSTEM[7]

MINIMUM HORIZONTAL DISTANCE IN CLEAR REQUIRED FROM	SURGE TANK (feet)	SUBSURFACE AND SUBSOIL IRRIGATION FIELD AND MULCH BED (feet)
Building structures[1]	5[2, 9]	2[3, 8]
Property line adjoining private property	5	5[8]
Water supply wells[4]	50	100
Streams and lakes[4]	50	50[5]
Sewage pits or cesspools	5	5
Sewage disposal field[10]	5	4[6]
Septic tank	0	5
On-site domestic water service line	5	5
Pressurized public water main	10	10[7]

For SI units: 1 foot = 304.8 mm

Notes:

[1] Including porches and steps, whether covered or uncovered, breezeways, roofed carports, roofed patios, carports, covered walks, covered driveways, and similar structures or appurtenances.

[2] The distance shall be permitted to be reduced to 0 feet for aboveground tanks where first approved by the Authority Having Jurisdiction.

[3] Reference to a 45 degree (0.79 rad) angle from the foundation.

[4] Where special hazards are involved, the distance required shall be increased as directed by the Authority Having Jurisdiction.

[5] These minimum clear horizontal distances shall apply between the irrigation or disposal field and the ocean mean higher high tide line.

[6] Add 2 feet (610 mm) for each additional foot of depth more than 1 foot (305 mm) below the bottom of the drain line.

[7] For parallel construction or crossings, approval by the Authority Having Jurisdiction shall be required.

[8] The distance shall be permitted to be reduced to 1½ feet (457 mm) for drip and mulch basin irrigation systems.

[9] The distance shall be permitted to be reduced to 0 feet for surge tanks of 75 gallons (284 L) or less.

[10] Where irrigation or disposal fields are installed in the sloping ground, the minimum horizontal distance between a part of the distribution system and the ground surface shall be 15 feet (4572 mm).

1503.5 Plot Plan Submission. No permit for a gray water system shall be issued until a plot plan with data satisfactory to the Authority Having Jurisdiction has been submitted and approved.

1503.6 Prohibited Location. Where there is insufficient lot area or inappropriate soil conditions for adequate absorption to prevent the ponding, surfacing, or runoff of the gray water, as determined by the Authority Having Jurisdiction, no gray water system shall be permitted. A gray water system is not permitted on a property in a geologically sensitive area as determined by the Authority Having Jurisdiction.

1503.7 Drawings and Specifications. The Authority Having Jurisdiction shall require the following information to be included with or in the plot plan before a permit is issued for a gray water system, or at a time during the construction thereof:

(1) Plot plan drawn to scale and completely dimensioned, showing lot lines and structures, direction and approximate slope of surface, location of present or proposed retaining walls, drainage channels, water supply lines, wells, paved areas and structures on the plot, number of bedrooms and plumbing fixtures in each structure, location of private sewage disposal system and expansion area or building sewer connecting to the public sewer, and location of the proposed gray water system.

(2) Details of construction necessary to ensure compliance with the requirements of this chapter, together with a full description of the complete installation, including installation methods, construction, and materials in accordance with the Authority Having Jurisdiction.

(3) Details for holding tanks shall include dimensions, structural calculations, bracings, and such other pertinent data as required.

(4) A log of soil formations and groundwater level as determined by test holes dug in proximity to proposed irrigation area, together with a statement of water absorption characteristics of the soil at the proposed site as determined by approved percolation tests.

    Exception: The Authority Having Jurisdiction shall permit the use of Table 1504.2 instead of percolation tests.

(5) Distance between the plot and surface waters such as lakes, ponds, rivers or streams, and the slope of the plot and the surface water, wherein close proximity.

1503.8 Procedure for Estimating Gray Water Discharge. Gray water systems shall be designed to distribute the total amount of estimated gray water on a daily basis. The water discharge for gray water systems shall be determined in accordance with Section 1503.8.1 or Section 1503.8.2.

1503.8.1 Single Family Dwellings and Multi-Family Dwellings. The gray water discharge for single family and multi-family dwellings shall be calculated by water use records, calculations of local daily per person interior water use, or the following procedure:

(1) The number of occupants of each dwelling unit shall be calculated as follows:

First bedroom	2 occupants
Each additional bedroom	1 occupant

(2) The estimated gray water flows of each occupant shall be calculated as follows:

Showers, bathtubs, and lavatories	25 gallons (95 L) per day/occupant
Laundry	15 gallons (57 L) per day/occupant

(3) The total number of occupants shall be multiplied by the applicable estimated gray water discharge as provided above and the type of fixtures connected to the gray water system.

1503.8.2 Commercial, Industrial, and Institutional Occupancies. The gray water discharge for commercial, industrial, and institutional occupancies shall be calculated by utilizing the procedure in Section 1503.8.1, water use records or other documentation to estimate gray water discharge.

1503.9 Gray Water System Components. Gray water system components shall comply with Section 1503.9.1 through Section 1503.9.7.

1503.9.1 Surge Tanks. Where installed, surge tanks shall be in accordance with the following:

(1) Surge tanks shall be constructed of solid, durable materials not subject to excessive corrosion or decay and shall be watertight. Surge tanks constructed of steel shall be approved by the Authority Having Jurisdiction, provided such tanks are in accordance with approved applicable standards.

(2) Each surge tank shall be vented in accordance with this code. The vent size shall be determined based on the total gray water fixture units as outlined in this code.

(3) Each surge tank shall have an access opening with lockable gasketed covers or approved equivalent to allow for inspection and cleaning.

(4) Each surge tank shall have its rated capacity permanently marked on the unit. Also, a sign stating GRAY WATER, DANGER — UNSAFE WATER shall be permanently marked on the holding tank.

(5) Each surge tank shall have an overflow drain. The overflow drains shall have permanent connections to the building drain or building sewer, upstream of septic tanks. The overflow drain shall not be equipped with a shutoff valve.

(6) The overflow drain pipes shall not be less in size than the inlet pipe. Unions or equally effective fittings shall be provided for piping connected to the surge tank.

(7) Surge tank shall be structurally designed to withstand anticipated earth or other loads. Surge tank covers shall be capable of supporting an earth load of not less than 300 pounds per square foot (lb/ft^2) (1465 kg/m^2) where the tank is designed for underground installation.

(8) Where a surge tank is installed underground, the system shall be designed so that the tank overflow will gravity drain to the existing sewer line or septic tank. The tank shall be protected against sewer line backflow by a backwater valve installed in accordance with this code.

(9) Surge tanks shall be installed on dry, level, well-compacted soil where underground or on a level 3 inch (76 mm) thick concrete slab where aboveground.

(10) Surge tanks shall be anchored to prevent against overturning where installed aboveground. Underground tanks shall be ballasted, anchored, or otherwise secured, to prevent the tank from floating out of the ground where empty. The combined weight of the tank and hold down system shall meet or exceed the buoyancy forces of the tank.

1503.9.2 Gray Water Pipe and Fitting Materials. Aboveground and underground building drainage and vent pipe and fittings for gray water systems shall comply with the requirements for aboveground and underground sanitary building drainage and vent pipe and fittings in this code. These materials shall extend not less than 2 feet (610 mm) outside the building.

1503.9.3 Subsoil Irrigation Field Materials. Subsoil irrigation field piping shall be constructed of perforated high-density polyethylene pipe, perforated ABS pipe, perforated PVC pipe, or other approved materials, provided that sufficient openings are available for distribution of the gray water into the trench area. Material, construction, and perforation of the pipe shall be in accordance with the appropriate absorption field drainage piping standards and shall be approved by the Authority Having Jurisdiction.

1503.9.4 Subsurface Irrigation Field and Mulch Basin Supply Line Materials. Materials for gray water piping outside the building shall be polyethylene or PVC. Drip feeder lines shall be PVC or polyethylene tubing.

1503.9.5 Valves. Valves shall be accessible.

1503.9.6 Trap. Gray water piping discharging into the surge tank or having a direct connection to the sanitary drain or sewer piping shall be downstream of an approved water seal type trap(s). Where no such trap(s) exists, an approved vented running trap shall be installed upstream of the connection to protect the building from possible waste or sewer gases.

1503.9.7 Backwater Valve. A backwater valve shall be installed on gray water drain connections to the sanitary drain or sewer.

1504.0 Subsurface Irrigation System Zones.

1504.1 General. Irrigation or disposal fields shall be permitted to have one or more valved zones. Each zone shall be of a size to receive the gray water anticipated in that zone.

1504.2 Required Area of Subsurface Irrigation Fields, Subsoil Irrigation Fields, and Mulch Basins. The minimum effective irrigation area of subsurface irrigation fields, subsoil irrigation fields, and mulch basins shall be determined by Table 1504.2 for the type of soil found in the excavation, based upon a calculation of estimated gray water discharge under Section 1503.8. For a subsoil irrigation field, the area shall be equal to the aggregate length of the perforated pipe sections within the valved zone multiplied by the width of the proposed subsoil irrigation field.

1504.3 Determination of Maximum Absorption Capacity. The irrigation field and mulch basin size shall be based on the maximum absorption capacity of the soil and determined using Table 1504.2. For soils not listed in Table 1504.2, the maximum absorption capacity for the proposed site shall be determined by percolation tests or another method acceptable to the Authority Having Jurisdiction. A gray water system shall not be permitted, where the percolation test shows the absorption capacity of the soil is unable to accommodate the maximum discharge of the proposed gray water irrigation system.

1504.4 Groundwater Level. No excavation for an irrigation field, disposal field, or mulch basin shall extend within 3 feet (914 mm) vertical of the highest known seasonal groundwater level, nor to a depth where gray water contaminates the groundwater or surface water. The applicant shall supply evidence of groundwater depth to the satisfaction of the Authority Having Jurisdiction.

TABLE 1504.2
DESIGN OF SIX TYPICAL SOILS

TYPE OF SOIL	MINIMUM SQUARE FEET OF IRRIGATION AREA PER 100 GALLONS OF ESTIMATED GRAY WATER DISCHARGE PER DAY	MAXIMUM ABSORPTION CAPACITY IN GALLONS PER SQUARE FOOT OF IRRIGATION/ LEACHING AREA FOR A 24-HOUR PERIOD
Coarse sand or gravel	20	5.0
Fine sand	25	4.0
Sandy loam	40	2.5
Sandy clay	60	1.7
Clay with considerable sand or gravel	90	1.1
Clay with small amounts of sand or gravel	120	0.8

For SI units: 1 square foot = 0.0929 m^2, 1 gallon per day = 0.000043 L/s

1504.5 Subsurface and Subsoil Irrigation Field Design and Construction. Subsurface and subsoil irrigation field design and construction shall be in accordance with Section 1504.5.1 through Section 1504.7.3. Where a gray water irrigation system design is predicated on soil tests, the subsurface or subsoil irrigation field or mulch basin shall be installed at the same location and depth as the tested area.

1504.5.1 Subsurface Irrigation Field. A subsurface irrigation field shall comply with Section 1504.5.2 through Section 1504.5.7.

1504.5.2 Minimum Depth. Supply piping, including drip feeders, shall be not less than 2 inches (51 mm) below finished grade and covered with mulch or soil.

1504.5.3 Filter. Not less than 140 mesh (105 microns) filter with a capacity of 25 gallons per minute (gpm) (1.58 L/s), or equivalent shall be installed. Where a filter backwash is installed, the backwash and flush discharge shall discharge into the building sewer or private sewage disposal system. Filter backwash and flush water shall not be used.

1504.5.4 Emitter Size. Emitters shall be installed in accordance with the manufacturer's installation instructions. Emitters shall have a flow path of not less than 1200 microns (μ) (1200 μm) and shall not have a coefficient of manufacturing variation (Cv) exceeding 7 percent. Irrigation system design shall be such that emitter flow variation shall not exceed 10 percent.

1504.5.5 Number of Emitters. The minimum number of emitters and the maximum discharge of each emitter in an irrigation field shall be in accordance with Table 1504.5.5.

1504.5.6 Controls. The system design shall provide user controls, such as valves, switches, timers, and other controllers, to rotate the distribution of gray water between irrigation zones.

1504.5.7 Maximum Pressure. Where pressure at the discharge side of the pump exceeds 20 pounds-force per square inch (psi) (138 kPa), a pressure-reducing valve able to maintain downstream pressure not exceeding 20 psi (138 kPa) shall be installed downstream from the pump and before an emission device.

TABLE 1504.5.5
SUBSURFACE IRRIGATION DESIGN CRITERIA FOR SIX TYPICAL SOILS

TYPE OF SOIL	MAXIMUM EMITTER DISCHARGE (gallons per day)	MINIMUM NUMBER OF EMITTERS PER GALLON OF ESTIMATED GRAY WATER DISCHARGE PER DAY* (gallons per day)
Sand	1.8	0.6
Sandy loam	1.4	0.7
Loam	1.2	0.9
Clay loam	0.9	1.1
Silty clay	0.6	1.6
Clay	0.5	2.0

For SI units: 1 gallon per day = 0.000043 L/s

* The estimated gray water discharge per day shall be determined in accordance with Section 1503.8 of this code.

1504.6 Mulch Basin Design and Construction. A mulch basin shall comply with Section 1504.6.1 through Section 1504.6.4.

1504.6.1 Single Family and Multi-Family Dwellings. The gray water discharge to a mulch basin is limited to single family and multi-family dwellings.

1504.6.2 Size. Mulch basins shall be of sufficient size to accommodate peak flow rates and distribute the total amount of estimated gray water on a daily basis without surfacing, ponding or runoff. Mulch basins shall have a depth of not less than 10 inches (254 mm) below finished grade. The mulch basin size shall be based on the maximum absorption capacity of the soil and determined using Table 1504.2.

1504.6.3 Minimum Depth. Gray water supply piping, including drip feeders, shall be not less than 2 inches (51 mm) below finished grade and covered with mulch.

1504.6.4 Maintenance. The mulch basin shall be maintained periodically to retain the required depth and area, and to replenish the required mulch cover.

1504.7 Subsoil Irrigation Field. Subsoil irrigation fields shall comply with Section 1504.7.1 through Section 1504.7.3.

1504.7.1 Minimum Pipe Size. Subsoil irrigation field distribution piping shall be not less than 3 inches (80 mm) diameter.

1504.7.2 Filter Material and Backfill. Filter material, clean stone, gravel, slag, or similar material acceptable to the Authority Having Jurisdiction, varying in size from ¾ of an inch (19.1 mm) to 2½ inches (64 mm) shall be placed in the trench to the depth and grade in accordance with Table 1504.7.3. The perforated section of subsoil irrigation field distribution piping shall be laid on the filter material in an approved manner. The perforated section shall then be covered with filter material to the minimum depth in accordance with Table 1504.7.3. The filter material shall then be covered with porous material to prevent the closure of voids with earth backfill. No earth backfill shall be placed over the filter material cover until after inspection and acceptance.

1504.7.3 Subsoil Irrigation Field Construction. Subsoil irrigation fields shall be constructed in accordance with Table 1504.7.3. Where necessary on sloping ground to prevent excessive line slopes, irrigation lines shall be stepped. The lines between each horizontal leaching section shall be made with approved watertight joints and installed on the natural or unfilled ground.

TABLE 1504.7.3
SUBSOIL IRRIGATION FIELD CONSTRUCTION

DESCRIPTION	MINIMUM	MAXIMUM
Number of drain lines per valved zone	1	–
Length of each perforated line	–	100 feet
Bottom width of trench	12 inches	18 inches
Spacing of lines, center to center	4 feet	–
Depth of earth covers of lines	10 inches	–
Depth of filter material cover of lines	2 inches	–
Depth of filter material beneath lines	3 inches	–
Grade of perforated lines level	level	3 inches per 100 feet

For SI units: 1 inch = 25.4 mm, 1 foot = 304.8 mm, 1 inch per foot = 83.3 mm/m

1504.8 Gray Water System Color and Marking Information. Pressurized gray water distribution systems shall be identified as containing nonpotable water in accordance with Section 601.3 of this code.

1504.9 Other Collection and Distribution Systems. Other collection and distribution systems shall be approved by the local Authority Having Jurisdiction, as allowed by Section 301.3 of this code.

1504.9.1 Higher Requirements. Nothing contained in this chapter shall be construed to prevent the Authority Having Jurisdiction from requiring compliance with higher requirements than those contained herein, where such higher requirements are essential to maintaining a safe and sanitary condition.

1504.10 Testing. Building drains and vents for gray water systems shall be tested in accordance with this code. Surge tanks shall be filled with water to the overflow line prior to and during the inspection. Seams and joints shall be left exposed, and the tank shall remain watertight. A flow test shall be performed through the system to the point of gray water discharge. Lines and components shall be watertight up to the point of the irrigation perforated and drip lines.

1504.11 Maintenance. Gray water systems and components shall be maintained in accordance with Table 1501.5.

1505.0 Reclaimed (Recycled) Water Systems.

1505.1 General. The provisions of this section shall apply to the installation, construction, alteration, and repair of reclaimed (recycled) water systems intended to supply uses such as water closets, urinals, trap primers for floor drains and floor sinks, aboveground and subsurface irrigation, industrial or commercial cooling or air conditioning and other uses approved by the Authority Having Jurisdiction.

1505.2 Permit. It shall be unlawful for a person to construct, install, alter, or cause to be constructed, installed, or altered a reclaimed (recycled) water system within a building or on premises without first obtaining a permit to do such work from the Authority Having Jurisdiction.

1505.2.1 Plumbing Plan Submission. No permit for a reclaimed (recycled) water system shall be issued until complete plumbing plans, with data satisfactory to the Authority Having Jurisdiction, have been submitted and approved.

1505.3 System Changes. No changes or connections shall be made to either the reclaimed (recycled) water system or the potable water system within site containing a reclaimed (recycled) water system without approval by the Authority Having Jurisdiction.

1505.4 Connections to Potable or Reclaimed (Recycled) Water Systems. Reclaimed (recycled) water systems shall have no connection to a potable water supply or alternate water source system. Potable water is permitted to be used as makeup water for a reclaimed (recycled) water storage tank provided the water supply inlet is protected by an air gap or reduced-pressure principle backflow preventer in accordance with this code.

1505.5 Initial Cross-Connection Test. A cross-connection test is required in accordance with Section 1502.3. Before the building is occupied or the system is activated, the installer shall perform the initial cross-connection test in the presence of the Authority Having Jurisdiction and other authorities having jurisdiction. The test shall be ruled successful by the Authority Having Jurisdiction before final approval is granted.

1505.6 Reclaimed (Recycled) Water System Materials. Reclaimed (recycled) water supply and distribution system materials shall comply with the requirements of this code for potable water supply and distribution systems unless otherwise provided for in this section.

1505.7 Reclaimed (Recycled) Water System Color and Marking Information. Reclaimed (recycled) water systems shall have a colored background and marking information in accordance with Section 601.3 of this code.

1505.8 Valves. Valves, except fixture supply control valves, shall be equipped with a locking feature.

1505.9 Hose Bibbs. Hose bibbs shall not be allowed on reclaimed (recycled) water piping systems located in areas accessible to the public. Access to reclaimed (recycled) water at points in the system accessible to the public shall be through a quick-disconnect device that differs from those installed on the potable water system. Hose bibbs supplying reclaimed (recycled) water shall be marked with the words: "CAUTION: NONPOTABLE RECLAIMED WATER, DO NOT DRINK," and the symbol in Figure 1505.9.

FIGURE 1505.9

1505.10 Required Appurtenances. The reclaimed (recycled) water system and the potable water system within the building shall be provided with the required appurtenances (e.g., valves, air/vacuum relief valves, etc.) to allow for deactivation or drainage as required for a cross-connection test in accordance with Section 1502.3.

1505.11 Same Trench as Potable Water Pipes. Reclaimed (recycled) water pipes shall be permitted to be run or laid in the same trench as potable water pipes with 12 inches (305 mm) minimum vertical and horizontal separation where both pipe materials are approved for use within a building. Where piping materials do not meet this requirement, the minimum horizontal separation shall be increased to 60 inches (1524 mm). The potable water piping shall be installed at an elevation above the reclaimed (recycled) water piping. Reclaimed (recycled) water pipes laid in the same trench or crossing building sewer or drainage piping shall be installed in accordance with this code for potable water piping.

1505.12 Signs. Signs in rooms and water closet tanks in buildings using reclaimed (recycled) water shall be in accordance with Section 1501.9 and Section 1501.9.1.

1505.13 Inspection and Testing. Reclaimed (recycled) water systems shall be inspected and tested in accordance with Section 1502.1.

1506.0 On-Site Treated Nonpotable Water Systems.

1506.1 General. The provisions of this section shall apply to the installation, construction, alteration, and repair of on-site treated nonpotable water systems intended to supply uses such as water closets, urinals, trap primers for floor drains and floor sinks, above and belowground irrigation, and other uses approved by the Authority Having Jurisdiction.

1506.2 Plumbing Plan Submission. No permit for an on-site treated nonpotable water system shall be issued until complete plumbing plans, with data satisfactory to the Authority Having Jurisdiction, have been submitted and approved.

1506.3 System Changes. No changes or connections shall be made to either the on-site treated nonpotable water system or the potable water system within a site containing an on-site treated nonpotable water system without approval by the Authority Having Jurisdiction.

1506.4 Connections to Potable or Reclaimed (Recycled) Water Systems. On-site treated nonpotable water systems shall have no connection to a potable water supply or reclaimed (recycled) water source system. Potable or reclaimed (recycled) water is permitted to be used as makeup water for a non-pressurized storage tank provided the makeup water supply is protected by an air gap in accordance with this code.

1506.5 Initial Cross-Connection Test. A cross-connection test is required in accordance with Section 1502.3. Before the building is occupied or the system is activated, the installer shall perform the initial cross-connection test in the presence of the Authority Having Jurisdiction and other authorities having jurisdiction. The test shall be ruled successful by the Authority Having Jurisdiction before final approval is granted.

1506.6 On-Site Treated Nonpotable Water System Materials. On-site treated nonpotable water supply, and distribution system materials shall comply with the requirements of this code for potable water supply and distribution systems unless otherwise provided for in this section.

1506.7 On-Site Treated Nonpotable Water Devices and Systems. Devices or equipment used to treat on-site treated nonpotable water to maintain the minimum water quality requirements determined by the Authority Having Jurisdiction shall be listed or labeled (third-party certified) by a listing agency (accredited conformity assessment body) or approved for the intended application. Devices or equipment used to treat on-site treated nonpotable water for use in the water closet and urinal flushing, surface irrigation, and similar applications shall comply with NSF 350 or approved by the Authority Having Jurisdiction.

1506.8 On-Site Treated Nonpotable Water System Color and Marking Information. On-site treated water systems shall have a colored background and marking information in accordance with Section 601.3 of this code.

1506.9 Design and Installation. The design and installation of on-site treated nonpotable systems shall be in accordance with Section 1506.9.1 through Section 1506.9.5.

1506.9.1 Listing Terms and Installation Instructions. On-site treated nonpotable water systems shall be installed in accordance with the terms of its listing and the manufacturer's installation instructions.

1506.9.2 Minimum Water Quality. On-site treated nonpotable water supplied to toilets or urinals or for other uses in which it is sprayed or exposed shall be disinfected. Acceptable disinfection methods shall include chlorination, ultraviolet sterilization, ozone, or other methods as approved by the Authority Having Jurisdiction. The minimum water quality for on-site treated nonpotable water systems shall meet the applicable water quality requirements for the intended applications as determined by the public health Authority Having Jurisdiction.

1506.9.3 Deactivation and Drainage. The on-site treated nonpotable water system and the potable water system within the building shall be provided with the required appurtenances (e.g., valves, air/vacuum relief valves, etc.) to allow for deactivation or drainage as required for a cross-connection test in accordance with Section 1502.3.

1506.9.4 Near Underground Potable Water Pipe. On-site treated nonpotable water pipes shall be permitted to be run or laid in the same trench as potable water pipes with a 12 inch (305 mm) minimum vertical and horizontal separation where both pipe materials are approved for use within a building. Where piping materials do not meet this requirement the minimum separation shall be increased to 60 inches (1524 mm). The potable water piping shall be installed at an elevation above the on-site treated nonpotable water piping.

1506.9.5 Required Filters. A filter permitting the passage of particulates no larger than 100 microns (100 μm) shall be provided for on-site treated nonpotable water supplied to water closets, urinals, trap primers, and drip irrigation system.

1506.10 Valves. Valves, except fixture supply control valves, shall be equipped with a locking feature.

1506.11 Signs. Signs in buildings using on-site treated nonpotable water shall comply with Section 1501.9 and Section 1501.9.1.

1506.12 Inspection and Testing. On-site treated nonpotable water systems shall be inspected and tested in accordance with Section 1502.1.

CHAPTER 16

NONPOTABLE RAINWATER CATCHMENT SYSTEMS

1601.0 General.

1601.1 Applicability. The provisions of this chapter shall apply to the installation, construction, alteration, and repair of nonpotable rainwater catchment systems.

1601.1.1 Allowable Use of Alternate Water. Where approved or required by the Authority Having Jurisdiction, rainwater shall be permitted to be used instead of potable water for the applications identified in this chapter.

1601.2 System Design. Rainwater catchment systems shall be designed in accordance with this chapter by a licensed plumbing contractor or registered design professional. Components, piping, and fittings used in a rainwater catchment system shall be listed.

Exceptions:

(1) A person registered or licensed to perform plumbing design work is not required to design rainwater catchment systems used for irrigation with a maximum storage capacity of 360 gallons (1363 L).

(2) A person registered or licensed to perform plumbing design work is not required to design rainwater catchment systems for single family dwellings where outlets, piping, and system components are located on the exterior of the building.

1601.3 Permit. It shall be unlawful for a person to construct, install, alter, or cause to be constructed, installed, or altered a rainwater catchment system in a building or on a premise without first obtaining a permit to do such work from the Authority Having Jurisdiction.

Exceptions:

(1) A permit is not required for exterior rainwater catchment systems used for outdoor drip and subsurface irrigation with a maximum storage capacity of 360 gallons (1363 L).

(2) A plumbing permit is not required for rainwater catchment systems for single family dwellings where outlets, piping, and system components are located on the exterior of the building. This does not exempt the need for permits where required for electrical connections, tank supports, or enclosures.

1601.4 Component Identification. System components shall be properly identified as to the manufacturer.

1601.5 Maintenance and Inspection. Rainwater catchment systems and components shall be inspected and maintained in accordance with Section 1601.5.1 through Section 1601.5.3.

1601.5.1 Frequency. Rainwater catchment systems and components shall be inspected and maintained in accordance with Table 1601.5 unless more frequent inspection and maintenance are required by the manufacturer.

TABLE 1601.5
MINIMUM ALTERNATE WATER SOURCE TESTING, INSPECTION, AND MAINTENANCE FREQUENCY

DESCRIPTION	MINIMUM FREQUENCY
Inspect and clean filters and screens, and replace (where necessary).	Every 3 months
Inspect and verify that disinfection, filters, and water quality treatment devices and systems are operational and maintaining minimum water quality requirements as determined by the Authority Having Jurisdiction.	In accordance with manufacturer's instructions and the Authority Having Jurisdiction.
Inspect and clear debris from rainwater gutters, downspouts, and roof washers.	Every 6 months
Inspect and clear debris from the roof or another aboveground rainwater collection surfaces.	Every 6 months
Remove tree branches and vegetation overhanging a roof or other aboveground rainwater collection surfaces.	As needed
Inspect pumps and verify operation.	After initial installation and every 12 months thereafter
Inspect valves and verify operation.	After initial installation and every 12 months thereafter
Inspect pressure tanks and verify operation.	After initial installation and every 12 months thereafter
Clear debris from and inspect storage tanks, locking devices, and verify operation.	After initial installation and every 12 months thereafter
Inspect caution labels and marking.	After initial installation and every 12 months thereafter
Cross-connection inspection and test.*	After initial installation and every 12 months thereafter
Test water quality of rainwater catchment systems required by Section 1603.5 to maintain a minimum water quality.	Every 12 months. After system renovation or repair.

* The cross-connection test shall be performed in the presence of the Authority Having Jurisdiction in accordance with the requirements of this chapter.

1601.5.2 Maintenance Log. A maintenance log for rainwater catchment systems is required to have a permit in accordance with Section 1601.3 and shall be maintained by the property owner and be available for inspection. The property owner or designated appointee shall ensure that a record of testing, inspection, and maintenance in accordance with Table 1601.5 is maintained in the log. The log will indicate the frequency of inspection and maintenance for each system.

1601.5.3 Maintenance Responsibility. The required maintenance and inspection of rainwater catchment systems shall be the responsibility of the property owner unless otherwise required by the Authority Having Jurisdiction.

1601.6 Operation and Maintenance Manual. An operation and maintenance manual for rainwater catchment systems required to have a permit in accordance with Section 1601.3, shall be supplied to the building owner by the system designer. The operating and maintenance manual shall include the following:

(1) Detailed diagram of the entire system and the location of system components.
(2) Instructions for operating and maintaining the system.
(3) Details on maintaining the required water quality as determined by the Authority Having Jurisdiction.
(4) Details on deactivating the system for maintenance, repair, or other purposes.
(5) Applicable testing, inspection, and maintenance frequencies in accordance with Table 1601.5.
(6) A method of contacting the manufacturer(s).

1601.7 Minimum Water Quality Requirements. The minimum water quality for rainwater catchment systems shall comply with the applicable water quality requirements for the intended application as determined by the Authority Having Jurisdiction. Water quality for nonpotable rainwater catchment systems shall comply with Section 1603.5.

Exceptions:

(1) Water treatment is not required for rainwater catchment systems used for aboveground irrigation with a maximum storage capacity of 360 gallons (1363 L).
(2) Water treatment is not required for rainwater catchment systems used for subsurface or drip irrigation.

1601.8 Material Compatibility. Rainwater catchment systems shall be constructed of materials that are compatible with the type of pipe and fitting materials, water treatment, and water conditions in the system.

1601.9 System Controls. Controls for pumps, valves, and other devices that contain mercury that come in contact with rainwater supply shall not be permitted.

1601.10 Separation Requirements. Underground rainwater catchment service piping shall be separated from the building sewer in accordance with Section 609.2. Treated nonpotable water pipes shall be permitted to be run or laid in the same trench as potable water pipes with a 12 inch (305 mm) minimum vertical and horizontal separation where both pipe materials are approved for use within a building. Where horizontal piping materials do not meet this requirement, the minimum separation shall be increased to 60 inches (1524 mm). The potable water piping shall be installed at an elevation above the treated nonpotable water piping.

1601.11 Abandonment. Rainwater catchment systems that are no longer in use, or fail to be maintained in accordance with Section 1601.5, shall be abandoned. Abandonment shall comply with Section 1601.11.1 and Section 1601.11.2.

1601.11.1 General. An abandoned system or part thereof covered under the scope of this chapter shall be disconnected from remaining systems, drained, plugged, and capped in an approved manner.

1601.11.2 Underground Tank. An underground water storage tank that has been abandoned or otherwise discontinued from use in a system covered under the scope of this chapter shall be completely drained and filled with earth, sand, gravel, concrete, or other approved material or removed in a manner satisfactory to the Authority Having Jurisdiction.

1601.12 Sizing. Unless otherwise provided for in this chapter, rainwater catchment piping shall be sized in accordance with Chapter 6 for sizing potable water piping.

1602.0 Nonpotable Rainwater Catchment Systems.

1602.1 General. The installation, construction, alteration, and repair of rainwater catchments systems intended to supply uses such as water closets, urinals, trap primers for floor drains and floor sinks, irrigation, industrial processes, water features, cooling tower makeup and other uses shall be approved by the Authority Having Jurisdiction. Rainwater catchment systems for collecting precipitation from rooftops shall comply with ARCSA/ASPE 63.

1602.2 Plumbing Plan Submission. No permit for a rainwater catchment system shall be issued until complete plumbing plans, with data satisfactory to the Authority Having Jurisdiction, have been submitted and approved.

1602.3 System Changes. No changes or connections shall be made to either the rainwater catchment system or the potable water system within a site containing a rainwater catchment system requiring a permit without approval by the Authority Having Jurisdiction.

1602.4 Connections to Potable or Reclaimed (Recycled) Water Systems. Rainwater catchment systems shall have no direct connection to a potable water supply or alternate water source system. Potable or reclaimed (recycled) water is permitted to be used as makeup water for a rainwater catchment system provided the potable or reclaimed (recycled) water supply connection is protected by an air gap or reduced-pressure principle backflow preventer in accordance with this code.

1602.5 Initial Cross-Connection Test. Where a portion of a rainwater catchment system is installed within a building, a cross-connection test is required in accordance with Section 1605.3. Before the building is occupied or the system is activated, the installer shall perform the initial cross-connection test in the presence of the Authority Having Jurisdiction and other authorities having jurisdiction. The test shall be ruled successful by the Authority Having Jurisdiction before final approval is granted.

1602.6 Sizing. The design and size of rainwater drains, gutters, conductors, and leaders shall comply with Chapter 11 of this code.

1602.7 Rainwater Catchment System Materials. Rainwater catchment system materials shall comply with Section 1602.7.1 through Section 1602.7.4.

1602.7.1 Water Supply and Distribution Materials. Rainwater catchment water supply and distribution materials shall comply with the requirements of this code for potable water supply and distribution systems unless otherwise provided for in this section.

1602.7.2 Rainwater Catchment System Drainage Materials. Materials used in rainwater catchment drainage systems, including gutters, downspouts, conductors, and leaders shall be in accordance with the requirements of this code for storm drainage.

1602.7.3 Storage Tanks. Rainwater storage tanks shall comply with Section 1603.6.

1602.7.4 Collections Surfaces. The collection surface shall be constructed of a hard, impervious material.

1602.8 Rainwater Catchment System Color and Marking Information. Rainwater catchment systems shall have a colored background in accordance with Section 601.3. Rainwater catchment systems shall be marked, in lettering in accordance with Section 601.3.3, with the words: "CAUTION: NONPOTABLE RAINWATER, DO NOT DRINK."

1602.9 Deactivation and Drainage for Cross-Connection Test. The rainwater catchment system and the potable water system within the building shall be provided with the required appurtenances (e.g., valves, air or vacuum relief valves, etc.) to allow for deactivation or drainage as required for a cross-connection test in accordance with Section 1605.3.

1603.0 Design and Installation.

1603.1 Rainwater Catchment Systems. The design and installation of nonpotable rainwater catchment systems shall be in accordance with Section 1603.2 through Section 1603.20.

1603.2 Outside Hose Bibbs. Outside hose bibbs shall be allowed on rainwater piping systems. Hose bibbs supplying rainwater shall be marked with the words: "CAUTION: NONPOTABLE RAINWATER, DO NOT DRINK" and in Figure 1603.2.

FIGURE 1603.2

1603.3 Construction. Rainwater storage tanks shall be constructed of solid, durable materials not subject to excessive corrosion or decay and shall be watertight.

1603.4 Rainwater Catchment Collection Surfaces. Rainwater shall be collected from roof surfaces or other manmade, aboveground collection surfaces.

1603.4.1 Other Surfaces. Natural precipitation collected from surface water runoff, vehicular parking surfaces, or manmade surfaces at or below grade shall be in accordance with the stormwater requirements for on-site treated nonpotable water systems in Section 1506.0.

1603.4.2 Prohibited Discharges. Overflows and bleed-off pipes from roof-mounted equipment and appliances shall not discharge onto roof surfaces that are intended to collect rainwater without prior approval from the Authority Having Jurisdiction.

1603.5 Minimum Water Quality. The minimum water quality for harvested rainwater shall meet the applicable water quality requirements for the intended applications as determined by the Authority Having Jurisdiction. In the absence of water quality requirements determined by the Authority Having Jurisdiction, the minimum treatment and water quality shall be in accordance with Table 1603.5.

Exception: No treatment is required for rainwater used for subsurface or nonsprinkled surface irrigation where the maximum storage volume is less than 360 gallons (1363 L).

1603.5.1 Treatment. If the quality of the tested water cannot consistently be maintained at the minimum levels specified in Table 1603.5, then the system shall be equipped with an appropriate treatment device meeting applicable NSF standards referenced in Chapter 17.

1603.6 Rainwater Storage Tanks. Rainwater storage tanks shall be constructed and installed in accordance with Section 1603.3 and Section 1603.7 through Section 1603.13.

1603.7 Location. Rainwater storage tanks shall be permitted to be installed above or below grade.

1603.8 Above Grade. Above grade, storage tanks shall be of an opaque material, approved for aboveground use in direct sunlight or shall be shielded from direct sunlight. Tanks shall be installed in an accessible location to allow for inspection and cleaning. The tank shall be installed on a foundation or platform that is constructed to accommodate loads in accordance with the building code.

1603.9 Below Grade. Rainwater storage tanks installed below grade shall be structurally designed to withstand anticipated earth or other loads. Holding tank covers shall be capable of supporting an earth load of not less than 300 pounds per square foot (lb/ft^2) (1465 kg/m^2) where the tank is designed for underground installation. Below grade rainwater tanks installed underground shall be provided with manholes. The manhole opening shall be not less than 20 inches (508 mm) in diameter and located not less than 4 inches (102 mm) above the surrounding grade. The surrounding grade shall be sloped away from the manhole. Underground tanks shall be ballasted, anchored, or otherwise secured, to prevent the tank from floating out of the ground where empty. The combined weight of the tank and hold down system shall meet or exceed the buoyancy force of the tank.

»

TABLE 1603.5
MINIMUM WATER QUALITY

APPLICATION	MINIMUM TREATMENT	MINIMUM WATER QUALITY
Car washing	Debris excluder or other approved means in accordance with Section 1603.18, and 100 microns in accordance with Section 1603.19 for drip irrigation.	N/A
Subsurface and drip irrigation	Debris excluder or other approved means in accordance with Section 1603.18, and 100 microns in accordance with Section 1603.19 for drip irrigation.	N/A
Spray irrigation where the maximum storage volume is less than 360 gallons	Debris excluder or other approved means in accordance with Section 1603.18, and disinfection in accordance with Section 1603.16.	N/A
Spray irrigation where the maximum storage volume is equal to or more than 360 gallons	Debris excluder or other approved means in accordance with Section 1603.18.	Escherichia coli: < 100 CFU/100 mL, and Turbidity: < 10 NTU
Urinal and water closet flushing, clothes washing, and trap priming	Debris excluder or other approved means in accordance with Section 1603.18, and 100 microns in accordance with Section 1603.19.	Escherichia coli: < 100 CFU/100 mL, and Turbidity: < 10 NTU
Ornamental fountains and other water features	Debris excluder or other approved means in accordance with Section 1603.18.	Escherichia coli: < 100 CFU/100 mL, and Turbidity: < 10 NTU
Cooling tower make-up water	Debris excluder or other approved means in accordance with Section 1603.18, and 100 microns in accordance with Section 1603.19.	Escherichia coli: < 100 CFU/100 mL, and Turbidity: < 10 NTU

For SI units: 1 micron = 1 μm, 1 gallon = 3.785 L

» **1603.10 Drainage and Overflow.** Rainwater storage tanks shall be provided with a means of draining and cleaning. The overflow drain shall not be equipped with a shutoff valve. The overflow outlet shall discharge in accordance with this code for storm drainage systems. Where discharging to the storm drainage system, the overflow drain shall be protected from backflow of the storm drainage system by a backwater valve or other approved method.

1603.10.1 Overflow Outlet Size. The overflow outlet shall be sized to accommodate the flow of the rainwater entering the tank and not less than the aggregate cross-sectional area of inflow pipes. «

1603.11 Opening and Access Protection. Rainwater tank openings shall be protected to prevent the entrance of insects, birds, or rodents into the tank. «

Rainwater tank access openings exceeding 12 inches (305 mm) in diameter shall be secured to prevent tampering and unintended entry by either a lockable device or other approved method.

1603.12 Marking. Rainwater tanks shall be permanently marked with the capacity and the language: "NONPOTABLE RAINWATER." Where openings are provided to allow a person to enter the tank, the opening shall be marked with the following language: "DANGER-CONFINED SPACE." «

1603.13 Storage Tank Venting. Where venting using drainage or overflow piping is not provided or is considered insufficient, a vent shall be installed on each tank. The vent shall extend from the top of the tank and terminate not less than 6 inches (152 mm) above grade and shall be not less than 1½ inches (40 mm) in diameter. The vent terminal shall be directed downward and covered with a ³⁄₃₂ of an inch (2.4 mm) mesh screen to prevent the entry of vermin and insects. «

1603.14 Pumps. Pumps serving rainwater catchment systems shall be listed. Pumps supplying water to water closets, urinals, and trap primers shall be capable of delivering not less than 15 pounds-force per square inch (psi) (103 kPa) residual pressure at the highest and most remote outlet served. Where the water pressure in the rainwater supply system within the building exceeds 80 psi (552 kPa), a pressure reducing valve reducing the pressure to 80 psi (552 kPa) or less to water outlets in the building shall be installed in accordance with this code. «

1603.15 Roof Drains. Primary and secondary roof drains, conductors, leaders, and gutters shall be designed and installed in accordance with this code. «

1603.16 Water Quality Devices and Equipment. Devices and equipment used to treat rainwater to maintain the minimum water quality requirements determined by the Authority Having Jurisdiction shall be listed or labeled (third-party certified) by a listing agency (accredited conformity assessment body) and approved for the intended application. «

1603.17 Freeze Protection. Tanks and piping installed in locations subject to freezing shall be provided with an approved means of freeze protection. «

1603.18 Debris Removal. The rainwater catchment conveyance system shall be equipped with a debris excluder or other approved means to prevent the accumulation of leaves, needles, other debris and sediment from entering the storage tank. Devices or methods used to remove debris or sediment shall be accessible and sized and installed in accordance with manufacturer's installation instructions. «

1603.19 Required Filters. A filter permitting the passage of particulates not larger than 100 microns (100 μm) shall be provided for rainwater supplied to water closets, urinals, trap primers, and drip irrigation system. «

» **1603.20 Roof Gutters.** Gutters shall maintain a minimum slope and be sized in accordance with Section 1103.3.

1604.0 Signs.

1604.1 General. Signs in buildings using rainwater shall be in accordance with Section 1604.2 and Section 1604.3.

1604.2 Commercial, Industrial, and Institutional Restroom Signs. A sign shall be installed in restrooms in commercial, industrial, and institutional occupancies using nonpotable rainwater for water closets, urinals, or both. Each sign shall contain ½ of an inch (12.7 mm) letters of a highly visible color on a contrasting background. The location of the sign(s) shall be such that the sign(s) shall be visible to users. The number and location of the signs shall be approved by the Authority Having Jurisdiction and shall contain the following text:

TO CONSERVE WATER, THIS BUILDING USES RAINWATER TO FLUSH TOILETS AND URINALS.

1604.3 Equipment Room Signs. Each equipment room containing nonpotable rainwater equipment shall have a sign posted with the following wording in 1 inch (25.4 mm) letters:

CAUTION NONPOTABLE RAINWATER, DO NOT DRINK. DO NOT CONNECT TO DRINKING WATER SYSTEM. NOTICE: CONTACT BUILDING MANAGEMENT BEFORE PERFORMING ANY WORK ON THIS WATER SYSTEM.

This sign shall be posted in a location that is visible to anyone working on or near rainwater equipment.

1605.0 Inspection and Testing.

1605.1 General. Rainwater catchment systems shall be inspected and tested in accordance with Section 1605.2 and Section 1605.3.

1605.2 Supply System Inspection and Test. Rainwater catchment systems shall be inspected and tested in accordance with the applicable provisions of this code for testing of potable water and storm drainage systems. Storage tanks shall be filled with water to the overflow opening for a period of 24 hours, and during the inspection, or by other means as approved by the Authority Having Jurisdiction. Seams and joints shall be exposed during the inspection and checked for watertightness.

1605.3 Annual Cross-Connection Inspection and Testing. An initial and subsequent annual inspection and test in accordance with Section 1602.5 shall be performed on both the potable and rainwater catchment water systems. The potable and rainwater catchment water systems shall be isolated from each other and independently inspected and tested to ensure there is no cross-connection in accordance with Section 1605.3.1 through Section 1605.3.4.

1605.3.1 Visual System Inspection. Prior to commencing the cross-connection testing, a dual system inspection shall be conducted by the Authority Having Jurisdiction and other authorities having jurisdiction as follows:

(1) Pumps, equipment, equipment room signs, and exposed piping in an equipment room shall be checked.

1605.3.2 Cross-Connection Test. The procedure for determining cross-connection shall be followed by the applicant in the presence of the Authority Having Jurisdiction and other authorities having jurisdiction to determine whether a cross-connection has occurred as follows:

(1) The potable water system shall be activated and pressurized. The rainwater catchment water system shall be shut down and completely drained.

(2) The potable water system shall remain pressurized for a minimum period of time specified by the Authority Having Jurisdiction while the rainwater catchment water system is empty. The minimum period the rainwater catchment water system is to remain depressurized shall be determined on a case-by-case basis, taking into account the size and complexity of the potable and rainwater catchment water distribution systems, but in no case shall that period be less than 1 hour.

(3) Fixtures, potable, and rainwater shall be tested and inspected for flow. Flow from a rainwater catchment water system outlet shall indicate a cross-connection. No flow from a potable water outlet shall indicate that it is connected to the rainwater system.

(4) The drain on the rainwater catchment water system shall be checked for flow during the test and at the end of the period.

(5) The potable water system shall then be completely drained.

(6) The rainwater catchment water system shall then be activated and pressurized.

(7) The rainwater catchment water system shall remain pressurized for a minimum period of time specified by the Authority Having Jurisdiction while the potable water system is empty. The minimum period the potable water system is to remain depressurized shall be determined on a case-by-case basis, but in no case shall that period be less than 1 hour.

(8) Fixtures, potable and rainwater catchment, shall be tested and inspected for flow. Flow from a potable water system outlet shall indicate a cross-connection. No flow from a rainwater catchment water outlet shall indicate that it is connected to the potable water system.

(9) The drain on the potable water system shall be checked for flow during the test and at the end of the period.

(10) Where there is no flow detected in the fixtures which would indicate a cross-connection, the potable water system shall be repressurized.

1605.3.3 Discovery of Cross-Connection. In the event that a cross-connection is discovered, the following procedure, in the presence of the Authority Having Jurisdiction, shall be activated immediately:

(1) Rainwater catchment piping to the building shall be shutdown at the meter, and the rainwater riser shall be drained.

(2) Potable water piping to the building shall be shutdown at the meter.

(3) The cross-connection shall be uncovered and disconnected.

(4) The building shall be retested following procedures listed in Section 1605.3.1 and Section 1605.3.2.

(5) The potable water system shall be chlorinated with 50 ppm chlorine for 24 hours.

(6) The potable water system shall be flushed after 24 hours, and a standard bacteriological test shall be performed. Where test results are acceptable, the potable water system shall be permitted to be recharged.

1605.3.4 Annual Inspection. An annual inspection of the rainwater catchment water system, following the procedures listed in Section 1605.3.1 shall be required. Annual cross-connection testing, following the procedures listed in Section 1605.3.2 shall be required by the Authority Having Jurisdiction, unless site conditions do not require it. In no event shall the test occur less than once in 4 years.

Alternate testing requirements shall be permitted by the Authority Having Jurisdiction.

CHAPTER 17
REFERENCED STANDARDS

1701.0 General.

1701.1 Standards. The standards listed in Table 1701.1 are referenced in various sections of this code and shall be considered part of the requirements of this document. The standards are listed herein by the standard number and effective date, the title, application and the section(s) of this code that references the standard. The application of the referenced standard(s) shall be as specified in Section 301.2.2. The promulgating agency acronyms referred to in Table 1701.1 are defined in a list found at the end of the tables.

TABLE 1701.1
REFERENCED STANDARDS

STANDARD NUMBER	STANDARD TITLE	APPLICATION	REFERENCED SECTION
49 CFR 192.281	Plastic Pipe	Plastic, Pipe	1208.6.7.2
49 CFR 192.283	Plastic Pipe: Qualifying Joining Procedures	Plastic, Pipe	1208.6.7.2
ARCSA/ASPE 63-2013	Rainwater Catchment Systems	Miscellaneous	1602.1
ASME A112.1.2-2012 (R2017)	Air Gaps in Plumbing Systems (For Plumbing Fixtures and Water-Connected Receptors)	Fittings	Table 603.2
ASME A112.1.3-2000 (R2015)	Air Gap Fittings for Use with Plumbing Fixtures, Appliances, and Appurtenances	Fittings	Table 603.2
ASME A112.3.1-2007 (R2017)	Stainless Steel Drainage Systems for Sanitary DWV, Storm, and Vacuum Applications, Above- and Below-Ground	Piping	418.1, Table 701.2, 705.7.2, 1102.1
ASME A112.3.4-2018/CSA B45.9-2018	Macerating Toilet Systems and Waste-Pumping Systems for Plumbing Fixtures	Fixtures	710.13
ASME A112.4.1-2009 (R2014)	Water Heater Relief Valve Drain Tubes	Appliances	608.5
ASME A112.4.2-2015/CSA B45.16-2015	Personal Hygiene Devices for Water Closets	Fixtures	411.4
ASME A112.4.4-2017	Plastic Push-Fit Drain, Waste, and Vent (DWV) Fittings	Fittings	Table 701.2
ASME A112.4.14-2017/CSA B125.14-2017	Manually Operated Valves for Use in Plumbing Systems	Valves	606.1
ASME A112.6.1M-1997 (R2017)	Floor-Affixed Supports for Off-the-Floor Plumbing Fixtures for Public Use	Fixtures	402.4
ASME A112.6.2-2017	Framing-Affixed Supports (Carriers) for Off-the-Floor Plumbing Fixtures	Fixtures	402.4
ASME A112.6.3-2016	Floor and Trench Drains	Fixtures	418.1
ASME A112.6.4-2003 (R2012)	Roof, Deck, and Balcony Drains	Fixtures	1102.1
ASME A112.6.7-2010 (R2015)	Sanitary Floor Sinks	Fixtures	421.1
ASME A112.6.9-2005 (R2015)	Siphonic Roof Drains	DWV Components	1106.3
ASME A112.14.1-2003 (R2017)	Backwater Valves	Valves	710.6
ASME A112.14.3-2018	Hydromechanical Grease Interceptors	Fixtures	1014.1
ASME A112.14.4-2001 (R2017)	Grease Removal Devices	Fixtures	1014.1
ASME A112.14.6-2010 (R2015)	FOG (Fats, Oils, and Greases) Disposal Systems	Fixtures	1015.2
ASME A112.18.1-2018/CSA B125.1-2018	Plumbing Supply Fittings	Fittings	408.3, 417.1, 417.2, 417.3, 417.4, 417.6, 603.5.19
ASME A112.18.2-2015/CSA B125.2-2015	Plumbing Waste Fittings	Fittings	404.1

TABLE 1701.1 (continued)
REFERENCED STANDARDS

STANDARD NUMBER	STANDARD TITLE	APPLICATION	REFERENCED SECTION
ASME A112.18.3-2002 (R2017)	Backflow Protection Devices and Systems in Plumbing Fixture Fittings	Backflow Protection	417.3, 417.4
ASME A112.18.6-2017/CSA B125.6-2017	Flexible Water Connectors	Piping	604.5, 604.13
ASME A112.18.9-2011 (R2017)	Protectors/Insulators for Exposed Waste and Supplies on Accessible Fixtures	Miscellaneous	403.3
ASME A112.19.1-2018/CSA B45.2-2018	Enameled Cast Iron and Enameled Steel Plumbing Fixtures	Fixtures	407.1, 408.1, 409.1, 415.1, 420.1
ASME A112.19.2-2018/CSA B45.1-2018	Ceramic Plumbing Fixtures	Fixtures	407.1, 408.1, 409.1, 410.1, 411.1, 412.1, 415.1, 420.1
ASME A112.19.3-2017/CSA B45.4-2017	Stainless Steel Plumbing Fixtures	Fixtures	407.1, 408.1, 409.1, 410.1, 411.1, 415.1, 420.1
ASME A112.19.5-2017/CSA B45.15-2017	Flush Valves and Spuds for Water Closets, Urinals, and Tanks	Fixtures	413.3
ASME A112.19.7-2012/CSA B45.10-2012 (R2017)	Hydromassage Bathtub Systems	Fixtures	409.1, 409.6
ASME A112.19.12-2014	Wall Mounted, Pedestal Mounted, Adjustable, Elevating, Tilting, and Pivoting Lavatory, Sink, and Shampoo Bowl Carrier Systems and Drain Waste Systems	Fixtures	407.1, 420.1
ASME A112.19.14-2013 (R2018)	Six-Liter Water Closets Equipped with a Dual Flushing Device	Fixtures	411.2.1
ASME A112.19.15-2012 (R2017)	Bathtubs/Whirlpool Bathtubs with Pressure Sealed Doors	Fixtures	409.1
ASME A112.19.19-2016	Vitreous China Nonwater Urinals	Fixtures	412.1, 412.1.2
ASME A112.36.2M-1991 (R2017)	Cleanouts	DWV Components	Table 707.2
ASME B1.20.1-2013	Pipe Threads, General Purpose, Inch	Joints	605.1.5, 605.2.3, 605.5.2, 605.12.3, 705.1.3, 705.3.4, 705.4.2, 705.6.3, 1208.6.9, 1322.5(2)
ASME B16.1-2015	Gray Iron Pipe Flanges and Flanged Fittings: Classes 25, 125, and 250	Fittings	1208.6.12.1
ASME B16.3-2016	Malleable Iron Threaded Fittings: Classes 150 and 300	Fittings	Table 604.1, Table 701.2
ASME B16.4-2016	Gray Iron Threaded Fittings: Classes 125 and 250	Fittings	Table 604.1
ASME B16.5-2017	Pipe Flanges and Flanged Fittings: NPS ½ through NPS 24 Metric/Inch	Fittings	1208.6.12.2(1)
ASME B16.12-2009 (R2014)	Cast Iron Threaded Drainage Fittings	Fittings	Table 701.2
ASME B16.15-2018	Cast Copper Alloy Threaded Fittings: Classes 125 and 250	Fittings	Table 604.1
ASME B16.18-2018	Cast Copper Alloy Solder Joint Pressure Fittings	Fittings	Table 604.1
ASME B16.20-2017	Metallic Gaskets for Pipe Flanges	Joints	1208.6.13.2
ASME B16.21-2016	Nonmetallic Flat Gaskets for Pipe Flanges	Joints	1208.6.13.3
ASME B16.22-2018	Wrought Copper and Copper Alloy Solder-Joint Pressure Fittings	Fittings	Table 604.1, 1321.1
ASME B16.23-2016	Cast Copper Alloy Solder Joint Drainage Fittings: DWV	Fittings	Table 701.2
ASME B16.24-2016	Cast Copper Alloy Pipe Flanges, Flanged Fittings, and Valves: Classes 150, 300, 600, 900, 1500, and 2500	Fittings	1208.6.12.3
ASME B16.26-2018	Cast Copper Alloy Fittings for Flared Copper Tubes	Fittings	Table 604.1

**TABLE 1701.1 (continued)
REFERENCED STANDARDS**

STANDARD NUMBER	STANDARD TITLE	APPLICATION	REFERENCED SECTION
ASME B16.29-2017	Wrought Copper and Wrought Copper Alloy Solder-Joint Drainage Fittings – DWV	Fittings	Table 701.2
ASME B16.34-2017	Valves-Flanged, Threaded, and Welding End	Valves	606.1
ASME B16.42-2016	Ductile Iron Pipe Flanges and Flanged Fittings: Classes 150 and 300	Fuel Gas Piping	1208.6.12.4
ASME B16.47-2017	Large Diameter Steel Flanges: NPS 26 through NPS 60 Metric/Inch	Fittings	1208.6.12.2(2)
ASME B16.50-2018	Wrought Copper and Copper Alloy Braze-Joint Pressure Fittings	Fittings	Table 604.1, 1321.1, 1321.11
ASME B16.51-2018	Copper and Copper Alloy Press-Connect Pressure Fittings	Fittings	Table 604.1
ASME B31.3-2016	Process Piping	Piping	1308.2(9)
ASME B36.10M-2018	Welded and Seamless Wrought Steel Pipe	Fuel Gas, Piping	1208.6.3.1
ASME BPVC Section VIII.1-2017	Rules for Construction of Pressure Vessels - Division 1	Miscellaneous	505.4, 1309.5(2), 1310.4(2), 1312.3(2)
ASME BPVC Section IX-2017	Welding, Brazing, and Fusing Qualifications - Qualification Standard for Welding, Brazing, and Fusing Procedures; Welders; Brazers; and Welding, Brazing, and Fusing Operators	Miscellaneous	1322.1.1, 1322.2.1, 1323.11
ASPE 45-2019	Siphonic Roof Drainage	Storm Drainage	1106.2
ASSE 1001-2017	Atmospheric Type Vacuum Breakers	Backflow Protection	Table 603.2
ASSE 1002/ASME A112.1002/CSA B125.12-2015	Anti-Siphon Fill Valves for Water Closet Tanks	Backflow Protection	413.3, Table 603.2
ASSE 1003-2009	Water Pressure Reducing Valves for Domestic Water Distribution Systems	Valves	608.2
ASSE 1004-2017	Commercial Dishwashing Machines	Backflow Protection	414.2
ASSE 1008-2006	Plumbing Aspects of Residential Food Waste Disposer Units	Appliances	419.1
ASSE 1010-2004	Water Hammer Arresters	Appliances	609.11
ASSE 1011-2017	Hose Connection Vacuum Breakers	Backflow Protection	Table 603.2
ASSE 1012-2009	Backflow Preventers with an Intermediate Atmospheric Vent	Backflow Protection	Table 603.2
ASSE 1013-2011	Reduced Pressure Principle Backflow Preventers and Reduced Pressure Principle Fire Protection Backflow Preventers	Backflow Protection	Table 603.2
ASSE 1014-2005	Backflow Prevention Devices for Hand-Held Shower	Backflow Protection	417.3
ASSE 1015-2011	Double Check Backflow Prevention Assemblies and Double Check Fire Protection Backflow Prevention Assemblies	Backflow Protection	Table 603.2
ASSE 1016-2017/ASME A112.1016-2017/CSA B125.16-2017	Automatic Compensating Valves for Individual Showers and Tub/Shower Combinations	Valves	408.3, 408.3.2(1)
ASSE 1018-2001	Trap Seal Primer Valves - Potable Water Supplied	Valves	1007.2
ASSE 1019-2011 (R2016)	Wall Hydrant with Backflow Protection and Freeze Resistance	Backflow Protection	Table 603.2
ASSE 1020-2004	Pressure Vacuum Breaker Assembly	Backflow Protection	Table 603.2
ASSE 1022-2017	Backflow Preventer for Beverage Dispensing Equipment	Backflow Protection	Table 603.2, 603.5.12
ASSE 1023-1979	Hot Water Dispensers Household Storage Type - Electrical	Appliances	417.6
ASSE 1024-2017	Dual Check Backflow Preventers	Backflow Protection	Table 603.2
ASSE 1035-2008	Laboratory Faucet Backflow Preventers	Backflow Protection	Table 603.2, 603.3.11
ASSE 1037-2015/ASME A112.1037-2015/CSA B125.37-2015	Pressurized Flushing Devices for Plumbing Fixtures	Backflow Protection	413.2

TABLE 1701.1 (continued)
REFERENCED STANDARDS

STANDARD NUMBER	STANDARD TITLE	APPLICATION	REFERENCED SECTION
ASSE 1044-2015	Trap Seal Primer - Drainage Types and Electric Design Types	DWV Components	1007.2
ASSE 1047-2011	Reduced Pressure Detector Fire Protection Backflow Prevention Assemblies	Backflow Protection	Table 603.2
ASSE 1048-2011	Double Check Detector Fire Protection Backflow Prevention Assemblies	Backflow Protection	Table 603.2
ASSE 1052-2016	Hose Connection Backflow Preventers	Backflow Protection	Table 603.2
ASSE 1053-2004	Dual Check Backflow Preventer Wall Hydrants – Freeze Resistant Type	Backflow Protection	Table 603.2
TIA TIA TIA TIA ASSE/IAPMO 1055-2020	Chemical Dispensers with Integral Backflow Protection	Backflow Protection	603.5.21
ASSE 1056-2013	Spill Resistant Vacuum Breaker Assemblies	Backflow Protection	Table 603.2
ASSE 1057-2012	Freeze Resistant Sanitary Yard Hydrants with Backflow Protection	Backflow Protection	Table 603.2
ASSE 1060-2017	Outdoor Enclosures for Fluid Conveying Components	Miscellaneous	603.4.7
ASSE 1061-2015	Push-Fit Fittings	Fittings	Table 604.1, 605.1.3.3, 605.2.1.1, 605.3.2.1, 605.9.3
ASSE 1062-2017	Temperature Actuated, Flow Reduction (TAFR) Valves for Individual Supply Fittings	Valves	408.3.2(5)
ASSE 1064-2006 (R2011)	Performance Requirements for Backflow Prevention Assembly Field Test Kits	Backflow Protection	603.4.2
ASSE 1069-2005	Automatic Temperature Control Mixing Valves	Valves	408.3.1, 408.3.2(2)
ASSE 1070-2015/ASME A112.1070-2015/CSA B125.70-2015	Water Temperature Limiting Devices	Valves	407.3(1), 408.3.2(3), 409.4(1), 410.3(1)
ASSE 1071-2012	Temperature Actuated Mixing Valves for Plumbed Emergency Equipment	Valves	416.2
ASSE 1079-2012	Dielectric Pipe Unions	Fittings	605.15, 605.16.1, 605.16.3
ASSE 1081-2014	Backflow Preventers with Integral Pressure Reducing Boiler Feed Valve and Intermediate Atmospheric Vent Style for Domestic and Light Commercial Water Distribution Systems	Backflow Protection	Table 603.2
ASSE 1084-2018	Water Heaters with Temperature Limiting Capacity	Appliances	407.3(2), 408.3.2(4), 409.4(2), 410.3(2)
ASSE 1085-2018	Water Heaters for Emergency Equipment	Appliances	416.2
ASSE 1087-2018	Commercial and Food Service Water Treatment Equipment Utilizing Drinking Water	Water Conditioning, Water Treatment	Table 611.1
ASSE Series 5000-2015	Cross-Connection Control Professional Qualifications Standard	Certification	603.2, 603.4.2
ASSE Series 6000-2015	Professional Qualifications Standard for Medical Gas Systems Personnel	Certification	1323.10
ASSE 6020-2015	Medical Gas Systems Inspectors	Miscellaneous	1324.5.4.7, 1324.5.6.2, 1324.5.6.5
ASSE 6030-2015	Medical Gas Systems Verifiers	Miscellaneous	1324.5.4.7, 1324.5.6.2, 1324.5.6.5, 1324.5.7.2
ASSE 6035-2015	Bulk Medical Gas Systems Verifiers	Miscellaneous	1324.5.7.3
ASSE Series 7000-2013	Residential Potable Water Fire Protection System Installers & Inspectors for One and Two Family Dwellings	Miscellaneous	612.1

TABLE 1701.1 (continued)
REFERENCED STANDARDS

STANDARD NUMBER	STANDARD TITLE	APPLICATION	REFERENCED SECTION
ASSE/IAPMO 12010-2018	Biological Pathogens Professional Qualifications Standard for Construction and Maintenance Personnel	Professional Qualifications	1303.9
ASSE/IAPMO 12030-2018	Waterborne Pathogens Professional Qualifications Standard for Construction and Maintenance Personnel	Professional Qualifications	1303.9
ASSE/IAPMO 12040-2018	Professional Qualifications Standard for Construction and Maintenance Personnel for Contamination/Infection Prevention Procedures to Protect Facility Occupants and Operations	Professional Qualifications	1303.9
ASSE Z359.1-2016	The Fall Protection Code	Miscellaneous	508.2.1.1
ASTM A53/A53M-2018	Pipe, Steel, Black and Hot-Dipped, Zinc-Coated, Welded and Seamless	Piping	Table 604.1, Table 701.2, 1208.6.3.1(1)
ASTM A74-2017	Cast Iron Soil Pipe and Fittings	Piping	301.2.4, Table 701.2
ASTM A106/A106M-2018	Seamless Carbon Steel Pipe for High-Temperature Service	Piping	1208.6.3.1(2)
ASTM A254/A254M-2012	Copper-Brazed Steel Tubing	Piping	1208.6.4.2
ASTM A268/A268M-2010 (R2016)	Seamless and Welded Ferritic and Martensitic Stainless Steel Tubing for General Service	Piping	1208.6.4.1(1)
ASTM A269/A269M-2015a	Seamless and Welded Austenitic Stainless Steel Tubing for General Service	Piping	Table 604.1, 1208.6.4.1(2), 1319.1(2)(a)
ASTM A312/A312M-2018a	Seamless, Welded, and Heavily Cold Worked Austenitic Stainless Steel Pipes	Piping	Table 604.1, 1208.6.3.1(3), 1319.1(2)(b), 1319.1(2)(c)
ASTM A403/A403M-2018a	Wrought Austenitic Stainless Steel Piping Fittings	Fittings	1319.1(2)(c)
ASTM A554-2016	Welded Stainless Steel Mechanical Tubing	Piping	Table 604.1
ASTM A778-2016	Welded, Unannealed Austenitic Stainless Steel Tubular Products	Piping	Table 604.1
ASTM A861-2004 (R2017)	High-Silicon Iron Pipe and Fittings	Piping	811.2
ASTM A888-2018a	Hubless Cast Iron Soil Pipe and Fittings for Sanitary and Storm Drain, Waste, and Vent Piping Applications	Piping	301.2.4, Table 701.2, Table 707.2
ASTM A1056-2012 (R2017)	Cast Iron Couplings used for Joining Hubless Cast Iron Soil Pipe and Fittings	Fittings	705.2.2
ASTM B32-2008 (R2014)	Solder Metal	Joints	605.1.4, 705.3.3
ASTM B42-2015a	Seamless Copper Pipe, Standard Sizes	Piping	Table 604.1
ASTM B43-2015	Seamless Red Brass Pipe, Standard Sizes	Piping	Table 604.1, Table 701.2
ASTM B75/B75M-2011	Seamless Copper Tube	Piping	Table 604.1, Table 701.2
ASTM B88-2016	Seamless Copper Water Tube	Piping	Table 604.1, 604.4, 903.2.3, 1208.6.4.3, 1319.1(1)(a)
ASTM B135/B135M-2017	Seamless Brass Tube	Piping	Table 604.1
ASTM B152/B152M-2013	Copper Sheet, Strip, Plate, and Rolled Bar	Miscellaneous	408.7.4
ASTM B210-2012	Aluminum and Aluminum-Alloy Drawn Seamless Tubes	Piping	1208.6.4.4
ASTM B241/B241M-2016	Aluminum and Aluminum-Alloy Seamless Pipe and Seamless Extruded Tube	Piping	1208.6.3.3, 1208.6.4.4
ASTM B251/B251M-2017	General Requirements for Wrought Seamless Copper and Copper-Alloy Tube	Piping	Table 604.1, Table 701.2
ASTM B280-2018	Seamless Copper Tube for Air Conditioning and Refrigeration Field Service	Piping	1208.6.4.3, 1319.1(1)(b)

TABLE 1701.1 (continued)
REFERENCED STANDARDS

STANDARD NUMBER	STANDARD TITLE	APPLICATION	REFERENCED SECTION
ASTM B302-2017	Threadless Copper Pipe, Standard Sizes	Piping	Table 604.1, Table 701.2
ASTM B306-2013	Copper Drainage Tube (DWV)	Piping	Table 701.2, 903.2.3
ASTM B447-2012a	Welded Copper Tube	Piping	Table 604.1
ASTM B813-2016	Liquid and Paste Fluxes for Soldering of Copper and Copper Alloy Tube	Joints	605.1.4, 705.3.3
ASTM B819-2018	Seamless Copper Tube for Medical Gas Systems	Piping	1318.4, 1318.5, 1319.1(1)(c), 1319.1.1
ASTM B828-2016	Making Capillary Joints by Soldering of Copper and Copper Alloy Tube and Fittings	Joints	605.1.4, 705.3.3
ASTM C4-2004 (R2018)	Clay Drain Tile and Perforated Clay Drain Tile	Piping	Table 1101.4.6
ASTM C425-2004 (R2018)	Compression Joints for Vitrified Clay Pipe and Fittings	Joints	705.8.1
ASTM C564-2014	Rubber Gaskets for Cast Iron Soil Pipe and Fittings	Joints	705.2.2
ASTM C700-2018	Vitrified Clay Pipe, Extra Strength, Standard Strength, and Perforated	Piping	Table 701.2, Table 1101.4.6
ASTM C1053-2000 (R2015)	Borosilicate Glass Pipe and Fittings for Drain, Waste, and Vent (DWV) Applications	Piping	811.2
ASTM C1173-2018	Flexible Transition Couplings for Underground Piping Systems	Fittings	705.10
ASTM C1277-2018	Shielded Couplings Joining Hubless Cast Iron Soil Pipe and Fittings	Fixtures	301.2.4, 705.2.2
ASTM C1460-2017	Shielded Transition Couplings for Use With Dissimilar DWV Pipe and Fittings Above Ground	Joints	705.10
ASTM C1461-2008 (R2017)	Mechanical Couplings Using Thermoplastic Elastomeric (TPE) Gaskets for Joining Drain, Waste, and Vent (DWV), Sewer, Sanitary, and Storm Plumbing Systems for Above and Below Ground Use	Joints	705.10
ASTM C1540-2018	Heavy-Duty Shielded Couplings Joining Hubless Cast Iron Soil Pipe and Fittings	Joints	705.2.2
ASTM C1563-2008 (R2017)	Gaskets for Use in Connection with Hub and Spigot Cast Iron Soil Pipe and Fittings for Sanitary Drain, Waste, Vent, and Storm Piping Applications	Joints	705.2.2
ASTM C1822-2015	Insulating Covers on Accessible Lavatory Piping	Miscellaneous	403.3
ASTM D1785-2015^{e1}	Poly (Vinyl Chloride) (PVC) Plastic Pipe, Schedules 40, 80, and 120	Piping	Table 604.1, Table 701.2
ASTM D2235-2004(R2016)	Solvent Cement for Acrylonitrile-Butadiene-Styrene (ABS) Plastic Pipe and Fittings	Joints	705.1.2
ASTM D2239-2012a	Polyethylene (PE) Plastic Pipe (SIDR-PR) Based on Controlled Inside Diameter	Piping	Table 604.1
ASTM D2241-2015	Poly (Vinyl Chloride) (PVC) Pressure-Rated Pipe (SDR Series)	Piping	Table 604.1
ASTM D2464-2015	Threaded Poly (Vinyl Chloride) (PVC) Plastic Pipe Fittings, Schedule 80	Fittings	Table 604.1
ASTM D2466-2017	Poly (Vinyl Chloride) (PVC) Plastic Pipe Fittings, Schedule 40	Fittings	Table 604.1
ASTM D2467-2015	Poly (Vinyl Chloride) (PVC) Plastic Pipe Fittings, Schedule 80	Fittings	Table 604.1
ASTM D2513-2018a	Polyethylene (PE) Gas Pressure Pipe, Tubing, and Fittings	Piping	1208.6.5, 1208.6.7.2, 1208.6.11.2, 1210.1.7.1(1)
ASTM D2564-2012 (R2018)	Solvent Cements for Poly (Vinyl Chloride) (PVC) Plastic Piping Systems	Joints	605.12.2, 705.6.2
ASTM D2609-2015	Plastic Insert Fittings for Polyethylene (PE) Plastic Pipe	Fittings	Table 604.1

TABLE 1701.1 (continued)
REFERENCED STANDARDS

STANDARD NUMBER	STANDARD TITLE	APPLICATION	REFERENCED SECTION
ASTM D2661-2014^{e1}	Acrylonitrile-Butadiene-Styrene (ABS) Schedule 40 Plastic Drain, Waste, and Vent Pipe and Fittings	Piping	Table 701.2, Table 707.2
ASTM D2665-2014	Poly (Vinyl Chloride) (PVC) Plastic Drain, Waste, and Vent Pipe and Fittings	Piping	Table 701.2, Table 707.2
ASTM D2680-2001 (R2014)	Acrylonitrile-Butadiene-Styrene (ABS) and Poly (Vinyl Chloride) (PVC) Composite Sewer Piping	Piping	Table 701.2
ASTM D2683-2014	Socket-Type Polyethylene Fittings for Outside Diameter-Controlled Polyethylene Pipe and Tubing	Fittings	Table 604.1
ASTM D2729-2017	Poly (Vinyl Chloride) (PVC) Sewer Pipe and Fittings	Piping	Table 701.2, Table 1101.4.6
ASTM D2737-2012a	Polyethylene (PE) Plastic Tubing	Piping, Plastic	Table 604.1
ASTM D2846/D2846M-2019	Chlorinated Poly (Vinyl Chloride) (CPVC) Plastic Hot- and Cold-Water Distribution Systems	Piping	Table 604.1, 605.2.2, 605.3.1
ASTM D3034-2016	Type PSM Poly (Vinyl Chloride) (PVC) Sewer Pipe and Fittings	Piping, Plastic	Table 701.2
ASTM D3035-2015	Polyethylene (PE) Plastic Pipe (DR-PR) Based on Controlled Outside Diameter	Piping	Table 604.1
ASTM D3138-2004 (R2016)	Solvent Cement for Transition Joints Between Acrylonitrile-Butadiene-Styrene (ABS) and Poly (Vinyl Chloride) (PVC) Non-Pressure Piping Components	Joints	705.9.4
ASTM D3139-1998 (R2011)	Joints for Plastic Pressure Pipes Using Flexible Elastomeric Seals	Joints	605.12.1
ASTM D3212-2007 (R2013)	Joints for Drain and Sewer Plastic Pipes Using Flexible Elastomeric Seals	Joints	705.1.1, 705.6.1
ASTM D3261-2016	Butt Heat Fusion Polyethylene (PE) Plastic Fittings for Polyethylene (PE) Plastic Pipe and Tubing	Fittings	Table 604.1
ASTM D4068-2017	Chlorinated Polyethylene (CPE) Sheeting for Concealed Water-Containment Membrane	Miscellaneous	408.7.2
ASTM D4551-2017	Poly (Vinyl Chloride) (PVC) Plastic Flexible Concealed Water-Containment Membrane	Miscellaneous	408.7.1
ASTM E84-2018b	Surface Burning Characteristics of Building Materials	Miscellaneous	701.2(2), 903.1(2), 1101.4
ASTM E119-2018c^{e1}	Fire Tests of Building Construction and Materials	Miscellaneous	1404.3, 1405.3
ASTM E814-2013a (R2017)	Fire Tests of Penetration Firestop Systems	Miscellaneous	1404.3, 1405.3
ASTM F409-2017	Thermoplastic Accessible and Replaceable Plastic Tube and Tubular Fittings	Piping, Plastic	404.1
ASTM F437-2015	Threaded Chlorinated Poly (Vinyl Chloride) (CPVC) Plastic Pipe Fittings, Schedule 80	Fittings	Table 604.1
ASTM F438-2017	Socket-Type Chlorinated Poly (Vinyl Chloride) (CPVC) Plastic Pipe Fittings, Schedule 40	Fittings	Table 604.1
ASTM F439-2013	Chlorinated Poly (Vinyl Chloride) (CPVC) Plastic Pipe Fittings, Schedule 80	Fittings	Table 604.1
ASTM F441/F441M-2015	Chlorinated Poly (Vinyl Chloride) (CPVC) Plastic Pipe, Schedules 40 and 80	Piping	Table 604.1
ASTM F442/F442M-2013^{e1}	Chlorinated Poly (Vinyl Chloride) (CPVC) Plastic Pipe (SDR-PR)	Piping	Table 604.1, 605.2.2
ASTM F493-2014	Solvent Cements for Chlorinated Poly (Vinyl Chloride) (CPVC) Plastic Pipe and Fittings	Joints	605.2.2, 605.3.1
ASTM F628-2012^{e3}	Acrylonitrile-Butadiene-Styrene (ABS) Schedule 40 Plastic Drain, Waste, and Vent Pipe with a Cellular Core	Piping	Table 701.2
ASTM F656-2015	Primers for Use in Solvent Cement Joints of Poly (Vinyl Chloride) (PVC) Plastic Pipe and Fittings	Joints	605.2.2, 605.3.1, 605.12.2, 705.6.2
ASTM F667/F667M-2016	3 through 24 in. Corrugated Polyethylene Pipe and Fittings	Piping, Plastic	Table 1101.4.6

TABLE 1701.1 (continued)
REFERENCED STANDARDS

STANDARD NUMBER	STANDARD TITLE	APPLICATION	REFERENCED SECTION
ASTM F714-2013	Polyethylene (PE) Plastic Pipe (DR-PR) Based on Outside Diameter	Piping	Table 701.2
ASTM F794-2003 (R2014)	Poly (Vinyl Chloride) (PVC) Profile Gravity Sewer Pipe and Fittings Based on Controlled Inside Diameter	Piping	Table 701.2
ASTM F876-2017	Crosslinked Polyethylene (PEX) Tubing	Piping	Table 604.1, 605.9.1
ASTM F877-2018a	Crosslinked Polyethylene (PEX) Hot- and Cold-Water Distribution Systems	Piping	Table 604.1
ASTM F891-2016	Coextruded Poly (Vinyl Chloride) (PVC) Plastic Pipe with a Cellular Core	Piping	Table 701.2
ASTM F894-2013	Polyethylene (PE) Large Diameter Profile Wall Sewer and Drain Pipe	Piping, Plastic	Table 701.2
ASTM F1055-2016a	Electrofusion Type Polyethylene Fittings for Outside Diameter Controlled Polyethylene and Crosslinked Polyethylene (PEX) Pipe and Tubing	Fittings	Table 604.1, 705.5.1.2
ASTM F1216-2016	Rehabilitation of Existing Pipelines and Conduits by the Inversion and Curing of a Resin-Impregnated Tube	Piping	715.3
ASTM F1281-2017	Crosslinked Polyethylene/Aluminum/Crosslinked Polyethylene (PEX-AL-PEX) Pressure Pipe	Piping	Table 604.1
ASTM F1282-2017	Polyethylene/Aluminum/Polyethylene (PE-AL-PE) Composite Pressure Pipe	Piping	Table 604.1
ASTM F1336-2015	Poly (Vinyl Chloride) (PVC) Gasketed Sewer Fittings	Fittings	Table 701.2
ASTM F1412-2016	Polyolefin Pipe and Fittings for Corrosive Waste Drainage Systems	Piping	811.2
ASTM F1488-2014 (R2019)	Coextruded Composite Pipe	Piping	Table 701.2
ASTM F1673-2010 (R2016)	Polyvinylidene Fluoride (PVDF) Corrosive Waste Drainage Systems	Piping	811.2
ASTM F1760-2016	Coextruded Poly(Vinyl Chloride) PVC Non-Pressure Plastic Pipe Having Reprocessed Recycled Content	Piping	Table 701.2
ASTM F1807-2018a	Metal Insert Fittings Utilizing a Copper Crimp Ring for SDR9 Cross-linked Polyethylene (PEX) Tubing and SDR9 Polyethylene of Raised Temperature (PE-RT) Tubing	Fittings	Table 604.1
ASTM F1866-2018	Poly (Vinyl Chloride) (PVC) Plastic Schedule 40 Drainage and DWV Fabricated Fittings	Fittings	Table 701.2
ASTM F1960-2018a	Cold Expansion Fittings with PEX Reinforcing Rings for Use with Cross-linked Polyethylene (PEX) and Polyethylene of Raised Temperature (PE-RT) Tubing	Fittings	Table 604.1
ASTM F1970-2012^{e1}	Special Engineered Fittings, Appurtenances or Valves for Use in Poly (Vinyl Chloride) (PVC) or Chlorinated Poly (Vinyl Chloride) (CPVC) Systems	Piping	Table 604.1, 606.1
ASTM F1973-2013 (R2018)	Factory Assembled Anodeless Risers and Transition Fittings in Polyethylene (PE) and Polyamide 11 (PA11) and Polyamide 12 (PA12) Fuel Gas Distribution Systems	Fuel Gas	1210.1.7.1(2)
ASTM F1974-2009 (R2015)	Metal Insert Fittings for Polyethylene/ Aluminum/Polyethylene and Crosslinked Polyethylene/Aluminum/Crosslinked Polyethylene Composite Pressure Pipe	Fittings	Table 604.1, 605.7.1, 605.10.1
ASTM F1986-2001 (R2011)	Multilayer Pipe Type 2, Compression Fittings, and Compression Joints for Hot and Cold Drinking-Water Systems	Fittings	Table 604.1
ASTM F2080-2018	Cold-Expansion Fittings with Metal Compression-Sleeves for Crosslinked Polyethylene (PEX) Pipe and SDR9 Polyethylene of Raised Temperature (PE-RT) Pipe	Fittings	Table 604.1
ASTM F2098-2015	Stainless Steel Clamps for Securing SDR9 Cross-linked Polyethylene (PEX) Tubing to Metal Insert and Plastic Insert Fittings	Fittings	Table 604.1

TABLE 1701.1 (continued)
REFERENCED STANDARDS

STANDARD NUMBER	STANDARD TITLE	APPLICATION	REFERENCED SECTION
ASTM F2159-2018a	Plastic Insert Fittings Utilizing a Copper Crimp Ring for SDR9 Cross-linked Polyethylene (PEX) Tubing and SDR9 Polyethylene of Raised Temperature (PE-RT) Tubing	Fittings	Table 604.1
ASTM F2389-2017a	Pressure-Rated Polypropylene (PP) Piping Systems	Piping	Table 604.1, 605.11.1, 606.1
ASTM F2434-2018	Metal Insert Fittings Utilizing a Copper Crimp Ring for SDR9 Cross-linked Polyethylene (PEX) Tubing and SDR9 Cross-linked Polyethylene/Aluminum/Cross-linked Polyethylene (PEX-AL-PEX) Tubing	Fittings	Table 604.1, 605.10.1
ASTM F2509-2015	Field-Assembled Anodeless Riser Kits for Use on Outside Diameter Controlled Polyethylene and Polyamide-11 (PA11) Gas Distribution Pipe and Tubing	Fuel Gas	1210.1.7.1(3)
ASTM F2561-2017	Rehabilitation of a Sewer Service Lateral and Its Connection to the Main Using a One Piece Main and Lateral Cured-in-Place Liner	Piping	715.3
ASTM F2599-2016	The Sectional Repair of Damaged Pipe by Means of an Inverted Cured-In-Place Liner	Piping	715.3
ASTM F2618-2015	Chlorinated Poly(Vinyl Chloride) CPVC Pipe and Fittings for Chemical Waste Drainage Systems	Piping	811.2
ASTM F2620-2013	Heat Fusion Joining of Polyethylene Pipe and Fittings	Joints	605.6.1.1, 605.6.1.3, 705.5.1.1, 705.5.1.3
ASTM F2735-2018	Plastic Insert Fittings for SDR9 Cross-linked Polyethylene (PEX) and Polyethylene of Raised Temperature (PE-RT) Tubing	Fittings	Table 604.1
ASTM F2769-2018	Polyethylene of Raised Temperature (PE-RT) Plastic Hot and Cold-Water Tubing and Distribution Systems	Piping, Fittings	Table 604.1
ASTM F2831-2012 (R2017)	Internal Non-Structural Epoxy Barrier Coating Material Used in Rehabilitation of Metallic Pressurized Piping Systems	Miscellaneous	320.1
ASTM F2855-2012	Chlorinated Poly (Vinyl Chloride)/Aluminum/Chlorinated Poly (Vinyl Chloride) (CPVC/AL/CPVC) Composite Pressure Tubing	Piping	Table 604.1, 605.3.1
ASTM F2945-2018	Polyamide 11 Gas Pressure Pipe, Tubing, and Fittings	Piping	1208.6.5
ASTM F3226/F3226M-2016^{e1}	Metallic Press-Connect Fittings for Piping and Tubing Systems	Fittings	Table 604.1
ASTM F3240-2017	Installation of Seamless Molded Hydrophilic Gaskets (SMHG) for Long-Term Watertightness of Cured-in-Place Rehabilitation of Main and Lateral Pipelines	Piping	715.3
AWS A5.8M/A5.8-2011-AMD 1	Filler Metals for Brazing and Braze Welding	Joints	605.1.1, 705.3.1, 1321.3
AWS A5.9/A5.9M-2017 (ISO 14343:2009 MOD)	Welding Consumables—Wire Electrodes, Strip Electrodes, Wires, and Rods for Arc Welding of Stainless and Heat Resisting Steels—Classification	Joints	605.13.2
AWS B2.2/B2.2M-2016	Brazing Procedure and Performance Qualification	Certification	1323.11
AWWA C110-2012	Ductile-Iron and Gray-Iron Fittings	Fittings	Table 604.1
AWWA C111-2017	Rubber-Gasket Joints for Ductile-Iron Pressure Pipe and Fittings	Joints	605.4.1, 605.4.2
AWWA C151-2017	Ductile-Iron Pipe, Centrifugally Cast	Piping	Table 604.1
AWWA C153-2011	Ductile-Iron Compact Fittings	Fittings	Table 604.1
AWWA C210-2015	Liquid-Epoxy Coatings and Linings for Steel Water Pipe and Fittings	Miscellaneous	604.9
AWWA C500-2009	Metal-Seated Gate Valves for Water Supply Service	Valves	606.1
AWWA C504-2015	Rubber-Seated Butterfly Valves	Valves	606.1
AWWA C507-2015	Ball Valves, 6 in. through 60 in. (150 mm through 1,500 mm)	Valves	606.1
AWWA C510-2017	Double Check-Valve Backflow Prevention Assembly	Backflow Protection	Table 603.2

TABLE 1701.1 (continued)
REFERENCED STANDARDS

STANDARD NUMBER	STANDARD TITLE	APPLICATION	REFERENCED SECTION
AWWA C511-2017	Reduced-Pressure Principle Backflow Prevention Assembly	Backflow Protection	Table 603.2
AWWA C900-2016	Polyvinyl Chloride (PVC) Pressure Pipe and Fabricated Fittings, 4 in. through 12 in. (100 mm through 300 mm)	Piping	Table 604.1
AWWA C901-2017	Polyethylene (PE) Pressure Pipe and Tubing, ¾ in. (19 mm) through 3 in. (76 mm), for Water Service	Piping	Table 604.1
AWWA C904-2016	Crosslinked Polyethylene (PEX) Pressure Tubing, ½ in. through 3 in. (13 mm through 76 mm), for Water Service	Piping	Table 604.1
AWWA C907-2017	Injection-Molded Polyvinyl Chloride (PVC) Pressure Fittings, 4 in. through 12 in. (100 mm through 300 mm), for Water, Wastewater, and Reclaimed Water Service	Fittings	Table 604.1
CGA G-4.1-2009	Cleaning Equipment for Oxygen Service	Miscellaneous	1318.2, 1321.8.7
CGA M-1-2018	Medical Gas Supply Systems at Health Care Facilities	Miscellaneous	1324.5.7.3
CGA V-5-2008 (R2013)	Diameter Index Safety System (Noninterchangeable Low Pressure Connections for Medical Gas Applications)	Connections	1315.5
CISPI 301-2017	Hubless Cast Iron Soil Pipe and Fittings for Sanitary and Storm Drain, Waste, and Vent Piping Applications	Piping, Ferrous	301.2.4, Table 701.2, Table 707.2
CISPI 310-2017	Couplings for Use in Connection with Hubless Cast Iron Soil Pipe and Fittings for Sanitary and Storm Drain, Waste, and Vent Piping Applications	Joints	301.2.4, 705.2.2
CSA B45.5-2017/IAPMO Z124-2017	Plastic Plumbing Fixtures (with Errata dated August 2017)	Fixtures	407.1, 408.1, 409.1, 411.1, 412.1, 420.1
CSA B45.8-2018/IAPMO Z403-2018	Terrazzo, Concrete, Composite Stone, and Natural Stone Plumbing Fixtures	Fixtures	407.1, 420.1
CSA B45.11-2017/IAPMO Z401-2017	Glass Plumbing Fixtures	Fixtures	407.1
CSA B45.12-2013/IAPMO Z402-2013	Aluminum and Copper Plumbing Fixtures	Fixtures	407.1, 408.1, 409.1, 420.1
CSA B64.1.1-2011 (R2016)	Atmospheric Vacuum Breakers (AVB)	Backflow Protection	Table 603.2
CSA B64.1.2-2011 (R2016)	Pressure Vacuum Breakers (PVB)	Backflow Protection	Table 603.2
CSA B64.2.1.1-2011 (R2016)	Hose Connection Dual Check Vacuum Breakers (HCDVB)	Backflow Protection	Table 603.2
CSA B64.4-2011 (R2016)	Reduced Pressure Principle (RP) Backflow Preventers	Backflow Protection	Table 603.2
CSA B64.4.1-2011 (R2016)	Reduced Pressure Principle Backflow Preventers for Fire Protection Systems (RPF)	Backflow Protection	Table 603.2
CSA B64.5-2011 (R2016)	Double Check Valve (DCVA) Backflow Preventers	Backflow Protection	Table 603.2
CSA B64.5.1-2011 (R2016)	Double Check Valve Backflow Preventers for Fire Protection Systems (DCVAF)	Backflow Protection	Table 603.2
CSA B79-2008 (R2018)	Commercial and Residential Drains and Cleanouts	Fixtures	418.1,Table 707.2
CSA B125.3-2018	Plumbing Fittings	Fittings	408.3.2(3), 409.4(1), 410.3(1)
CSA B125.5/IAPMO Z2600-2011 (R2016)	Flexible Water Connectors with Excess Flow Shut-off Device	Miscellaneous	604.5
CSA B137.1-2017	Polyethylene (PE) Pipe, Tubing, and Fittings for Cold-Water Pressure Services	Piping	Table 604.1
CSA B137.5-2017	Crosslinked Polyethylene (PEX) Tubing Systems for Pressure Applications	Piping	Table 604.1
CSA B137.6-2017	Chlorinated Polyvinylchloride (CPVC) Pipe, Tubing, and Fittings for Hot- and Cold-Water Distribution Systems	Piping, Fittings	Table 604.1
CSA B137.9-2017	Polyethylene/Aluminum/Polyethylene (PE-AL-PE) Composite Pressure-Pipe Systems	Piping	Table 604.1

TABLE 1701.1 (continued)
REFERENCED STANDARDS

STANDARD NUMBER	STANDARD TITLE	APPLICATION	REFERENCED SECTION
CSA B137.10-2017	Crosslinked Polyethylene/Aluminum/Crosslinked Polyethylene (PEX-AL-PEX) Composite Pressure-Pipe Systems	Piping	Table 604.1
CSA B137.11-2017	Polypropylene (PP-R) Pipe and Fittings for Pressure Applications	Piping	Table 604.1, 605.11.1
CSA B137.18-2017	Polyethylene of Raised Temperature Resistance (PE-RT) Tubing Systems for Pressure Applications	Piping, Fittings	Table 604.1
CSA B181.3-2018	Polyolefin and Polyvinylidene Fluoride (PVDF) Laboratory Drainage Systems	Piping	811.2
CSA B481-2012 (R2017)	Grease Interceptors	Fixtures	1014.1
CSA LC 1-2018	Fuel Gas Piping Systems Using Corrugated Stainless Steel Tubing (same as CSA 6.26)	Fuel Gas	1208.6.4.5, 1210.4.1(4), 1211.3
CSA LC 4a-2013 (R2017)	Press-Connect Metallic Fittings for Use in Fuel Gas Distribution Systems (same as CSA 6.32a)	Fuel Gas	1208.6.10.1, 1208.6.10.2, 1208.6.10.3, 1210.4.1(3)
CSA Z21.10.1-2017	Gas Water Heaters, Volume I, Storage Water Heaters with Input Ratings of 75,000 Btu Per Hour or Less (same as CSA 4.1)	Fuel Gas, Appliances	Table 501.1(1)
CSA Z21.10.3-2017	Gas-Fired Water Heaters, Volume III, Storage Water Heaters with Input Ratings Above 75,000 Btu Per Hour, Circulating and Instantaneous (same as CSA 4.3)	Fuel Gas, Appliances	Table 501.1(1)
CSA Z21.22-2015	Relief Valves for Hot Water Supply Systems (same as CSA 4.4)	Valves	607.5, 608.7
CSA Z21.24-2015	Connectors for Gas Appliances (same as CSA 6.10)	Fuel Gas	1212.1(3), 1212.2
CSA Z21.41-2014	Quick-Disconnect Devices for Use with Gas Fuel Appliances (same as CSA 6.9)	Fuel Gas	1212.7
CSA Z21.54-2014	Gas Hose Connectors for Portable Outdoor Gas-Fired Appliances (same as CSA 8.4)	Fuel Gas	1212.3.2
CSA Z21.69-2015	Connectors for Moveable Gas Appliances (same as CSA 6.16)	Fuel Gas	1212.1.1
CSA Z21.75-2016	Connectors for Outdoor Gas Appliances and Manufactured Homes (same as CSA 6.27)	Fuel Gas	1212.1(4)
CSA Z21.80a-2012	Line Pressure Regulators (same as CSA 6.22a)	Fuel Gas	1208.8.1, 1208.8.4(1)
CSA Z21.90-2015	Gas Convenience Outlets and Optional Enclosures (same as CSA 6.24)	Fuel Gas	1212.8
CSA Z21.93-2017	Excess Flow Valves for Natural Gas and Propane Gas with Pressures up to 5 psig (same as CSA 6.30)	Fuel Gas	1209.1
DOE-STD-3020-2015	HEPA Filters Used by DOE Contractors	Miscellaneous	1312.4(3)
EPA/625/R-04/108-2004	Guidelines for Water Reuse	Miscellaneous	1501.7
IAPMO IGC 78-2018	Drain, Waste and Vent (DWV) Internal Cleanout Fittings	DWV Components	Table 707.2
IAPMO IGC 115-2013^{e1}	Automatic Water Leak Detection Devices	Miscellaneous	606.9
IAPMO IGC 127-2018	Combined Hand-Washing Systems	Fixtures	407.1, 420.1
IAPMO IGC 154-2019	Shower and Tub/Shower Enclosures, Bathtubs with Glass Pressure-Sealed Doors, and Shower/Steam Panels	Fixtures	408.1
IAPMO IGC 224-2018	ABS, PVC and Cast Iron DWV Test Fitting with Integral Cleanout	DWV Components	Table 707.2
IAPMO IGC 322-2018	Alkaline Water – Drinking Water Treatment Units	Miscellaneous	611.1.1
IAPMO IGC 349-2018	Electronic Plumbing Supply System Integrity Protection Devices	Miscellaneous	606.9
IAPMO IGC 352-2018	Diverter Valves for Diversion of Rainwater or Storm Water for Use in Alternate Nonpotable Water Source Systems	Valves	1503.2.4
IAPMO PS 59-2016a^{e1}	Wastewater Diverter Valves and Diversion Systems	Fittings	1503.2.2
IAPMO PS 65-2002	Airgap Units for Water Conditioning Equipment Installation	Backflow Protection	611.2

TABLE 1701.1 (continued)
REFERENCED STANDARDS

STANDARD NUMBER	STANDARD TITLE	APPLICATION	REFERENCED SECTION
IAPMO PS 90-2014	Elastomeric Test Caps, Cleanout Caps, and Combination Test Caps/Shielded Couplings	DWV Components	Table 707.2
IAPMO PS 117-2017	Press and Nail Connections	Fittings	Table 604.1
IAPMO Z124.5-2013e1	Plastic Toilet Seats	Appurtenance	411.3
IAPMO Z601-2018	Scale Reduction Devices	Water Conditioning, Water Treatment	611.1.2
IAPMO Z1001-2016	Prefabricated Gravity Grease Interceptors	Fixtures	1014.3.4
IAPMO Z1033-2015	Flexible PVC Hoses and Tubing for Pools, Hot Tubs, Spas, and Jetted Bathtubs	Tubing	409.6.1
IAPMO Z1088-2013	Pre-Pressurized Water Expansion Tanks	Miscellaneous	608.3
IAPMO Z1157-2014e1	Ball Valves	Valves	606.1
ICC A117.1-2017	Accessible and Usable Buildings and Facilities	Miscellaneous	403.2, 408.6
ISEA Z358.1-2014	Emergency Eyewash and Shower Equipment	Miscellaneous	416.1, 416.2
MSS SP-58-2018	Pipe Hangers and Supports – Materials, Design, Manufacture, Selection, Application, and Installation	Miscellaneous	1210.3.5, 1323.4.1
MSS SP-67-2017	Butterfly Valves	Valves	606.1
MSS SP-70-2011	Gray Iron Gate Valves, Flanged and Threaded Ends	Valves	606.1
MSS SP-71-2018	Gray Iron Swing Check Valves, Flanged and Threaded Ends	Valves	606.1
MSS SP-72-2010a	Ball Valves with Flanged or Butt-Welding Ends for General Service	Valves	606.1
MSS SP-78-2011	Gray Iron Plug Valves, Flanged and Threaded Ends	Valves	606.1
MSS SP-80-2013	Bronze Gate, Globe, Angle, and Check Valves	Valves	606.1
MSS SP-110-2010	Ball Valves Threaded, Socket-Welding, Solder Joint, Grooved and Flared Ends	Valves	606.1
MSS SP-122-2017	Plastic Industrial Ball Valves	Valves	606.1
NFPA 13D-2019	Installation of Sprinkler Systems in One- and Two-Family Dwellings and Manufactured Homes	Miscellaneous	612.1, 612.5.3.1
NFPA 30A-2018	Motor Fuel Dispensing Facilities and Repair Garages	Miscellaneous	507.14.2
NFPA 31-2016	Installation of Oil-Burning Equipment	Fuel Gas, Appliances	505.3, 1201.1
NFPA 51-2018	Design and Installation of Oxygen-Fuel Gas Systems for Welding, Cutting, and Allied Processes	Fuel Gas	507.9
NFPA 54/Z223.1-2018	National Fuel Gas Code	Fuel Gas	Chapter 5, Chapter 12
NFPA 58-2017	Liquefied Petroleum Gas Code	Fuel Gas	1208.5(7), 1208.6.7.3, 1208.6.11.4, 1212.11
NFPA 70-2017	National Electrical Code	Miscellaneous	1210.12.5(2), 1211.2.4, 1211.7, 1317.1(11), 1323.3.1
NFPA 88A-2019	Parking Structures	Miscellaneous	507.14.1
NFPA 99-2018	Health Care Facilities Code	Miscellaneous	1301.3, 1309.13(2), 1317.1(9), 1324.5.9.4, 1327.1
NFPA 211-2019	Chimneys, Fireplaces, Vents, and Solid Fuel-Burning Appliances	Fuel Gas, Appliances	509.5.2, 509.5.3, 509.5.6.1, 509.5.6.3
NFPA 409-2016	Aircraft Hangars	Miscellaneous	507.15
NFPA 780-2017	Installation of Lightning Protection Systems	Fuel Gas	1211.5
NFPA 1192-2018	Recreational Vehicles	Fuel Gas	1202.3(18)
NSF 3-2017	Commercial Warewashing Equipment	Appliances	414.1

TABLE 1701.1 (continued)
REFERENCED STANDARDS

STANDARD NUMBER	STANDARD TITLE	APPLICATION	REFERENCED SECTION
NSF 14-2018	Plastics Piping System Components and Related Materials	Miscellaneous	301.2.3, 604.1
NSF 42-2018	Drinking Water Treatment Units – Aesthetic Effects	Appliances	Table 611.1
NSF 44-2018	Residential Cation Exchange Water Softeners	Appliances	Table 611.1
NSF 53-2017	Drinking Water Treatment Units-Health Effects	Appliances	Table 611.1
NSF 55-2018	Ultraviolet Microbiological Water Treatment Systems	Appliances	Table 611.1
NSF 58-2017	Reverse Osmosis Drinking Water Treatment Systems	Appliances	611.2, Table 611.1
NSF 61-2018	Drinking Water System Components – Health Effects	Miscellaneous	415.1, 417.1, 604.1, 604.9, 606.1, 607.2, 608.2, 609.8.2, Table 611.1
NSF 62-2018	Drinking Water Distillation Systems	Appliances	Table 611.1
NSF 350-2017a	Onsite Residential and Commercial Water Reuse Treatment Systems	Miscellaneous	1501.7, 1506.7
NSF 359-2018	Valves for Crosslinked Polyethylene (PEX) Water Distribution Tubing Systems	Valves	606.1
PDI G-101-2017	Testing and Rating Procedure for Hydro Mechanical Grease Interceptors with Appendix of Installation and Maintenance	DWV Components	1014.1
PDI G-102-2010	Testing and Certification for Grease Interceptors with FOG Sensing and Alarm Devices	Certification	1014.1
PDI-WH 201-2017	Water Hammer Arresters	Water Supply Components	609.11
UL 17-2008	Vent or Chimney Connector Dampers for Oil-Fired Appliances (with revisions through September 25, 2013)	Fuel Gas, Vent Dampers	509.14.1
UL 103-2010	Factory-Built Chimneys for Residential Type and Building Heating Appliances (with revisions through March 15, 2017)	Fuel Gas, Appliances	509.5.1, 509.5.1.1
UL 174-2004	Household Electric Storage Tank Water Heaters (with revisions through December 15, 2016)	Appliances	Table 501.1(1)
UL 263-2011	Fire Tests of Building Construction and Materials (with revisions through March 2, 2018)	Miscellaneous	1404.3, 1405.3
UL 378-2006	Draft Equipment (with revisions through September 17, 2013)	Fuel Gas, Appliances	509.3.3, 509.14.1
UL 399-2017	Drinking Water Coolers (with revisions through August 29, 2018)	Fixtures	415.1
UL 430-2015	Waste Disposers (with revisions through February 23, 2018)	Appliances	419.1
UL 441-2016	Gas Vents (with revisions through July 27, 2016)	Fuel Gas, Vents	509.1
UL 467-2013	Grounding and Bonding Equipment	Miscellaneous	1211.2.5
UL 499-2014	Electric Heating Appliances (with revisions through February 23, 2017)	Appliances	417.6, Table 501.1(1)
UL 641-2010	Type L Low-Temperature Venting Systems (with revisions through April 23, 2018)	Fuel Gas	509.1
UL 651-2011	Schedule 40, 80, Type EB and A Rigid PVC Conduit and Fittings (with revisions through June 15, 2016)	Piping	1208.6.6
UL 723-2018	Test for Surface Burning Characteristics of Building Materials	Miscellaneous	701.2(2), 903.1(2), 1101.4
UL 732-2018	Oil-Fired Storage Tank Water Heaters (with revisions through August 9, 2018)	Fuel Gas, Appliances	Table 501.1(1)
UL 749-2018	Household Dishwashers	Appliances	414.1
UL 778-2016	Motor-Operated Water Pumps (with revisions through January 17, 2019)	Appliances	1101.14
UL 921-2016	Commercial Dishwashers (with revisions through September 20, 2017)	Appliances	414.1

TABLE 1701.1 (continued)
REFERENCED STANDARDS

STANDARD NUMBER	STANDARD TITLE	APPLICATION	REFERENCED SECTION
UL 959-2010	Medium Heat Appliance Factory-Built Chimneys (with revisions through June 12, 2014)	Fuel Gas, Appliances	509.5.1
UL 1453-2016	Electric Booster and Commercial Storage Tank Water Heaters (with revisions through May 18, 2018)	Appliances	Table 501.1(1)
UL 1479-2015	Fire Tests of Penetration Firestops	Miscellaneous	1404.3, 1405.3
UL 1738-2010	Venting Systems for Gas-Burning Appliances, Categories II, III, and IV (with revisions through November 7, 2014)	Fuel Gas, Appliances	509.4.1, 509.4.2, 509.4.3
UL 1777-2015	Chimney Liners	Chimney Liners	509.5.3(2)
UL 2523-2009	Solid Fuel-Fired Hydronic Heating Appliances, Water Heaters, and Boilers (with revisions through March 16, 2018)	Fuel Gas, Appliances	Table 501.1(1)
UL 2561-2016	1400 Degree Fahrenheit Factory-Built Chimneys (with revisions through April 19, 2018)	Fuel Gas, Appliances	509.5.1

1701.2 Standards, Publications, Practices, and Guides. The standards, publications, practices, and guides listed in Table 1701.2 are not referenced in other sections of this code. The application of the referenced standards, publications, practices, and guides shall be as specified in Section 301.2.2. The promulgating agency acronyms are found at the end of the tables.

TABLE 1701.2
STANDARDS, PUBLICATIONS, PRACTICES, AND GUIDES

DOCUMENT NUMBER	DOCUMENT TITLE	APPLICATION
AHAM FWD-1-2016	Food Waste Disposers	Appliances
ASCE 25-2016	Earthquake-Actuated Automatic Gas Shutoff Devices	Fuel Gas
ASHRAE/IES 90.1-2016	Energy Standard for Buildings Except Low-Rise Residential Buildings	Miscellaneous
ASHRAE/IES 90.2-2018	Energy-Efficient Design of Low-Rise Residential Buildings	Miscellaneous
ASHRAE 188-2018	Legionellosis: Risk Management for Building Water Systems	Risk Management
ASHRAE Guideline 12-2000	Minimizing the Risk of Legionellosis Associated with Building Water Systems	Risk Management
ASME A13.1-2015	Scheme for the Identification of Piping Systems	Piping
ASME A112.4.3-1999 (R2015)	Plastic Fittings for Connecting Water Closets to the Sanitary Drainage System	Fittings
ASME A112.19.10-2017	Retrofit Dual Flush Devices for Water Closets	Fixtures
ASME A112.21.3M-1985 (R2017)	Hydrants for Utility and Maintenance Use	Valves
ASME B1.20.3-1976 (R2018)	Dryseal Pipe Threads (Inch)	Joints
ASME B16.33-2012 (R2017)	Manually Operated Metallic Gas Valves for Use in Gas Piping Systems Up to 175 psi (Sizes NPS ½ through NPS 2)	Valves
ASME B16.39-2014	Malleable Iron Threaded Pipe Unions: Classes 150, 250 and 300	Fittings
ASME B16.40-2013	Manually Operated Thermoplastic Gas Shutoffs and Valves in Gas Distribution Systems	Valves
ASME B31.1-2018	Power Piping	Piping
ASME B36.19M-2018	Stainless Steel Pipe	Piping, Ferrous
ASME BPVC Section IV-2015	Rules for Construction of Heating Boilers	Miscellaneous
ASSE 1017-2009	Temperature Actuated Mixing Valves for Hot Water Distribution Systems	Valves
ASSE 1032-2004 (R2011)	Dual Check Valve Type Backflow Preventers for Carbonated Beverage Dispensers, Post Mix Type	Backflow Protection
ASSE 1066-1997	Individual Pressure Balancing In-Line Valves for Individual Fixture Fittings	Valves
ASSE 1082-2018	Water Heaters with Integral Temperature Control Devices for Hot Water Distribution Systems	Appliances
ASTM A48/A48M-2003 (R2016)	Gray Iron Castings	Piping, Ferrous
ASTM A126-2004 (R2014)	Gray Iron Castings for Valves, Flanges, and Pipe Fittings	Piping, Ferrous
ASTM A377-2018	Ductile Iron Pressure Pipe	Piping, Ferrous
ASTM A479/A479M-2018	Stainless Steel Bars and Shapes for Use in Boilers and Other Pressure Vessels	Piping, Ferrous
ASTM A536-1984 (R2014)	Ductile Iron Castings	Piping, Ferrous
ASTM A733-2016	Welded and Seamless Carbon Steel and Austenitic Stainless Steel Pipe Nipples	Piping, Ferrous
ASTM A1045-2010 (R2017)	Flexible Poly (Vinyl Chloride) (PVC) Gaskets used in Connection of Vitreous China Plumbing Fixtures to Sanitary Drainage Systems	Piping, Plastic
ASTM B29-2014	Refined Lead	Joints
ASTM B370-2012	Copper Sheet and Strip for Building Construction	Miscellaneous
ASTM B687-1999 (R2016)	Brass, Copper, and Chromium-Plated Pipe Nipples	Piping, Copper Alloy
ASTM C14-2015a	Nonreinforced Concrete Sewer, Storm Drain, and Culvert Pipe	Piping, Non-Metallic

TABLE 1701.2 (continued)
STANDARDS, PUBLICATIONS, PRACTICES, AND GUIDES

DOCUMENT NUMBER	DOCUMENT TITLE	APPLICATION
ASTM C412-2015	Concrete Drain Tile	Piping, Non-Metallic
ASTM C443-2012 (R2017)	Joints for Concrete Pipe and Manholes, Using Rubber Gaskets	Joints
ASTM C444-2017	Perforated Concrete Pipe	Piping, Non-Metallic
ASTM C478-2018	Circular Precast Reinforced Concrete Manhole Sections	Miscellaneous
ASTM C1227-2013	Precast Concrete Septic Tanks	DWV Components
ASTM C1440-2017	Thermoplastic Elastomeric (TPE) Gasket Materials for Drain, Waste, and Vent (DWV), Sewer, Sanitary and Storm Plumbing Systems	Joints
ASTM D1784-2011	Rigid Poly (Vinyl Chloride) (PVC) Compounds and Chlorinated Poly (Vinyl Chloride) (CPVC) Compounds	Piping, Plastic
ASTM D2321-2018	Underground Installation of Thermoplastic Pipe for Sewers and Other Gravity-Flow Applications	Piping, Plastic
ASTM D2517-2018	Reinforced Epoxy Resin Gas Pressure Pipe and Fittings	Piping, Plastic
ASTM D2657-2007 (R2015)	Heat Fusion Joining of Polyolefin Pipe and Fittings	Joints
ASTM D2774-2012	Underground Installation of Thermoplastic Pressure Piping	Piping, Plastic
ASTM D2855-2015	Two-Step (Primer and Solvent Cement) Method of Joining Poly (Vinyl Chloride) (PVC) or Chlorinated Poly (Vinyl Chloride) (CPVC) Pipe and Piping Components with Tapered Sockets	Joints
ASTM D3122-2015	Solvent Cement for Styrene-Rubber (SR) Plastic Pipe and Fittings	Joints
ASTM D3311-2017	Drain, Waste, and Vent (DWV) Plastic Fittings Patterns	Fittings
ASTM F402-2018	Safe Handling of Solvent Cements, Primers, and Cleaners Used for Joining Thermoplastic Pipe and Fittings	Joints
ASTM F480-2014	Thermoplastic Well Casing Pipe and Couplings Made in Standard Dimension Ratios (SDR), SCH 40 and SCH 80	Piping, Plastic
ASTM F810-2012 (R2018)	Smoothwall Polyethylene (PE) Pipe for Use in Drainage and Waste Disposal Absorption Fields	Piping, Plastic
ASTM F949-2015	Poly (Vinyl Chloride) (PVC) Corrugated Sewer Pipe with a Smooth Interior and Fittings	Piping, Plastic
ASTM F1476-2007 (R2013)	Performance of Gasketed Mechanical Couplings for Use in Piping Applications	Joints
ASTM F1499-2017	Coextruded Composite Drain, Waste, and Vent Pipe (DWV)	Piping, Plastic
ASTM F1743-2017	Rehabilitation of Existing Pipelines and Conduits by Pulled-in-Place Installation of Cured-in-Place Thermosetting Resin Pipe (CIPP)	Piping, Plastic
ASTM F1924-2012	Plastic Mechanical Fittings for Use on Outside Diameter Controlled Polyethylene Gas Distribution Pipe and Tubing	Fittings
ASTM F1948-2015	Metallic Mechanical Fittings for Use on Outside Diameter Controlled Thermoplastic Gas Distribution Pipe and Tubing	Fittings
ASTM F2165-2013	Flexible Pre-Insulated Piping	Piping, Plastic
ASTM F2206-2014	Fabricated Fittings of Butt-Fused Polyethylene (PE)	DWV Components
ASTM F2306/F2306M-2018	12 to 60 in. [300 to 1500 mm] Annular Corrugated Profile-Wall Polyethylene (PE) Pipe and Fittings for Gravity-Flow Storm Sewer and Subsurface Drainage Applications	Piping, Plastic
AWS B2.4-2012	Welding Procedure and Performance Qualification for Thermoplastics	Joints, Certification
AWWA C203-2015	Coal-Tar Protective Coatings and Linings for Steel Water Pipe	Miscellaneous
AWWA C213-2015	Fusion-Bonded Epoxy Coatings and Linings for Steel Water Pipe and Fittings	Miscellaneous
AWWA C215-2016	Extruded Polyolefin Coatings for Steel Water Pipe	Miscellaneous
AWWA C606-2015	Grooved and Shouldered Joints	Joints
CGA C-9-2013	Standard Color Marking of Compressed Gas Containers for Medical Use	Miscellaneous
CGA S-1.3-2008	Pressure Relief Device Standards-Part 3-Stationary Storage Containers for Compressed Gases	Fuel Gas
CGA V-1-2013	Compressed Gas Cylinder Valve Outlet and Inlet Connections	Valves
CSA A257-2014	Concrete Pipe and Manhole Sections	Piping

TABLE 1701.2 (continued)
STANDARDS, PUBLICATIONS, PRACTICES, AND GUIDES

DOCUMENT NUMBER	DOCUMENT TITLE	APPLICATION
CSA B55.2-2015	Drain Water Heat Recovery Units	Miscellaneous
CSA B64.7-2011 (R2016)	Laboratory Faucet Vacuum Breakers (LFVB)	Backflow Protection
CSA B66-2016	Design, Material, and Manufacturing Requirements for Prefabricated Septic Tanks and Sewage Holding Tanks	DWV Components
CSA B128.1-2006/B128.2-2006 (R2016)	Design and Installation of Non-Potable Water Systems/Maintenance and Field Testing of Non-Potable Water Systems	Miscellaneous
CSA B242-2005 (R2016)	Groove- and Shoulder-Type Mechanical Pipe Couplings	Fittings
CSA B356-2010 (R2015)	Water Pressure Reducing Valves for Domestic Water Supply Systems	Valves
CSA G401-2014	Corrugated Steel Pipe Products	Piping, Ferrous
CSA Z21.12b-1994 (R2015)	Draft Hoods	Fuel Gas, Appliances
CSA Z21.13-2017	Gas-Fired Low-Pressure Steam and Hot Water Boilers (same as CSA 4.9)	Fuel Gas, Appliances
CSA Z21.15b-2013 (R2014)	Manually Operated Gas Valves for Appliances, Appliance Connector Valves, and Hose End Valves (same as CSA 9.1b)	Fuel Gas
CSA Z21.81a-2007 (R2015)	Cylinder Connection Devices (same as CSA 6.25a)	Fuel Gas
CSA Z21.86-2016	Vented Gas-Fired Space Heating Appliances (same as CSA 2.32)	Fuel Gas, Appliances
CSA Z83.11-2016	Gas Food Service Equipment (same as CSA 1.8)	Fuel Gas, Appliances
CSA Z317.1-2016	Special Requirements for Plumbing Installations in Health Care Facilities	Miscellaneous
Energy Star-2012 (version 2.0)	Program Requirements Product Specification for Commercial Dishwashers	Appliances
Energy Star-2018 (version 8.0)	Program Requirements Product specification for Clothes Washers (effective February 5, 2018)	Appliances
Energy Star-2016 (version 6.0)	Program Requirements for Residential Dishwashers	Appliances
EPA/600/R-12/618-2012	Guidelines for Water Reuse	Miscellaneous
EPA WaterSense-2007	High-Efficiency Lavatory Faucet Specification	Fixtures
EPA WaterSense-2009	Specification for Flushing Urinals	Fixtures
EPA WaterSense-2014	Specification for Tank-Type Toilets	Fixtures
IAPMO IGC 67-2014[e1]	Specialized ABS and PVC DWV Fittings	DWV Components
IAPMO IGC 109-2017	Water Distribution Manifolds	Valves
IAPMO IGC 183-2016	Oil/Water Separators and Coalescing Plate Separators	DWV Components
IAPMO IGC 193-2010	Safety Plates, Plate Straps, Notched Plates and Safety Collars	Miscellaneous
IAPMO IGC 226-2006a	Drinking Water Fountains With or Without Chiller or Heater	Fixtures
IAPMO IGC 244-2015a	Tub and Shower Flow-Reduction Systems	Valves
IAPMO IGC 267-2015[e1]	Hydrants Without Integral Backflow Preventers	Valves
IAPMO IGC 276-2011	Bundled Expanded Polystyrene Synthetic Aggregate Units	DWV Components
IAPMO IGC 315-2016	Water Manifold Systems	Fittings
IAPMO IGC 327-2016	Flexible Metallic Expansion Joints for Pressure Systems	Joints
IAPMO IGC 330-2017	Recirculating Shower Systems	Fixtures
IAPMO IGC 332-2017a	Hydronic Radiators	Miscellaneous
IAPMO PS 1-2007	Tank Risers	DWV Components
IAPMO PS 23-2006a	Dishwasher Drain Airgaps	Backflow Protection
IAPMO PS 25-2002	Metallic Fittings for Joining Polyethylene Pipe for Water Service and Yard Piping	Joints
IAPMO PS 34-2003	Encasement Sleeve for Potable Water Pipe and Tubing	Piping

TABLE 1701.2 (continued)
STANDARDS, PUBLICATIONS, PRACTICES, AND GUIDES

DOCUMENT NUMBER	DOCUMENT TITLE	APPLICATION
IAPMO PS 36-2014[e1]	Lead-Free Sealing Compounds for Threaded Joints	Joints
IAPMO PS 37-1990	Black Plastic PVC or PE Pressure-Sensitive Corrosion Preventive Tape	Miscellaneous
IAPMO PS 42-2013[e1]	Pipe Alignment and Secondary Support Systems	Miscellaneous
IAPMO PS 50-2010	Flush Valves with Dual Flush Device for Water Closets or Water Closet Tank with an Integral Flush Valves with a Dual Flush Device	Fixtures
IAPMO PS 51-2016	Expansion Joints and Flexible Expansion Joints for DWV Piping Systems	Joints
IAPMO PS 52-2009[e1]	Pump/Dose, Sumps and Sewage Ejector Tanks with or without a Pump	DWV Components
IAPMO PS 53-2016a	Grooved Mechanical Pipe Couplings and Grooved Fittings	Joints
IAPMO PS 54-2015	Metallic and Plastic Utility Boxes	Miscellaneous
IAPMO PS 57-2002	PVC Hydraulically Actuated Diaphragm Type Water Control Valves	Valves
IAPMO PS 63-2014	Plastic Leaching Chambers	DWV Components
IAPMO PS 64-2012a[e1]	Roof Pipe Flashings	Miscellaneous
IAPMO PS 66-2015	Dielectric Fittings	Fittings
IAPMO PS 67-2010	Early-Closure Replacement Flappers or Early-Closure Replacement Flapper with Mechanical Assemblies	Fixtures
IAPMO PS 69-2006	Bathwaste and Overflow Assemblies with Tub Filler Spout	DWV Components
IAPMO PS 72-2007[e1]	Valves with Atmospheric Vacuum Breakers	Valves
IAPMO PS 73-2015	Dental Liquid-Ring Vacuum Pumps	Miscellaneous
IAPMO PS 76-2012a	Trap Primers for Fill Valves and Flushometer Valves	DWV Components
IAPMO PS 79-2005	Multiport Electronic Trap Primer	DWV Components
IAPMO PS 80-2008	Clarifiers	DWV Components
IAPMO PS 81-2006	Precast Concrete Seepage Pit Liners and Covers	DWV Components
IAPMO PS 82-1995	Fiberglass (Glass Fiber Reinforced Thermosetting Resin) Fittings	Fittings
IAPMO PS 85-1995	Tools for Mechanically Formed Tee Connections in Copper Tubing	Miscellaneous
IAPMO PS 86-1995	Rainwater Diverter Valve for Non-Roofed Area Slabs	DWV Components
IAPMO PS 89-1995	Soaking and Hydrotherapy (Whirlpool) Bathtubs with Hydraulic Seatlift	Fixtures
IAPMO PS 90-2014	Elastomeric Test Caps, Cleanout Caps, and Combination Test Caps/Shielded Couplings	DWV Components
IAPMO PS 91-2005a	Plastic Stabilizers for Use with Plastic Closet Bends	DWV Components
IAPMO PS 92-2013[e1]	Heat Exchangers and Indirect Water Heaters	Miscellaneous
IAPMO PS 94-2012[e1]	Insulated Protectors for P-Traps, Supply Stops and Risers	Miscellaneous
IAPMO PS 95-2018[e1]	Pipe Support Hangers and Hooks	DWV Components
IAPMO PS 98-1996	Prefabricated Fiberglass Church Baptisteries	Fixtures
IAPMO PS 100-1996	Porous Filter Protector for Sub-Drain Weep Holes	DWV Components
IAPMO PS 101-1997	Suction Relief Valves	Valves
IAPMO PS 104-1997	Pressure Relief Connection for Dispensing Equipment	Valves
IAPMO PS 105-1997	Polyethylene Distribution Boxes	DWV Components
IAPMO PS 106-2015[e1]	Tileable Shower Receptors and Shower Kits	Fixtures
IAPMO PS 110-2006a	PVC Cold Water Compression Fittings	Fittings
IAPMO PS 111-1999	PVC Cold Water Gripper Fittings	Fittings
IAPMO PS 112-1999	PVC Plastic Valves for Cold Water Distribution Systems Outside a Building and CPVC Plastic Valves for Hot and Cold Water Distribution Systems	Valves
IAPMO PS 113-2010	Hydraulically Powered Household Food Waste Disposers	Appliances
IAPMO PS 114-1999[e1]	Remote Floor Box Industrial Water Supply, Air Supply, Drainage	Miscellaneous

TABLE 1701.2 (continued)
STANDARDS, PUBLICATIONS, PRACTICES, AND GUIDES

DOCUMENT NUMBER	DOCUMENT TITLE	APPLICATION
IAPMO PS 115-2007	Hot Water On-Demand or Automatic Activated Hot Water Pumping Systems	Miscellaneous
IAPMO PS 116-1999	Hot Water Circulating Devices Which Do Not Use a Pump	Miscellaneous
IAPMO PS 119-2012a[e3]	Water-Powered Sump Pumps	Miscellaneous
IAPMO Z124.7-2013	Prefabricated Plastic Spa Shells	Fixtures, Swimming Pools, Spas, and Hot Tubs
IAPMO Z124.8-2013[e2]	Plastic Liners for Bathtubs and Shower Receptors	Fixtures
IAPMO Z1000-2013	Prefabricated Septic Tanks	DWV Components
MSS SP-25-2018	Standard Marking System for Valves, Fittings, Flanges, and Unions	Miscellaneous
MSS SP-42-2013	Corrosion-Resistant Gate, Globe, Angle, and Check Valves with Flanged and Butt Weld Ends (Classes 150, 300, & 600)	Piping, Ferrous
MSS SP-44-2016 (R2017)	Steel Pipeline Flanges	Fittings
MSS SP-83-2018	Class 3000 and 6000 Pipe Unions, Socket Welding and Threaded (Carbon Steel, Alloy Steel, Stainless Steels, and Nickel Alloys)	Joints
MSS SP-104-2018	Wrought Copper, Solder-Joint Pressure Fittings	Fittings
MSS SP-106-2012	Cast Copper Alloy Flanges and Flanged Fittings: Class 125, 150, and 300	Fittings
MSS SP-109-2018	Weld-Fabricated, Copper Solder-Joint Pressure Fittings	Fittings
MSS SP-123-2013	Non-Ferrous Threaded and Solder-Joint Unions for Use with Copper Water Tube	Joints
NFPA 13R-2019	Installation of Sprinkler Systems in Low-Rise Residential Occupancies	Miscellaneous
NFPA 80-2019	Fire Doors and Other Opening Protectives	Miscellaneous
NFPA 501A-2017	Fire Safety Criteria for Manufactured Home Installations, Sites, and Communities	Miscellaneous
NFPA 1981-2013	Open-Circuit Self-Contained Breathing Apparatus (SCBA) for Emergency Services	Miscellaneous
NFPA 1989-2019	Breathing Air Quality for Emergency Services Respiratory Protection	Miscellaneous
NFPA 5000-2018	Building Construction and Safety Code	Miscellaneous
NSF 2-2018	Food Equipment	Appliances
NSF 4-2016	Commercial Cooking, Rethermalization, and Powered Hot Food Holding and Transportation Equipment	Appliances
NSF 5-2016	Water Heaters, Hot Water Supply Boilers, and Heat Recovery Equipment	Appliances
NSF 12-2018	Automatic Ice Making Equipment	Appliances
NSF 18-2016	Manual Food and Beverage Dispensing Equipment	Appliances
NSF 29-2017	Detergent and Chemical Feeders for Commercial Spray-Type Dishwashing Machines	Appliances
NSF 40-2018	Residential Wastewater Treatment Systems	DWV Components
NSF 41-2018	Non-Liquid Saturated Treatment Systems	DWV Components
NSF 46-2018	Evaluation of Components and Devices Used in Wastewater Treatment Systems	DWV Components
NSF 169-2016	Special Purpose Food Equipment and Devices	Appliances
PSAI Z4.1-2016	For Sanitation – In Places of Employment: Minimum Requirements	Miscellaneous
SAE J512-1997	Automotive Tube Fittings	Fittings
SAE J1670-2008	Type "F" Clamps for Plumbing Applications	Joints
TCNA A118.10-2014	Load Bearing, Bonded, Waterproof Membranes for Thin-Set Ceramic Tile and Dimension Stone Installation	Miscellaneous
Title 49, Code of Federal Regulations, Part 192	Transportation of Natural and Other Gas by Pipeline: Minimum Federal Standards	Miscellaneous
UL 70-2001	Septic Tanks, Bituminous-Coated Metal	DWV Components
UL 80-2007	Steel Tanks for Oil-Burner Fuels and Other Combustible Liquids (with revisions through January 16, 2014)	Fuel Gas
UL 144-2012	LP-Gas Regulators (with revisions through November 5, 2014)	Fuel Gas

**TABLE 1701.2 (continued)
STANDARDS, PUBLICATIONS, PRACTICES, AND GUIDES**

DOCUMENT NUMBER	DOCUMENT TITLE	APPLICATION
UL 252-2017	Compressed Gas Regulators (with revisions through August 10, 2018)	Fuel Gas
UL 296-2017	Oil Burners (with revisions through November 29,2017)	Fuel Gas, Appliances
UL 404-2010	Gauges, Indicating Pressure, for Compressed Gas Service (with revisions through February 11, 2015)	Fuel Gas
UL 429-2013	Electrically Operated Valves	Valves
UL 536-2014	Flexible Metallic Hose	Fuel Gas
UL 563-2009	Ice Makers (with revisions through August 30, 2018)	Appliances
UL 569-2013	Pigtails and Flexible Hose Connectors for LP-Gas (with revisions through July 28, 2017)	Fuel Gas
UL 726-1995	Oil-Fired Boiler Assemblies (with revisions through October 9, 2013)	Fuel Gas, Appliances
UL 1206-2003	Electric Commercial Clothes-Washing Equipment (with revisions through January 7, 2019)	Appliances
UL 1331-2005	Station Inlets and Outlets (with revisions through May 12, 2017)	Medical Gas
UL 1795-2016	Hydromassage Bathtubs (with revisions through December 8, 2017)	Fixtures
UL 1951-2011	Electric Plumbing Accessories (with revisions through August 25, 2017)	Miscellaneous
UL 2157-2018	Electric Clothes Washing Machines and Extractors	Appliances
WQA S-300-2000	Point-of-Use Low-Pressure Reverse Osmosis Drinking Water Systems	Appliances

ABBREVIATIONS IN TABLE 1701.1 AND TABLE 1701.2

AHAM	Association of Home Appliance Manufacturers, 1111 19th Street, NW, Suite 402, Washington, DC 20036.
ANSI	American National Standards Institute, Inc., 25 W. 43rd Street, 4th Floor, New York, NY 10036.
ARCSA	American Rainwater Catchment Systems Association, 6101 Long Prairie Road, Suite 744, PMB 251, Flower Mound, TX 75028.
ASCE	American Society of Civil Engineers, 1801 Alexander Bell Drive, Reston, VA 20191-4400.
ASHRAE	American Society of Heating, Refrigerating, and Air Conditioning Engineers, Inc., 1791 Tullie Circle, NE, Atlanta, GA 30329-2305.
ASME	American Society of Mechanical Engineering, Two Park Avenue, New York, NY 10016-5990.
ASPE	American Society of Plumbing Engineers, 6400 Shafer Court, Suite 350, Rosemont, IL 60018.
ASSE	American Society of Sanitary Engineering, 18927 Hickory Creek Drive, Suite 220, Mokena, IL 60448.
ASTM	ASTM International, 100 Barr Harbor Drive, West Conshohocken, PA 19428-2959.
AWS	American Welding Society, 8669 NW 36 Street, #130 Miami, FL 33166-6672.
AWWA	American Water Works Association, 6666 W. Quincy Avenue, Denver, CO 80235.
CFR	Code of Federal Regulations, U.S Government Publishing Office, 723 North Capitol Street, N.W. Washington, DC 20401-001
CGA	Compressed Gas Association, 14501 George Carter Way, Suite 103, Chantilly, VA 20151.
CISPI	Cast-Iron Soil Pipe Institute, 2401 Fieldcrest Drive, Mundelein, IL 60060.
CSA	Canadian Standards Association, 178 Rexdale Boulevard, Toronto, Ontario, Canada, M9W 1R3.
DOE	U.S. Department of Energy, Forrestal Building, 1000 Independence Avenue, SW, Washington, DC 20585.
e1	An editorial change since the last revision or reapproval.
ENERGY STAR	1200 Pennsylvania Avenue, N.W., Washington, D.C. 20460.
EPA WATERSENSE	U.S. Environmental Protection Agency, Office of Wastewater Management (4204M), 1200 Pennsylvania Avenue, N.W., Washington, D.C. 20460.
IAPMO	International Association of Plumbing and Mechanical Officials, 4755 E. Philadelphia Street, Ontario, CA 91761.
ICC	International Code Council, 500 New Jersey Avenue, NW, 6th Floor, Washington, DC 20001.
ISEA	International Safety Equipment Association, 1901 N. Moore Street, Suite 808, Arlington, VA 22209-1762.
ISO	International Organization for Standardization, 1 ch. de la Voie-Creuse, Casa Postale 56, CH-1211 Geneva 20, Switzerland.
MSS	Manufacturers Standardization Society of the Valve and Fittings Industry, 127 Park Street, NE, Vienna, VA 22180.
NFPA	National Fire Protection Association, 1 Batterymarch Park, Quincy, MA 02169-7471.
NSF	NSF International, 789 N. Dixboro Road, Ann Arbor, MI 48105.
PDI	Plumbing and Drainage Institute, 800 Turnpike Street, Suite 300, North Andover, MA 01845.
PSAI	Portable Sanitation Association International, 2626 E. 82nd Street, Suite 175, Bloomington, MN 55425.
SAE	Society of Automotive Engineers, 400 Commonwealth Drive, Warrendale, PA 15096.
TCNA	Tile Council of North America, Inc. 100 Clemson Research Blvd., Anderson, SC 29625.
UL	Underwriters Laboratories, Inc., 333 Pfingsten Road, Northbrook, IL 60062.
WQA	Water Quality Association, 4151 Naperville Road, Lisle, IL 60532-3696.

APPENDICES

The appendices are intended to supplement the provisions of the installation requirements of this code. The definitions in Chapter 2 are also applicable to the appendices.

TABLE OF CONTENTS

Page

Appendix A
Recommended Rules for Sizing the Water Supply System 289

Appendix B
Explanatory Notes on Combination Waste and Vent Systems 303

Appendix C
Alternate Plumbing Systems 305

Appendix D
Sizing Storm Water Drainage Systems 313

Appendix E
Manufactured/Mobile Home Parks and Recreational Vehicle Parks 321

Appendix F
Firefighter Breathing Air Replenishment Systems 331

Appendix G
Sizing of Venting Systems 335

Appendix H
Private Sewage Disposal Systems 345

Appendix I
Installation Standards 355

Appendix J
Combination of Indoor and Outdoor Combustion and Ventilation Opening Design 393

Appendix K
Potable Rainwater Catchment Systems 395

Appendix L
Sustainable Practices 399

Appendix M
Peak Water Demand Calculator 415

Appendix N
Impact of Water Temperature on the Potential for Scalding and Legionella Growth 421

APPENDIX A

RECOMMENDED RULES FOR SIZING THE WATER SUPPLY SYSTEM

A 101.0 General.

A 101.1 Applicability. This appendix provides a general procedure for sizing a water supply system. Because of the variable conditions encountered, it is impractical to lay down definite detailed rules of procedure for determining the sizes of water supply pipes in an appendix, which shall necessarily be limited in length. For an adequate understanding of the problems involved, refer to Water-Distributing Systems for Buildings, Report BMS 79 of the National Bureau of Standards; and Plumbing Manual, Report BMS 66, also published by the National Bureau of Standards.

A 102.0 Preliminary Information.

A 102.1 Daily Service Pressure. Obtain the necessary information regarding the minimum daily service pressure in the area where the building is to be located.

A 102.2 Water Meter. Where the building supply is to be metered, obtain information regarding friction loss relative to the rate of flow of meters in the range of sizes likely to be used. Friction-loss data is capable of being obtained from most manufacturers of water meters. Friction losses for disk-type meters shall be permitted to be obtained from Chart A 102.2.

A 102.3 Local Information. Obtain available local information regarding the use of different kinds of pipe with respect both to durability and to decrease in capacity with the length of service in the particular water supply.

A 103.0 Demand Load.

A 103.1 Supply Demand. Estimate the supply demand for the building main, the principal branches and risers of the system by totaling the fixture units on each, Table A 103.1, and then by reading the corresponding ordinate from Chart A 103.1(1) or Chart A 103.1(2), whichever is applicable.

A 103.2 Continuous Supply Demand. Estimate continuous supply demands in gallons per minute (gpm) (L/s) for lawn sprinklers, air conditioners, etc., and add the sum to the total demand for fixtures. The result is the estimated supply demand of the building supply.

A 104.0 Permissible Friction Loss.

A 104.1 Residual Pressure. Decide what is the desirable minimum residual pressure that shall be maintained at the highest fixture in the supply system. Where the highest group of fixtures contains flushometer valves, the residual pressure for the group shall be not less than 15 pounds-force per square inch (psi) (103 kPa). For flush tank supplies, the available residual pressure shall be not less than 8 psi (55 kPa).

A 104.2 Elevation. Determine the elevation of the highest fixture or group of fixtures above the water (street) main. Multiply this difference in elevation by 0.43. The result is the loss of static pressure in psi (kPa).

CHART A 102.2
FRICTION LOSSES FOR DISK-TYPE WATER METERS

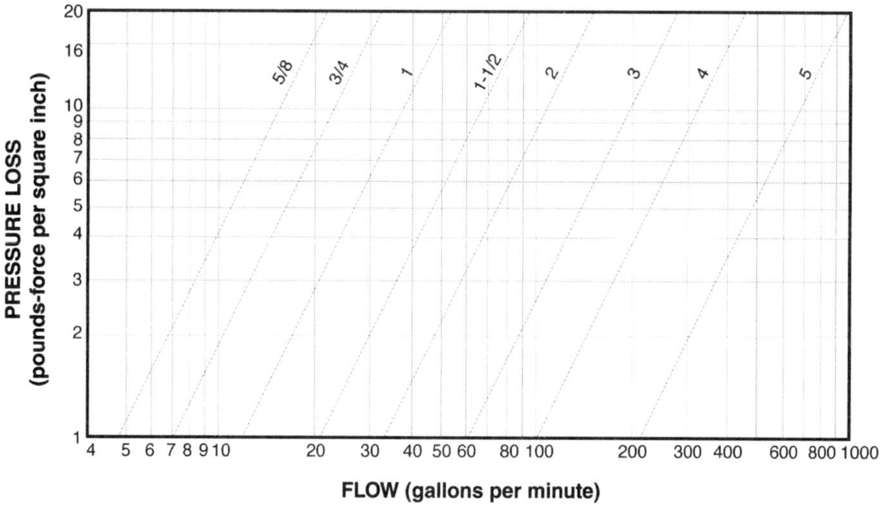

For SI units: 1 inch = 25 mm, 1 pound-force per square inch = 6.8947 kPa, 1 gallon per minute = 0.06 L/s

A 104.3 Available Pressure. Subtract the sum of loss in static pressure and the residual pressure to be maintained at the highest fixture from the average minimum daily service pressure. The result will be the pressure available for friction loss in the supply pipes, where no water meter is used. Where a meter is to be installed, the friction loss in the meter for the estimated maximum demand should also be subtracted from the service pressure to determine the pressure loss available for friction loss in the supply pipes.

A 104.4 Developed Length. Determine the developed length of pipe from the water (street) main to the highest fixture. Where close estimates are desired, compute with the aid of Table A 104.4(1), Table A 104.4(2), or Table A 104.4(3), whichever is applicable, the equivalent length of pipe for fittings in the line from the water (street) main to the highest fixture and add the sum to the developed length. The pressure available for friction loss in psi (kPa), divided by the developed lengths of pipe from the water (street) main to the highest fixture, times 100, will be the average permissible friction loss per 100 feet (30 480 mm) length of pipe.

A 105.0 Size of Building Supply.

A 105.1 Diameter. Knowing the permissible friction loss per 100 feet (30 480 mm) of pipe and the total demand, the diameter of the building supply pipe shall be permitted to be obtained from Chart A 105.1(1), Chart A 105.1(2), Chart A 105.1(3), Chart A 105.1(4) Chart A 105.1(5), Chart A 105.1(6), or Chart A 105.1(7), whichever is applicable. The diameter of pipe on or next above the coordinate point corresponding to the estimated total demand and the permissible friction loss will be the size needed up to the first branch from the building supply pipe.

A 105.2 Copper and Copper Alloy Piping. Where copper tubing or copper alloy pipe is to be used for the supply piping and where the character of the water is such that slight changes in the hydraulic characteristics are expected, Chart A 105.1(1) shall be permitted to be used.

A 105.3 Hard Water. Chart A 105.1(2) shall be used for ferrous pipe with the most favorable water supply in regards to corrosion and caking. Where the water is hard or corrosive, Chart A 105.1(3) or Chart A 105.1(4) will be applicable. For extremely hard water, it will be advisable to make additional allowances for the reduction of the capacity of hot-water lines in service.

A 106.0 Size of Principal Branches and Risers.

A 106.1 Size. The required size of branches and risers shall be permitted to be obtained in the same manner as the building supply, by obtaining the demand load on each branch or riser and using the permissible friction loss computed in Section A 104.0.

A 106.2 Branches. Where fixture branches to the building supply are sized for the same permissible friction loss per 100 feet (30 480 mm) of pipe as the branches and risers to the highest level in the building and lead to the inadequate water supply to the upper floor of a building, one of the following shall be provided:

(1) Selecting the sizes of pipe for the different branches so that the total friction loss in each lower branch is approximately equal to the total loss in the riser, including both friction loss and loss in static pressure.

(2) Throttling each such branch using a valve until the preceding balance is obtained.

(3) Increasing the size of the building supply and risers above the minimum required to meet the maximum permissible friction loss.

A 106.3 Water Closets. The size of branches and mains serving flushometer tanks shall be consistent with sizing procedures for flush tank water closets.

A 107.0 General.

A 107.1 Velocities. Velocities shall not exceed 10 feet per second (ft/s) (3 m/s), except as otherwise approved by the Authority Having Jurisdiction.

A 107.2 Pressure-Reducing Valves. Where a pressure-reducing valve is used in the building supply, the developed length of supply piping and the permissible friction loss shall be computed from the building side of the valve.

A 107.3 Fittings. The allowances in Table A 104.4(1) for fittings are based on non-recessed threaded fittings. For recessed threaded fittings and streamlined soldered fittings, one-half of the allowances given in the table will be ample.

TABLE A 103.1
WATER SUPPLY FIXTURE UNITS (WSFU) AND MINIMUM FIXTURE BRANCH PIPE SIZES[3]

APPLIANCES, APPURTENANCES, OR FIXTURES[2]	MINIMUM FIXTURE BRANCH PIPE SIZE[1,4] (inches)	PRIVATE	PUBLIC	ASSEMBLY[6]
Bathtub or Combination Bath/Shower (fill)	½	4.0	4.0	–
¾ inch Bathtub Fill Valve	¾	10.0	10.0	–
Bidet	½	1.0	–	–
Clothes Washer	½	4.0	4.0	–
Dental Unit, cuspidor	½	–	1.0	–
Dishwasher, domestic	½	1.5	1.5	–
Drinking Fountain or Water Cooler	½	0.5	0.5	0.75
Hose Bibb	½	2.5	2.5	–
Hose Bibb, each additional[7]	½	1.0	1.0	–
Lavatory	½	1.0	1.0	1.0
Lawn Sprinkler, each head[5]	–	1.0	1.0	–
Mobile Home, each (minimum)	–	12.0	–	–
Sinks	–	–	–	–
Bar	½	1.0	2.0	–
Clinical Faucet	½	–	3.0	–
Clinical Flushometer Valve with or without faucet	1	–	8.0	–
Kitchen, domestic	½	1.5	1.5	–
Laundry	½	1.5	1.5	–
Service or Mop Basin	½	1.5	3.0	–
Washup, each set of faucets	½	–	2.0	–
Shower per head	½	2.0	2.0	–
Urinal, 1.0 GPF Flushometer Valve	¾	3.0	4.0	5.0
Urinal, greater than 1.0 GPF Flushometer Valve	¾	4.0	5.0	6.0
Urinal, flush tank	½	2.0	2.0	3.0
Wash Fountain, circular spray	¾	–	4.0	–
Water Closet, 1.6 GPF Gravity Tank	½	2.5	2.5	3.5
Water Closet, 1.6 GPF Flushometer Tank	½	2.5	2.5	3.5
Water Closet, 1.6 GPF Flushometer Valve	1	5.0	5.0	8.0
Water Closet, greater than 1.6 GPF Gravity Tank	½	3.0	5.5	7.0
Water Closet, greater than 1.6 GPF Flushometer Valve	1	7.0	8.0	10.0

For SI units: 1 inch = 25 mm

Notes:

1 Size of the cold branch pipe, or both the hot and cold branch pipes.

2 Appliances, appurtenances, or fixtures not included in this table shall be permitted to be sized by reference to fixtures having a similar flow rate and frequency of use.

3 The listed fixture unit values represent their total load on the cold water building supply. The separate cold water and hot water fixture unit value for fixtures having both cold and hot water connections shall be permitted to be three-quarters of the listed total value of the fixture.

4 The listed minimum supply branch pipe sizes for individual fixtures are the nominal (I.D.) pipe size.

5 For fixtures or supply connections likely to impose continuous flow demands, determine the required flow in gallons per minute (gpm) (L/s) and add it separately to the demand in gpm (L/s) for the distribution system or portions thereof.

6 Assembly [Public Use (see Table 422.1)].

7 Reduced fixture unit loading for additional hose bibbs is to be used where sizing total building demand and for pipe sizing where more than one hose bibb is supplied by a segment of water distribution pipe. The fixture branch to each hose bibb shall be sized by 2.5 fixture units.

CHART A 103.1(1)
ESTIMATE CURVES FOR DEMAND LOAD

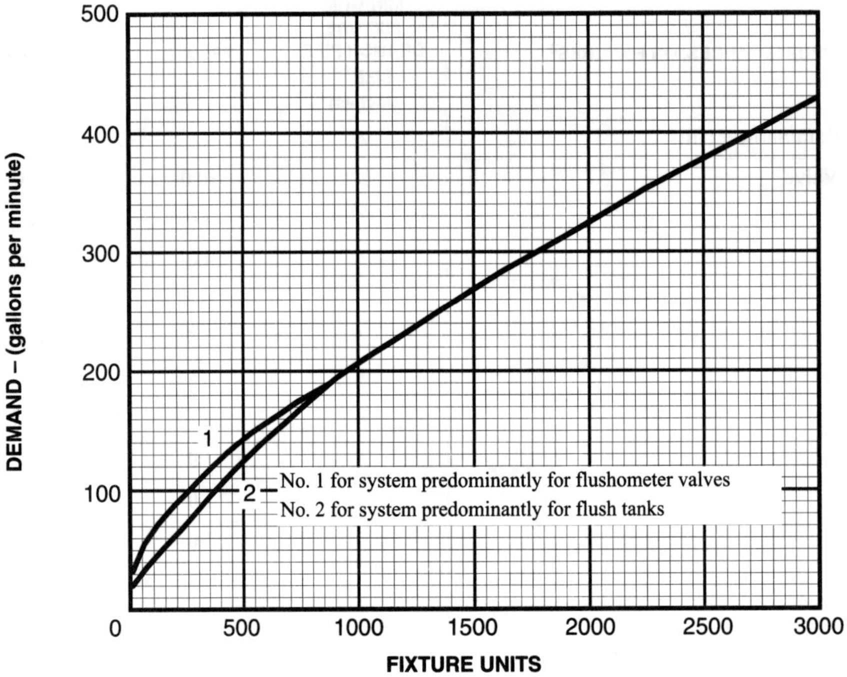

For SI units: 1 gallon per minute = 0.06 L/s

CHART A 103.1(2)
ENLARGED SCALE DEMAND LOAD

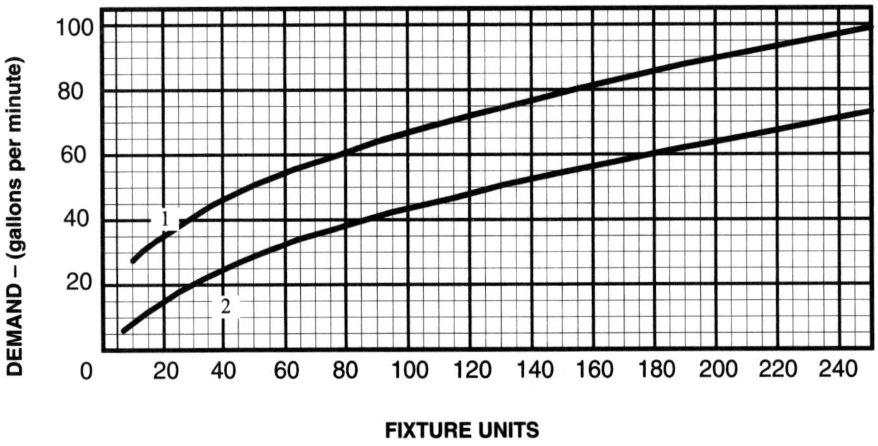

For SI units: 1 gallon per minute = 0.06 L/s

TABLE A 104.4(1)
ALLOWANCE IN EQUIVALENT LENGTH OF PIPE FOR FRICTION LOSS IN VALVES AND THREADED FITTINGS*
EQUIVALENT LENGTH OF PIPE FOR VARIOUS FITTINGS

DIAMETER OF FITTING (inches)	90° STANDARD ELBOW (feet)	45° STANDARD ELBOW (feet)	90° STANDARD TEE (feet)	COUPLING OR STRAIGHT RUN OF TEE (feet)	GATE VALVE (feet)	GLOBE VALVE (feet)	ANGLE VALVE (feet)
3/8	1.0	0.6	1.5	0.3	0.2	8	4
1/2	2.0	1.2	3.0	0.6	0.4	15	8
3/4	2.5	1.5	4.0	0.8	0.5	20	12
1	3.0	1.8	5.0	0.9	0.6	25	15
1 1/4	4.0	2.4	6.0	1.2	0.8	35	18
1 1/2	5.0	3.0	7.0	1.5	1.0	45	22
2	7.0	4.0	10.0	2.0	1.3	55	28
2 1/2	8.0	5.0	12.0	2.5	1.6	65	34
3	10.0	6.0	15.0	3.0	2.0	80	40
4	14.0	8.0	21.0	4.0	2.7	125	55
5	17.0	10.0	25.0	5.0	3.3	140	70
6	20.0	12.0	30.0	6.0	4.0	165	80

For SI units: 1 inch = 25 mm, 1 foot = 304.8 mm, 1 degree = 0.017 rad

* Allowances are based on nonrecessed threaded fittings. Use one-half the allowances for recessed threaded fittings or streamlined solder fittings.

TABLE A 104.4(2)
EQUIVALENT LENGTH OF COPPER TUBE SIZE CPVC PIPE FOR VARIOUS FITTINGS

DIAMETER OF FITTING (inches)	90 DEGREE ELBOW (feet)	45 DEGREE ELBOW (feet)	COUPLING OR STRAIGHT RUN OF TEE (feet)	90 DEGREE STANDARD TEE (feet)
1/2	1.6	0.8	1.0	3.1
3/4	2.1	1.1	1.4	4.1
1	2.6	1.4	1.7	5.3
1 1/4	3.5	1.8	2.3	6.9
1 1/2	4.0	2.1	2.7	8.1
2	5.2	2.8	3.5	10.3

For SI units: 1 inch = 25 mm, 1 foot = 304.8 mm

TABLE A 104.4(3)
EQUIVALENT LENGTH OF SCHEDULE 40 AND 80 CPVC PIPE FOR VARIOUS FITTINGS

DIAMETER OF FITTING (inches)	90 DEGREE ELBOW (feet)	45 DEGREE ELBOW (feet)	COUPLING OR STRAIGHT RUN OF TEE (feet)	90 DEGREE STANDARD TEE (feet)
1/2	1.5	0.8	1.0	4.0
3/4	2.0	1.1	1.4	5.0
1	2.5	1.4	1.7	6.0
1 1/4	3.8	1.8	2.3	7.0
1 1/2	4.0	2.1	2.7	8.0
2	5.7	2.6	4.3	12.0

For SI units: 1 inch = 25 mm, 1 foot = 304.8 mm

CHART A 105.1(1)

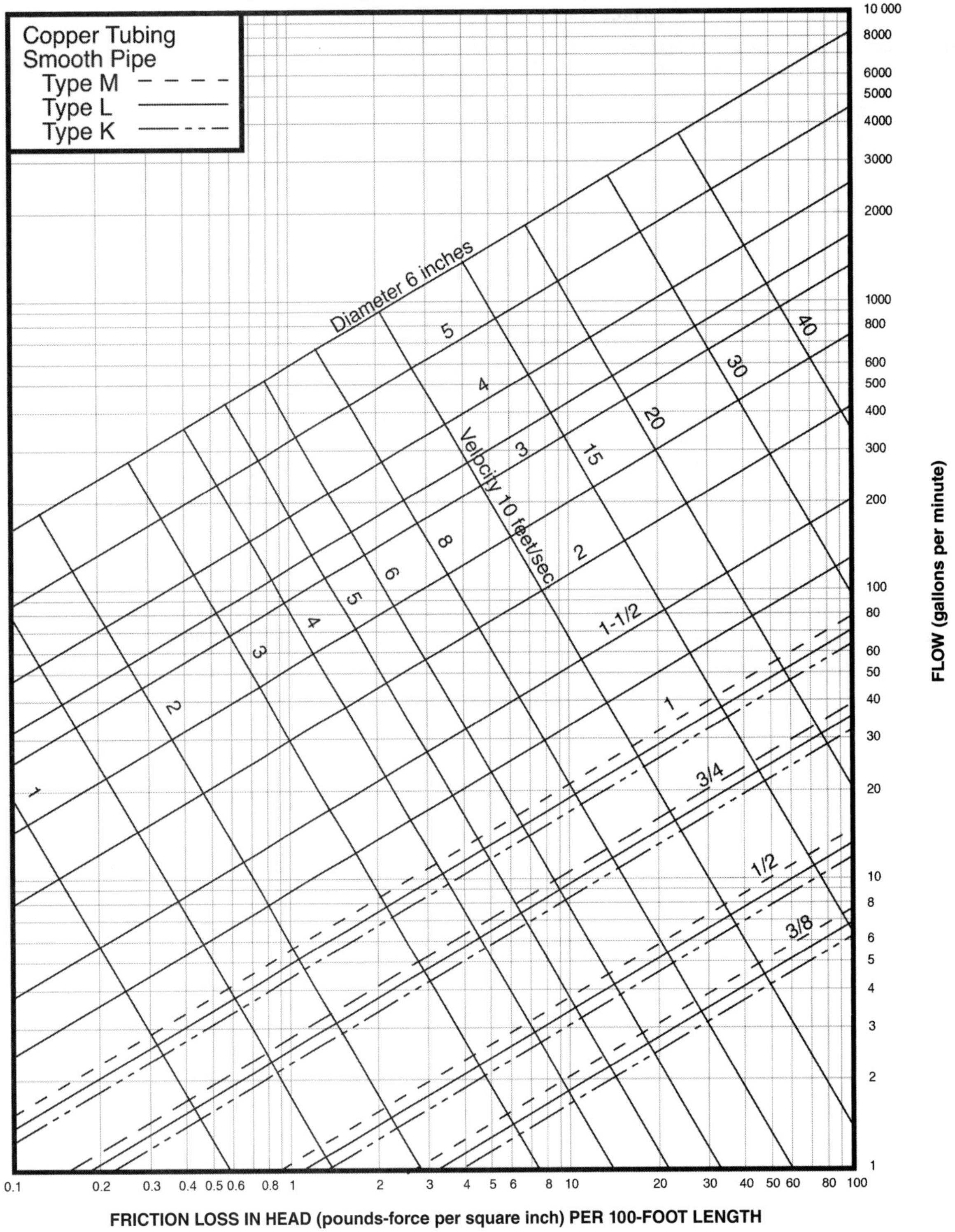

For SI units: 1 inch = 25 mm, 1 gallon per minute = 0.06 L/s, 1 pound-force per square inch = 6.8947 kPa, 1 foot = 304.8 mm, 1 foot per second = 0.3048 m/s

CHART A 105.1(2)

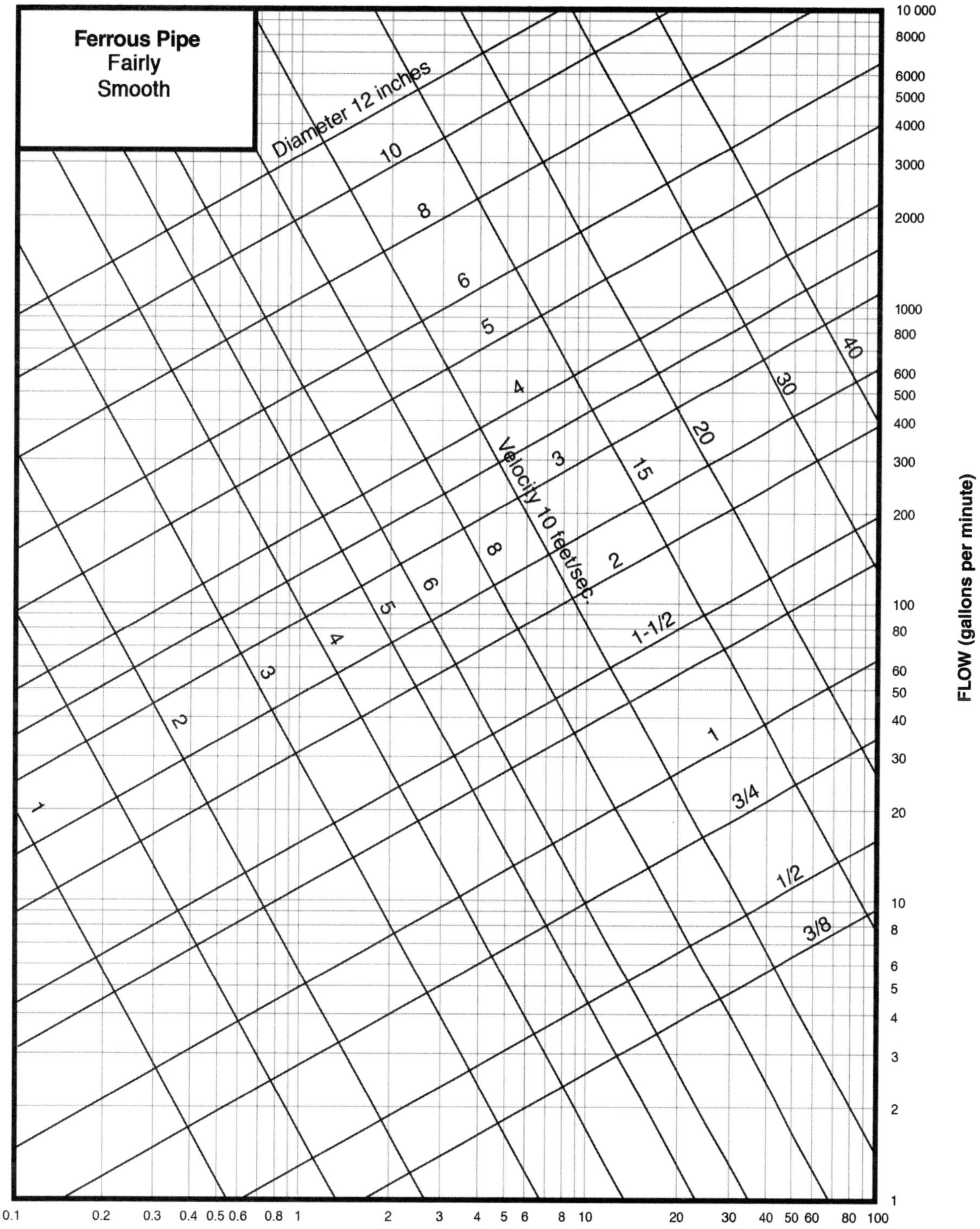

For SI units: 1 inch = 25 mm, 1 gallon per minute = 0.06 L/s, 1 pound-force per square inch = 6.8947 kPa, 1 foot = 304.8 mm, 1 foot per second = 0.3048 m/s

CHART A 105.1(3)

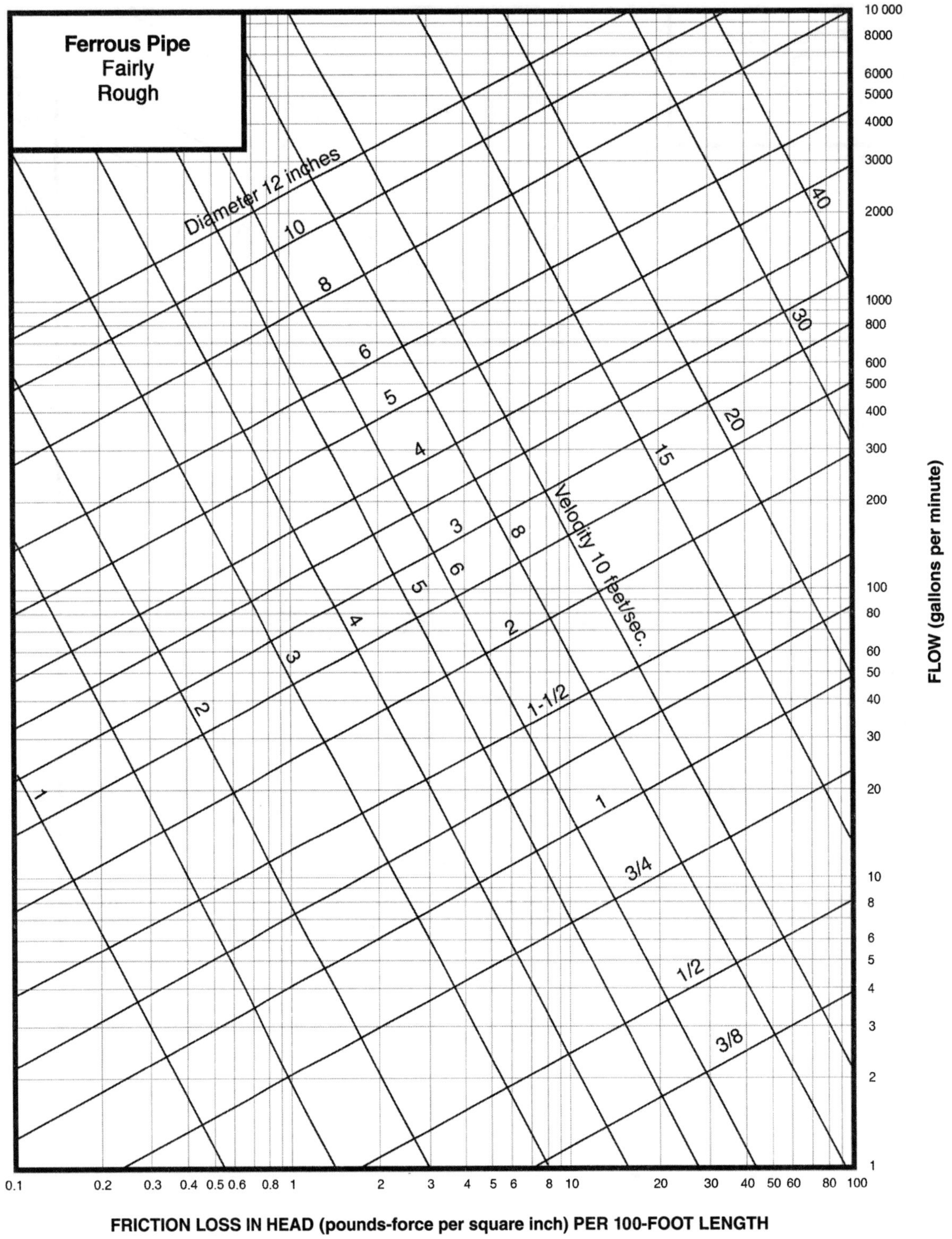

For SI units: 1 inch = 25 mm, 1 gallon per minute = 0.06 L/s, 1 pound-force per square inch = 6.8947 kPa, 1 foot = 304.8 mm, 1 foot per second = 0.3048 m/s

CHART A 105.1(4)

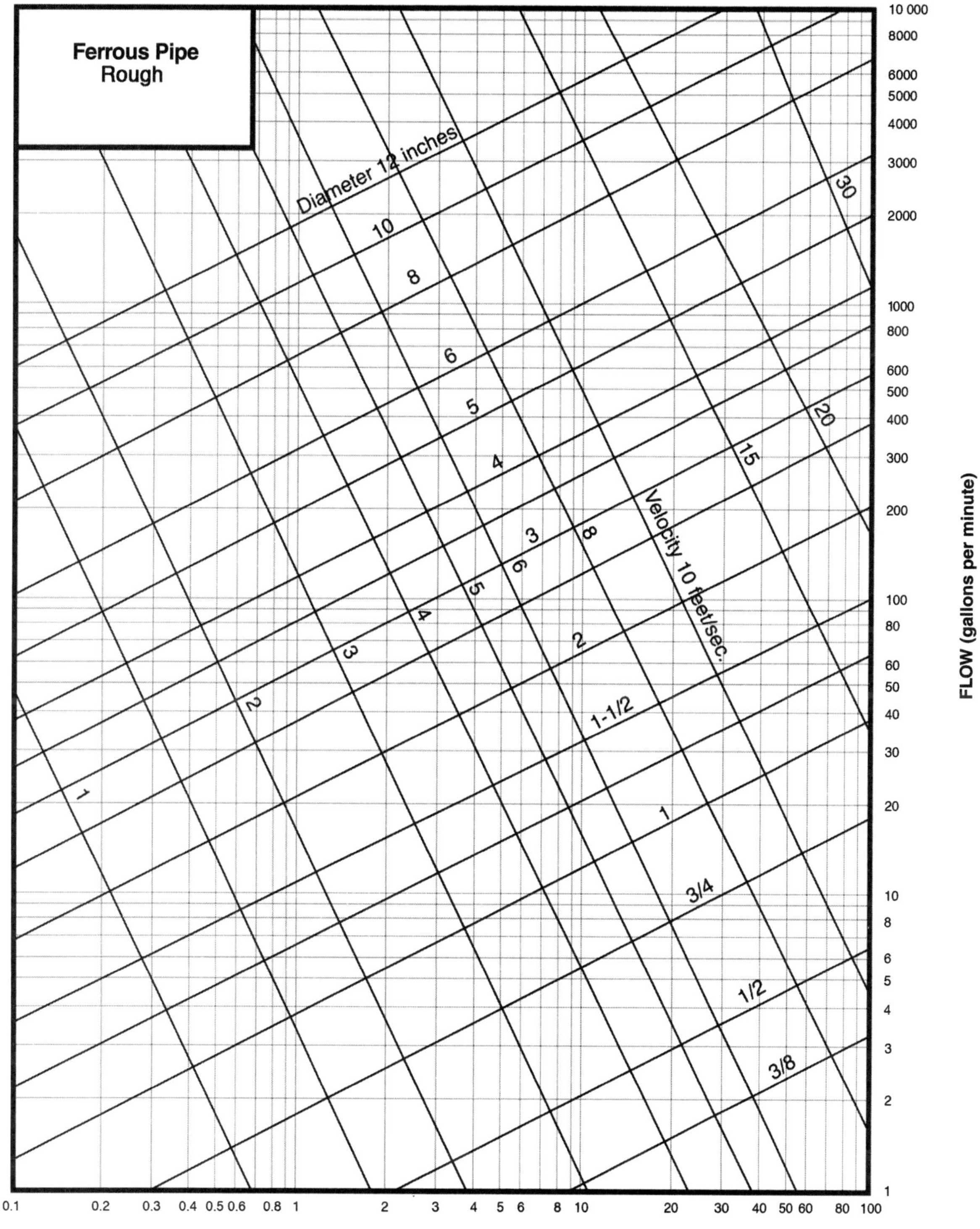

For SI units: 1 inch = 25 mm, 1 gallon per minute = 0.06 L/s, 1 pound-force per square inch = 6.8947 kPa, 1 foot = 304.8 mm, 1 foot per second = 0.3048 m/s

CHART A 105.1(5)

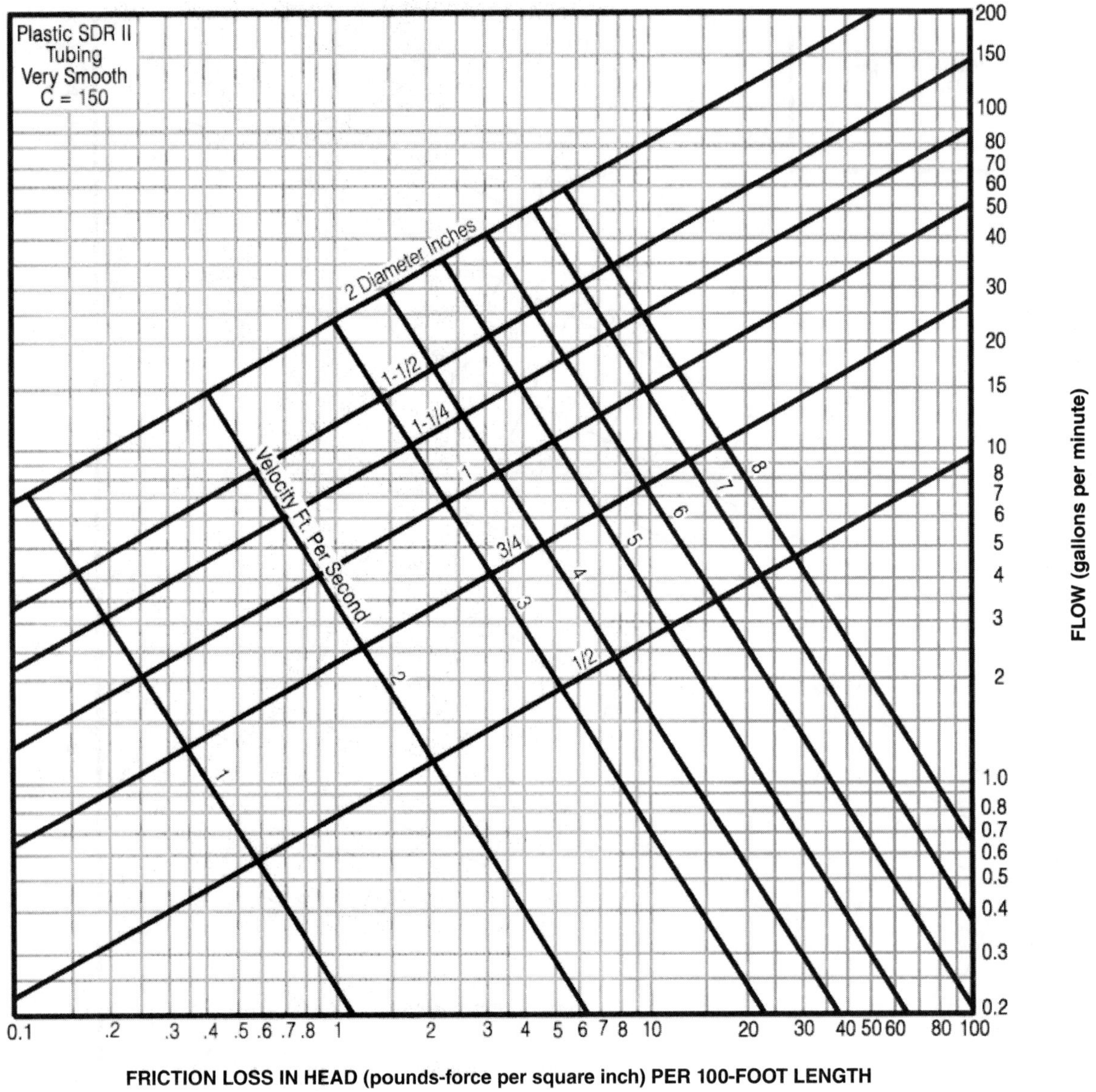

For SI units: 1 inch = 25 mm, 1 gallon per minute = 0.06 L/s, 1 pound-force per square inch = 6.8947 kPa, 1 foot = 304.8 mm, 1 foot per second = 0.3048 m/s

CHART A 105.1(6)

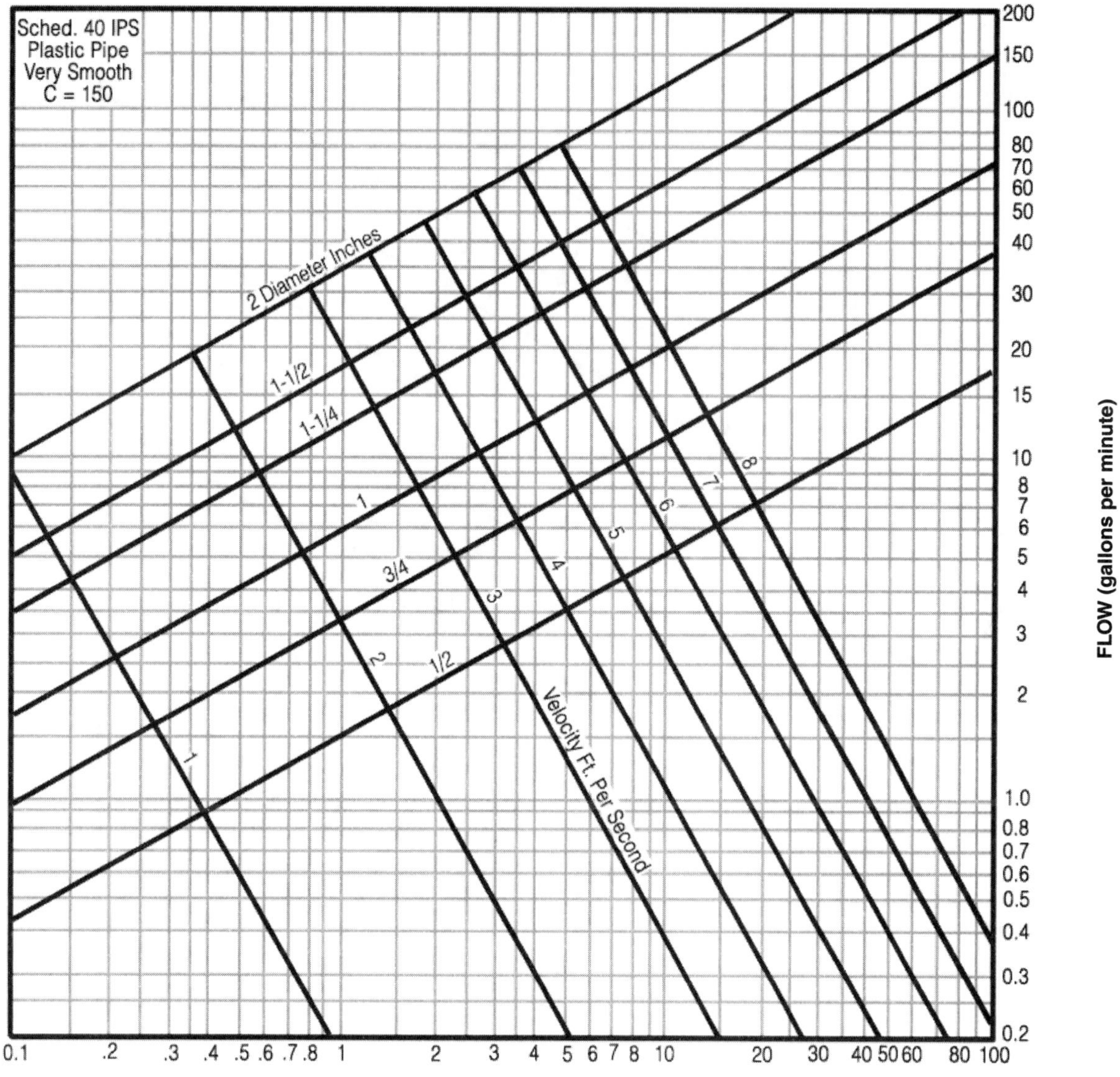

For SI units: 1 inch = 25 mm, 1 gallon per minute = 0.06 L/s, 1 pound-force per square inch = 6.8947 kPa, 1 foot = 304.8 mm, 1 foot per second = 0.3048 m/s

CHART A 105.1(7)

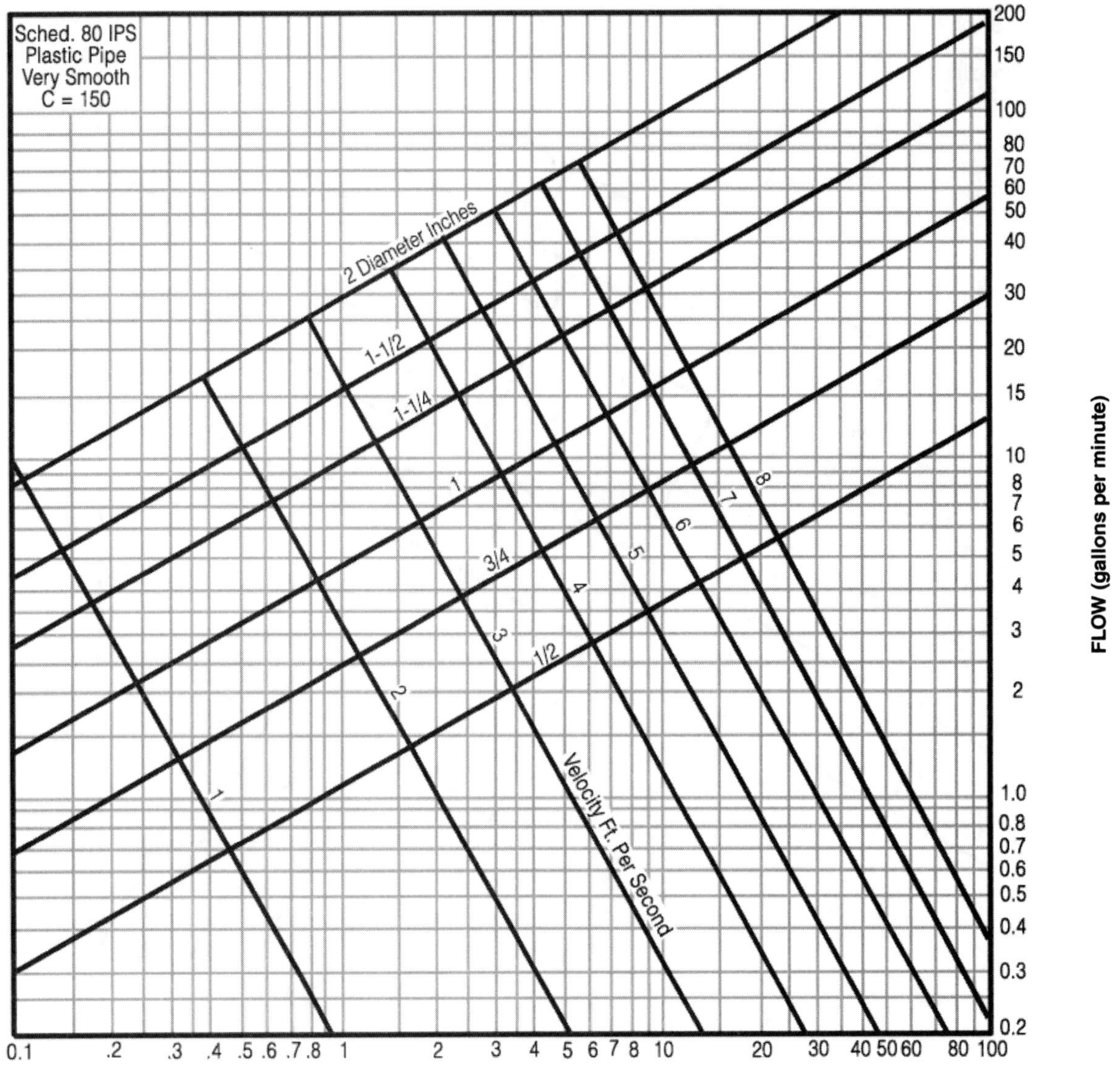

For SI units: 1 inch = 25 mm, 1 gallon per minute = 0.06 L/s, 1 pound-force per square inch = 6.8947 kPa, 1 foot = 304.8 mm, 1 foot per second = 0.3048 m/s

A 108.0 Sizing.

A 108.1 Example. Assume an office building of four stories and basement; pressure on the building side of the pressure-reducing valve of 55 psi (379 kPa) (after an allowance for reduced pressure falloff at peak demand); an elevation of highest fixture above the pressure-reducing valve of 45 feet (13 716 mm); a developed length of pipe from the pressure-reducing valve to the most distant fixture of 200 feet (60 960 mm); and fixtures to be installed with flush valves for water closets and stall urinals as follows:

Where the pipe material and water supply are such that Chart A 105.1(2) applies, the required diameter of the building supply is 3½ inches (90 mm), and the required diameter of the branch to the hot-water heater is 1½ inches (40 mm).

The sizes of the various branches and risers shall be permitted to be determined in the same manner as the size of the building supply or the branch to the hot-water system, by estimating the demand for the riser or branch from Chart A 103.1(1) or Chart A 103.1(2) and applying the total demand estimate from the branch, riser, or section thereof to the appropriate flowchart.

A 108.1 EXAMPLE

FIXTURE UNITS AND ESTIMATED DEMANDS							
BUILDING SUPPLY DEMAND					BRANCH TO HOT WATER SYSTEM		
KIND OF FIXTURES	NUMBER OF FIXTURES	FIXTURE UNIT DEMAND	TOTAL UNITS	BUILDING SUPPLY DEMAND (gallons per minute)	NUMBER OF FIXTURES	FIXTURE UNIT DEMAND CALCULATION	DEMAND (gallons per minute)
Water Closets	130	8.0	1040	–	–	–	–
Urinals	30	4.0	120	–	–	–	–
Showerheads	12	2.0	24	–	12	12 x 2 x ¾ = 18	–
Lavatories	100	1.0	100	–	100	100 x 1 x ¾ = 75	–
Service Sinks	27	3.0	81	–	27	27 x 3 x ¾ = 61	–
Total	–	–	1365	252	–	154	55

For SI units: 1 gallon per minute = 0.06 L/s, 1 pound-force per square foot = 6.8947 kPa

Allowing for 15 psi (103 kPa) at the highest fixture under the maximum demand of 252 gallons per minute (15.90 L/s), the pressure available for friction loss is found by the following:

$$55 - [15 + (45 \times 0.43)] = 20.65 \text{ psi } (142.38 \text{ kPa})$$

The allowable friction loss per 100 feet (30 480 mm) of the pipe is, therefore:

$$100 \times 20.65 \div 200 = 10.32 \text{ psi } (71.15 \text{ kPa})$$

APPENDIX B

EXPLANATORY NOTES ON COMBINATION WASTE AND VENT SYSTEMS

(See Section 910.0 for specific limitations)

B 101.0 General.

B 101.1 Applicability. This appendix provides general guidelines for the design and installation of a combination waste and vent system.

B 101.2 General Requirements. Combination waste and vent systems, as outlined in Section 910.0 of this code, cover the horizontal wet venting of a series of traps using a common waste and vent pipe. Pipe sizes not less than two pipe sizes larger than those required for a conventional system are designed to maintain a wetted perimeter or flow line low enough in the waste pipe to allow adequate air movement in the upper portion, thus balancing the system. Sinks, lavatories, and other fixtures that rough in above the floor, shall not be permitted on a combination waste and vent system, which, at best, is merely an expedient designed to be used in locations where it would be structurally impractical to provide venting in a conventional manner.

Combination waste and vent systems are intended primarily for extensive floor or shower drain installations where separate venting is not practical, for floor sinks in markets, demonstration or work tables in school buildings, or for similar applications where the fixtures are not adjacent to walls or partitions. Due to its oversize characteristics, such a waste system is not self-scouring and, consequently, care shall be exercised as to the type of fixtures connected to it and the location of cleanouts. Given its grease-producing potential, restaurant kitchen equipment shall not be connected to a combination waste and vent system.

B 101.3 Caution. Caution shall be exercised to exclude appurtenances delivering large quantities or surges of water (such as pumps, sand interceptors, etc.) from combination waste and vent systems so that adequate venting will be maintained. Small fixtures with a waste-producing potential of less than 7½ gallons per minute (gpm) (0.47 L/s) shall be permitted to be safely assigned a loading value of one unit. Long runs shall be laid at the minimum permissible slope to keep tailpieces as short as possible. Tailpieces shall not exceed 2 feet (610 mm) in length, which shall necessitate slopes up to 45 degrees (0.79 rad) (see definition of horizontal pipe) on some branches.

B 101.4 Pneumatics. It is essential that the pneumatics of such a system be properly engineered, as the air pressure within the line shall at all times balance that of outside atmosphere to prevent either trap seal loss or air locking between traps. Long mains shall be provided with additional relief vents located at intervals not exceeding 100 feet (30 480 mm). Each such relief vent shall equal not less than one-half of the inside cross-sectional area of the drain pipe served.

B 101.5 Trap Sizes. Trap sizes are required to be equivalent to the branches they serve (two pipe sizes larger than normal), and tailpieces between fixtures or floor drains and such traps shall be reduced to normal size.

B 101.6 Layout Drawings. Duplicate layout drawings of each such proposed piping system shall be presented to the Authority Having Jurisdiction and approval obtained before an installation is made. Complicated layouts shall be checked by qualified personnel.

B 101.6.1 Example of Sizing. A floor drain normally requires a 2 inch (50 mm) trap and waste. On a combination waste and vent system, both trap and waste shall be increased two pipe sizes (through 2½ inches and 3 inches) (65 mm and 80 mm), which would make the trap 3 inches (80 mm). Pipe sizes recognized for this purpose are 2 inches, 2½ inches, 3 inches, 3½ inches, 4 inches, 4½ inches, 5 inches, 6 inches, etc. (50 mm, 65 mm, 80 mm, 90 mm, 100 mm, 115 mm, 125 mm, 150 mm, etc.). The tailpiece between the floor drain and its trap shall be 2 inches (50 mm) (or normal size) to ensure that the amount of wastewater entering the trap partially fills the waste branch. A 3 inch (80 mm) floor drain would thus require a 4 inch (100 mm) trap, and a 4 inch (100 mm) floor drain would require a 5 inch (125 mm) trap for the reasons previously stated.

WHERE IN DOUBT, CHECK WITH YOUR LOCAL Authority Having Jurisdiction.

APPENDIX C

ALTERNATE PLUMBING SYSTEMS

C 101.0 General.

C 101.1 Applicability. The intent of this appendix is to provide clarification of procedures for the design and approval of engineered plumbing systems, alternate materials, and equipment not specifically covered in other parts of the code.

C 101.2 Provisions. The provisions of this appendix apply to the design, installation, and inspection of an engineered plumbing system, alternate material, and equipment.

C 101.3 Authority Having Jurisdiction. The Authority Having Jurisdiction has the right to require descriptive details of an engineered plumbing system, alternate material, or equipment including pertinent technical data to be filed.

C 101.4 Standards and Specifications. Components, materials, and equipment shall comply with standards and specifications listed in Chapter 17 of this code and other national consensus standards applicable to plumbing systems and materials.

C 101.5 Alternate Materials and Equipment. Where such standards and specifications are not available, alternate materials and equipment shall be approved in accordance with the provisions of Section 301.3 of this code.

C 201.0 Definitions.

C 201.1 General. For the purposes of this code, these definitions shall apply to this appendix:

Branch Interval. A length of soil or waste stack corresponding in general to a story height, but in no case less than 8 feet (2438 mm), within which the horizontal branches from one floor or story of the building are connected to the stack.

Engineered Plumbing System. A system designed for a specific building project with drawings and specifications indicating plumbing materials to be installed, all as prepared by a registered design professional.

C 301.0 Engineered Plumbing Systems.

C 301.1 Inspection and Installation. In other than one- and two-family dwellings, the designer of the system is to provide periodic inspection of the installation on a schedule approved by the Authority Having Jurisdiction. Prior to the final approval, the designer shall verify to the Authority Having Jurisdiction that the installation is in accordance with the approved plans, specifications, and data and such amendments to it. The designer shall certify to the Authority Having Jurisdiction that the installation is in accordance with the applicable engineered design criteria.

C 301.2 Owner Information. The designer of the system shall provide the building owner with information concerning the system, considerations applicable for subsequent modifications to the system, and maintenance requirements as applicable.

C 302.0 Water Heat Exchangers.

C 302.1 Protection from Contamination. Heat exchangers used for heat transfer, heat recovery, or solar heating shall protect the potable water system from being contaminated by the heat-transfer medium.

C 302.2 Single-Wall Heat Exchangers. Single-wall heat exchangers shall comply with the following requirements:

(1) The heat-transfer medium is either potable water or contains essentially nontoxic transfer fluids having a toxicity rating or class of 1 (see Section 207.0).

(2) The pressure of the heat-transfer medium is maintained at less than the normal minimum operating pressure of the potable water system.

Exception: Steam in accordance with Section C 302.2(1) above.

(3) The equipment is permanently labeled to indicate that only additives recognized as safe by the FDA shall be used in the heat-transfer medium.

C 302.3 Alternate Designs. Other heat exchanger designs shall be permitted where approved by the Authority Having Jurisdiction.

C 303.0 Fixture Unit Values for Private or Private Use Bathroom Groups.

C 303.1 Fixtures. Table C 303.2 and Table C 303.3 reflect the fixture unit loads for the fixtures in bathrooms as groups, rather than as individual fixtures. Such fixtures include water closets, lavatories, and bathtubs or showers. The tables reflect diversity in the use of fixtures within a bathroom and between multiple bathrooms.

C 303.2 Water Supply Fixture Unit Values. The listed water supply fixture unit values in Table C 303.2 reflect the load of entire bathroom groups on the cold water building supply. Individual hot and cold water branch piping to the fixtures shall be permitted to be sized in accordance with Chapter 6 and Appendix A.

C 303.3 Drainage Fixture Unit Values. The listed drainage fixture unit values in Table C 303.3 reflect a load of entire bathroom groups on the sanitary drainage system. Where fixtures within bathrooms connect to different branches of the drainage system, the fixture unit values for the individual fixtures shall be used, as listed in Table 702.1 of this code.

C 304.0 Drainage System Sizing.

C 304.1 Drainage Fixture Units. Drainage fixture unit values shall be sized in accordance with Section 702.0 and Table 702.1.

TABLE C 303.2
WATER SUPPLY FIXTURE UNITS (WSFU) FOR BATHROOM GROUPS[1, 2]

	PRIVATE USE BATHROOM GROUP		SERVING 3 OR MORE PRIVATE USE BATHROOM GROUPS	
	COLD	HOT[3]	COLD	HOT
Bathroom Groups Having up to 1.6 GPF Gravity-Tank Water Closets				
Half-Bath or Powder Room	3.5	0.8	2.5	0.5
1 Bathroom Group	5.0	2.5	3.5	1.8
1½ Bathrooms	6.0	2.5	–	–
2 Bathrooms	7.0	3.5	–	–
2½ Bathrooms	8.0	3.6	–	–
3 Bathrooms	9.0	4.5	–	–
Each Additional ½ Bath	0.5	0.1	–	–
Each Additional Bathroom Group	1.0	0.5	–	–
Bathroom Groups Having up to 1.6 GPF Pressure-Tank Water Closets				
Half-Bath or Powder Room	3.5	0.8	2.5	0.5
1 Bathroom Group	5.0	2.5	3.5	1.8
1½ Bathrooms	6.0	2.5	–	–
2 Bathrooms	7.0	3.5	–	–
2½ Bathrooms	8.0	3.6	–	–
3 Bathrooms	9.0	4.5	–	–
Each Additional ½ Bath	0.5	0.1	–	–
Each Additional Bathroom Group	1.0	0.5	–	–
Bathroom Group (1.6 GPF Flushometer Value)	6.0	2.5	4.0	1.7
Kitchen Group (Sink and Dishwasher)	2.0	2.0	1.5	1.5
Laundry Group (Sink and Clothes Washer)	5.0	5.0	3.0	3.0

Notes:

[1] A bathroom group, for this table, consists of one water closet, up to two lavatories, and either one bathtub or one shower.

[2] A half-bath or powder room, for this table, consists of one water closet and one lavatory.

[3] Multi-unit dwellings with individual water heaters use the same WSFU as for individual dwellings.

C 304.2 Size of Building Drain and Building Sewer. The maximum number of drainage fixture units allowed on the building drain or building sewer of a given size shall be in accordance with Table C 304.2. The size of a building drain or building sewer serving a water closet shall be not less than 3 inches (80 mm).

C 304.3 Size of Horizontal Branch or Vertical Stack. The maximum number of drainage fixture units allowed on a horizontal branch or vertical soil or waste stack of a given size shall be in accordance with Table C 304.3. Stacks shall be sized based on the total accumulated connected load at each story or branch interval.

C 304.3.1 Horizontal Stack Offsets. Horizontal stack offsets shall be sized in accordance with Table C 304.2 as required for building drains.

C 304.3.2 Vertical Stack Offsets. Vertical stack offsets shall be sized in accordance with Table C 304.3 as required for stacks.

C 304.4 Horizontal Stack Offset and Horizontal Branch Connections. Horizontal branch connections shall not connect to a horizontal stack offset or within 2 feet (610 mm) above or below the offset where such horizontal offset is located more than four branch intervals below the top of the stack.

C 401.0 Vent System Sizing.

C 401.1 Size of Vents. The size of vent piping shall be determined from the developed length and the total number of drainage fixture units connected in accordance with Table C 401.1. Vents shall be not less than one-half the required size of the drainage pipe size served as determined by Table C 304.3 for horizontal fixture branches and stacks nor less than 1¼ inches (32 mm) in diameter. The drainage system shall be vented by not less than one vent pipe which shall be not less than one-half the size of the required building drain and which shall extend from the building drain or extension of building drain to the outdoors. Vents shall be installed in accordance with Chapter 9.

C 401.2 Vent Stack. A vent stack shall be required for a drainage stack that extends five or more branch intervals above the building drain or horizontal branch. The developed length of the vent stack shall be measured from the lowest connection of a branch vent to the termination outdoors.

C 401.3 Branch Vents. Where branch vents exceed 40 feet (12 192 mm) in developed length, such vent shall be increased by one pipe size for the entire developed length of the vent pipe.

C 401.4 Venting Horizontal Offsets. Drainage stacks with horizontal offsets shall be vented where five or more branch

TABLE C 303.3
DRAINAGE FIXTURE UNIT VALUES (DFU) FOR BATHROOM GROUPS[1, 2]

	PRIVATE USE BATHROOM GROUP	SERVING 3 OR MORE PRIVATE USE BATHROOM GROUP
Bathroom Groups having 1.6 GPF Gravity-Tank Water Closets		
Half-Bath or Powder Room	3.0	2.0
1 Bathroom Group	5.0	3.0
1½ Bathrooms	6.0	–
2 Bathrooms	7.0	–
2½ Bathrooms	8.0	–
3 Bathrooms	9.0	–
Each Additional ½ Bath	0.5	–
Each Additional Bathroom Group	1.0	–
Bathroom Groups having 1.6 GPF Pressure-Tank Water Closets		
Half-Bath or Powder Room	3.5	2.5
1 Bathroom Group	5.5	3.5
1½ Bathrooms	6.5	–
2 Bathrooms	7.5	–
2½ Bathrooms	8.5	–
3 Bathrooms	9.5	–
Each Additional ½ Bath	0.5	–
Each Additional Bathroom Group	1.0	–
Bathroom Groups having 3.5 GPF Gravity-Tank Water Closets		
Half-Bath or Powder Room	3.0	2.0
1 Bathroom Group	6.0	4.0
1½ Bathrooms	8.0	–
2 Bathrooms	10.0	–
2½ Bathrooms	11.0	–
3 Bathrooms	12.0	–
Each Additional ½ Bath	0.5	–
Each Additional Bathroom	1.0	–
Bathroom Group (1.6 GPF Flushometer Valve)	3.0	–
Bathroom Group (3.5 GPF Flushometer Valve)	4.0	–

Notes:

[1] A bathroom group, for this table, consists of not more than one water closet, up to two lavatories, and either one bathtub or one shower.

[2] A half-bath or powder room, for this table, consists of one water closet and one lavatory.

intervals are located above the offset. The upper and lower section of the horizontal offset shall be vented in accordance with Section C 401.4.1 and Section C 401.4.2.

C 401.4.1 Venting Upper Section. The vent for the upper section of the stack shall be vented as a separate stack with a vent stack connection installed at the base of the drainage stack. Such vent stack shall connect below the lowest horizontal branch or building drain. Where vent stack connects to the building drain, the connection shall be located downstream of the drainage stack and within a distance of 10 times the diameter of the drainage stack.

C 401.4.2 Venting Lower Section. The vent for the lower section of the stack shall be vented by a yoke vent connecting between the offset and the next lower horizontal branch using a wye-branch fitting. The size of the yoke vent and connection shall be not less in diameter than the required size for the vent serving the drainage stack. The yoke vent connection shall be permitted to be a vertical extension of the drainage stack.

C 501.0 Vacuum Drainage Systems.

C 501.1 General. This section regulates the design and installation provisions for vacuum waste drainage systems. Plans for vacuum waste drainage systems shall be submitted to the Authority Having Jurisdiction for approval and shall be considered an engineered designed system. Such plans shall be prepared by a registered design professional to perform plumbing design work. Details are necessary to ensure compliance with the requirements of this section, together with a full description of the complete installation including quality, grade of materials, equipment, construction, and methods of assembly and installation. Components, materials, and equipment shall comply with standards and specifications listed in Chapter 17 of this code or approved by the Authority Having Jurisdiction and other national consensus standards applicable to plumbing systems and materials. Where such standards and specifications are not available, alternate materials and equipment shall be approved in accordance with Section 301.3.

TABLE C 304.2
BUILDING DRAINS AND BUILDING SEWERS[1]

DIAMETER OF PIPE (inches)	MAXIMUM NUMBER OF DRAINAGE FIXTURE UNITS FOR SANITARY BUILDING DRAINS AND RUNOUTS FROM STACKS			
	SLOPE (inches per foot)			
	1/16	1/8	1/4	1/2
2	–	–	21	26
2½	–	–	24	31
3	–	20	42[2]	50[2]
4	–	180	216	250
5	–	390	480	575
6	–	700	840	1000
8	1400	1600	1920	2300
10	2500	2900	3500	4200
12	3900	4600	5600	6700
15	7000	8300	10 000	12 000

For SI units: 1 inch = 25 mm, 1 inch per foot = 83.3 mm/m

Notes:

1 On-site sewers that serve more than one building shall be permitted to be sized according to the current standards and specifications of the administrative authority for public sewers.

2 A maximum of two water closets or two bathroom groups, except in single-family dwellings, where a maximum of three water closets or three bathroom groups shall be permitted to be installed.

TABLE C 304.3
HORIZONTAL FIXTURE BRANCHES AND STACKS

DIAMETER OF PIPE (inches)	MAXIMUM NUMBER OF DRAINAGE FIXTURE UNITS			
	HORIZONTAL FIXTURE BRANCH[1]	ONE STACK OF THREE OR FEWER BRANCH INTERVALS	STACKS WITH MORE THAN THREE BRANCH INTERVALS	
			TOTAL FOR STACK	TOTAL AT ONE BRANCH INTERVAL
1½	3	4	8	2
2	6	10	24	6
2½	12	20	42	9
3	20[2]	48[2]	72[2]	20[2]
4	160	240	500	90
5	360	540	1100	200
6	620	960	1900	350
8	1400	2200	3600	600
10	2500	3800	5600	1000
12	3900	6000	8400	1500
15	7000	6000	8400	1500

For SI units: 1 inch = 25 mm

Notes:

1 Does not include branches of the building drain.

2 A maximum of two water closets or bathroom groups within each branch interval or more than six water closets or bathroom groups on the stack.

C 501.2 System Design. Vacuum waste drainage systems shall be designed and installed in accordance with the manufacturer's installation instructions. A vacuum waste drainage system shall include a vacuum generating system, waste collection center, piping network, vacuum valve, and control components used to isolate the vacuum piping network from atmospheric pressure and to collect waste at its point of origin. Where a vacuum system provides the only means of sanitation, the duplicate vacuum generating equipment set to operate automatically shall be installed to allow the system to continue in operation during periods of maintenance.

C 501.2.1 Vacuum Generating System. The vacuum generating station shall include vacuum pumps to create a constant vacuum pressure within the piping network and storage tanks. The operation of pumps, collection tanks, and alarms shall be automated by controls. The vacuum pumps shall be activated on demand and accessible for repair or replacement. The vent from the vacuum pump shall be provided for vacuum pump air exhaust and shall be of a size capable of handling the total air volume of the vacuum pump.

C 501.2.2 Waste Collection Center or Storage Tanks. Vacuum collection center or storage tanks shall be of such capacity as to provide storage of waste to prevent fouling of the system. Such collection or storage tank shall be capable of withstanding 150 percent of the

TABLE C 401.1
SIZE AND LENGTH OF VENTS

SIZE OF SOIL OR WASTE STACK (inches)	FIXTURE UNITS CONNECTED	DIAMETER OF VENT REQUIRED (inches)								
		1¼	1½	2	2½	3	4	5	6	8
		MAXIMUM LENGTH OF VENT (feet)								
1½	8	50	150	—	—	—	—	—	—	—
2	12	30	75	200	—	—	—	—	—	—
2	20	26	50	150	—	—	—	—	—	—
2½	42	—	30	100	300	—	—	—	—	—
3	10	—	30	100	100	600	—	—	—	—
3	30	—	—	60	200	500	—	—	—	—
3	60	—	—	50	80	400	—	—	—	—
4	100	—	—	35	100	260	1000	—	—	—
4	200	—	—	30	90	250	900	—	—	—
4	500	—	—	20	70	180	700	—	—	—
5	200	—	—	—	35	80	350	1000	—	—
5	500	—	—	—	30	70	300	900	—	—
5	1100	—	—	—	20	50	200	700	—	—
6	350	—	—	—	25	50	200	400	1300	—
6	620	—	—	—	15	30	125	300	1100	—
6	960	—	—	—	—	24	100	250	1000	—
6	1900	—	—	—	—	20	70	200	700	—
8	600	—	—	—	—	—	50	150	500	1300
8	1400	—	—	—	—	—	40	100	400	1200
8	2200	—	—	—	—	—	30	80	350	1100
8	3600	—	—	—	—	—	25	60	250	800
10	1000	—	—	—	—	—	—	75	125	1000
10	2500	—	—	—	—	—	—	50	100	500
10	3800	—	—	—	—	—	—	30	80	350
10	5600	—	—	—	—	—	—	25	60	250

For SI units: 1 inch = 25 mm, 1 foot = 304.8 mm

rated vacuum (negative pressure) created by the vacuum source without leakage or collapse. Waste collection center or storage tanks shall be accessible for adjustment, repair, or replacement.

C 501.2.3 Piping Network. The piping network shall be under a continuous vacuum and shall be designed to withstand 150 percent of the vacuum (negative pressure) created by the vacuum source within the system without leakage or collapse. Sizing the piping network shall be in accordance with the manufacturer's instructions. The water closet outlet fitting shall connect with a piping network having not less than an 1½ inch (40 mm) nominal inside diameter.

C 501.2.4 Vacuum Interface Valve. A closed vacuum interface valve shall be installed to separate the piping network vacuum from atmospheric pressure. A control device shall open the vacuum interface valve where a signal is generated to remove waste from the plumbing fixture.

C 501.2.5 Control Components. Where a pneumatic signal is generated at the controller, a vacuum from the system to open the extraction valve shall be designed to operate where vacuum pressure exists to remove the accumulated waste. Each tank shall incorporate a level indicator switch that automatically controls the discharge pump and warns of malfunction or blockage as follows:

(1) Start discharge.

(2) Stop discharge.

(3) Activate an audible alarm where the level of effluent is usually high.

(4) Warning of system shutdown where the tank is full.

C 501.3 Fixtures. Fixtures utilized in a vacuum waste drainage system shall be in accordance with referenced standards listed in Chapter 17. Components shall be of corrosion resistant materials. The water closet outlet shall be able to pass a 1 inch (25.4 mm) diameter ball and shall have a smooth, impervious surface. The waste outlet and passages shall be free of obstructions, recesses, or chambers that are capable of permitting fouling. The mechanical valve and its seat shall be of such materials and design to provide a leak-free connection where at atmospheric pressure or under vacuum. The flushing mechanism shall be so designed as to ensure proper cleansing of the interior surfaces during the flushing cycle at a minimum operating flow rate. Mechanical seal mechanisms shall withdraw completely from the path of the waste discharge during the flushing operation. Each mechanical seal vacuum water closet shall be equipped with a listed vacuum breaker. The vacuum breaker shall be mounted with the critical level or marking not less than 1 inch (25.4 mm) above the flood-level rim of the fixture. Vacuum breakers shall be installed on the discharge side of the last control valve in the potable water supply line and shall be located to be protected from physical damage and contamination.

C 501.4 Drainage Fixture Units. Drainage fixture units shall be determined by the manufacturer's instructions. The pump discharge load from the collector tanks shall be in accordance with this appendix.

C 501.5 Water Supply Fixture Units. Water supply fixture units shall be determined by the manufacturer's instructions.

C 501.6 Materials. Materials used for water distribution pipe and fittings shall be in accordance with Table 604.1. Materials used for aboveground drainage shall be in accordance with Table 701.2 and shall have a smooth bore, and be constructed of non-porous material.

C 501.7 Traps and Cleanouts. Traps and cleanouts shall be installed in accordance with Chapter 7 and Chapter 10.

C 501.8 Testing. The entire vacuum waste system shall be subjected to a vacuum test of 29 inches of mercury (98 kPa) or not less than the working pressure of the system for 30 minutes. The system shall be gastight and watertight at all points. Verification of test results shall be submitted to the Authority Having Jurisdiction.

C 501.9 Manufacturer's Instructions. Manufacturer's instructions shall be provided to provide information regarding safe and proper operating instructions whether or not as part of the condition of listing to determine compliance. Such instructions shall be submitted and approved by the Authority Having Jurisdiction.

C 601.0 Single-Stack Vent System.

C 601.1 Where Permitted. Single-stack venting shall be designed by a registered design professional as an engineered design. A drainage stack shall be permitted to serve as a single-stack vent system where sized and installed in accordance with Section C 601.2 through Section C 601.9. The drainage stack and branch piping in a single-stack vent system shall provide for the flow of liquids, solids, and air without the loss of fixture trap seals.

C 601.2 Stack Size. Drainage stacks shall be sized in accordance with Table C 601.2. Not more than two water closets shall be permitted to discharge to a 3 inch (80 mm) stack. Stacks shall be uniformly sized based on the total connected drainage fixture unit load, with no reductions in size.

C 601.2.1 Stack Vent. The drainage stack vent shall have a stack vent of the same size terminating to the outdoors.

C 601.3 Branch Size. Horizontal branches connecting to a single-stack vent system shall be sized in accordance with Table 703.2.

Exceptions:

(1) Not more than one water closet within 18 inches (457 mm) of the stack horizontally shall be permitted on a 3 inch (80 mm) horizontal branch.

(2) A water closet within 18 inches (457 mm) of a stack horizontally and one other fixture with up to 1½ inch (40 mm) fixture drain size shall be permitted on a 3 inch (80 mm) horizontal branch where connected to the stack through a sanitary tee.

C 601.4 Length of Horizontal Branches. Water closets shall be not more than 4 feet (1219 mm) horizontally from the stack.

Exception: Water closets shall be permitted to be up to 8 feet (2438 mm) horizontally from the stack where connected to the stack through a sanitary tee.

C 601.4.1 Other Fixtures. Fixtures other than water closets shall be not more than 12 feet (3658 mm) horizontally from the stack.

C 601.4.2 Length of Vertical Piping. The length of a vertical piping from a fixture trap to a horizontal branch shall not be considered in computing the fixture's horizontal distance from the stack.

C 601.5 Maximum Vertical Drops from Fixtures. Vertical drops from fixture traps to horizontal branch piping shall be one size larger than the trap size, but not less than 2 inches (50 mm) in diameter. Vertical drops shall be 4 feet (1219 mm) maximum length. Fixture drains that are not increased in size, or have a vertical drop exceeding 4 feet (1219 mm) shall be individually vented.

C 601.6 Additional Venting Required. Additional venting shall be provided where more than one water closet is on a horizontal branch and where the distance from a fixture trap to the stack exceeds the limits in Section C 601.4. Where additional venting is required, the fixture(s) shall be vented by individual vents, common vents, wet vents, circuit vents, or a combination waste and vent pipe. The dry vent extensions for the additional venting shall connect to a branch vent, vent stack, stack vent, or be extended outdoors and terminate to the open air.

C 601.7 Stack Offsets. Where there are no fixture drain connections below a horizontal offset in a stack, the offset does not need to be vented. Where there are fixture drain connections below a horizontal offset in a stack, the offset shall be vented. There shall be no fixture connections to a stack within 2 feet (610 mm) above and below a horizontal offset.

C 601.8 Separate Stack Required. Where stacks are more than two stories high, a separate stack shall be provided for the fixtures on the lower two stories. The stack for the lower two stories shall be permitted to be connected to the branch of the building drain that serves the stack for the upper stories at a point that is not less than 10 pipe diameters downstream from the base of the upper stack.

C 601.9 Sizing Building Drains and Sewers. In a single-stack vent system, the building drain and branches thereof shall be sized in accordance with Table 703.2, and the building sewer shall be sized in accordance with Table 717.1.

TABLE C 601.2
SINGLE STACK SIZE*

	MAXIMUM CONNECTED DRAINAGE FIXTURE UNITS		
STACK SIZE (inches)	STACKS LESS THAN 75 FEET IN HEIGHT	STACK 75 FEET TO LESS THAN 160 FEET IN HEIGHT	STACK 160 FEET OR GREATER IN HEIGHT
3	24	NP	NP
4	225	24	NP
5	480	225	24
6	1015	480	225
8	2320	1015	480
10	4500	2320	1015
12	8100	4500	2320
15	13 600	8100	4500

For SI units: 1 inch = 25 mm, 1 foot = 304.8 mm
* NP = Not permitted

APPENDIX D

SIZING STORM WATER DRAINAGE SYSTEMS

D 101.0 General.

D 101.1 Applicability. This appendix provides general guidelines for the sizing of storm water drainage systems based on maximum rates of rainfall for various cities. The rainfall rates in Table D 101.1 shall be permitted to be used for design unless higher values are established locally.

D 102.0 Sizing by Flow Rate.

D 102.1 General. Storm drainage systems shall be permitted to be sized by storm water flow rates, using the gallons per minute per square foot [(L/s)/m^2] of rainfall listed in Table D 101.1 for the local area. Multiplying the listed gallons per minute per square foot [(L/s)/m^2] by the roof area being drained (in square feet) (m^2) by each inlet produces the gallons per minute (gpm) (L/s) of required flow for sizing each drain inlet. The flow rates shall be permitted to be added to determine the flows in each of the drainage systems. Required pipe sizes for various flow rates are listed in Table 1103.1 and Table 1103.2.

D 103.0 Sizing by Roof Area.

D 103.1 General. Storm drainage systems shall be permitted to be sized using the roof area served by each of the drainage systems. Maximum allowable roof areas with various rainfall rates are listed in Table 1103.1 and Table 1103.2, along with the required pipe sizes. By using this method, it shall be permitted to interpolate between two listed rainfall rate columns (inches per hour) (mm/h). To determine the allowable roof area for a listed pipe size at a listed slope, divide the allowable square feet (m^2) of the roof for a 1 inch per hour (in/h) (25.4 mm/h) rainfall rate by the listed rainfall rate for the local area. For example, the allowable roof area for a 6 inch (150 mm) drain at ⅛ inch per foot (10.4 mm/m) slope with a rainfall rate of 3.2 in/h (81 mm/h) is 21 400/3.2 = 6688 square feet (621.3 m^2).

D 104.0 Capacity of Rectangular Scuppers.

D 104.1 General. Table D 104.1 lists the discharge capacity of rectangular roof scuppers of various widths with various heads of water. The maximum allowable level of water on the roof shall be obtained from the registered design professional, based on the design of the roof.

TABLE D 101.1
MAXIMUM RATES OF RAINFALL FOR VARIOUS CITIES*

STATES AND CITIES	STORM DRAINAGE 60-MINUTE DURATION, 100-YEAR RETURN	
	inches per hour	gallons per minute per square foot
ALABAMA	–	–
Birmingham	3.7	0.038
Huntsville	3.3	0.034
Mobile	4.5	0.047
Montgomery	3.8	0.039
ALASKA	–	–
Aleutian Islands	1.0	0.010
Anchorage	0.6	0.006
Bethel	0.8	0.008
Fairbanks	1.0	0.010
Juneau	0.6	0.006
ARIZONA	–	–
Flagstaff	2.3	0.024
Phoenix	2.2	0.023
Tucson	3.0	0.031
ARKANSAS	–	–
Eudora	3.8	0.039
Ft. Smith	3.9	0.041
Jonesboro	3.5	0.036
Little Rock	3.7	0.038

TABLE D 101.1
MAXIMUM RATES OF RAINFALL FOR VARIOUS CITIES* (continued)

STATES AND CITIES	STORM DRAINAGE 60-MINUTE DURATION, 100-YEAR RETURN	
	inches per hour	gallons per minute per square foot
CALIFORNIA	–	–
Eureka	1.5	0.016
Lake Tahoe	1.3	0.014
Los Angeles	2.0	0.021
Lucerne Valley	2.5	0.026
Needles	1.5	0.016
Palmdale	3.0	0.031
Redding	1.5	0.016
San Diego	1.5	0.016
San Francisco	1.5	0.016
San Luis Obispo	1.5	0.016
COLORADO	–	–
Craig	1.5	0.016
Denver	2.2	0.023
Durango	1.8	0.019
Stratton	3.0	0.031
CONNECTICUT	–	–
Hartford	2.8	0.029
New Haven	3.0	0.031
DELAWARE	–	–
Dover	3.5	0.036
Rehobeth Beach	3.6	0.037
DISTRICT OF COLUMBIA	–	–
Washington	4.0	0.042
FLORIDA	–	–
Daytona Beach	4.0	0.042
Ft. Myers	4.0	0.042
Jacksonville	4.3	0.045
Melbourne	4.0	0.042
Miami	4.5	0.047
Palm Beach	5.0	0.052
Tampa	4.2	0.044
Tallahassee	4.1	0.043
GEORGIA	–	–
Atlanta	3.5	0.036
Brunswick	4.0	0.042
Macon	3.7	0.038
Savannah	4.0	0.042
Thomasville	4.0	0.042
HAWAII	–	–
Rainfall rates in the Hawaiian Islands vary from 1½ inches per hour to 8 inches per hour, depending on location and elevation. Consult local data.		
IDAHO	–	–
Boise	1.0	0.010
Idaho Falls	1.2	0.012
Lewiston	1.0	0.010
Twin Falls	1.1	0.011

TABLE D 101.1
MAXIMUM RATES OF RAINFALL FOR VARIOUS CITIES* (continued)

STATES AND CITIES	STORM DRAINAGE 60-MINUTE DURATION, 100-YEAR RETURN	
	inches per hour	gallons per minute per square foot
ILLINOIS	–	–
Chicago	2.7	0.028
Harrisburg	3.1	0.032
Peoria	2.9	0.030
Springfield	3.0	0.031
INDIANA	–	–
Evansville	3.0	0.031
Indianapolis	2.8	0.029
Richmond	2.7	0.028
South Bend	2.7	0.028
IOWA	–	–
Council Bluffs	3.7	0.038
Davenport	3.0	0.031
Des Moines	3.4	0.035
Sioux City	3.6	0.037
KANSAS	–	–
Goodland	3.5	0.036
Salina	3.8	0.039
Topeka	3.8	0.039
Wichita	3.9	0.041
KENTUCKY	–	–
Bowling Green	2.9	0.030
Lexington	2.9	0.030
Louisville	2.8	0.029
Paducah	3.0	0.031
LOUISIANA	–	–
Monroe	3.8	0.039
New Orleans	4.5	0.047
Shreveport	4.0	0.042
MAINE	–	–
Bangor	2.2	0.023
Kittery	2.4	0.025
Millinocket	2.0	0.021
MARYLAND	–	–
Baltimore	3.6	0.037
Frostburg	2.9	0.030
Ocean City	3.7	0.038
MASSACHUSETTS	–	–
Adams	2.6	0.027
Boston	2.7	0.028
Springfield	2.7	0.028
MICHIGAN	–	–
Detroit	2.5	0.026
Grand Rapids	2.6	0.027
Kalamazoo	2.7	0.028

TABLE D 101.1
MAXIMUM RATES OF RAINFALL FOR VARIOUS CITIES* (continued)

STATES AND CITIES	STORM DRAINAGE 60-MINUTE DURATION, 100-YEAR RETURN	
	inches per hour	gallons per minute per square foot
Sheboygan	2.1	0.022
Traverse City	2.2	0.023
MINNESOTA	–	–
Duluth	2.6	0.027
Grand Forks	2.5	0.026
Minneapolis	3.0	0.031
Worthington	3.4	0.035
MISSISSIPPI	–	–
Biloxi	4.5	0.047
Columbus	3.5	0.036
Jackson	3.8	0.039
MISSOURI	–	–
Independence	3.7	0.038
Jefferson City	3.4	0.035
St. Louis	3.2	0.033
Springfield	3.7	0.038
MONTANA	–	–
Billings	1.8	0.019
Glendive	2.5	0.026
Great Falls	1.8	0.019
Missoula	1.3	0.014
NEBRASKA	–	–
Omaha	3.6	0.037
North Platte	3.5	0.036
Scotts Bluff	2.8	0.029
NEVADA	–	–
Las Vegas	1.5	0.016
Reno	1.2	0.012
Winnemucca	1.0	0.010
NEW HAMPSHIRE	–	–
Berlin	2.2	0.023
Manchester	2.5	0.026
NEW JERSEY	–	–
Atlantic City	3.4	0.035
Paterson	3.0	0.031
Trenton	3.2	0.033
NEW MEXICO	–	–
Albuquerque	2.0	0.021
Carlsbad	2.6	0.027
Gallup	2.1	0.022
NEW YORK	–	–
Binghamton	2.4	0.025
Buffalo	2.3	0.024

TABLE D 101.1
MAXIMUM RATES OF RAINFALL FOR VARIOUS CITIES* (continued)

STATES AND CITIES	STORM DRAINAGE 60-MINUTE DURATION, 100-YEAR RETURN	
	inches per hour	gallons per minute per square foot
New York City	3.1	0.032
Schenectady	2.5	0.026
Syracuse	2.4	0.025
NORTH CAROLINA	–	–
Asheville	3.2	0.033
Charlotte	3.4	0.035
Raleigh	4.0	0.042
Wilmington	4.4	0.046
NORTH DAKOTA	–	–
Bismarck	2.7	0.028
Fargo	2.9	0.030
Minot	2.6	0.027
OHIO	–	–
Cincinnati	2.8	0.029
Cleveland	2.4	0.025
Columbus	2.7	0.028
Toledo	2.6	0.027
Youngstown	2.4	0.025
OKLAHOMA	–	–
Boise City	3.4	0.035
Muskogee	4.0	0.042
Oklahoma City	4.1	0.043
OREGON	–	–
Medford	1.3	0.014
Ontario	1.0	0.010
Portland	1.3	0.014
PENNSYLVANIA	–	–
Erie	2.4	0.025
Harrisburg	2.9	0.030
Philadelphia	3.2	0.033
Pittsburgh	2.5	0.026
Scranton	2.8	0.029
RHODE ISLAND	–	–
Newport	3.0	0.031
Providence	2.9	0.030
SOUTH CAROLINA	–	–
Charleston	4.1	0.043
Columbia	3.5	0.036
Greenville	3.3	0.034
SOUTH DAKOTA	–	–
Lemmon	2.7	0.028
Rapid City	2.7	0.028
Sioux Falls	3.4	0.035

TABLE D 101.1
MAXIMUM RATES OF RAINFALL FOR VARIOUS CITIES* (continued)

STATES AND CITIES	STORM DRAINAGE 60-MINUTE DURATION, 100-YEAR RETURN	
	inches per hour	gallons per minute per square foot
TENNESSEE	–	–
Knoxville	3.1	0.032
Memphis	3.5	0.036
Nashville	3.0	0.031
TEXAS	–	–
Corpus Christi	4.6	0.048
Dallas	4.2	0.044
El Paso	2.0	0.021
Houston	4.6	0.048
Lubbock	3.3	0.034
San Antonio	4.4	0.046
UTAH	–	–
Bluff	2.0	0.021
Cedar City	1.5	0.016
Salt Lake City	1.3	0.014
VERMONT	–	–
Bennington	2.5	0.026
Burlington	2.3	0.024
Rutland	2.4	0.025
VIRGINIA	–	–
Charlottesville	3.4	0.035
Norfolk	4.0	0.042
Richmond	4.0	0.042
Roanoke	3.3	0.034
WASHINGTON	–	–
Seattle	1.0	0.010
Spokane	1.0	0.010
Walla Walla	1.0	0.010
WEST VIRGINIA	–	–
Charleston	2.9	0.030
Martinsburg	3.0	0.031
Morgantown	2.7	0.028
WISCONSIN	–	–
Green Bay	2.5	0.026
Lacrosse	2.9	0.030
Milwaukee	2.7	0.028
Wausau	2.5	0.026
WYOMING	–	–
Casper	1.9	0.020
Cheyenne	2.5	0.026
Evanston	1.3	0.014
Rock Springs	1.4	0.015

For SI units: 1 inch per hour = 25.4 mm/h, 1 gallon per minute per square foot = 0.679 [(L/s)/m^2]

* The rainfall rates in this table are based on U.S. Weather Bureau Technical Paper No. 40, Chart 14: 100-Year 60-Minute Rainfall (inches).

TABLE D 104.1
DISCHARGE FROM RECTANGULAR SCUPPERS (gallons per minute)[1, 2, 3, 4]

WATER HEAD (inches)	WIDTH OF SCUPPER (inches)					
	6	12	18	24	30	36
½	6	13	19	25	32	38
1	17	35	53	71	89	107
1½	31	64	97	130	163	196
2	–	98	149	200	251	302
2½	–	136	207	278	349	420
3	–	177	271	364	458	551
3½	–	–	339	457	575	693
4	–	–	412	556	700	844

For SI units: 1 inch = 25.4 mm, 1 gallon per minute = 0.06 L/s

Notes:

[1] Table D 104.1 is based on discharge over a rectangular weir with end contractions.
[2] The head is the depth of water above the bottom of the scupper opening.
[3] The height of the scupper opening shall be not less than two times the design head.
[4] Coordinate the allowable head of water with the structural design of the roof.

APPENDIX E

MANUFACTURED/MOBILE HOME PARKS AND RECREATIONAL VEHICLE PARKS

E 101.0 Manufactured/Mobile Home Park.

E 101.1 Applicability. The manufactured home park plumbing and drainage systems shall be designed and installed in accordance with the requirements of this appendix and the requirements of this code.

E 101.2 Construction Documents. Before plumbing or sewage disposal facilities are installed or altered in a manufactured home park, duplicate construction documents shall be filed and proper permits obtained from the department or departments having jurisdiction. Plans shall show in detail:

(1) Plot plan of the park drawn to scale, indicating elevations, property lines, driveways, existing or proposed buildings, and the sizes of manufactured home lots.

(2) Complete specification and piping layout of proposed plumbing systems or alteration.

(3) Complete specification and layout of proposed sewage disposal system or alteration.

(4) The nature and extent of the work proposed, showing clearly that such work will conform to the provisions of this code.

E 201.0 Definitions.

E 201.1 General. For purposes of this chapter, the following definitions shall apply:

Manufactured/Mobile Home. A structure transportable in one or more sections, which in the traveling mode is 8 feet (2438 mm) or more in width and 40 feet (12 192 mm) or more in length or, where erected on-site, is 320 square feet (29.73 m^2) or more, and which is built on a permanent chassis, and designed to be used as a dwelling with or without a permanent foundation where connected to the required utilities. It includes the plumbing, heating, air-conditioning, and electrical systems contained therein. For further clarification of definition, see Federal Regulation 24 CFR.

Manufactured/Mobile Home Accessory Building or Structure. A building or structure that is an addition to or supplements the facilities provided to a manufactured home. It is not a self-contained, separate, habitable building or structure. Examples are awnings, cabanas, ramadas, storage structures, carports, fences, windbreaks, or porches.

Manufactured/Mobile Home Lot. A portion of a manufactured home park designed for the accommodation of one manufactured home and its accessory buildings or structures for the exclusive use of the occupants.

Manufactured/Mobile Home Park. A parcel (or contiguous parcels) of land that has been so designated and improved that it contains two or more manufactured home lots available to the general public for the placement thereon of a manufactured home for occupancy.

Recreational Vehicle (RV). A vehicular-type unit primarily designed as temporary living quarters for recreational, camping, travel, or seasonal use that either has its own motive power or is mounted on or towed by another vehicle. The basic entities are camping trailer, fifth-wheel trailer, motor home, park trailer, travel trailer, and truck camper.

Recreational Vehicle Park. A plot of land upon which two or more recreational vehicle sites are located, established or maintained for occupancy by recreational vehicles of the general public as temporary living quarters for recreation or vacation purpose.

Recreational Vehicle Site. Within a recreational vehicle park, a plot of ground intended for the accommodation of a recreational vehicle, a tent, or another individual camping unit on a temporary basis.

E 301.0 Manufactured/Mobile Home Park Drainage System Construction.

E 301.1 General. A drainage system shall be provided in manufactured home parks for conveying and disposing of sewage. Where feasible, the connection shall be made to a public system. New improvements shall be designed, constructed, and maintained in accordance with applicable laws and regulations. Where the drainage lines of the manufactured home park are not connected to a public sewer, the Authority Having Jurisdiction shall approve sewage disposal facilities prior to construction.

E 301.2 Underground Drainage System Location, Size, and Slope. Drainage (sewage) collection lines shall be located in trenches at an approved depth to be free of breakage from traffic or other movements and shall be separated from the park water supply system as specified in this code. Drainage (sewage) lines shall have a minimum size and slope as specified in Table E 301.2(1) and Table E 301.2(2).

TABLE E 301.2(1)
DRAINAGE PIPE DIAMETER AND NUMBER OF FIXTURE UNITS ON DRAINAGE SYSTEM

SIZE OF DRAINAGE (inches)	MAXIMUM NUMBER OF FIXTURE UNITS
2*	8
3	35
4	256
5	428
6	720
8	2640
10	4680
12	8200

For SI units: 1 inch = 25 mm
* Except for six unit fixtures

TABLE E 301.2(2)
MINIMUM GRADE AND SLOPE OF DRAINAGE PIPE

PIPE SIZE	SLOPE (per 100 feet)	PIPE SIZE	SLOPE (per 100 feet)
inches	inches	inches	inches
2	25	6	8
3	25	8	4
4	15	10	3½
5	11	12	3

For SI units: 1 inch = 25 mm, 1 inch per foot = 83.3 mm/m

E 301.2.1 Inlet, System, and Lateral Sizing. Each manufactured home lot drainage inlet shall be assigned a waste loading value of 12 drainage fixture units, and each park drainage system shall be sized in accordance with Table E 301.2(1) or as provided herein. Drainage laterals shall be not less than 3 inches (80 mm) in diameter.

E 301.2.2 Engineered Design. A park drainage system that exceeds the fixture unit loading of Table E 301.2(1) or in which the grade and slope of drainage pipe do not meet the minimum specified in Table E 301.2(2) shall be designed by a registered design professional.

E 301.2.3 Materials. Pipe and fittings installed underground in manufactured home park drainage systems shall be of a material approved for the purpose. Manufactured home lot drainage inlets and extensions to grade shall be of a material approved for underground use within a building.

E 301.3 Lot Drainage Inlet. Provision shall be made for plugging or capping the sewage drain inlet where a manufactured home does not occupy the lot. Surface drainage shall be diverted away from the inlet. The rim of the inlet shall extend to a maximum of 4 inches (102 mm) aboveground elevation.

E 301.3.1 Location. Each lot drainage inlet shall be located in the third rear section and within 4 feet (1219 mm) of the proposed location of the manufactured home.

E 301.3.2 Materials. Materials used for drainage connections between a manufactured home, and the lot drainage inlet shall be semi-rigid, corrosion resistant, nonabsorbent, and durable. The inner surface shall be smooth.

E 301.4 Drain Connector. A manufactured home shall be connected to the lot drainage inlet using a drain connector consisting of approved pipe not less than Schedule 40, approved fittings and connectors, and not less in size than the manufactured home drainage outlet. An approved cleanout shall be provided between the manufactured home and the lot drainage inlet. The fitting connected to the lot drainage inlet shall be a directional fitting to discharge the flow into the drainage inlet.

E 301.4.1 Grade and Gastightness. A drain connector shall be installed or maintained with a grade not less than ¼ inch per foot (20.8 mm/m). A drain connector shall be gastight and no longer than necessary to make the connection between the manufactured home outlet and the drain inlet on the lot. A flexible connector shall be permitted to be used at the lot drainage inlet area. Each lot drainage inlet shall be capped gastight where not in use.

E 302.0 Manufactured/Mobile Home Park Water Supply.

E 302.1 Potable Water Supply. An accessible and approved supply of potable water shall be provided in each manufactured home park. Where a public supply of water of approved quantity, quality, and pressure is available at or within the boundary of the park site, the connection shall be made to it and its supply used exclusively. Where an approved public water supply is not available, a private water supply system shall be developed and used as approved by the Authority Having Jurisdiction.

E 302.2 Water Service Outlet. Each manufactured home lot shall be provided with a water service outlet delivering potable water. The water service outlet riser shall be not less than ¾ of an inch (20 mm) nominal pipe size and capable of delivering 12 water supply fixture units.

E 302.2.1 Connection. A manufactured home shall be connected to the park water service outlet by a flexible connector, such as copper or copper alloy or other approved material not less than ¾ of an inch nominal (20 mm) interior diameter.

E 302.2.2 Water Supply Fixture Units. Park water distribution systems shall be designed to deliver a minimum of 12 water supply fixture units to each lot and installed with materials in accordance with Chapter 6, Appendix A, or both of this code.

E 302.2.3 Pressure. Each manufactured home park water distribution system shall be so designed and maintained as to provide a pressure of not less than 20 pounds-force per square inch (psi) (138 kPa) at each manufactured home lot at maximum operating conditions.

E 302.2.4 Location. Each lot water service outlet shall be located in the third rear section and within 4 feet (1219 mm) of the proposed location of the manufactured home.

E 302.3 Shutoff Valve. A separate water shutoff valve shall be installed in each water service outlet at each manufactured home lot. Where a listed backflow protective device is installed, the service shutoff shall be located on the supply side of such device.

E 302.4 Backflow Preventer. Where a condition exists in the plumbing of a manufactured home that creates a cross-connection, a listed backflow preventer shall be installed in the water service line to the manufactured home at or near the water service outlet. Where a hose bibb or outlet is installed on the supply outlet riser in addition to the service connector, a listed backflow preventer shall be installed on each additional outlet.

E 302.5 Pressure-Relief Valve. Where it is required to install a backflow preventer at the manufactured home lot service outlet, a listed pressure-relief valve shall be installed in the water service line on the discharge side of the backflow preventer. Pressure-relief valves shall be set to release at a pressure at a maximum of 150 psi (1034 kPa). Pressure-relief valves shall discharge toward the ground. Backflow preventers and pressure-relief valves shall be not less than 12 inches (305 mm) above the ground.

E 302.6 Mechanical Protection. Park water service outlets, backflow preventers, and pressure-relief valves shall be protected from damage by vehicles or other causes. Such protection shall be permitted to consist of posts, fencing, or other permanent barriers.

E 302.7 Water-Conditioning Equipment. A permit shall be obtained from the Authority Having Jurisdiction before installing water-conditioning equipment on a manufactured home lot. Approval of the park operator is required on applications for a permit to install such equipment. Where the water-conditioning equipment is of the regenerating type, and the park drainage system discharges into a public sewer, approval of the sanitary district or agency having jurisdiction over the public sewer is required.

E 302.7.1 Approval. Regenerating water-conditioning equipment shall be listed and labeled by an approved listing agency.

E 302.7.2 Installation. Regenerating units shall discharge the effluent of regeneration into a trap not less than 1½ inches (40 mm) in diameter connected to the manufactured home park drainage system. An approved air gap shall be installed on the discharge line a minimum of 12 inches (305 mm) above the ground.

E 302.8 Testing. Installations shall be tested and inspected in accordance with Chapter 3 of this code.

E 401.0 Fuel Supply.

E 401.1 Fuel Gas Piping Systems. All fuel gas piping systems serving manufactured homes, accessory buildings, or structures and communities shall be designed and constructed in accordance with any applicable provisions of NFPA 54 and NFPA 58. NFPA 31 shall apply to oil fuel-burning systems and shall conform to the criteria of the Authority Having Jurisdiction. [NFPA 501A:4.1.1.1 – 4.1.1.2]

E 401.2 Gas Supply Connections. Gas supply connections at sites, where provided from an underground gas supply piping system, shall be located and arranged to permit attachment to a manufactured home occupying the site.

For the installation of liquefied petroleum gas (LP-Gas) storage systems, the provisions of NFPA 58 shall be followed. [NFPA 501A:4.1.2.1 – 4.1.2.2]

E 401.3 Location of Gas Supply Connection. The gas supply to the manufactured home shall be located within 4 feet (1219 mm) of the manufactured home stand.

Exception: The requirement of Section E 401.3 shall not apply to gas supply connections for manufactured homes located on all-weather wood, concrete, or concrete block foundation systems or on foundations constructed in accordance with the local building code or, in the absence of a local code, with a recognized model building code. [NFPA 501A:4.1.3]

E 401.4 Recreational Vehicle Park Fuel Gas Equipment and Installations. Fuel gas equipment and installations shall comply with this appendix, except as otherwise permitted or required by this code.

E 402.0 Single and Multiple Manufactured Home Site Fuel Supply Systems.

E 402.1 Underground Installations. Underground gas piping system installations shall comply with any applicable building code, Section E 402.2 and Section E 402.2.1. [NFPA 501A:4.2.1]

E 402.2 Open-Ended Gastight Conduit. Underground gas piping shall not be installed beneath that portion of a manufactured home site reserved for the location of a manufactured home or manufactured home accessory building or structure unless installed in the open-ended gastight conduit of Section E 402.2.1. [NFPA 501A:4.2.1.1]

E 402.2.1 Requirements. The open-ended gastight conduit shall conform to the requirements in the following:

(1) The conduit shall be not less than Schedule 40 pipe that is approved for underground installation beneath buildings.

(2) The interior diameter of the conduit shall be not less than ½ of an inch (15 mm) larger than the outside diameter of the gas piping.

(3) The conduit shall extend to a point not less than 4 inches (102 mm) beyond the outside wall of the manufactured home or accessory building or structure, and the outer ends shall not be sealed.

(4) Where the conduit terminates within a manufactured home or accessory building or structure, it shall be accessible, and the space between the conduit and the gas piping shall be sealed to prevent leakage of gas into the building. [NFPA 501A:4.2.1.2 – 4.2.1.2.4]

E 402.3 Shutoff Valve. Each manufactured home site shall have a listed gas shutoff valve installed upstream of the manufactured home site gas outlet. The gas shutoff valve shall be located on the outlet riser at a height of not less than 6 inches (152 mm) above grade. A gas shutoff valve shall not be located under any manufactured home. The outlet shall be equipped with a cap or plug to prevent discharge of gas whenever the manufactured home site outlet is not connected to a manufactured home. [NFPA 501A:4.2.2.1 – 4.2.2.4]

Exception: Gas shutoff valves shall conform to Section E 402.3, except for manufactured homes located on foundations constructed in accordance with the local building code or, in the absence of a local code, with a recognized model building code. [NFPA 501A:4.2.2]

E 402.4 Gas Meters. Where installed, gas meters shall be supported by a post or bracket placed on a firm footing or other means providing equivalent support and shall not depend on the gas outlet riser for support. [NFPA 501A:4.2.3.1]

E 402.4.1 Location of Meters. Each gas meter shall be installed in an accessible location and shall be provided with unions or other fittings so that the meter can be removed easily and placed in an upright position. Meters shall not be installed in unventilated or inaccessible locations or closer than 3 feet (914 mm) to sources of ignition. [NFPA 501A:4.2.3.2 – 4.2.3.2.2]

E 402.4.2 Meter Shutoff Valve or Cock. All gas meter installations shall be provided with shutoff valves or cocks located adjacent to and on the inlet side of the meters. In the case of a single meter installation utilizing an LP-Gas container, the container service valve shall be permitted to be used in lieu of the shutoff valve or cock. All gas meter installations shall be provided with test tees located adjacent to and on the outlet side of the meters. [NFPA 501A:4.2.4.1 – 4.2.4.3]

E 403.0 Multiple Manufactured Home Site Fuel Distribution and Supply Systems.

E 403.1 Manufactured Home Community LP-Gas Supply Systems. Where 10 or more customers are served by one LP-Gas supply system, the installation of the gas supply system shall be in accordance with 49 CFR 192. Other types of liquefied petroleum gas supply systems and the storage and handling of LP-Gas shall be in accordance with NFPA 58 (See also Section E 403.10). [NFPA 501A:4.3.2.1 – 4.3.2.2]

E 403.2 Required Gas Supply. The minimum hourly volume of gas required at each manufactured home site outlet or any section of the manufactured home community gas piping system shall be calculated as shown in Table E 403.2.

In extreme climate areas, additional capacities other than those in Table E 403.2 shall be considered. [NFPA 501A:4.3.4.1 – 4.3.4.2]

TABLE E 403.2
DEMAND FACTORS FOR USE IN CALCULATING GAS PIPING SYSTEMS IN MANUFACTURED HOME COMMUNITIES
[NFPA 501A: TABLE 4.3.4.1]

NUMBER OF MANUFACTURED HOME SITES	BTU/H PER MANUFACTURED HOME SITE
1	125 000
2	117 000
3	104 000
4	96 000
5	92 000
6	87 000
7	83 000
8	81 000
9	79 000
10	77 000
11–20	66 000
21–30	62 000
31–40	58 000
41–60	55 000
Over 60	50 000

For SI units: 1000 British thermal units per hour = 0.293 kW

E 403.3 Size. The size of each section of a gas piping system shall be determined in accordance with NFPA 54 or by other standard engineering methods acceptable to the Authority Having Jurisdiction. [NFPA 501A:4.3.5.1]

E 403.4 Pressure. Where all connected appliances are operated at their rated capacity, the gas supply pressure shall be not less than 7 inches water column (1.7 kPa). The gas supply pressure shall not exceed 14 inches water column (3.5 kPa). [NFPA 501A:4.3.5.2]

E 403.5 Metal Gas Piping. Metal gas pipe shall be standard-weight wrought iron or steel (galvanized or black), yellow brass containing not more than 75 percent copper, or internally tinned or treated copper of iron pipe size. Galvanizing shall not be considered protection against corrosion.

Seamless copper or steel tubing shall be permitted to be used with gases not corrosive to such material. Steel tubing shall comply with ASTM A254. Copper tubing shall comply with ASTM B88 (Type K or Type L) or ASTM B280. Copper tubing (unless tin-lined) shall not be used if the gas contains more than an average of 0.3 grains of hydrogen sulfide per 100 standard cubic feet (scf) of gas (0.7 mg/100 L). [NFPA 501A:4.3.6.1.1 – 4.3.6.1.6]

E 403.6 Protection Coatings for Metal Gas Piping. All buried or submerged metallic gas piping shall be protected from corrosion by approved coatings or wrapping materials. All gas pipe protective coatings shall be approved types, shall be machine applied, and shall conform to recognized standards. Field wrapping shall provide equivalent protection and is restricted to those short sections and fittings that are necessarily stripped for threading or welding. Risers shall be coated or wrapped to a point at least 6 inches (152 mm) above ground. [NFPA 501A:4.3.6.2.1 – 4.3.6.2.4]

E 403.7 Plastic Piping. Plastic piping shall only be used underground and shall meet the requirements of ASTM D2513 or ASTM D2517, as well as the design pressure and design limitations of 49 CFR (Section 192.123), and shall otherwise conform to the installation requirements thereof. [NFPA 501A:4.3.6.3]

E 403.8 Gas Piping Installation. All gas piping installed below ground level shall have a minimum earth cover of 18 inches (457 mm) and shall be installed with at least 12 inches (305 mm) of clearance in any direction from any other underground utility system. [NFPA 501A:4.3.7.1]

E 403.8.1 Metallic Gas Piping. All metallic gas piping systems shall be installed in accordance with approved plans and specifications, including provisions for cathodic protection. Each cathodic protection system shall be designed and installed to conform to the provisions of 49 CFR 192. [NFPA 501A:4.3.7.2.1, 4.3.7.2.2]

E 403.8.2 Cathodic Protection. Where the cathodic protection system is designed to protect only the gas piping system, the gas piping system shall be electrically isolated from all other underground metallic systems or installations. Where only the gas piping system is cathodically protected against corrosion, a dielectric fitting shall be used in the manufactured home gas connection to insulate the manufactured home from the underground gas piping system. [NFPA 501A:4.3.7.2.3, 4.3.7.2.4]

E 403.8.3 Underground Metallic Systems. Where a cathodic protection system is designed to provide all underground metallic systems and installations with protection against corrosion, all such systems and installations shall be electrically bonded together and protected as a whole. [NFPA 501A:4.3.7.2.5]

E 403.8.4 Plastic Gas Piping. Plastic gas piping shall be used only underground and shall be installed with an electrically conductive wire for locating the pipe. The wire used to locate the plastic pipe shall be copper, not smaller in size than 18 AWG, with insulation approved for direct burial. Every portion of a plastic gas piping system consisting of metallic pipe shall be cathodically protected against corrosion. [NFPA 501A:4.3.7.3.1 – 4.3.7.3.3]

E 403.9 Gas Piping System Shutoff Valve. An accessible and identifiable shutoff valve controlling the flow of gas to the entire manufactured home community gas piping system shall be installed in a location acceptable to the Authority Having Jurisdiction and near the point of connection to the service piping or to the supply connection of an LP-Gas container. [NFPA 501A:4.3.7.4]

E 403.10 Liquefied Petroleum Gas Equipment. LP-Gas equipment shall be installed in accordance with the applicable provisions of NFPA 58. [NFPA 501A:4.3.8]

E 403.11 Oil Supply. The following three methods of supplying oil to an individual manufactured home site shall be permitted:

(1) Supply from an outside underground tank (see Section E 404.6).

(2) Supply from a centralized oil distribution system designed and installed in accordance with accepted engineering practices and in compliance with NFPA 31.

(3) Supply from an outside aboveground tank (see Section E 404.6). [NFPA 501A:4.3.9]

E 403.12 Minimum Oil Supply Tank Size. Oil supply tanks shall have a minimum capacity equal to 20 percent of the average annual oil consumption. [NFPA 501A:4.3.10]

E 403.13 Oil Supply Connections. Oil supply connections at manufactured home sites, where provided from a centralized oil distribution system, shall be located and arranged to permit attachment to a manufactured home utilizing the stand. [NFPA 501A:4.3.11.1] The installation of such facilities shall comply with the following requirements:

(1) The main distribution pipeline shall be permitted to be connected to a tank or tanks having an aggregate capacity not exceeding 20 000 gallons (75 708 L) at a point below the liquid level.

(2) Where this piping is so connected, a readily accessible internal or external shutoff valve shall be installed in the piping as close as practicable to the tank.

(3) If external and aboveground, the shutoff valve and its tank connections shall be made of steel.

(4) Connections between the tank(s) and the main pipeline shall be made with double swing joints or flexible connectors, or shall otherwise be arranged to permit the tank(s) to settle without damaging the system.

(5) If located aboveground, the connections specified in Section E 403.13(4) shall be located within the diked area.

(6) A readily accessible and identified manual shutoff valve shall be installed either inside or outside of the structure in each branch supply pipeline that enters a building, mobile home, travel trailer, or other structure. If outside, the valve shall be protected from weather and damage. If inside, the valve shall be located directly adjacent to the point at which the supply line enters the structure.

(7) A device shall be provided in the supply line at or ahead of the point where it enters the interior of the structure that will automatically shut off the oil supply, if the supply line between this device and the appliance is broken. This device shall be located on the appliance side of the manual shutoff valve required in Section E 403.13(6) and shall be solidly supported and protected from damage.

(8) Means shall be provided to limit the oil pressure at the appliance inlet to a maximum gauge pressure of 3 pounds-force per square inch gauge (psig) (21 kPa). If a pressure-reducing valve is used, it shall be a type approved for the service.

(9) A device shall be provided that will automatically shut off the oil supply to the appliance if the oil pressure at the appliance inlet exceeds a gauge pressure of 8 psig (55 kPa). This device shall not be required under either of the following conditions:

   (a) Where the distribution system is supplied from a gravity tank and the maximum hydrostatic head of oil in the tank is such that the oil pressure at the appliance inlet will not exceed a gauge pressure of 8 psig (55 kPa).

   (b) Where a means is provided to automatically shut off the oil supply if the pressure-regulating device provided in accordance with Section E 403.13(8) fails to regulate the pressure as required.

(10) Only appliances equipped with primary safety controls specifically listed for the appliance shall be connected to a centralized oil distribution system. [NFPA 31:9.2.10 – 9.2.15]

E 404.0 Fuel Supply Systems Installation.

E 404.1 Flexible Gas Connector. Except for manufactured homes located on an all-weather wood, concrete, or concrete block foundation system or on a foundation constructed in accordance with the local building code or, in the absence of a local code, with a recognized model building code, each gas supply connector shall be listed for outside manufactured home use, shall be not more than 6 feet (1829 mm) in length, and shall have a capacity rating to supply the connected load. [NFPA 501A:4.4.1]

E 404.2 Use of Approved Pipe and Fittings of Extension. Where it is necessary to extend a manufactured home inlet to permit connection of the 6 foot (1829 mm) listed connector to the site gas outlet, the extension shall be of approved materials of the same size as the manufactured home inlet and shall be adequately supported at no more than 4 foot (1219 mm) intervals to the manufactured home. [NFPA 501A:4.4.2]

E 404.3 Mechanical Protection. All gas outlet risers, regulators, meters, valves, and other exposed equipment shall be protected against accidental damage. [NFPA 501A:4.4.3]

E 404.4 Special Rules on Atmospherically Controlled Regulators. Atmospherically controlled regulators shall be installed in such a manner that moisture cannot enter the regulator vent and accumulate above the diaphragm. Where the regulator vent is obstructed due to snow and icing conditions, shields, hoods, or other suitable devices shall be provided to guard against closing of the vent opening. [NFPA 501A:4.4.4.1 – 4.4.4.2]

E 404.5 Fuel Gas Piping Test. The manufactured home fuel gas piping system shall be tested only with air before it is connected to the gas supply. The manufactured home gas piping system shall be subjected to a pressure test with all appliance shutoff valves in their closed positions. [NFPA 501A:4.4.5]

E 404.5.1 Procedures. The fuel gas piping test shall consist of air pressure at not less than 10 inches water column or more than 14 inches water column (2.5 kPa to 3.5 kPa). The fuel gas piping system shall be isolated from the air pressure source and shall maintain this pressure for not less than 10 minutes without perceptible leakage. Upon satisfactory completion of the fuel gas piping test, the appliance valves shall be opened, and the gas appliance connectors shall be tested with soapy water or bubble solution while under the pressure remaining in the piping system. Solutions used for testing for leakage shall not contain corrosive chemicals. Pressure shall be measured with a manometer, slope gauge, or gauge that is calibrated in either water inch (mm) or psi (kPa) with increments of either $\frac{1}{10}$ inch (2.5 mm) or $\frac{1}{10}$ psi (0.7 kPa) gauge, as applicable. Upon satisfactory completion of the fuel gas piping test, the manufactured home gas supply connector shall be installed, and the connections shall be tested with soapy water or bubble solution. [NFPA 501A:4.4.5.1.1 – 4.4.5.1.6]

E 404.5.2 Warning. The following warning shall be supplied to the installer:

WARNING: Do not overpressurize the fuel gas piping system. Damage to valves, regulators, and appliances can occur due to pressurization beyond the maximums specified. [NFPA 501A:4.4.5.2]

E 404.5.3 Vents. Gas appliance vents shall be visually inspected to ensure that they have not been dislodged in transit and are connected securely to the appliance. [NFPA 501A:4.4.5.3]

E 404.6 Oil Tanks. Oil tank capacities shall comply with the following:

(1) No more than one 660 gallon (2498 L) tank or two tanks with an aggregate capacity of 660 gallons (2498 L) or less shall be connected to one oil-burning appliance.

(2) Two supply tanks, where used, shall be cross-connected and provided with a single fill and single vent, as described in NFPA 31 and shall be on a common slab and rigidly secured, one to the other.

(3) Tanks having a capacity of 660 gallons (2498 L) or less shall be securely supported by rigid, noncombustible supports to prevent settling, sliding, or lifting. [NFPA 501A:4.4.6]

E 404.6.1 Installation. Oil supply tanks shall be installed in accordance with the applicable provisions of NFPA 31. [NFPA 501A:4.4.6.1]

E 404.6.2 Capacity. A tank with a capacity no larger than 60 gallons (227 L) shall be permitted to be a DOT-5 shipping container (drum), and so marked, or a tank meeting the provisions of UL 80. Tanks other than DOT-5 shipping containers having a capacity of not more than 660 gallons (2498 L) shall meet the provisions of UL 80. Pressure tanks shall be built in accordance with Section VIII, Pressure Vessels of the ASME Boiler and Pressure Vessel Code. [NFPA 501A:4.4.6.2.1 – 4.4.6.2.2]

E 404.6.3 Location. Tanks, as described in Section E 404.6 and Section E 404.6.2, that are adjacent to buildings shall be located not less than 10 feet (3048 mm) from a property line that is permitted to be built upon. [NFPA 501A:4.4.6.3]

E 404.6.4 Vent. Tanks with a capacity no larger than 660 gallons (2498 L) shall be equipped with an open vent no smaller than $1\frac{1}{2}$ inch (40 mm) iron pipe size; tanks with a 500 gallon (1892 L) or less capacity shall have a vent of $1\frac{1}{4}$ inch (32 mm) iron pipe size. [NFPA 501A:4.4.6.4]

E 404.6.5 Liquid Level. Tanks shall be provided with a means of determining the liquid level. [NFPA 501A:4.4.6.5]

E 404.6.6 Fill Opening. The fill opening shall be a size and in a location that permits filling without spillage. [NFPA 501A:4.4.6.6]

E 405.0 Manufactured Home Accessory Building Fuel Supply Systems.

E 405.1 General. Fuel gas supply systems installed in a manufactured home accessory building or structure shall comply with the applicable provisions of NFPA 54 and NFPA 58. Fuel oil supply systems shall comply with the applicable provisions of NFPA 31. [NFPA 501A:4.5.1 – 4.5.2]

E 406.0 Community Building Fuel Supply Systems in Manufactured Home Communities.

E 406.1 Fuel Gas Piping and Equipment Installations. Fuel gas piping and equipment installed within a permanent building in a manufactured home community shall comply with nationally recognized appliance and fuel gas piping codes and standards adopted by the Authority Having Jurisdiction. Where the state or other political subdivision does not assume jurisdiction, such fuel gas piping and equipment installations shall be designed and installed in accordance with the applicable provisions of NFPA 54 or NFPA 58. [NFPA 501A:4.6.1.1 – 4.6.1.2]

E 406.2 Oil Supply Systems. Oil burning equipment and installation within a manufactured home community shall be designed and constructed in accordance with the applicable codes and standards adopted by the Authority Having Jurisdiction. Where the state or other political subdivision does not assume jurisdiction, such installations shall be designed and constructed in accordance with the applicable provisions of NFPA 31. [NFPA 501A:4.6.2.1 – 4.6.2.2]

E 406.3 Oil-Burning Equipment and Installation. Oil burning equipment and installation within a building constructed in a manufactured home community in accordance with the local building code or a nationally recognized building code shall comply with nationally recognized codes and standards adopted by the Authority Having Jurisdiction. Where the state or other political subdivision does not assume jurisdiction, such oil-burning equipment and installations shall be designed and installed in accordance with the applicable provisions of NFPA 31. [NFPA 501A:4.6.3.1 – 4.6.3.2]

E 406.4 Inspections and Tests. Inspections and tests for fuel gas piping shall be made in accordance with Chapter 1 and Chapter 12 of this code.

E 501.0 Recreational Vehicle Parks.

E 501.1 Plumbing Systems. Plumbing systems shall be installed in accordance with the plumbing codes of the Authority Having Jurisdiction and with this appendix.

E 501.2 Toilet Facilities. Water closets and urinals shall be provided at one or more locations in a recreational vehicle park. They shall be of convenient access and shall be located within a 500 foot (152 m) radius from a recreational vehicle site not provided with an individual sewer connection.

E 501.2.1 Signage. Facilities for males and females shall be appropriately marked.

E 501.2.2 Interior Finish. The interior finish of walls shall be moisture resistant to a height of not less than 4 feet (1219 mm) to facilitate washing and cleaning.

E 501.2.3 Receptacle. Each toilet room for women shall be provided with a receptacle for sanitary napkins. The receptacle shall be of durable, impervious, readily cleanable material, and shall be provided with a lid.

E 501.2.4 Ceiling Height and Doors. A toilet facility shall have a ceiling height of not less than 7 feet (2134 mm) and, unless the artificial light is provided, a window or skylight area equal to not less than 10 percent of the floor area shall be provided.

Doors to the exterior shall open outward, be self-closing, and shall be visually screened using a vestibule or wall to prevent a direct view of the interior where the exterior doors are open. Such screening shall not be required on single toilet units.

E 501.2.5 Ventilation. A toilet facility shall have permanent, non-closable, screened opening(s), having a total area not less than 5 percent of the floor area and opening directly to the exterior to provide proper ventilation. A listed exhaust fan(s), vented to the exterior, the rating of which in cubic feet per minute (L/s) is not less than 25 percent of the total volume of the room(s) served, shall be considered as meeting the requirements of this section. Openable windows and vents to the outside shall be provided with flyproof screens of not less than number 16 mesh.

E 501.3 Water Closets. Not less than one water closet shall be provided for each sex up to the first 25 sites. For each additional 25 sites not provided with sewer connections, an additional water closet shall be provided.

E 501.3.1 Application. Water closets shall be of an approved, elongated bowl type and shall be provided with seats with open fronts.

E 501.3.2 Compartment. Each water closet shall be in a separate compartment and shall be provided with a latched door for privacy. A holder or dispenser for toilet paper shall be provided. Dividing walls or partitions shall be not less than 5 feet (1524 mm) high and shall be separated from the floor by a space not exceeding 12 inches (305 mm).

E 501.3.3 Size. Water closet compartments shall be not less than 30 inches (762 mm) in width [no water closet shall be set closer than 15 inches (381 mm) from its center to a side wall] and shall be not less than 30 inches (762 mm) of clear space in front of each water closet.

E 501.4 Lavatories. Where water-supplied water closets are provided, an equal number of lavatories shall be provided for up to six water closets. One additional lavatory shall be provided for each two water closets where more than six water closets are required. Each lavatory basin shall have a piped supply of potable water and shall drain into the drainage system.

E 501.5 Urinals. Where separate facilities are provided for men and women, urinals shall be acceptable for not more than one-third of the water closets required in the men's facilities, except that one urinal shall be permitted to be used to replace a water closet in a minimum park. Individual stall or wall-hung types of urinals shall be installed. Floor-type trough units shall be prohibited.

E 501.6 Floors and Drains. The floors shall be constructed of material impervious to water and shall be easily cleanable. A building having water-supplied water closets shall be provided with a floor drain in the toilet room. This drain shall be provided with means to protect the trap seal in accordance with this code.

E 501.7 Shower Size. Each shower, where provided, shall have a floor area of 36 inches by 36 inches (914 mm by 914 mm), shall be capable of encompassing a 30 inch (762 mm) diameter circle and shall be of the individual type. The shower area shall be visually screened from view with a minimum floor area of 36 by 36 inches (914 mm by 914 mm) per shower. Each shall be provided with individual dressing areas screened from view and shall contain a minimum of one clothing hook and stool (or bench area).

E 501.7.1 Drainage Connection. Each shower area shall be designed to minimize the flow of water into the dressing area and shall be connected to the drainage system using a properly trapped and vented inlet. Each such area shall have an impervious, skid-resistant surface; wooden racks (duck boards) over shower floors shall be prohibited.

E 501.8 Drinking Fountains. Where provided, drinking fountains shall be in accordance with the requirements of this code.

E 502.0 Recreational Vehicle Park Potable Water Supply and Distribution.

E 502.1 Quality. The supply or supplies of water shall comply with the potable water standards of the state, local health authority or, in the absence thereof, with the Drinking Water Standard of the Federal Environmental Protection Agency.

E 502.2 Sources. Water approved by a regulating agency shall be acceptable. Where an approved public water supply system is available, it shall be used. Where the park has its own water supply system, the components of the system shall be approved. A water supply system that is used on a seasonal basis shall be provided with means for draining.

E 502.3 Prohibited Connections. The potable water supply shall not be connected to a nonpotable or unapproved water supply, nor be subjected to backflow or backsiphonage.

E 502.4 Supply. The water supply system shall be designed and constructed in accordance with the following:

(1) A minimum of 25 gallons (95 L) per day per site for sites without individual water connections.

(2) A minimum of 50 gallons (189 L) per day per site for sites with individual water connections.

(3) A minimum of 50 gallons (189 L) per day per site where water-supplied water closets are provided in restrooms.

E 502.5 Pressure and Volume. Where water is distributed under pressure to an individual site, the water supply system shall be designed to provide a minimum flow pressure of not less than 20 psi (138 kPa) with a minimum flow of 2 gallons per minute (gpm) (0.1 L/s) at an outlet. The pressure shall not exceed 80 psi (552 kPa).

E 502.6 Outlets. Water outlets shall be convenient to access and, where not piped to individual recreational vehicle sites, shall not exceed 300 feet (91 m) from a site. Provisions shall be made to prevent accumulation of standing water or the creation of muddy conditions at each water outlet.

E 502.7 Storage Tanks. Water storage tanks shall be constructed of impervious materials, protected against contamination, and provided with locked, watertight covers. Overflow or ventilation openings shall be down-facing and provided with a corrosion-resistant screening of not less than number 24 mesh to prevent the entrance of insects and vermin. Water storage tanks shall not have direct connections to sewers.

E 503.0 Recreational Vehicle Park Water Connections for Individual Recreational Vehicles.

E 503.1 Location. Where provided, the water connections for potable water to individual recreational vehicle sites shall be located on the left rear half of the site (left side of recreational vehicle) within 4 feet (1219 mm) of the stand.

E 503.2 Water Riser Pipe. Each potable water connection shall consist of a water riser pipe that is equipped with a threaded male spigot located not less than 12 inches (305 mm) but not more than 24 inches (610 mm) above grade level for the attachment of a standard water hose. The water riser pipe shall be protected from physical damage in accordance with this code. This connection shall be equipped with a listed antisiphon backflow prevention device.

E 504.0 Recreational Vehicle Park Drainage System.

E 504.1 Where Required. An approved drainage system shall be provided in recreational vehicle parks for conveying and disposing of sewage. Where available, parks shall be connected to a public sewer system.

E 504.2 Location. Sewer lines shall be located to prevent damage from vehicular traffic.

E 504.3 Materials. Pipe and fittings installed in the drainage system shall be of material listed, approved, and installed in accordance with this code.

E 504.4 Pipe Sizes. The minimum diameters of drainage laterals, branches, and mains serving recreational vehicle sites shall be in accordance with Table E 504.4.

TABLE E 504.4
PIPE SIZES

MAXIMUM NUMBER OF RECREATIONAL VEHICLE STANDS SERVED	MINIMUM PIPE SIZES (inches)
5	3
36	4
71	5
120	6
440	8

For SI units: 1 inch = 25 mm

E 504.5 Cleanouts. Cleanouts shall be provided in accordance with Chapter 7 of this code.

E 504.6 Drainage Inlet. Where provided, the site drainage system inlet connections for individual recreational vehicles shall be located to prevent damage by the parking of recreational vehicles or automobiles and shall consist of a sewer riser extending vertically to grade. The minimum diameter of the sewer riser pipe shall be not less than 3 inches (80 mm) in diameter, and shall be provided with a 4 inch (100 mm) inlet or not less than a 3 inch (80 mm) female fitting.

E 504.6.1 Location. Where provided, the sewer inlet to individual recreational vehicle sites shall be located on the left rear half of the site (left side of the recreational vehicle) within 4 feet (1219 mm) of the stand.

E 504.6.2 Protection. The sewer riser pipe shall be firmly imbedded in the ground and protected against damage from movement. It shall be provided with a tight-fitting plug or cap, which shall be secured by a durable chain (or equivalent) to prevent loss.

E 505.0 Recreational Vehicle Park Sanitary Disposal Stations.

E 505.1 Where Required. One recreational vehicle sanitary disposal station shall be provided for each 100 recreational vehicle sites, or part thereof, which are not equipped with individual drainage system connections.

E 505.2 Access. Each station shall be level and convenient of access from the service road and shall provide easy ingress and egress for recreational vehicles.

E 505.3 Construction. Unless other approved means are used, each station shall have a concrete slab with the drainage system inlet located to be on the road (left) side of the recreational vehicle. The slab shall be not less than 3 feet by 3 feet (914 mm by 914 mm), not less than 3½ inches (89 mm) thick and properly reinforced. The slab surface is to be troweled to a smooth finish and sloped from each side inward to a drainage system inlet.

The drainage system inlet shall consist of a 4 inch (102 mm), self-closing, foot-operated hatch of approved material with the cover milled to fit tight. The hatch body shall be set in the concrete of the slab with the lip of the opening flush with its surface to facilitate the cleansing of the slab with water. The hatch shall be properly connected to a drainage system inlet, which shall discharge to an approved sanitary sewage disposal facility.

E 505.4 Flushing Device. Where the recreational vehicle park is provided with a piped water supply system, means for flushing the recreational vehicle holding tank and the sanitary disposal station slab shall be provided that consists of a piped supply of water under pressure, terminating in an outlet located and installed to prevent damage by automobiles or recreational vehicles. The flushing device shall consist of a properly supported riser terminating not less than 2 feet (610 mm) above the ground surface, with a ¾ of an inch (20 mm) valved outlet adaptable for a flexible hose.

The water supply to the flushing device shall be protected from backflow using a listed vacuum breaker or backflow prevention device located downstream from the last shutoff valve.

Adjacent to the flushing arrangement shall be posted a sign of durable material not less than 2 feet by 2 feet (610 mm by 610 mm) in size. Inscribed thereon in clearly legible letters shall be the following:

"DANGER – NOT TO BE USED FOR DRINKING OR DOMESTIC PURPOSES."

E 506.0 Recreational Vehicle Park Water Supply Stations.

E 506.1 Potable Watering Stations. A potable watering station, where provided for filling recreational vehicle potable water tanks, shall be located not less than 50 feet (15 240 mm) from a sanitary disposal station. Where such is provided, adjacent to the potable water outlet there shall be a posted sign of durable material not less than 2 feet by 2 feet (610 mm by 610 mm) in size. Inscribed thereon in clear, legible letters on a contrasting background shall be:

"POTABLE WATER. NOT TO BE USED FOR FLUSHING WASTE TANKS."

The potable water shall be protected from backflow using a listed vacuum breaker located downstream from the last shutoff valve.

APPENDIX F

FIREFIGHTER BREATHING AIR REPLENISHMENT SYSTEMS

F 101.0 General.

F 101.1 Applicability. This chapter covers minimum requirements for the installation of firefighter breathing air replenishment systems.

F 201.0 Definitions.

F 201.1 General. For purposes of this chapter, the following definitions shall apply:

High-Rise Building. A building where the floor of an occupiable story is greater than 75 feet (22 860 mm) above the lowest level of fire department vehicle access. [NFPA 5000:3.3.68.10]

Interior Cylinder Fill Panels. Lockable interior panels that provide firefighters the ability to regulate breathing air pressure and refill self-contained breathing apparatus (SCBA) cylinders.

Interior Cylinder Fill Stations and Enclosures. Freestanding fill containment stations that provide firefighters the ability to regulate breathing air pressure and refill SCBA cylinders.

Open-Circuit Self-Contained Breathing Apparatus. An SCBA in which exhalation is vented to the atmosphere and not rebreathed. [NFPA 1981:3.3.34]

Self-Contained Breathing Apparatus (SCBA). An atmosphere-supplying respirator that supplies a respirable air atmosphere to the user from a breathing air source that is independent of the ambient environment and designed to be carried by the user. [NFPA 1981:3.3.46]

For this appendix, where this term is used without a qualifier, it indicates an open-circuit self-contained breathing apparatus or combination SCBA/SARs. For this appendix, combination SCBA/SARs are encompassed by the terms self-contained breathing apparatus or SCBA.

Welding Procedure Specification (WPS). A written qualified welding procedure prepared to provide direction for making production welds to code requirements. [ASME B31.1:100.2]

F 301.0 System Components.

F 301.1 General. Firefighter breathing air replenishment systems shall contain, as a minimum, the following components:

(1) Exterior fire department connection panel
(2) Interior fire department air fill panel or station
(3) Interconnected piping distribution system
(4) Pressure monitoring switch

F 401.0 Required Installations.

F 401.1 General. A firefighter air system shall be installed in the following buildings:

(1) High-rise buildings.
(2) Underground structures that are three or more floors below grade with an area greater than 20 000 square feet (1858 m^2).
(3) Large area structures with an area greater than 200 000 square feet (18 580 m^2) and where the travel distance from the building centerline to the closest exit is greater than 500 feet (152 m), such as warehouses, manufacturing complexes, malls, or convention centers.
(4) Underground transportation or pedestrian tunnels exceeding 500 feet (152 m) in length.

F 501.0 Exterior Fire Department Connection Panel and Enclosure.

F 501.1 Purpose. The exterior fire department connection panel shall provide the fire department's mobile air operator access to the system and shall be compatible with the fire department's mobile air unit.

F 501.2 Number of Panels. Each building or structure shall have a minimum of two panels.

F 501.3 Location. Each panel shall be attached to the building or on a remote monument at the exterior of the building with a minimum 6 foot (1829 mm) radius and 180-degree (3.14 rad) clear, unobstructed access to the front of the panel. The panel shall be weather-resistant or secured inside of a weather-resistant enclosure. The panel shall be located on opposite sides of the building within 50 feet (15 240 mm) of an approved roadway or driveway, or other locations approved by the Authority Having Jurisdiction.

F 501.4 Construction. The fire department connection panel shall be installed in a metal cabinet constructed of not less than 18-gauge carbon steel or equivalent. The cabinet shall be provided with a coating or other means to protect the cabinet from corrosion.

F 501.5 Vehicle Protection. Where the panel is located in an area subject to vehicle traffic, impact protection shall be provided.

F 501.6 Enclosure Marking. The front of the enclosure shall be marked, "FIREFIGHTER AIR SYSTEM." The lettering shall be in a color that contrasts with the enclosure front and in letters that are not less than 2 inches (51 mm) high with a 3/8 of an inch (9.5 mm) brush stroke.

F 501.7 Enclosure Components. The exterior fire department connection panel shall contain the necessary gauges, isolation valves, pressure-relief valves, pressure-regulating valves, check valves, tubing, fittings, supports, connectors, adapters, and other necessary components as required to allow the fire department's mobile air unit to connect and augment the system with a constant source of breathing air. Each fire department connection panel shall contain not less than two inlet air connections.

F 501.8 Pressure-Relief Valve. Pressure-relief valves shall be installed downstream of the pressure regulator inlet. The relief valve shall meet the requirements of CGA S-1.3 and shall not be field adjustable. The relief valve shall have a set-to-open pressure not exceeding 1.1 times the design pressure of the system. Pressure-relief valve discharge shall terminate so that the exhaust air stream cannot impinge upon personnel in the area. Valves, plugs, or caps shall not be installed in the discharge of a pressure-relief valve. Where discharge piping is used, the end shall not be threaded.

F 501.9 Security. The fire department connection panel enclosure shall be locked by an approved means.

F 601.0 Interior Cylinder Fill Panels.

F 601.1 Cabinet Requirements. Each cylinder fill panel shall be installed in a metal cabinet constructed of not less than 18-gauge carbon steel or equivalent. The depth of the cabinet shall not create an exit obstruction where installed in building stairwells. Except for the shutoff valve, pressure gauges, fill hoses and ancillary components; no system components shall be visible and shall be contained behind a not less than an 18-gauge interior panel.

F 601.2 Clearance and Access. The panel shall be located not less than 36 inches (914 mm) but not more than 60 inches (1524 mm) above the finished floor or a stairway landing. Clear, unobstructed access shall be provided to each panel.

F 601.3 Door. The door shall be arranged such that where the door is open, it does not reduce the required exit width or create an obstruction in the path of egress.

F 601.4 Cabinet Marking. The front of each cylinder fill panel shall be marked, "FIREFIGHTER AIR SYSTEM." The lettering shall be in a color that contrasts with the cabinet front and in letters that are not less than 2 inches (51 mm) high with a $^3/_8$ of an inch (9.5 mm) brush stroke.

F 601.5 Cabinet Components. The cabinet shall be of a size to allow for the installation of the components in Section F 601.5.1.

> **F 601.5.1 Cylinder Fill Panel.** The cylinder fill panel shall contain the gauges, isolation valves, pressure-relief valves, pressure-regulating valves, check valves, tubing, fittings, supports, connectors, hoses, adapters, and other components to refill SCBA cylinders.

F 601.6 Cylinder Filling Hose. The design of the cabinet shall provide a means for storing the hose to prevent kinking. Where the hose is coiled, the brackets shall be installed so that the hose bend radius is maintained at 4 inches (102 mm) or greater. Fill hose connectors for connection to SCBA cylinders shall comply with the requirements of CGA V-1, number 346 or 347. No other SCBA cylinder fill connections shall be permitted.

F 601.7 Security. Each panel cover shall be maintained and locked by an approved means.

F 701.0 Interior Cylinder Fill Stations and Enclosures.

F 701.1 Location. The location of the closet or room for each air fill station shall be approved by the Authority Having Jurisdiction. Where approved by the Authority Having Jurisdiction, space shall be permitted to be utilized for other firefighting purposes. The door to each room enclosing the air filling station enclosure shall be readily accessible at all times. Not less than a 6 foot (1829 mm) radius and 180-degree (3.14 rad) clear, unobstructed access to the front of the air filling station shall be provided. The enclosure shall have emergency lighting installed in accordance with NFPA 70.

F 701.2 Security. Each air fill station shall be installed within a lockable enclosure, closet, or room by an approved means. Access to fill equipment and controls shall be restricted to authorized personnel by key or other means.

F 701.3 Components. The air fill station shall contain the gauges, isolation valves, pressure-relief valves, pressure-regulating valves, check valves, tubing, fittings, supports, connectors, hoses, adapters, and other components to refill SCBA cylinders.

F 701.4 Cylinder Filling Hose. Where hoses are used, the design of the cabinet shall provide a means for storing the hose to prevent kinking. Where the hose is coiled, the brackets shall be installed so that the hose bend radius is maintained at 4 inches (102 mm) or greater. Fill hose connectors for connection to SCBA cylinders shall comply with the requirements of CGA V-1, no. 346 or 347. For high-pressure SCBA cylinders of 4500 pounds-force per square inch (psi) (31 026 kPa), no. 347 connectors shall be used. For low-pressure SCBA cylinders of 3000 psi (20 684 kPa) and 2200 psi (15 168 kPa), no. 346 connectors shall be used. No other SCBA cylinder fill connections shall be permitted.

F 701.5 Enclosure and Air Filling Station Marking. Each enclosure, closet, or room shall be marked, "FIREFIGHTERS AIR SYSTEM." The lettering shall be in a color that contrasts with the cabinet front and in letters that are not less than 2 inches (51 mm) high with a $^3/_8$ of an inch (9.5 mm) brush stroke.

F 801.0 Materials.

F 801.1 General. Pressurized components shall be compatible for use with high-pressure breathing air equipment and self-contained breathing air apparatus. Pressurized breathing air components shall be rated for not less than a working pressure of 5000 psi (34 474 kPa).

F 801.2 Tubing. Tubing shall be stainless steel in accordance with ASTM A269, or other approved materials that are compatible with breathing air at the system pressure. Routing of tubing and bends shall be such as to protect the tubing from mechanical damage.

F 801.3 Fittings. Fittings shall be constructed of stainless steel in accordance with ASTM A479, or other approved materials that are compatible with breathing air at the system pressure.

F 801.4 Prohibited Materials. The use of nonmetallic materials, carbon steel, iron pipe, malleable iron, high-strength gray iron, or alloy steel shall be prohibited for breathing air pipe and tubing materials.

F 801.5 Pressure Monitoring Switch. An electric low-pressure monitoring switch shall be installed in the piping system to monitor the air pressure. The pressure switch shall transmit a supervisory signal to the central alarm monitoring station where the pressure of the breathing air system is less than 80 percent of the system operating pressure. Activation of the pressure switch shall also activate an audible alarm and visual strobe located at the building annunciator panel. A weather-resistant sign shall be provided in conjunction with the audible alarm stating, "FIREFIGHTER AIR SYSTEM – LOW AIR PRESSURE ALARM." Where not part of a building annunciator panel, the lettering shall be in a contrasting color, and the letters shall be not less than 2 inches (51 mm) high with a ⅜ of an inch (9.5 mm) brush stroke.

F 801.6 Isolation Valve. A system isolation valve shall be installed downstream of each air fill station and shall be located in the panel or within 3 feet (914 mm) of the station. The isolation valve shall be marked with its function in letters that are not less than $^3/_{16}$ of an inch (4.8 mm) high with a $^1/_{16}$ of an inch (1.6 mm) brush stroke.

F 901.0 System Requirements.

F 901.1 Protection. Components of the firefighter breathing air replenishment system installed in a building or structure shall be protected by not less than a 2-hour fire-resistive construction. Components shall be protected from physical damage.

F 901.2 Markings. Components shall be clearly identified using stainless steel or plastic labels or tags indicating their function. This shall include not less than all fire department connection panels, air fill stations, air storage system, gauges, valves, air connections, air outlets, enclosures, and doors.

F 901.3 Tubing Markings. Tubing shall be clearly marked, "FIREFIGHTERS AIR SYSTEM" and "HIGH-PRESSURE BREATHING AIR" using signs or self-adhesive labels. Signs shall be 1 inch (25.4 mm) high and shall be secured to the tubing. Signs shall be made of copper alloy, stainless steel, or plastic and engraved with ⅜ of an inch (9.5 mm) letters with a $^1/_{16}$ of an inch (1.6 mm) stroke lettering. Signs or labels shall be placed at not less than 20 foot (6096 mm) intervals and at each fitting, whether the tubing is concealed or in plain view. Tubing shall have a sign or label at an accessible point.

F 901.4 Support. Pipe and tubing shall be supported at intervals not less than that shown in Table 313.3 of this code. Pipe and tubing shall be supported in accordance with Section 313.0 of this code.

F 1001.0 Design Criteria.

F 1001.1 Fill Time. The system shall be designed to fill, at the most remote fill station or panel, not less than two 66 cubic foot (ft^3) (1.87 m^3) compressed breathing air cylinders to a pressure not to exceed 4500 psi (31 026 kPa) simultaneously in 3 minutes or less. Where greater capacity is required, the Authority Having Jurisdiction shall specify the required system capacity.

F 1001.2 Fill Panels or Stations Location. Cylinder fill panels or stations shall be installed in the interior of buildings in accordance with Section F 1001.2.1 through Section F 1001.2.3.

F 1001.2.1 High-Rise Buildings. An interior cylinder fill panel or station shall be installed commencing on the third floor and every third floor thereafter above grade. For underground floors in buildings with more than five underground floors, an interior cylinder fill panel or station shall be installed commencing on the third floor below grade and every three floors below grade thereafter, except for the bottom-most floor.

F 1001.2.2 Underground Structures. For underground floors in buildings with more than five underground floors, an interior cylinder fill panel or station shall be installed commencing on the third floor below grade and every three floors below grade thereafter, except for the bottom-most floor.

F 1001.2.3 Installation Locations. The specific location or locations on each floor shall be approved by the Authority Having Jurisdiction.

F 1101.0 System Assembly Requirements.

F 1101.1 General. The system shall be an all-welded system except where the tubing joints are readily accessible and at the individual air fill panels or stations. Where mechanical high-pressure tube fittings are used, they shall be approved for the type of materials to be joined and rated for the maximum pressure of the system.

F 1101.2 Welding Requirements. Prior to and during the welding of sections of tubing, a continuous, regulated dry nitrogen or argon purge at 3 psi (21 kPa) shall be maintained to eliminate contamination with products of the oxidation or welding flux. The purge shall commence not less than 2 minutes prior to welding operations and continue until the welded joint is at ambient temperature. Welding procedures shall comply with the following requirements:

(1) Qualification of the WPS to be used, and of the performance of welders and operators, is required.

(2) No welding shall be done if there is impingement of rain, snow, sleet, or high wind on the weld area.

(3) Tack welds permitted to remain in the finished weld shall be made by a qualified welder. Tack welds made by an unqualified welder shall be removed. Tack welds that remain shall be made with an electrode and WPS, which is the same as or equivalent to the electrode and WPS to be used for the first pass. The stopping and starting ends shall be prepared by grinding or other means so that they are capable of being satisfactorily incorporated into the final weld. Tack welds that have cracked shall be removed.

(4) CAUTION: Arc strikes outside the area of the intended weld shall be avoided on a base metal. [ASME B31.1:127.4.1]

F 1101.3 Prevention of Contamination. The system components shall not be exposed to contaminants, including but not limited to, oils, solvents, dirt, and construction materials. Where contamination of system components has occurred, the affected component shall not be installed in the system.

F 1201.0 System Acceptance and Certification.

F 1201.1 Static Pressure Testing. Following fabrication, assembly, and installation of the piping distribution system, exterior connection panel, and interior cylinder fill panels, the Authority Having Jurisdiction shall witness the pneumatic testing of the complete system at a test pressure of not less than 7500 psi (51 711 kPa) using oil-free dry air, nitrogen, or argon. A pneumatic test of not less than 24 hours shall be performed. During this test, all fittings, joints, and system components shall be inspected for leaks. A solution compatible with the system component materials shall be used on each joint and fitting. Defects in the system or leaks detected shall be documented on an inspection report, repaired or replaced. As an alternate, a pressure-decay test in accordance with ASME B31.3 shall be permitted.

F 1201.2 Low-Pressure Switch Test. Upon successful completion of the 24 hour static pressure test, the system's low-pressure monitoring switch shall be calibrated to not less than 3000 psi (20 684 kPa) descending, and tested to verify that the signal is annunciated at the building's main fire alarm panel and by means of an audible alarm and visual strobe located in a visible location.

F 1201.3 Compatibility Check. Each air fill panel and station, and each exterior fire department connection panel shall be tested for compatibility with the fire department's SCBA fill fittings.

F 1201.4 Material Certifications. The pipe or tubing material certifications shall be provided to the Authority Having Jurisdiction.

F 1201.5 Air Sampling. Before the system is placed into service, a minimum of two samples shall be taken from separate air fill panels and submitted to an independent certified gas analysis laboratory to verify the system's cleanliness and that the air is in accordance with the following requirements for breathing air:

(1) Breathing air shall have oxygen content not less than 19.5 percent and not greater than 23.5 percent by volume.

(2) Breathing air shall not have a concentration of carbon monoxide exceeding 5.0 parts per million (ppm) by volume.

(3) Breathing air shall not have a concentration of carbon dioxide exceeding 1000 ppm by volume.

(4) Breathing air shall not have a concentration of condensed oil and particulate exceeding 7.2 E-11 pounds per cubic inch (lb/in^3) (2.0 mg/m^3) at 72°F (22°C) and 30 in Hg (102 kPa).

(5) Where breathing air supply for respirators is stored at pressures exceeding 15 bars (1500 kPa), the breathing air shall not have a concentration of water exceeding 24 ppm by volume.

(6) Breathing air shall not have a nonmethane volatile organic compound (VOC) content exceeding 25 ppm as methane equivalents.

(7) Breathing air shall not have a pronounced or unusual odor.

(8) Breathing air shall have a concentration of nitrogen not less than 75 percent and not more than 81 percent. [NFPA 1989:5.6]

The written report of the analysis shall be submitted to the Authority Having Jurisdiction, documenting that the breathing air is in accordance with this section.

F 1201.5.1 Air Quality Analysis. During the period of air quality analysis, the air fill panel inlet shall be secured so that no air is capable of being introduced into the system and each air fill panel shall be provided with a sign stating, "AIR QUALITY ANALYSIS IN PROGRESS, DO NOT FILL OR USE ANY AIR FROM THIS SYSTEM." This sign shall be not less than 8½ inches (216 mm) by 11 inches (279 mm) with not less than 1 inch (25.4 mm) lettering.

F 1201.6 Annual Air Sampling. The breathing air within the system shall be sampled and certified annually and inspected in accordance with the procedure in Section F 1201.5.

F 1201.7 Final Proof Test. The Authority Having Jurisdiction shall witness the filling of two empty 66 cubic foot (ft^3) (1.87 m^3) capacity SCBA cylinders in 3 minutes or less, using compressed air supplied by fire department equipment connected to the exterior fire department connection panel. The SCBA cylinders shall be filled at the air fill panel or station farthest from the exterior fire department connection panel. Following this, not less than two air samples shall then be taken from separate air filling stations and submitted to an independent certified gas analyst laboratory to verify the system's cleanliness and that the air is in accordance with the requirements of NFPA 1989. The written report shall be provided to the Authority Having Jurisdiction certifying that the air analysis is in accordance with the above requirements.

APPENDIX G

SIZING OF VENTING SYSTEMS

(The content of this Appendix is based on Annex F of NFPA 54)

G 101.0 General.

G 101.1 Applicability. This appendix provides general guidelines for sizing venting systems serving appliances equipped with draft hoods, Category I appliances, and appliances listed for use with Type B vents.

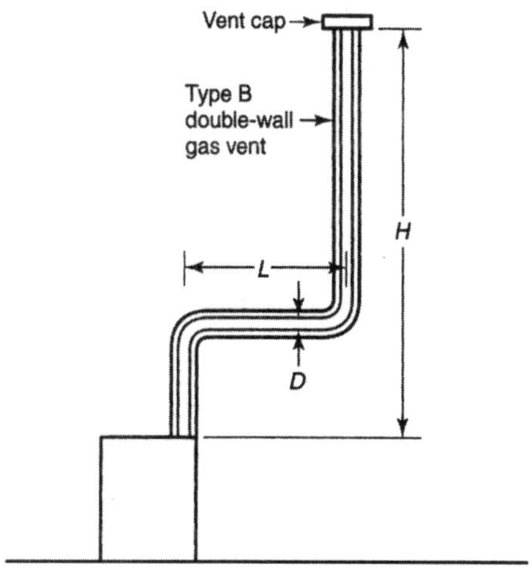

Table 510.1.2(1) is used when sizing a Type B double-wall gas vent connected directly to the appliance.

Note: The appliance can be either Category I draft hood-equipped or fan-assisted type.

FIGURE G 101.2(1)
TYPE B DOUBLE-WALL VENT SYSTEM SERVING A SINGLE APPLIANCE WITH A TYPE B DOUBLE-WALL VENT

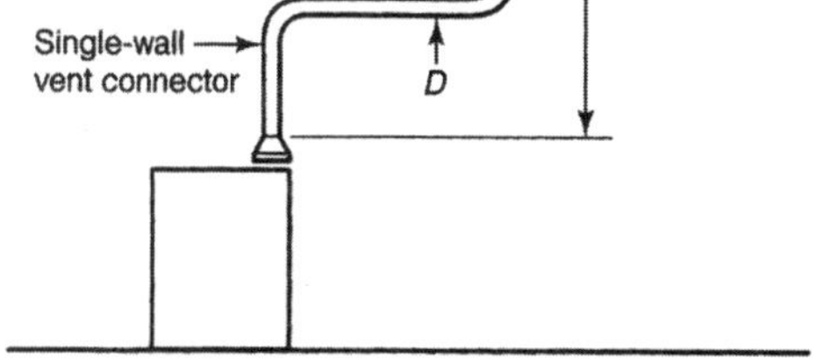

Table 510.1.2(2) is used when sizing a single-wall metal vent connector attached to a Type B double-wall gas vent.

Note: The appliance can be either Category I draft hood-equipped or fan-assisted type.

FIGURE G 101.2(2)
TYPE B DOUBLE-WALL VENT SYSTEM SERVING A SINGLE APPLIANCE WITH A SINGLE-WALL METAL VENT CONNECTOR

G 101.2 Examples Using Single Appliance Venting Tables. See Figure G 101.2(1) through Figure G 101.2(14).

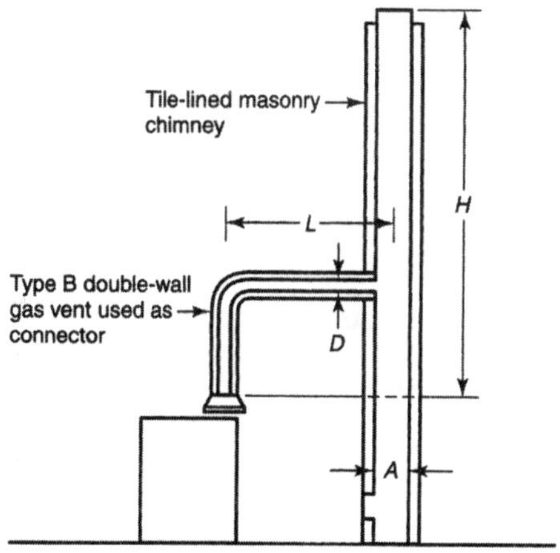

Table 510.1.2(3) is used when sizing a Type B double-wall gas vent connector attached to a tile-lined masonry chimney.

Notes:
1. *A* is the equivalent cross-sectional area of the tile liner.
2. The appliance can be either Category I draft hood-equipped or fan-assisted type.

FIGURE G 101.2(3)
VENT SYSTEM SERVING A SINGLE APPLIANCE WITH A MASONRY CHIMNEY AND A TYPE B DOUBLE-WALL VENT CONNECTOR

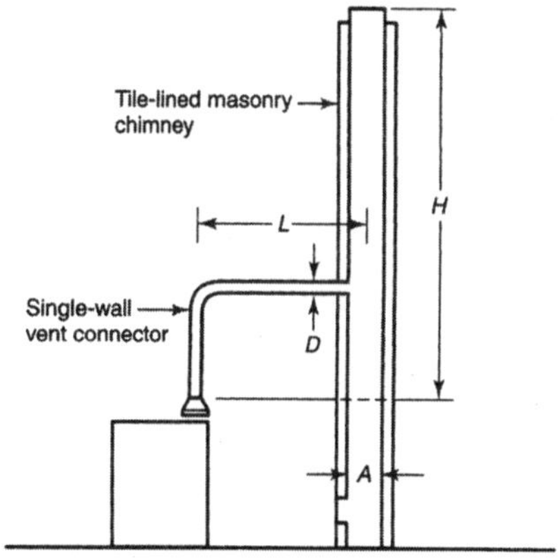

Table 510.1.2(4) is used when sizing a single-wall vent connector attached to a tile-lined masonry chimney.

Notes:
1. *A* is the equivalent cross-sectional area of the tile liner.
2. The appliance can be either Category I draft hood-equipped or fan-assisted type.

FIGURE G 101.2(4)
VENT SYSTEM SERVING A SINGLE APPLIANCE USING A MASONRY CHIMNEY AND A SINGLE-WALL METAL VENT CONNECTOR

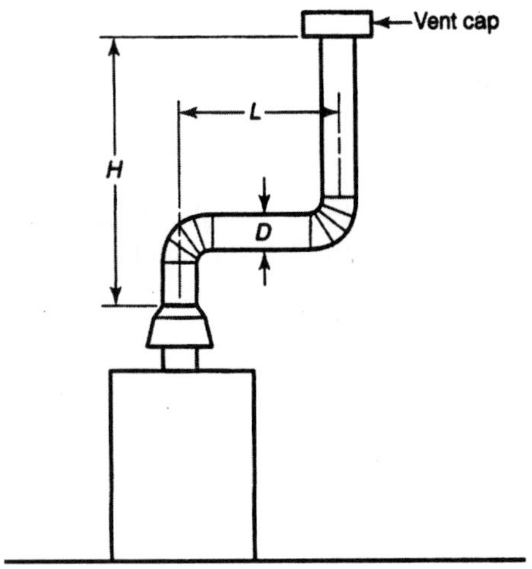

Asbestos cement Type B or single-wall metal vent serving a single draft hood-equipped appliance. [See Table 510.1.2(5)]

FIGURE G 101.2(5)
ASBESTOS CEMENT TYPE B OR SINGLE-WALL METAL VENT SYSTEM SERVING A SINGLE DRAFT HOOD-EQUIPPED APPLIANCE

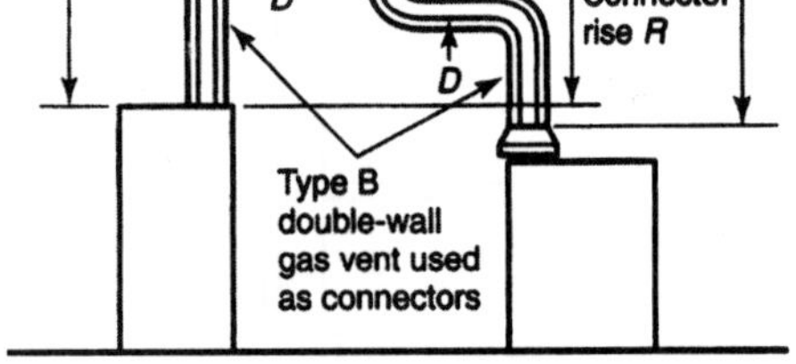

Table 510.2(1) is used when sizing Type B double-wall gas vent connectors attached to a Type B double-wall common vent.

Note: Each appliance can be either Category I draft hood-equipped or fan-assisted type.

FIGURE G 101.2(6)
VENT SYSTEM SERVING TWO OR MORE APPLIANCES WITH TYPE B DOUBLE-WALL VENT AND TYPE B DOUBLE-WALL VENT CONNECTORS

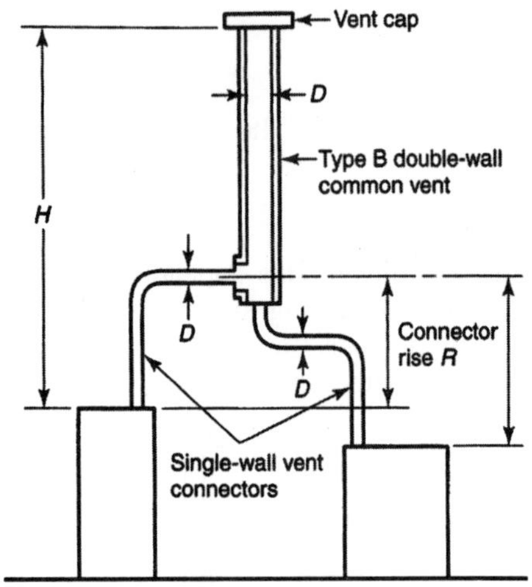

Table 510.2(2) is used when sizing single-wall vent connectors attached to a Type B double-wall common vent.

Note: Each appliance can be either Category I draft hood-equipped or fan-assisted type.

FIGURE G 101.2(7)
VENT SYSTEM SERVING TWO OR MORE APPLIANCES WITH TYPE B DOUBLE-WALL VENT AND SINGLE-WALL METAL VENT CONNECTORS

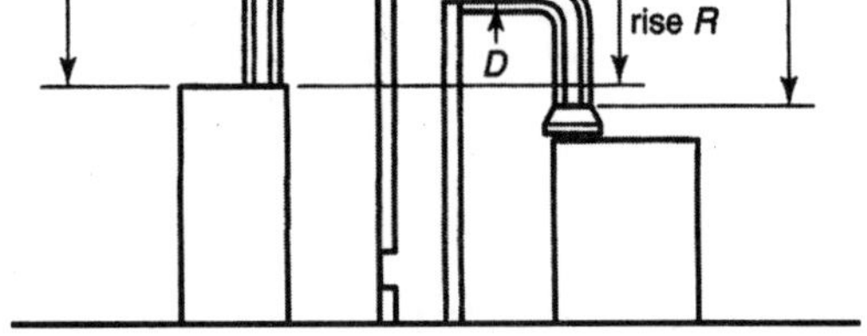

Table 510.2(3) is used when sizing Type B double-wall vent connectors attached to a tile-lined masonry chimney.

Notes:

1. *A* is the equivalent cross-sectional area of the tile liner.
2. Each appliance can be either Category I draft hood-equipped or fan-assisted type.

FIGURE G 101.2(8)
MASONRY CHIMNEY SERVING TWO OR MORE APPLIANCES WITH TYPE B DOUBLE-WALL VENT CONNECTORS

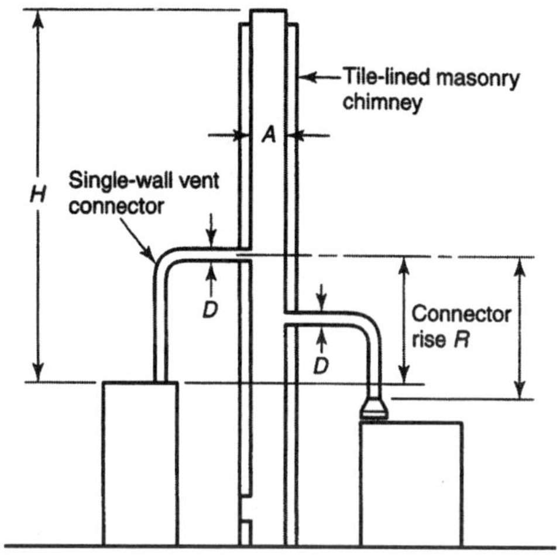

Table 510.2(4) is used when sizing single-wall metal vent connectors attached to a tile-lined masonry chimney.

Notes:

1. *A* is the equivalent cross-sectional area of the tile liner.
2. Each appliance can be either Category I draft hood-equipped or fan-assisted type.

FIGURE G 101.2(9)
MASONRY CHIMNEY SERVING TWO OR MORE APPLIANCES WITH SINGLE-WALL METAL VENT CONNECTORS

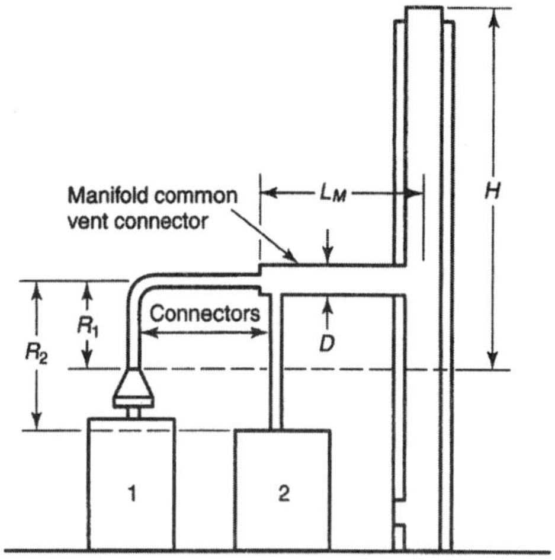

Example: Manifolded common vent connector *LM* can be no greater than 18 times the common vent connector manifold inside diameter; that is, a 4 inch (102 mm) inside diameter common vent connector manifold should not exceed 72 inches (1829 mm) in length. [See Section 510.2.3]

Note: This is an illustration of a typical manifolded vent connector. Different appliance, vent connector, or common vent types are possible. [See Section 510.2]

FIGURE G 101.2(11)
USE OF MANIFOLDED COMMON VENT CONNECTOR

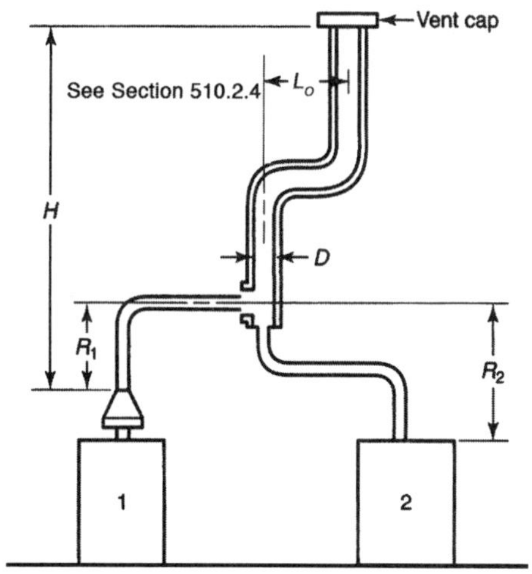

Asbestos cement Type B or single-wall metal pipe vent serving two or more draft hood-equipped appliances. [See Table 510.2(5)]

FIGURE G 101.2(10)
ASBESTOS CEMENT TYPE B OR SINGLE-WALL METAL VENT SYSTEM SERVING TWO OR MORE DRAFT HOOD-EQUIPPED APPLIANCES

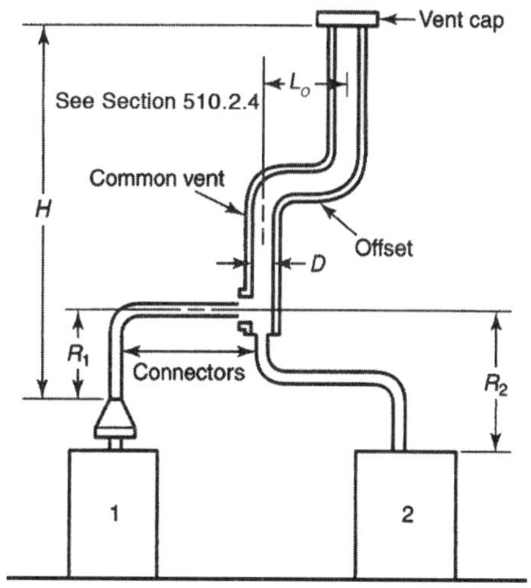

Example: Offset common vent

Note: This is an illustration of a typical offset vent. Different appliance, vent connector, or vent types are possible. [See Section 510.1 and Section 510.2]

FIGURE G 101.2(12)
USE OF OFFSET COMMON VENT

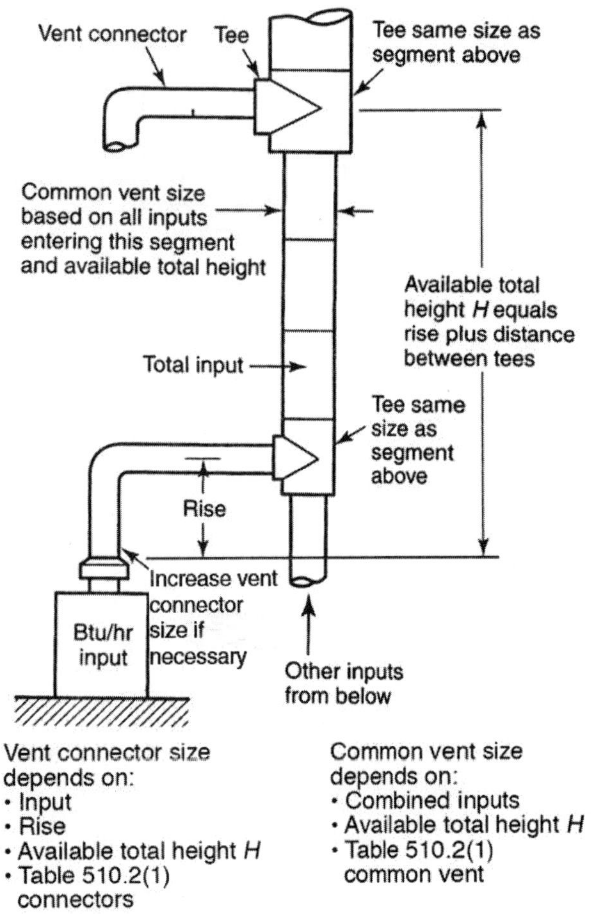

FIGURE G 101.2(13)
MULTISTORY GAS VENT DESIGN PROCEDURE FOR EACH SEGMENT OF SYSTEM

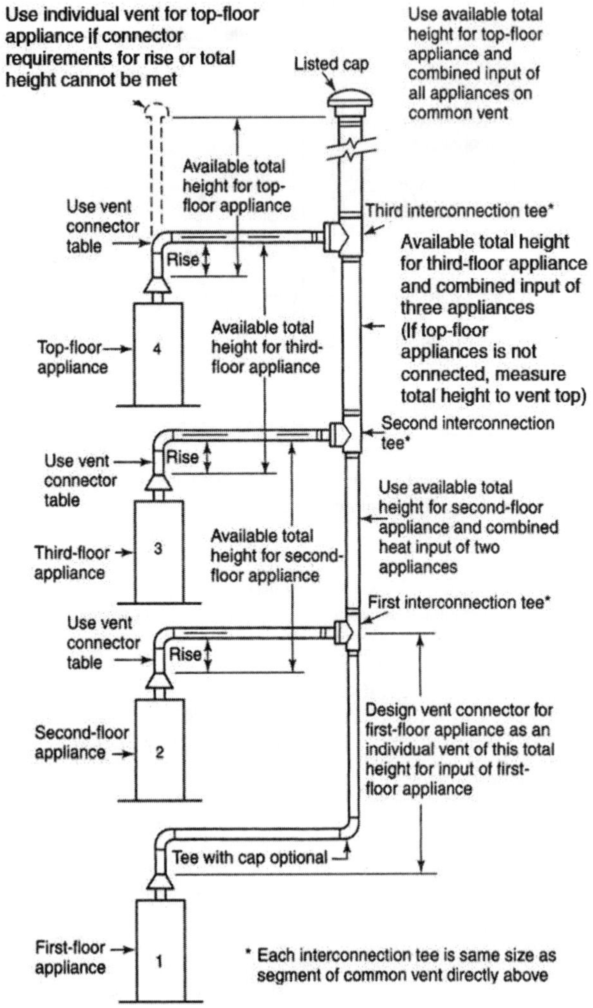

FIGURE G 101.2(14)
PRINCIPLES OF DESIGN OF MULTISTORY VENTS USING VENT CONNECTOR AND COMMON VENT DESIGN TABLES

G 101.3 Example 1: Single Draft Hood-Equipped Appliance. An installer has a 120 000 British thermal units per hour (Btu/h) (35 kW) input appliance with a 5 inch (127 mm) diameter draft hood outlet that needs to be vented into a 10 foot (3048 mm) high Type B vent system. What size vent shall be used assuming: (1) a 5 foot (1524 mm) lateral single-wall metal vent connector is used with two 90 degree (1.57 rad) elbows or (2) a 5 foot (1524 mm) lateral single-wall metal vent connector is used with three 90 degree (1.57 rad) elbows in the vent system? (See Figure G 101.3)

Solution:

Table 510.1.2(2) should be used to solve this problem because single-wall metal vent connectors are being used with a Type B vent, as follows:

(1) Read down the first column in Table 510.1.2(2) until the row associated with a 10 foot (3048 mm) height and 5 foot (1524 mm) lateral is found. Read across this row until a vent capacity greater than 120 000 Btu/h (35 kW) is located in the shaded columns labeled NAT Max for draft hood-equipped appliances. In this case, a 5 inch (127 mm) diameter vent has a capacity of 122 000 Btu/h (35.7 kW) and can be used for this application.

(2) If three 90 degree (1.57 rad) elbows are used in the vent system, the maximum vent capacity listed in the tables must be reduced by 10 percent. This implies that the 5 inch (127 mm) diameter vent has an adjusted capacity of only 110 000 Btu/h (32 kW). In this case, the vent system must be increased to 6 inches (152 mm) in diameter. See the following calculations:

122 000 Btu/h (35.7 kW) x 0.90 = 110 000 Btu/h (32 kW) for 5 inch (127 mm) vent

From Table 510.1.2(2), select 6 inch (152 mm) vent.

186 000 Btu/h (54.5 kW) x 0.90 = 167 000 Btu/h (49 kW)

This figure is greater than the required 120 000 Btu/h (35 kW). Therefore, use a 6 inch (152 mm) vent and connector where three elbows are used.

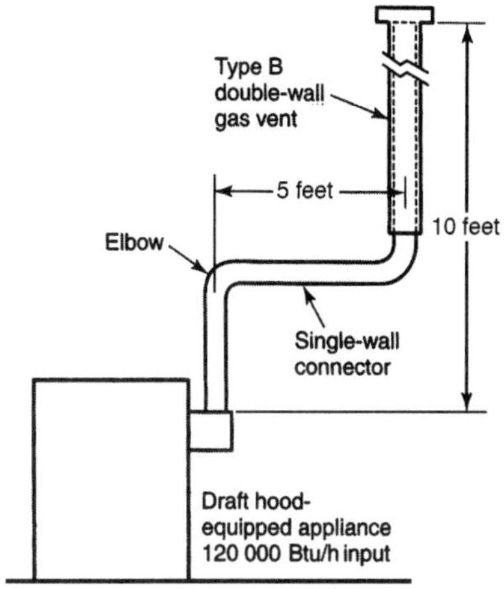

For SI units: 1 foot = 304.8 mm, 1000 British thermal units per hour = 0.293 kW

FIGURE G 101.3
SINGLE DRAFT HOOD-EQUIPPED APPLIANCE – EXAMPLE 1

G 101.4 Example 2: Single Fan-Assisted Appliance. An installer has an 80 000 Btu/h (23.4 kW) input fan-assisted appliance that must be installed using 10 feet (3048 mm) of lateral connector attached to a 30 foot high (9144 mm) Type B vent. Two 90-degree (1.57 rad) elbows are needed for the installation. Can a single-wall metal vent connector be used for this application? (See Figure G 101.4)

Solution:

Table 510.1.2(2) refers to the use of single-wall metal vent connectors with Type B vent. In the first column find the row associated with a 30 foot (9144 mm) height and a 10 foot (3048 mm) lateral. Read across this row, looking at the FAN Min and FAN Max columns, to find that a 3 inch (76 mm) diameter single-wall metal vent connector is not recommended. Moving to the next larger size single-wall connector [4 inch (102 mm)] we find that a 4 inch (102 mm) diameter single-wall metal connector has a recommended minimum vent capacity of 91 000 Btu/h (26.7 kW) and a recommended maximum vent capacity of 144 000 Btu/h (42.2 kW). The 80 000 Btu/h (23.4 kW) fan-assisted appliance is outside this range, so the conclusion is that a single-wall metal vent connector cannot be used to vent this appliance using 10 feet (3048 mm) of lateral for the connector.

However, if the 80 000 Btu/h (23.4 kW) input appliance could be moved to within 5 feet (1524 mm) of the vertical vent, a 4 inch (102 mm) single-wall metal connector could be used to vent the appliance. Table 510.1.2(2) shows the acceptable range of vent capacities for a 4 inch (102 mm) vent with 5 feet (1524 mm) of lateral to be between 72 000 Btu/h (21.1 kW) and 157 000 Btu/h (46 kW).

If the appliance cannot be moved closer to the vertical vent, a Type B vent could be used as the connector material. In this case, Table 510.1.2(1) shows that, for a 30 foot (9144 mm) high vent with 10 feet (3048 mm) of lateral, the acceptable range of vent capacities for a 4 inch (102 mm) diameter vent attached to a fan-assisted appliance is between 37 000 Btu/h (10.8 kW) and 150 000 Btu/h (44 kW).

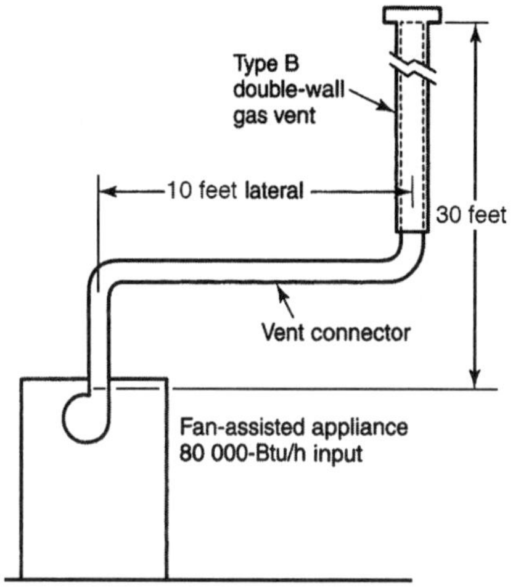

For SI units: 1 foot = 304.8 mm, 1000 British thermal units per hour = 0.293 kW

FIGURE G 101.4
SINGLE FAN-ASSISTED APPLIANCE – EXAMPLE 2

G 101.5 Example 3: Interpolating Between Table Values. An installer has an 80 000 Btu/h (23.4 kW) input appliance with a 4 inch (102 mm) diameter draft hood outlet that needs to be vented into a 12 foot (3658 mm) high Type B vent. The vent connector has a 5 foot (1524 mm) lateral length and is also Type B. Can this appliance be vented using a 4 inch (102 mm) diameter vent?

Solution:

Table 510.1.2(1) is used in the case of an all Type B Vent system. However, Table 510.1.2(1) does not have an entry for a height of 12 feet (3658 mm), and interpolation must be used. Read down the 4 inch (102 mm) diameter NAT Max column to the row associated with 10 foot (3048 mm) height and 5 foot (1524 mm) lateral to find the capacity value of 77 000 Btu/h (22.6 kW). Read further down to the 15 foot (4572 mm) height, 5 foot (1524 mm) lateral row to find the capacity value of 87 000 Btu/h (25.5 kW). The difference between the 15 foot (4572 mm) height capacity value and the 10 foot (3048 mm) height capacity value is 10 000 Btu/h (3 kW). The capacity for a vent system with a 12 foot (3658 mm) height is equal to the capacity for a 10 foot (3048 mm) height plus two-fifths of the difference between the 10 foot (3048 mm) and 15 foot (4572 mm) height values, or 77 000 Btu/h (22.6 kW) + ⅖ x 10 000 Btu/h (3 kW) = 81 000 Btu/h (23.7 kW). Therefore, a 4 inch (102 mm) diameter vent can be used in the installation.

G 101.6 Example 4: Common Venting Two Draft Hood-Equipped Appliances. A 35 000 Btu/h (10.3 kW) water heater is to be common vented with a 150 000 Btu/h (44 kW) furnace, using a common vent with a total height of 30 feet (9144 mm). The connector rise is 2 feet (610 mm) for the water heater with a horizontal length of 4 feet (1219 mm). The connector rise for the furnace is 3 feet (914 mm) with a horizontal length of 8 feet (2438 mm). Assume single-wall metal connectors will be used with Type B vent. What size connectors and combined vent should be used in this installation? (See Figure G 101.6)

Solution:

Table 510.2(2) should be used to size single-wall metal vent connectors attached to Type B vertical vents. In the vent connector capacity portion of Table 510.2(2), find the row associated with a 30 foot (9144 mm) vent height. For a 2 foot (610 mm) rise on the vent connector for the water heater, read the shaded columns for draft hood-equipped appliances to find that a 3 inch (76 mm) diameter vent connector has a capacity of 37 000 Btu/h (10.8 kW). Therefore, a 3 inch (76 mm) single-wall metal vent connector can be used with the water heater. For a draft hood-equipped furnace with a 3 foot (914 mm) rise, read across the appropriate row to find that a 5 inch (127 mm) diameter vent connector has a maximum capacity of 120 000 Btu/h (35 kW) (which is too small for the furnace) and a 6 inch (152 mm) diameter vent connector has a maximum vent capacity of 172 000 Btu/h (50 kW). Therefore, a 6 inch (152 mm) diameter vent connector should be used with the 150 000 Btu/h (44 kW) furnace. Because both vent connector, horizontal lengths are less than the maximum lengths listed in Section 510.2.1; the table values can be used without adjustments.

In the common vent capacity portion of Table 510.2(2), find the row associated with a 30 foot (9144 mm) vent height and read over to the NAT + NAT portion of the 6 inch (152 mm) diameter column to find a maximum combined capacity of 257 000 Btu/h (75 kW). Since the two appliances total only 185 000 Btu/h (54 kW), a 6 inch (152 mm) common vent can be used.

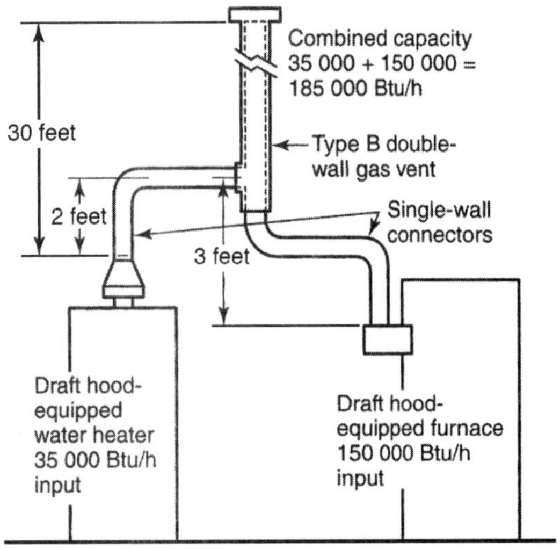

For SI units: 1 foot = 304.8 mm, 1000 British thermal units per hour = 0.293 kW

FIGURE G 101.6
COMMON VENTING TWO DRAFT HOOD-EQUIPPED APPLIANCES – EXAMPLE 4

G 101.7 Example 5(a): Common Venting a Draft Hood-Equipped Water Heater with a Fan-Assisted Furnace into a Type B Vent. In this case, a 35 000 Btu/h (10.3 kW) input draft hood-equipped water heater with a 4 inch (102 mm) diameter draft hood outlet, 2 feet (610 mm) of connector rise, and 4 feet (1219 mm) of horizontal length is to be common vented with a 100 000 Btu/h (29 kW) fan-assisted furnace with a 4 inch (102 mm) diameter flue collar, 3 feet (914 mm) of connector rise, and 6 feet (1829 mm) of horizontal length. The common vent consists of a 30 foot (9144 mm) height of Type B vent. What are the recommended vent diameters for each connector and the common vent? The installer would like to use a single-wall metal vent connector. (See Figure G 101.7)

Solution:

Water Heater Vent Connector Diameter. Since the water heater vent connector, horizontal length of 4 feet (1219 mm) is less than the maximum value listed in Table 510.2(2), the venting table values can be used without adjustments. Using the Vent Connector Capacity portion of Table 510.2(2), read down the Total Vent Height (*H*) column to 30 feet (9144 mm) and read across the 2 feet (610 mm) Connector Rise (*R*) row to the first Btu/h rating in the NAT Max column that is equal to or greater than the water heater input rating. The table shows that a 3 inch (76 mm) vent connector has a maximum input rating of 37 000 Btu/h (10.8 kW). Although this rating is greater than the water heater input rating, a 3 inch (76 mm) vent connector is prohibited by Section 510.2.18. A 4 inch (102 mm) vent connector has a maximum input rating of 67 000 Btu/h (19.6 kW) and is equal to the draft hood outlet diameter. A 4 inch (102 mm) vent connector is selected. Since the water heater is equipped with a draft hood, there are no minimum input rating restrictions.

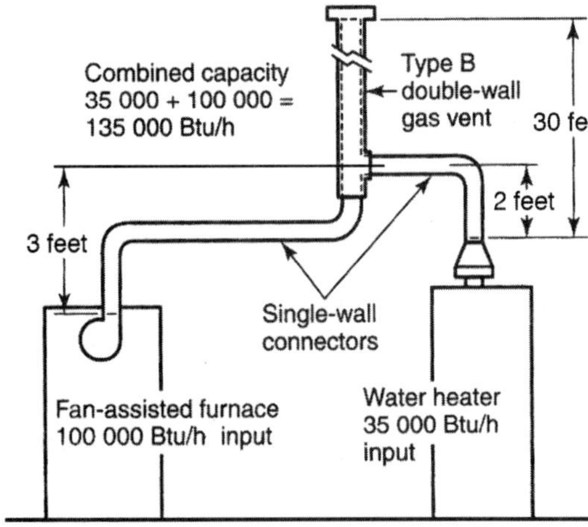

For SI units: 1 foot = 304.8 mm, 1000 British thermal units per hour = 0.293 kW

FIGURE G 101.7
COMMON VENTING A DRAFT HOOD-EQUIPPED WATER HEATER WITH A FAN-ASSISTED FURNACE INTO A TYPE B DOUBLE-WALL COMMON VENT – EXAMPLE 5(a)

Furnace Vent Connector Diameter. Using the Vent Connector Capacity portion of Table 510.2(2), read down the Total Vent Height (*H*) column to 30 feet (9144 mm) and across the 3 feet (914 mm) Connector Rise (*R*) row. Because the furnace has a fan-assisted combustion system, find the first FAN Max column with a Btu/h rating greater than the furnace input rating. The 4 inch (102 mm) vent connector has a maximum input rating of 119 000 Btu/h (34.9 kW) and a minimum input rating of 85 000 Btu/h (24.9 kW).

The 100 000 Btu/h (29 kW) furnace in this example falls within this range, so a 4 inch (102 mm) connector is adequate. Because the furnace vent connector horizontal length of 6 feet (1829 mm) is less than the maximum value listed in Section 510.2.1; the venting table values can be used without adjustment. If the furnace had an input rating of 80 000 Btu/h (23.4 kW), a Type B vent connector would be needed in order to meet the minimum capacity limit.

Common Vent Diameter. The total input to the common vent is 135 000 Btu/h (40 kW). Using the Common Vent Capacity portion of Table 510.2(2), read down the Total Vent Height (*H*) column to 30 feet (9144 mm) and across this row to find the smallest vent diameter in the FAN + NAT column that has a Btu/h rating equal to or greater than 135 000 Btu/h (40 kW). The 4 inch (102 mm) common vent has a capacity of 132 000 Btu/h (39 kW) and the 5 inch (127 mm) common vent has a capacity of 202 000 Btu/h (59 kW). Therefore, the 5 inch (127 mm) common vent should be used in this example.

Summary: In this example, the installer can use a 4 inch (102 mm) diameter, single-wall metal vent connector for the water heater and a 4 inch (102 mm) diameter, single-wall metal vent connector for the furnace. The common vent should be a 5 inch (127 mm) diameter Type B vent.

G 101.8 Example 5(b): Common Venting into an Interior Masonry Chimney. In this case, the water heater and fan-assisted furnace of Example 5(a) are to be common-vented into a clay-tile-lined masonry chimney with a 30 foot (9144 mm) height. The chimney is not exposed to the outdoors below the roof line. The internal dimensions of the clay tile liner are nominally 8 inches (203 mm) by 12 inches (305 mm). Assuming the same vent connector heights, laterals, and materials found in Example 5(a), what are the recommended vent connector diameters, and is this an acceptable installation?

Solution:

Table 510.2(4) is used to size common venting installations involving single-wall connectors into masonry chimneys.

Water Heater Vent Connector Diameter. Using Table 510.2(4), Vent Connector Capacity, read down the Total Vent Height (*H*) column to 30 feet (9144 mm), and read across the 2 feet (610 mm) Connector Rise (*R*) row to the first Btu/h rating in the NAT Max column that is equal to or greater than the water heater input rating. The table shows that a 3 inch (76 mm) vent connector has a maximum input of only 31 000 Btu/h (9 kW), while a 4 inch (102 mm) vent connector has a maximum input of 57 000 Btu/h (16.7 kW). A 4 inch (102 mm) vent connector must therefore be used.

Furnace Vent Connector Diameter. Using the Vent Connector Capacity portion of Table 510.2(4), read down the Total Vent Height (*H*) column to 30 feet (9144 mm) and across the 3 feet (914 mm) Connector Rise (*R*) row. Because the furnace has a fan-assisted combustion system, find the first FAN Max column with a Btu/h rating greater than the furnace input rating. The 4 inch (102 mm) vent connector has a maximum input rating of 127 000 Btu/h (37 kW) and a minimum input rating of 95 000 Btu/h (27.8 kW). The 100 000 Btu/h (29 kW) furnace in this example falls within this range, so a 4 inch (102 mm) connector is adequate.

Masonry Chimney. From Table G 101.8, the equivalent area for a nominal liner size of 8 inches (203 mm) by 12 inches (305 mm) is 63.6 square inches (0.041 m^2). Using Table 510.2(4), Common Vent Capacity, read down the FAN + NAT column under the Minimum Internal Area of Chimney value of 63 to the row for 30 foot (9144 mm) height to find a capacity value of 739 000 Btu/h (217 kW). The combined input rating of the furnace and water heater, 135 000 Btu/h (40 kW), is less than the table value, so this is an acceptable installation.

Subsection 510.2.17 requires the common vent area to be no greater than seven times the smallest listed appliance categorized vent area, flue collar area, or draft hood outlet area. Both appliances in this installation have 4 inch (102 mm) diameter outlets. From Table G 101.8, the equivalent area for an inside diameter of 4 inches (102 mm) is 12.2 square inches (0.008 m^2). Seven times 12.2 equals 85.4, which is greater than 63.6, so this configuration is acceptable.

G 101.9 Example 5(c): Common Venting into an Exterior Masonry Chimney. In this case, the water heater and fan-assisted furnace of Examples 5(a) and 5(b) are to be common-vented into an exterior masonry chimney. The chimney height, clay-tile-liner dimensions, and vent connector heights and laterals are the same as in Example 5(b). This system is being installed in Charlotte, North Carolina. Does this exterior masonry chimney need to be relined? If so, what corrugated metallic liner size is recommended? What vent connector diameters are recommended? (See Table G 101.8 and Figure 510.1.10)

Solution:

According to Section 510.2.20, Type B vent connectors are required to be used with exterior masonry chimneys. Use Table 510.2(8) and Table 510.2(9) to size FAN+NAT common venting installations involving Type B double-wall connectors into exterior masonry chimneys.

The local 99 percent winter design temperature needed to use Table 510.2(8) and Table 510.2(9) can be found in ASHRAE Handbook – Fundamentals. For Charlotte, North Carolina, this design temperature is 19°F (-7.2°C).

TABLE G 101.8
MASONRY CHIMNEY LINER DIMENSIONS WITH CIRCULAR EQUIVALENTS*

NOMINAL LINER SIZE (inches)	INSIDE DIMENSIONS OF LINER (inches)	INSIDE DIAMETER OR EQUIVALENT DIAMETER (inches)	EQUIVALENT AREA (square inches)
4 x 8	2½ x 6½	4.0	12.2
		5.0	19.6
		6.0	28.3
		7.0	38.3
8 x 8	6¾ x 6¾	7.4	42.7
		8.0	50.3
8 x 12	6½ x 10½	9.0	63.6
		10.0	78.5
12 x 12	9¾ x 9¾	10.4	83.3
		11.0	95.0
12 x 16	9½ x 13½	11.8	107.5
		12.0	113.0
		14.0	153.9
16 x 16	13¼ x 13¼	14.5	162.9
		15.0	176.7
16 x 20	13 x 17	16.2	206.1
		18.0	254.4
20 x 20	16½ x 16¾	18.2	260.2
		20.0	314.1
20 x 24	16½ x 20½	20.1	314.2
		22.0	380.1
24 x 24	20¼ x 20¼	22.1	380.1
		24.0	452.3
24 x 28	20¼ x 24¼	24.1	456.2
28 x 28	24¼ x 24¼	26.4	543.3
		27.0	572.5
30 x 30	25½ x 25½	27.9	607.0
		30.0	706.8
30 x 36	25½ x 31½	30.9	749.9
		33.0	855.3
36 x 36	31½ x 31½	34.4	929.4
		36.0	1017.9

For SI units: 1 inch = 25.4 mm, 1 square inch = 0.000645 m^2

* When liner sizes differ dimensionally from those shown in this table, equivalent diameters can be determined from published tables for square and rectangular ducts of equivalent carrying capacity or by other engineering methods.

Chimney Liner Requirement. As in Example 5(b), use the 63 square inches (0.04 m^2) Internal Area columns for this size clay tile liner. Read down the 63 square inches (0.04 m^2) column of Table 510.2(8) to the 30 foot (9144 mm) height row to find that the combined appliance maximum input is 747 000 Btu/h (218.9 kW). The combined input rating of the appliances in this installation, 135 000 Btu/h (40 kW), is less than the maximum value, so this criterion is satisfied. Table 510.2(9), at a 19°F (-7.2°C) design temperature, and at the same vent height and internal area used earlier, shows that the minimum allowable input rating of a space-heating appliance is 470 000 Btu/h (137.7 kW). The furnace input rating of 100 000 Btu/h (29 kW) is less than this minimum value. So this criterion is not satisfied, and an alternative venting design needs to be used, such as a Type B vent shown in Example 5(a) or a listed chimney liner system shown in the remainder of the example.

According to Section 510.2.19, Table 510.2(1) or Table 510.2(2) is used for sizing corrugated metallic liners in masonry chimneys, with the maximum common vent capacities reduced by 20 percent. This example will be continued assuming Type B vent connectors.

Water Heater Vent Connector Diameter. Using Table 510.2(1) Vent Connector Capacity, read down the total Vent Height (*H*) column to 30 feet (9144 mm), and read across the 2 feet (610 mm) Connector Rise (*R*) row to the first Btu/hour rating in the NAT Max column that is equal to or greater than the water heater input rating. The table shows that a 3 inch (76 mm) vent connector has a maximum capacity of 39 000 Btu/h (11.4 kW). Although this rating is greater than the water heater input rating, a 3 inch (76 mm) vent connector is prohibited by Section 510.2.20. A 4 inch (102 mm) vent connector has a maximum input rating of 70 000 Btu/h (20.5 kW) and is equal to the draft hood outlet diameter. A 4 inch (102 mm) vent connector is selected.

Furnace Vent Connector Diameter. Using Table 510.2(1), Vent Connector Capacity, read down the total Vent Height (*H*) column to 30 feet (9144 mm), and read across the 3 feet (914 mm) Connector Rise (*R*) row to the first Btu/h rating in the FAN MAX column that is equal to or greater than the furnace input rating. The 100 000 Btu/h (29 kW) furnace in this example falls within this range, so a 4 inch (102 mm) connector is adequate.

Chimney Liner Diameter. The total input to the common vent is 135 000 Btu/h (40 kW). Using the Common Vent Capacity portion of Table 510.2(1), read down the total Vent Height (*H*) column to 30 feet (9144 mm) and across this row to find the smallest vent diameter in the FAN + NAT column that has a Btu/h rating greater than 135 000 Btu/h (40 kW). The 4 inch (102 mm) common vent has a capacity of 138 000 Btu/h (40.4 kW). Reducing the maximum capacity by 20 percent results in a maximum capacity for a 4 inch (102 mm) corrugated liner of 110 000 Btu/h (32 kW), less than the total input of 135 000 Btu/h (40 kW). So a larger liner is needed. The 5 inch (127 mm) common vent capacity listed in Table

510.2(1) is 210 000 Btu/h (62 kW), and after reducing by 20 percent is 168 000 Btu/h (49.2 kW). Therefore, a 5 inch (127 mm) corrugated metal liner should be used in this example.

Single Wall Connectors. Once it has been established that relining the chimney is necessary, Type B double-wall vent connectors are not specifically required. This example could be redone using Table 510.2(2) for single-wall vent connectors. For this case, the vent connector and liner diameters would be the same as found for Type B double-wall connectors.

APPENDIX H

PRIVATE SEWAGE DISPOSAL SYSTEMS

H 101.0 General.

H 101.1 Applicability. This appendix provides general guidelines for the materials, design, and installation of private sewage disposal systems.

H 101.2 General Requirements. Where permitted by Section 713.0, the building sewer shall be permitted to be connected to a private sewage disposal system in accordance with the provisions of this appendix. The type of system shall be determined on the basis of location, soil porosity, and groundwater level, and shall be designed to receive all sewage from the property. The system, except as otherwise approved, shall consist of a septic tank with effluent discharging into a subsurface disposal field, into one or more seepage pits, or into a combination of subsurface disposal field and seepage pits. The Authority Having Jurisdiction shall be permitted to grant exceptions to the provisions of this appendix for permitted structures that have been destroyed due to fire or natural disaster, and that cannot be reconstructed in compliance with these provisions provided that such exceptions are the minimum necessary.

H 101.3 Quantity and Quality. Where the quantity or quality of the sewage is such that the above system cannot be expected to function satisfactorily for commercial, agricultural, and industrial plumbing systems; for installations where appreciable amounts of industrial or indigestible wastes are produced; for occupancies producing abnormal quantities of sewage or liquid waste; or where grease interceptors are required by other parts of this code, the method of sewage treatment and disposal shall be first approved by the Authority Having Jurisdiction. Special sewage disposal systems for minor, limited, or temporary uses shall be first approved by the Authority Having Jurisdiction.

H 101.4 Septic Tank and Disposal Field Systems. Disposal systems shall be designed to utilize the most porous or absorptive portions of the soil formation. Where the groundwater level extends to within 12 feet (3658 mm) or less of the ground surface or where the upper soil is porous, and the underlying stratum is rock or impervious soil, a septic tank and disposal field system shall be installed.

H 101.5 Flood Hazard Areas. Disposal systems shall be located outside of flood hazard areas.

Exception: Where suitable sites outside of flood hazard areas are not available, disposal systems shall be permitted to be located in flood hazard areas on sites where the effects of inundation under conditions of the design flood are minimized.

H 101.6 Design. Private sewage disposal systems shall be so designed that additional seepage pits or subsurface drain fields, equivalent to not less than 100 percent of the required original system, shall be permitted to be installed where the original system cannot absorb all the sewage. No division of the lot or erection of structures on the lot shall be made where such division or structure impairs the usefulness of the 100 percent expansion area.

H 101.7 Capacity. No property shall be improved more than its capacity to absorb sewage effluent properly by the means provided in this code.

Exception: The Authority Having Jurisdiction shall be permitted to, at its discretion, approve an alternate system.

H 101.8 Location. No private sewage disposal system, or part thereof, shall be located in any lot other than the lot that is the site of the building or structure served by such private sewage disposal system, nor shall any private sewage disposal system or part thereof be located at any point having less than the minimum distances indicated in Table H 101.8.

Nothing contained in this code shall be construed to prohibit the use of all or part of an abutting lot to provide additional space for a private sewage disposal system or part thereof where proper cause, transfer of ownership, or change of boundary not in violation of other requirements has been first established to the satisfaction of the Authority Having Jurisdiction. The instrument recording such action shall constitute an agreement with the Authority Having Jurisdiction, which shall clearly state and show that the areas so joined or used shall be maintained as a unit during the time they are so used. Such agreement shall be recorded in the office of the County Recorder as part of the conditions of ownership of said properties and shall be binding on heirs, successors, and assigns to such properties. A copy of the instrument recording such proceedings shall be filed with the Authority Having Jurisdiction.

H 101.9 Building Permit. Where there is insufficient lot area or improper soil conditions for sewage disposal for the building or land use proposed, and the Authority Having Jurisdiction so finds, no building permit shall be issued, and no private sewage disposal shall be permitted. Where space or soil conditions are critical, no building permit shall be issued until engineering data, and test reports satisfactory to the Authority Having Jurisdiction have been submitted and approved.

H 101.10 Additional Requirements. Nothing contained in this appendix shall be construed to prevent the Authority Having Jurisdiction from requiring compliance with additional requirements than those contained herein, where such additional requirements are essential to maintaining a safe and sanitary condition.

H 101.11 Alternate Systems. Alternate systems shall be permitted to be used by special permission of the Authority Having Jurisdiction after being satisfied with their adequacy. This authorization is based on extensive field and test data from conditions similar to those at the proposed site or requires such additional data as necessary to assure that the alternate system will produce continuous and long-range results at the proposed site, not less than equivalent to systems which are specifically authorized.

Where demonstration systems are to be considered for installation, conditions for installation, maintenance, and monitoring at each such site shall first be established by the Authority Having Jurisdiction.

Approved aerobic systems shall be permitted to be substituted for conventional septic tanks provided the Authority Having Jurisdiction is satisfied that such systems will produce results not less than equivalent to septic tanks, whether their aeration systems are operating or not.

TABLE H 101.8
LOCATION OF SEWAGE DISPOSAL SYSTEM

MINIMUM HORIZONTAL DISTANCE	BUILDING SEWER	SEPTIC TANK	DISPOSAL FIELD	SEEPAGE PIT OR CESSPOOL
Building or structures[1]	2 feet	5 feet	8 feet	8 feet
Property line adjoining private property	Clear[2]	5 feet	5 feet	8 feet
Water supply wells	50 feet[3]	50 feet	100 feet	150 feet
Streams and other bodies of water	50 feet	50 feet	100 feet[7]	150 feet[7]
Trees	–	10 feet	–	10 feet
Seepage pits or cesspools[8]	–	5 feet	5 feet	12 feet
Disposal field[8]	–	5 feet	4 feet[4]	5 feet
On-site domestic water service line	1 foot[5]	5 feet	5 feet	5 feet
Distribution box	–	–	5 feet	5 feet
Pressure public water main	10 feet[6]	10 feet	10 feet	10 feet

For SI units: 1 foot = 304.8 mm

Notes:

1 Including porches and steps, whether covered or uncovered, breezeways, roofed porte cocheres, roofed patios, carports, covered walks, covered driveways, and similar structures or appurtenances.

2 See Section 312.3.

3 Drainage piping shall clear domestic water supply wells by not less than 50 feet (15 240 mm). This distance shall be permitted to be reduced to not less than 25 feet (7620 mm) where the drainage piping is constructed of materials approved for use within a building.

4 Plus 2 feet (610 mm) for each additional 1 foot (305 mm) of depth more than 1 foot (305 mm) below the bottom of the drain line. (See Section H 601.0)

5 See Section 720.0.

6 For parallel construction – For crossings, approval by the Health Department shall be required.

7 These minimum clear horizontal distances shall also apply to disposal fields, seepage pits, and the mean high-tide line.

8 Where disposal fields, seepage pits, or both are installed in sloping ground, the minimum horizontal distance between any part of the leaching system and ground surface shall be 15 feet (4572 mm).

H 201.0 Capacity of Septic Tanks.

H 201.1 General. The liquid capacity of septic tanks shall comply with Table H 201.1(1) and Table H 201.1(4) as determined by the number of bedrooms or apartment units in dwelling occupancies and the estimated waste/sewage design flow rate or the number of plumbing fixture units as determined from Table 702.1 of this code, whichever is greater in other building occupancies. The capacity of any one septic tank and its drainage system shall be limited to the soil structure classification in Table H 201.1(2), and as specified in Table H 201.1(3).

TABLE H 201.1(1)
CAPACITY OF SEPTIC TANKS[1, 2, 3, 4]

SINGLE-FAMILY DWELLINGS - NUMBER OF BEDROOMS	MULTIPLE DWELLING UNITS OR APARTMENTS - ONE BEDROOM EACH	OTHER USES: MAXIMUM FIXTURE UNITS SERVED PER TABLE 702.1	MINIMUM SEPTIC TANK CAPACITY (gallons)
1 or 2	–	15	750
3	–	20	1000
4	2 units	25	1200
5 or 6	3	33	1500
–	4	45	2000
–	5	55	2250
–	6	60	2500
–	7	70	2750
–	8	80	3000
–	9	90	3250
–	10	100	3500

For SI units: 1 gallon = 3.785 L

Notes:

1 Extra bedroom, 150 gallons (568 L) each.

2 Extra dwelling units over 10: 250 gallons (946 L) each.

3 Extra fixture units over 100: 25 gallons (94.6 L) per fixture unit.

4 Septic tank sizes in this table include sludge storage capacity and the connection of domestic food waste disposers without further volume increase.

TABLE H 201.1(2)
DESIGN CRITERIA OF FIVE TYPICAL SOILS

TYPE OF SOIL	REQUIRED SQUARE FEET OF LEACHING AREA PER 100 GALLONS	MAXIMUM ABSORPTION CAPACITY IN GALLONS PER SQUARE FEET OF LEACHING AREA FOR A 24 HOUR PERIOD
Coarse sand or gravel	20	5.0
Fine sand	25	4.0
Sandy loam or sandy clay	40	2.5
Clay with considerable sand or gravel	90	1.1
Clay with small amount of sand or gravel	120	0.8

For SI units: 1 square foot = 0.0929 m^2, 1 gallon = 3.785 L, 1 gallon per square foot = 40.7 L/m^2

**TABLE H 201.1(3)
LEACHING AREA SIZE BASED ON SEPTIC TANK CAPACITY**

REQUIRED SQUARE FEET OF LEACHING AREA PER 100 GALLONS SEPTIC TANK CAPACITY (square feet per 100 gallons)	MAXIMUM SEPTIC TANK SIZE ALLOWABLE (gallons)
20–25	7500
40	5000
90	3500
120	3000

For SI units: 1 square foot per 100 gallons = 0.000245 m^2/L, 1 gallon = 3.785 L

**TABLE H 201.1(4)
ESTIMATED WASTE/SEWAGE FLOW RATES[1, 2, 3]**

TYPE OF OCCUPANCY	GALLONS PER DAY
Airports (per employee)	15
Airports (per passenger)	5
Auto washers – check with equipment manufacturer	–
Bowling alleys – with snack bar only (per lane)	75
Campground – with central comfort station (per person)	35
Campground – with flush toilets - no showers (per person)	25
Camps (day) – no meals served (per person)	15
Camps (summer and seasonal camps) – (per person)	50
Churches – sanctuary (per seat)	5
Churches – with kitchen waste (per seat)	7
Dance halls – (per person)	5
Factories – no showers (per employee)	25
Factories – with showers (per employee)	35
Factories – with cafeteria (per employee)	5
Hospitals – (per bed)	250
Hospitals – kitchen waste only (per bed)	25
Hospitals – laundry waste only (per bed)	40
Hotels – no kitchen waste (per bed)	60
Institutions – resident (per person)	75
Nursing home – (per person)	125
Rest home – (per person)	125
Laundries – self-service with minimum 10 hours per day (per wash cycle)	50
Laundries – commercial check with manufacturer's specification	–
Motel (per bed space)	50
Motel – with kitchen (per bed space)	60
Offices – (per employee)	20
Parks – mobile homes (per space)	250
Parks (picnic) – with toilets only (per parking space)	20
Parks (recreational vehicles) – without water hook-up (per space)	75
Parks (recreational vehicles) – with water and sewer hook-up (per space)	100
Restaurants – cafeteria (per employee)	20
Restaurants – with toilet waste (per customer)	7
Restaurants – with kitchen waste (per meal)	6
Restaurants – with kitchen waste disposable service (per meal)	2
Restaurants – with garbage disposal (per meal)	1
Restaurants – with cocktail lounge (per customer)	2
Schools staff and office (per person)	20
Schools – elementary (per student)	15
Schools – intermediate and high (per student)	20
Schools – with gym and showers (per student)	5

TABLE H 201.1(4) (continued)
ESTIMATED WASTE/SEWAGE FLOW RATES[1, 2, 3]

TYPE OF OCCUPANCY	GALLONS PER DAY
Schools – with cafeteria (per student)	3
Schools (boarding) – total waste (per person)	100
Service station – with toilets for 1st bay	1000
Service station – with toilets for each additional bay	500
Stores – (per employee)	20
Stores – with public restrooms (per 10 square feet of floor space)	1
Swimming pools – public (per person)	10
Theaters – auditoriums (per seat)	5
Theaters – with drive-in (per space)	10

For SI units: 1 square foot = 0.0929 m^2, 1 gallon per day = 3.785 L/day

Notes:

1 Sewage disposal systems sized using the estimated waste/sewage flow rates shall be calculated as follows:
(a) Waste/sewage flow, up to 1500 gallons per day (5678 L/day)
Flow x 1.5 = septic tank size
(b) Waste/sewage flow, over 1500 gallons per day (5678 L/day)
Flow x 0.75 + 1125 = septic tank size
(c) Secondary system shall be sized for total flow per 24 hours.

2 See Section H 201.1.

3 Because of the many variables encountered, it is not possible to set absolute values for waste/sewage flow rates for all situations. The designer should evaluate each situation and, where figures in this table need modification; they should be made with the concurrence of the Authority Having Jurisdiction.

H 301.0 Area of Disposal Fields and Seepage Pits.

H 301.1 General. The minimum effective absorption area in disposal fields in square feet (m^2), and in seepage pits in square feet (m^2) of sidewall, shall be predicated on the required septic tank capacity of gallons (liters), estimated waste/sewage flow rate, or whichever is greater, and shall be in accordance with Table H 201.1(2) as determined by the type of soil found in the excavation, and shall be as follows:

(1) Where disposal fields are installed, not less than 150 square feet (13.9 m^2) of trench bottom shall be provided for each system exclusive of any hard pan, rock, clay, or other impervious formations. Sidewall area more than the required 12 inches (305 mm) and not exceeding 36 inches (914 mm) below the leach line shall be permitted to be added to the trench bottom area where computing absorption areas.

(2) Where leaching beds are permitted instead of trenches, the area of each such bed shall be not less than 50 percent greater than the tabular requirements for trenches. Perimeter sidewall area more than the required 12 inches (305 mm) and not exceeding 36 inches (914 mm) below the leach line shall be permitted to be added to the trench bottom area where computing absorption areas.

(3) No excavation for a leach line or leach bed shall be located within 5 feet (1524 mm) of the water table nor to a depth where sewage is capable of contaminating the underground water stratum that is usable for domestic purposes.

Exception: In areas where the records or data indicate that the groundwaters are grossly degraded, the 5 foot (1524 mm) separation requirement shall be permitted to be reduced by the Authority Having Jurisdiction. The applicant shall supply evidence of groundwater depth to the satisfaction of the Authority Having Jurisdiction.

(4) The minimum effective absorption area in any seepage pit shall be calculated as the excavated sidewall area below the inlet exclusive of any hardpan, rock, clay, or other impervious formations. The minimum required area of porous formation shall be provided in one or more seepage pits. No excavation shall extend within 10 feet (3048 mm) of the water table nor to a depth where sewage is capable of contaminating underground water stratum that is usable for domestic purposes.

Exception: In areas where the records or data indicate that the groundwaters are grossly degraded, the 10 foot (3048 mm) separation requirement shall be permitted to be reduced by the Authority Having Jurisdiction.

The applicant shall supply evidence of groundwater depth to the satisfaction of the Authority Having Jurisdiction.

(5) Leaching chambers that comply with IAPMO PS 63 and bundled expanded polystyrene synthetic aggregate units that comply with IAPMO IGC 276 shall be sized using the required area calculated using Table H 201.1(2) with a 0.70 multiplier.

H 401.0 Percolation Test.

H 401.1 Pit Sizes. Where practicable, disposal field and seepage pit sizes shall be computed from Table H 201.1(2). Seepage pit sizes shall be computed by percolation tests unless use of Table H 201.1(2) is approved by the Authority Having Jurisdiction.

H 401.2 Absorption Qualities. The absorption qualities of seepage pits and questionable soils other than those listed in Table H 201.1(2), the proposed site, shall be subjected to percolation tests acceptable to the Authority Having Jurisdiction.

H 401.3 Absorption Rates. Where a percolation test is required, no private disposal system shall be permitted to serve a building where that test shows the absorption capacity of the soil is less than 0.83 gallons per square foot (gal/ft^2) (33.8 L/m^2) or more than 5.12 gal/ft^2 (208.6 L/m^2) of leaching area per 24 hours. Where the percolation test shows an absorption rate greater than 5.12 gal/ft^2 (208.6 L/m^2) per 24 hours, a private disposal system shall be permitted where the site does not overlie groundwaters protected for drinking water supplies, a minimum thickness of 2 feet (610 mm) of the native soil below the entire proposed system is replaced by loamy sand, and the system design is based on percolation tests made in the loamy sand.

H 501.0 Septic Tank Construction.

H 501.1 Plans. Plans for septic tanks shall be submitted to the Authority Having Jurisdiction for approval. Such plans shall show dimensions, reinforcing, structural calculations, and such other pertinent data as required.

H 501.2 Design. Septic tank design shall be such as to produce a clarified effluent consistent with accepted standards and shall provide adequate space for sludge and scum accumulations.

H 501.3 Construction. Septic tanks shall be constructed of solid, durable materials not subject to excessive corrosion or decay and shall be watertight.

H 501.4 Compartments. Septic tanks shall have not less than two compartments unless otherwise approved by the Authority Having Jurisdiction. The inlet compartment of any septic tank shall be not less than two-thirds of the total capacity of the tank, nor less than 500 gallons (1892 L) liquid capacity, and shall be not less than 3 feet (914 mm) in width and 5 feet (1524 mm) in length. Liquid depth shall be not less than 2½ feet (762 mm) nor more than 6 feet (1829 mm). The secondary compartment of a septic tank shall have a capacity of not less than 250 gallons (946 L) and a capacity not exceeding one-third of the total capacity of such tank. In septic tanks having a 1500 gallon (5678 L) capacity, the secondary compartment shall be not less than 5 feet (1524 mm) in length.

H 501.5 Access. Access to each septic tank shall be provided by not less than two manholes 20 inches (508 mm) in minimum dimension or by an equivalent removable cover slab. One access manhole shall be located over the inlet, and one access manhole shall be located over the outlet. Where a first compartment exceeds 12 feet (3658 mm) in length, an additional manhole shall be provided over the baffle wall.

H 501.6 Pipe Opening Sizes. The inlet and outlet pipe openings shall not be larger in size than the connecting sewer pipe. The vertical leg of round inlet and outlet fittings shall not be less in size than the connecting sewer pipe nor less than 4 inches (100 mm) in diameter. A baffle-type fitting shall have the equivalent cross-sectional area of the connecting sewer pipe and not less than a 4 inch (102 mm) horizontal dimension where measured at the inlet and outlet pipe inverts.

H 501.7 Pipe Extension. The inlet and outlet pipe or baffle shall extend 4 inches (102 mm) above and not less than 12 inches (305 mm) below the water surface. The invert of the inlet pipe shall be at a level not less than 2 inches (51 mm) above the invert of the outlet pipe.

H 501.8 Free Vent Area. Inlet and outlet pipe fittings or baffles and compartment partitions shall have a free vent area equal to the required cross-sectional area of the house sewer or private sewer discharging therein to provide free ventilation above the water surface from the disposal field or seepage pit through the septic tank, house sewer, and stack to the outer air.

H 501.9 Sidewalls. The sidewalls shall extend not less than 9 inches (229 mm) above the liquid depth. The cover of the septic tank shall be not less than 2 inches (51 mm) above the back vent openings.

H 501.10 Partitions and Baffles. Partitions or baffles between compartments shall be of solid, durable material and shall extend not less than 4 inches (102 mm) above the liquid level. The transfer port between compartments shall be a minimum size equivalent to the tank inlet, but in no case less than 4 inches (102 mm) in size, shall be installed in the inlet compartment side of the baffle so that the entry into the port is placed 65 percent to 75 percent in the depth of the liquid. Wooden baffles are prohibited.

H 501.11 Structural Design. The structural design of septic tanks shall comply with the following requirements:

(1) Each such tank shall be structurally designed to withstand all anticipated earth or other loads. Septic tank covers shall be capable of supporting an earth load of not less than 500 pounds per square foot (lb/ft^2) (2441 kg/m^2) where the maximum coverage does not exceed 3 feet (914 mm).

(2) In flood hazard areas, tanks shall be anchored to counter buoyant forces during conditions of the design flood. The vent termination and service manhole of the tank shall be not less than 2 feet (610 mm) above the design flood elevation or fitted with covers designed to prevent the inflow of floodwater or the outflow of the contents of the tanks during conditions of the design flood.

H 501.12 Manholes. Septic tanks installed under concrete or blacktop paving shall have the required manholes accessible by extending the manhole openings to grade in a manner acceptable to the Authority Having Jurisdiction.

H 501.13 Materials. The materials used for constructing a septic tank shall be in accordance with the following:

(1) Materials used in constructing a concrete septic tank shall be in accordance with applicable standards in Chapter 17.

(2) The minimum wall thickness of a steel septic tank shall be number 12 U.S. gauge (0.109 of an inch) (2.77 mm), and each such tank shall be protected from corrosion both externally and internally by an approved bituminous coating or by other acceptable means.

(3) Septic tanks constructed of alternate materials shall be permitted to be approved by the Authority Having Jurisdiction where in accordance with approved applicable standards. Wooden septic tanks shall be prohibited.

H 501.14 Prefabricated Septic Tanks. Prefabricated septic tanks shall comply with the following requirements:

(1) Manufactured or prefabricated septic tanks shall comply with approved applicable standards and be approved by the Authority Having Jurisdiction. Prefabricated bituminous coated septic tanks shall comply with UL 70.

(2) Independent laboratory tests and engineering calculations certifying the tank capacity and structural stability shall be provided as required by the Authority Having Jurisdiction.

H 601.0 Disposal Fields.

H 601.1 Distribution Lines. Distribution lines shall be constructed of clay tile laid with open joints, perforated clay pipe, perforated bituminous fiber pipe, perforated high-density polyethylene pipe, perforated ABS pipe, perforated PVC pipe, or other approved materials, provided that approved openings are available for distribution of the effluent into the trench area.

H 601.2 Filter Material. Before placing filter material or drain lines in a prepared excavation, smeared or compacted surfaces shall be removed from trenches by raking to a depth of 1 inch (25.4 mm) and the loose material removed. Clean stone, gravel, slag, or similar filter material acceptable to the Authority Having Jurisdiction, varying in size from ¾ of an inch to 2½ inches (19.1 mm to 64 mm), shall be placed in the trench to the depth and grade required by this section. Drain pipe shall be placed on filter material in an approved manner. The drain lines shall then be covered with filter material to the minimum depth required by this section, and this material covered with untreated building paper, straw, or similar porous material to prevent the closure of voids with earth backfill. No earth backfill shall be placed over the filter material cover until after inspection and acceptance.

Exception: Listed or approved plastic leaching chambers and bundled expanded polystyrene synthetic aggregate units shall be permitted to be used in lieu of pipe and filter material. Chamber and bundled expanded polystyrene synthetic aggregate unit installations shall follow the rules for disposal fields, where applicable, and shall be in accordance with the manufacturer's instructions.

H 601.3 Grade Board. A grade board staked in the trench to the depth of filter material shall be utilized where the distribution line is constructed with drain tile or a flexible pipe material that will not maintain alignment without continuous support.

H 601.4 Seepage Pits. Where seepage pits are used in combination with disposal fields, the filter material in the trenches shall terminate not less than 5 feet (1524 mm) from the pit excavation, and the line extending from such points to the seepage pit shall be approved pipe with watertight joints.

H 601.5 Distribution Boxes. Where two or more drain lines are installed, an approved distribution box of sufficient size to receive lateral lines shall be installed at the head of each disposal field. The invert of outlets shall be level, and the invert of the inlet shall be not less than 1 inch (25.4 mm) above the outlets. Distribution boxes shall be designed to ensure equal flow and shall be installed on a level concrete slab in natural or compacted soil.

H 601.6 Laterals. Laterals from a distribution box to the disposal field shall be approved pipe with watertight joints. Multiple disposal field laterals, where practicable, shall be of uniform length.

H 601.7 Connections. Connections between a septic tank and a distribution box shall be laid with approved pipe with watertight joints on natural ground or compacted fill.

H 601.8 Dosing Tanks. Where the quantity of sewage exceeds the amount that is permitted to be disposed of in 500 lineal feet (152.4 m) of leach line, a dosing tank shall be used. Dosing tanks shall be equipped with an automatic siphon or pump that discharges the tank once every 3 or 4 hours. The tank shall have a capacity equal to 60 to 75 percent of the interior capacity of the pipe to be dosed at one time. Where the total length of pipe exceeds 1000 lineal feet (305 m), the dosing tank shall be provided with two siphons or pumps dosing alternately and each serving one-half of the leach field.

H 601.9 Construction. Disposal fields shall be constructed in accordance with Table H 601.9.

Minimum spacing between trenches or leaching beds shall be not less than 4 feet (1219 mm) plus 2 feet (610 mm) for each additional foot (305 mm) of depth more than 1 foot (305 mm) below the bottom of the drain line. Distribution drain lines in leaching beds shall be not more than 6 feet (1829 mm) apart on centers, and no part of the perimeter of the leaching bed shall exceed 3 feet (914 mm) from a distribution drain line. Disposal fields, trenches, and leaching beds shall not be paved over or covered by concrete or a material that is capable of reducing or inhibiting a possible evaporation of sewer effluent.

H 601.10 Joints. Where necessary on sloping ground to prevent excessive line slope, leach lines or leach beds shall be stepped. The lines between each horizontal section shall be made with watertight joints and shall be designed, so each horizontal leaching trench or bed shall be utilized to the maximum capacity before the effluent shall pass to the next lower leach line or bed. The lines between each horizontal leaching section shall be made with approved watertight joints and installed on the natural or unfilled ground.

TABLE H 601.9
GENERAL DISPOSAL FIELD REQUIREMENTS

ELEMENT	MINIMUM	MAXIMUM
Number of drain lines per field	1	-
Length of each line	-	100 feet
Bottom width of trench	18 inches	36 inches
Spacing of lines, center-to-center	6 feet	-
Depth of earth cover of lines (preferred 18 inches)	12 inches	-
Grade of lines	level	3 inches per 100 feet
Filter material under drain lines	12 inches	-
Filter material over drain lines	2 inches	-

For SI units: 1 inch = 25.4 mm, 1 foot = 304.8 mm, 1 inch per foot = 83.3 mm/m

H 701.0 Seepage Pits.

H 701.1 Capacity. The capacity of seepage pits shall be based on the quantity of liquid waste discharging thereinto and on the character and porosity of the surrounding soil, and shall be in accordance with Section H 301.0 of this appendix.

H 701.2 Multiple Installations. Multiple seepage pit installations shall be served through an approved distribution box or be connected in series using a watertight connection laid on undistributed or compacted soil. The outlet from the pit shall have an approved vented leg fitting extending not less than 12 inches (305 mm) below the inlet fitting.

H 701.3 Construction. A seepage pit shall be circular in shape and shall have an excavated diameter of not less than 4 feet (1219 mm). Each such pit shall be lined with approved-type whole new hard-burned clay brick, concrete brick, concrete circular-type cesspool blocks, or other approved materials. Approval shall be obtained before construction for any pit having an excavated diameter greater than 6 feet (1829 mm).

H 701.4 Lining. The lining in a seepage pit shall be laid on a firm foundation. Lining materials shall be placed tight together and laid with joints staggered. Except in the case of approved-type precast concrete circular sections, no brick or block shall be greater in height than its width and shall be laid flat to form not less than a 4 inch (102 mm) wall. Brick or block greater than 12 inches (305 mm) in length shall have chamfered matching ends and be scored to provide for seepage. Excavation voids behind the brick, block, or concrete liner shall have not less than 6 inches (152 mm) of clean $^3/_4$ of an inch (19.1 mm) gravel or rock.

H 701.5 Brick and Block. Brick or block used in seepage pit construction shall have a compressive strength of not less than 2500 pounds per square inch (lb/in^2) (1 757 674 kg/m^2).

H 701.6 Sidewall. A seepage pit shall have a minimum sidewall (not including the arch) of 10 feet (3048 mm) below the inlet.

H 701.7 Arch and Dome. The arch or dome of a seepage pit shall be permitted to be constructed in one of three ways:

(1) Approved-type hard-burned clay brick or solid concrete brick or block laid in cement mortar.

(2) Approved brick or block laid dry. In both of the above methods, an approved cement mortar covering of not less than 2 inches (51 mm) in thickness shall be applied, said covering to extend not less than 6 inches (152 mm) beyond the sidewalls of the pit.

(3) Approved-type one or two-piece reinforced concrete slabs of not less than 2500 lb/in^2 (1 757 674 kg/m^2) minimum compressive strength, not less than 5 inches (127 mm) thick, and designed to support an earth load of not less than 400 pounds per square foot (lb/ft^2) (1953 kg/m^2). Each such cover shall be provided with a 9 inch (229 mm) minimum inspection hole with plug or cover and shall be coated on the underside with an approved bituminous or other nonpermeable protective compound.

H 701.8 Location. The top of the arch or cover shall be not less than 18 inches (457 mm) but not exceed 4 feet (1219 mm) below the surface of the ground.

H 701.9 Inlet Fitting. An approved vented inlet fitting shall be provided in the seepage pit so arranged as to prevent the inflow from damaging the sidewall.

Exception: Where using a one- or two-piece concrete slab cover inlet, fitting shall be permitted to be a one-fourth bend fitting discharging through an opening in the top of the slab cover. On multiple seepage pit installations, the outlet fittings shall comply with Section H 701.2 of this appendix.

H 801.0 Cesspools.

H 801.1 Limitations. A cesspool shall be considered as a temporary expedient pending the construction of a public sewer; as an overflow facility where installed in conjunction with an existing cesspool; or as a means of sewage disposal for limited, minor, or temporary uses, where first approved by the Authority Having Jurisdiction.

H 801.2 Septic Tanks. Where it is established that a public sewer system will be available in less than 2 years, and soil and groundwater conditions are favorable to cesspool disposal, cesspools without septic tanks shall be permitted to be installed for single-family dwellings or for other limited uses where first approved by the Authority Having Jurisdiction.

H 801.3 Construction. Each cesspool, where permitted, shall be in accordance with the construction requirements set forth in Section H 701.0 of this appendix for seepage pits and shall have a sidewall (not including arch) of not less than 20 feet (6096 mm) below the inlet, provided, however, that where a strata of gravel or equally pervious material of 4 feet (1219 mm) in thickness is found, the depth of such sidewall shall not exceed 10 feet (3048 mm) below the inlet.

H 801.4 Existing Installations. Where overflow cesspools or seepage pits are added to existing installations, the effluent shall leave the existing pit through an approved vented leg extending not less than 12 inches (305 mm) downward into such existing pit and having its outlet flow line not less than 6 inches (152 mm) below the inlet. The pipe between pits shall be laid with approved watertight joints.

H 901.0 Commercial or Industrial Special Liquid-Waste Disposal.

H 901.1 Interceptor. Where liquid wastes contain excessive amounts of grease, garbage, flammable wastes, sand, or other ingredients that affect the operation of a private sewage disposal system, an interceptor for such wastes shall be installed.

H 901.2 Installation. Installation of such interceptors shall comply with Section 1009.0 of this code, and their location shall comply with Table H 101.8 of this appendix.

H 901.3 Sampling Box. A sampling box shall be installed where required by the Authority Having Jurisdiction.

H 901.4 Design and Structural Requirement. Interceptors shall be of approved design and be not less than two compartments. Structural requirements shall comply with Section H 501.0 of this appendix.

H 901.5 Location. Interceptors shall be located as close to the source as possible and be accessible for servicing. Necessary manholes for servicing shall be at grade level and be gastight.

H 901.6 Waste Discharge. Waste discharge from interceptors shall be permitted to be connected to a septic tank or other primary system or be disposed into a separate disposal system.

H 901.7 Design Criteria. A formula shall be permitted to be adapted to other types of occupancies with similar wastes. (See Chart H 901.7)

H 1001.0 Inspection and Testing.

H 1001.1 Inspection. Inspection requirements shall comply with the following:

(1) Applicable provisions of Section 105.0 of this code and this appendix shall be required. Plans shall be required in accordance with Section 103.3 of this code.

(2) System components shall be properly identified as to manufacturer. Septic tanks or other primary systems shall have the rated capacity permanently marked on the unit.

(3) Septic tanks or other primary systems shall be installed on dry, level, well-compacted soil.

(4) Where design is predicated on soil tests, the system shall be installed at the same location and depth as the tested area.

H 1001.2 Testing. Testing requirements shall comply with the following:

(1) Septic tanks or other primary components shall be filled with water to flow line before requesting an inspection. Seams or joints shall be left exposed (except the bottom), and the tank shall remain watertight.

(2) A flow test shall be performed through the system to the point of effluent disposal. All lines and components shall be watertight. Capacities required air space, and fittings shall comply with the provisions outlined in this appendix.

H 1101.0 Abandoned Sewers and Sewage Disposal Facilities.

H 1101.1 Plugged and Capped. An abandoned building (house) sewer, or part thereof, shall be plugged or capped in an approved manner within 5 feet (1524 mm) of the property line.

H 1101.2 Fill Material. A cesspool, a septic tank, or a seepage pit that has been abandoned or has been discontinued otherwise from further use, or to which no waste or soil pipe from a plumbing fixture is connected, shall have the sewage removed therefrom and be completely filled with the earth, sand, gravel, concrete, or other approved material.

H 1101.3 Filling Requirements. The top cover or arch over the cesspool, septic tank, or seepage pit shall be removed before filling, and the filling shall not extend above the top of the vertical portions of the sidewalls or above the level of any outlet pipe until inspection has been called and the cesspool, septic tank, or seepage pit has been inspected. After such inspection, the cesspool, septic tank, or seepage pit shall be filled to the level of the top of the ground.

H 1101.4 Owner. No person owning or controlling a cesspool, septic tank, or seepage pit on the premises of such person or in that portion of any public street, alley, or other

CHART H 901.7
RECOMMENDED DESIGN CRITERIA

GREASE AND GARBAGE, COMMERCIAL KITCHENS										
Number of meals per peak hour	x	Waste flow rate[1]	x	Retention time[2]	x	Storage factor[3]	=	Interceptor size (liquid capacity)		
SAND-SILT OIL, AUTO WASHERS										
Number of vehicles per hour	x	Waste flow rate[1]	x	Retention time[2]	x	Storage factor[3]	=	Interceptor size (liquid capacity)		
SILT-LINT GREASE, LAUNDRIES, LAUNDROMATS										
Number of machines	x	2 cycles per hour	x	Waste flow rate[1]	x	Retention time[2]	=	Storage factor[3]	=	Interceptor size (liquid capacity)

Notes:

[1] For waste flow rate see Table H 201.1(4).

[2] Retention Times:
(a) Kitchen (commercial) – with dishwasher, garbage disposal, or both = 2.5 hours
(b) Kitchen (single service) – with garbage disposal = 1.5 hours
(c) Auto Washers (sand-silt oil) = 2.0 hours
(d) Laundries/Laundromats = 2.0 hours

[3] Storage Factors:
(a) Kitchen (commercial) – with 8 hours operation = 1
(b) Kitchen (commercial) – with 16 hours operation = 2
(c) Kitchen (commercial) – with 24 hours operation = 3
(d) Kitchen (single service) = 1.5
(e) Auto Washers (sand-silt oil) – with self service = 1.5
(e) Auto Washers (sand-silt oil) – with employee operated = 2
(d) Laundries/Laundromats – with rock filter = 1.5 hours

public property abutting such premises shall fail, refuse, or neglect to be in accordance with the provisions of this section or upon receipt of notice so to be in accordance with the Authority Having Jurisdiction.

H 1101.5 Permittee. Where disposal facilities are abandoned consequent to connecting any premises with the public sewer, the permittee making the connection shall fill all abandoned facilities in accordance with the Authority Having Jurisdiction within 30 days from the time of connecting to the public sewer.

H 1201.0 Drawings and Specifications.

H 1201.1 General. The Authority Having Jurisdiction, Health Officer, or other department having jurisdiction shall be permitted to require the following information before a permit is issued for a private sewage disposal system or at a time during the construction thereof:

(1) Plot plan drawn to scale, completely dimensioned, showing direction and approximate slope of surface, location of present or proposed retaining walls, drainage channels, water supply lines or wells, paved areas and structures on the plot, number of bedrooms or plumbing fixtures in each structure, and location of the private sewage disposal system with relation to lot lines and structures.

(2) Details of construction necessary to ensure compliance with the requirements of this appendix together with a full description of the complete installation including quality, kind, and grade of materials, equipment, construction, workmanship, and methods of assembly and installation.

(3) A log of soil formations and groundwater levels as determined by test holes dug in close proximity to a proposed seepage pit or disposal field, together with a statement of water absorption characteristics of the soil at the proposed site, as determined by approved percolation tests.

APPENDIX I

INSTALLATION STANDARDS

CONTENTS

Page

IAPMO IS 26-2019^{e2}
Trenchless Insertion of Polyethylene (PE) Pipe for Sewer Laterals....357

IAPMO IS 31-2014^{e2}
Installation Standard for PEX Tubing Systems for Hot- and Cold Water Distribution....361

IAPMO IS 33-2019^{e1}
Thrust Blocking for Rubber Gasketed and Solvent Cement Joints....369

TCNA Handbook-2018
For Ceramic, Glass, and Stone Tile Installation....377

"The information contained in this appendix is not part of this American National Standard (ANS) and has not been processed in accordance with ANSI's requirements for an ANS. As such, this appendix may contain material that has not been subjected to public review or a consensus process. In addition, it does not contain requirements necessary for conformance to the standard."

The following IAPMO Installation Standard is included here for the convenience of the users of the Uniform Plumbing Code. It is not considered as a part of the Uniform Plumbing Code unless formally adopted as such. This Installation Standard is an independent, stand-alone document published by the International Association of Plumbing and Mechanical Officials and is printed herein by the expressed written permission of IAPMO.

TRENCHLESS INSERTION OF POLYETHYLENE (PE) PIPE FOR SEWER LATERALS

IAPMO IS 26-2019^{e2}

1 Scope

1.1 General

This standard shall govern the trenchless installation of polyethylene (PE) pipe for use in sanitary and storm sewers. The installed pipe shall comply with the requirements of the Uniform Plumbing Code (UPC™) published by the International Association of Plumbing and Mechanical Officials (IAPMO) as to grade and connections to existing pipe and shall also comply with this standard. This standard specifies requirements for the installation of the trenchless insertion of polyethylene (PE) pipe for use in sanitary and storm sewers.

1.2 Terminology

In this Standard,

(a) "shall" is used to express a requirement, i.e., a provision that the user is obliged to satisfy to comply with the Standard;

(b) "should" is used to express a recommendation, but not a requirement;

(c) "may" is used to express an option or something permissible within the scope of the Standard; and

(d) "can" is used to express a possibility or a capability.

Notes accompanying sections of the Standard do not specify requirements or alternative requirements; their purpose is to separate explanatory or informative material from the text. Notes to tables and figures are considered part of the table or figure and can be written as requirements.

2 Reference Publications

This Standard refers to the following publications, and where such reference is made, it shall be to the current edition of those publications, including all amendments published thereto.

ASTM International

ASTM D3261

Standard Specification for Butt Heat Fusion Polyethylene (PE) Plastic Fittings for Polyethylene (PE) Plastic Pipe and Tubing

ASTM F714

Standard Specification for Polyethylene (PE) Plastic Pipe (DR-PR) Based on Outside Diameter

ASTM F894

Standard Specification for Polyethylene (PE) Large Diameter Profile Wall Sewer and Drain Pipe

ASTM F1055

Standard Specification for Electrofusion Type Polyethylene Fittings for Outside Diameter Controlled Polyethylene and Crosslinked Polyethylene (PEX) Pipe and Tubing

ASTM F2620

Standard Practice for Heat Fusion Joining of Polyethylene Pipe and Fittings

3 Abbreviations

The following abbreviations apply in this Standard:

PE — Polyethylene

HDPE — High Density Polyethylene

4 General

4.1 Product Requirements

Polyethylene (PE) sewer pipe or tubing and fitting joining methods shall be installed in accordance with the manufacturer's installation instructions and comply with ASTM F714, ASTM F894 the applicable nationally recognized standard.

4.2 HDPE Materials

HDPE Extra High Molecular Weight 3408 SDR 17 Pipe Socket-Type PE Fittings for Outside Diameter Controlled.

Note: *The HDPE 3408 SDR 17 pipe used in this process was selected because of its ability to retain its circular shape even when bent on a 1.2 m (4 ft) radius during and after installation.*

4.3 Protection of Pipe

Pipe shall be stored in a way to protect it from mechanical damage (slitting, puncturing, etc.). It shall be stored under cover to keep it clean and avoid long term exposure to sunlight. Exposure to sunlight during normal construction periods is acceptable.

4.4 Joining Methods

PE joints shall be made in accordance with the manufacturer's installation instructions. PE pipe shall be joined to other pipe materials by an approved listed adapter or transition fittings listed for the specific transition intended.

4.4.1 Butt-Fusion Joints [2021 UPC 705.5.1.1]

Butt-fusion joints for PE pipe shall be installed in accordance with ASTM F2620 and shall be made by heating the prepared ends of two pipes, pipe and fitting, or two fittings by holding ends against a heated element. The heated element shall be removed when the required melt or times are obtained, and heated ends shall be placed together with applied force. Do not disturb the joint until cooled to ambient temperature.

4.4.2 Electro-Fusion Joints [2021 UPC 705.5.1.2]

Electro-fusion joints shall be heated internally by a conductor at the interface of the joint. Fittings shall comply with ASTM F1055 for the performance requirements of polyethylene electro-fusion fittings. The specified electro-fusion cycle used to form the joint requires consideration of the properties of the materials being joined, the design of the fitting being used, and the environmental conditions. Align and restrain fitting to pipe to prevent movement and apply electric current to the fitting. Turn off the current when the required time has elapsed to heat the joint. Do not disturb the joint until cooled to ambient temperature.

4.4.3 Socket-Fusion Joints [2021 UPC 705.5.1.3]

Socket fusion joints shall be installed in accordance with ASTM F2620 and shall be made by simultaneously heating the outside surface of a pipe end and the inside of a fitting socket. Where the required melt is obtained, the pipe and fitting shall be joined by inserting one into the other with applied force. Do not disturb the joint until cooled to ambient temperature.

4.5 Trenchless Installation of Sewers

The trenchless installation of sewers will be as follows:

(a) Preliminary Steps

(i) Inspect the inside of the sewer line using a camera and recording device to ascertain the line condition.

(ii) Mark the details revealed by the video inspection including:

1. The ground surface to show the location of the lateral tie of the city wye.
2. The line location with an arrow in the street pointing back at the lateral.
3. The property denoting the lateral location.
4. The locations of the proposed excavations.

(iii) Obtain utility line identification service contact information and all applicable permits.

(b) Excavation

In addition to the above markings, the local utility companies will mark utilities. Considerations are soil density; clearance from obstacles, utilities, and structures; location of bends, and water service locations. Excavations and shoring shall be in accordance with jurisdictional safety requirements.

(c) Set Up

Fuse the proper length of polyethylene pipe in accordance with ASTM F2620, or ASTM D3261 and fuse the end to a small length that is attached to the pulling head. A rod pusher cable is pushed through the damaged host pipe and attached to the pulling cable, which is then drawn through the pipe. The clevis end of the cable is attached to the pulling head. The pulling equipment is then set up according to the manufacturer's instructions.

(d) Pulling

Pull the pulling head through. Once the pull is done, complete the connection to the existing piping.

4.6 Cleanouts

4.6.1 Plug

Each cleanout fitting for cast-iron pipe shall consist of a cast-iron or brass body and an approved plug. Each cleanout for galvanized wrought-iron, galvanized steel, copper, or brass pipe shall consist of a brass plug as specified in Table 1, or a standard weight brass cap, or an approved ABS or PVC plastic plug, or an approved stainless-steel cleanout or plug. Plugs shall have raised square heads or approved countersunk rectangular slots.

4.6.2 Approved

Each cleanout fitting and each cleanout plug, or cap shall be of an approved type.

4.6.3 Watertight and Gastight

Cleanouts shall be designed to be watertight and gastight.

5 Testing and Inspection Requirements

5.1 Media

The piping of the building sewer shall be tested with water. The Authority Having Jurisdiction shall be permitted to require the removal of cleanouts, etc., to ascertain whether the pressure has reached all parts of the system.

5.2 Water Test

The system shall be tested by plugging the end of the building sewer at its points of connection to the public sewer or private sewage disposal system and completely filling the building sewer with water from the lowest to the highest point thereof.

5.3 Inspections

The completed piping shall be internally inspected by camera unless waived by the Administrative Authority.

TABLE 1
CLEANOUTS
(See Section 4.6.1)

NPS	SIZE OF CLEANOUT (inches)	THREADS PER INCH
1.5	1.5	11.5
2	1.5	11.5
2.5	2.5	8
3	2.5	8
4 & larger	3.5	8

Note: *For SI units: 1 inch = 25 mm*

Uniform Plumbing Code References

The following sections of the 2021 Uniform Plumbing Code apply.

105.2	Required Inspections
105.2.2	Other Inspections
105.3	Testing of Systems
301.2	Minimum Standards
309.0	Workmanship
312.0	Protection of Piping, Materials, and Structures
314.0	Trenching, Excavation, and Backfill
701.3	Drainage Fittings
705.5.1.1	Butt Fusion Joints
705.5.1.2	Electro-Fusion Joints
705.5.1.3	Socket-Fusion Joints
723.0	Building Sewer Test

The following standards from Tables 1701.1 and 1701.2 of the 2021 Uniform Plumbing Code apply.

Table 1701.1 Standards

ASTM D2239	Polyethylene (PE) Plastic Pipe (SIDR-PR) Based on Controlled Inside Diameter
ASTM D2683	Socket-Type Polyethylene Fittings for Outside Diameter-Controlled Polyethylene Pipe and Tubing
ASTM D3261	Butt Heat Fusion Polyethylene (PE) Plastic Fittings for Polyethylene (PE) Plastic Pipe and Tubing
ASTM F714	Polyethylene (PE) Plastic Pipe (DR-PR) Based on Outside Diameter
ASTM F894	Polyethylene (PE) Large Diameter Profile Wall Sewer and Drain Pipe
ASTM F1055	Electrofusion Type Polyethylene Fittings for Outside Diameter Controlled Polyethylene and Crosslinked Polyethylene (PEX) Pipe and Tubing
ASTM F2620	Heat Fusion Joining of Polyethylene Pipe and Fittings

Table 1701.2 Standards

ASTM D2657	Heat Fusion Joining of Polyolefin Pipe and Fittings
IAPMO PS 25	Metallic Fittings for Joining Polyethylene Pipe for Water Service and Yard Piping

"The information contained in this appendix is not part of this American National Standard (ANS) and has not been processed in accordance with ANSI's requirements for an ANS. As such, this appendix may contain material that has not been subjected to public review or a consensus process. In addition, it does not contain requirements necessary for conformance to the standard."

The following IAPMO Installation Standard is included here for the convenience of the users of the Uniform Plumbing Code. It is not considered as a part of the Uniform Plumbing Code unless formally adopted as such. This Installation Standard is an independent, stand-alone document published by the International Association of Plumbing and Mechanical Officials and is printed herein by the expressed written permission of IAPMO.

INSTALLATION STANDARD FOR PEX TUBING SYSTEMS FOR HOT- AND COLD-WATER DISTRIBUTION

IAPMO IS 31-2014e2

1.0 Scope.

1.1 General.

1.1.1 This Standard specifies requirements for the installation of SDR 9 CTS crosslinked polyethylene (PEX) tubing and fittings, including cold-expansion, crimp, press, and mechanical compression fittings, intended for hot- and cold-water distribution systems within buildings.

1.1.2 This Standards applies to

(a) SDR 9 CTS PEX tubing complying with ASTM F876 and pressure-rated in accordance with PPI TR-3; and

(b) PEX fitting systems complying with

(i) ASTM F877, for mechanical compression fittings and metal or plastic insert fittings with stainless steel press sleeves;

(ii) ASTM F1807 or ASTM F2159, for metal or plastic insert fittings with copper crimp rings;

(iii) ASTM F1960, for cold expansion fittings with PEX reinforced rings; or

(iv) ASTM F2080, for cold expansion fittings with metal compression sleeves.

1.2 Terminology.

In this Standard,

(a) "shall" is used to express a requirement, i.e., a provision that the user is obliged to satisfy to comply with the Standard;

(b) "should" is used to express a recommendation, but not a requirement;

(c) "may" is used to express an option or something permissible within the scope of the Standard; and

(d) "can" is used to express a possibility or a capability.

Notes accompanying sections of the Standard do not specify requirements or alternative requirements; their purpose is to separate explanatory or informative material from the text. Notes to tables and figures are considered part of the table or figure and can be written as requirements.

1.3 Amendments.

Proposals for amendments to this Standard will be processed in accordance with the standards-writing procedures of IAPMO.

2.0 Reference Publications.

This Standard refers to the following publications, and where such reference is made, it shall be to the current edition of those publications, including all amendments published thereto.

ASTM F876	Standard Specification for Crosslinked Polyethylene (PEX) Tubing
ASTM F877	Standard Specification for Crosslinked Polyethylene (PEX) Hot- and Cold-Water Distribution Systems
ASTM F1807	Standard Specification for Metal Insert Fittings Utilizing a Copper Crimp Ring for SDR9 Cross-linked Polyethylene (PEX) Tubing and SDR9 Polyethylene of Raised Temperature (PE-RT) Tubing
ASTM F1960	Standard Specification for Cold Expansion Fittings with PEX Reinforcing Rings for Use with Cross-linked Polyethylene (PEX) Tubing
ASTM F2080	Standard Specification for Cold-Expansion Fittings With Metal Compression-Sleeves for Cross-Linked Polyethylene (PEX) Pipe

ASTM F2159	Standard Specification for Plastic Insert Fittings Utilizing a Copper Crimp Ring for SDR9 Cross-linked Polyethylene (PEX) Tubing and SDR9 Polyethylene of Raised Temperature (PE-RT) Tubing
ASTM F2657	Standard Test Method for Outdoor Weathering Exposure of Crosslinked Polyethylene (PEX) Tubing
AWWA C904	Cross-Linked Polyethylene (PEX) Pressure Tubing, ½ In. (12 mm) Through 3 In. (76 mm) for Water Service
IAPMO/ ANSI UPC-1	Uniform Plumbing Code
PPI TR-3	Policies and Procedures for Developing Hydrostatic Design Basis (HDB), Pressure Design Basis (PDB), Strength Design Basis (SDB), and Minimum Required Strength (MRS) Ratings for Thermoplastic Tubing Materials or Tubing

3.0 Abbreviations.

The following abbreviations apply in this Standard:

CTS — copper tube size
HDPE — high density polyethylene
IC — insulation contact
NTS — nominal tubing size
PEX — crosslinked polyethylene
SDR — standard dimension ratio
UV — ultraviolet light

4.0 General.

4.1 Tubing.

4.1.1 PEX tubing can be

(a) pigmented throughout (i.e., with color);
(b) non-pigmented (e.g., translucent or natural); or
(c) coated with a pigmented layer.

4.1.2 PEX tubing is typically available in NTS-¼ to NTS-3.

4.1.3 Before installation, the installer shall review the tubing markings and verify that

(a) the standard designation(s) of the fittings to which the tube can be joined to is included in the markings;
(b) it bears a certification mark from an accredited certification organization; and
(c) pressure and temperature ratings meet or exceed that of the intended end-use.

4.2 Fittings.

4.2.1 Cold-Expansion Fittings.

Cold-expansion fittings typically

(a) are made of brass, stainless steel, or sulfone;
(b) consist of an insert and a PEX reinforcing ring; and
(c) are available in NTS-⅜ to NTS-3.

4.2.2 Crimp or Press Insert Fittings.

Crimp or press insert fittings typically

(a) are made of brass, stainless steel, or sulfone;
(b) consist of an insert and a copper crimp ring or a stainless steel press ring
(c) are available in NTS-⅜ to NTS-2.

4.2.3 Compression Fittings.

Compression (i.e., transition) fittings typically

(a) are made of brass; and
(b) consist of
  (i) a nut, a compression ring, and an insert; or
  (ii) an O-ring brass insert with a compression sleeve
(c) are available in NTS-¼ to NTS-3.

4.3 Installation.

Only fittings systems marked on the tubing shall be used for installation with that particular tubing.

4.4 Tools.

Tools and tool accessories (e.g., tool heads) used for the installation of PEX tubing systems shall be in accordance with the manufacturer's specifications and written instructions.

4.5 Tubing Protection.

4.5.1 Abrasion.

PEX tubing passing through drilled or notched metal studs or metal joists, or hollow-shell masonry walls shall be protected from abrasion by elastomeric or plastic sleeves or grommets.

4.5.2 Puncture.

Steel-plate protection shall be installed in accordance with the local plumbing code.

5.0 Handling.

5.1 Receiving.

When receiving PEX tubing shipments, the receiver shall inspect and inventory each shipment, ensuring that there has been no loss or damage. In addition:

(a) At the time of unloading, the markings of all tubing, fittings, and accessories shall be verified to ensure that all items have been manufactured in accordance with the applicable product Standard and appropriately certified.

(b) An overall examination of the shipment shall be made. If the shipment is intact, ordinary inspection while unloading shall be sufficient to ensure that the items have arrived in good condition.

(c) If the load has shifted, has broken packaging, or shows evidence of rough treatment, each item shall be carefully inspected for damage.

(d) The total quantities of each shipment (e.g., tubing, gaskets, fittings, and accessories) shall be checked against shipping records.

(e) Any damaged or missing items shall be noted on the delivery slip. The carrier shall be notified immediately and a claim made in accordance with its instructions.

(f) No damaged material shall be disposed of. The carrier shall recommend the procedure to follow.

(g) Shortages and damaged materials are normally not reshipped without request. If replacement material is needed, it shall be reordered from the manufacturer, the distributor, or a manufacturer's representative.

5.2 Storage and UV Exposure.

5.2.1 PEX tubing and fittings shall be stored indoors and in its original packaging until the time of installation. Appropriate precautions to protect the tubing from damage, impact, and punctures shall be taken.

5.2.2 Accumulative exposure time to UV radiation during storage and installation shall not exceed the UV exposure limits recommended by the manufacturer or specified in ASTM F876.

Note: *ASTM F876 has four categories for UV-resistance, ranging from untested to 6 months of continuous exposure, as listed in the material designation code.*

5.3 Exposure to Heat.

5.3.1 PEX tubing and fittings shall not be exposed to open flames.

5.3.2 PEX tubing shall not be exposed to temperatures exceeding 93°C (200°F).

5.4 Exposure to Chemicals.

5.4.1 Chemical compatibility (e.g., with common construction materials) shall be verified with the manufacturer prior to direct contact.

5.4.2 In general, petroleum- or solvent-based chemicals (e.g., paints, greases, pesticides, or sealants) shall not be allowed to come in direct contact with PEX tubing or fittings.

6.0 Thermal Expansion and Contraction.

6.1 Horizontal Tubing Runs.

Thermal expansion and contraction forces on suspended horizontal runs of PEX tubing that can experience a 22°C (40°F) or greater change in temperature (operating temperature compared to ambient temperature) shall be controlled by a means of mitigating temperature-induced stresses to other parts of the water distribution system. Means for controlling thermal expansion and contraction include

(a) loops;

(b) offsets;

(c) arms with rigid anchor points; and

(d) supporting the tubing with continuous runs of CTS support channels with

  (i) rigid anchor points installed every 20 m (65 ft); and

  (ii) proper strapping (e.g., 27 kg (60 lb) straps or equivalent) spaced 1 m (3 ft) and rated for the maximum temperature and UV exposure of the PEX tubing application.

6.2 Vertical Tubing Runs.

Thermal expansion and contraction forces on vertical runs of PEX tubing that pass through more than one floor and can experience a 22°C (40°F) or greater change in temperature (operating temperature compared to ambient temperature) shall be controlled by installing

(a) a riser clamp at the top of every other floor; and

(b) mid-story guides to maintain the alignment of the vertical tubing.

Note: *Installing riser clamps isolates expansion and contraction to two-floor intervals allowing the PEX tubing to naturally compensate for the expansion and contraction.*

6.3 Clearance.

Adequate clearance shall be provided between PEX tubing and the building structure (e.g., using bored holes and sleeves) to allow for free longitudinal movement of the tubing.

6.4 Expansion Arms and Expansion Loops.

6.4.1 Expansion Arms (See Figure 1).

6.4.1.1 Expansion arms shall be installed as illustrated in Figure 1.

6.4.1.2 The minimum length of expansion arms shall be calculated using the following equation:

$$LB = C \times \sqrt{(D \times \Delta L)}$$

where

LB = length of flexible arm

C = material constant (12 for PEX)

D = nominal outside diameter of tubing

ΔL = thermal expansion length

6.4.2 Expansion Loops (See Figure 2).

6.4.2.1 Expansion loops shall be installed at the mid-point between anchors, as illustrated in Figure 2.

6.4.2.2 The minimum length of expansion loops shall be calculated using the equation in Section 6.4.1.2; however, the distance *LB* shall be divided into three sections, as illustrated in Figure 2, where

$L1 = LB \div 5$; and

$L2 = L1 \times 2$

7.0 Hangers and Supports.

7.1 Vertical Tubing.

Vertical PEX tubing shall

(a) be supported at each floor or as specified by the water-distribution system designer to allow for expansion and contraction; and

(b) have mid-story guides.

7.2 Horizontal Tubing.

Unless otherwise authorized by the authority having jurisdiction, suspended horizontal runs of PEX tubing

(a) NTS-1 and smaller shall be supported every 0.8 m (32 in), unless continuously supported by metallic CTS or V channels that

  (i) are supported at intervals not exceeding 1.8 m (6 ft);

  (ii) have a maximum cantilever, measured from the support to the end of the CTS support channel, of 0.5 m (1.5 ft); and

(b) NTS-1¼ and larger shall be supported every 1.2 m (4 ft), unless continuously supported by metallic CTS or V channels that

  (i) are supported at intervals not exceeding 2.4 m (8 ft); and

  (ii) have a maximum cantilever, measured from the support to the end of the CTS support channel, of 0.5 m (1.5 ft).

7.3 Anchors.

Anchors shall be

(a) used to restrict PEX tubing movement;

(b) made of materials that provide rigidity to the support system and utilize pipe clamps designed for plastic tubing capable of restraining the tubing; and

(c) installed in accordance with Figures 1 or 2, as applicable (i.e., anchor distances and size of arms and offsets).

Note: *Anchors are typically installed every 20 m (65 ft). See Section 6.*

8.0 Joints and Connections.

8.1 Assembly Procedure.

The procedure for making joints shall be as specified by the manufacturer.

8.2 Concealed Joints.

PEX tubing systems manufactured in accordance with the applicable standards referenced in Section 2 are deemed manufactured joints and may be installed in concealed spaces without the need for access panels.

9.0 Clearances.

9.1 Gas Vents.

Except for double-wall B-vents, which require a 25 mm (1 in) clearance, the clearance between gas appliance vents and PEX tubing shall be at least 150 mm (6 in).

9.2 Recessed Light Fixtures.

Except when the PEX tubing is protected with fiberglass or closed-cell insulation or the recessed light is IC-rated, the clearance between recessed light fixtures and PEX tubing shall be at least 300 mm (12 in).

9.3 Fluorescent Lighting.

When in direct view of the light source, the clearance between fluorescent lighting and PEX tubing shall be at least 1.5 m (5 ft). If the minimum clearance cannot be achieved, the PEX tubing shall be protected with a UV-blocking sleeve.

10.0 Other Considerations.

10.1 Hot-Work Joints.

Hot-work joints (e.g., soldering, brazing, welding, and fusion-welding) shall be

(a) made at least 500 mm (18 in) from PEX tubing in the same water line; and

(b) performed prior to completing the PEX joints.

10.2 Bending Radius.

10.2.1 The free (unsupported) bending radius for PEX tubing, measured at the outside of the bend, shall be not less than six times the actual outside diameter of the tubing, unless otherwise specified by the PEX manufacturer. Supports should be used to facilitate rigid bends and to alleviate stress on PEX joints when bends are needed in close proximity to such joints.

10.2.2 Tighter bends may be used when the PEX tubing is uniformly bent (supported) around a curved bracket or other rigid fixture. In this case, the minimum outside radius of the supported bend shall be as specified by the PEX manufacturer.

10.3 Directional Fittings.

Directional fittings (e.g., 90° and 45° elbows) should only be installed where necessary.

Note: *The flexible nature of PEX tubing allows for sweeping bends resulting in less fittings and joints.*

10.4 Direct Burial.

PEX tubing and fittings may be used in direct burial applications when allowed in the manufacturer's written installation instructions.

Note: *AWWA C904 should be consulted for water service applications.*

10.5 Fire-Resistive Construction.

Manufacturer's installation instructions shall be consulted prior to installation of PEX tubing in fire resistive constructions. PEX tubing penetrating a wall or floor-and-ceiling fire-rated assembly shall include a means of passive fire protection in accordance with the local codes.

10.6 Sizing and Flow Velocities.

10.6.1 PEX tubing shall be sized in accordance with IAPMO/ANSI UPC 1.

Note: *Potable water piping sizing is addressed in Chapter 6 and Appendix A of IAPMO/ANSI UPC.*

10.6.2 The tubing manufacturer's pressure-loss data should be referenced when using Appendix A of IAPMO/ANSI UPC 1. In absence of such data, Figures 3 and 4 shall be used.

10.6.3 Flow velocities through the water distribution system, used for calculating flush tank and flush valve fixture units depending on the tubing sizes (see Table 1), shall not exceed

(a) 3.0 m/s (10 ft/s) for cold-water distribution systems; and

(b) 2.4 m/s (8 ft/s) for hot-water distribution systems.

Note: *The flow velocities in Items (a) and (b) account for the increased velocities through the fittings.*

10.6.4 Hot-water recirculation systems shall

(a) be balanced to maintain adequate system temperatures; and

(b) have flow velocities that do not exceed 0.6 m/s (2 ft/s) (see Table 2); and

(c) use only PEX tubing designated for hot, chlorinated water recirculation systems and rated for the maximum percentage of time during which the system is intended to be operated at elevated temperatures, in accordance with ASTM F876.

10.7 Installation Testing.

Installation of PEX water distribution systems may be tested with air when

(a) expressly allowed in the written instructions of the manufacturers of all plastic pipe and fittings installed at the time the PEX piping system is being tested; and

(b) compressed air or other gas testing is not prohibited by the authority having jurisdiction.

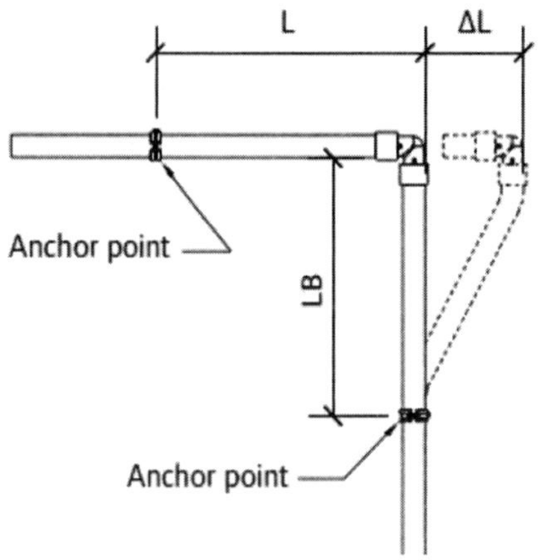

Note:

$LB = C \times \sqrt{(D \times \Delta L)}$

where

LB = *length of flexible arm*

C = *material constant (12 for PEX)*

D = *nominal outside diameter of tubing*

ΔL = *thermal expansion length*

FIGURE 1
EXPANSION ARMS
(See Sections 6.4.1 and 7.3)

TABLE 1
CALCULATION OF FLUSH TANK AND FLUSH VALVE FIXTURE UNITS
(See Section 10.6.3)

NOMINAL TUBING SIZE	FLOW VELOCITY: 3.0 m/s (10 ft/s)			FLOW VELOCITY: 2.4 m/s (8 ft/s)		
	FLOW VOLUME, L/min (gpm)	FLUSH TANK FIXTURE UNITS	FLUSH VALVE FIXTURE UNITS	FLOW, L/min (gpm)	FLUSH TANK FIXTURE UNITS	FLUSH VALVE FIXTURE UNITS
½	20.8 (5.5)	6	—	16.7 (4.4)	4	—
¾	41.6 (11.0)	15	—	33.3 (8.8)	11	—
1	68.9 (18.2)	26	—	55.3 (14.6)	20	—
1¼	103.0 (27.2)	46	10	82.5 (21.8)	33	5
1½	143.5 (37.9)	77	24	114.7 (30.3)	54	13
2	246.1 (65.0)	200	91	196.8 (52.0)	135	52
3	533.0 (140.8)	590	495	426.2 (112.6)	443	310

TABLE 2
TUBING SIZES, FLOWS, AND FRICTION LOSSES FOR HOT-WATER RECIRCULATION SYSTEMS
(See Section 10.6.4)

NOMINAL TUBING SIZE	FLOW VELOCITY m/s (ft/s)	FLOW VOLUME L/min (gpm)	FRICTION LOSSES AT 49°C (120°F) kPa/m (psi/ft)
½	0.6 (2)	4.2 (1.1)	0.4411 (0.0195)
¾	0.6 (2)	8.3 (2.2)	0.2850 (0.0126)
1	0.6 (2)	13.6 (3.6)	0.2081 (0.0092)
1¼	0.6 (2)	20.4 (5.4)	0.1629 (0.0072)
1½	0.6 (2)	28.4 (7.5)	0.1335 (0.0059)
2	0.6 (2)	48.8 (12.9)	0.0950 (0.0042)

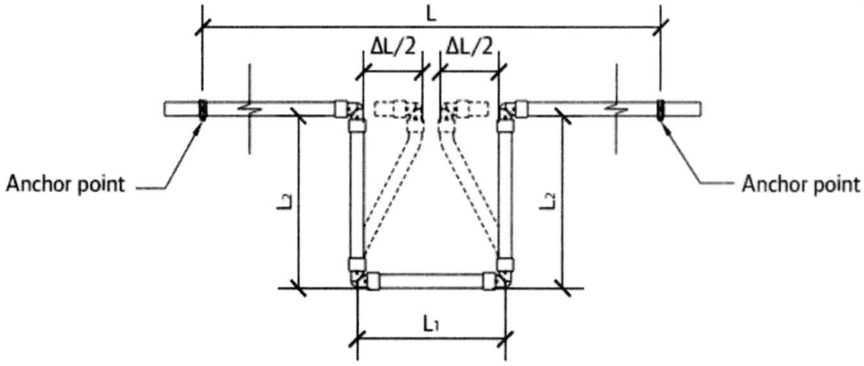

Note:

LB shall be calculated as specified in Figure 2 and divided into three sections, as follows:

LB = L1 + (2 × L2)

where

L1 = LB ÷ 5; and

L2 = L1 × 2.

FIGURE 2
EXPANSION LOOPS
(See Sections 6.4.2 and 7.3)

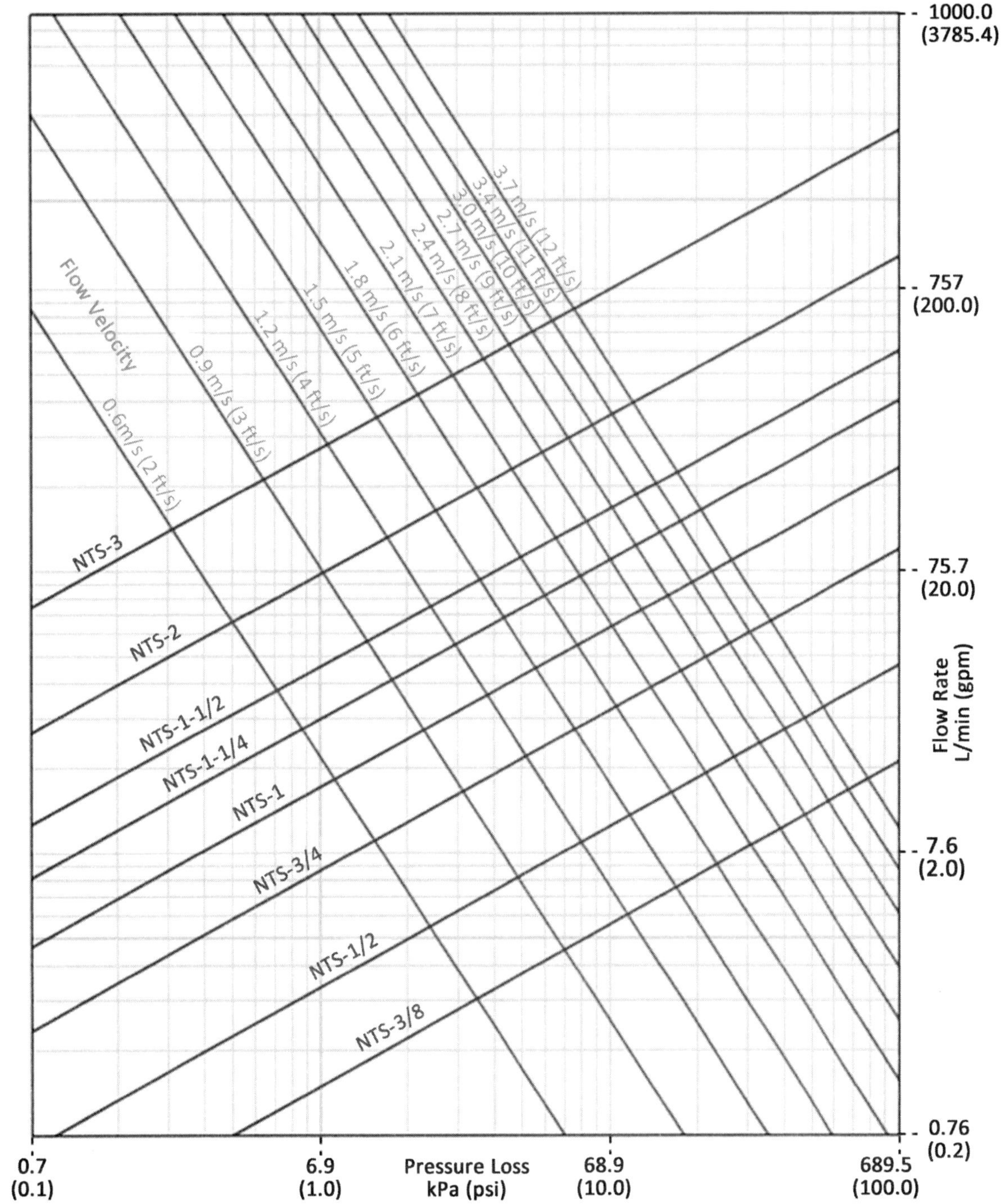

FIGURE 3
PRESSURE LOSS OF PEX TUBING AT 16 °C (60°F)
(See Section 10.6.2)

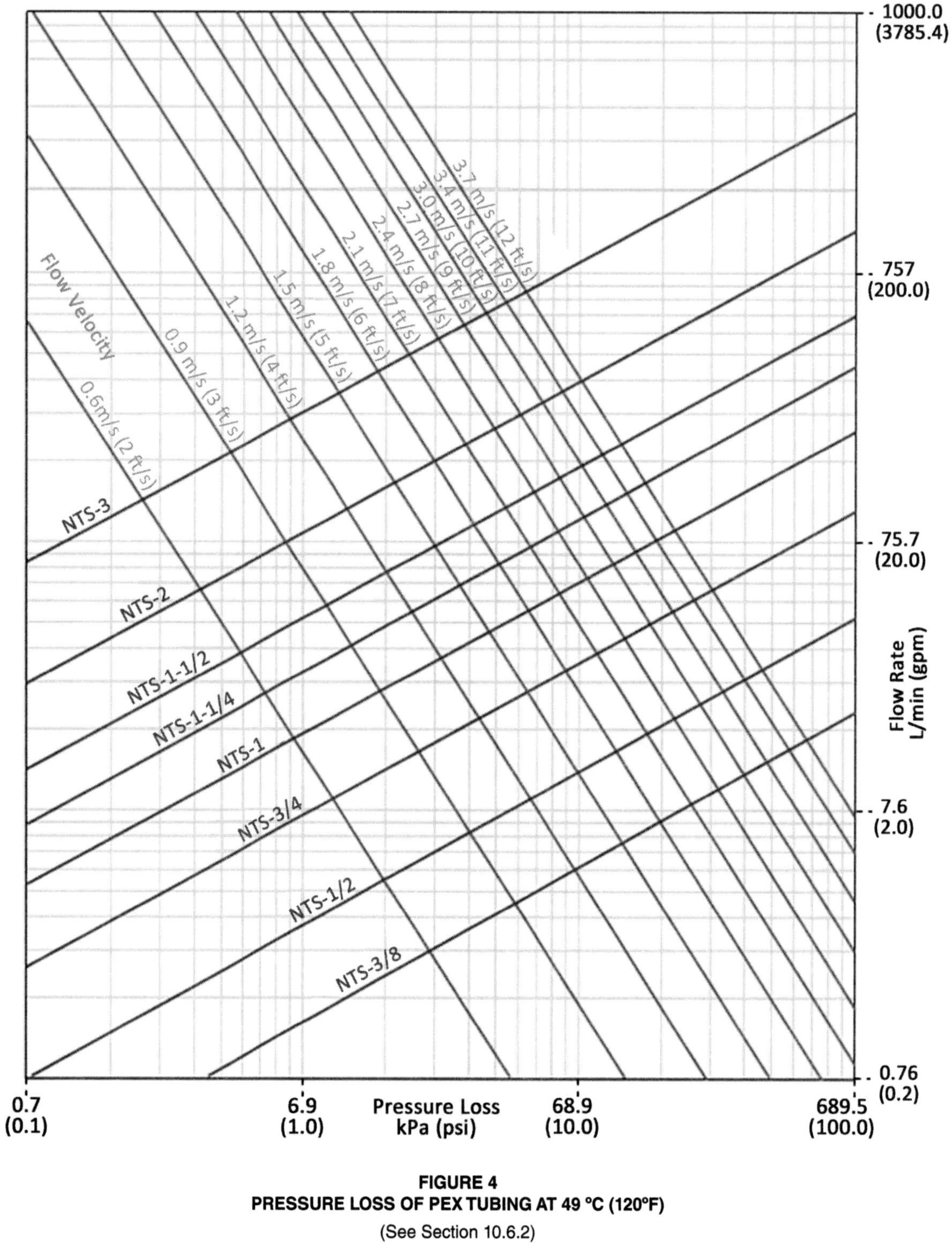

FIGURE 4
PRESSURE LOSS OF PEX TUBING AT 49 °C (120°F)

(See Section 10.6.2)

"The information contained in this appendix is not part of this American National Standard (ANS) and has not been processed in accordance with ANSI's requirements for an ANS. As such, this appendix may contain material that has not been subjected to public review or a consensus process. In addition, it does not contain requirements necessary for conformance to the standard."

The following IAPMO Installation Standard is included here for the convenience of the users of the Uniform Plumbing Code. It is not considered as a part of the Uniform Plumbing Code unless formally adopted as such. This Installation Standard is an independent, stand-alone document published by the International Association of Plumbing and Mechanical Officials and is printed herein by the expressed written permission of IAPMO.

THRUST BLOCKING FOR RUBBER GASKETED AND SOLVENT CEMENT JOINTS

IAPMO IS 33-2019[e1]

1 Scope

1.1 General

1.1.1 This Standard specifies requirements for the installation of thrust blocking for ductile iron pipe with elastomeric gasketed joints and fittings and PVC piping with solvent cemented or elastomeric gasketed joints intended for cold water building supply and yard piping.

1.1.2 Thrust blocks covered by this standard prevent separation of joints and pipe movement by transferring the resultant thrust force at a bend to the undisturbed soil behind the thrust block. The bearing strength of the soil is expressed in pounds per square foot. Therefore, the area behind the thrust block must engage enough soil area to resist the resultant thrust force at a change in direction.

1.2 Terminology

In this Standard,

(a) "shall" is used to express a requirement, i.e., a provision that the user is obliged to satisfy to comply with the Standard;

(b) "should" is used to express a recommendation, but not a requirement;

(c) "may" is used to express an option or something permissible within the scope of the Standard; and

(d) "can" is used to express a possibility or a capability.

Notes accompanying sections of the Standard do not specify requirements or alternative requirements; their purpose is to separate explanatory or informative material from the text. Notes to tables and figures are considered part of the table or figure and can be written as requirements.

1.3 Amendments

Proposals for amendments to this Standard will be processed in accordance with the standards-writing procedures of IAPMO.

2 Reference Publications

This Standard refers to the following publications, and where such reference is made, it shall be to the current edition of those publications, including all amendments published thereto.

IAPMO/ANSI UPC-1	Uniform Plumbing Code

3 Abbreviations

The following abbreviations apply in this Standard:

PVC — polyvinyl chloride

4 General

4.1 Thrust Blocking for Rubber Gasket Joints

In lines with rubber gasketed joints, thrust blocks shall be installed at all locations including:

(a) Changes in direction, as at tees and bends

(b) Changes in size, as at reducers

(c) Stops, as at dead ends

(d) Valves, where thrusts may be expected.

Note: *See examples in Figure 1.*

4.2 Sizing

Piping shall be sized in accordance with the applicable code or jurisdictional requirements.

Note: *For example, Appendix A, "Recommended Rules for Sizing the Water Supply System", or Section 610.0, "Size of Potable Water Piping" of IAPMO/ANSI UPC-1 may be use for sizing pipe.*

4.3 Flow Velocity

Flow velocity shall not exceed 8 fps (2.4 m/s) for PVC pipe or 10 fps (3m/s) for other pipe materials.

4.4 Deflection

Elastomeric gasketed pipe may be deflected in accordance with the manufacturer's recommendations provided that it shall not be permanently staked or blocked to maintain this deflection.

4.5 Maximum Working Pressure

The pipe and fitting system shall be designed for the maximum working pressure in accordance with Table 1.

5 Thrust Blocking Sizing and Calculations

5.1 Sizes

Thrust block sizes shall be based on the maximum line pressure, pipe size and the safe bearing load of the soil.

5.2 Calculation Examples

5.2.1 Thrust

5.2.1.1 Example 1

From Table 2, *Thrust at Fittings in pounds at 100 psi of Water Pressure*, the thrust for a NPS 4 Dead End Tee at a pressure of 100 psi is 1620 lb.

5.2.1.2 Example 2

The thrust for an NPS 4, Dead End Tee, at a pressure of 150 psi is 2430 lb or 1.5 times the thrust found in Table 2.

The calculation is as follows:

(a) The thrust from Table 2, for an NPS 4, Dead End Tee, at 100 psi, is 1,620 lb.

(b) The actual pressure of the system is 150 psi.

(c) Divide the actual pressure by the table pressure or 150 psi/100 psi is 1.5.

(d) Multiply the thrust from Table 2, for an NPS 4, Dead End Tee, at 100 psi, times the ratio of the actual/table pressure, or 1.5 times 1620 lb for a thrust of 2430 lb.

5.2.1.3 Example 3

From Table 3, *Thrust at Fittings in Newtons at 689 kPa of Water Pressure*, the thrust for a DN 125, 90° Bend, at a pressure of 689 kPa is 15,757 N.

5.2.1.4 Example 4

The thrust for a DN 125, 90° Bend, at a pressure of 861 kPa is 19,469 N or 1.25 times the force in Table 3.

The calculation is as follows

(a) The thrust from Table 3, for a DN 127, 90° Bend at a pressure of 689 kPa, is 15,575 N.

(b) The actual pressure of the system is 861 kPa.

(c) Divide the actual pressure by the table pressure or 861 kPa/689 kPa or 1.25.

(d) Multiply the thrust from Table 2, for a DN 125, 90° Bend, at a pressure of 689 kPa times, the ratio of the actual/table pressure, or 1.25 times 15,757 N for a thrust of 19,469 N.

Note: *1000 N is equal to 1 kN for a thrust of 19.469 kN.*

5.2.2 Thrust Block Bearing Area

The thrust block bearing area is determined by dividing the thrust by the safe bearing load of the soil.

Note: *See examples in Figure 2*

5.2.2.1 Example 5

From Table 4, *Safe Bearing Loads of Various Soils* the safe bearing load of sand is 2000 lb/ft^2.

5.2.2.2 Example 6

The thrust block bearing area for an NPS 4, Dead End Tee, at a pressure of 100 psi in sand is 0.81 ft^2.

The calculation is as follows

(a) From Table 2, *Thrust at Fittings in pounds at 100 psi of Water Pressure*, the thrust for a 4 in Dead End Tee at a pressure of 100 psi is 1620 lb.

(b) From Table 4 the safe bearing load of sand is 2000 lb/ft^2.

(c) Divide the thrust by the safe bearing load of sand, or 1620 lb/2000 lb/ft^2 for a thrust block bearing area of 0.81 ft^2.

5.2.2.3 Example 7

The thrust block bearing area for an NPS 4, Dead End Tee at a pressure of 150 psi in sand is 1.22 ft^2.

The calculation is as follows:

(a) The thrust for an NPS 4, Dead End Tee, at a pressure of 150 psi was found in Example 2 to be 2430 lb.

(b) The soil type is sand and the safe bearing load of sand from Table 4 is 2000 lb/ft^2.

(c) The required bearing area of the thrust block is determined by dividing the thrust by the safe bearing load or 2430 lb/2000 lb/ft^2 for a thrust block bearing area of 1.22 ft^2.

5.2.2.4 Example 8

The thrust block bearing area for a DN 125, 90° Bend, at a pressure of 861 kPa in Soft Clay is 2.82 m^2

The calculation is as follows:

(a) The thrust for a DN 125, 90° Bend at a pressure of 861 kPa was found in Example 4 to be 19,469 N.

(b) The soil condition is soft clay and the safe bearing load of soft clay from Table 4 is 48 kPa or 48,000 N/m^2.

Note: *1 kPa = 1 kN/m^2 = 1000 N/m^2*

(c) The required block bearing area is determined by dividing the thrust by the safe bearing load or 19,469 N/48,000 N/ m^2 for a thrust block bearing area of 0.4 m^2.

6 Testing Requirements

6.1 Rubber Gasketed Joints

Properly sized thrust blocks, either permanent or temporary, shall be installed at all required points before testing. When concrete thrust blocks are installed, wait at least 24 hours before pressure testing.

6.2 Solvent Cement Joints

The entire system shall be purged before testing to eliminate all solvent cement vapors and air. The system shall not be pressurized until the joints have cured (set) at least as long as recommended by the manufacturer. If the manufacturers recommendation is not available the cure times in Table 5 shall apply. Systems with solvent cement joints shall be pressure tested filled with water or other fluid. *CAUTION: Water Test Only*.

7 Identification

A label shall be fastened to the main electrical meter panel stating, "This structure has a nonmetallic water service".

TABLE 1
MAXIMUM WORKING PRESSURE FOR PIPE AND FITTING SYSTEM DESIGN
(See Section 4.5)

MAXIMUM WORKING PRESSURE kPa (psi)	SIZES		PIPE	FITTINGS
	NPS	DN		
1103 kPa (160 psi)	1/2 to 8	15 to 200	SDR 26	Schedule 40
				Schedule 80
1379 kPa (200 psi)	1/2 to 4	15 to 100	SDR 21	Schedule 40
	1/2 to 8	15 to 200		Schedule 80
1724 kPa (250 psi)	1/2 to 3	15 to 80	SDR 17	Schedule 40
	1/2 to 8	15 to 200		Schedule 80
2172 kPa (315 psi)	1/2 to 1-1/2	15 to 40	SDR 13.5	Schedule 40
	1/2 to 4	15 to 100		Schedule 80
1103 kPa (160 psi)	5 to 8	125 to 200	Schedule 40	Schedule 40
1517 kPa (220 psi)	2 to 4	50 to 100	Schedule 40	Schedule 40
				Schedule 80
2206 kPa (320 psi)	1/2 to 1-1/2	15 to 40	Schedule 40	Schedule 40
				Schedule 80
1103 kPa (160 psi)	5 to 8	125 to 200	Schedule 80	Schedule 40
1517 kPa (220 psi)	2 to 4	50 to 100	Schedule 80	Schedule 40
2206 kPa (320 psi)	1/2 to 1-1/2	15 to 40	Schedule 80	Schedule 40
1724 kPa (250 psi)	5 to 8	125 to 200	Schedule 80	Schedule 80
2206 kPa (320 psi)	1/2 to 4	15 to 100	Schedule 80	Schedule 80

TABLE 2
THRUST AT FITTINGS IN POUNDS AT 100 psi OF WATER PRESSURE
(See Sections 5.2.1.1, 5.2.1.2, and 5.2.2.2)

PIPE SIZE NPS	90° BENDS	45° BENDS	22-1/2° BENDS	DEAD ENDS AND TEES
1-1/2	415	225	115	295
2	645	350	180	455
2-1/2	935	510	260	660
3	1395	755	385	985
3-1/2	1780	962	495	1260
4	2,295	1245	635	1620
5	3,500	1900	975	2,490
6	4,950	2,710	1385	3,550
8	8,300	4,500	2,290	5,860
10	12,800	6,900	3,540	9,050
12	18,100	9,800	5,000	12,800

TABLE 3
THRUST AT FITTINGS IN NEWTONS (N) AT 689 kPa OF WATER PRESSURE
(See Sections 5.2.1.3, and 5.2.1.4)

PIPE SIZE DN	90° BENDS	45° BENDS	22-1/2° BENDS	DEAD ENDS AND TEES
40	1847	1000	515	1315
50	2870	1560	800	2025
65	4160	2270	1160	3940
80	6210	3360	1715	4385
90	7925	4280	2205	5610
100	10,215	5540	2815	7210
125	15,575	8455	4340	11,080
150	22,030	12,060	6165	15,800
200	36,935	20,025	10,190	26,080
250	56,960	30,705	15,755	40,275
300	80,545	43,610	22,250	56,960

TABLE 4
SAFE BEARING LOADS OF VARIOUS SOILS
(See Sections 5.2.2.1, and 5.2.2.3 and 5.2.2.4)

	Safe Bearing Load	
SOIL	**lbs/ft^2 (psf)**	**kN/m^2 (kPa)**
Mulch, Peat, etc.	0	0
Soft Clay	1000	48
Sand	2000	96
Sand and Gravel	3000	143
Sand and Gravel Cement with Clay	4000	191
Hard Shale	10,000	478

TABLE 5
MINIMUM CURE TIME, IN HOURS (h)*
TEST PRESSURE FOR PIPE
(See Section 6.2)

	DN 15 TO 32 (NPS 1/2 TO 1-1/4)		DN 40 TO 80 (NPS 1-1/2 TO 3)		DN 80 TO 200 (NPS 3-1/2 TO 8)	
	DN 15 (NPS 1/2)	DN 32 (NPS 1-1/2)	DN 40 (NPS 1-1/2)	DN 80 (NPS 1-1/2)	DN 90 (NPS 3-1/2)	DN 200 (NPS 8)
TEMPERATURE RANGE DURING CURE PERIOD	UP TO 1240 kPa (180 PSI)	ABOVE 1240 TO 2549 kPa (180 TO 370 psi)	UP TO 1240 kPa (180 PSI)	ABOVE 1240 TO 2549 kPa (180 TO 370 psi)	UP TO 1240 kPa (180 PSI)	ABOVE 1240 TO 2549 kPa (180 TO 370 psi)
60°F - 100°F (16°C-38°C)	1 h	6 h	2 h	12 h	6 h	24 h
40°F-60°F (4°C-16°C)	2 h	12 h	4 h	24 h	12 h	48 h
10°F- 40°F (-12°C+4°C)	8 h	48 h	16 h	96 h	48 h	8 days

*If gaps or loose fits are encountered in the system, double these cure times.

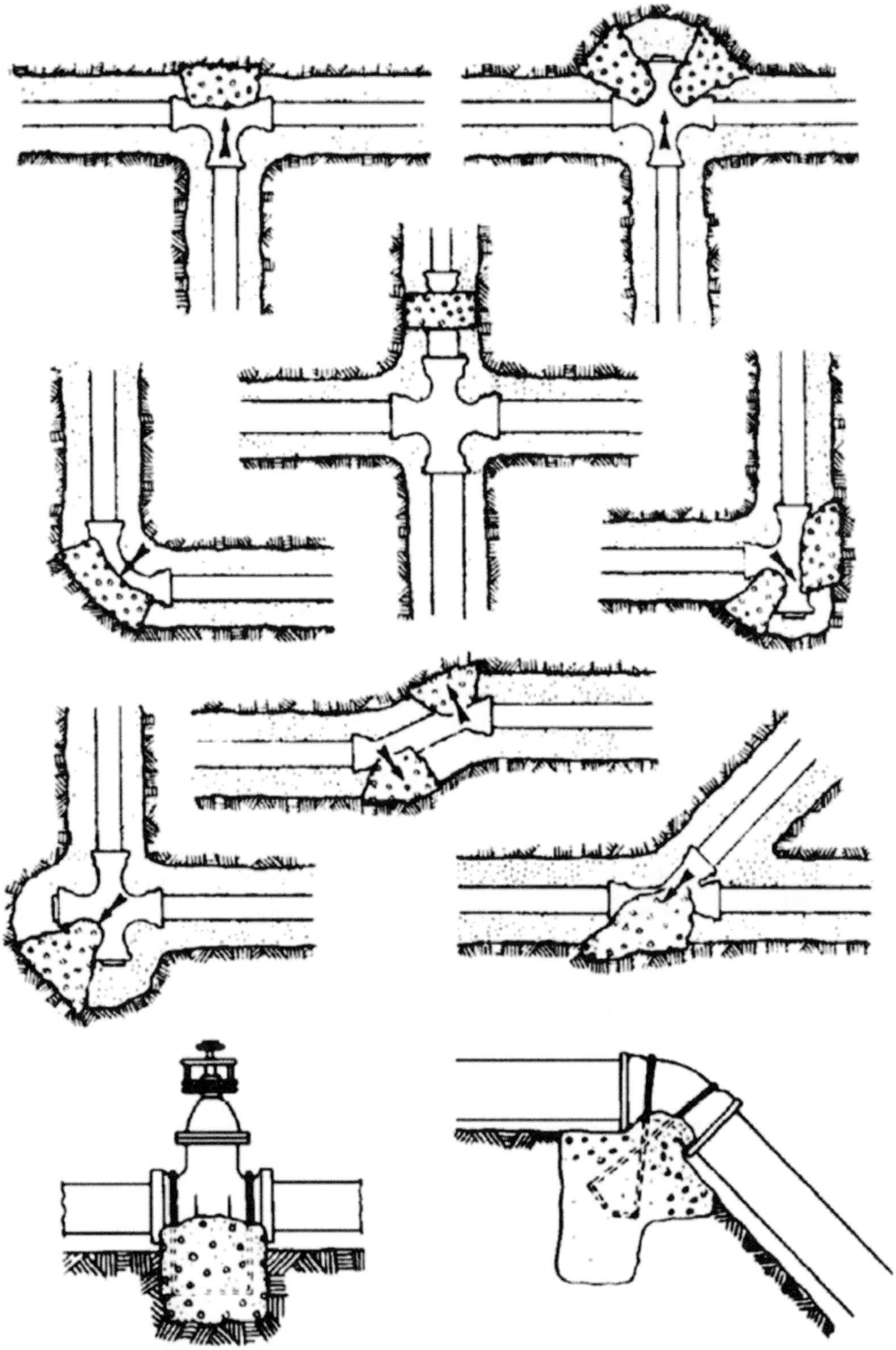

FIGURE 1
LOCATION OF THRUST BLOCKS
(STANDARD AND METRIC COMBINED)
A COMPARISON OF THRUST-BLOCK AREAS
(See Section 4.1)

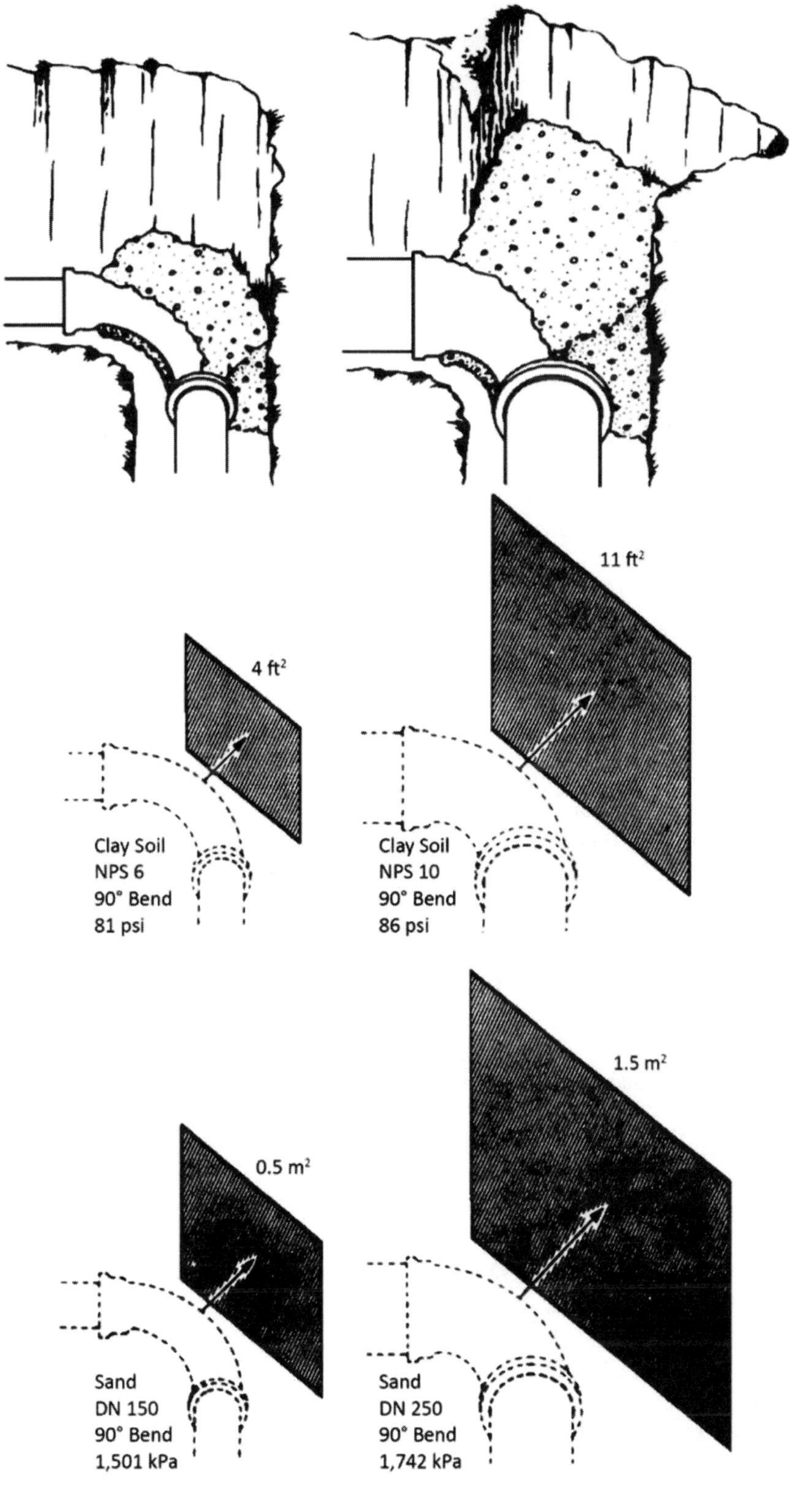

FIGURE 2
THRUST BLOCK BEARING AREA
(See Section 5.2.2)

Uniform Plumbing Code References for Ductile Iron and PVC Pressure Building Supply and Exterior Cold-Water Piping

The following sections of the 2021 Uniform Plumbing Code apply to ductile iron and PVC pressure building supply and exterior cold-water piping.

Chapter 3	**General Regulations**
309.0	Workmanship
312.0	Protection of Piping, Materials, and Structures
313.0	Hangers and Supports
314.0	Trenching, Excavation, and Backfill

Chapter 6	**Water Supply and Distribution**
604.0	Materials (water piping)
Table 604.1	Materials for Building Supply and Water Distribution Piping and Fittings
605.4	Ductile Iron Pipe and Joints
605.12	PVC Plastic Pipe and Joints
605.16.2	Plastic Pipe to Other Materials
609.0	Installation, Testing, Unions, and Location

Abbreviations

IAPMO	International Association of Plumbing and Mechanical Officials
UPC	Uniform Plumbing Code published by IAPMO
AWWA	American Water Works Association
CSA	Canadian Standards Association

Applicable Standards.

ASME B16.4	Gray Iron Threaded Fittings Classes 125 and 250
ASTM D1785	Poly (Vinyl Chloride) (PVC) Plastic Pipe, Schedules 40, 80, and 120
ASTM D2241	Poly (Vinyl Chloride) (PVC) Pressure-Rated Pipe (SDR Series)
ASTM D2466	Poly (Vinyl Chloride) (PVC) Plastic Pipe Fittings, Schedule 40
ASTM D2467	Poly (Vinyl Chloride) (PVC) Plastic Pipe Fittings, Schedule 80
ASTM F1970	Special Engineered Fittings, Appurtenances or Valves for use in Poly (Vinyl Chloride) (PVC) or Chlorinated Poly (Vinyl Chloride) (CPVC) Systems
AWWA C110	Ductile Iron and Gray-Iron Fittings
AWWA C111	Rubber-Gasket Joints for Ductile-Iron Pressure Pipe and Fittings
AWWA C151	Ductile-Iron Pipe, Centrifugally Cast
AWWA C153	Ductile-Iron Compact Fitting
AWWA C900	Polyvinyl Chloride (PVC) Pressure Pipe and Fabricated Fittings, 4 in. through 12 in. (100 mm through 300 mm)
AWWA C904	Crosslinked Polyethylene (PEX) Pressure Tubing, ½ in. through 3 In. (13 mm through 76 mm), for Water Service
AWWA C907	Injection-Molded Polyvinyl Chloride (PVC) Pressure Fittings, 4 In, through 12 in (100 mm through 300 mm) for Water, Wastewater and Reclaimed Water Service
IAPMO/ ANSI UPC-1	Uniform Plumbing Code

The information contained in this appendix is not part of this American National Standard (ANS) and has not been processed in accordance with ANSI's requirements for an ANS. As such, this appendix may contain material that has not been subjected to public review or a consensus process. In addition, it does not contain requirements necessary for conformance to the standard.

The following Sections B414-18, B417-18, B421C-18, B422C-18, SR614-18, TR418-18, TR420-18, are excerpted in their entirety from the *2018 Handbook for the Installation of Ceramic, Glass, and Stone Tile Installations*, with permission from the Tile Council of North America, Inc.

In addition to these, the TCNA *Handbook*, includes 200+ methods for tile and stone installation based on various installation requirements and application types, as well as product selection guides, field and installation requirements, and guidelines for wet areas. For a complete copy, visit www.TCNAtile.com.

TCNA HANDBOOK-2018

FOR CERAMIC, GLASS, AND STONE TILE INSTALLATION

NOTICE OF DISCLAIMER AND LIMITATION OF LIABILITY

This publication is provided as a guide, not a product, and all of the information it contains is provided 'as is' without warranty of any kind, whether express or implied. All implied warranties, including, without limitation, implied warranties of merchantability, fitness for a particular purpose, and non-infringement, are hereby expressly disclaimed.

Publisher has not performed any independent analysis in connection with any of the information contained herein, and expressly disclaims any obligation to obtain and include information other than that presented herein resulting from the handbook committee consensus process. This information does not purport to address safety issues or applicable regulatory requirements associated with its use. It is the responsibility of the user of this information to review any applicable codes and other regulations and any site specific conditions in connection with the use of this information. Publisher expressly makes no representations or warranties regarding use of this information and compliance with any applicable statute, rule or regulation.

The reader is expressly warned to consider and adopt all safety precautions appropriate for the activities herein and to avoid all potential hazards. Such precautions are generally not listed herein and are outside the scope of this document.

Under no circumstances will publisher be liable to any person or business entity for any damages, including without limitation any and all direct, indirect, special, incidental, consequential, or exemplary damages, resulting, in whole or in part, from any use of, reference to, or reliance upon this publication, even if advised of the possibility of such damages. The foregoing limitation of liability is a fundamental element of the use of this information and the information would not be offered by the publisher without such limitation.

B414-18

- **Wood or Metal Studs**
- **Mortar Bed Walls**
- **Mortar Bed Floor**
- **Ceramic Tile, Glass Tile**

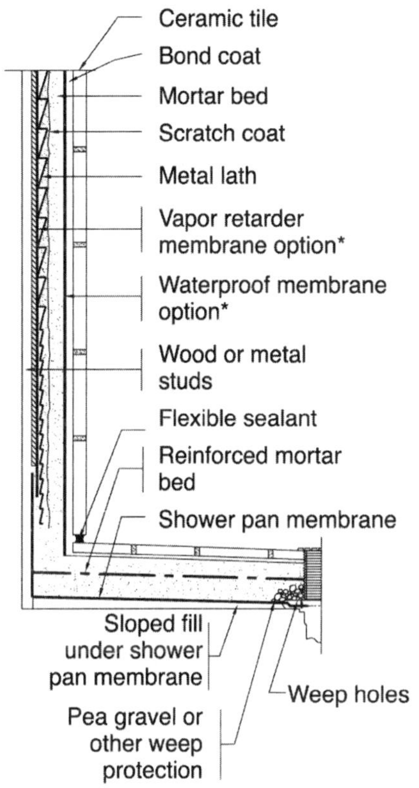

* Use of a membrane on walls is required. See membrane options.

Shower receptors, curbs, seats, etc., must be properly waterproofed and installed to avoid water damage to adjacent building materials. See common shower configurations section.

Wall membrane laps over shower pan.

Recommended Uses

- For showers that do not have prefabricated receptors.
- For areas where wall flatness is critical, such as when tiles with any edge longer than 15" are specified.

Environmental Exposure Classifications

- Res1, 2, 3, 5; Com1, 2, 3, 5.
- For Res4 and Com4, see SR614.
- For installations that may be exposed to staining, specify tile and grout suitable for exposure. Consult product manufacturers; see also "Product Selection Guides."
- For installations that may be exposed to mild chemical attack, specify epoxy grout and tile suitable for exposure.

  For greater resistance to chemical exposure, also specify an epoxy bonding material. Consult product manufacturers; see also "Product Selection Guides."

Limitations

- Maximum mortar bed thickness (walls)—1" for metal studs, 1½" for wood studs.
- Maximum stud spacing 16" on center.

Membrane Options

- A waterproof membrane (A118.10) or vapor retarder membrane (A108.02-3.8) must be specified for walls to prevent moisture intrusion and protect adjacent building materials. Specifier shall indicate if complete waterproofing of walls is required, including treatment at termination points.
- If a waterproof membrane is applied over the mortar bed walls, membrane over framing members may not be required.
- Check with membrane manufacturer for suitability for applicable conditions, as not all membranes are suitable for steam, high-temperature and/or chemical exposure, or exterior use.
- When glass tile is used, consult glass tile manufacturer for membrane options and recommendations.

Requirements

- Wood studs—dry and well-braced, minimum depth 3½".
- Metal studs—well-braced; 20 gauge (0.033") or heavier; minimum depth 3½" for residential applications or 3⅝" for commercial applications.
- Mortar bed thickness—¾" minimum to 1" maximum (metal studs) or 1½" maximum (wood studs)
- Membrane behind mortar bed, when used, must lap over shower pan membrane.
- Slope shower pan membrane ¼" per foot to weep holes in drain.
- Turn shower pan membrane up walls a minimum of 3" above shower curb (6" above floor in showers without curbs).
- Surround drain with pea gravel or other weep protection to prevent mortar from blocking weep holes.

Materials

- **Multiple options exist for membranes, mortars, grouts, and other materials and *must be clearly specified* to be included. If not specifically indicated, optional materials are not included and mortar/grout choice defaults to minimum performance specification indicated. Consider each system component and intended use to determine minimum requirements and to specify options.**
- Ceramic tile—ANSI A137.1.
- Glass tile, when used—ANSI A137.2; see also "Glass Tile Selection and Installation Guide," and consult tile manufacturer for environmental exposure classification recommendations. Not all glass tiles are suitable.

- Cementitious grout—ANSI A118.6 or better or ISO CG1 or better. When glass tile is used, specify grout designated by tile and grout manufacturers.
- Epoxy grout, when used—ANSI A118.3 or ISO RG.
- Cementitious bond coat—portland cement paste on a mortar bed that is still workable. For a cured mortar bed, follow recommendations below to select appropriate bonding mortar:
  - When a waterproof membrane is not used—ANSI A118.1 or better or ISO C1 or better.
  - When a waterproof membrane is used—ANSI A118.4 or better or ISO C2S1 or better unless ANSI A118.1 or ISO C1 is recommended by membrane manufacturer.
  - When porcelain tile is used—ANSI A118.4 or better or ISO C2 or better.
  - When glass tile is used, specify mortar designated by tile and mortar manufacturers. Bond coat color will impact the final appearance of translucent glass tile. Specifier shall confirm bond coat color is acceptable.
- Epoxy bond coat, when used—ANSI A118.3 or ISO R1 or better.
  - When glass tile is used, specify epoxy bond coat designated by tile and bond coat manufacturers. Bond coat color will impact the final appearance of translucent glass tile. Specifier shall confirm bond coat color is acceptable.
- Vapor retarder membrane, when used—ANSI A108.02-3.8.
- Waterproof membrane, when used—ANSI A118.10.
- Mortar bed, metal lath, and cleavage membrane—ANSI A108.1A.
- Flexible mildew-resistant sealant—ASTM C920.
- Shower pan membrane—local building code.
- Metal studs—ASTM C645.

Materials for Green/Sustainable Design

- See "Green Building Standards and Green Product Selection Guide" and consult manufacturers and suppliers for product sustainability and contribution to green building design.
- Consider specifying tile and installation materials that meet ANSI A138.1, the *American National Standard Specifications for Sustainable Ceramic Tiles, Glass Tiles, and Tile Installation Materials*.

Preparation by Other Trades

- Wall framing shall meet the general framing requirements of ANSI A108.11-4.0–4.3.
- Apply blocking between the studs to support the shower pan membrane.

Movement Joint (architect must specify type of joint and show location and details on drawings)

- Movement joints—mandatory according to EJ171.
- When glass tile is used, adhere to more frequent placement recommendations within the ranges listed in EJ171.

Installation Specifications

- Shower pan membrane—ANSI A108.01-3.6.
- Tile—ANSI A108.1A, .1B, or .1C. A108.1B required if waterproof membrane or epoxy bond coat to be used.
- Glass tile—ANSI A108.14, .15, .16, or manufacturer's directions.
- Cementitious grout—ANSI A108.10.
- Epoxy mortar/grout—ANSI A108.6.
- Waterproof membrane—ANSI A108.13.
- Movement Joints—EJ171 and ASTM C1193.

Notes

- Test shower pan membrane and drainage fitting for leaks before commencing tilework.
- Materials adversely affected by moisture in areas immediately adjacent to showers, tubs, and roman tubs should be properly protected.
- A sloped portland cement mortar fill or approved preformed slope may be used under shower pan membrane when subfloor is not sloped to drain.
- All horizontal surfaces, for example shower seats, sills, curbs, etc., must slope towards drain or other surface sloped toward drain. Where present, waterproofing also must be sloped.
- When glass tile is used, see "Glass Tile Selection and Installation Guide, " and consult manufacturer for recommendations and requirements.
- For curbless shower receptor, see B421C and B422C.

For the following standards listed, refer to the 2018 TCNA Handbook for Ceramic, Glass, and Stone Tile Installation.

a. B441-18 Wood or Metal Studs • Backer Board • Mortar Bed Walls (One Coat Method) • Mortar Bed Floor • Ceramic Tile

b. B415-18 Wood or Metal Studs • Cement Backer Board or Fiber-Cement Backer Board Walls • Mortar Bed Floor • Ceramic Tile, Glass Tile

c. B420-18 Wood or Metal Studs • Coated Glass Mat Water-Resistant Gypsum Backer Board Walls • Mortar Bed Floor • Ceramic Tile, Glass Tile

d. B426-18 Wood or Metal Studs • Cementitious-Coated Extruded Foam Backer Board Walls • Mortar Bed Floor • Ceramic Tile, Glass Tile

e. B431-18 Wood or Metal Studs • Fiber-Reinforced Water-Resistant Gypsum Backer Board Walls • Mortar Bed Floor • Ceramic Tile, Glass Tile

f. B421-18 Solid Backing • Bonded Waterproof Membrane • Ceramic Tile, Glass Tile

g. B422-18 Solid Backing • Bonded Waterproof Membrane • Integrated Bonding Flange • Ceramic Tile, Glass Tile

B417-18

- **Concrete Tank**
- **Mortar Bed**

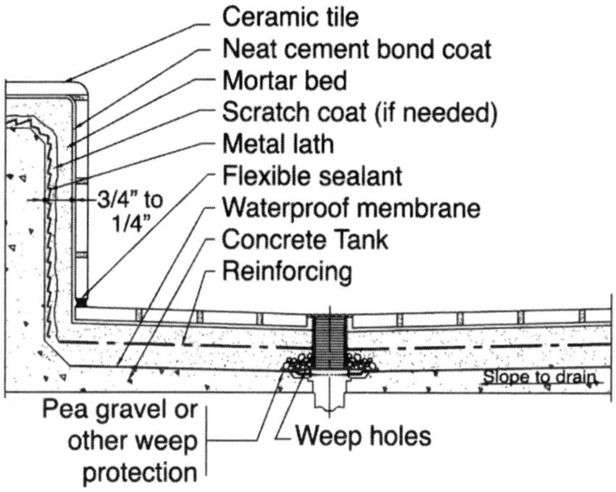

- **Wood Form**
- **Mortar Bed**

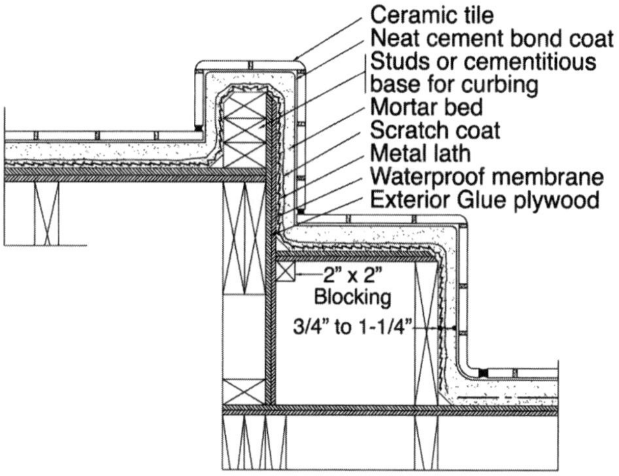

- **Concrete Curb**
- **Mortar Bed**

- **Wood Curb**
- **Mortar Bed**

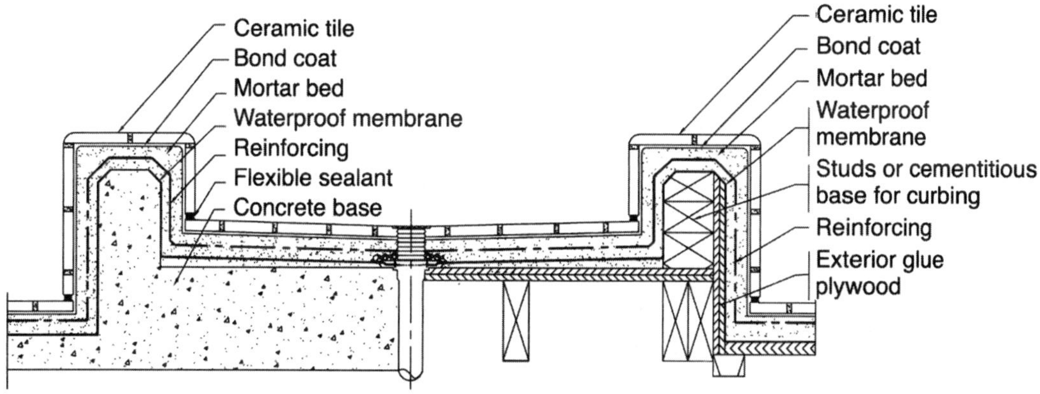

Note: This detail reflects both concrete and wood substrates. Specify one or the other.

Requirements

- Waterproof membrane—slope membrane ¼" per foot to weep holes in drain.
- Wood framing, when used, should be pressure treated and designed to resist deflection and movement.

Preparation by Other Trades

- Test tank, membrane, and drainage fittings for leaks before starting tilework.

Materials for Green/Sustainable Design

- See "Green Building Standards and Green Product Selection Guide" and consult manufacturers and suppliers for product sustainability and contribution to green building design.
- Consider specifying tile and installation materials that meet ANSI A138.1, the *American National Standard Specifications for Sustainable Ceramic Tiles, Glass Tiles, and Tile Installation Materials*.

Installation Methods

- Attach metal lath only above water line.
- Floor—follow F121.
- Walls—follow W221.

Installation Specifications

- Tile—ANSI A108.1A.
- Grout—ANSI A108.10.
- Movement Joints—EJ171 and ASTM C1193.

B421C-18

- **Solid Backing (wall)**
- **Concrete Substrate (floor)**
- **Bonded Waterproof Membrane**
- **Ceramic Tile, Glass Tile**

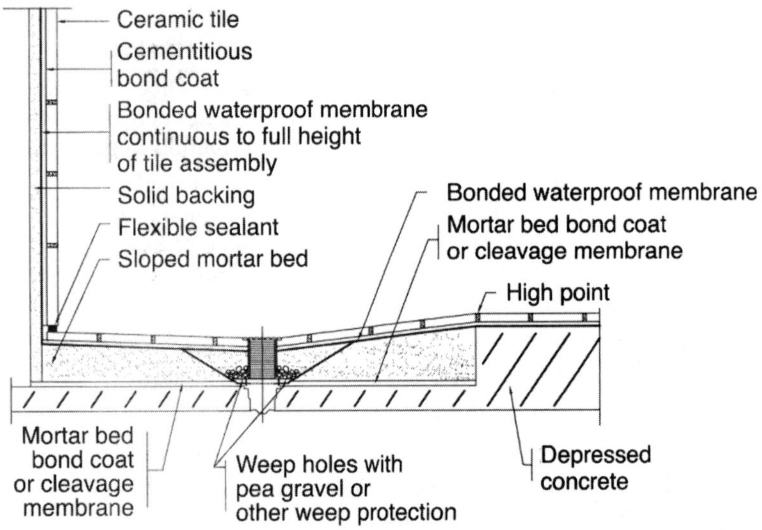

Shower receptors, curbs, seats, etc., must be properly waterproofed and installed to avoid water damage to adjacent building materials.
See common shower configuration section.

For mortar beds over 65 s.f., wire reinforcing is required.

Recommended Uses

- For construction of a shower without a curb.

Environmental Exposure Classifications

- Res1, 2, 3; Com1, 2
- May be suitable for Com3, Res5, and Com5 as determined by membrane and backing material manufacturers.
- For Res4 and Com4, see SR613 and SR614.
- For installations that may be exposed to staining, specify tile and grout suitable for exposure. Consult product manufacturers; see also "Product Selection Guides. "
- For installations that may be exposed to mild chemical attack, specify epoxy grout and tile suitable for exposure. Consult product manufacturers; see also "Product Selection Guides. "

Limitations

- Facilitates construction of a curbless shower without adding floor height at restroom entryway, but does not inherently comply with Americans with Disabilities Act (ADA) Standards for Accessible Design. Follow ADA where ADA compliance is required or desired.
- Refer to appropriate wall method for applicable limitations based on type of backing used—W201, W202E, W202I, W211, W221, W222, W231/W241, W243, W244C, W244E, W244F, W245, W246, and W247.

Requirements

- When glass tile is used, consult glass tile manufacturer for tile suitability over non-absorptive surface.
- Bonded waterproof membrane must be continuous, including at changes in plane. Follow membrane manufacturer's requirements for corners, seaming, and overlap.
- Waterproof membrane inside shower area must extend to full height of tile assembly. Also, the floor and wall waterproofing must continue outside the immediate shower area one foot beyond the high point of the floor, but not beyond the tiled area. Additional waterproofing of floor and/or walls outside the shower area may be needed to effectively contain and evacuate shower water and splash water and to protect building materials. When additional waterproofing of floor and/or walls outside the shower area is desired or required, the building design professional must specify all areas to be waterproofed and indicate membrane termination points.
- A secondary drain may be required outside the immediate shower area to facilitate evacuation of shower water and splash water that is not contained in the shower area due to curbless design. When required or desired, building design professional must specify type and location of secondary drain.
- High point of the floor must be outside the shower area, i.e., beyond shower door or curtain, to facilitate evacuation of shower water and splash water that is not contained in the shower area due to curbless design. Location of the high point of the floor is especially critical when secondary drain is not included outside the shower area.
- Surround drain with pea gravel or other weep protection to prevent mortar from blocking weep holes.
- Refer to appropriate wall method for applicable requirements based on type of backing used.

- Slope mortar bed ¼" per foot toward drain and follow membrane manufacturer's instructions for connecting membrane to drain.

Materials

- **Multiple options exist for membranes, mortars, grouts, and other materials and *must be clearly specified* to be included. If not specifically indicated, optional materials are not included and mortar/grout choice defaults to minimum performance specification indicated. Consider each system component and intended use to determine minimum requirements and to specify options.**
- Ceramic tile—ANSI A137.1.
- Glass tile, when used—ANSI A137.2; see also "Glass Tile Selection and Installation Guide," and consult tile manufacturer for environmental exposure classification recommendations. Not all glass tiles are suitable.
- Cementitious grout—ANSI A118.6 or better or ISO CG1 or better. When glass tile is used, specify grout designated by tile and grout manufacturers.
- Epoxy grout, when used—ANSI A118.3 or ISO RG.
- Cementitious bond coat:
  - ANSI A118.4 or better or ISO C2S1 or better unless ANSI A118.1 or ISO C1 is recommended by membrane manufacturer.
  - When glass tile is used, specify mortar designated by tile and mortar manufacturers. Bond coat color will impact the final appearance of translucent glass tile. Specifier shall confirm bond coat color is acceptable.
- Waterproof membrane—ANSI A118.10 and recommended by manufacturer of membrane for use over backing type in intended application.
- Flexible mildew-resistant sealant—ASTM C920.

Materials for Green/Sustainable Design

- See "Green Building Standards and Green Product Selection Guide" and consult manufacturers and suppliers for product sustainability and contribution to green building design.
- Consider specifying tile and installation materials that meet ANSI A138.1, the *American National Standard Specifications for Sustainable Ceramic Tiles, Glass Tiles, and Tile Installation Materials.*

Preparation by Other Trades/Backer Board Installers

- Depressed slab by others. Depressed area must extend beyond shower area to facilitate a finished floor installation that effectively evacuates shower water and splash water, which are often difficult to contain within a curbless shower.
- Refer to appropriate wall method for applicable preparations by other trades and backer board installers.
- Maximum allowable variation in the tile substrate—for tiles with all edges shorter than 15", maximum allowable variation is ¼" in 10' from the required plane, with no more than 1⁄16" variation in 12" when measured from the high points in the surface. For tiles with at least one edge 15" in length, maximum allowable variation is ⅛" in 10' from the required plane, with no more than 1⁄16" variation in 24" when measured from the high points in the surface.
- Center backer board end or edge joints on framing and stagger joints in adjacent rows so four corners do not come together within the same plane. Space panel ends and edges in accordance with manufacturer's recommendations.

Movement Joint (architect must specify type of joint and show location and details on drawings)

- Movement joints—mandatory according to EJ171.
- When glass tile is used, adhere to more frequent placement recommendations within the ranges listed in EJ171.

Installation Specifications

- Waterproof membrane—ANSI A108.13.
- Tile—ANSI A108.5.
- Glass tile—manufacturer's directions.
- Cementitious grout—ANSI A108.10.
- Epoxy grout—ANSI A108.6.
- Movement Joints—EJ171 and ASTM C1193.

Notes

- Test shower pan membrane/waterproof membrane and drainage fitting for leaks before commencing tilework.
- Materials adversely affected by moisture in areas immediately adjacent to showers, tubs, and roman tubs should be properly protected.
- All horizontal surfaces, for example shower seats, sills, curbs, etc., must slope towards drain or other surface sloped toward drain. Where present, waterproofing also must be sloped.
- When glass tile is used, see "Glass Tile Selection and Installation Guide, " and consult manufacturer for recommendations and requirements.
- Follow applicable plumbing and building codes.

B422C-18

- **Solid Backing (wall)**
- **Concrete Substrate (floor)**
- **Bonded Waterproof Membrane**
- **Ceramic Tile, Glass Tile**

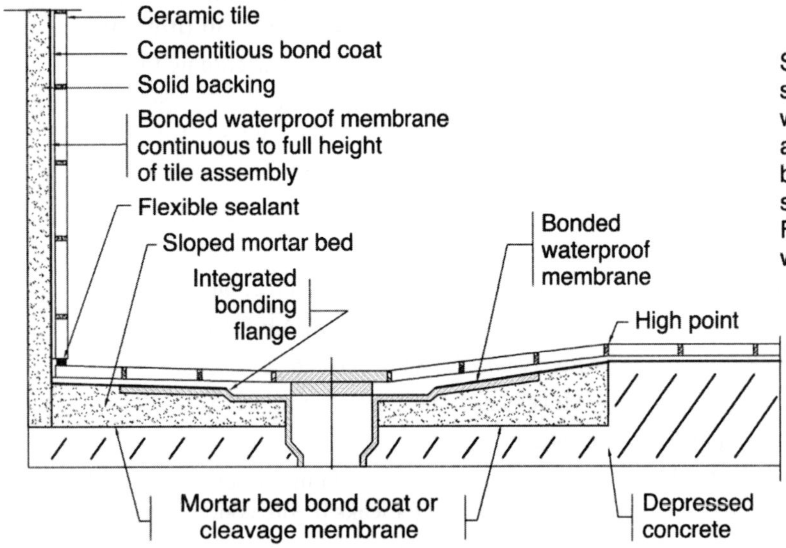

Shower receptors, curbs, seats, etc., must be properly waterproofed and installed to avoid water damage to adjacent building materials. See common shower configurations section. For mortar beds over 65 s.f., wire reinforcing is required.

Recommended Uses

- For construction of a shower without a curb.

Environmental Exposure Classifications

- Res1, 2, 3; Com1, 2
- May be suitable for Com3, Res5, and Com5 as determined by membrane and backing material manufacturers.
- For Res4 and Com4, see SR613 and SR614.
- For installations that may be exposed to staining, specify tile and grout suitable for exposure. Consult product manufacturers; see also "Product Selection Guides. "
- For installations that may be exposed to mild chemical attack, specify epoxy grout and tile suitable for exposure. Consult product manufacturers; see also "Product Selection Guides."

Limitations

- Facilitates construction of a curbless shower without adding floor height at restroom entryway, but does not inherently comply with Americans with Disabilities Act (ADA) Standards for Accessible Design. Follow ADA where ADA compliance is required or desired.
- Refer to appropriate wall method for applicable limitations based on type of backing used—W201, W202E, W202I, W211, W221, W222, W231/W241, W243, W244C, W244E, W244F, W245, W246, and W247.

Requirements

- When glass tile is used, consult glass tile manufacturer for tile suitability over non-absorptive surface.
- Bonded waterproof membrane must be continuous, including at changes in plane. Follow membrane manufacturer's requirements for corners, seaming, and overlap.
- Waterproof membrane inside shower area must extend to full height of tile assembly. Also, the floor and wall waterproofing must continue outside the immediate shower area one foot beyond the high point of the floor, but not beyond the tiled area. Additional waterproofing of floor and/or walls outside the shower area may be needed to effectively contain and evacuate shower water and splash water and to protect building materials. When additional waterproofing of floor and/or walls outside the shower area is desired or required, the building design professional must specify all areas to be waterproofed and indicate membrane termination points.
- A secondary drain may be required outside the immediate shower area to facilitate evacuation of shower water and splash water that is not contained in the shower area due to curbless design. When required or desired, building design professional must specify type and location of secondary drain.
- High point of the floor must be outside the shower area, i.e., beyond shower door or curtain, to facilitate evacuation of

shower water and splash water that is not contained in the shower area due to curbless design. Location of the high point of the floor is especially critical when secondary drain is not included outside the shower area.

- Refer to appropriate wall method for applicable requirements based on type of backing used.
- Slope mortar bed ¼" per foot toward drain and follow membrane and integrated bonding flange manufacturer's instructions for connecting membrane to integrated bonding flange.

Materials

- **Multiple options exist for membranes, mortars, grouts, and other materials and *must be clearly specified* to be included. If not specifically indicated, optional materials are not included and mortar/grout choice defaults to minimum performance specification indicated. Consider each system component and intended use to determine minimum requirements and to specify options.**
- Ceramic tile—ANSI A137.1.
- Glass tile, when used—ANSI A137.2; see also "Glass Tile Selection and Installation Guide," and consult tile manufacturer for environmental exposure classification recommendations. Not all glass tiles are suitable.
- Cementitious grout—ANSI A118.6 or better or ISO CG1 or better. When glass tile is used, specify grout designated by tile and grout manufacturers.
- Epoxy grout, when used—ANSI A118.3 or ISO RG.
- Cementitious bond coat:
  - ANSI A118.4 or better or ISO C2S1 or better unless ANSI A118.1 or ISO C1 is recommended by membrane manufacturer.
  - When glass tile is used, specify mortar designated by tile and mortar manufacturers. Bond coat color will impact the final appearance of translucent glass tile. Specifier shall confirm bond coat color is acceptable.
- Waterproof membrane—ANSI A118.10 and recommended by manufacturer of membrane for use over backing type in intended application.
- Flexible mildew-resistant sealant—ASTM C920.

Materials for Green/Sustainable Design

- See "Green Building Standards and Green Product Selection Guide" and consult manufacturers and suppliers for product sustainability and contribution to green building design.
- Consider specifying tile and installation materials that meet ANSI A138.1, the *American National Standard Specifications for Sustainable Ceramic Tiles, Glass Tiles, and Tile Installation Materials.*

Preparation by Other Trades/Backer Board Installers

- Depressed slab by others. Depressed area must extend beyond shower area to facilitate a finished floor installation that effectively evacuates shower water and splash water, which are often difficult to contain within a curbless shower.
- Refer to appropriate wall method for applicable preparations by other trades and backer board installers.
- Maximum allowable variation in the tile substrate—for tiles with all edges shorter than 15", maximum allowable variation is ¼" in 10' from the required plane, with no more than ¹⁄₁₆" variation in 12" when measured from the high points in the surface. For tiles with at least one edge 15" in length, maximum allowable variation is ⅛" in 10' from the required plane, with no more than ¹⁄₁₆" variation in 24" when measured from the high points in the surface.
- Center backer board end or edge joints on framing and stagger joints in adjacent rows so four corners do not come together within the same plane. Space panel ends and edges in accordance with manufacturer's recommendations.

Movement Joint (architect must specify type of joint and show location and details on drawings)

- Movement joints—mandatory according to EJ171.
- When glass tile is used, adhere to more frequent placement recommendations within the ranges listed in EJ171.

Installation Specifications

- Waterproof membrane—ANSI A108.13.
- Tile—ANSI A108.5.
- Glass tile—manufacturer's directions.
- Cementitious grout—ANSI A108.10.
- Epoxy grout—ANSI A108.6.
- Movement Joints—EJ171 and ASTM C1193.

Notes

- Test shower pan membrane/waterproof membrane and drainage fitting for leaks before commencing tilework.
- Materials adversely affected by moisture in areas immediately adjacent to showers, tubs, and roman tubs should be properly protected.
- All horizontal surfaces, for example shower seats, sills, curbs, etc., must slope towards drain or other surface sloped toward drain. Where present, waterproofing also must be sloped.
- When glass tile is used, see "Glass Tile Selection and Installation Guide, " and consult manufacturer for recommendations and requirements.
- Follow applicable plumbing and building codes.

Mortar Curb*

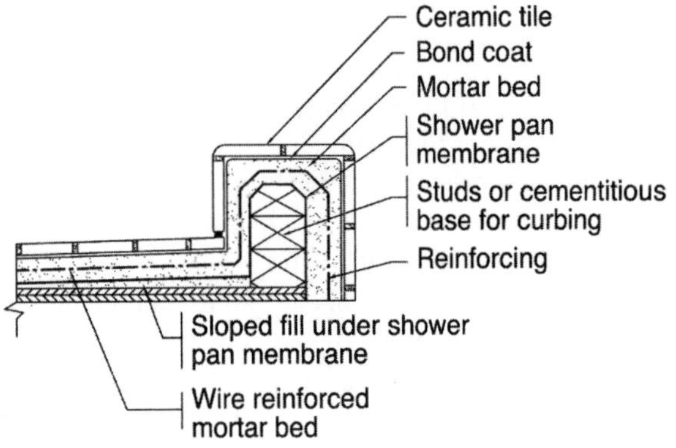

Preformed Curb*

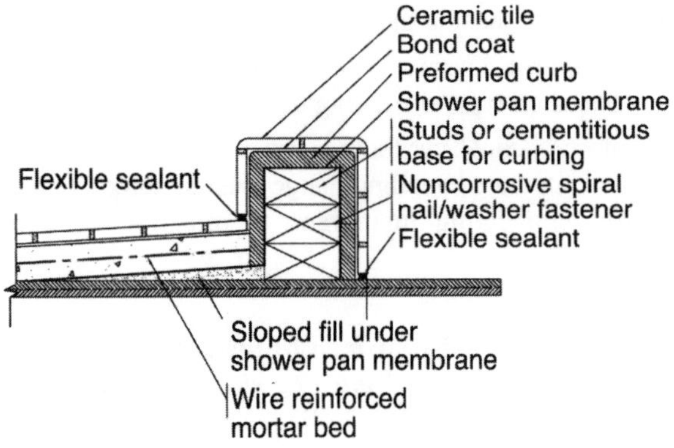

*Note: Construct curb such that membrane on top of curb (shower pan membrane or bonded waterproof membrane) is sloped toward drain.

Solid Curb*

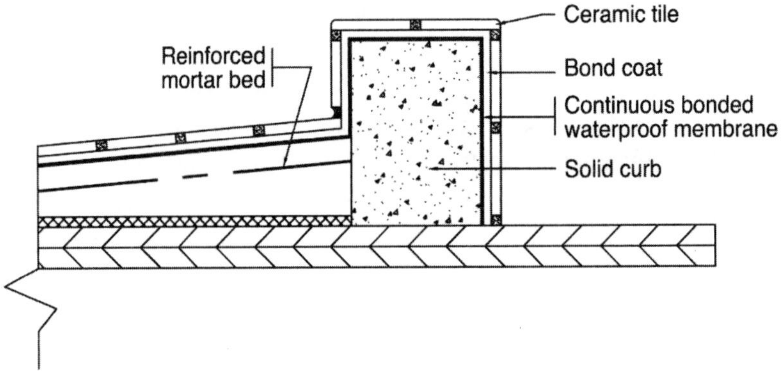

Alternate Receptor Base

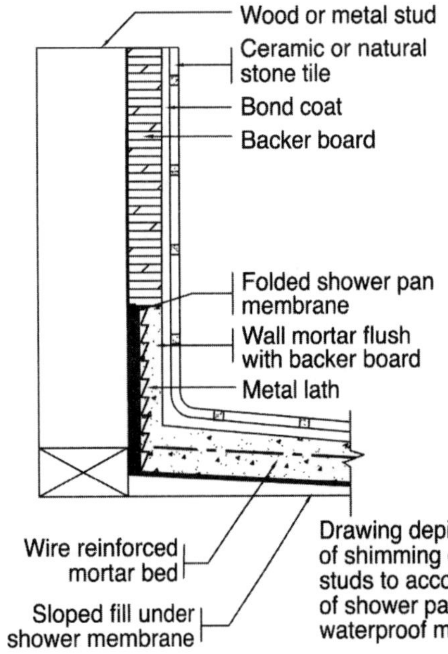

Drawing depicts use of wall mortar in lieu of shimming out backer board or notching studs to accommodate thickness of folds of shower pan membrane. Vapor retarder or waterproof membrane as required (now shown).

Shower Pan Membrane at Curb/Jamb

1. Construct curb such that waterproof membrane on top side of curb is sloped toward the drain
2. Fasteners - min. 3" above curb
3. Shower pan membrane is sloped min. 1/4" per foot to drain, folded up walls min. 3" above curb, and extends to floor outside of shower.

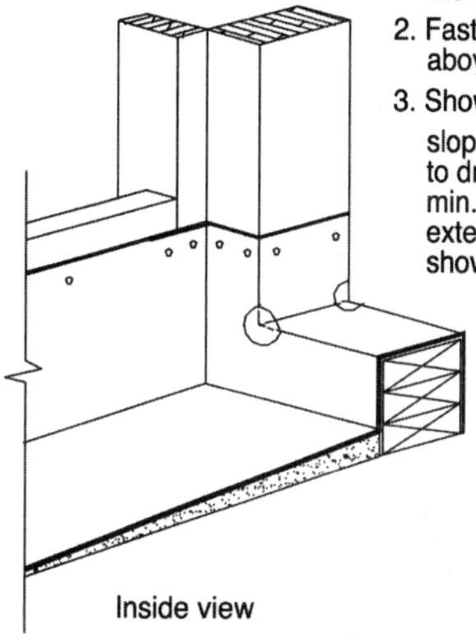

Inside view

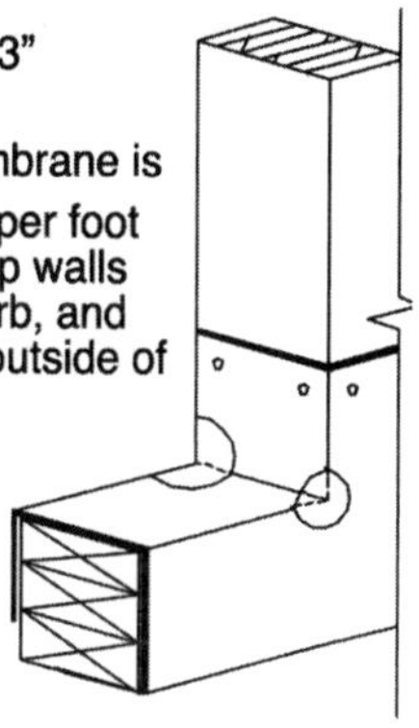

Outside view

Backer Board Installation Over Shower Pan Membrane

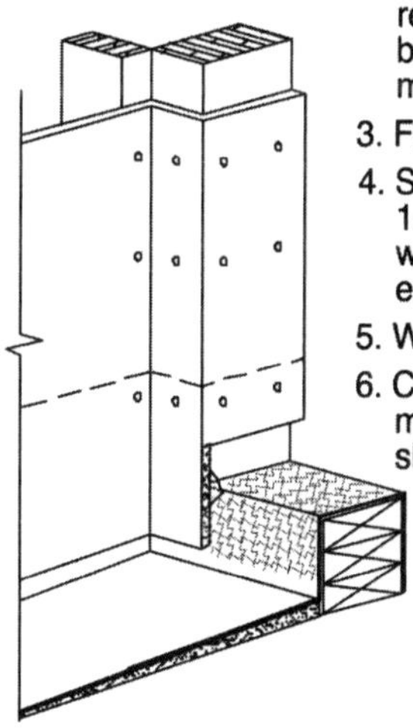

1. Space, tape, seal, etc., backer board per manufacturer's requirements, including for bottom edge of board.
2. Because of varying requirements, vapor retarder behind backer board not shown. Include and install vapor retarder when and as required by building code and backer board manufacturer.
3. Fasteners - Min. 3" above curb
4. Shower pan membrane is sloped 1/4" per foot to drain, folded up walls min. 3" above curb, and extends to floor outside of shower.
5. Wire mesh for curb - Per ANSI
6. Construct curb such that waterproof membrane on top side of curb is sloped toward the drain.

Typical Shower Seat for Backer Board Showers

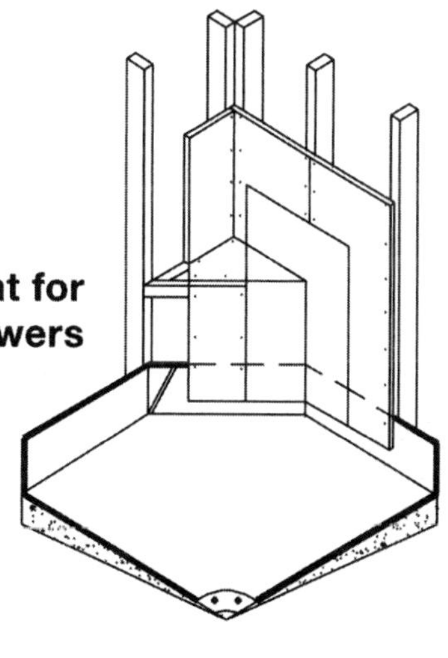

1. Construct shower seat such that bonded waterproof membrane on backer board on top side of shower seat is sloped min. 1/4" per foot to drain, and extend bonded waterproof membrane beyond seat/wall intersection, and below top edge of shower pan membrane.
2. Shower pan membrane is sloped min. 1/4" per foot to drain and folded up walls.
3. Space, tape, seal, etc., backer board per manufacturer's requirements, including for bottom edge of board.
4. Because of varying requirements, vapor retarder behind backer board now shown. Include and install vapor retarder when and as required by building code and backer board manufacturer.

SR614-18

- **Wood or Metal Studs**
- **Mortar Bed or Cement Backer Board Walls**
- **Mortar Bed Floor**
- **Ceramic Tile**

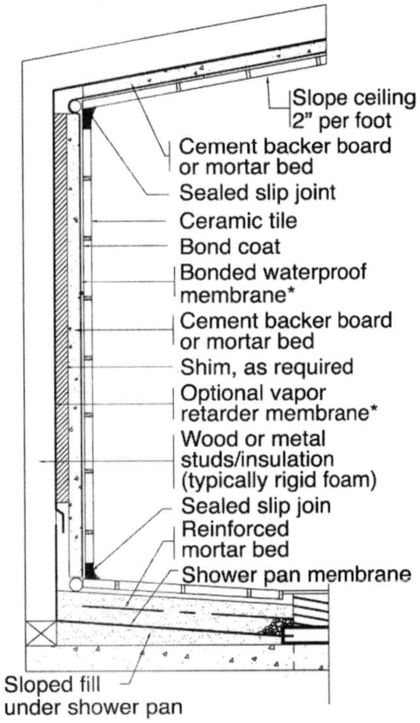

Shower receptors, curbs, seats, etc., must be properly detailed and installed to avoid water damage to adjacent building materials. See common shower configurations section.

* For continuous use steam showers: If bonded waterproof membrane has a water vapor permeance greater than 0.5 perms, a secondary vapor retarder membrane is required. See requirements. When used, vapor retarder membrane must overlap into shower membrane.

Configuration of base with integrated bonding flange

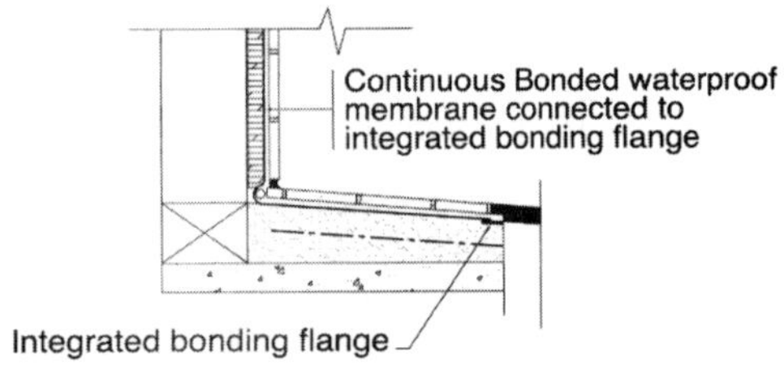

Recommended Uses

- For steam showers and steam rooms framed with wood or metal studs. Specify mortar bed walls option when wall flatness is critical, such as when tiles with any edge longer than 15" are specified.

Environmental Exposure Classifications

- Res4.
- For installations that may be exposed to staining, specify tile and grout suitable for exposure. Consult product manufacturers; see also Notes and "Product Selection Guides."

Limitations

- Duration of use as a steam shower or steam room as determined by membrane and cement backer board (when used) manufacturers; lower membrane water vapor permeance decreases water vapor transmission.
- Maximum stud spacing 16" on center.

Requirements

- Steam rooms are highly specialized applications. Design and installation are critical to avoid damage to adjoining materials from vapor migration. Design criteria must include consideration of necessary insulation and temperature and humidity differential.
- Use a tile contractor knowledgeable in steam applications and experienced with the materials specified.
- Bonded waterproof membrane (sheet, liquid, and trowel-on) must be continuous and must adequately limit vapor transmission into adjacent spaces and building materials, according to intended duration of use as a steam shower.
- Steam showers designed for continuous use applications should specify a low perm waterproof membrane (a waterproof membrane meeting ANSI A118.10 and with a water vapor permeance rating of 0.5 perms or less when tested per ASTM E96 Procedure E, tested at 90% relative humidity). When a waterproof membrane with a water vapor permeance rating greater than 0.5 perms is specified, a vapor retarder behind the wall assembly is required, and vapor retarder must have a water vapor permeance rating of 0.1 perm or less when tested per ASTM E96 Procedure A, tested at 50% relative humidity. Consult waterproof membrane manufacturer for water vapor permeance rating and vapor retarder requirements.
- Specifier shall indicate how waterproofing and vapor retarding is to be achieved, including details for membrane penetrations such as penetrations for plumbing, lighting fixtures, fasteners, etc. Specifier shall also indicate where and how to waterproof curbs and jambs and where membrane terminates. Area outside steam shower door is a wet area and should be treated accordingly.
- Check with membrane manufacturer for suitability for applicable conditions, as not all membranes are suitable for steam, high temperature and/or chemical exposure.
- Some waterproof membrane manufacturers require use of a vapor retarder membrane in addition to the waterproof membrane. Consult membrane manufacturer for requirements. When used, vapor retarder membrane must weather lap itself and lap into the shower pan membrane.
- If vapor retarder membrane is required, integrated bonding flange cannot be used.
- Follow waterproof membrane manufacturers' directions for interface between drain and membrane(s).
- Design professional to specify adequate insulation on walls and ceilings to reduce condensation. Consult insulation manufacturer for application suitability.
- Seal all membrane penetrations with appropriate sealant according to membrane manufacturer's requirements.

- Specify slip joints at change in plane, such as where walls and ceiling meet.
- Slope ceiling minimum 2" per foot to avoid condensate dripping onto occupants; sloping ceiling from center can minimize rundown on walls.
- When mortar bed walls, studs must provide firm anchorage for metal lath.
- When mortar bed walls, mortar bed wall thickness—¾" minimum to 1½" maximum.
- When mortar bed walls, cut lath at all slip joints.
- Slope shower pan membrane minimum ¼" per foot to weep holes in drain.
- Turn shower pan membrane up walls a minimum of 3" above shower curb (6" above floor in showers without curbs).
- Surround drain with pea gravel or other weep protection to prevent mortar from blocking weep holes. (Does not apply if integrated bonding flange is used.)
- Wood studs, when used—dry and well braced, minimum depth 3½".
- Metals studs, when used—well-braced; 20 gauge (0.033") or heavier; minimum depth 3½" for residential applications or 3⅝" for commercial applications.

Preparation by Backer Board Installers

- Maximum allowable variation in the tile substrate—for tiles with all edges shorter than 15", maximum allowable variation is ¼" in 10' from the required plane, with no more than ¹⁄₁₆" variation in 12" when measured from the high points in the surface. For tiles with at least one edge 15" in length, maximum allowable variation is ¼" in 10' from the required plane, with no more than ¹⁄₁₆" variation in 24" when measured from the high points in the surface.
- Horizontal joints—⅛" spacing filled solid and taped with latex-portland cement mortar and 2" alkali-resistant glass fiber mesh tape.
- Vertical joints—fill any space and tape with latex-portland cement mortar and 2" alkali-resistant glass fiber mesh tape.
- Corners—leave space between backer units. Tape joints using skim coat of latex-portland cement mortar, but do not fill.
- Center backer board end or edge joints on framing and stagger joints in adjacent rows so four corners do not come together within the same plane. Space panel ends and edges in accordance with manufacturer's recommendations.

Materials

- **Multiple options exist for membranes, mortars, grouts, and other materials and *must be clearly specified* to be included. If not specifically indicated, optional materials are not included and mortar/grout choice defaults to minimum performance specification indicated. Consider each system component and intended use to determine minimum requirements and to specify options.**
- Ceramic tile—ANSI A137.1 and recommended by manufacturer for use in steam showers.
- Cementitious grout—ANSI A118.6 or better or ISO CG1 or better.
- Epoxy grout, when used—ANSI A118.3 or ISO RG.
- Cementitious bond coat:
  - ANSI A118.15 or better or ISO C2S1 or better, unless ANSI A118.1 or ISO C1 is recommended by membrane manufacturer.
- Cement backer board, when used—ANSI A118.9 or ASTM C1325 (Type A).
- Fasteners—noncorrosive and nonoxidizing.
- Hot dipped fasteners meeting ASTM F2329-05 required in wet areas.
- 2" alkali-resistant glass fiber mesh tape.
- Waterproof membrane—ANSI A118.10 and recommended by membrane manufacturer for use in specific application. For continuous use applications, see Requirements for water vapor permeance.
- Vapor retarder membrane, when used—recommended by manufacturer for use in specific application.
- Metal studs, when used—ASTM C645.
- Shower pan membrane—ANSI A118.10, ASTM D4068, or D4551 and meeting applicable building code.
- Mortar bed and reinforcing—ANSI A108.1A.
- Flexible mildew-resistant sealant—ASTM C920 and recommended by manufacturer for use in steam showers.

Materials for Green/Sustainable Design

- See "Green Building Standards and Green Product Selection Guide" and consult manufacturers and suppliers for product sustainability and contribution to green building design.
- Consider specifying tile and installation materials that meet ANSI A138.1, the *American National Standard Specifications for Sustainable Ceramic Tiles, Glass Tiles, and Tile Installation Materials*.

Preparation by Other Trades

- Wall framing shall meet the general framing requirements of ANSI A108.11-4.0-4.3.

Movement Joint (architect must show type of joint and show location and details on drawings)

- Movement joints—mandatory according to EJ171.
- Slip joints at changes in plane, such as where walls and ceiling meet.

Installation Specifications

- Shower pan membrane—ANSI A108.01-3.6.
- Mortar bed and reinforcing—ANSI A108.1A, .1B, and .1C.
- Tile—ANSI A108.5.
- Cementitious grout—A108.10.
- Epoxy grout—A108.6.
- Waterproof membrane and slip joint—ANSI A108.13 and membrane manufacturer's slip joint directions.
- Movement Joints—EJ171 and ASTM C1193.

Notes

- Use of softened water in steam showers and steam rooms helps reduce grout and tile staining due to iron and/or hard water. Such stains may require harsh chemicals for removal. Select products suitable for water type and maintenance practices that will be used.
- Standard grouts will need to be periodically maintained over the life of the steam shower.
- Steam unit design must take into consideration the affect of Moisture Vapor Transmission (MVT) on opposite side of steam unit walls. MVT can cause efflorescence and can affect paints and other adhered finishes.

TR418-18

- **Mortar Bed Walls**
- **Mortar Bed Floor**

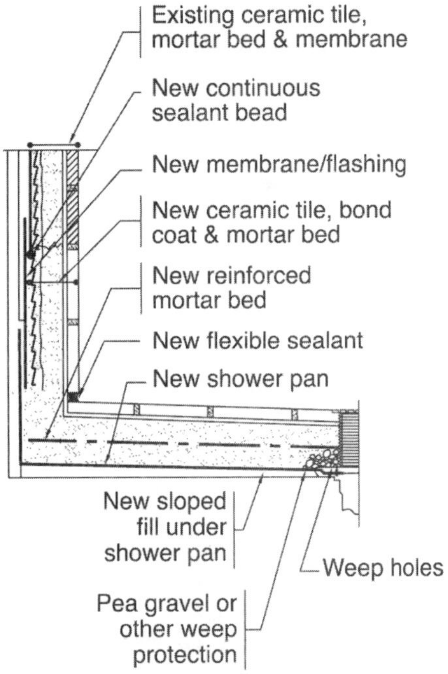

Recommended Uses

- For use where old shower pan has failed.

Requirements

- Remove existing shower receptor, shower pan, and wall tile, as required, to install new shower pan.
- Replace any damaged wall and floor substrate materials.
- New shower pan—slope ¼" per foot to weep holes in drain.
- New shower pan to turn up wall a minimum of 3" above curb (6" above floor in showers without curbs).
- New wall membrane/flashing (ANSI A108.02-3.8) placed behind existing wall membrane, out over new shower pan, and fastened to studs.
- Place continuous bead of sealant on existing wall mortar with new wall mortar brought up tight against it.

Materials for Green/Sustainable Design

- See "Green Building Standards and Green Product Selection Guide" and consult manufacturers and suppliers for product sustainability and contribution to green building design.
- Consider specifying tile and installation materials that meet ANSI A138.1, the *American National Standard Specifications for Sustainable Ceramic Tiles, Glass Tiles, and Tile Installation Materials*.

Notes

- See B414 for complete specifications.

TR420-18

- **Backer Board Walls**
- **Mortar Bed Floor**

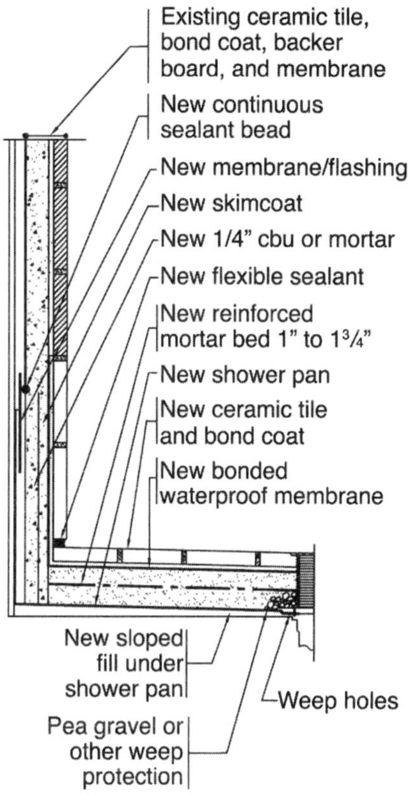

Recommended Uses

- For use where old shower pan has failed.

Requirements

- Remove existing shower receptor, shower pan, and wall tile, as required, to install new shower pan.
- Replace any damaged wall and floor substrate materials.
- New shower pan—slope ¼" per foot to weep holes in drain.
- New shower pan to turn up wall a minimum of 3" above curb (6" above floor in showers without curbs).
- New wall membrane/flashing (ANSI A108.02-3.8) placed behind existing wall membrane, out over new shower pan, and fastened to studs.
- Place continuous bead of sealant on existing wall substrate material with new wall substrate material brought up tight against it.

Materials for Green/Sustainable Design

- See "Green Building Standards and Green Product Selection Guide" and consult manufacturers and suppliers for product sustainability and contribution to green building design.
- Consider specifying tile and installation materials that meet ANSI A138.1, the *American National Standard Specifications for Sustainable Ceramic Tiles, Glass Tiles, and Tile Installation Materials*

Notes

- See B415, B420, B426, or B431 for complete specifications based on wall backing type.

APPENDIX J

COMBINATION OF INDOOR AND OUTDOOR COMBUSTION AND VENTILATION OPENING DESIGN

(The content of this Appendix is based on Annex I of NFPA 54)

J 101.0 General.

J 101.1 Applicability. This appendix provides general guidelines for the sizing of combination indoor and outdoor combustion and ventilation air openings.

J 101.2 Example of Combination Indoor and Outdoor Combustion Air Opening. Determine the required combination of indoor and outdoor combustion air opening sizes for the following appliance installation example.

Example Installation: A fan-assisted furnace and a draft-hood-equipped water heater with the following inputs are located in a 15 foot by 30 foot (4572 mm by 9144 mm) basement with an 8 foot (2438 mm) ceiling. No additional indoor spaces can be used to help meet the appliance combustion air needs.

Fan-Assisted Furnace Input: 100 000 British thermal units per hour (Btu/h) (29 kW)

Draft Hood-Equipped Water Heater Input: 40 000 Btu/h (11.7 kW)

Solution:

(1) Determine the total available room volume. Appliance room volume:

15 feet by 30 feet (4572 mm by 9144 mm) with an 8 foot (2438 mm) ceiling = 3600 cubic feet (101.94 m^3)

(2) Determine the total required volume. The standard method to determine combustion air is used to calculate the required volume. The combined input for the appliances located in the basement is calculated as follows:

100 000 Btu/h (29 kW) + 40 000 Btu/h (11.7 kW) = 140 000 Btu/h (41 kW)

The standard method requires that the required volume be determined based on 50 cubic feet per 1000 Btu/h (4.83 m^3/kW). Using Table J 101.2, the required volume for a 140 000 Btu/h (41 kW) water heater is 7000 cubic feet (198.22 m^3).

Conclusion: The indoor volume is insufficient to supply combustion air since the total of 3600 cubic feet (101.94 m^3) does not meet the required volume of 7000 cubic feet (198.22 m^3). Therefore, additional combustion air shall be provided from the outdoors.

(3) Determine ratio of the available volume to the required volume:

$$\frac{3600 \text{ cubic feet}}{7000 \text{ cubic feet}} = 0.51$$

(4) Determine the reduction factor to be used to reduce the full outdoor air opening size to the minimum required based on the ratio of indoor spaces:

1.00 – 0.51 (from Step 3) = 0.49

(5) Determine the single outdoor combustion air opening size as though all combustion air is to come from outdoors. In this example, the combustion air opening directly communicates with the outdoors:

$$\frac{140\,000 \text{ Btu/h}}{3000 \text{ British thermal units per square inch (Btu/in}^2\text{)}} = 47 \text{ square inches } (0.03 \text{ m}^2)$$

(6) Determine the minimum outdoor combustion air opening area:

Outdoor opening area = 0.49 (from Step 4) x 47 square inches (0.03 m^2) = 23 square inches (0.01 m^2)

Section 506.5.3(3) requires the minimum dimension of the air opening should not be less than 3 inches (76 mm).

TABLE J 101.2
STANDARD METHOD: REQUIRED VOLUME, ALL APPLIANCES*
[NFPA 54: TABLE A.9.3.2.1]

APPLIANCE INPUT (Btu/h)	REQUIRED VOLUME (cubic feet)
5000	250
10 000	500
15 000	750
20 000	1000
25 000	1250
30 000	1500
35 000	1750
40 000	2000
45 000	2250
50 000	2500
55 000	2750
60 000	3000
65 000	3250
70 000	3500
75 000	3750
80 000	4000
85 000	4250
90 000	4500
95 000	4750
100 000	5000
105 000	5250
110 000	5500
115 000	5750
120 000	6000
125 000	6250
130 000	6500
135 000	6750
140 000	7000
145 000	7250
150 000	7500
160 000	8000
170 000	8500
180 000	9000
190 000	9500
200 000	10 000
210 000	10 500
220 000	11 000
230 000	11 500
240 000	12 000
250 000	12 500
260 000	13 000
270 000	13 500
280 000	14 000
290 000	14 500
300 000	15 000

For SI units: 1000 British thermal units per hour = 0.293 kW, 1 cubic foot = 0.0283 m^3

APPENDIX K

POTABLE RAINWATER CATCHMENT SYSTEMS

K 101.0 General.

K 101.1 Applicability. The provisions of this appendix shall apply to the installation, construction, alteration, and repair of potable rainwater catchment systems.

K 101.2 System Design. Potable rainwater catchment systems in accordance with this appendix shall be designed by a registered design professional or person deemed competent by the Authority Having Jurisdiction to perform potable rainwater catchment system design work.

K 101.3 Permit. It shall be unlawful for a person to construct, install, or alter, or cause to be constructed, installed, or altered a potable rainwater catchment systems in a building or on a premise without first obtaining a permit to do such work from the Authority Having Jurisdiction.

K 101.3.1 Plumbing Plan Submission. No permit for a rainwater catchment system requiring a permit shall be issued until complete plumbing plans, with data satisfactory to the Authority Having Jurisdiction, have been submitted and approved. No changes or connections shall be made to either the rainfall catchment or the potable water system within a site containing a rainwater catchment water system without approval by the Authority Having Jurisdiction.

K 101.3.2 System Changes. No changes or connections shall be made to either the rainwater catchment system or the potable water system within a site containing a rainwater catchment system requiring a permit without approval by the Authority Having Jurisdiction.

K 101.4 Product and Material Approval. System components shall be properly identified as to the manufacturer.

K 101.4.1 Plumbing Materials and Systems. Pipe, pipe fittings, traps, fixtures, material, and devices used in a potable rainwater system shall be listed or labeled (third-party certified) by a listing agency (accredited conformity assessment body) and shall be in accordance with approved applicable recognized standards referenced within this code, and shall be free from defects. Unless otherwise provided for in this appendix, materials, fixtures, or devices used or entering into the construction of plumbing systems, or parts thereof shall be submitted to the Authority Having Jurisdiction for approval.

K 101.5 Maintenance and Inspection. Potable rainwater catchment systems and components shall be inspected and maintained in accordance with Section K 101.5.1 through Section K 101.5.3.

K 101.5.1 Frequency. Potable rainwater catchment systems and components shall be inspected and maintained in accordance with Table K 101.5.1 unless more frequent inspection and maintenance are required by the manufacturer.

K 101.5.2 Maintenance Log. A maintenance log for potable rainwater catchment systems shall be maintained by the property owner and be available for inspection. The property owner or designated appointee shall ensure that a record of testing, inspection, and maintenance in accordance with Table K 101.5.1 is maintained in the log. The log will indicate the frequency of inspection, and maintenance of each system. A record of the required water quality tests shall be retained for not less than 2 years.

K 101.5.3 Maintenance Responsibility. The required maintenance and inspection of potable rainwater catchment systems shall be the responsibility of the property owner unless otherwise required by the Authority Having Jurisdiction.

K 101.6 Operation and Maintenance Manual. An operation and maintenance manual for potable rainwater catchment systems shall be supplied to the building owner by the system designer. The operating and maintenance manual shall include the following:

(1) Detailed diagram of the entire system and the location of system components.

(2) Instructions for operating and maintaining the system.

(3) Details on maintaining the required water quality as determined by the Authority Having Jurisdiction.

(4) Details on deactivating the system for maintenance, repair, or other purposes.

(5) Applicable testing, inspection, and maintenance frequencies in accordance with Table K 101.5.1.

(6) A method of contacting the manufacturer(s).

K 101.7 Minimum Water Quality Requirements. The minimum water quality for potable rainwater catchment systems shall comply with the applicable water quality requirements as determined by the Authority Having Jurisdiction. In the absence of water quality requirements, the guidelines EPA/625/R-04/108 contains recommended water reuse guidelines to assist regulatory agencies develop, revise, or expand alternate water source water quality standards.

K 101.8 Material Compatibility. In addition to the requirements of this appendix, potable rainwater catchment systems shall be constructed of materials that are compatible with the type of pipe and fitting materials and water conditions in the system.

K 101.9 System Controls. Controls for pumps, valves, and other devices that contain mercury that come in contact with the water supply shall not be permitted.

K 102.0 Connection.

K 102.1 General. No water piping supplied by a potable rainwater catchment system shall be connected to a source of supply without the approval of the Authority Having Jurisdiction, Health Department, or other department having jurisdiction.

TABLE K 101.5.1
MINIMUM POTABLE RAINWATER CATCHMENT SYSTEM TESTING, INSPECTION, AND MAINTENANCE FREQUENCY

DESCRIPTION	MINIMUM FREQUENCY
Inspect and clean filters and screens, and replace (where necessary).	Every 3 months
Inspect and verify that disinfection, filters and water quality treatment devices and systems are operational. Perform water quality tests in accordance with the Authority Having Jurisdiction.	In accordance with the manufacturer's instructions, and the Authority Having Jurisdiction.
Perform applicable water quality tests to verify compliance with Section K 104.3.	Every 3 months
Perform a water quality test for E. Coli, Total Coliform, and Heterotrophic bacteria. For a system where 25 different people consume water from the system over a 60 day period, a water quality test for cryptosporidium shall be performed.	After initial installation and every 12 months thereafter, or as directed by the Authority Having Jurisdiction.
Inspect and clear debris from rainwater gutters, downspouts, and roof washers.	Every 6 months
Inspect and clear debris from the roof or other aboveground rainwater collection surface.	Every 6 months
Remove tree branches and vegetation overhanging roof or other aboveground rainwater collection surface.	As needed
Inspect pumps and verify operation.	After initial installation and every 12 months thereafter
Inspect valves and verify operation.	After initial installation and every 12 months thereafter
Inspect pressure tanks and verify operation.	After initial installation and every 12 months thereafter
Clear debris and inspect storage tanks, locking devices, and verify operation.	After initial installation and every 12 months thereafter
Inspect caution labels and marking.	After initial installation and every 12 months thereafter

K 102.2 Connections to Public or Private Potable Water Systems. Potable rainwater catchment systems shall have no direct connection to a public or private potable water supply or alternate water source system. Potable water from a public or private potable water system is permitted to be used as makeup water to the rainwater storage tank provided the public, or private potable water supply connection is protected by an air gap or reduced-pressure principle backflow preventer in accordance with this code.

K 102.3 Backflow Prevention. The potable rainwater catchment system shall be protected against backflow in accordance with this code.

K 103.0 Potable Rainfall Catchment System Materials.

K 103.1 Collections Surfaces. The collection surface for potable applications shall be constructed of a hard, impervious material and shall be approved for potable water use. Roof coatings, paints, and liners shall comply with NSF Protocol P151.

K 103.1.1 Prohibited. Roof paints and coatings with lead, chromium, or zinc shall not be permitted. Wood roofing material and lead flashing shall not be permitted.

K 103.2 Rainwater Catchment System Drainage Materials. Materials used in rainwater catchment drainage systems, including gutters, downspouts, conductors, and leaders shall be in accordance with the requirements of this code for storm drainage.

K 103.3 Storage Tanks. Rainwater storage shall comply with Section K 105.0.

K 103.4 Water Supply and Distribution Materials. Potable rainwater supply and distribution materials shall comply with the requirements of this code for potable water supply and distribution systems.

K 104.0 Design and Installation.

K 104.1 Collection Surfaces. Rainwater shall be collected from a roof or other cleanable aboveground surfaces specifically designed for rainwater catchment. A rainwater catchment system shall not collect rainwater from:

(1) Vehicular parking surfaces

(2) Surface water runoff

(3) Bodies of standing water

K 104.2 Prohibited Discharges. Overflows, condensate, and bleed-off pipes from roof-mounted equipment and appliances shall not discharge onto roof surfaces that are intended to collect rainwater.

K 104.3 Minimum Water Quality. Upon initial system startup, the quality of the water for the intended application shall be verified at the point(s) of use as determined by the Authority Having Jurisdiction in accordance with Section K 104.3.1 and Section K 104.3.2. Water quality maintenance shall be in accordance with Section K 104.3.3.

K 104.3.1 Private Potable Water System. In the absence of water quality requirements determined by the Authority Having Jurisdiction, the minimum water quality for a private potable water system at the point of use shall comply with Table K 104.3.1.

TABLE K 104.3.1
MINIMUM WATER QUALITY

Escherichia coli (fecal coliform)	Non-detectable
Protozoan Cysts	Non-detectable
Viruses	Non-detectable
Turbidity	<0.3 NTU

K 104.3.2 Public Use Occupancies. The minimum water quality for a potable water system for public use occupancies at the point of use and testing procedures shall comply with the Environmental Protection Agency (EPA) Safe Drinking Water Act for a public water system.

K 104.3.3 Maintenance. Normal system maintenance shall require system testing for Escherichia coli (fecal coliform) and turbidity every 3 months in accordance with Table K 104.3.3. Upon failure of the fecal coliform test, the system shall be re-commissioned involving cleaning, and retesting in accordance with Section K 104.3. Testing for viruses and cysts shall occur once after 3 months of initial operation and once every 12 months thereafter.

Exception: Upon failure of the virus or cyst test, the tests will be repeated every 3 months until the tests results are negative for two consecutive tests.

TABLE K 104.3.3
MINIMUM SYSTEM MAINTENANCE REQUIREMENTS

Escherichia coli (fecal coliform)	Non-detectable
Turbidity	<0.3 NTU

K 104.4 Water Quality Devices and Equipment. Devices and equipment used to treat rainwater to maintain the minimum water quality requirements determined by the Authority Having Jurisdiction shall be listed or labeled (third-party certified) by a listing agency (accredited conformity assessment body) and approved for the intended application.

K 104.4.1 Filtration Devices. Potable water filters shall comply with NSF 53 and shall be installed in accordance with the manufacturer's installation instructions.

K 104.4.2 Disinfection Devices. Chlorination, ozone, ultraviolet, or other disinfection methods approved by the Authority Having Jurisdiction, or the product is listed and certified according to a microbiological reduction performance standard for drinking water, shall be used to treat harvested rainwater to meet the required water quality permitted. The disinfection devices and systems shall be installed in accordance with the manufacturer's installation instructions and the conditions of listing. Disinfection devices and systems shall be located downstream of the storage tank.

K 104.4.3 Filtration and Disinfection Systems. Filtration and disinfection systems shall be located after the water storage tank. Where a chlorination system is installed, it shall be installed upstream of filtration systems. Where an ultraviolet disinfection system is installed, a filter not more than 5 microns (5 μm) shall be installed upstream of the disinfection system.

K 104.5 Overhanging Tree Branches and Vegetation. Tree branches and vegetation shall not be located over the roof or other aboveground rainwater collection surface. Where existing tree branch and vegetation growth extends over the rainwater collection surface, it shall be removed in accordance with Section K 101.5.

K 105.0 Rainwater Storage Tanks.

K 105.1 General. Rainwater storage tanks shall be installed in accordance with Section K 105.2 through Section K 105.10.

K 105.2 Construction. Rainwater storage tanks shall be constructed of solid, durable materials not subject to excessive corrosion or decay and shall be watertight. Storage tanks shall be approved by the Authority Having Jurisdiction for potable water applications, provided such tanks are in accordance with approved applicable standards.

K 105.3 Location. Rainwater storage tanks shall be permitted to be installed above or below grade.

K 105.3.1 Above Grade. Above grade, storage tanks shall be of an opaque material, approved for aboveground use in direct sunlight, or shall be shielded from direct sunlight. Tanks shall be installed in an accessible location to allow for inspection and cleaning. The tank shall be installed on a foundation or platform that is constructed to accommodate loads in accordance with the building code.

K 105.3.2 Below Grade. Rainwater storage tanks installed below grade shall be structurally designed to withstand anticipated earth or other loads. Holding tank covers shall be capable of supporting an earth load of not less than 300 pounds per square foot (lb/ft^2) (1465 kg/m^2) where the tank is designed for underground installation. Below grade rainwater tanks installed underground shall be provided with manholes. The manhole opening shall be not less than 20 inches (508 mm) in diameter and located not less than 4 inches (102 mm) above the surrounding grade. The surrounding grade shall be sloped away from the manhole. Underground tanks shall be ballasted, anchored, or otherwise secured, to prevent the tank from floating out of the ground where empty. The combined weight of the tank and hold down system shall meet or exceed the buoyancy force of the tank.

K 105.4 Drainage and Overflow. Rainwater storage tanks shall be provided with a means of draining and cleaning. The overflow drain shall not be equipped with a shutoff valve. The overflow outlet shall discharge in accordance with this code for storm drainage systems. Where discharging to the storm drainage system, the overflow drain shall be protected from backflow of the storm drainage system by a backwater valve or other approved method.

K 105.4.1 Overflow Outlet Size. The overflow outlet shall be sized to accommodate the flow of the rainwater entering the tank and not less than the aggregate cross-sectional area of the inflow pipes.

K 105.5 Animals and Insects. Rainwater tank openings to the atmosphere shall be protected to prevent the entrance of insects, birds, or rodents into the tank.

K 105.6 Human Access. Rainwater tank access openings exceeding 12 inches (305 mm) in diameter shall be secured to prevent tampering and unintended entry by either a lockable device or other approved method.

K 105.7 Exposure to Sunlight. Rainwater tank openings shall not be exposed to direct sunlight.

K 105.8 Inlets. A device or arrangement of fittings shall be installed at the inlet of the tank to prevent rainwater from disturbing sediment as it enters the tank.

K 105.9 Primary Tank Outlets. The primary tank outlet shall be located not less than 4 inches (102 mm) above the bottom of the tank, or shall be provided with a floating inlet to draw water from the cistern just below the water surface.

K 105.10 Storage Tank Venting. Where venting using drainage or overflow piping is not provided or is considered insufficient, a vent shall be installed on each tank. The vent shall extend from the top of the tank and terminate not less than 6 inches (152 mm) above grade and shall be not less than 1½ inches (40 mm) in diameter. The vent terminal shall be directed downward and covered with a $\frac{3}{32}$ of an inch (2.4 mm) mesh screen to prevent the entry of vermin and insect.

K 105.11 Pumps. Pumps serving rainwater catchment systems shall be listed for potable water use. Pumps supplying water to water closets, urinals, and trap primers shall be capable of delivering not less than the minimum residual pressure required by the highest and most remote outlet served. Where the water pressure in the rainwater supply system within the building exceeds 80 psi (552 kPa), a pressure reducing valve reducing the pressure to 80 psi (552 kPa) or less to water outlets in the building shall be installed in accordance with this code.

K 105.12 Roof Drains. Primary and secondary roof drains, conductors, leaders, overflows, and gutters shall be designed and installed in accordance with this code.

K 106.0 Freeze Protection.

K 106.1 General. Tanks and piping installed in locations subject to freezing shall be provided with an approved means of freeze protection.

K 106.2 Roof Washer or Pre-Filtration System. Collected rainwater shall pass through a roof washer or pre-filtration system before the water enters the rainwater storage tank. Roof washer systems shall comply with Section K 106.2.1 through Section K 106.2.4.

K 106.2.1 Size. The roof washer shall be sized to direct rainwater containing debris that has accumulated on the collection surface away from the storage tank. ARCSA/ASPE 63 contains additional guidance on acceptable methods of sizing roof washers.

K 106.2.2 Debris Screen. The inlet to the roof washer shall be provided with a debris screen or other approved means that protects the roof washer from the intrusion of debris and vermin. Where the debris screen is installed, the debris screen shall be corrosion resistant and shall have openings not larger than ½ of an inch (12.7 mm).

K 106.2.3 Drain Discharge. Water drained from the roof washer, or pre-filter shall be diverted away from the storage tank and discharged to a disposal area that does not cause property damage or erosion. Roof washer drainage shall not drain over a public way.

K 106.2.4 Automatic Drain. Roof washing systems shall be provided with an automatic means of self-draining between rain events.

K 106.3 Roof Gutters. Gutters shall maintain a minimum slope and be sized in accordance with this code.

K 106.4 Drains, Conductors, and Leaders. The design and size of rainwater drains, conductors, and leaders shall comply with this code.

K 106.5 Size of Potable Water Piping. Potable rainwater system distribution piping shall be sized in accordance with this code for sizing potable water piping.

K 107.0 Cleaning.

K 107.1 General. The interior surfaces of tanks and equipment shall be clean before they are put into service.

K 108.0 Supply System Inspection and Test.

K 108.1 General. Rainwater catchment systems shall be inspected and tested in accordance with the applicable provisions of this code for testing of potable water and storm drainage systems. Storage tanks shall be filled with water to the overflow opening for 24 hours, and during the inspection, or by other means as approved by the Authority Having Jurisdiction. Seams and joints shall be exposed during the inspection and checked for water tightness.

APPENDIX L
SUSTAINABLE PRACTICES

L 101.0 General.

L 101.1 Applicability. The purpose of this appendix is to provide a comprehensive set of technically sound provisions that encourage sustainable practices and works towards enhancing the design and construction of plumbing systems that result in a positive long-term environmental impact. This appendix is not intended to circumvent the health, safety, and general welfare requirements of this code.

L 101.2 Definition of Terms. For the purposes of this code, the definitions in Section L 201.0 shall apply to this appendix.

No attempt is made to define ordinary words, which are used in accordance with their established dictionary meanings, except where a word has been used loosely, and it is necessary to define its meaning as used in this appendix to avoid misunderstanding.

The definitions of terms are arranged alphabetically according to the first word of the term.

L 201.0 Definitions.

Catch Can Test. Method to measure the precipitation rate of an irrigation system by placing catchment containers at various random positions in the irrigation zone for a prescribed amount of time during irrigation application. The volumes of water in the containers are measured, averaged, and calculated to determine precipitation rate. Tests are conducted using irrigation industry accepted practices.

Combination Ovens. A device that combines the function of hot air convection (oven mode) and saturated and superheated steam heating (steam mode), or both, to perform steaming, baking, roasting, rethermalizing, and proofing of various food products. In general, the term combination oven is used to describe this type of equipment, which is self-contained. The combination oven is also referred to as a combination oven/steamer, combi or combo.

Energy Star. A joint program of the U.S. Environmental Protection Agency and the U.S. Department of Energy. Energy Star is a voluntary program designed to identify and promote energy-efficient products and practices.

Evapotranspiration (ET). The water transpired from vegetation, evaporated from the soil, water, and plant surfaces. Evapotranspiration rates are values expressed in inches (mm) per unit of time (day, week, month, or year). Evapotranspiration rates vary by components of weather conditions, including insolation, humidity, temperatures and wind, and time of year.

Food Steamers (Steam Cookers). A cooking appliance wherein heat is imparted to food in a closed compartment by direct contact with steam. The compartment can be at or above atmospheric pressure. The steam can be static or circulated.

Gang Showers. Shower compartments designed and intended for use by multiple persons simultaneously.

Hydrozone. A grouping of plants with similar water requirements that are irrigated by the same irrigation zone.

Irrigation Control System. An irrigation control system consists of a combination of a programmable controller using one or more inputs or sensors that, in combination, estimate or measure the availability of moisture for plants in order to operate an irrigation system, in such a manner that the system replenishes water as needed while minimizing excess water use. A properly programmed irrigation control system requires initial site specific set-up and will make irrigation schedule adjustments, including run times and required cycles throughout the irrigation season without human intervention.

Irrigation Demand. The amount of water not supplied by natural precipitation that is needed to maintain landscape plant life in good condition. Irrigation demand is calculated by subtracting natural effective precipitation from the ET rate adjusted by the landscape coefficient, which includes the functional purpose and desired quality of the plant being irrigated.

Irrigation Emission Device. The various landscape irrigation equipment terminal fittings or outlets that emit water for irrigating vegetation in a landscape.

Irrigation Zone. The landscape area that is irrigated by a set of landscape irrigation emission devices installed on the same water supply line downstream of a single valve.

Kitchen and Bar Sink Faucets. A faucet that discharges into a kitchen or bar sinks in domestic or commercial installations. Supply fittings that discharge into other type sinks, including clinical sinks, floor sinks, service sinks and laundry trays are not included.

Lavatory. (1) A basin or vessel for washing. (2) A plumbing fixture, as defined in (1), especially placed for use in personal hygiene. Principally not used for laundry purposes and never used for food preparation, or utensils, in food services. (3) A fixture designed for the washing of the hands and face. Sometimes called a wash basin.

Lavatory Faucet. A faucet that discharges into a lavatory basin in a domestic or commercial installation.

Low Application Rate Irrigation. A means of irrigation using low precipitation rate sprinkler heads or low flow emitters in conjunction with cycling irrigation schedules to apply water at a rate less than the soil absorption rate.

Low Flow Emitter. Low-flow irrigation emission device designed to dissipate water pressure and discharge a small uniform flow or trickle of water at a constant flow rate. To be classified as a low flow emitter: drip emitters shall discharge water at less than 4 gallons (15 L) per hour per emitter; micro-spray, micro-jet, and misters shall discharge water at a maximum of 30 gallons (114 L) per hour per nozzle.

Low Precipitation Rate Sprinkler Heads. Landscape irrigation emission devices or sprinkler heads with a maximum precipitation rate of 1 inch per hour (25.4 mm/h) over the applied irrigation area.

Maintenance. The upkeep of property or equipment by the owner of the property in compliance with the requirements of this appendix.

Metering Faucet. A self-closing faucet that dispenses a specific volume of water for each actuation cycle. The volume or cycle duration can be fixed or adjustable.

Modified Evapotranspiration. Numeric values, expressed in inches/hour (in/h), of evapotranspiration rates, derived by altering ETo rates by applying factors of specific needs of the vegetation and local climate conditions. Modified evapotranspiration rates are used as a factor in estimating the irrigation water needs of landscapes. Common usage includes reference evapotranspiration as the base rate, modified by coefficients or factors for specific plant types and densities.

Multi-Occupant Spaces. Indoor spaces used for presentations and training, including classrooms and conference rooms.

Precipitation Rate. The sprinkler head application rate of water applied to landscape irrigation zone, measured as inches per hour (mm/h). Precipitation rates of sprinkler heads are calculated according to the flow rate, pattern, and spacing of the sprinkler heads.

Pre-Rinse Spray Valve. A handheld device for use with commercial dishwashing and ware washing equipment that sprays water on dishes, flatware, and other food service items for the purpose of removing food residue before cleaning and sanitizing the items.

Recirculation System. A system of hot water supply and return piping with shutoff valves, balancing valves, circulating pumps, and a method of controlling the circulating system.

Reference Evapotranspiration (ETo). Numeric value, expressed in inches/hour (in/h), calculated as the water necessary to produce maximum biomass based upon a cool-season turf grass 4 inches to 6 inches (102 mm to 152 mm) tall. Common sources for obtaining local reference evapotranspiration rates are local agriculture extension services, state departments of agriculture, water agencies, irrigation professionals, the United States Geological Survey, and internet websites.

Reverse Osmosis Reject Water. Water that does not pass through a membrane of a reverse osmosis system.

Run Out. The developed length of pipe that extends away from the circulating loop system to a fixture(s).

Self Closing Faucet. A faucet that closes itself after the actuation or control mechanism is deactivated. The actuation or control mechanism can be mechanical or electronic.

Single Occupant Spaces. Private offices, workstations in open offices, reception workstations, and ticket booths.

Soil Absorption Rate. The rate of the soil's ability to allow water to percolate or infiltrate the soil and be retained in the root zone of the soil expressed as inches per hour (mm/h).

Sprinkler Head. Landscape irrigation emission device discharging water in the form of sprays or rotating streams, not including low flow emitters.

Storage Tank. The central component of the rainwater, stormwater, or dry weather runoff catchment system. Also known as a cistern or rain barrel.

Stormwater. Natural precipitation that has contacted a surface at grade or below grade and has not been put to beneficial use.

Stormwater Catchment System. A system that collects and stores stormwater for beneficial use.

Submeter. A meter installed subordinate to a site meter. Also known as a dedicated meter.

WaterSense. A voluntary program of the U.S., Environmental Protection Agency, designed to identify and promote water-efficient products and practices.

Water Closet. A fixture with a water-containing receptor that receives liquid and solid body waste and on actuation conveys the waste through an exposed integral trap into a drainage system. Also referred to as a toilet.

Water Factor (WF). A measurement and rating of appliance water efficiency, most often used for residential and light commercial clothes washers, as follows:

> **Water Factor (WF), Clothes Washer.** The quantity of water in gallons used to complete a full wash and rinse cycle per measured cubic foot capacity of the clothes container.

L 301.0 General Regulations.

L 301.1 Installation. Plumbing systems covered by this appendix shall be installed in accordance with this code, other applicable codes, and the manufacturer's installation instructions.

L 301.2 Qualifications. Where permits are required, the Authority Having Jurisdiction shall have the authority to require contractors, installers, or service technicians to demonstrate competency. Where determined by the Authority Having Jurisdiction, the contractor, installer or service technician shall be licensed to perform such work.

L 302.0 Disposal of Liquid Waste.

L 302.1 Disposal. It shall be unlawful for a person to cause, suffer, or permit the disposal of sewage, human excrement, or other liquid wastes, in a place or manner, except through and by means of an approved drainage system, installed and maintained in accordance with the provisions of this code.

L 302.2 Connections to Plumbing System Required. Equipment and appliances, used to receive or discharge liquid wastes or sewage, shall be connected properly to the drainage system of the building or premises, in accordance with the requirements of this code.

L 303.0 Abandonment.

L 303.1 General. An abandoned system or part thereof covered under the scope of this appendix shall be disconnected from remaining systems, drained, plugged, and capped in an approved manner.

L 401.0 Water Conservation and Efficiency.

L 401.1 Scope. The provisions of this section establish the means of conserving potable and nonpotable water used in and around a building.

L 402.0 Water-Conserving Plumbing Fixtures and Fittings.

L 402.1 General. The maximum water consumption of fixtures and fixture fittings shall comply with the flow rates specified in Table L 402.1, and Section L 402.2 through Section L 402.10.

TABLE L 402.1
MAXIMUM FIXTURE AND FIXTURE FITTINGS FLOW RATES

FIXTURE TYPE	FLOW RATE
Showerheads	2.0 gpm at 80 psi[1]
Kitchen faucets residential[4]	1.8 gpm at 60 psi
Lavatory faucets residential	1.5 gpm at 60 psi
Lavatory faucets other than residential	0.5 gpm at 60 psi
Metering faucets	0.25 gallons/cycle
Metering faucets for wash fountains	One 0.25 gallons/cycle fixture fitting for each 20 inches rim space
Wash fountains	One 2.2 gpm at 60 psi fixture fitting for each 20 inches rim space
Water Closets	1.28 gallons/flush[2]
Urinals	0.5 gallons/flush[3]
Commercial Pre-Rinse Spray Valves	1.3 gpm at 60 psi

For SI units: 1 gallon per minute = 0.06 L/s, 1 pound-force per square inch = 6.8947 kPa, 1 inch = 25.4 mm, 1 gallon = 3.785 L

Notes:

1 For multiple showerheads serving one shower compartment see Section L 402.6.1.

2 Shall be listed to EPA WaterSense Tank-Type Toilet Specification.

3 Shall be listed to EPA WaterSense Flushing Urinal Specification. Nonwater urinals shall comply with specifications listed in Section L 402.3.1.

4 See Section L 402.4.

L 402.2 Water Closets. No water closet shall have an effective flush volume exceeding 1.28 gallons per flush (gpf) (4.8 Lpf).

L 402.2.1 Gravity, Pressure Assisted, and Electro-Hydraulic Tank Type Water Closets. Gravity, pressure assisted, and electro-hydraulic tank-type water closets shall have a maximum effective flush volume of not more than 1.28 gallons (4.8 Lpf) of water per flush in accordance with ASME A112.19.2/CSA B45.1 or ASME A112.19.14 and shall be listed to the EPA WaterSense Specification for Tank-Type Toilets. The effective flush volume for dual flush toilets is defined as the composite, average flush volume of two reduced flushes and one full flush.

L 402.2.2 Flushometer-Valve Activated Water Closets. Flushometer-valve activated water closets shall have a maximum flush volume of not more than 1.28 gallons (4.8 Lpf) of water per flush in accordance with ASME A112.19.2/CSA B45.1 and shall be listed to the EPA WaterSense Specification for Flushometer-Valve Water Closets.

L 402.3 Urinals. Urinals shall have a maximum flush volume of not more than 0.5 gallon (1.9 Lpf) of water per flush in accordance with ASME A112.19.2/CSA B45.1 or CSA B45.5/IAPMO Z124. Flushing urinals shall be listed to the EPA WaterSense Flushing Urinal Specification.

L 402.3.1 Nonwater Urinals. Nonwater urinals shall comply with ASME A112.19.3/CSA B45.4, ASME A112.19.19, or CSA B45.5/IAPMO Z124. Nonwater urinals shall be cleaned and maintained in accordance with the manufacturer's instructions after installation. Where nonwater urinals are installed, they shall have a water distribution line roughed-in to the urinal location at a height not less than 56 inches (1422 mm) to allow for the installation of an approved backflow prevention device in the event of a retrofit. Such water distribution lines shall be installed with shutoff valves located as close as possible to the distributing main to prevent the creation of dead ends. Where nonwater urinals are installed, not less than one water supplied fixture rated at not less than 1 drainage fixture unit (DFU) shall be installed upstream on the same drain line to facilitate drain line flow and rinsing.

L 402.4 Residential Kitchen Faucets. The maximum flow rate of residential kitchen faucets shall not exceed 1.8 gallons per minute (gpm) (6.8 L/m) at 60 pounds-force per square inch (psi) (414 kPa). Kitchen faucets are permitted to temporarily increase the flow above the maximum rate, but not to exceed 2.2 gpm (8.3 L/m) at 60 psi (414 kPa), and shall revert to a maximum flow rate of 1.8 gpm (6.8 L/m) at 60 psi (414 kPa) upon valve closure.

L 402.5 Lavatory Faucets. The maximum water flow rate of faucets shall comply with Section L 402.5.1 and Section L 402.5.2.

L 402.5.1 Lavatory Faucets in Residences, Apartments, and Private Bathrooms in Lodging Facilities, Hospitals, and Patient Care Facilities. The flow rate for lavatory faucets installed in residences, apartments, and private bathrooms in lodging, hospitals, and patient care facilities (including skilled nursing and long-term care facilities) shall not exceed 1.5 gpm (5.7 L/m) at 60 psi (414 kPa) in accordance with ASME A112.18.1/CSA B125.1 and shall be listed to the EPA WaterSense High Efficiency Lavatory Faucet Specification.

L 402.5.2 Lavatory Faucets in Other Than Residences, Apartments, and Private Bathrooms in Lodging Facilities. Lavatory faucets installed in bathrooms of buildings or occupancies other than those specified in Section L 402.5.1 shall be in accordance with Section L 402.5.2(1) or Section L 402.5.2(2).

(1) The flow rate shall not exceed 0.5 gpm (1.9 L/m) at 60 psi (414 kPa) in accordance with ASME A112.18.1/CSA B125.1.

(2) Metering faucets shall deliver not more than 0.25 gallons (0.95 L) of water per cycle.

L 402.6 Showerheads. Showerheads shall not exceed 2.0 gpm (7.6 L/m) at 80 psi (552 kPa), and shall be listed to ASME A112.18.1/CSA B125.1 and the EPA WaterSense Specification for Showerheads.

L 402.6.1 Multiple Showerheads Serving One Shower Compartment. The total allowable flow rate of water from multiple showerheads flowing at a given time, with or without a diverter, including rain systems, waterfalls, bodysprays, and jets, shall not exceed 2.0 gpm (7.6 L/m) per shower compartment, where the floor area of the shower compartment is less than 1800 square inches (1.161 m^2). For each increment of 1800 square inches (1.161 m^2) of floor area after that or part thereof, additional showerheads are allowed, provided the total flow rate of water from flowing devices shall not exceed 2.0 gpm (7.6 L/m) for each such increment.

Exceptions:

(1) Gang showers in nonresidential occupancies. Singular showerheads or multiple shower outlets serving one showering position in gang showers shall not have more than 2.0 gpm (7.6 L/m) total flow.

(2) Where provided, shower compartments required for persons with disabilities in accordance with Chapter 17 shall not have more than 4.0 gpm (15 L/m) total flow, where one outlet is the hand shower.

L 402.6.2 Bath and Shower Diverters. The rate of leakage out of the tub spout of bath and shower diverters while operating in the shower mode shall not exceed 0.1 gpm (0.4 L/m) in accordance with ASME A112.18.1/CSA B125.1.

L 402.6.3 Shower Valves. Shower valves shall comply with the temperature control performance requirements of ASSE 1016 or ASME A112.18.1/CSA B125.1 where tested at 2.0 gpm (7.6 L/m).

L 402.7 Recirculating Shower Systems. Recirculating shower systems shall comply with IAPMO IGC 330.

» **L 402.8 Commercial Pre-Rinse Spray Valves.** The flow rate for a pre-rinse spray valve installed in a commercial kitchen to remove food waste from cookware and dishes before cleaning shall not be more than 1.28 gpm (4.8 L/m) at 60 psi (414 kPa). Where pre-rinse spray valves with maximum flow rates of 1.0 gpm (3.8 L/m) or less are installed, the static pressure shall be not less than 30 psi (207 kPa). Commercial kitchen pre-rinse spray valves shall be equipped with an integral automatic shutoff. Pre-rinse spray valves shall be listed to the EPA WaterSense Commercial Pre-rinse Spray Valve Specification.

» **L 402.9 Emergency Safety Showers and Eye Wash Stations.** Emergency safety showers and emergency eyewash stations shall not be limited to their water supply flow rates.

» **L 402.10 Drinking Fountains.** Drinking fountains shall be self-closing.

L 403.0 Appliances.

L 403.1 Dishwashers. Residential and commercial dishwashers shall comply with the Energy Star program requirements.

L 403.2 Clothes Washers. Residential clothes washers shall comply with the Energy Star program requirements. Commercial clothes washers shall comply with Energy Star program requirements, where such requirements exist.

L 404.0 Occupancy Specific Water Efficiency Requirements.

L 404.1 Commercial Food Service. Commercial food service facilities shall comply with the water efficiency requirements in Section L 404.2 through Section L 404.7.5.

L 404.2 Ice Makers. Ice makers shall be air cooled and shall be in accordance with Energy Star for commercial ice machines.

L 404.3 Food Steamers. Boilerless type steamers shall consume not more than 2.0 gallons (7.6 L) per compartment. Boiler type steamers shall not consume more than 1.5 gallons (5.7 L) per pan per hour.

L 404.4 Combination Ovens. Combination ovens shall not use water in the convection mode except when utilizing a moisture nozzle for food products in the oven. The total amount of water used by the moisture nozzle in the convection mode shall not exceed a half a gallon per hour per oven cavity. When operating in the steamer mode, combination ovens shall not consume more than 1.5 gallons per hour (gph) (5.7 L/h) per pan.

L 404.5 Grease Interceptors. Grease interceptor maintenance procedures shall not include post-pumping/cleaning refill using potable water. Refill shall be by connected appliance accumulated discharge only.

L 404.6 Dipper Well Faucets. Where dipper wells are installed, the water supply to a dipper well shall have a shutoff valve and flow control. The flow of water into a dipper well shall be limited by not less than one of the following methods:

(1) Water flow shall not exceed the water capacity of the dipper well in one minute at a supply pressure of 60 psi (414 kPa), and the maximum flow shall not exceed 2.2 gpm (8.3 L/m) at a supply pressure of 60 psi (414 kPa). The water capacity of a dipper well shall be the maximum amount of water that the fixture can hold before water flows into the drain.

(2) The volume of water dispensed into a dipper well in each activation cycle of a self-closing fixture fitting shall not exceed the water capacity of the dipper well, and the maximum flow shall not exceed 2.2 gpm (8.3 L/m) at a supply pressure of 60 psi (414 kPa).

L 404.7 Food Waste Devices. Where installed, food waste devices shall be in accordance with Section L 404.7.1 through Section L 404.7.5.

L 404.7.1 Pulpers and Mechanical Strainers. The water use for pulpers or mechanical strainers shall not exceed 3 gpm (11.4 L/m). A flow restrictor shall be installed on the water supply to limit the water flow.

L 404.7.2 Food Waste Disposers. The water use for the food waste grinder shall not exceed 8 gpm (30.3 L/m) under full load condition and 1 gpm (3.8 L/m) under no-load condition. Flow restrictors shall be installed on the water supply to limit the water flow rate to a maximum of 8 gpm (30.3 L/m). A load sensing device shall be installed to monitor current demand and regulate water flow.

L 404.7.3 Time Out and Shut Off. Pulpers, mechanical strainers, and food waste disposers shall have a time out system with push button to reactivate. The maximum allowable run time cycle shall be 10 minutes.

L 404.7.4 Sink Drain Outlets. Where a strainer or basket is installed, they shall be readily removable.

L 404.7.5 Strainer Baskets. Strainer (scrapper) baskets shall either fit over a sink compartment or be attached to a drain system. The strainer baskets shall be readily removable for emptying.

›› **L 404.8 Medical and Laboratory Facilities.** Medical and laboratory facilities shall comply with the water efficiency requirements in Section L 404.9 through Section L 404.11.

›› **L 404.9 Steam Sterilizers.** Controls shall be installed to limit the discharge temperature of condensate or water from steam sterilizers to 140°F (60°C) or less. A venturi-type vacuum system shall not be utilized with vacuum sterilizers.

›› **L 404.10 X-Ray Film Processing Units.** Processors for X-ray film exceeding 6 inches (152 mm) in any dimension shall be equipped with water recycling units.

›› **L 404.11 Exhaust Hood Liquid Scrubber Systems.** Liquid scrubber systems for exhaust hoods and ducts shall be of the recirculation type. Liquid scrubber systems for perchloric acid exhaust hoods and ducts shall be equipped with a timer-controlled water recirculation system. The collection sump for perchloric acid exhaust systems shall be designed to drain automatically after the wash down process has completed.

L 405.0 Leak Detection and Control.

L 405.1 General. Where installed, leak detection and control devices shall comply with IAPMO IGC 115 or IAPMO IGC 349. Leak detection and control devices help protect property from water damage and also conserve water by shutting off the flow when leaks are detected.

L 406.0 Fountains and Other Water Features.

L 406.1 Use of Alternate Water Source for Special Water Features. Special water features such as ponds and water fountains shall be provided with reclaimed (recycled) water, rainwater, or on-site treated nonpotable water where the source and capacity are available on the premises and approved by the Authority Having Jurisdiction.

L 407.0 Meters.

L 407.1 Required. A water meter shall be required for buildings connected to a public water system, including municipally supplied reclaimed (recycled) water. In other than single-family houses, multifamily structures of three stories or fewer above grade, and modular houses, a separate meter or submeter shall be installed in the following locations:

(1) The water supply for irrigated landscape with an accumulative area exceeding 2500 square feet (232.3 m^2).

(2) The water supply to a water-using process where the consumption exceeds 1000 gallons per day (gal/d) (0.0438 L/s), except for manufacturing processes.

(3) The water supply to each building on a property with multiple buildings where the water consumption exceeds 500 gals/d (0.021 L/s).

(4) The water supply to an individual tenant space on a property where one or more of the following applies:

   (a) Water consumption exceeds 500 gals/d (0.021 L/s) for that tenant.

   (b) Tenant space is occupied by a commercial laundry, cleaning operation, restaurant, food service, medical office, dental office, laboratory, beauty salon, or barbershop.

   (c) Total building area exceeds 50 000 square feet (4645 m^2).

(5) The makeup water supplies to a swimming pool.

L 407.2 Approval. Dedicated meters, other than water utility meters shall be approved by the Authority Having Jurisdiction for the intended use.

L 407.3 Consumption Data. A means of communicating water consumption data from submeters to the water consumer shall be provided.

L 407.4 Access. Meters and submeters shall be accessible.

L 408.0 Condensate Recovery.

L 408.1 General. Condensate is permitted to be used as onsite treated nonpotable water when collected, stored, and treated in accordance with Section 1506.0.

L 408.1.1 Condensate Drainage Recovery. Condensate from air-conditioning, boiler and steam systems used to supply water for non-potable water systems shall be in accordance with Section 1506.0.

L 409.0 Water-Powered Sump Pumps.

L 409.1 General. Sump pumps powered by potable or reclaimed (recycled) water pressure shall be used as an emergency backup pump. The water-powered pump shall be equipped with a battery powered alarm having a minimum rating of 85 dBa at 10 feet (3048 mm). Water-powered pumps shall have a water efficiency factor of pumping at least 1.4 gallons (5.3 L) of water to a height of 10 feet (3048 mm) for every gallon of water used to operate the pump, measured at a water pressure of 60 psi (414 kPa). Pumps shall be labeled as to the gallons of water pumped per gallon of potable water consumed.

Water-powered stormwater sump pumps shall be equipped with a reduced pressure principle backflow prevention assembly.

L 410.0 Water Softeners and Treatment Devices.

L 410.1 Water Softeners. Water softeners shall be listed to NSF 44. Water softeners shall have a rated salt efficiency exceeding 3400 grains (gr) (0.222 kg) of total hardness exchange per pound (0.5 kg) of salt, based on sodium chloride (NaCl) equivalency, and shall not generate more than 4 gallons (15.1 L) of water per 1000 grains (0.0647 kg) of hardness removed during the service cycle.

L 410.2 Water Softener Limitations. In residential buildings, where the supplied potable water hardness is equal to or less than 8 grains per gallon (gr/gal) (137 mg/L) measured as total calcium carbonate equivalents, water softening equipment that discharges water into the wastewater system during the service cycle shall not be allowed, except as required for medical purposes.

L 410.3 Point-of-Use Reverse Osmosis Water Treatment Systems. Reverse osmosis water treatment systems installed in residential occupancies shall be equipped with automatic shutoff valves to prevent discharge when there is no call for producing treated water. Reverse osmosis water treatment systems shall be listed in accordance with NSF 58.

L 411.0 Landscape Irrigation Systems.

L 411.1 General. Where landscape irrigation systems are installed, they shall be in accordance with Section L 411.2 through Section L 411.14. Requirements limiting the amount or type of plant material used in landscapes shall be established by the Authority Having Jurisdiction.

Exception: Plants grown for food production.

L 411.2 Backflow Protection. Potable water and supplies to landscape irrigation systems shall be protected from backflow in accordance with this code and the Authority Having Jurisdiction.

L 411.3 Use of Alternate Water Sources for Landscape Irrigation. Where available by pre-existing treatment, storage, or distribution network, and where approved by the Authority Having Jurisdiction, alternative water source(s) shall be utilized for landscape irrigation. Where adequate capacity and volumes of pre-existing alternative water sources are available, the irrigation system shall be designed to use a minimum of 75 percent of alternative water for the annual irrigation demand before supplemental potable water is used.

L 411.4 Irrigation Control Systems. Where installed as part of a landscape irrigation system, irrigation control systems shall:

(1) Automatically adjust the irrigation schedule to respond to plant water needs determined by weather or soil moisture conditions.

(2) Utilize sensors to suspend irrigation during a rainfall.

(3) Utilize sensors to suspend irrigation where adequate soil moisture is present for plant growth.

(4) Have the capability to program multiple and different run times for each irrigation zone to enable cycling of water applications and durations to mitigate water flowing off of the intended irrigation zone.

(5) The site-specific settings of the irrigation control system affecting the irrigation and shall be posted at the control system location. The posted data, where applicable to the settings of the controller, shall include:

- (a) Precipitation rate for each zone.
- (b) Plant evapotranspiration coefficients for each zone.
- (c) Soil absorption rate for each zone.
- (d) Rain sensor settings.
- (e) Soil moisture setting.
- (f) Peak demand schedule including run times for each zone and the number of cycles to mitigate runoff and monthly adjustments or percentage.

L 411.5 Low Flow Irrigation. Irrigation zones using low flow irrigation emitters shall comply with ASABE/ICC 802 Landscape Irrigation Sprinkler and Emitter Standard and shall be equipped with filters sized according to the manufacturer's recommendation for the specific low flow emitter, and with a pressure regulator installed upstream of the irrigation emission devices as necessary to reduce the operating water pressure in accordance with the manufacturers' equipment requirements.

L 411.6 Mulched Planting Areas. Only low flow emitters are allowed to be installed in mulched planting areas with vegetation taller than 12 inches (305 mm).

L 411.7 System Performance Requirements. The landscape irrigation system shall be designed and installed to:

(1) Prevent irrigation water from runoff out of the irrigation zone.

(2) Prevent water in the supply line drainage from draining out between irrigation events.

(3) Not allow irrigation water to be applied onto or enter non-targeted areas including adjacent property and vegetation areas, adjacent hydrozones not requiring the irrigation water to meet its irrigation demand, non-vegetative areas, impermeable surfaces, roadways, and structures.

L 411.8 Narrow or Irregularly Shaped Landscape Areas. Narrow or irregularly shaped landscape areas, less than 4 feet (1219 mm) in any direction across opposing boundaries, shall not be irrigated by an irrigation emission device except low flow emitters.

L 411.9 Sloped Areas. Where soil surface rises more than 1 foot (305 mm) per 4 feet (1219 mm) of length, the irrigation zone system average precipitation rate shall not exceed 0.75 inches (19.1 mm) per hour as verified through either of the following methods:

(1) Manufacturer documentation that the precipitation rate for the installed sprinkler head does not exceed 0.75 inches (19.1 mm) per hour where the sprinkler heads are installed not closer than the specified radius and where the water pressure of the irrigation system is not more than the manufacturer's recommendations.

(2) Catch can test in accordance with the requirements of the Authority Having Jurisdiction and where emitted water volume is measured with a minimum of six catchment containers at random places within the irrigation zone for a minimum of 15 minutes to determine the average precipitation rate, expressed as inches per hour (mm/h).

L 411.10 Sprinkler Head Installations. All installed sprinkler heads shall comply with ASABE/ICC 802 or other approved standard(s).

L 411.10.1 Sprinkler Heads in Common Irrigation Zones. Sprinkler heads installed in irrigation zones served by a common valve shall be limited to applying water to plants with similar irrigation needs, and shall have matched precipitation rates (identical inches of water application per hour as rated or tested, plus or minus 5 percent).

L 411.10.2 Sprinkler Head Pressure Regulation. Sprinkler heads shall utilize pressure regulating devices (as part of an irrigation system or integral to the sprinkler head) to maintain manufacturer's recommended operating pressure for each sprinkler and nozzle type.

L 411.10.3 Pop-up Type Sprinkler Heads. Where pop-up type sprinkler heads are installed, the sprinkler heads shall rise to a height above vegetation level and of not less than 4 inches (102 mm) above the soil level where emitting water.

L 411.11 Irrigation Zone Performance Criteria. Irrigation zones shall be designed and installed to ensure the average precipitation rate of the sprinkler heads over the irrigated area does not exceed 1 inch per hour (25.4 mm/h) as verified through either of the following methods:

(1) Manufacturer's documentation that the precipitation rate for the installed sprinkler head does not exceed 1 inch per hour (25.4 mm/h) where the sprinkler heads are installed not closer that the specified radius and where the water pressure of the irrigation system is not more than the manufacturer's recommendations.

(2) Catch can test in accordance with the requirements of the Authority Having Jurisdiction and where emitted water volume is measured with a minimum of six catchment containers at random places within the irrigation zone for a minimum of 15 minutes to determine the average precipitation rate, expressed as inches per hour (mm/h).

L 411.12 Depth of Irrigation Pipe. Irrigation pipe downstream from the backflow preventer shall be buried at a minimum depth according to Section L 411.12.1 and Section L 411.12.2.

L 411.12.1 Landscape Areas. Irrigated landscaped areas not exceeding 10 000 square feet (929 m^2) shall have irrigation main lines buried a minimum of 12 inches (305 mm) and irrigation lateral lines buried a minimum of 8 inches (203 mm). Irrigated landscaped areas greater than 10 000 square feet (929 m^2) shall have irrigation main lines buried a minimum of 18 inches (457 mm) and irrigation lateral lines buried a minimum of 12 inches (305 mm).

L 411.12.2 Vehicular Surfaces. Irrigation pipe installed under vehicular paving and pervious pavers, including landscaped fire lanes, shall be sleeved with a minimum of one 1-inch pipe (25 mm) size greater than the irrigation pipe and buried at a minimum depth of 24 inches (610 mm) in all cases.

L 411.13 Backfill. All excavation for irrigation pipe installation shall be backfilled in thin layers to 12 inches (305 mm) with clean earth, which shall not contain stones, boulders, cinderfill, frozen earth, construction debris, or other materials that would damage or break the piping. Fill shall be properly compacted. Suitable precautions shall be taken to ensure permanent stability for pipe laid in filled or made ground.

L 411.14 Qualifications. The Authority Having Jurisdiction shall have the authority to require landscape irrigation contractors, installers, or designers to demonstrate competency. Where required by the Authority Having Jurisdiction, the contractor, installer, or designer shall be certified to perform such work.

L 412.0 Trap Seal Protection.

L 412.1 Water Supplied Trap Primers. Water supplied trap primers shall be electronic or pressure activated and shall use not more than 30 gallons (114 L) per year per drain. Where an alternate water source, as defined by this code, is used for fixture flushing or other uses in the same room, the alternate water source shall be used for the trap primer water supply.

Exception: Flushometer tailpiece trap primers in accordance with IAPMO PS 76.

L 412.2 Drainage Type Trap Seal Primer Devices. Drainage type trap seal primer devices shall not be limited in the amount of water they discharge.

L 413.0 Vehicle Wash Facilities.

L 413.1 Automatic. The maximum make-up water use for automobile washing shall not exceed 40 gallons (151 L) per vehicle for in-bay automatic car washes and 35 gallons (132 L) for conveyor and express type car washes.

L 413.2 Self-Service. Spray wands and foamy brushes shall use not more than 3.0 gpm (0.19 L/s).

L 413.3 Reverse Osmosis. Spot-free reverse osmosis discharge (reject) water shall be recycled.

L 413.4 Towel Ringers. Towel ringers shall have a positive shutoff valve. Spray nozzles shall be replaced annually.

Exception: Bus and large commercial vehicle washes are exempt from the requirements of this section.

L 501.0 Water Heating Design, Equipment, and Installation.

L 501.1 Scope. The provisions of this section shall establish the means of conserving potable and nonpotable water and energy associated with the generation and use of hot water in a building. This includes provisions for the hot water distribution system, which is the portion of the potable water distribution system between a water heating device and the plumbing fixtures, including dedicated return piping and appurtenances to the water heating device in a recirculation system.

L 501.2 Insulation. Hot water supply and return piping shall be thermally insulated. The wall thickness of the insulation shall be equal to the nominal diameter of the pipe up to 2 inches (50 mm). The wall thickness shall be not less than 2 inches (51 mm) for nominal pipe diameters exceeding 2 inches (50 mm). The conductivity of the insulation [k-factor (Btu•in/(h•ft^2•°F))], measured radially, shall not be more than 0.28 [Btu•in/(h•ft^2•°F)] [0.04 W/(m•K)]. Hot water piping to be insulated shall be installed such that insulation is continuous. Pipe insulation shall be installed to within ¼ of an inch (6.4 mm) of appliances, appurtenances, fixtures, structural members, or a wall where the pipe passes through to connect to a fixture within 24 inches (610 mm).

Exceptions:

(1) Where the hot water pipe is installed in a wall that is not of a width to accommodate the pipe and insulation, the insulation thickness shall be permitted to have the maximum thickness that the wall is capable of accommodating and not less than ½ of an inch (12.7 mm) thick.

(2) Hot water supply piping exposed under sinks, lavatories, and similar fixtures.

L 501.2.1 Pipe Supports. Pipe supports shall be installed on the outside of the pipe insulation.

Exception: Vertical supports, and horizontal and vertical anchors shall be installed on the pipe inside the pipe insulation.

L 501.3 Recirculation Systems. Recirculation systems shall comply with Section L 501.3.1 and Section L 501.3.2.

L 501.3.1 For Low-Rise Residential Buildings. Circulating hot water systems shall be arranged so that the circulating pump(s) are capable of being turned off (automatically or manually) where the hot water system is not in operation. [ASHRAE 90.2-2007:7.2]

L 501.3.2 For Pumps Between Boilers and Storage Tanks. Where used to maintain storage tank water temperature, recirculating pumps shall be equipped with controls limiting operation to a period from the start of the heating cycle to a maximum of 5 minutes after the end of the heating cycle. [ASHRAE 90.1:7.4.4.4]

L 501.4 Recirculation Pump Controls. Pump controls shall include on-demand activation or time clocks combined with temperature sensing. Time clock controls for pumps shall not let the pump operate more than 15 minutes every hour. Temperature sensors shall stop circulation where the temperature set point is reached and shall be located on the circulation loop at or near the last fixture. The pump, pump controls, and temperature sensors shall be accessible. Pump operation shall be limited to the building's hours of operation.

L 501.5 Temperature Maintenance Controls. Systems designed to maintain usage temperatures in hot-water pipes, such as recirculating hot water systems or heat trace, shall be equipped with automatic time switches or other controls that are capable of being set to switch off the usage temperature maintenance system during extended periods where hot water is not required. [ASHRAE 90.1:7.4.4.2]

L 501.6 System Balancing. Systems with multiple recirculation zones shall be balanced to distribute hot water uniformly, or they shall be operated with a pump for each zone. The circulation pump controls shall comply with the provisions of Section L 501.4.

L 501.7 Flow Balancing Valves. Flow balancing valves shall be a factory preset automatic flow control valve, a flow regulating valve, or a balancing valve with memory stop.

L 501.8 Air Elimination. Provision shall be made for the elimination of air from the return system.

L 501.9 Gravity or Thermosyphon Systems. Gravity or thermosyphon systems are prohibited.

L 502.0 Service Hot Water – Low-Rise Residential Buildings.

L 502.1 General. The service water heating system for single-family houses, multi-family structures of three stories or fewer above grade, and modular houses shall comply with Section L 502.2 through Section L 502.7.3. The service water heating system of all other buildings shall comply with Section L 503.0.

L 502.2 Water Heaters and Storage Tanks. Residential-type water heaters, pool heaters, and unfired water heater storage tanks shall comply with the minimum performance requirements specified by federal law.

Unfired storage water heating equipment shall have a heat loss through the tank surface area of less than 6.5 British thermal units per square foot hour [Btu/(ft^2•h)] (20.5 W/m^2). [ASHRAE 90.2-2007:7.1]

L 502.3 Recirculation Systems. Recirculation systems shall comply with the provisions of Section L 501.3.

L 502.4 Central Water Heating Equipment. Service water heating equipment (central systems) that do not fall under the requirements for residential-type service water heating equipment addressed in Section L 502.2 shall comply with the applicable requirements for service water-heating equipment found in Section L 503.0. [ASHRAE 90.2-2007:7.3]

L 502.5 Insulation. Insulation for hot water and return piping shall comply with the provisions of Section L 501.2.

L 502.6 Hard Water. Where water has hardness equal to or exceeding 9 grains per gallon (gr/gal) (154 mg/L) measured as total calcium carbonate equivalents, the water supply line to water heating equipment in new one- and two-family dwellings shall be roughed-in to allow for the installation of water treatment equipment.

L 502.7 Maximum Volume of Hot Water. The maximum volume of water contained in hot water distribution pipes shall be in accordance with Section L 502.7.1 or Section L 502.7.2. The water volume shall be calculated using Table L 502.7.

L 502.7.1 Maximum Volume of Hot Water Without Recirculation or Heat Trace. The maximum volume of water contained in hot water distribution pipe between the water heater and any fixture fitting shall not exceed 32 ounces (oz) (946 mL). Where a fixture fitting shutoff valve (supply stop) is installed ahead of the fixture fitting, the maximum volume of water is permitted to be calculated between the water heater and the fitting shutoff valve (supply stop).

L 502.7.2 Maximum Volume of Hot Water with Recirculation or Heat Trace. The maximum volume of water contained in the branches between the recirculation loop or electrically heat traced pipe, and the fixture fitting shall not exceed 16 oz (473 mL). Where a fixture fitting shutoff valve (supply stop) is installed ahead of the fixture fitting, the maximum volume of water is permitted to be calculated between the recirculation loop or electrically, heat traced pipe and the fixture fitting shutoff valve (supply stop).

TABLE L 502.7
WATER VOLUME FOR DISTRIBUTION PIPING MATERIALS*

NOMINAL SIZE (inch)	OUNCES OF WATER PER FOOT LENGTH OF PIPING												
	COPPER M	COPPER L	COPPER K	CPVC CTS SDR 11	CPVC SCH 40	PEX-AL- PEX	PE-AL- PE	CPVC SCH 80	PEX CTS SDR 9	PE-RT SDR 9	PP SDR 6	PP SDR 7.3	PP SDR 11
⅜	1.06	0.97	0.84	NA	1.17	0.63	0.63	NA	0.64	0.64	0.91	1.09	1.24
½	1.69	1.55	1.45	1.25	1.89	1.31	1.31	1.46	1.18	1.18	1.41	1.68	2.12
¾	3.43	3.22	2.90	2.67	3.38	3.39	3.39	2.74	2.35	2.35	2.23	2.62	3.37
1	5.81	5.49	5.17	4.43	5.53	5.56	5.56	4.57	3.91	3.91	3.64	4.36	5.56
1¼	8.70	8.36	8.09	6.61	9.66	8.49	8.49	8.24	5.81	5.81	5.73	6.81	8.60
1½	12.18	11.83	11.45	9.22	13.20	13.88	13.88	11.38	8.09	8.09	9.03	10.61	13.47
2	21.08	20.58	20.04	15.79	21.88	21.48	21.48	19.11	13.86	13.86	14.28	16.98	21.39

For SI units: 1 ounce = 29.573 mL
* NA: Not Applicable

Exception: Whirlpool bathtubs or bathtubs that are not equipped with a shower are exempted from the requirements of Section L 502.7.

L 502.7.3 Hot Water System Submeters. Where a hot water pipe from a circulation loop or electric heat trace line is equipped with a submeter, the hot water distribution system downstream of the submeter shall have either an end-of-line hot water circulation pump or shall be electrically heat traced. The maximum volume of water in a branch from the circulation loop or electric heat trace line downstream of the submeter shall not exceed 16 oz (473 mL).

Where there is no circulation loop or electric heat traced line downstream of the submeter, the submeter shall be located within 2 feet (610 mm) of the central hot water system; or the branch line to the submeter shall be circulated or heat traced to within 2 feet (610 mm) of the submeter. The maximum volume from the submeter to each fixture shall not exceed 32 oz (946 mL).

The circulation pump controls shall comply with the provisions of Section L 501.4.

L 503.0 Service Hot Water – Other Than Low-Rise Residential Buildings.

L 503.1 General. The service hot water, other than single-family houses, multifamily structures of three stories or fewer above grade, and modular houses shall comply with this section.

L 503.1.1 New Buildings. Service water-heating systems and equipment shall comply with the requirements of this section as described in Section L 503.2. [ASHRAE 90.1:7.1.1.1]

L 503.1.2 Additions to Existing Buildings. Service water heating systems and equipment shall comply with the requirements of this section.

Exception: Where the service water-heating to an addition is provided by existing service water-heating systems and equipment, such systems and equipment shall not be required to be in accordance with this appendix. However, new systems or equipment installed shall be in accordance with specific requirements applicable to those systems and equipment. [ASHRAE 90.1:7.1.1.2]

L 503.1.3 Alterations to Existing Buildings. Building service water-heating equipment installed as a direct replacement for existing building service water-heating equipment shall be in accordance with the requirements of Section L 503.0 applicable to the equipment being replaced. New and replacement piping shall comply with Section L 503.3.3.

Exception: Compliance shall not be required where there is insufficient space or access to meet these requirements. [ASHRAE 90.1:7.1.1.3]

L 503.2 Compliance Path(s). Compliance shall be achieved in accordance with the requirements of Section L 503.1, Section L 503.3, Section L 503.4, and Section L 503.5. [ASHRAE 90.1:7.2.1]

L 503.2.1 Energy Cost Budget Method. Projects using the energy cost budget method of ASHRAE 90.1 for demonstrating compliance with the standard shall be in accordance with the requirements of Section L 503.3 in conjunction with the energy cost budget method of ASHRAE 90.1. [ASHRAE 90.1:7.2.2]

L 503.3 Mandatory Provisions. The mandatory provisions of Section L 503.3.1 through Section L 503.3.7 shall be followed.

L 503.3.1 Load Calculations. Service water-heating system design loads for the purpose of sizing systems and equipment shall be determined in accordance with manufacturer's published sizing guidelines or accepted engineering standards and handbooks acceptable to the adopting authority (e.g., ASHRAE Handbook – HVAC Applications). [ASHRAE 90.1:7.4.1]

L 503.3.2 Equipment Efficiency. Water-heating equipment, hot-water supply boilers used solely for heating potable water, pool heaters, and hot water storage tanks shall comply with the criteria listed in Table L 503.3.2. Where multiple criteria are listed, all criteria shall be met. The omission of minimum performance requirements for certain classes of equipment does not preclude the use of such equipment where appropriate. Equipment not listed in Table L 503.3.2 has no minimum performance requirements.

Exceptions: Water heaters and hot-water supply boilers having more than 140 gallons (530 L) of storage capacity are not required to meet the standby loss (SL) requirements of Table L 503.3.2 where:

(1) The tank surface is thermally insulated to R-12.5.

(2) A standing pilot light is not installed.

(3) Gas- or oil-fired storage water heaters have a flue damper or fan-assisted combustion. [ASHRAE 90.1:7.4.2]

L 503.3.3 Insulation. The following piping shall be insulated in accordance with Table L 503.3.3:

(1) Recirculating system piping, including the supply and return piping of a circulating tank type water heater.

(2) The first 8 feet (2438 mm) of outlet piping for a constant temperature nonrecirculating storage system.

(3) The first 8 feet (2438 mm) of branch piping connecting to recirculated, heat-traced, or impedance heated piping.

(4) The inlet piping between the storage tank and a heat trap in a nonrecirculating storage system.

(5) Piping that is externally heated (such as heat trace or impedance heating). [ASHRAE 90.1:7.4.3]

L 503.3.4 Hot Water System Design. Hot water systems shall comply with Section L 503.3.4(1) and Section L 503.3.4(2).

(1) Recirculation systems shall comply with the provisions of Section L 501.3.

(2) The maximum volume of water contained in hot water distribution lines between the water heater and the fixture stop or connection to showers, kitchen faucets, and lavatories shall be determined in accordance with Section L 502.7.

L 503.3.5 Service Water Heating System Controls. Service water heating system controls shall comply with Section L 503.3.5(1) and Section L 503.3.5(2).

(1) Temperature controls shall be provided that allows for storage temperature adjustment from 120°F (49°C) or lower to a maximum temperature compatible with the intended use.

Exception: Where the manufacturer's installation instructions specify a higher minimum thermostat setting to minimize condensation and resulting corrosion. [ASHRAE 90.1:7.4.4.1]

(2) Temperature controlling means shall be provided to limit the maximum temperature of water delivered from lavatory faucets in public facility restrooms to 110°F (43°C). [ASHRAE 90.1:7.4.4.3]

L 503.3.6 Pools. Pool heating systems shall comply with Section L 503.3.6(1) through Section L 503.3.6(3).

(1) Pool heaters shall be equipped with a readily accessible ON/OFF switch to allow shutting off the heater without adjusting the thermostat setting. Pool heaters fired by natural gas shall not have continuously burning pilot lights. [ASHRAE 90.1:7.4.5.1]

(2) Heated pools shall be equipped with a vapor retardant pool cover on or at the water surface. Pools heated to more than 90°F (32°C) shall have a pool cover with a minimum insulation value of R-12.

Exception: Pools that are deriving over 60 percent of the energy for heating from site-recovered energy or solar energy. [ASHRAE 90.1:7.4.5.2]

(3) Time switches shall be installed on swimming pool heaters and pumps.

Exceptions:

(1) Where public health standards require 24-hour pump operation.

(2) Where pumps are required to operate solar and waste heat recovery pool heating systems. [ASHRAE 90.1:7.4.5.3]

L 503.3.7 Heat Traps. Vertical pipe risers serving storage water heaters and storage tanks not having integral heat traps and serving a nonrecirculating system shall have heat traps on both the inlet and outlet piping as close as practical to the storage tank. A heat trap is a means to counteract the natural convection of heated water in a vertical pipe run. The means is either of the following:

(1) A device specifically designed for the purpose or an arrangement of tubing that forms a loop of 360 degrees (6.28 rad).

(2) Piping that, from the point of connection to the water heater (inlet or outlet) includes a length of piping directed downward before connection to the vertical piping of the supply water or hot-water distribution system, as applicable. [ASHRAE 90.1:7.4.6]

L 503.4 Prescriptive Path. The prescriptive path for space or water heating efficiency shall comply with Section L 503.4.1 through Section L 503.4.5.

L 503.4.1 Space Heating and Water Heating. The use of a gas-fired or oil-fired space heating boiler system, otherwise in accordance with Section L 503.0, to provide the total space heating and service water heating for a building is allowed where one of the following conditions is met:

(1) The single space-heating boiler, or the component of a modular or multiple boiler system that is heating the service water, has a standby loss in Btu/h (kW) not exceeding $(13.3 \times pmd + 400)/n$, where (pmd) is the probable maximum demand in gallons per hour, determined in accordance with the procedures described in generally accepted engineering standards and handbooks, and (n) is the fraction of the year where the outdoor daily mean temperature exceeds 64.9°F (18.28°C).

The standby loss is to be determined for a test period of 24 hours duration while maintaining a boiler water temperature of not less than 90°F (32°C) above ambient, with an ambient temperature between 60°F (16°C) and 90°F (32°C). For a boiler with a modulating burner, this test shall be conducted at the lowest input.

TABLE L 503.3.2
PERFORMANCE REQUIREMENTS FOR WATER-HEATING EQUIPMENT
MINIMUM EFFICIENCY REQUIREMENTS
[ASHRAE 90.1: TABLE 7.8]

EQUIPMENT TYPE	SIZE CATEGORY (INPUT)	SUBCATEGORY OR RATING CONDITION	PERFORMANCE REQUIRED[1]	TEST PROCEDURE[2,3]
Electric table top water heaters	≤12 kW	Resistance ≥20 gal	See footnote 7	—
Electric water heaters	≤12 kW[5]	Resistance ≥20 gal	See footnote 7	—
	>12 kW[5]	Resistance ≥20 gal	$0.3 + 27\sqrt{V_m}$ %/h	Section G.2 of CSA Z21.10.3
	≤24 Amps and ≤250 Volts	Heat Pump	See footnote 7	—
Gas storage water heaters	≤75 000 Btu/h	≥20 gal	See footnote 7	—
	>75 000 Btu/h[6]	<4000 (Btu/h)/gal	80% E_t $(Q/800 + 110\sqrt{V})$ SL, Btu/h	Sections G.1 and G.2 of CSA Z21.10.3
Gas instantaneous water heaters	>50 000 Btu/h and <200 000 Btu/h	≥4000 (Btu/h)/gal and <2 gal	See footnote 7	—
	≥200 000 Btu/h[4,6]	≥4000 (Btu/h)/gal and <10 gal	80% E_t	Sections G.1 and G.2 of CSA Z21.10.3
	≥200 000 Btu/h[6]	≥4000 (Btu/h)/gal and ≥10 gal	80% E_t $(Q/799 + 16.6\sqrt{V})$ SL, Btu/h	
Oil storage water heaters	≤105 000 Btu/h	≥20 gal	0.59-0.0005V EF	—
	>105 000 Btu/h	<4000 (Btu/h)/gal	80% E_t $(Q/800 + 110\sqrt{V})$ SL, Btu/h	Sections G.1 and G.2 of CSA Z21.10.3
Oil instantaneous water heaters	≤210 000 Btu/h	≥4000 (Btu/h)/gal and <2 gal	See footnote 7	—
	>210 000 Btu/h	≥4000 (Btu/h)/gal and <10 gal	80% E_t	Sections G.1 and G.2 of CSA Z21.10.3
	>210 000 Btu/h	≥4000 (Btu/h)/gal and ≥10 gal	78% E_t $(Q/800 + 110\sqrt{V})$ SL, Btu/h	
Hot-water supply boilers, gas and oil[6]	≥300 000 Btu/h and <12 500 000 Btu/h	≥4000 (Btu/h)/gal and <10 gal	80% E_t	Sections G.1 and G.2 of CSA Z21.10.3
Hot-water supply boilers, gas[6]	—	≥4000 (Btu/h)/gal and ≥10 gal	80% E_t $(Q/800 + 110\sqrt{V})$ SL, Btu/h	
Hot-water supply boilers, oil	—	≥4000 (Btu/h)/gal and ≥10 gal	78% E_t $(Q/800 + 110\sqrt{V})$ SL, Btu/h	
Pool heaters, oil and gas	All	—	See footnote 7	ASHRAE 146
Heat pump pool heaters	All	50°F db 44.2°F wb Outdoor air 80.0°F entering water	4.0 COP	AHRI 1160
Unfired storage tanks	All	—	R-12.5	(none)

For SI units: 1 gallon = 3.785 L, 1000 British thermal units per hour = 0.293 kW, °C = (°F-32)/1.8

Notes:

[1] Thermal efficiency (*Et*) is a minimum requirement, while standby loss (SL) is maximum Btu/h (kW) based on a 70°F (21°C) temperature difference between stored water and ambient requirements. In the SL equation, V is the rated volume in gallons and Q is the nameplate input rate in Btu/h (kW). Vm is the measured volume in the tank in gallons.

[2] ASHRAE 90.1 contains a complete specification, including the year version, of the referenced test procedure.

[3] Section G.1 is titled "Test Method for Measuring Thermal Efficiency" and Section G.2 is titled "Test Method for Measuring Standby Loss."

[4] Instantaneous water heaters with input rates below 200 000 Btu/h (58.6 kW) shall be in accordance with these requirements where the water heater is designed to heat water to temperatures of 180°F (82°C) or higher.

[5] Electric water heaters with input rates less than 40 946 Btu/h (12 kW) shall be in accordance with these requirements where the water heater is designed to heat water to temperatures of 180°F (82°C) or higher.

[6] Refer to Section L 503.4.3 for additional requirements for gas storage and instantaneous water heaters and gas hot water supply boilers.

[7] In the U.S., the efficiency requirements for water heaters or gas pool heaters in this category or subcategory are specified by the U.S. Department of Energy. Those requirements and applicable test procedures are found in the Code of Federal Regulations 10 CFR Part 430.

TABLE L 503.3.3
MINIMUM PIPING INSULATION THICKNESS FOR HEATING AND HOT-WATER SYSTEMS
(STEAM, STEAM CONDENSATE, HOT-WATER HEATING, AND DOMESTIC WATER SYSTEMS)
[ASHRAE 90.1: TABLE 6.8.3-1]

FLUID OPERATING TEMPERATURE RANGE AND USAGE (°F)	INSULATION CONDUCTIVITY		≥NOMINAL PIPE SIZE OR TUBE SIZE (inches)				
	CONDUCTIVITY Btu•inch/(h•ft²•°F)	MEAN RATING TEMPERATURE (°F)	<1	1 to <1½	1½ to <4	4 to <8	≥8
			INSULATION THICKNESS (inches)				
>350	0.32 to 0.34	250	4.5	5.0	5.0	5.0	5.0
251 to 350	0.29 to 0.32	200	3.0	4.0	4.5	4.5	4.5
201 to 250	0.27 to 0.30	150	2.5	2.5	2.5	3.0	3.0
141 to 200	0.25 to 0.29	125	1.5	1.5	2.0	2.0	2.0
105 to 140	0.22 to 0.28	100	1.0	1.0	1.5	1.5	1.5

For SI units: °C=(°F-32)/1.8, 1 British thermal unit inch per hour square foot degree Fahrenheit = [0.1 W/(m•K)], 1 inch = 25 mm

Notes:

1 For insulation outside the stated conductivity range, the minimum thickness (T) shall be determined as follows:
$T = r\{(1 + t/r)^{K/k} - 1\}$ Where:
T = minimum insulation thickness (inches) (mm).
r = actual outside radius of pipe (inches) (mm).
t = insulation thickness listed in this table for applicable fluid temperature and pipe size.
K = conductivity of alternate material at mean rating temperature indicated for the applicable fluid temperature [Btu•in/(h•ft²•°F)] [W/(m•K)].
k = the upper value of the conductivity range listed in this table for the applicable fluid temperature.

2 These thicknesses are based on energy efficiency considerations only. Additional insulation is sometimes required relative to safety issues or surface temperature.

3 For piping 1½ inches (40 mm) or less, and located in partitions within conditioned spaces, reduction of insulation thickness by 1 inch (25.4 mm) shall be permitted before thickness adjustment required in Footnote 1, but not a thickness less than 1 inch (25.4 mm).

4 For direct-buried heating and hot water system piping, reduction of insulation thickness by 1½ inch (38 mm) shall be permitted before thickness adjustment required in Footnote 1, but not a thickness less than 1 inch (25.4 mm).

5 Table L 503.3.3 is based on steel pipe. Non-metallic pipes, Schedule 80 thickness or less shall use the table values. For other non-metallic pipes having a thermal resistance more than that of steel pipe, reduced insulation thicknesses shall be permitted where documentation is provided showing that the pipe with the proposed insulation has no more heat transfer per foot (mm) than a steel pipe of the same size with the insulation thickness shown in Table L 503.3.3.

(2) It is demonstrated to the satisfaction of the Authority Having Jurisdiction that the use of a single heat source will consume less energy than separate units.

(3) The energy input of the combined boiler and water heater system is less than 150 000 British thermal units per hour (Btu/h) (44 kW). [ASHRAE 90.1:7.5.1]

L 503.4.2 Service Water Heating Equipment. Service water-heating equipment used to provide the additional function of space heating as part of a combination (integrated) system shall satisfy stated requirements for the service water-heating equipment. [ASHRAE 90.1:7.5.2]

L 503.4.3 Buildings with High-Capacity Service Water Heating Systems. New buildings with gas service hot-water systems with a total installed gas water-heating input capacity of 1 000 000 Btu/h (293 kW) or more, shall have gas service water-heating equipment with a thermal efficiency (E_t) of not less than 90 percent. Multiple units of gas water-heating equipment shall be permitted to comply with this requirement where the water-heating input provided by the equipment, with thermal efficiency (E_t) of more or less than 90 percent, provides an input capacity-weighted average thermal efficiency of not less than 90 percent.

Exceptions:

(1) Where 25 percent of the annual service water-heating requirement is provided by site-solar or site-recovered energy.

(2) Water heaters installed in individual dwelling units.

(3) Individual gas water heaters with input capacity, not more than 100 000 Btu/h (29.3 kW). [ASHRAE 90.1:7.5.3]

L 503.4.4 Heat Recovery for Service Water Heating. Condenser heat recovery systems shall be installed for heating or preheating of service hot water provided the following are true:

(1) The facility operates 24 hours a day.

(2) The total installed heat rejection capacity of the water-cooled systems exceeds 6 000 000 Btu/h (1758 kW) of heat rejection.

(3) The design service water-heating load exceeds 1 000 000 Btu/h (293 kW). [ASHRAE 90.1:6.5.6.2.1]

L 503.4.5 Capacity. The required heat recovery system shall have the capacity to provide the smaller of:

(1) Sixty percent of the peak heat-rejection load at design conditions.

(2) Preheat of the peak service hot-water draw to 85°F (29°C).

Exceptions:

(1) Facilities that employ condenser heat recovery for space heating with a heat recovery design exceeding 30 percent of the peak water-cooled condenser load at design conditions.

(2) Facilities that provide 60 percent of their service water heating from site-solar or site-recovered energy or other sources. [ASHRAE 90.1:6.5.6.2.2]

L 503.5 Submittals. The Authority Having Jurisdiction shall require submittal of compliance documentation and supplemental information in accordance with Section 104.3.1 of this code.

L 504.0 Solar Water Heating Systems.

L 504.1 General. The erection, installation, alteration, addition to, use or maintenance of solar water heating systems shall be in accordance with this section and the Uniform Solar Energy and Hydronics Code.

L 504.2 Annual Inspection and Maintenance. Solar energy systems that utilize a heat transfer fluid shall annually be inspected unless inspections are required on a more frequent basis by the solar energy system manufacturer.

L 505.0 Hard Water.

L 505.1 Softening and Treatment. Where water has a hardness equal to or exceeding 10 gr/gal (171 mg/L) measured as total calcium carbonate equivalents, the water supply line to water heating equipment and the circuit of boilers shall be softened or treated to prevent accumulation of limescale and consequent reduction in energy efficiency.

L 506.0 Drain Water Heat Exchangers.

L 506.1 General. Drain water heat exchangers shall comply with IAPMO PS 92. The heat exchanger shall be accessible.

L 507.0 Heat Recovery from Steam Boiler Blowdown.

L 507.1 General. Where heat recovery can be used beneficially to heat boiler makeup water or for other purposes, boiler blowdown from steam boilers exceeding 15 psi (103 kPa) and 3.4 E+06 Btu/h (996 kW) shall be directed to a heat recovery system that reduces the temperature of the blowdown discharge to below 140°F (60°C) without using tempering water.

L 601.0 Installer Qualifications.

L 601.1 Scope. The provisions of this section address minimum qualifications of installers of plumbing and mechanical systems covered within the scope of this appendix.

L 602.0 Qualifications.

L 602.1 General. Where permits are required, the Authority Having Jurisdiction shall have the authority to require contractors, installers, or service technicians to demonstrate competency. Where determined by the Authority Having Jurisdiction, the contractor, installer, or service technician shall be licensed to perform such work.

L 701.0 Method of Calculating Water Savings.

L 701.1 Purpose. The purpose of this section is to provide a means of estimating the water savings where installing plumbing and fixture fittings that use less water than the maximum required by the Energy Policy Act of 1992 and 2005 and this code.

L 701.2 Calculation of Water Savings. Table L 701.2(1) and Table L 701.2(2) shall be permitted to be used to establish a water use baseline in calculating the amount of water saved as a result of using plumbing fixtures and fixture fittings that use less water than the required maximum. Water use is determined by the following equation:

Water use = (flow rate or consumption) x (duration) x (occupants) x (daily uses)

TABLE L 701.2(1)
WATER USE BASELINE[5]

FIXTURE TYPE	MAXIMUM FLOW-RATE CONSUMPTION[2]	DURATION	ESTIMATED DAILY USES PER PERSON	OCCUPANTS[3, 4]
Showerheads	2.5 gpm at 80 psi	8 minutes	1	–
Private or Private Use Lavatory Faucets	2.2 gpm at 60 psi	0.25 minutes	4	–
Residential Kitchen Faucets	2.2 gpm at 60 psi	4 minutes	1	–
Wash Fountains	One 2.2 gpm at 60 psi fixture fitting for every 20 inches rim space	–	–	–
Lavatory Faucets in other than Residences, Apartments, and Private Bathrooms in Lodging Facilities	0.5 gpm	0.25 minutes	4	–
Metering Faucets	0.25 gallons/cycle	–	3	–
Metering Faucets for Wash Fountains	One 0.25 gallon per cycle fixture fitting for every 20 inches rim space	–	–	–
Water Closets	1.6 gallons per flush	1 flush	1 male[1]	–
			3 female	–
Urinals	1.0 gallons per flush	1 flush	2 male	–

For SI units: 1 gallon per minute = 0.06 L/s, 1 pound-force per square inch = 6.8947 kPa, 1 gallon = 3.785 L, 1 inch = 25.4 mm

Notes:

1 The daily use number shall be increased to three where urinals are not installed in the room.

2 The maximum flow rate or consumption is from the Energy Policy Act.

3 For residential occupancies, the number of occupants shall be based on two persons for the first bedroom and one additional person for each additional bedroom.

4 For nonresidential occupancies, refer to Table 422.1 for occupant load factors.

5 Where determining calculations, assume one use per person for metering or self-closing faucets.

TABLE L 701.2(2)
WATER SAVINGS CALCULATOR[1, 2, 3]

NONRESIDENTIAL BUILDINGS					
FIXTURE TYPE	**CONSUMPTION (gallons per minute)**	**DAILY USES**	**DURATION (minutes)**	**OCCUPANTS**	**DAILY WATER USES (gallons)**
1.6 gpf (gallons per flush) toilet - male	1.6	1	1	150	240
1.6 gpf toilet - female	1.6	3	1	150	720
1.0 gpf urinal - male	1	2	1	150	300
Commercial lavatory faucet - 0.5 gpm	0.5	3	0.25	300	113
Kitchen sink - 2.2 gpm	2.2	1	0.25	300	165
Showerhead - 2.5 gpm	2.5	0.1	8	300	600
				Total Daily Volume	2138
				Annual Work Days	260
				Total Annual Usage	555 750

NONRESIDENTIAL BUILDINGS					
FIXTURE TYPE	**CONSUMPTION (gallons per minute)**	**DAILY USES**	**DURATION (minutes)**	**OCCUPANTS**	**DAILY WATER USES (gallons)**
1.6 gpf toilet - male	1.28	1	1	150	192
1.6 gpf toilet - female	1.28	3	1	150	576
1.0 gpf urinal - male	0.5	2	1	150	150
Commercial lavatory faucet - 0.5 gpm	0.5	3	0.25	300	113
Kitchen sink - 2.2 gpm	2.2	1	0.25	300	165
Showerhead - 2.5 gpm	2.5	0.1	8	300	600
				Total Daily Volume	1796
				Annual Work Days	260
				Total Annual Usage	466 830
				Annual Savings	88 920
				% Reduction	-16.0 percent

For SI units: 1 gallon per minute = 0.06 L/s, 1 gallon = 3.785 L

Notes:

1 Consumption values shown as underlined reflect the maximum consumption values associated with the provisions called out in the IAPMO Green Plumbing & Mechanical Code Supplement.

2 Where metering faucets are used, insert the flow rate of the faucet in the "Consumption" column and insert the cycle time in the "Duration" column (assume 1 cycle per use).

3 To determine estimated savings, insert occupant values (same as Baseline) and consumption values based on fixtures and fixture fittings installed.

TABLE L 701.2(2) (continued)
WATER SAVINGS CALCULATOR[1, 2]

NONRESIDENTIAL BUILDINGS					
FIXTURE TYPE	**CONSUMPTION (gallons per minute)**	**DAILY USES**	**DURATION (minutes)**	**OCCUPANTS**	**DAILY WATER USES (gallons)**
1.6 gpf toilets	1.6	5	1	4	32
Lavatory faucet - 2.2 gpm	2.2	8	0.25	4	18
Kitchen sink - 2.2 gpm	2.2	6	0.25	4	13
Showerhead - 2.5 gpm	2.5	0.75	8	4	60
				Total Daily Volume	123
				Annual Work Days	44 822

NONRESIDENTIAL BUILDINGS					
FIXTURE TYPE	**CONSUMPTION (gallons per minute)**	**DAILY USES**	**DURATION (minutes)**	**OCCUPANTS**	**DAILY WATER USES (gallons)**
1.6 gpf toilet - male	1.28	5	1	4	26
Lavatory faucet - 1.5 gpm	1.5	8	0.25	4	12
Kitchen sink - 2.2 gpm	2.2	6	0.25	4	13
Showerhead - 2.5 gpm	2.5	0.75	8	4	60
				Total Daily Volume	111
				Annual Usage	40 442
				Annual Savings	4380
				% Reduction	-9.8 percent

For SI units: 1 gallon per minute = 0.06 L/s, 1 gallon = 3.785 L

Notes:

1 Consumption values shown as underlined reflect the maximum consumption values associated with the provisions called out in the IAPMO Green Plumbing & Mechanical Code Supplement.

2 To determine estimated savings, insert occupant values (same as Baseline) and consumption values based on fixtures and fixture fittings installed.

Notes and instructions for Table L 701.2(2):

Table L 701.2(2) is an example of a calculator that is capable of helping estimate water savings in residential and nonresidential structures. The "Duration" of use and "Daily Uses" values that appear in the table are estimates and based on previous studies. The first example shown below is a commercial office building with 300 occupants, 150 females, and 150 males. The second example is a 3 bedroom residential building. To obtain and use a working copy of this calculator, follow the download and use instructions below.

Instructions for download:

1. Go to the IAPMO website at www.iapmogreen.org to download the water savings calculator. The calculator is a Microsoft Office Excel file (1997 or later), your computer must be capable of running MS Excel.
2. Follow the instructions for downloading and running the file.

Instructions for use:

1. In the Baseline Case section, insert the number of total occupants, male occupants and female occupants that apply to the building in the "Occupants" column. Unless specific gender ratio values are provided, assume a 50/50 gender ratio.
2. Copy and paste these same values in the "Occupants" column of the Calculator section.
3. In the Calculator section, insert the consumption values (flow rates in gpm or gallons per flush or per cycle) in the "Consumption" column.
4. Estimated water savings regarding percent savings versus baseline values, gallons per day and gallons per year will be automatically calculated.

APPENDIX M

PEAK WATER DEMAND CALCULATOR

M 101.0 General.

M 101.1 Applicability. This appendix provides a method for estimating the demand load for the building water supply and principal branches for single- and multi-family dwellings with water-conserving plumbing fixtures, fixture fittings, and appliances.

M 102.0 Demand Load.

M 102.1 Water-Conserving Fixtures. Plumbing fixtures, fixture fittings, and appliances shall not exceed the design flow rate in Table M 102.1.

TABLE M 102.1
DESIGN FLOW RATE FOR WATER-CONSERVING PLUMBING FIXTURES AND APPLIANCES IN RESIDENTIAL OCCUPANCIES

FIXTURE AND APPLIANCE	MAXIMUM DESIGN FLOW RATE (gallons per minute)
Bar Sink	1.5
Bathtub	5.5
Bidet	2.0
Clothes Washer*	3.5
Combination Bath/Shower	5.5
Dishwasher*	1.3
Kitchen Faucet	2.2
Laundry Faucet (with aerator)	2.0
Lavatory Faucet	1.5
Shower, per head	2.0
Water Closet, 1.28 GPF Gravity Tank	3.0

For SI units: 1 gallon per minute = 0.06 L/s
* Clothes washers and dishwashers shall have an energy star label.

M 102.2 Water Demand Calculator. The estimated design flow rate for the building supply and principal branches and risers shall be determined by the IAPMO Water Demand Calculator available for download at http://www.iapmo.org/WEStand/Pages/WaterDemandCalculator.aspx

M 102.3 Meter and Building Supply. To determine the design flow rate for the water meter and building supply, enter the total number of indoor plumbing fixtures and appliances for the building in Column [B] of the Water Demand Calculator and run Calculator. See Table M 102.3 for an example.

M 102.4 Fixture Branches and Fixture Supplies. To determine the design flow rate for fixture branches and risers, enter the total number of plumbing fixtures and appliances for the fixture branch or riser in Column [B] of the Water Demand Calculator and run Calculator. The flow rate for one fixture branch and one fixture supply shall be the design flow rate of the fixture according to Table M 102.1.

M 102.5 Continuous Supply Demand. Continuous supply demands in gallons per minute (gpm) for lawn sprinklers, air conditioners, hose bibbs, etc., shall be added to the total estimated demand for the building supply as determined by Section M 102.3. Where there is more than one hose bibb installed on the plumbing system, the demand for only one hose bibb shall be added to the total estimated demand for the building supply. Where a hose bibb is installed on a fixture branch, the demand of the hose bibb shall be added to the design flow rate for the fixture branch as determined by Section M 102.4.

M 102.6 Other Fixtures. Fixtures not included in Table M 102.1 shall be added in Rows 12 through 14 in the Water Demand Calculator as Other Fixture. The probability of use and flow rate for Other Fixtures shall be added by selecting the comparable probability of use and flow rate from Columns [C] and [E].

M 102.7 Size of Water Piping per Appendix A. Except as provided in Section M 102.0 for estimating the demand load for single- and multi-family dwellings, the size of each water piping system shall be determined in accordance with the procedure set forth in Appendix A. After determining the permissible friction loss per 100 feet (30 480 mm) of pipe in accordance with Section A 104.0 and the demand flow in accordance with the Water Demand Calculator, the diameter of the building supply pipe, branches and risers shall be obtained from Chart A 105.1(1) through Chart A 105.1(7), whichever is applicable, in accordance with Section A 105.0 and Section A 106.0. Velocities shall be in accordance with Section A 107.0. Appendix I (IS 31), Figure 3 and Figure 4 shall be permitted when sizing PEX systems.

M 102.7.1 Minimum Fixture Branch Size. The minimum fixture branch size shall be $^{1}/_{2}$ inch (15 mm) in diameter.

TABLE M 102.3
WATER DEMAND CALCULATOR EXAMPLE

	[A] FIXTURE	[B] ENTER NUMBER OF FIXTURES	[C] PROBABILITY OF USE (%)	[D] ENTER FIXTURE FLOW RATE (GPM)	[E] MAXIMUM RECOMMENDED FIXTURE FLOW RATE (GPM)
1	Bar Sink	0	2.0	1.5	1.5
2	Bathtub	0	1.0	5.5	5.5
3	Bidet	0	1.0	2.0	2.0
4	Clothes Washer	1	5.5	3.5	3.5
5	Combination Bath/Shower	1	5.5	5.5	5.5
6	Dishwasher	1	0.5	1.3	1.3
7	Kitchen Faucet	1	2.0	2.2	2.2
8	Laundry Faucet	0	2.0	2.0	2.0
9	Lavatory Faucet	1	2.0	1.5	1.5
10	Shower, per head	0	4.5	2.0	2.0
11	Water Closet, 1.28 GPF Gravity Tank	1	1.0	3.0	3.0
12	Other Fixture 1	0	0.0	0.0	6.0
13	Other Fixture 2	0	0.0	0.0	6.0
14	Other Fixture 3	0	0.0	0.0	6.0
Total Number of Fixtures		6		RESET	RUN WATER DEMAND CALCULATOR
99th Percentile Demand Flow =		8.5 GPM			

For SI units: 1 gallon per minute = 0.66 L/s, 1 gallon = 3.785 L

M 102.8 Examples Illustrating Use of Water Demand Calculator with Appendix A.

Example 1: Indoor Water Use Only – Use the information given below to find the pipe size for the building supply to a residential building with six indoor fixtures as shown in Figure 1 [Pipe Section 4].

Given Information:

Type of construction:	Residential, one-bathroom	Friction loss per 100 ft (30 480 mm):	15 psi (103 kPa)
Type of pipe material:	L-copper	Maximum velocity:	10 ft/s (3.05 m/s)
Fixture number/type:	1 combination bath/shower 1 dishwasher	1 kitchen faucet 1 WC	1 lavatory faucet 1 clothes washer

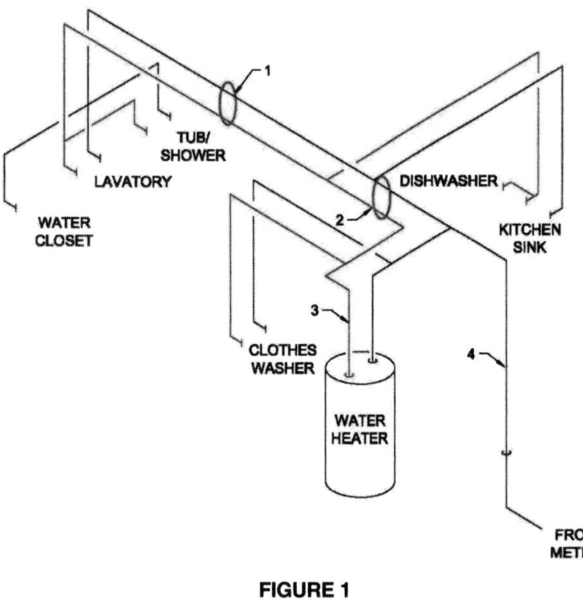

FIGURE 1
RESIDENTIAL BUILDING WITH SIX INDOOR FIXTURES

Solution: Step 1 of 2 – Find Demand Load for the Building Supply.

The Water Demand Calculator [WDC] in Figure 2 is used to determine the demand load expected from indoor water use. The WDC has white-shaded cells and light gray-shaded cells. The values in the light gray cells are derived from a national survey of indoor water use at homes with efficient fixtures and cannot be changed.

The white-shaded cells accept input from the designer. For instance, fixture counts from the given information are entered in Column [B]; the corresponding recommended fixture flow rates are already provided in Column [D]. The flow rates in Column [D] may be reduced only if the manufacturer specifies a lower flow rate for the fixture. Column [E] establishes the upper limits for the flow rates entered into Column [D]. Clicking the Run Water Demand Calculator button gives 8.5 gpm (0.54 L/s) as the estimated indoor water demand for the whole building. This result appears in the dark gray box of the WDC in Figure 2.

[A] FIXTURE		[B] ENTER NUMBER OF FIXTURES	[C] PROBABILITY OF USE (%)	[D] ENTER FIXTURE FLOW RATE (GPM)	[E] MAXIMUM RECOMMENDED FIXTURE FLOW RATE (GPM)
1	Bar Sink	0	2.0	1.5	1.5
2	Bathtub	0	1.0	5.5	5.5
3	Bidet	0	1.0	2.0	2.0
4	Clothes Washer	1	5.5	3.5	3.5
5	Combination Bath/Shower	1	5.5	5.5	5.5
6	Dishwasher	1	0.5	1.3	1.3
7	Kitchen Faucet	1	2.0	2.2	2.2
8	Laundry Faucet	0	2.0	2.0	2.0
9	Lavatory Faucet	1	2.0	1.5	1.5
10	Shower, per head	0	4.5	2.0	2.0
11	Water Closet, 1.28 GPF Gravity Tank	1	1.0	3.0	3.0
12	Other Fixture 1	0	0.0	0.0	6.0
13	Other Fixture 2	0	0.0	0.0	6.0
14	Other Fixture 3	0	0.0	0.0	6.0
Total Number of Fixtures		6		RESET	RUN WATER DEMAND CALCULATOR
99th Percentile Demand Flow =		8.5 GPM			

For SI units: 1 gallon per minute = 0.66 L/s, 1 gallon = 3.785 L

FIGURE 2
WATER DEMAND CALCULATOR FOR INDOOR USE AT HOME WITH SIX EFFICIENT FIXTURES (EXAMPLE 1).

Solution: Step 2 of 2 – Determine the Pipe Size of the Building Supply.

Chart A 105.1(1) for copper piping systems (from Appendix A of the UPC, shown in Figure 3) is used to determine the pipe size, based on given friction loss, given maximum allowable pipe velocity, given pipe material and the demand load computed in Step 1. In Figure 3, the intersection of the given friction loss (15 psi) (103 kPa) and the maximum allowable pipe velocity (10 ft/s) (3.05 m/s) is labeled point A. The vertical line that descends from point A to the base of the chart intersects four nominal sizes for L-copper pipe. These intersection points are labeled B, C, D, E and correspond to pipe sizes of 1 inch (25 mm), 3/4 inch (20 mm), 1/2 inch (15 mm) and 3/8 inch (10 mm), respectively. A horizontal line from points B, C, D, E to the right-hand side of the chart gives maximum flow rates of 24 gpm (1.5 L/s), 12 gpm (0.757 L/s), 4.5 gpm (0.28 L/s), and 2.3 gpm (0.145 L/s), respectively. These results are summarized in Table 1 which shows that a 3/4 inch (20 mm) type L copper line is the minimum size that can convey the peak water demand of 8.5 gpm (0.54 L/s).

TABLE 1
PIPE SIZE OPTIONS FOR BUILDING SUPPLY

POINT IN FIGURE 3	PIPE DIAMETER (INCH)	MAXIMUM FLOW (GPM)	OK FOR BUILDING SUPPLY*
E	3/8	2.3	No
D	1/2	4.5	No
C	3/4	12	Yes
B	1	24	Yes

For SI units: 1 inch = 25 mm, 1 gallon per minute = 0.06 L/s
* For Building in Examples 1, 2, 3, and 4.

CHART A 105.1(1)

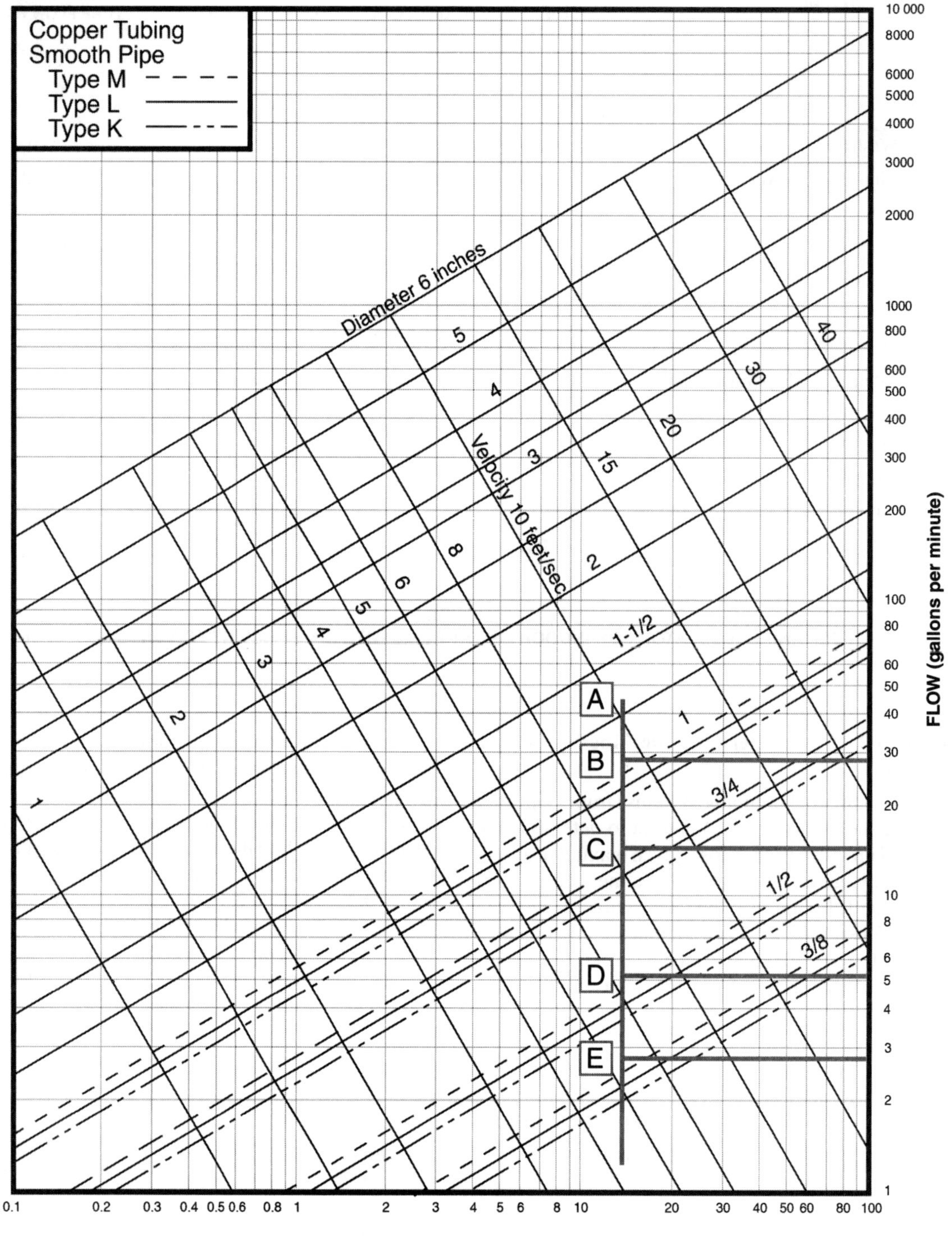

FIGURE 3
CHART A 105.1(1) FOR FINDING PIPE SIZE

For SI units: 1 foot = 304.8 mm, 1 gallon per minute = 0.06 L/s), 1 pound-force per square inch = 6.8947 kPa

Example 2: Indoor and Outdoor Water Use – Find the pipe size for the building supply [Figure 1, Pipe Section 4] if the building in Example 1 adds two outdoor fixtures (hose bibb, each with a fixture flow of 2.0 gpm) (0.13 L/s).

Solution: Step 1 of 2 – Find Demand Load for the Building Supply.

The WDC has been developed exclusively for peak indoor water use which can be viewed as a high-frequency short duration process. Because fixtures for outdoor water use may operate continuously for very long periods, they are not included in the WDC. To account for water use from one or more outdoor fixtures, add the demand of the single outdoor fixture with the highest flowrate to the calculated demand for indoor water use. With two hose bibbs, the demand of only one hose bibb is included. Hence, in this example, the total demand for the whole house is 8.5 gpm (0.54 L/s) + 2.0 gpm (0.13 L/s) = 10.5 gpm 0.662 L/s).

Solution: Step 2 of 2 – Determine the Pipe Size of the Building Supply.

Table 1 shows that at 10.5 gpm (0.662 L/s) the building supply shall be 3/4 inch (20 mm) in diameter.

Example 3: Indoor, Outdoor and Other Fixture Water Use – Find the pipe size for the water supply [Figure 1, Pipe Section 4] if the building in Example 2 adds a kitchen pot filler and a dog bath each with a faucet flow rate of 5.5 gpm (0.35 L/s).

Solution: Step 1 of 2 – Find Demand Load for the Building Supply.

The kitchen pot filler and dog bath are not listed in Column [A] of the WDC. To accommodate cases such as this, the WDC provides up to three additional rows for "Other Fixtures". Enter the kitchen pot filler and dog bath in Column [A] of the WDC and enter the fixture count for each in Column [B]. Find an indoor fixture that has a similar probability of use in Column [C] and add that to the column. Finally, enter the flow rate of the kitchen pot filler and dog bath in Column [D]. The estimated indoor water demand for the whole building is 11 gpm, as shown in the WDC in Figure 4.

As illustrated in Example 2, the hose bibb will increase the total demand for the whole house to 13 gpm (0.820 L/s).

Note that a reset button is provided to clear any numbers in Column [B] from a previous calculation.

Solution: Step 2 of 2 – Determine the Pipe Size of the Building Supply.

Table 1 shows that at 13 gpm (0.820 L/s) the building supply shall be 1inch (25 mm) in diameter.

Example 4: Sizing Branches and Risers – For individual hot and cold branches, repeat Steps 1 and 2. For example, for the hot water branch at the water heater [Figure 1, Pipe Section 3], enter all the fixtures and appliances that use hot water into the Water Demand Calculator (toilets will be excluded) as seen in Figure 5. Use the calculated demand load to find the pipe size in Step 2. Table 1 shows that at 7.7 (0.49 L/s) gpm, the hot water branch shall be 3/4 inch (20 mm) in diameter.

For each additional hot and cold branch [Figure 1, Pipe Sections 1 and 2], enter the number of fixtures and appliances served by that branch into the WDC and use that demand in Step 2 to determine the branch size. If the branch serves a hose bibb, add the demand of the hose bibb to the calculated demand flow for the branch. As discussed in Example 2, the hose bibb is not to be entered into WDC, since the Calculator is for indoor uses only.

When there is only one fixture or appliance served by a fixture branch, the demand flow shall not exceed the fixture flow rate in Column [E] of the Water Demand Calculator. The fixture flow rate would be used in Step 2 to determine the size of the fixture branch and supply.

[A] FIXTURE		[B] ENTER NUMBER OF FIXTURES	[C] PROBABILITY OF USE (%)	[D] ENTER FIXTURE FLOW RATE (GPM)	[E] MAXIMUM RECOMMENDED FIXTURE FLOW RATE (GPM)
1	Bar Sink	0	2.0	1.5	1.5
2	Bathtub	0	1.0	5.5	5.5
3	Bidet	0	1.0	2.0	2.0
4	Clothes Washer	1	5.5	3.5	3.5
5	Combination Bath/Shower	1	5.5	5.5	5.5
6	Dishwasher	1	0.5	1.3	1.3
7	Kitchen Faucet	1	2.0	2.2	2.2
8	Laundry Faucet	0	2.0	2.0	2.0
9	Lavatory Faucet	1	2.0	1.5	1.5
10	Shower, per head	0	4.5	2.0	2.0
11	Water Closet, 1.28 GPF Gravity Tank	1	1.0	3.0	3.0
12	Pot Filler	1	2.0	5.5	6.0
13	Dog Bath	1	1.0	5.5	6.0
14	Other Fixture 3	0	0.0	0.0	6.0
Total Number of Fixtures		8		RESET	RUN WATER DEMAND CALCULATOR
99th Percentile Demand Flow =		11.0 GPM			

For SI units: 1 gallon per minute = 0.66 L/s, 1 gallon = 3.785 L

FIGURE 4
WATER DEMAND CALCULATOR TO ACCOMMODATE OTHER FIXTURES (EXAMPLE 3).

[A] FIXTURE		[B] ENTER NUMBER OF FIXTURES	[C] PROBABILITY OF USE (%)	[D] ENTER FIXTURE FLOW RATE (GPM)	[E] MAXIMUM RECOMMENDED FIXTURE FLOW RATE (GPM)
1	Bar Sink	0	2.0	1.5	1.5
2	Bathtub	0	1.0	5.5	5.5
3	Bidet	0	1.0	2.0	2.0
4	Clothes Washer	1	5.5	3.5	3.5
5	Combination Bath/Shower	1	5.5	5.5	5.5
6	Dishwasher	1	0.5	1.3	1.3
7	Kitchen Faucet	1	2.0	2.2	2.2
8	Laundry Faucet	0	2.0	2.0	2.0
9	Lavatory Faucet	1	2.0	1.5	1.5
10	Shower, per head	0	4.5	2.0	2.0
11	Water Closet, 1.28 GPF Gravity Tank	0	1.0	3.0	3.0
12	Other Fixture 1	0	0.0	0.0	6.0
13	Other Fixture 2	0	0.0	0.0	6.0
14	Other Fixture 3	0	0.0	0.0	6.0
Total Number of Fixtures		5		RESET	RUN WATER DEMAND CALCULATOR
99th Percentile Demand Flow =		7.7 GPM			

For SI units: 1 gallon per minute = 0.66 L/s, 1 gallon = 3.785 L

FIGURE 5
WATER DEMAND CALCULATOR FOR THE HOT WATER BRANCH (EXAMPLE 4).

APPENDIX N

IMPACT OF WATER TEMPERATURE ON THE POTENTIAL FOR SCALDING AND LEGIONELLA GROWTH

N 101.0 General.

N 101.1 Applicability. This appendix provides guidelines on the impact of water temperature in minimizing both scalding and Legionella growth potential associated with occupiable commercial, institutional, multi-unit residential, and industrial building plumbing systems.

This Appendix shall not include single-family residential buildings. This appendix shall not be considered a risk management guidance document for scalding or Legionella.

Note: Published documents which address Legionella risk management include ASHRAE 188, ASHRAE Guideline 12, or as required by the Authority Having Jurisdiction.

There are additional factors associated with the potential for scalding and Legionella growth other than temperature.

For scalding potential, other factors include, but are not limited to, user age, health, body part, length of contact time, and water source.

For Legionella growth potential other factors include, but are not limited to, water source and plumbing system: size, design, circulation rate, water age, disinfectant residual, piping material and component complexity.

N 102.0 Definitions.

N 102.1 Cold Water. Water at a temperature less than 77°F (25°C).

N 102.2 Tepid Cold Water. Water at a temperature not less than 77°F (25°C) and less than 85°F (29°C).

N 102.3 Tepid Water. Water at a temperature not less than 85°F (29°C) and less than 110°F (43°C).

N 102.4 Warm Water. Water at a temperature not less than 110°F (43°C) and less than 120°F (49°C).

N 102.5 Tempered Hot Water. Water at a temperature not less than 120°F (49°C) and less than 130°F (54°C).

N 102.6 Hot Water. Water at a temperature not less than 130°F (54°C) and less than 140°F (60°C).

N 102.7 Very Hot Water. Water at a temperature not less than 140°F (60°C) and less than 160°F (71°C).

N 102.8 Disinfecting Hot Water. Water at a temperature not less than 160°F (71°C).

N 102.9 Legionella Growth Potential. The likelihood that Legionella bacteria will reproduce.

N 102.10 Scald Potential. The likelihood of burning the skin.

N 103.0 Building Water System Design Documentation.

N 103.1 Required Documentation. Construction documents shall be required for new construction, renovation, refurbishment, replacement, or repurposing of an occupiable building water system and shall be submitted to the Authority Having Jurisdiction.

N 104.0 Potential Exposure.

N 104.1 Legionella Growth Potential. The Authority Having Jurisdiction shall have the authority to require documentation to address Legionella growth potential, where water temperatures in a water distribution system are within ranges shown in Table N 104.1 that pose a Legionella growth potential.

N 104.2 Scald Potential. Where the water distribution system's water temperature(s) range poses a scald potential in accordance with Table N 104.1, protection shall be provided in accordance with Chapter 4.

N 105.0 Disinfection.

N 105.1 Disinfection Documentation. Where required by the Authority Having Jurisdiction, documentation for disinfection of all building water systems shall be provided by the registered design professional in the construction documents.

Methods for new construction and any repaired system disinfection shall include, but not be limited to, the chlorination methods and procedures for flushing and disinfection in accordance with Section 609.10.

Other or alternative water treatment methods for disinfection shall include, but not be limited to, one of the following methods:

(1) Copper-silver ionization methods and procedures, including the following documentation.
   - (a) Copper and silver ionization concentrations shall be included in the documentation.
   - (b) Methods and documentation for monitoring ion levels.
   - (c) Electrode cleaning cycles and methods shall be reported.

(2) Ultraviolet light methods shall include the following documentation:
   - (a) Locations of ultraviolet light units.
   - (b) Cleaning cycles and methods of the quartz sleeves and housing shall be documented.

TABLE N 104.1
CORRELATION BETWEEN TEMPERATURE RANGES, LEGIONELLA, AND SCALD POTENTIAL

WATER DESCRIPTION	TEMPERATURE (°F)	SCALD POTENTIAL[1]	LEGIONELLA GROWTH POTENTIAL[2]
Cold	<77	None	Minimal
Tepid Cold	≥77 and <85	None	Low
Tepid	≥85 and <110	None Hyperthermia is possible after long exposure in a bathtub or whirlpool tub.	High
Warm	≥110 and <120	Minimal At 111°F, greater than 220 minutes for second-degree burn.	Moderate
Tempered Hot	≥120 and <130	Low At 120°F, greater than 5 minutes for second-degree burn, and 10 minutes to third-degree burn; At 124°F, two minutes for second-degree burn, and 4 minutes, 10 seconds for third-degree burn.	Low
Hot	≥130 and <140	Moderate to High At 130°F, 18 seconds for second-degree burn, and 30 seconds for third-degree burn.	None
Very Hot	≥140 and <160	High At 140°F, three seconds for second-degree burn, and 5 seconds for third-degree burn; At 150°F, instant for second-degree burn, and less than two seconds for third-degree burn; At 158°F, instant for second-degree burn, and less than a second for third-degree burn.	None
Disinfecting Hot	≥160	Immediate	None

For SI units: °C = (°F-32)/1.8

Notes:

1 The infant, elderly, and infirmed have a higher potential for scalding at temperatures lower than listed.

2 Temperature ranges reported are experimentally determined in a laboratory setting in the absence of a realistic microbial community. Legionella can survive for longer periods of time at temperatures higher and lower than the growth temperature ranges indicated due to changes in their metabolic state and/or protection from thermal disinfection within biofilm or amoeba host organisms.

USEFUL TABLES

CONVERSION TABLES

Note: The information contained in these tables are not part of this American National Standard (ANS) and have not been processed in accordance with ANSI's requirements for an ANS. As such, these tables may contain material that has not been subjected to public review or a consensus process. In addition, they do not contain requirements necessary for conformance to the standard.

UNIT CONVERSIONS

MULTIPLY	BY	TO OBTAIN
Acres	43 560	Square feet (ft^2)
Acre-feet	43 560	Cubic feet (ft^3)
Acre-feet	325 851	Gallons (U. S. liquid)
Atmosphere (atm)	76.0	Centimeters of mercury (0°C)
Atmosphere (standard) (atm)	76.0	Centimeters of mercury (0°C)
Atmosphere (standard) (atm)	33.90	Feet of water (4°C)
Atmosphere (standard) (atm)	29.92	Inches of mercury
Atmosphere (standard) (atm)	101.32501	Kilopascals (kPa)
Atmosphere (standard) (atm)	14.70	Pounds-force/square inch (lbf/in^2)
Barrels	42	Gallons (gal)
Barrels	158.9873	Liters (L)
British thermal units (Btu)	1055.055	Joules (J)
British thermal units/hour (Btu/h)	0.000293	Kilowatts (kW)
British thermal units/hour (Btu/h)	0.293	Watts (W)
British thermal unit hour (Btu/h)	1.055056	Kilojoules (kJ)
British thermal unit hour (Btu/h)	1055.056	Joule (J)
British thermal units/minute (Btu/min)	12.97	Foot pounds-force/second (ft•lbf/s)
British thermal units/minute (Btu/min)	0.02358	Horsepower (hp) (international)
British thermal unit per hour square foot degree Fahrenheit [Btu/(h•ft^2•°F)]	5.678263	Watt per square meter kelvin [W/(m^2•K)]
British thermal unit inch per hour foot degree Fahrenheit [Btu•in/(h•ft•°F)]	0.14442279	Watt per meter kelvin [W/(m•K)]
Candelas per square inch (cd/in^2)	1.550003×10^3	Candela per square meters (cd/m^2)
Celsius (°C)	°C + 273.15	Kelvin (k)
Celsius (°C)	1.8 x °C + 32	Fahrenheit (°F)
Centimeters (cm)	0.3937	Inches (in)
Centimeters of mercury (0°C)	0.01316	Atmosphere (standard)
Centimeters of mercury (0°C)	0.446	Feet of water (4°C)
Centimeters of mercury (0°C)	27.84	Pounds-force/square feet (lbf/ft^2)
Centimeters of mercury (0°C)	0.1934	Pounds-force/square inch (lbf/in^2)
Circumference	6.283	Radians (rad)
Cubic feet (ft^3)	1728	Cubic inches (in^3)
Cubic feet (ft^3)	0.0283	Cubic meters (m^3)
Cubic feet (ft^3)	0.03704	Cubic yards (yd^3)
Cubic feet (ft^3)	7.48052	Gallons (U.S. liquid) (gal)
Cubic feet (ft^3)	28.32	Liters (L)
Cubic feet (ft^3)	29.92	Quarts (U.S. liquid)
Cubic feet/hour (ft^3/h)	0.0283	Cubic meters/hour (m^3/h)

UNIT CONVERSIONS (continued)

MULTIPLY	BY	TO OBTAIN
Cubic feet/minute (ft^3/min)	0.000472	Cubic meters/second (m^3/s)
Cubic feet/minute (ft^3/min)	0.1247	Gallons/second
Cubic feet/minute (ft^3/min)	0.47194	Liters/second (L/s)
Cubic feet/second (ft^3/s)	646 316.89	Gallons/day
Cubic feet/second (ft^3/s)	448.831	Gallons/minute (gpm)
Cubic inches (in^3)	1.64×10^{-5}	Cubic meters (m^3)
Cubic inches (in^3)	0.01639	Liters (L)
Cubic meters (m^3)	264.17	Gallons (U.S. liquid)
Cubic yards (yd^3)	27	Cubic feet (ft^3)
Cubic yards (yd^3)	0.76455	Cubic meters (m^3)
Cubic yards (yd^3)	201.97	Gallons (U.S. liquid)
Degrees (deg)	0.0174	Radians (rad)
Fahrenheit (°F)	(°F-32)/1.8	Celsius (°C)
Feet (ft)	0.3048	Meters (m)
Feet (ft)	304.8	Millimeters (mm)
Feet of water (4°C)	0.0295	Atmosphere (standard) (atm)
Feet of water (4°C)	0.8827	Inches of mercury (0°C)
Feet of water (4°C)	62.43	Pounds-force/square feet
Feet of water (4°C)	0.4335	Pounds-force/square inch
Feet/minute (ft/min)	0.01667	Feet/second (ft/s)
Feet/minute (ft/min)	0.01136	Miles/hour (mi/h)
Feet/second (ft/s)	0.3048	Meters/second (m/s)
Feet/second (ft/s)	0.6818	Miles/hour (mi/h)
Feet/second (ft/s)	0.01136	Miles/minute (mi/min)
Foot lambert (fL)	3.426259	Candela per square meter (cd/m^2)
Foot pounds-force (ft•lbf)	1.355	Joules (J)
Foot pounds-force/minute (ft•lbf/min)	2.260×10^{-5}	Kilowatts (kW)
Foot pounds-force/second (ft•lbf/s)	1.356×10^{-3}	Kilowatts (kW)
Gallons (U.S. liquid) (gal)	231	Cubic inches (in^3)
Gallons (U.S. liquid) (gal)	0.003785	Cubic meters (m^3)
Gallons (U.S. liquid) (gal)	0.1337	Cubic feet (ft^3)
Gallons (U.S. liquid) (gal)	3.785	Liters (L)
Gallons (U.S. liquid) (gal)	4	Quarts (U.S. liquid)
Gallons/day	4.3×10^{-5}	Liters/second
Gallons/minute (gal/min) (gpm)	8.0208	Cubic feet/hour (ft^3/h)
Gallons/minute (gal/min) (gpm)	0.003785412	Cubic meters/minute (m^3/min)
Gallons/minute (gal/min) (gpm)	0.00223	Cubic feet/second (ft^3/s)
Gallons/minute (gal/min) (gpm)	6.309020×10^{-5}	Cubic meters/second (m^3/s)
Gallons/minute (gal/min) (gpm)	0.06309	Liters/second (L/s)
Grains (gr)	0.00006479	Kilograms (kg)
Horsepower (hp)	0.7457	Kilowatts (kW)
Horsepower-hours	$2.684 \times 10^{+6}$	Joules (J)
Horsepower-hours	0.7457	Kilowatt-hours (kW•h)
Inch (water column at 60°F)	0.24884	Kilopascals (kPa)
Inches (in)	2.54	Centimeters (cm)

UNIT CONVERSIONS (continued)

MULTIPLY	BY	TO OBTAIN
Inches (in)	25.4	Millimeters (mm)
Inches/hour	25.4	Millimeters/hour (mm/h)
Inches of mercury (0°C)	0.03342	Atmosphere (standard) (atm)
Inches of mercury (0°C)	1.133	Feet of water (4°C)
Inches mercury (0°C)	3.3863	Kilopascals (kPa)
Inches of mercury (0°C)	0.4912	Pounds-force/square inch
Inches of water (4°C)	0.002458	Atmosphere (standard) (atm)
Inches of water (4°C)	0.07356	Inches of mercury (0°C)
Inches of water (4°C)	5.202	Pounds-force/square feet
Inches of water (4°C)	0.03613	Pounds-force/square inch
Joules (J)	9.480×10^{-4}	British thermal units (Btus)
Joules (J)	0.7376	Foot-pounds
Joules (J)	2.778×10^{-4}	Watt-hours
Kelvin (K)	°K - 273.15	Celsius (°C)
Kilograms (kg)	2.2046	Pounds (lbs)
Kilograms (kg)	1.102×10^{-3}	Tons (short)
Kilopascals (kPa)	0.145038	Pounds-force/square inch
Kilometers (km)	0.6214	Miles (statute)
Kilometers/hour (km/h)	0.6214	Miles/hour (mi/h)
Kilowatts (kW)	3412.14	British thermal units/hour (Btus/hour)
Kilowatts (kW)	1.341	Horsepower (hp)
Kilowatt-hours	3413	British thermal units (Btus)
Kilowatt-hours	$2.655 \times 10^{+6}$	Foot-pounds (ft•lbs)
Kilowatt-hours	$3.6 \times 10^{+6}$	Joules (J)
Kip (1000 lbf)	4.448222	Kilonewtons (kN)
Kip-foot	1.35671	Kilonewton meters (kN•m)
Kips per square inch (kip/in^2)	6.8947	Megapascals (MPa)
Lambert (la)	3.183099×10^3	Candela per square meter (cd/m^2)
Liters (L)	0.03531	Cubic feet (ft^3)
Liters (L)	61.02	Cubic inches (in^3)
Liters (L)	0.001	Cubic meters (m^3)
Liters (L)	0.2642	Gallons (U.S. liquid)
Lumens per square foot (lm/ft^2) (footcandle)	10.76391	Lux (lx)
Meters (m)	3.281	Feet (ft)
Meters (m)	39.37	Inches (in)
Meters (m)	1.094	Yards
Meters/second (m/s)	3.281	Feet/second (ft/s)
Meters/second (m/s)	2.237	Miles/hour (mi/h)
Miles (mi)	5280	Feet (ft)
Miles (statute)	1.609	Kilometers (km)
Miles/hour (mi/h)	88	Feet/minute (ft/min)
Miles/hour (mi/h)	1.467	Feet/second (ft/s)
Miles/hour (mi/h)	1.609344	Kilometers/hour (km/h)
Miles/hour (mi/h)	26.82	Meters/minute (m/min)
Miles/hour (mi/h)	0.44704	Meters/second (m/s)

UNIT CONVERSIONS (continued)

MULTIPLY	BY	TO OBTAIN
Millimeters (mm)	0.1	Centimeters (cm)
Millimeters (mm)	0.03937	Inches (in)
Millimeters (mm)	0.001	Meters (m)
Minutes (min)	2.908882×10^{-4}	Radians (rads)
Ounces/square inch (oz/in^2)	43.94185	Kilograms/square meter (kg/m^2)
Ounces/square feet (oz/ft^2)	0.03051517	Kilograms/square meter (kg/m^2)
Ounces (oz)	0.02834	Kilograms (kg)
Ounces (oz)	29.57353	Millimeters (mL)
Pints	0.4731765	Liters (L)
Pound-force feet (lbf•ft)	1.355818	Newton meters (N•m)
Pound-force inch (lbf•in)	0.1129848	Newton meters (N•m)
Pound-force per foot (lbf/ft)	14.5939	Newton meters (N•m)
Pound-force per square foot (lbf/ft^2)	47.88026	Pascal (Pa)
Pound-force per square inch (lbf/in^2)	6.89476	Kilopascals (kPa)
Pound-force per inch (lbf/in)	175.1268	Newton meters (N•m)
Pounds/cubic inch (lb/in^3)	2.767990×10^{4}	Kilograms/cubic meter (kg/m^3)
Pounds/cubic yard (lb/yd^3)	0.5932764	Kilograms/cubic meter (kg/m^3)
Pounds (lb)	0.45359	Kilograms (kg)
Pounds/cubic foot (lb/ft^3)	16.0184	Kilograms/cubic meter (kg/m^3)
Pounds/foot (lb/ft)	1.4881	Kilograms/meters (kg/m)
Pounds/square inch (lb/in^2)	703.1	Kilograms-force/square meter (kg/m^2)
Pounds/square foot (lb/ft^2)	4.882427	Kilograms-force/square meter (kg/m^2)
Pounds-force (lbf)	4.4482	Newtons (N)
Pounds-force/square inch (psi)	0.06805	Atmosphere (standard) (atm)
Pounds-force/square inch (psi)	2.307	Feet of water (4°C)
Pounds-force/square inch (psi)	2.036	Inches of mercury (0°C)
Pounds-force/square inch (psi)	6.89476	Kilopascals (kPa)
Quarts (U.S. dry) (dry qt)	67.20	Cubic inches (in^3)
Quarts (U.S. liquid) (liq qt)	57.75	Cubic inches (in^3)
Quarts (liquid)	0.9463529	Liters (L)
Radians (rads)	57.30	Degrees (deg)
Seconds (s)	4.848137×10^{-6}	Radians (rads)
Square acre	0.404687	Square kilometers (km^2)
Square feet (ft^2)	144	Square inches (in^2)
Square feet (ft^2)	0.0929	Square meters (m^2)
Square inches (in^2)	645.16	Square millimeters (mm^2)
Square inches (in^2)	0.000645	Square meters (m^2)
Square meters (m^2)	1550	Square inches (in^2)
Square miles (mi^2)	640	Acres
Square miles (mi^2)	2.589988	Square kilometers (km^2)
Square millimeters (mm^2)	1.550×10^{-3}	Square inches (in^2)
Square yards (yd^2)	9	Square feet (ft^2)
Square yards (yd^2)	0.8361274	Square meters (m^3)
Temperature (°C) + 17.28	1.8	Temperature (°F)
Temperature (°F) – 32	$5/9$	Temperature (°C)

UNIT CONVERSIONS (continued)

MULTIPLY	BY	TO OBTAIN
Ton-force (tonf) (2000 lbf)	8.896443	Kilonewtons (kN)
Ton-force foot (tonf•f/ft)	2.71342	Kilonewton meters (kN•m)
Ton-force per square foot (tonf/ft^2)	95.7605	Kilopascals (kPa)
Ton-force per square inch (tonf/in^2)	13.7895	Megapascals (MPa)
Ton (metric)	1000	Kilograms (kg)
Tons (long) (2240 lbs)	1016.047	Kilograms (kg)
Tons (short)	2000	Pounds (lbs)
Watts	3.4121	British thermal units per hour (Btus/hour)
Watts	1.341×10^{-3}	Horsepower (hp)
Yards (y)	0.9144	Meters (m)

SI SYMBOLS AND PREFIXES

SI PREFIXES				
MULTIPLICATION FACTOR			PREFIX	SYMBOL
1 000 000 000 000 000 000	=	E+18	exa	E
1 000 000 000 000 000	=	E+15	peta	P
1 000 000 000 000	=	E+12	tera	T
1 000 000 000	=	E+09	giga	G
1 000 000	=	E+06	mega	M
1 000	=	E+03	kilo	k
100	=	E+02	hecto	h
10	=	E+01	deka	da
0.1	=	E-01	deci	d
0.01	=	E-02	centi	c
0.001	=	E-03	milli	m
0.000 001	=	E-06	micro	µ
0.000 000 001	=	E-09	nano	n
0.000 000 000 001	=	E-12	pico	p
0.000 000 000 000 001	=	E-15	femto	f
0.000 000 000 000 000 001	=	E-18	atto	a

FLOW IN GALLONS PER MINUTE CONVERTED TO FIXTURE UNITS FOR GRAVITY FLUSH TANKS AND FLUSHOMETERS

These tables may be interpolated

FLOW GPM	FIXTURE UNITS	
	FLUSH VALVE TANKS	FLUSHOMETER VALVES
1	0	—
2	1	—
3	3	—
4	4	—
5	6	—
6	7	—
7	8	—
8	10	—
9	12	—
10	13	—
11	15	—
12	16	—
13	18	—
14	20	—
15	21	—
16	23	—
17	24	—
18	26	—
19	28	—
20	30	—
21	32	—
22	34	5
23	36	6
24	39	7
25	42	8
26	44	9
27	46	10
28	49	11
29	51	12
30	54	13
31	56	14
32	58	15
33	60	16
34	63	18
35	66	20
36	69	21
37	74	23
38	78	25
39	83	26
40	86	28
41	90	30
42	95	31

FLOW GPM	FIXTURE UNITS	
	FLUSH VALVE TANKS	FLUSHOMETER VALVES
43	99	33
44	103	35
45	107	37
46	111	39
47	115	42
48	119	44
49	123	46
50	127	48
51	130	50
52	135	52
53	141	54
54	146	57
55	151	60
56	155	63
57	160	66
58	165	69
59	170	73
60	175	76
62	185	82
64	195	88
66	205	95
68	215	102
70	225	108
72	236	116
74	245	124
76	254	132
78	264	140
80	275	148
82	284	158
84	294	168
86	305	176
88	315	186
90	325	195
92	337	205
94	348	214
96	359	223
98	370	234
100	380	245
105	406	270
110	431	295
115	455	329
120	479	365

FLOW GPM	FIXTURE UNITS	
	FLUSH VALVE TANKS	FLUSHOMETER VALVES
125	506	396
130	533	430
135	559	460
140	585	490
145	611	521
150	638	559
155	665	596
160	692	631
165	719	666
170	748	700
175	778	739
180	809	775
185	840	811
190	874	850
200	945	931
210	1018	1009
220	1091	1091
230	1173	1173
240	1254	1254
250	1335	1335
260	1418	1418
270	1500	1500
280	1598	1598
290	1668	1668
300	1755	1755
310	1845	1845
320	1926	1926
330	2018	2018
340	2110	2110
350	2204	2204
360	2298	2298
370	2388	2388
380	2480	2480
390	2575	2575
400	2670	2670
410	2765	2765
420	2862	2862
430	2960	2960
440	3060	3060
450	3150	3150
500	3620	3620

AREAS AND CIRCUMFERENCES OF CIRCLES

DIAMETER		CIRCUMFERENCE		AREA	
Inches	mm	Inches	mm	Inches²	mm²
1/8	6	0.40	10	0.01227	8.0
1/4	8	0.79	20	0.04909	31.7
3/8	10	1.18	30	0.11045	71.3
1/2	15	1.57	40	0.19635	126.7
3/4	20	2.36	60	0.44179	285.0
1	25	3.14	80	0.7854	506.7
1 1/4	32	3.93	100	1.2272	791.7
1 1/2	40	4.71	120	1.7671	1140.1
2	50	6.28	160	3.1416	2026.8
2 1/2	65	7.85	200	4.9087	3166.9
3	80	9.43	240	7.0686	4560.4
4	100	12.55	320	12.566	8107.1
5	125	15.71	400	19.635	12 667.7
6	150	18.85	480	28.274	18 241.3
7	175	21.99	560	38.485	24 828.9
8	200	25.13	640	50.265	32 428.9
9	225	28.27	720	63.617	41 043.1
10	250	31.42	800	78.540	50 670.9

EQUAL PERIPHERIES

S = 0.7854 D

D = 1.2732 S

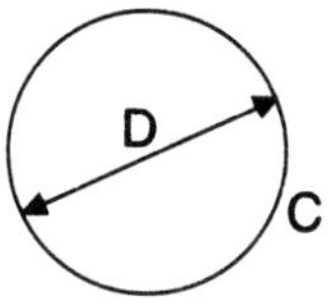

S = 0.8862 D

D = 1.1284 S

S = 0.2821 C

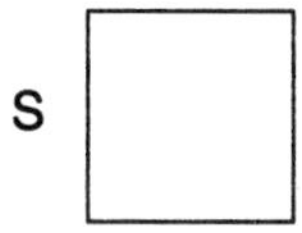

EQUAL AREAS

Area of square (S') =
1.2732 x area of circle

Area of square (S) =
0.6366 x area of circle

$C = \pi D = 2\pi R$

$C = 3.5446 \sqrt{area}$

$D = 0.3183\ C = 2R$

$D = 1.1283 \sqrt{area}$

$Area = \pi R^2 = 0.7854\ D^2$

$Area = 0.07958\ C^2 = \frac{\pi D^2}{4}$

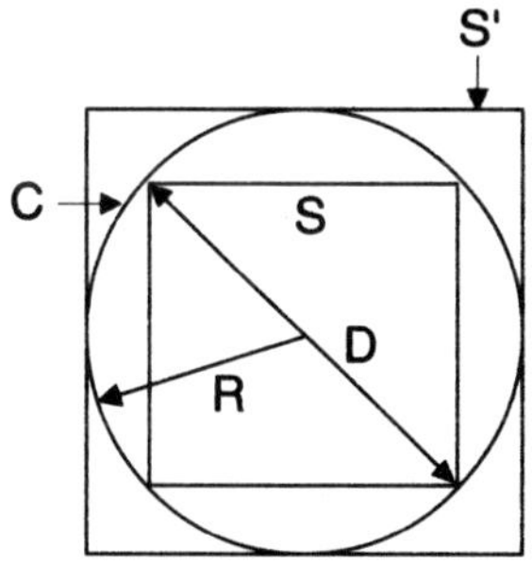

$\pi = 3.1416$

FRICTION LOSS PER FOOT LENGTH
[SCHEDULE 80 PVC EXCEEDING CHART A105.1(7)]

The following formula should be used for where Schedule 80 IPS CPVC sizes 2½" through 10" exceed Chart A 105.1(7):

Head loss formula: $H_L = 0.2083\ (100/C)\ 1.852 \text{ x } Q\ 1.852\ /\ d_I\ 4.8655$

Where:

H_L = frictional head loss (feet of water per 100 feet)
C = Hazen-Williams factor (150 for CPVC)
Q = flow rate (gal/min)
d_I = inside diameter of pipe (inches)

Note: head loss in feet of water per 100 feet can be multiplied by 0.4335 to obtain pressure drop in psi

Velocity formula: $V_W = 0.4085\ Q\ /\ d_I^2$

Where: V_W = velocity of water (feet per second)

FLOW IN PARTLY FILLED (ONE-HALF FULL) PIPES
(BASED ON MANNING'S FORMULA WITH n = .012)

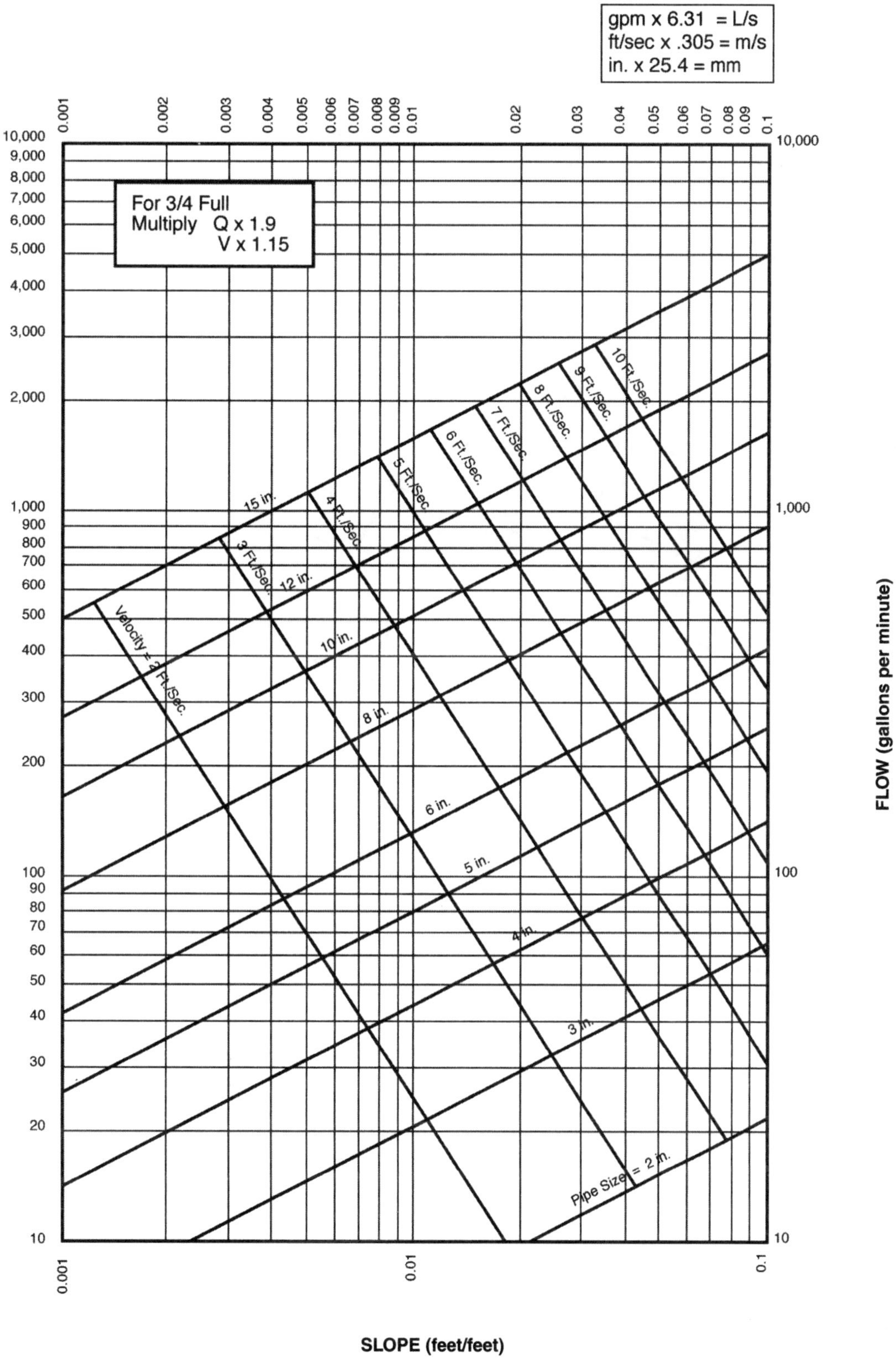

FLOW IN PARTLY FILLED (FULL) PIPES
(BASED ON MANNING'S FORMULA WITH n = .012)

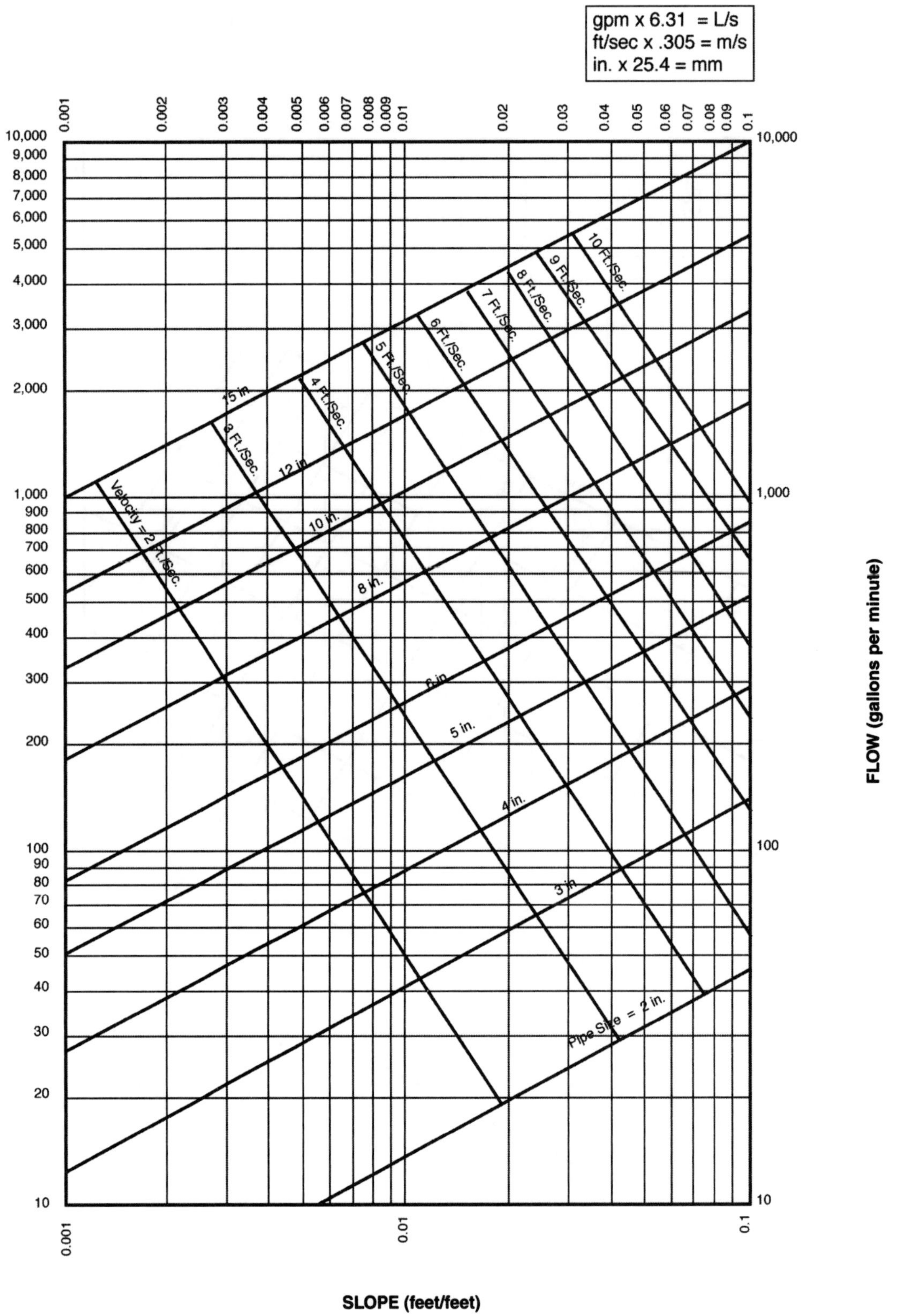

INDEX

– A –

ABANDONED
Grease interceptors 1014.3.7
Sewers and sewage disposal .....722.0, H 1101.0
ACCESS PANELS 317.1, 402.10
ACCESS TO
Appliances 507.26, 508.3, 508.4
Backwater valves 710.6, 1101.6.4, 1503.2.3, 1503.9.7
Chemical waste system piping 811.4
Cleanouts 707.8, 719.5
Drips 1210.7.1
Emergency eyewash and shower equipment 416.4
Flexible connectors 604.5, 1212.5.3
Flushometers 413.2
Fuel gas meters .....1208.7.1, E 402.4.1
Fuel gas piping 1210.3.3.1, 1210.4.4
Fuel gas pressure regulators 1208.8.2, 1210.5.1
Fullway valves 606.2, 710.6
Grease interceptors 1014.3.4
Indirect waste receptors 804.1
Interceptors .....1009.3, 1009.5, 1016.4, H 901.5
Lapped flanges 1208.6.12.7
Manholes .....H 501.12
Meters and submeters .....L 407.4
Plumbing fixtures 402.1
Pool heaters on/off switch .....L 503.3.6(1)
Rainwater storage tanks .....1603.4, K 105.3.1
Septic tanks .....H 501.5
Shutoff valves .....606.3, 606.6, 1210.9, 1210.9.1, 1210.9.2, 1210.9.4, 1212.5.3, 1212.6, 1327.2, E 403.9, E 403.13(2), E 403.13(6)
Sumps 710.10, 810.3
Toilet facilities 422.4
Trap seal primers 1007.1
Vacuum pumps .....C 501.2.1
Valves for nonpotable application 1503.9.5
Vent connectors, fuel gas 509.10.10
Water meters and submeters .....L 407.4
Water pressure regulators and strainers 608.2
ACCESSIBLE, DEFINITION 203.0
ACCESSIBLE, READILY, DEFINITION 203.0
ACCESSIBLE PLUMBING FACILITIES
For persons with disabilities 403.2
Protection against contact 403.3
Where required 403.1
ACRYLONITRILE-BUTADIENE-STYRENE (ABS) PIPE .....203.0, Table 313.3, 701.2, Table 701.2, 705.1 705.9.4, 903.1, 1101.4
AIR GAPS
Backflow prevention .....603.2, Table 603.2, 603.3.1, Table 603.3.1, 603.4.4
Definition 203.0
Discharge from drinking water treatment units 611.2
AIR TANK 710.11
ALTERATIONS AND REPAIRS
Fuel gas piping 1203.1, 1213.1.2, 1213.4.2
Health care facilities 1301.5
Plumbing systems 102.4, 301.2.5
ALTERNATE MATERIALS, DESIGN, AND METHODS 301.3
ALTERNATE WATER SOURCE
Definition 203.0
For nonpotable applications .....Chapter 15
For special water features .....L 406.1
Identification of piping 601.3.3
Nonpotable rainwater catchment system .....Table 1601.5, 1602.4, K 102.2
APPLIANCES
Automatically operated 509.3.4, 509.5.7.1
Categorized vent diameter/area, definition 203.0
Connections to building fuel gas piping 1212.0
Definition 203.0
Direct-vent 206.0, 509.2.6, 509.5.10.1, 509.8.2, 509.12
In attics 508.4
Indirect waste type 806.0, 807.0
In under-floor spaces 508.4
On roofs 508.0
Vent Connectors 509.10, 510.2.11
Venting of 509.0

APPLICATION FOR PERMIT .104.3
APPROVED, DEFINITION .203.0
APPROVED TESTING AGENCY, DEFINITION .203.0
AREA DRAINS .203.0, 1101.15
ASBESTOS-CEMENT PIPE .Table 510.1.2(5), Table 510.2(5)
ASPIRATORS .203.0, 603.5.9, 1303.5
AUTHORITY HAVING JURISDICTION .103.0, 203.0
AUTHORITY TO
- Disconnect gas .1206.0
- Plumbing system .106.5
- Render gas service .1205.0

AUTOMATIC
- Drains .K 106.2.4
- Flushometer valves .413.2
- Pressure relief valve .608.4
- Pumps .612.7.2, 1017.1, 1101.6.2, C 501.2.5, H 601.8
- Seat cover dispensers .411.3
- Shutoff .420.3, 504.6
- Siphon, septic tanks .H 601.8
- Sprinkler systems .603.5.14
- Sumps and receiving tanks .710.9
- Vent dampers, fuel gas .509.14, 509.14.1

AUTO WASH RACKS .1011.0

– B –

BACKFILLING .314.4, 314.4.1, 1210.1.2.1, 1323.5.1, 1323.5.4, 1323.5.5, 1504.7.2, H 601.2
BACKFLOW
- Definition. 204.0
- Prevention . 602.2, 602.3
- Prevention devices or assembly . 603.1, Table 603.2, 603.3

BACKFLOW CONNECTION .204.0, 603.0
BACKFLOW PREVENTERS
- Approved standards .Table 603.2
- Definition .204.0
- For hot water .603.4.5
- Integral .417.2, 417.3, 417.4, 603.4.6
- Prohibited locations .603.4.9
- Protection from damage .E 302.6
- Required .603.2

BACKFLOW PREVENTION .602.3, 603.3
BACK-TO-BACK FIXTURES .704.2
BACKWATER VALVES
- Definition .204.0
- Where Required.710.6

BALLCOCKS .Table 603.2, 603.5.2
BAPTISTRIES .406.1
BATHROOM, HALF, DEFINITION .204.0
BATHROOM GROUP .204.0, 908.2, C 303.0
BATHTUB, PROHIBITED TYPE .405.3
BATHTUBS AND WHIRLPOOL BATHTUBS .409.0
BIDETS .402.5, 410.0
BLOWOFF CONDENSERS .810.1, Table 810.1
BOARD OF APPEALS .107.0
BOILER BLOWOFF
- Definition .204.0
- Discharge requirements .810.1

BONDING JUMPER
- Bonding of CSST fuel gas piping .1211.2
- Definition .204.0

BOTTLING ESTABLISHMENTS .1013.0
BUILDING DRAIN
- Definition .204.0
- Existing .102.4.1
- Material .Table 701.2
- Sizing .703.0, Table 703.2, 710.5, C 304.2, Table C 304.2
- Testing .712.0

BUILDING DRAIN (STORM), DEFINITION .204.0
BUILDING SEWERS
- Definition .204.0
- Existing .102.4.1, 715.3
- Location .721.0
- Material .Table 701.2, 715.1, 720.0
- Sizing .710.5, 717.0, Table 717.1, C 304.2
- Testing .723.0

BUILDING SUPPLY .204.0, Chapter 6
BURRED ENDS .309.3, 609.1, 701.3.1, 903.3, 1208.6.8, 1321.7.2

– C –

CAPACITY OF
- Breathing air cylinders .F 1001.1
- Graywater systems .1503.2.1
- Interceptors .1014.2.1
- Irrigation field and mulch basin .1503.2.1, 1504.2, Table 1504.2, 1504.3, 1504.6.2
- Medical air compressors .1310.3.1, 1311.2
- Rectangular scuppers .D 104.0
- Sewage ejectors or pumps .710.3
- Sump pump .1101.6.2
- Water heaters .501.1, Table 501.1(1)

CAST-IRON PIPE
- Approved standards .301.2.4

Discharge line from ejectors 710.4
Drainage piping 701.2
Fittings 301.2.3, 604.6, 1208.6.10.5
Flanges 1208.6.12.1
Joining 705.2, 705.10.1
Prohibited use 316.1, 1208.6.3
CAULKED JOINTS 705.2.1
CAULKING FERRULES .... 701.6, Table 701.6
CERTIFICATION
Backflow assembly testers 205.0, 603.2
Medical gas or medical vacuum installers 1323.10
Medical gas or medical vacuum systems 1306.0
Pipe welder, definition 225.0
Residential fire sprinklers installers 612.1
CESSPOOL .... H 801.1
Abandoned .... 722.0, H 1101.0
Connections to sewer prohibited 714.3
Definition 205.0
CHECK VALVES
In drainage systems 710.4, 710.12.1, 710.13.2, 1004.2
In firefighter breathing air replenishment systems .... F 501.7, F 601.5.1, F 701.3
In fuel gas systems 1208.8.1
In medical gas and medical vacuum systems .... 1311.5(2), 1313.5
In water supply systems .... Table 603.2, 603.3, 608.3
CHEMICAL RESISTANT GLASS PIPE 811.2
CHEMICAL WASTES 205.0, 811.0
CHIMNEYS 509.5
Cleanouts 509.5.6.2, 509.5.9
Decorative Shrouds 509.5.1.1, 509.5.4.3
Definition 205.0
Existing 509.5.6, 509.5.6.3
Factory-built 509.5.1
Masonry 509.5.3
Metal 509.5.2
Size 509.5.5
Support 509.5.8
Termination 509.5.4
CHLORINATED POLYVINYL CHLORIDE (CPVC) PIPE OR TUBING
Approved standards, water supply .... Table 604.1
Definition 205.0
Joining and connections .... (see Joints and connections)
Residential sprinklers .... 612.4.2, Table 612.5.3.2(6), Table 612.5.3.2(7)
CHLORINATED POLYVINYL CHLORIDE/ALUMINUM/CHLORINATED POLYVINYL CHLORIDE (CPVC-AL-CPVC) PIPE OR TUBING
Approved standards, water supply .... Table 604.1
Joining and connections .... (see Joints and connections)
CIRCUIT VENTING 911.0
CLARIFIERS 205.0, 1009.0
CLEANOUT PLUGS OR CAPS .... 707.1 – 707.3, 707.13, 1107.2
CLEANOUTS
Access 707.8, 719.5, 801.3.2
Building sewer 719.0
Change of direction 707.4, 719.1, 801.3.2, 803.3, 1002.3
Clearances 707.8, 707.9
Condensate drain 814.3.1
Cover plate 707.8
Direction of flow 707.5, 719.4
Horizontal drains 707.4, 719.1
Manholes 719.6
Plugs 707.1, 707.2, 707.12
Prohibited use of 316.1
Size .... 707.1, Table 707.1
Soil or drain pipes 317.1
Where required .... 707.4, 719.1, 909.1, 910.6, 1002.3, 1008.1, 1017.1, 1101.13.1, E 301.4
CLEARANCE
Cleanouts 707.8, 707.9
Fuel gas vents .... 509.7.3.4, Table 509.7.3.4(1), Table 509.7.3.4(2)
Water heaters 504.3.1, 504.3.2
CLEAR WATER WASTE
Definition 205.0
Discharge 812.1
CLINICAL SINKS 205.0, 413.1, 1303.7
CLOSET BENDS 310.3, 402.6.1
CLOSET RINGS 402.6.1
CLOSET SCREWS 402.6.2
CODE, DEFINITION 205.0
COMBINATION WASTE AND VENT SYSTEMS .... 205.0, 910.0, Appendix B
COMBUSTIBLE MATERIAL, DEFINITION 205.0
COMBUSTIBLE PIPING INSTALLATIONS 1404.0
COMBUSTION AIR 506.0
COMMON VENT, FUEL GAS 509.6.3, 509.10.2.2, 509.10.2.3, 509.10.3.1, 509.10.7.2, 510.2.3, 510.2.9, 510.2.15, 510.2.16

COMPRESSION JOINTS 605.7.1.1, 605.10.1.1
705.2.2, 719.6, 1208.6.11.3
CONCEALED SLIP JOINTS 402.10
CONCEALING OF WORK PROHIBITED
BEFORE INSPECTION 105.1, 105.2.1.1, 105.2.1.2,
503.2, 1203.2
CONDENSATE WASTE
AND CONTROL . . . . .814.0 – 814.6
CONDENSATE PUMPS 814.1.1
CONDENSERS AND SUMPS 810.0
CONDUCTORS,
STORM WATER 205.0, 1101.12.1, 1101.13.1,
1101.16, 1103.0, 1602.6,
1602.7.2, 1603.11, K 103.2,
K 104.5, K 106.4
CONTAMINATION 205.0, 602.2, 602.3,
603.1, 811.1, 1321.8.8,
1321.10.4, 1321.10.7,
C 302.1, F 1101.3
CONTINUOUS VENT, DEFINITION 205.0
CONTINUOUS WASTES
Definition 205.0
Materials 407.5, 701.4
CONTROL VALVES 408.3, 408.9,
606.5 – 606.7, 1303.1
COOLING WATER 808.0
COPPER OR COPPER ALLOY
PIPE OR TUBING
Fuel gas 1208.6.3.2, 1208.6.4.2,
Table 1215.2(7) – Table 1215.2(13),
E 403.5
Joining and connections . . . .(see Joints
and connections)
Medical gas and
medical vacuum 1318.4,
1319.1, 1323.4.2
Penetrating framing members 312.9
Prohibited use of 811.1, 1208.6.3.2,
1208.6.4
Standard weight 302.1
Underground drainage and vent 903.2
Water supply 604.3, 604.4,
A 105.2, E 302.2.1
CRITICAL CARE AREA,
DEFINITION 205.0
CRITICAL LEVEL . . . . .205.0, Table 603.2,
603.5.1, 603.5.2, 603.5.5,
603.5.13, C 501.3
CROSS-CONNECTION CONTROL 603.0
CROSS-LINKED POLYETHYLENE (PEX)
PIPING OR TUBING
Approved standards, water supply . . . . Table 604.1
Definition 218.0
Flexible connectors 604.13
Joining and connections . . . . (see Joints
and connections)
Residential sprinklers 612.4.2,
Table 612.5.3.2(8),
Table 612.5.3.2(9)
CROSS-LINKED POLYETHYLENE-ALUMINUM-
CROSSLINKED POLYETHYLENE (PEX-AL-PEX)
PIPING OR TUBING
Approved standards, water supply . . . . Table 604.1
Definition 218.0
Flexible connectors 604.13
Joining and connections . . . . (see Joints
and connections)
Residential sprinklers 612.4.2

– D –

DESIGN FLOOD ELEVATION 206.0, 301.4
DEVELOPED LENGTH
Bar and fountain sink trap 801.4
Definition 206.0
DWV piping 707.4, 710.13.2
Indirect waste piping 801.3.1
Trap arm .1002.2, Table 1002.2
Water supply piping . . . . .610.9, A 104.4, A 107.2
Wet vent 908.1
DIELECTRIC INSULATORS .507.1
DISHWASHING MACHINES 414.0
Air gap 807.3
Approved standards 414.1
Backflow protection 414.2
Discharge, drainage 414.3, 807.3, 1014.3.2
Drainage discharge, prohibited 419.2, 1014.1.3
Drainage fixture unit (DFU) . . . . .Table 702.1
Energy efficiency . . . . .L 403.1
Flexible connector length 604.12
Water supply
fixture units (WSFU) . . . . .Table 610.3, Table A 103.1,
Table C 303.2
DISPOSAL FIELDS . . . . .1504.1, 1504.4, Appendix H
DOSING TANKS . . . . .H 601.8
DRAINAGE SYSTEM
Cleanouts 707.0, 719.0
Change in direction 706.0
Damage to public sewer or private
sewage disposal system 305.1, 714.0
Definition 206.0
Fixture units . . . . .Table 702.1
Independent systems 311.0
Indirect waste . . . . .Chapter 8
Joints and connections 705.0

Materials .701.2, Table 701.2, 715.0
Obstruction of flow .310.5
Sizing .702.0, Table 702.1, 703.0, Table 703.2, 717.0, Table 717.1
Slope of piping .708.1
Sumps and ejectors .710.0
Testing .712.0, 723.0
Waste detrimental to .306.0, 811.1, 1012.1

DRINKING FOUNTAINS .206.0, 415.0, 809.0, E 501.8

DUCTILE IRON PIPE .Table 604.1, 605.4

DURHAM SYSTEM, DEFINITION .206.0

– E –

EFFECTIVE GROUND-FAULT CURRENT PATH, DEFINITION .207.0

EFFECTIVE OPENING .207.0, Table 603.3.1

EJECTORS .710.2 – 710.5, 710.9 – 710.12

EMERGENCY EYEWASH AND SHOWER EQUIPMENT .416.0

EXCAVATIONS .314.0, 1203.2, 1504.4, H 301.1, H 701.4

EXCESSIVE WATER PRESSURE .608.2

EXPANSION AND CONTRACTION OF PIPE, PROVISION FOR .312.2

EXPANSION TANKS .608.2, 608.3

– F –

F RATING .208.0, 1404.3, 1405.3, 1406.4

FAUCETS
Approved standards 417.1, 417.4, L 402.5.1
Maximum flow rate 407.2, 420.2, L 402.4, L 402.5
Metering 407.2.2, 407.4, Table L 402.1, L 402.5.2, Table L 701.2(1)
Residential kitchen L 402.4
Self closing 407.4
Separate controls for hot and cold water 417.5
Sink faucets 420.2
Temperature limitation 407.3, 408.3, 409.4, 410.3, 416.2
With pull out spouts 417.4

FIRE SPRINKLERS, RESIDENTIAL .612.0

FIREFIGHTER BREATHING AIR REPLENISHMENT SYSTEMS .Appendix F

FITTINGS
Branch .419.2
Cleanout .707.1, 707.2
Dielectric .605.15, 605.16.1, 605.16.3, 705.10.2, 1322.6.1, E 403.8.2
Drainage system .Table 701.2, 701.3, 705.0, 706.0, E 504.3
Firefighter breathing air replenishment systems .F 801.3
Fixture .402.9, 403.2, 417.0, 603.5.19, L 402.0
Fuel gas piping .1208.6.5, 1208.6.10, 1208.6.11, 1210.4.1
Fuel gas vent .509.3.6(2), 509.4.2, 510.2.8, 510.2.9
Gray water .1503.9.2
Inlet, for fixture connections .704.1, 704.2
Lead content .604.2, 604.2.1
Material of .604.0, Table 604.1, Table 701.2, 701.3, 903.0, 1208.6, 1208.6.10, 1208.6.10.5
Medical gas and medical vacuum .1320.2, 1320.2.1, 1321.1
Plumbing vent .903.0
Prohibited .310.1
Screwed .310.8
Water supply system .604.0, Table 604.1, 605.0, A 107.3

FIXTURE .(see Plumbing fixtures)
Branch, definition .208.0
Connections .704.0
Drain, definition .208.0
Flanged connection .402.6
Future installations .402.11
Plumbing (see Plumbing fixtures)
Units, drainage (DFU) 702.1, Table 702.1, C 303.1, Table C 303.3
Units, water supply (WSFU) 610.3, Table 610.3, 610.10, Table 610.10, A 103.1, Table A 103.1, C 303.1, Table C 303.2

FLAGPOLING OF VENTS .906.3

FLAMMABLE VAPOR OR FUMES .208.0, 507.12

FLASHING .701.5, 906.5, 906.6, 1102.3, K 103.1.1

FLOOD RESISTANT CONSTRUCTION
Coastal high hazard area .205.0, 301.4.1
Design flood elevation .206.0
Flood hazard area .208.0
Flood hazard resistance .301.4

FLOOD-LEVEL RIM .208.0, 801.3.2

FLOODED, DEFINITION .208.0

FLOOR DRAINS .418.0
Connected into a trap .1006.1, 1007.1
Directly connected to drainage system .1007.1
Location .418.3, 418.4, 1017.1
Required .408.7, 704.3, 1017.1, E 501.6

FLUSH TANKS .208.0, 402.4, 413.1, 413.3, 413.5, 601.2
FLUSH VALVES, DEFINITION .208.0
FLUSHOMETER TANKS .208.0, 411.2, 413.1, 603.5.3
FLUSHOMETER VALVES .208.0, 413.1, 413.2, 601.2, 603.5.1, 610.10, Table 610.10, L 402.2.2
FLUX .604.1, 604.2, 605.1.1, 605.1.4, 705.3.1, 705.3.3, 1321.9, 1321.9.2, 1321.11.1
FOG DISPOSAL SYSTEMS .208.0, 1015.0
FOOD-HANDLING EQUIPMENT AND FIXTURES .801.3
FOOD WASTE DISPOSER .419.0, 714.4, 1001.2, 1014.1.3
FREEZE PROTECTION .312.6, 603.4.7, 603.5.7, 608.2, 1005.1 1210.1.4, 1210.7.1, 1323.2, 1603.17, K 106.0
FROST OR SNOW CLOSURE .906.7
FUEL GAS PIPING .Chapter 12
- Cathodic protection .E 403.8.1 – E 403.8.4
- Change in direction .1210.6
- Concealed location .1210.4
- Detection of leaks .1213.4
- Electrical bonding and grounding .1211.0
- Inspection and testing .1203.0, 1213.0
- Installation .1210.0
- Interconnections .1208.3
- Joints and fittings .1208.6.5, 1208.6.11 – 1208.6.12.4, 1213.2.1, 1213.2.2
- Materials .1208.6
- Obstruction of flow .1210.10
- Purging .1213.6
- Sizing .1208.4, 1214.0, 1215.0, Table 1215.2(1) – Table 1215.2(36), E 403.3
- Supports .1210.3.5, Table 1210.3.5.1
- Underground installation .1208.6.3.3, 1208.6.4.4, 1210.1, E 401.2, E 403.7, E 403.8.4

FUEL GAS VENTS
- Appliances not requiring vents .509.2.1
- Caps .509.6.1(6), 509.7.4(3)
- Clearance .509.7.3.4, Table 509.7.3.4(1), Table 509.7.3.4(2)
- Dampers .509.13, 509.14, 509.14.1
- Direct-vent .509.2.6
- General requirements .509.0
- Insulation shield .509.6.1.1
- Integral .509.2.7
- Listing .509.1
- Mechanical draft .509.3.3 – 509.3.3.5
- Obstruction of flow .509.15
- Plastic piping .509.4.1
- Special .509.4.3
- Termination .509.6.1, 509.7.2, 509.8
- Wall penetrations .509.8

FUEL GAS VENT, SIZING
- Category I appliances .509.6.2.1, 510.0
- Multiple appliances .510.2, Table 510.2(1) – Table 510.2(9)
- Multistory .509.6.3, 509.6.3.2, 510.2.13 – 510.2.16
- Single appliance .509.7.4(1)(b), 509.10.2.1, 510.1, 510.1.2, Table 510.1.2(1) – Table 510.1.2(6)

– G –

GANG OR GROUP SHOWERS .209.0, 408.3, 408.8
GARBAGE GRINDER .(see Food waste disposer)
GRADE
- Building sewer .718.1
- Definition .209.0
- Horizontal drainage .708.0

GRAVITY DRAINAGE REQUIRED .709.0
GRAY WATER, DEFINITION .209.0
GRAY WATER DIVERTER VALVE, DEFINITION .209.0
GRAY WATER SYSTEMS .601.3.3, Table 1501.5, 1501.5.2, 1501.6, 1502.0
GREASE INTERCEPTORS .1014.0
- Approved standards .1014.1
- Definition .209.0
- Capacity .1014.2.1, Table 1014.2.1
- Gravity-type .1014.3, Table 1014.3.6
- Hydromechanical-type .1014.2
- Not required .1014.1.3
- Required .1010.1, 1014.1, H 101.3

GRINDER PUMPS .710.12

– H –

HANDHELD SHOWERS .417.3
HANGERS .(see Supports)
HARMFUL WASTES .(see Wastes, detrimental or harmful)
HEALTH AND SAFETY .101.3, 102.5
HEAT EXCHANGERS .505.4.1, 603.5.4, C 302.0

HEAT-FUSION WELD JOINTS 210.0, 605.6.1, 605.11.1, 1208.6.11.2
HORIZONTAL BRANCH, DEFINITION 210.0
HORIZONTAL PIPE, DEFINITION 210.0
HORIZONTAL-TO-HORIZONTAL DRAINAGE FITTINGS 706.3
HORIZONTAL-TO-VERTICAL DRAINAGE FITTINGS 706.2
HOT-MOPPED SHOWER PANS . . . . . .(see Shower, pans)
HOT WATER SUPPLY PIPING . . . . .601.2, 609.1, 609.4, 609.12, 610.3, 610.12.1, L 501.0, L 502.0, L 503.0
HOUSE DRAINS . . . .(see Building drain)
HOUSE SEWERS . . . .(see Building sewer)

– I –

IDENTIFICATION
- Cast-iron soil pipe 701.2
- Copper tubing 604.4
- Fuel gas piping 1208.7.4
- Medical gas and medical vacuum systems . . . . .1318.5, 1323.13 – 1323.13.2, 1323.14, 1323.15
- Piping, fittings, traps, and fixtures 301.2.1
- Water, nonpotable and potable 601.3

INCREASERS AND REDUCERS 316.0
INDIRECT-FIRED WATER HEATERS 211.0, 505.4, 603.5.4, 608.7
INDIRECT WASTE . . . . .Chapter 8
- Air gap or break 801.2
- Appliances 807.0
- Approval required 802.0
- Aspirators 603.5.9, 1303.5
- Chemical wastes 811.0
- Cleanouts 801.3.2, 803.3
- Clear water wastes 812.0
- Condensate 814.1, 814.1.1, 814.5
- Food handling fixtures .801.3.3
- High temperature discharge 810.1
- Interceptor 902.1
- Receptors 801.4, 804.0
- Sterile equipment 801.6, 806.0, 1303.4.1
- Swimming pools 813.0
- Traps 801.3.2, 803.3, 814.5
- Vents 803.3
- Where required 801.2

INDIVIDUAL VENTS . . . . .211.0, Table 703.2, 904.1, 908.2.1, C 601.6
INDUSTRIAL WASTES
- Access 811.4
- Approval required 811.1
- Definition 211.0
- Disposal 306.0, 811.7
- Materials 811.2, 811.3, 903.2.2
- Pretreatment required 306.1, 811.1
- Prohibited material 811.1, 903.2.2
- Vents 811.6

INSANITARY
- Defective systems 105.3.1
- Definition 211.0
- Health and safety 102.5

INSPECTION AND TESTING .105.0
- Alternative engineering design . . . . .301.3.1 – 301.3.1.2, 301.5.6
- Final 105.2
- Reinspection 105.2.6
- Required .105.1, 105.2, 1203.0, 1213.1, 1321.12, 1324.0, 1406.0, 1501.5, Table 1501.5, 1502.0, 1601.5, Table 1601.5, 1605.0
- Rough-in 105.2, 1203.3.1
- Scheduling of 105.2.3
- Testing 105.3

INSPECTOR . . . .(see Authority Having Jurisdiction)
INSULATION
- Domestic hot water supply . . . . .609.11, L 501.2
- Fuel gas vent 509.6.1.1
- Penetrating firestop system 1404.5, 1405.7
- Pool covers . . . . .L 503.3.6
- Service hot water . . . . .L 503.3.3, Table L 503.3.3
- Tracer wire .604.10.1, E 403.8.4

INTERCEPTING HOPPERS OR SINKS . . .(see Receptors)
INTERCEPTOR (CLARIFIER)
- Abandoned 1014.3.7
- Auto wash racks 1011.0
- Bottling establishments 1013.0
- Cleanouts, location 707.7
- Commercial and industrial laundries 1012.0
- Definition 211.0
- Grease 1014.0
- Industrial 1009.0
- Oil and flammable liquid 1017.0
- Packing establishments .1010.0
- Sand 1016.0
- Slaughterhouses 1010.0
- Vents 902.1

INTERMIXING OF METALS PROHIBITED 310.6, 604.1

INVERT, DEFINITION .211.0
IRON PIPE SIZE (IPS) PIPE .302.0
ISLAND FIXTURE, VENTING .909.0

– J –

JOINTS AND CONNECTIONS .315.0, 605.0, 705.0, 1208.6.11 – 1208.6.13.5, 1320.0
- ABS .705.1
- Asbestos cement pipe .Table 510.1.2(5), Table 510.2(5)
- Brazed joint .212.0, 605.1.1, 609.3, 705.3.1, 705.10.1, 705.10.2, 1208.6.11.1, 1208.6.11.2, 1309.4, 1321.1
- Cast-iron piping .604.6, 701.2, 705.2, 705.10.1
- Caulked joint .705.2.1
- Compression joint .605.6.2, 605.7.1, 605.10.1.1, 605.11.2, 705.2.2, 1208.6.11.3
- Copper or copper alloy pipe or tubing .605.1, 705.3, 705.10.1, 705.10.2
- CPVC piping .605.2
- CPVC-AL-CPVC piping or tubing .605.3
- Ductile iron piping .605.4
- Expansion joint .705.9.2, 1101.5, 1213.2.2
- Flared .605.1.2, 1208.6.10.4
- Galvanized steel pipe .605.5, 705.4
- Ground joints .705.9.3
- Heat-fusion joints .210.0, 605.6.1, 605.11.1, 1208.6.11.2
- Lead content .604.2
- Mechanical joints .605.1.3, 605.2.1, 605.4.1, 605.5.1, 605.6.2, 605.7.1, 605.9.2, 605.10.1, 605.11.2, 605.12.1, 605.13.1, 605.16.3, 705.1.1, 705.2.2, 705.3.2, 705.4.1, 705.6.1, 705.7.1, 705.8.1, 705.10.4, 1208.6.11.3
- PE pipe or tubing .605.6, 1208.6.5
- PE-AL-PE pipe or tubing .605.7
- PE-RT tubing .605.8
- PEX tubing .605.9
- PEX-AL-PEX pipe or tubing .605.10
- Plastic piping, joints, and fittings .605.2 – 605.3.2.1, 605.6 – 605.12.3, 705.1.1 – 705.1.3, 705.5 – 705.5.1.3, 705.6 – 705.6.3, 705.10.3, 1208.6.11, 1210.1.7.1
- PP piping .605.11
- Prohibited fittings and practices .310.0, 315.2, 604.2, 1321.8.4, 1321.12.3, 1322.7
- PVC piping .605.12, 705.6, 705.9.4
- Screwed joint .310.8, 701.3.1
- Slip joint .221.0, 402.10, 605.14, 705.9.1, 1003.2
- Soldered joint .212.0, 605.1.4, 705.3.3, A 107.3
- Solvent cement plastic pipe .605.2.2, 605.3.1, 605.12.2, 705.1.2, 705.6.2, 705.9.4
- Stainless steel pipe .605.13, 605.16.3, 705.7, 705.10.4
- Threaded joint .605.1.5, 605.2.3, 605.5.2, 605.11.3, 605.12.3, 605.16.1, 705.1.3, 705.3.4, 705.4.2, 705.6.3, 705.10.2, 1208.6.9, 1208.6.10.1, 1208.6.10.5, 1322.5, A 107.3
- Tightness .105.3
- Transition joint .705.9.4
- Various materials .605.16, 705.10, 1210.1.3.8
- Vitrified clay .705.8
- Welded joint .212.0, 605.13.2, 705.7.2, 1208.6.10.1, 1322.1, F 1101.2

– L –

LABELED .214.0, 301.2
LAUNDRY TUBS .1001.2
LAVATORIES .407.0, Table 702.1, E 501.4
LAVATORIES IN SETS .214.0, Table 702.1
LAWN SPRINKLERS .Table 610.3, Table A 103.1
LEAD
- Allowable lead contents .604.2, 604.2.1
- Bends and traps .701.5, 906.6
- Closet rings .402.6.1
- Flashing .1102.3.1
- Roof drains .1102.1
- Sheet .408.7.3, 701.5, 906.6
- Water supply .604.2, 604.2.1

LEADERS .214.0, 1101.4.3, 1101.12.1, 1101.13.1, 1101.15, 1101.16, 1101.16.1, 1103.0, Table 1103.1

LIQUEFIED PETROLEUM GAS (LP-GAS) .214.0, 1202.1, 1208.5.1, 1208.6.7.3, 1208.6.11.4, 1212.11, E 401.2, E 403.1, E 403.10
LIQUID WASTES .214.0, 303.0, L 302.0
LISTED .214.0, 301.2, 402.6.2, 501.1
LISTING AGENCY .214.0, 301.2
LOCATION OF
Backwater valves .710.1, 710.3, 710.4, 710.6, 1006.0, 1101.6.4, 1503.2.3, 1503.9.1, 1503.9.7
Fixtures .308.0, 402.0
Fuel gas meters .1208.7 – 1208.7.3, E 402.4 – E 402.4.2
Fuel gas valves .1210.9 – 1210.9.4, 1212.5.3, 1212.6, E 402.3
Interceptors and separators .1009.3, 1009.5
Medical gas and medical vacuum system valves .1314.0
Plumbing system or sewer on lot .307.0, 721.1, Table 721.1
Private sewage disposal system .H 101.8
Water heaters .504.1, 504.3 – 504.6, 506.3, 507.2, 507.5, 507.10, 507.12 – 507.15, 507.25, 507.26, 507.27
LOT, DEFINITION .214.0
LOW HAZARD .214.0, Table 603.2

– M –

MACERATING TOILET SYSTEMS .215.0, 710.13
MAIN, DEFINITION .215.0
MAINTENANCE .102.2, 102.3, 1009.6, 1014.1.2, 1501.5, Table 1501.5, 1501.6, 1504.6.4, 1504.11, 1601.5, Table 1601.5, K 101.5, K 101.5.1, Table K 104.3.3, L 501.5, L 504.2
MALLEABLE IRON .604.7, Table 604.1, Table 701.2
MANUFACTURED/MOBILE HOME PARKS AND RECREATIONAL VEHICLE PARKS .Appendix E
MARKINGS .(see Identification)
MATERIALS
Alternate .301.3, H 501.13
Approval required .301.2
Approved standards .301.2.2
Backwater, gate, or ball valves .710.6
Building drains .Table 701.2
Building sewers .Table 701.2, 715.1, 720.0
Cesspools .H 801.0
Cleanouts .707.1
Drainage piping, joints, and fittings .701.2, Table 701.2, 701.3, 705.0
Fuel gas piping, joints, and fittings .1208.6 – 1208.6.2, 1208.6.3 – 1208.6.5, 1208.6.10, 1208.6.11, 1208.6.12
Marking of .301.2.1
Medical gas and medical vacuum .1318.0
Potable water piping, joints, and fittings .604.0, 605.0
Rainwater catchment system .1602.7
Septic tanks .H 501.3
Storm drainage pipe .1101.4
Subsoil drain .1101.4.6
Sumps or receiving tanks .710.8
Traps .1003.1
Vents (DWV) .Table 701.2, 903.0
MAY, DEFINITION .215.0
MEAT PACKING AND SLAUGHTERHOUSES .1010.0
MEDICAL GAS AND MEDICAL VACUUM SYSTEMS .Chapter 13
Alarms .1324.4
Aspirators .1303.5
Category 1 .205.0
Category 2 .205.0
Category 3 .205.0
Category 4 .205.0
Cleaning .1318.2
Construction documents .1304.3
Definitions .215.0
Installation requirements .1304.0, 1323.0
Joints and connections .1320.0
Labeling and identification .1323.13
Materials .1318.0
Medical air systems .1312.0
Medical vacuum systems .1319.0
Pipe sizing .1323.1, Table 1323.1.4(1) – Table 1323.1.4(6)
Pressure-regulating equipment .1308.0, 1321.10.2
Shutoff valves .1318.1
Station outlets/inlets .1305.3, 1314.10, 1315.0
Sterilizers and bedpan steamers .1303.4
System certification .1306.0
Testing and inspection .1324.0
Warning system .1317.0
METALS, INTERMIXING PROHIBITED .310.6, 604.1
METERING FAUCETS .407.2.2
MINIMUM STANDARDS .301.2

MULCH, DEFINITION 215.0
MULCH BASIN .... 215.0, 1504.0 – 1504.6.4

– N –

NATURAL GAS, SPECIFIC GRAVITY 1214.1
NITROGEN, NF, DEFINITION 216.0
NONCOMBUSTIBLE PIPING INSTALLATIONS 1405.0
NONPOTABLE WATER
Alternate water source .... Chapter 15
Identification 601.3
Rainwater catchment system .... Chapter 16
Unlawful connections to .... 603.5.11, E 502.3
NONWATER URINALS .... 412.1.1, L 402.3.1
NUISANCE 102.5, 216.0

– O –

OBSTRUCTIONS OF FLOW 310.5, 315.2
OFFSETS
Definition 217.0
Drainage .... 706.4, C 304.3.1, C 304.3.2, C 304.4, C 401.4, C 601.7
Medical gas 1320.2
Prohibited use 402.6.3
Venting (fuel gas) 509.6.2.2, 510.1.2, 510.2.4, 510.2.16
Venting (plumbing) 910.5
OIL AND FLAMMABLE INTERCEPTORS 1017.0
ON-SITE TREATED NONPOTABLE WATER SYSTEMS 217.0, 601.3.3, 1506.0
OPEN TRENCH WORK 314.3
OPENING, EFFECTIVE, DEFINITION 207.0
OVERFLOWS 404.2, 413.5, 607.4, 1603.10, 1603.10.1

– P –

PATIENT CARE SPACE, DEFINITION 218.0
PENALTIES 106.0
PENETRATION FIRESTOP SYSTEMS 218.0, 1404.3, 1405.3
PERCOLATION TESTS .... 1504.3, H 401.0
PERMITS, REQUIRED .... 104.1, 502.1, 1501.3, 1601.3, E 101.2, E 302.7, H 101.9, K 101.3
PERMITS FEES .... 104.5, Table 104.5
PIPING
Drainage, changes in direction 706.0
Drainage, existing sewers 715.3
Drainage, horizontal slope 708.0, 718.1
Drainage, material .... 701.2, Table 701.2, 715.1, 720.1
Exposed to sunlight, ABS and PVC 906.1
Fuel gas 1208.0, 1210.0
Hangers and supports .... 313.0, Table 313.3, 1210.3.5, 1210.3.5.1, Table 1210.3.5.1, 1323.4 – 1323.4.5 Table 1323.1.4(1) – Table 1323.4.4
Identification 601.3
Indirect waste 803.0
Joints and connections .... (see Joints and connections)
Medical gas, material 1318.0
Plumbing vent, material .... Table 701.2, 903.0
Protection of .... 312.0, 718.3, 1210.1.3 – 1210.1.5
Rehabilitation of 320.0
Sprinkler system 612.4, 612.5
Storm drainage, materials .... 1101.4 – 1101.4.6
Trenching, excavation, and backfill 314.0
Underground 314.4.1, 1210.1.6
Water supply, material .... 604.0, Table 604.1
PITS, SEEPAGE .... 221.0, 714.3, 716.1, 722.2 – 722.4, Appendix H
PLUGS, CAPS, CLEANOUT 707.1, 707.2
PLUMBING
Additions, alterations, renovations or repairs 102.4
Appliances .... 218.0, 304.1, 310.5, 313.1, 417.2, 603.4.6, 606.5, Table 610.3, Table 702.1, 801.7, 803.3, 806.1, 807.0, 808.1, Table A 103.1, L 302.2
Appurtenances .... 218.0, 304.1, 313.1, 603.4.6, Table 610.3, Table 702.1, 814.2.1, 1505.10, 1506.9.3, 1602.9, Table A 103.1, L 501.2
Connected to sewer or private disposal systems .... 304.1, 713.1, L 302.2
Definition 218.0
Inspection of 105.0
Official .... (see Authority Having Jurisdiction)
System, definition 218.0
Tests 105.3, 105.3.1
Uncovered for inspection 105.2.1
Vent, definition 218.0
Vent system, definition 218.0
PLUMBING FIXTURES
Access 402.1, 413.2, 416.4, 422.4

Accessibility 403.1, 403.2, 408.5, 408.6
Alternate materials 301.3, 406.3
Approved standards . . . . .301.2.1, 301.2.2, 408.1, 409.1, 410.1, 411.1, 412.1, 415.1, 416.1, 417.1 – 417.4, 418.1, 420.1
Back-to-back 704.2
Below curb or sewer level 710.1, 710.2
Clearance from wall, toilet, urinal 402.5
Connections, drainage 304.1, 406.1, 408.4, 409.2, 415.3, 416.5, 419.2, 420.4, 704.0
Control valves . . . . .408.3, 408.9, 601.2, 606.5 – 606.7, 1303.1, 1505.8, 1506.10
Definition 218.0
Distance from traps . . . . .1001.2, 1002.2, Table 1002.2
Fixture units . . . . .610.3, Table 610.4, 610.7 – 610.10, Table 610.10, 702.0, Table 702.1, Table 702.1(1), Table 702.2, 703.0, Table 703.2, Table A 103.1, Table C 303.2, Table C 303.3, C 501.4, C 501.5
Floor drains . . . . .408.7, 414.3, 418.0, 704.3, 801.3.2, 1003.1, 1006.0, 1007.1, 1014.3.2, 1016.1, 1017.1, 1101.11, 1505.1, 1506.1, 1602.1, B 101.5, B 101.6.1, Table C 304.2, Table C 304.3, E 501.6
Improper location of 308.0
Indirect wastes 804.1
Installation of . . . . .402.0, 405.3, 408.3, 408.7, 409.3, 409.6, 412.1.1, 413.1 – 413.3, 413.5, 415.4, 416.3, 418.3
Islands 909.0
Joints 402.2
Material of . . . . .401.0, 406.2 – 406.4, 408.7
Minimum number 422.0
Must be connected to sewer 304.1, 713.1
Must be provided with trap 1001.2
Overflows 404.0
Prohibited 310.5, 405.0, 910.7
Quality 401.2
Setting 402.5, 402.6
Size of waste for . . . . .702.0, Table 702.1, 703.0, Table 703.2
Special fixtures 406.0
Strainers and connections 407.5, 408.4
Supply fittings 402.7
Waste connections 407.5, 408.4, 409.2, 415.3, 416.5, 419.2, 420.4
Water supply required 601.2

POLLUTION . . . . .218.0, 603.1, Table 603.2, 714.5

POLYVINYL CHLORIDE (PVC) PIPE OR TUBING
Approved standards, water supply and DWV . . . . .Table 604.1, Table 701.2
Definition 218.0
Exposed to sunlight 906.1
Flexible tubing for whirlpool bathtubs 409.6.1
Joining and connections . . . .(see Joints and connections)
Lining for shower receptors 408.7
Roof drains 1102.1

POLYETHYLENE (PE) PIPE OR TUBING
Approved standards, water supply and DWV . . . . .Table 604.1, Table 701.2
Definition 218.0
Fuel gas . . . . .1208.6.5, Table 1215.2(19) – Table 1215.2(23), Table 1215.2(34) – Table 1215.2(36)
Joining and connections . . . .(see Joints and connections)
Roof drains 1102.1

POLYETHYLENE-ALUMINUM-POLYETHYLENE (PE-AL-PE) PIPE OR TUBING
Approved standards, water supply . . . . .Table 604.1
Definition 218.0
Joining and connections . . . .(see Joints and connections)

POLYETHYLENE OF RAISED TEMPERATURE (PE-RT) PIPE OR TUBING
Approved standards, water supply . . . . .Table 604.1
Definition 218.0
Flexible connectors 604.13
Joining and connections . . . .(see Joints and connections)

POLYPROPYLENE (PP) PIPE OR TUBING
Approved standards, water supply . . . . .Table 604.1
Chemical waste 811.2
Definition 218.0
Joining and connections . . . .(see Joints and connections)
Roof drains 1102.1

POLYVINYLIDENE FLUORIDE (PVDF) PIPE OR TUBING
Chemical waste 811.2
Definition 218.0

POTABLE RAINWATER CATCHMENT SYSTEMS . . . . .Appendix K
POTABLE WATER
Connections . . . . .603.4.4, 603.5.6, 603.5.7, 603.5.10 – 603.5.12, 603.5.14, 603.5.20, 801.5, 1303.5
Definition 218.0
Devices or assemblies 603.2
Disinfection 609.10
Protection from heat transfer medium . . . . .505.4.1, C 302.2
Heating . . . . .501.1, L 507.0
Identification 601.3
Joints and fittings 604.2, 604.6, 605.0
Materials . . . . .604.1, Table 604.1
Outlets 603.5.17
Protection of 602.0, 603.0
Sizing piping . . . . .610.0, Appendix A
Supply required .601.2, E 302.1
Tanks 607.0
Unlawful connection to . . . . .602.0, 603.5.11, 1503.3, 1505.4, 1506.4, 1602.4, E 502.3, K 102.2
Valves 606.0
PRE-RINSE SPRAY VALVES . . . . .420.3, L 201.0
PRESSED FITTINGS 605.1.3, 605.1.3.2, 605.13.1, 605.16.1, 705.7.1
PRESSURE DRAINAGE CONNECTIONS 805.0
PRESSURE, EXCESSIVE, WATER 608.2
PRESSURE-LOCK-TYPE CONNECTION, DEFINITION 218.0
PRESSURE LOSSES, WATER 610.2
PRESSURE REGULATORS
Firefighter breathing air replenishment systems . . . . .F 501.8
Fuel gas 507.20, 507.21, 507.22, 1202.1, 1208.8, 1210.9, 1215.1.3
Medical gas and medical vacuum 1321.10.2
Water 608.2
PRESSURE-RELIEF VALVES
Definition 224.0
Differential, backflow prevention . . . . .603.3.7, Table 603.2
Discharge 608.5
Firefighter breathing air replenishment systems . . . . .F 501.1, F 501.8, F 701.3
Medical gas and medical vacuum 1308.2, 1308.3, 1310.4, 1316.1
Water Heaters 504.4, 504.6, 505.2
Water Supply . . . . .608.2 – 608.4, E 302.5
PRIMARY SETTLING TANKS 902.1
PRIVATE OR PRIVATE USE, DEFINITION 218.0
PRIVATE SEWAGE
Capacity of system . . . . .H 101.7, H 201.0, H 701.1
Construction .H 501.3, H 601.9, H 701.3, H 801.3
Definition 218.0
Design . . . . .H 101.6, H 501.2, H 501.11, H 901.4, H 901.7
Disposal systems . . . . .218.0, 713.2, Appendix H
General requirements .H 101.2
Inspections 105.0
Location . . . . .721.0, H 101.8, Table H 101.8
PRIVATE WATER SUPPLY . . . . .602.2, 602.4, E 302.1
PROHIBITED
Appliances 310.5
Concealed gas piping 1210.4.2, 1212.3.1, 1212.5.3, 1323.6
Concealment of defects .309.2
Concealment of work before inspection 105.0, 503.2, 1203.0, 1406.1
Connections
Sewer 714.3
Waste 310.3, 404.2, 603.4.4, 805.1, 807.3, 810.1
Water . . . . .602.0, 603.4.4, E 502.3
Copper tubing, use of 903.2.2, 1208.6.4.2
Damage to drainage system 305.1, 312.3, 312.6
Devices 310.5, 1210.10
Direct drainage .603.4.4, 805.1, 807.3, 810.1
Discharge . . . . .509.1.2, 714.2, 1603.4.2, K 104.2
Fittings 310.0, 315.2, 604.2
Fixtures 310.5, 405.0
Food waste disposal, connection to 1014.1.3
Fuel gas piping 1208.6.4.3, 1210.3.4
Industrial waste disposal .306.2
Installations 312.2, 602.1
Joints and connections 315.2, 605.1.4, 605.6.2, 1321.12.3, 1322.7
Locations 415.4, 1210.3.4, 1503.6
Materials . . . . .F 801.4, H 501.10, H 501.13, K 103.1.1, L 503.1.1
Medical gas and medical vacuum piping 1323.3.1
Oxygen 1213.1.6

Pipes in concrete or masonry 312.2
Practices ....102.4.1, 104.1, 106.0, 303.1, 305.1, 309.2, 310.0, 312.2, 312.6, 502.1, 602.0, 714.1, 1101.3, 1204.3, 1205.0, 1206.3, 1501.3, K 101.3, L 302.1
Receptors, location of 804.1
Systems 310.4
Traps 1004.0
Trough type urinals ....405.2, E 501.5
Urinals and water closets on a combination waste and vent system 910.7
Vent termination 906.3
Water piping 312.6, 604.2, 720.1
PROPERTY LINE, TERMINATION OF VENT 906.2
PROTECTION FROM
Corrosion ....312.1, 312.4, 402.3, 402.6.2, 402.6.3, 506.9, 507.5, 508.1.2, 509.5.3, 509.7, 509.10.1, 509.10.1.2, 509.10.1.3, 603.5.14.1, 603.5.14.2, 609.3, 710.4, 710.8, 810.2, 811.2, 814.1, 1004.1, 1208.6.4.4, 1208.7.1.1, 1210.1.3, 1210.3, 1210.4.4, 1310.4, 1323.2, 1323.3, 1503.9.1, 1603.3, C 501.3, E 301.3.2, E 403.5, E 403.6, E 403.8.2, E 403.8.4, E 502.7, F 501.4, H 501.3, H 501.13, K 105.2, K 106.2.2
Decay ....1503.9.1, 1603.3, H 501.3, K 105.2
Freezing 312.6, 603.4.7, 1210.1.4, 1210.7.1, 1603.17,
Seismic 507.2, 1323.4.5
PROTECTION OF
Fuel gas meters 1208.7.1, 1208.7.3
Piping materials and structures 312.0, 314.4, 1210.1
Potable water supply 603.0
Trap seals 1002.0, 1005.1, 1007.0
Water heaters ....504.3 – 504.6, 505.2, 507.2, 507.12 – 507.15
PUBLIC OR PUBLIC USE, DEFINITION 218.0
PUBLIC SEWER, DEFINITION 218.0
PUBLIC SHOWER FLOORS 408.8
PUSH FIT FITTINGS 218.0, 605.1.3, 605.1.3.3, 605.2.1

– Q –

QUALITY AND WEIGHT OF MATERIALS, GENERAL REGULATIONS 301.2.1, 301.2.2, 301.3, 301.5
QUALITY OF FIXTURES 401.2
QUICK-DISCONNECT DEVICE, DEFINITION 219.0

– R –

RAINWATER, DEFINITION 220.0
RAINWATER CATCHMENT SYSTEMS ....220.0, 601.3.3, Chapter 16, Appendix K
RAINWATER DIVERSION VALVES 1503.2.4
RAINWATER STORAGE TANK, DEFINITION 220.0
REAMING PIPES ....309.3, 605.1.1 – 605.1.2, 605.1.3.2, 605.1.3.3, 605.1.4, 609.1, 701.3.1, 705.3.1, 705.3.3, 903.3
RECEIVING TANKS OR SUMPS ....710.2, 710.3, 710.7 – 710.10, 710.13.1
RECEPTORS
Definition 220.0
Indirect waste 414.3, 801.2, 801.3.2, 801.3.3, 801.4, 801.6, 801.7, 804.0, 807.1, 810.4, 814.5, 902.2, 1303.4.1
Shower ....408.1, 408.5 – 408.7
RECLAIMED (RECYCLED) WATER SYSTEMS 220.0, 601.3.3, 1505.0
REDUCED PRESSURE DETECTOR FIRE PROTECTION BACKFLOW ASSEMBLY ....Table 603.2, 603.3.9, 603.5.14
REDUCERS 316.0
REFRIGERATORS 801.3
REGULATORS
Backpressure 507.9
Draft ....506.1.2, 509.12.3 – 509.12.5, 509.15
Firefighter breathing air ....F 501.8
Fuel gas supply ....507.20 – 507.22, 1202.1, 1208.3, 1208.8 – 1208.12, 1213.1.5, 1214.1, 1215.1.3, E 404.0, E 404.4, E 404.5.2
Medical gas and medical vacuum ...1308.3, 1321.10.2
Water supply 608.2

RELIEF VENTS . . . . .907.0, 911.3, B 101.4
Connections 805.1
Definition 220.0
Interceptors 1009.4
REMOTE OUTLET, DEFINITION 220.0
REPAIRING
Defective gas pipe, prohibited 1208.6.8
REPAIRS AND ALTERATIONS TO PLUMBING AND DRAINAGE SYSTEMS 102.4
REQUIRED
Air gaps or air breaks 801.2
Appurtenances 1505.10, 1506.9.3, 1602.9
Area of subsurface irrigation fields and mulch basins 1504.2
Drainage or sewer connections 304.1, 713.0
Filters 1506.9.5, 1603.19
Flushing devices 413.1
Fuel gas pipe size 1215.0
Fuel gas supply 1214.0
Gravity drainage 709.0
Hot and cold water 601.2.1
Inspections .1406.0, 1502.3.4, 1605.3.4
Low-pressure cutoff 609.9
Medical gas and medical vacuum pipe size 1323.1
Permits 104.0
Plans 103.3
Plumbing fixtures . . . . .422.0, Table 422.1
Tests . . . . .1502.2 – 1502.3.4, 1504.10, 1505.5, 1506.5, 1602.5
Vents 901.2
RESPONSIBLE FOR TESTING 105.2.5
RESTAURANT, GREASE INTERCEPTORS 1014.1
RETESTING 105.3.2
RISERS
Fuel gas . . . . .1208.6.7, 1210.1.7, 1210.4.4, E 403.6, E 404.3
Medical gas and medical vacuum 1323.3, 1323.13.2
Sewer . . . . .E 504.6, E 504.6.2
Water . . . . .220.0, 408.10, 610.1, 1502.3.3, A 103.1, A 106.1, A 106.2, A 108.1, E 302.1, E 302.4, E 503.2, E 505.4, L 503.3.7
RODENTPROOFING 312.12
ROOF DRAINS . . . . .220.0, 1101.12.1, 1101.12.2, 1101.15, 1102.0, Table 1103.1, 1106.3, 1107.2.1, K 105.12
ROOF DRAINAGE . . . . .1101.12, Appendix D
Combined 1101.12.2.2.2
Controlled-flow 1105.0
Primary 1101.12.1
Secondary 1101.12.2
Separate piping 1101.12.2.2.1
ROOF GUTTERS .1101.12.1, 1103.3, Table 1103.3, 1602.6, 1602.7.2, 1603.15, 1603.20, K 103.2, K 103.2, K 105.12, K 106.3
ROOF, TERMINATION OF VENTS ABOVE 509.6.1, 905.4, 906.1, 906.3, 906.5, 906.7
ROOF WASHERS . . . . .220.0, K 106.2
ROUGHING-IN, DEFINITION .220.0

– S –

SADDLE, SLEEVE, BAND OUTLETS, PROHIBITED 310.1
SAFETY AND HEALTH 102.5
SAFETY DEVICES . . . . .504.4 – 504.6, 505.2, 608.0
SAND INTERCEPTORS 1016.0
Cleanout for 707.7
Construction and size 1016.3
Definition 221.0
Design 1009.3
Discharge 1009.7, 1016.1
Location 1009.5
Maintenance 1009.6
Required 1009.1, 1016.1
Vent, required and not required 902.1, 1002.1, 1016.4
Water seals 1009.3, 1016.4
SANITARY TEES, PERMISSIBLE USE . . . . .706.2, C 601.3, C 601.4
SCREW JOINTS 1205.2, 1206.3
SDR, DEFINITION 221.0
SEAL, TRAP . . . . .222.0, 315.1, 412.1.1, 413.4, 901.3, 912.2, 1005.0, 1007.0, C 601.1, E 501.6
SEAM, WELDED, DEFINITION 221.0
SECONDARY INTERCEPTORS 902.1
SECONDARY TEMPERATURE LIMITING DEVICE 505.2
SEDIMENT TRAPS 1210.7.2, 1212.9
SEEPAGE PITS . . . . .H 701.0
Abandoned 722.2
Area of disposal . . . . .H 301.0
Capacity . . . . .H 701.1
Combination with disposal fields . . . . .H 601.4
Connection to sewer prohibited 714.3

Construction . . . . . H 701.3
Definition 221.0
Location . . . . . 721.0, H 101.8, Table H 101.8
Markings 716.1
Multiple installations . . . . . H 701.2
Soil absorption . . . . . Table H 201.1(2), H 301.1
SELF-CLEANING TRAPS 1003.1
SEPARATE PLUMBING AND DRAINAGE SYSTEMS 311.0
SEPARATE VENTING 811.6, 906.3
SEPARATORS 1009.0
SEPTIC TANKS
Abandoned 722.2
Capacity . . . . . H 201.0, Table H 201.1(1)
Connections . . . . . H 601.7
Connections to sewer prohibited 714.3
Construction . . . . . H 501.0
Definition 221.0
Location of . . . . . H 101.8, Table H 101.8
Markings 716.1
Vents . . . . . H 501.8
SERVICE, AUTHORITY TO RENDER GAS 1205.0
SETTING FIXTURES 402.5
SETTLING TANKS 902.1
SEWAGE
Definition 221.0
Disposal of . . . . . 303.0, 709.0, 713.0, 714.0, 1014.1, 1503.2, E 301.1, E 504.1, L 302.1
SEWAGE DISPOSAL SYSTEMS, PRIVATE . . . . . Appendix H
Abandoned 722.2
Commercial food waste grinders 714.4
Design . . . . . H 101.6
Holding tank 714.5
Inspection and testing . . . . . H 1001.0
Location . . . . . 307.0, 721.0, H 101.8, Table H 101.8
Materials 715.0
Vents 904.1
SEWAGE EJECTOR 221.0, 710.2, 710.3, 710.10
SEWAGE PUMP 221.0, 710.2, 710.3, 710.13
SEWERS . . . . . Chapter 7
Abandoned 722.0
Adjacent to walls . . . . . 312.3, 715.1, 718.3, Table 721.1
Available 713.2, 713.4
Building, definition 204.0
Building, material of . . . . . 701.0, Table 701.2, 715.0
Cleanouts 719.0
Connection to, required 304.0, 713.0
Construction of . . . . . 715.0 – 719.0
Damage to 305.1, 306.1, 714.1, 718.3, 811.1, 1009.1
Depth of 312.3, 718.3, 720.1
Easements for 721.2
Ejectors 710.2, 710.3, 710.10
Existing 102.4.1, 713.6, 715.3
Grade and support of 708.0, 718.0
Gray water . . . . . 1503.0 – 1503.9.7
Independent system required 311.0
Inspections 105.0
Joints and connections 705.0
Location . . . . . 307.0, 312.3, 718.3, 720.0, 721.0, Table 721.1
Materials . . . . . 701.0, Table 701.2, 715.0, 720.1
Mobile home and recreational vehicle parks . . . . . E 301.0, E 504.0
Must be on same lot as building 307.1, 721.1
Private sewage disposal systems 713.0, 714.0
Prohibited connections . . . . . 603.4.4, 714.2, 714.3, E 502.7
Reclaimed (recycled) water 1505.11
Required 709.1, 713.0
Size of building . . . . . Table 702.1, 703.2, Table 703.2, 710.5, 717.0, Table 717.1, C 304.2, Table C 304.2, C 601.9
Storm water 1101.3, 1101.16.2
Support 718.2
Testing of 723.0
Vents 904.1
Water pipe in trench 609.2, 720.0
SHALL, DEFINITION 221.0
SHEET LEAD 408.7, 408.7.3, 701.5, 906.6
SHIELDED COUPLINGS . . . . . 221.0, Table 313.3, 705.2.2
SHOCK ARRESTER, DEFINITION 221.0
SHORT TURN ¼ BEND, PROHIBITED 310.5, 706.0
SHOULD, DEFINITION 221.0
SHOWERS
Approved standards 408.1
Control valves 408.3, 601.2
Finished curb or threshold 408.5
Gang . . . . . 209.0, 408.3.1, 408.3.2, L 201.0, L 402.6.1
Mobile home and recreational vehicle parks . . . . . E 501.7
Pans 408.7
Public floors 408.8
Receptors 408.5, 408.6, 408.7
Stall or compartment 408.6
Tests 408.7.5
Valve and head location 408.9

Valves .417.2, 417.3
Waste outlet or drains .317.1, 408.4
Water consumption .408.2, L 402.6

SHUTOFF VALVES
Fuel gas .507.8, 509.5.7.1, 1208.13, 1210.3.3.1, 1210.8.2, 1210.11, 1212.3.1, 1212.3.2, 1212.5.3, 1212.6, 1212.7, 1212.9, 1213.2.5, 1213.6.1.4, E 402.3, E 402.4.2, E 403.0, E 403.9, E 403.13, E 404.5
Medical gas and medical vacuum .1314.6, 1314.7, 1314.8, 1314.11, 1327.2, 1310.3.2, 1312.2
Sewage .710.12.1, 710.13.2
Water .504.6, 606.2, 606.3, 608.4, 609.9, E 302.3, L 402.3.1, L 404.6, L 410.3, L 413.4

SINGLE-FAMILY DWELLING, DEFINITION .221.0

SINGLE-STACK VENT SYSTEMS .C 601.0

SINKS .420.0
Approved standards .420.1
Clinical .205.0, 413.1, 1303.7
Floor .421.0, B 101.2
Island .909.1
Laundry .420.4
Prohibited types .405.3, B 101.2
Service .420.4
Special use .406.2
Three-compartment .1001.2

SIZE OF
Building sewers .703.2, 717.0, Table 717.1
Cesspools .Appendix H
Chimneys .509.5.5
Circuit venting .911.0
Combination waste and vent piping .910.0, Appendix B
Combustion air openings .506.3 – 506.5
Condensate pipe .814.3, Table 814.3
Disposal fields .Appendix H
Drainage piping .Table 702.1, 703.0, Table 703.2, C 304.0, E 302.2, E 504.4
FOG disposal systems .1015.3
Fuel gas piping systems .1208.4, 1214.0, 1215.0, E 403.2, Table E 403.2
Fuel gas vents .509.6.2, 509.7.4, 509.10.2, 510.0
Flushometer valves and tanks .610.10, 610.11
Gray water .1503.8
Grease interceptors .1014.2.1, Table 1014.2.1, 1014.3.6, Table 1014.3.6
Hanger rods .313.6
Indirect waste pipes .801.3.1, 803.3
Leaders, conductors, and storm drains .1103.0, Table 1103.1, Table 1103.2, Table 1103.3
Louvers, grilles, and screens .506.8
Medical gas and medical vacuum pipe .1323.1.1
Mulch basins .1504.6.2
Oil tanks .E 403.12
Pipe cleanouts .Table 707.1, 707.10
Rainwater catchment systems .1602.6, K 106.5
Reclaimed (recycled) water piping .1502.6
Roof drainage .1101.12, Table 1103.2, Table 1103.3, 1105.1, 1603.16, Appendix D, K 106.4
Sand interceptors .1016.3
Seepage pits and disposal fields .H 301.0
Septic tanks .H 201.0
Single-stack vent systems .C 601.0
Subsoil irrigation field piping .1504.7
Traps .702.1, Table 702.1, 910.4, 1003.3, 1006.1, 1101.15.2
Vacuum drainage systems .C 501.2.1
Vent pipes .Table 703.2, 710.10, 810.1, Table 810.1, 904.0, 908.1.1, 908.1.2, 908.2.2, C 401.0
Water heaters .501.1, Table 501.1(1)
Water piping .610.0, Table 610.3, Table 610.4, Table A 103.1, Appendix A, E 302.2
Water softener connections .611.4
Wet venting .908.0

SLAUGHTERHOUSES .1010.0

SLIP JOINTS .221.0, 402.10, 509.10.9, 605.14, 705.9.1, 705.9.3, 1003.2

SLOPE .(see Grade)

SOIL PIPE .221.0, 310.2, 310.4, 402.6.1, 402.6.3, Table 701.2, 701.2(6), 722.2, 1002.4

SOIL POROSITY .H 101.2

SOLDER, BUSHINGS .701.7, Table 701.7

SOLDER JOINTS .212.0, 309.2, Table 313.3, 402.6, 605.1.3.1, 605.1.4, 605.16.1, 705.3.3, 705.10.1, 705.10.2

SPACING
Pipe supports .313.0, Table 313.3, 1210.3.5, 1210.3.5.1, Table 1210.3.5.1, 1323.4, Table 1323.4.4, 1327.6, Table 1327.6

Subsoil irrigation fields 1504.7.3
Trenching and leaching beds . . . . . H 601.9
SPECIAL FIXTURES 406.0
SPECIAL VENTING 909.0
SPECIAL WASTE PIPE . . . . . (see Industrial wastes)
SPECIAL WASTES
Acid 811.2
Chemical 811.0
Definition 221.0
Industrial 811.0
High temperatures . . . . . 810.1, Table 810.1
Laundries, slaughterhouses, bottling establishments, and auto wash racks . . . . . 711.0, 1010.0 – 1013.0
SPECIFIC GRAVITY, NATURAL GAS 1214.1
SPECIFICATIONS FOR MATERIALS . . . . . 301.2, Table 1701.1
STACKS . . . . . 221.0, 317.1, 706.2, 711.1, 906.1, 907.0, 911.3, 1101.16.2, C 304.3, C 304.4, C 401.0, C 601.0
STAINLESS STEEL PIPE . . . . . Table 313.3, Table 604.1, 605.13, Table 701.2, 705.7
STANDPIPE, CLOTHES WASHER 804.1
STATION INLETS AND OUTLETS . . . . . 221.0, 1305.3, Table 1305.3, Table 1323.1.5(1), 1323.1.2, 1315.0
STEAM EXHAUST CONDENSER OR BLOW OFF . . . . . 810.0, Table 810.1
STERILE EQUIPMENT 806.0
STORAGE TANKS
Alternate water sources 1503.3, 1505.4, 1506.4
Hot water . . . . . 505.2, 608.6, 608.7, L 502.2, L 503.3.2, L 503.3.7
Potable water . . . . . E 502.7
Rainwater . . . . . 1602.7.3, 1603.0, K 102.2, K 103.3, K 105.0
Vacuum drainage . . . . . C 501.2.1
STORM DRAINAGE SYSTEMS . . . . . Chapter 11, Appendix D
STORM DRAINS 221.0, 1101.4.4, 1101.6, 1101.11, 1101.12, 1101.13, 1101.15, 1103.0, 1104.1
STORM SEWERS 221.0, 715.3, 1101.2, 1101.4.5, 1101.12.2.2.2, 1101.13.1, 1103.2, 1104.1
STRAINERS AND CONNECTIONS 407.5, 408.4, 810.4, 1012.1, 1102.2
S-TRAP PROHIBITED 1004.1
STREET MAIN; (SEE WATER MAIN) . . . . . 225.0, 608.1, 608.3, 610.7, Table 721.1, Table 1503.4, A 104.2, A 104.4, Table H 101.8
STRUCTURAL CONDITION SAFETY 312.2, 312.11
SUBMITTAL OF PLANS . . . . . 103.3, 104.3, 104.3.2, 104.3.3, 301.5.3, 811.1, 910.2, 1105.1, 1304.3, 1505.2.1, 1602.2, C 301.1, C 501.1, E 101.2, H 501.1, K 101.3.1
SUBSOIL DRAINS . . . . . 221.0, 1101.4.6, Table 1101.4.6, 1101.6, 1101.8, 1101.9
SUBSOIL IRRIGATION FIELDS . . . . . 221.0, 1503.2.1, Table 1503.4, 1503.9.3, 1504.2, 1504.5, 1504.7, Table 1504.7.3
SUBSTITUTE MATERIALS 301.3
SUBSURFACE DISPOSAL FIELDS . . . . . 1504.0, H 101.2
SUBSURFACE IRRIGATION FIELDS . . . . . 221.0, 1503.2.1, 1504.2 – 1504.4, 1504.5
SUDS RELIEF 711.0
SUMPS
Boiler room, installation 810.1, 810.3
Building, backwater devices 710.1, 710.3, 710.4, 710.6
Building drainage systems below sewer level 710.2
Condensers . . . . . 810.0, Table 810.1
Construction 810.2
Definition 221.0
Macerating toilet systems 710.13.1
Pump connection permitted . . . . . 710.2 – 710.5, 710.9, 805.1, 1101.6.2, 1101.7
Pump discharges, conversion to fixture units 702.3, 710.5
Sewage 710.2
Sizing, sewage ejector and pumps 710.3
Storm drainage 1101.6.2, 1101.6.3, 1101.7, 1101.10, 1101.11, 1101.14
Vents . . . . . 710.7, 710.10, 710.13.3, 810.1, Table 810.1
Water supply inlets 603.5.5
SUPPLY, WATER
Fittings, fixtures 402.7

Fixture, definition 208.0
Fixture units . . . . .610.3, Table 610.3, 610.7 – 610.10, Table 610.10, Table A 103.1, C 303.2, C 501.5, E 302.1, E 302.2, E 302.2.2, L 402.3.1
Hot water . . . . .601.2.1, 602.1, 603.4.5, 603.5.10, 608.7, 609.1, 609.4, 610.3, Table 610.3, 610.12.1, 610.12.2, Table A 103.1, A 105.3, Table C 303.2, L 502.5, L 502.7, L 502.7.3
Materials .604.0, Table 604.1, 1503.9.2 – 1503.9.4, 1505.6, 1506.6, 1602.7, K 103.4
Nonpotable water 601.3, 603.5.11, 1506.6
Pipe, size of . . . . .610.0, Table 610.3, Table 610.4, Table A 103.1, Table C 303.2, Appendix M
Potable water, protection . . . . .409.5, 410.2, 412.2, 414.2, 419.3, 602.0, 603.0, 1303.5, 1503.3, 1505.4, 1506.4, 1602.4, E 502.3, K 101.9, K 102.2, L 411.2
Pressure . . . . .608.0, 1603.14, A 104.3, E 302.2.3, E 502.5, K 105.11
Quality . . . . .601.2, 1501.7, 1506.7, 1506.9.2, 1601.7, 1603.5, 1603.16, E 302.1, K 101.7, K 104.3, K 104.4
Quantity . . . . .413.1 – 413.4, 416.2, 419.3, 610.3, E 302.1
Risers . . . . .408.10, E 302.2
Tanks, water 603.5.5, 608.1
Testing 609.4
Valves 606.0

SUPPORTS . . . . .221.0, 313.0, Table 313.3, 402.4, 507.3, 507.3.1, 507.4, 507.19, 509.5.8, 509.6.4, 509.7.5, 509.10.8, 609.1, 718.2, 906.4, 1208.7.2, 1208.15.1, 1210.3, 1210.3.5, Table 1210.3.5.1, 1323.4, 1323.4.4, Table 1323.4.4, E 402.4, E 404.2, F 901.4, H 601.3, L 501.2

SURFACE WATER, DISCHARGE 306.2, 714.2, 811.1, 1101.1, 1101.11

SURGE TANKS 221.0

SWIMMING POOLS, SPAS, AND HOT TUBS . . . . .603.5.5, 603.5.20, 813.0, L 407.1, L 503.3.6

SYSTEMS
Chimney venting 509.5.5, 509.6.2.4
Circuit venting 911.0
Combination waste and vent . . . . .205.0, 910.0, Appendix B
Defective, testing of 105.3.1
Dialysis water 603.5.18.1
Direct-vent 509.6.1, 509.8.2
Durham, definition 206.0
Engineered . . . . .509.10.7.1, 912.0, C 201.1
Exhaust 507.22, 509.3.4, 509.6.1
Fire protection (protection from) 603.5.14
Firefighter breathing air replenishment . . . . .Appendix F
Firestop . . . . .Chapter 14
FOG disposal 1015.0
Fuel gas piping . . . . .507.18, Chapter 12, E 401.0 – E 406.4
Fuel gas venting . . . . .509.0, Table 509.4, 510.0
Gray water 601.3.3, 1502.0
Independent 311.0
Lightning protection 1211.5
Macerating toilet 710.13
Mechanical draft 509.3.3, 509.8, 509.10.3.2, 509.10.6
Medical gas and medical vacuum . . . . .319.0, Chapter 13
Natural draft venting 509.6.2.1
On-site treated nonpotable 601.3.3, 1506.0
Plumbing vent, definition .218.0
Plumbing venting . . . . .710.7, Chapter 9, C 401.0
Potable water (definition) . . . . .218.0, Appendix A, E 302.1 – E 302.8
Private sewage disposal . . . . .713.2, 713.3, 713.5, 714.0, 716.1, 723.1, 904.1, Appendix H
Pure water process 603.5.18
Rainwater catchment . . . . . . .220.0, 601.3.3, Chapter 16, Appendix K, L 501.0
Reclaimed (recycled) 601.3.3, 1503.3, 1505.0, 1506.4, 1602.4
Sanitary drainage . . . . .Chapter 7, E 301.0 – E 301.4.1, E 504.1 – E 504.5
Single-stack vent . . . . .C 601.0
Sprinkler 612.0

Storm water drainage . . . . Chapter 11, Appendix D
Ultraviolet water treatment . . . . . . Table 611.1, 1506.9.2, K 104.4.2, N 105.1
Vacuum drainage . . . . C 501.0
Ventilation 506.1.3, 506.7.3
Wet vent 908.2.5

– T –

T RATING 222.0, 1404.3, 1405.3, 1406.4
TAILPIECES
Connection 414.3, 419.2, 910.5, 910.6
Definition 222.0
Maximum length of 910.5, 1001.2
Sizes . . . . 407.5, 408.4, 409.2, 420.4, B 101.3, B 101.5, B 101.6.1
TANKS
Dosing . . . . H 601.8
Expansion 608.2, 608.3, 812.1
Flush . . . . 208.0, 402.4, 413.1, 601.2, A 104.1
Flushometer . . . . 610.11, A 106.3
Hot water storage . . . . 608.3, 608.6, 608.7, L 502.2, L 503.3.2
Materials 710.8, 810.2
Oil . . . . 1017.1, E 403.12, E 403.13, E 404.6
Rainwater storage . . . . 1602.7.3, 1603.6, K 102.2, K 103.3, K 105.0
Reclaimed water storage 1505.4
Septic . . . . 714.3, 716.1, 722.2 – 722.4, Appendix H
Sump or receiving . . . . 710.2, 710.7 – 710.10, 810.2, 1101.7
Surge . . . . 221.0, 1503.2.1, Table 1503.4 1503.9.1, 1504.10
Urinal 603.5.2
Water closet . . . . 404.2, 405.1, 601.2, 603.5.2, 603.5.3, L 402.2
Water supply . . . . 608.1, 608.6, 608.7, E 502.7, E 506.1
TEMPERATURE AND PRESSURE-RELIEF VALVES 504.6, 505.2
TEMPERATURE-LIMITING DEVICES 407.3, 408.3, 409.4, 410.3, 504.5
TEMPORARY USE OF GAS 1207.0
TERMINATION OF VENTS . . . . 509.6.1, 509.8 509.8.2, 906.0, H 501.11
TESTS
Air 105.3, 105.3.1, 609.4, 712.3, 723.1, 1107.2.2, 1107.2.3, 1213.1.6
Alternate materials and methods 301.3
Building sewer 102.4.1, 105.2.1, 105.2.1.2, 723.0
Exceptions 105.3, 1107.2.3
Final 712.1
Fuel gas piping 1213.0
Gauges 318.0, 1203.3.2
Percolation . . . . 1504.3, H 401.0
Required 105.3
Responsibility for 105.2.5
Shower receptors 408.7.5
Storm drainage 1106.0
Testing agency 301.3.1.1, 505.4, 610.2, 1406.4
Water 609.4, 712.2, 1107.2.1, 1107.2.3
Water supply systems 105.2.1.1, 609.4
THREADED JOINTS 605.1.5, 605.2.3, 605.5.2, 605.11.3, 605.12.3, 605.16.1, 705.1.3, 705.3.4, 705.4.2, 705.6.3, 705.10.2, 1208.6.9, 1208.6.10.1, 1322.5
THREE SINGLE COMPARTMENT SINKS 1001.2
TOILETS . . . . (see Water closets)
TOILETS, FAMILY OR ASSISTED - USE FACILITY 422.1.1
TRACER WIRE 604.10.1, 1210.1.7.2
TRANSITION GAS RISER 222.0, 1210.4.4
TRANSITION JOINTS 705.9.4
TRAPS . . . . Chapter 10
Arms . . . . 222.0, Table 702.1, 707.14, 1002.2, Table 1002.2, 1002.3
Bar and fountain sinks 801.4, 902.2
Building 1008.0
Bottle 1004.1
Condensers and sumps 810.1
Crown weir 222.0, 409.6, 905.5, 1001.2, 1002.2, 1002.4
Definition 222.0
Drum 1004.1
Floor drains 704.3, 1003.1, 1006.0, 1007.1
Food storage areas 418.4, 801.3.2
For three-compartment sinks or three lavatories 1001.2
From fixture, distance of 1001.2
From vent, distance of . . . . 1002.2, Table 1002.2
Gray water system 1503.9.6
Grease interceptors 1014.1.1
Indirect wastes 803.3
Integral 1001.2

Island sinks .909.1
Materials .807.2, 906.6, 1003.1
Minimum size .702.1, Table 702.1, 1003.3
Nonwater urinals .412.1.1
Prohibited types .1004.0
Required .1001.0, 1101.15
Seal .222.0, 412.1.1, 912.2, 1005.0, 1007.1, C 601.1, E 501.6
Seal primer .222.0, 1007.1
Separate for each fixture .1001.2
Slip joints .705.9.1, 1003.2
Standpipe receptors .804.1
Storm drains and leaders .1101.15
Unions .315.1
Unit loading .702.0, Table 702.1, Table 703.2
Vacuum drainage system .C 501.0
Vents for .Table 703.2, 810.1, 901.0, 905.2, 905.5, 908.2.1, 909.1, 1002.0, Table 1002.2

TRENCHING, EXCAVATION, AND BACKFILLING .314.0, 1203.2, 1210.1.1 – 1210.1.2.1, 1323.5.1

TROUGH URINALS, PROHIBITED .405.2

TUBING, COPPER .Table 604.1, 604.3, 604.4, 605.1, 609.1, 609.3, 701.2, Table 701.2, 903.2, 1208.6.4.3, A 105.2, E 403.5

TUNNELING .314.2

TYPE B GAS VENTS .224.0, 509.1, Table 509.4, 509.5.5, 509.6.1, 509.6.2.1, 509.6.2.2, 509.6.3.2, Table 509.7.3.4(1), 509.7.3.5, 509.10.1.1, 509.10.1.2, 509.10.7.2, 509.10.12, 510.1.6, 510.1.10, 510.2.2, 510.2.15, 510.2.19, 510.2.20, 510.2.22, Appendix G

TYPE BW GAS VENTS .224.0, Table 509.4, 509.6.1,

TYPE L GAS VENTS .224.0, Table 509.4, 509.6.1, 509.6.2.2, Table 509.7.3.4(1), 509.10.1.1, 509.10.1.2, 509.10.12

– U –

UNDERGROUND
Backflow prevention devices .603.4.9
Building storm drains .1101.4.4
Drainage piping .701.2, Table 701.2, 720.0
Fuel gas piping .1203.2, 1208.6.3.3, 1208.6.4.4, 1210.1, E 401.2, E 402.2, E 403.7, E 403.8, E 403.8.4
Gray water piping .1503.9.2
Medical gas and medical vacuum piping .1323.2, 1323.5.1
Stop-and-waste valves .603.5.17
Storm water piping .1101.4, 1101.4.2
Tanks .1503.9.1, 1601.11.2, 1603.9, E 403.11, K 105.0
Unions .315.1
Vent material .903.1, 903.2
Water piping .604.3, 609.9, 1506.9.4

UNIONS
Dielectric .605.15, 605.16.1, 605.16.3
Fuel gas piping .1212.6, E 404.2
Location .315.1, 609.5, 710.6
Medical gas and medical vacuum piping .1322.7
Sump pump .1101.6.2
Where allowed .315.1

UNIT LOADING OF TRAPS .702.1, Table 702.1(1)

UNITS, FIXTURE .(see Fixture, units)

UNLAWFUL CONNECTIONS .404.2, 602.0, 1205.2

UNSANITARY, DEFINITION .223.0

URINALS .402.5, 402.6, 405.2, 412.1, 413.0, 418.3, 601.2, 603.5.1, 603.5.2, 707.4, 710.3, 910.7, 1003.1, 1014.1, 1014.3.2.1, 1501.9, 1505.1, 1506.1, 1506.9.2, 1506.9.5, 1602.1, 1603.14, 1603.19, 1604.2, E 501.2, E 501.5, K 105.11 L 402.3

USE OF ALTERNATE MATERIAL .301.3

USER OUTLET, DEFINITION .223.0

– V –

VACUUM BREAKERS .224.0, 412.2, 601.3.4, Table 603.2, 603.3.2, 603.3.5, 603.3.6, 603.4.4, 603.5.1, 603.5.5, 603.5.6, 603.5.7, 603.5.9, 603.5.13, 603.5.15, 603.5.16, 1303.5, C 501.3, E 505.4, E 506.1

VACUUM DRAINAGE SYSTEMS .C 501.0

VACUUM RELIEF VALVES .224.0, 504.6, 608.7, 1602.9

VACUUM SYSTEM – LEVEL 1, DEFINITION .224.0

VALIDITY OF CODE .101.5

VALVES
Backwater .204.0, 710.1, 710.3, 710.6, 1006.1, 1503.2.3, 1503.9.1, 1503.9.7, 1603.10, K 105.4
Ball 606.1, 606.4, 710.4, 710.6
Bathtub filler 409.5
Check .603.5.10, 608.3, 710.4, 710.12.1, 710.13.2, 1004.2, 1208.12.1, 1309.11, 1311.5, 1313.5, F 701.3
Combination stop-and-waste 603.5.17
Combination temperature
and pressure-relief 505.2, 608.3
Combination thermostatic/
pressure balancing 205.0, 408.3
Control . . . . .408.9, 601.2, 606.5, 606.6, 1303.1, 1306.2, C 501.3
Diaphragm-type 507.22, 1208.8.4, 1208.12
Excess flow 207.0, 1209.0
Flow balancing . . . . .L 501.7
Flush, definition 208.0
Flushometer . . . . .208.0, 411.2.2, 413.1, 413.2, 601.2, 603.5.1, 610.10, Table 610.10, A 104.1, L 402.2.2
Fullway 606.2, 710.6
Gate 606.1, 606.4, 710.3, 710.4, 710.6
Gray water diverter 209.0, 1503.2.2
Installation 310.7
Isolation . . . . .224.0, F 501.7, F 601.5.1, F 701.3, F 801.6
Mixing 417.5
Pre-rinse spray .420.3, L 201.0, Table L 402.1, L 402.8
Pressure-balancing 218.0, 408.3
Pressure-reducing . . . . .612.5.3.2.2, 612.7.2, 1504.5.7, 1603.14, A 107.2, E 403.13, F 701.3, K 105.11
Pressure-relief . . . . .504.4, 504.6, 505.2, 507.9, 607.5, 608.2 – 608.6, 1208.11.4, 1306.2 1308.2, 1308.3, 1309.5, 1310.3, 1310.4, 1315.15, 1316.1, 1324.4, 1324.5, 1327.7.1, E 302.5, E 302.6, F 501.8, F 701.3
Quick-acting 609.11
Riser 224.0, 1314.8, 1323.14.4
Service . . . . .224.0, 1314.9, 1323.14.5, E 402.4.2
Shutoff . . . . .504.6, 606.2, 606.3, 606.6, 608.4, 609.9, 612.4.3, 612.6, 710.12.1, 1202.1, 1203.3.2, 1208.12, 1208.13, 1210.3.3.1, 1210.8.2, 1210.9 – 1210.9.4, 1210.11, 1210.12.5, 1212.3.1, 1212.3.2, 1212.5.3, 1212.6, 1212.7, 1213.2.5, 1213.6.1.4, 1310.3.2, 1312.2, 1314.6, 1323.14, 1327.2, 1503.9.1, E 302.3, E 402.3, E 402.4.2, E 403.0, E 403.9, E 403.13, E 404.5, E 505.4, E 506.1, F 601.1, K 105.4, L 402.3.1, L 404.6, L 410.3, L 413.4
Source 224.0, 1314.6, 1314.7, 1323.14.2, 1324.5.4.2
Temperature relief 504.5, 504.6, 505.2
Thermostatic 222.0, 408.3
Vacuum interface . . . . .C 501.2.4
Vacuum relief 224.0, 504.6, 607.5, 608.7, 1505.10, 1506.9.3, 1602.9
Zone 224.0, 1314.4
VENT, PLUMBING . . . . .Chapter 9
Below trap weir .905.5, 1002.4
Branch, definition 204.0
Buildings over 10 stories .907.1
Changes in direction of .903.3
Chemical 811.2, 811.5, 811.6
Combination waste and
vent system design . . . . .205.0, 910.0, Appendix B
Connections . . . . .310.2, 805.0, Table 810.1, 905.1, 905.2, 908.1.2, 908.2.1, 911.2, 1503.2.2
Continuous, definition 205.0
Definition 224.0
Distance of trap from . . . . .1002.2, Table 1002.2
Double fitting for 704.2, 905.6
Engineered 912.0
Expansion joints 705.9.2, 1101.5, 1213.2.2
Fittings . . . . .701.2, Table 701.2, 903.1
Fixtures, back-to-back 704.2, 905.6
Flagpoling, prohibited 906.3
Flashing for 312.8, 701.5, 906.1, 906.5, 906.6
Frost or snow closure 906.7
Grade of 905.1, 905.3, 911.4
Height above fixture 905.3
Height above grade 906.4

Horizontal . . . . .Table 703.2, 904.2, 905.3
Indirect waste 803.3, 805.1, 902.1
Individual, definition 211.0
Interceptors 902.1, 1009.4, 1014.1.1, 1014.2.2, 1016.4, 1017.1
Island sinks and fixtures 909.0
Length of . . . . .Table 702.1, Table 703.2, 904.0
Material for .701.0, Table 701.2, 811.2, 903.0, 1503.9.2
Method of installation 310.5, 903.3, 906.2, 906.4
Not required 902.0
Not to be used as soil or waste pipe 310.4
Opening location 905.5, 1002.4
Parallel 907.1
Piping, support of .Table 313.3
Prohibited 310.4, 903.1, 903.2.2
Relief 220.0, 805.0, 907.0, 911.3, 1009.4
Required 901.2, 904.1
Required separately on every trap 901.2, 1002.1, 1002.2
Sizing . . . . .Table 702.1, 703.0, Table 703.2, 904.0, 908.1.1, 908.1.2, 908.2.2, 910.4, C 401.0, Table C 401.1, C 601.2
Special 909.0
Stack . . . . .224.0, 906.1, 907.0, C 401.2, C 601.2
Sump . . . . .710.7, 710.10, 710.13.3, 810.1, Table 810.1
System, definition 224.0
Termination 905.4, 906.0
Testing 712.0, 1504.10
Trap seal protection 901.3, 1002.0
Undiminished in size through roof 905.4, 907.1
Unit loading . . . . .Table 703.2
Wet, Horizontal 908.2
Wet, Vertical 908.1
Yoke 227.0, 907.0

VENTILATING HOODS 509.2.4, 509.3.4
VERTICAL-TO-HORIZONTAL DRAINAGE FITTINGS 706.4
VERTICAL WET VENTING . . . .(see Wet vents)
VIOLATION, NOTICES OF 106.2
VITRIFIED CLAY PIPE . . . . .701.2, Table 701.2, 705.8, 811.2, Table 1101.4.6

– W –

WALL, EXTENSION THROUGH 312.1, 509.6.1, 509.8
WALL-HUNG FIXTURES, INSTALLATION OF .402.4, 402.6
WASH RACKS, AUTO 1011.0
WASTES
Anesthetic gas disposal .1301.2
Chemical .205.0, 811.0, 903.2.2
Clear water 205.0, 812.0
Condensate 807.2, 814.0
Continuous 205.0, 407.5, 408.4, 409.2, 419.2
Cooling water 808.0
Definition 225.0
Detrimental or harmful . . . . .305.1, 306.1, 714.1, 811.1, 1009.1, H 901.1
Flammable . . . . .1009.1, H 901.1
Indirect . . . . .Chapter 8, 1303.4.1
Industrial 211.0, 306.0, 801.1, 811.1, 903.2.2
Liquid . . . . .214.0, 303.0, 304.0, 710.2, 811.1, 1006.1, 1009.1, H 101.3, H 701.1, H 901.1, L 302.0
Materials, pipe 701.1, 811.2
Pipe, definition 225.0
Pretreatment required 811.1, 1014.1
Process 1013.1
Special, definition 221.0
Swimming pools 813.1
WATER
Backflow prevention . . . . .Table 603.2, 603.3
Cross contamination . . . . .601.2, 602.2, 603.1 – 603.3.10
Distribution piping material . . . . .604.1, Table 604.1
Distribution system design 601.1
Excessive pressure 608.2
Hammer 609.11
Identification 601.3
Lead content 604.2, 604.2.1
Nonpotable .603.5.11, 1506.6, 1506.9.5, L 408.1
Rainwater, material 1101.4
Reclaimed 1602.4
Storm water . . . .(see Storm drainage)
Supply piping material . . . . .604.1, Table 604.1
Treatment units 611.0
WATER CLOSETS 411.0
Approved standards 411.1
Clearance 402.5
Compartments . . . . .E 501.3.2, E 501.3.3
Flanged fixture Connection 402.6
Floor-mounted, back-outlet 402.6.3
Seats 411.3

Manufactured/Mobile home ....E 501.2, E 501.3, E 501.3.1, E 501.6
Prohibited 405.1, 910.7
Water consumption ....411.2, L 402.2
WATER-CONSERVING FIXTURES AND FITTINGS ....L 402.0
WATER HAMMER ARRESTORS 225.0, 609.11
WATER HEATERS ....Chapter 5
Access 504.3.1, 507.13.2, 507.26, 508.4
Approval requirements 501.1
Approved standards ......Table 501.1(1), 505.3, 505.4
Clearances..................... 504.3 – 504.3.2, 508.1, 508.2.1
Combustion air required 506.0
Connectors 604.13
Definition 225.0
Garages 507.13
Inspections 503.2
Installation 507.0
Location ....504.1, 507.12 – 507.15
Oil-burning and other 505.0
Permits required 502.0
Pressure-relief valves ....504.4, 504.6, 505.2, 608.2 – 608.5
Protection ....504.3 – 504.6, 505.2, 507.2, 507.8, 507.10, 507.14.1, 507.25
Relief valve discharge 608.5
Seismic provisions 507.2
Temperature and pressure-relief valves 504.6, 505.2
Temperature-limiting devices 504.5
Vents 509.0, 510.0
WATER SOFTENERS................ 611.4, Table 611.4
WATER TESTS
Building sewers 723.0
Drainage and vent systems 105.3, 712.2
Storm drainage 1107.2.1, 1107.2.3
Water supply systems 105.2.1.1, 105.3, 609.4
WATERPROOFING AND WATERTIGHT
Building sewer 723.1
Cleanouts 707.3
Connections 105.3
Drain pans 507.5
Fixtures 402.2, 408.5, 408.7, 418.2
Flashings 312.8, 906.5, 1102.3
Interceptors 1016.3
Joints ....105.3, 1504.7.3, H 601.4, H 601.6, H 601.7, H 601.10, H 801.4
Openings 312.8, 312.10.2, 317.1
Sump or receiving tanks 710.2, 710.8
Tanks ....714.5, 1503.9.1, 1504.10, 1603.3, H 501.3, H 1001.2, K 105.2
Vacuum drainage systems ....C 501.8
WATER/WASTEWATER UTILITY, DEFINITION 225.0
WEEP HOLES, SUBDRAIN OF SHOWER STALLS 408.7
WEIGHT OF FLASHING, LEAD 701.5, 906.6, 1102.3.1
WEIGHTS AND QUALITY OF MATERIALS 301.2, 401.2, 604.1, 701.2, 906.6, 1102.3.1
WELDED JOINTS ....212.0, 605.13.2, 705.7.2, 1208.6.10.1, 1208.6.10.3, 1322.1, F 1101.2
WELDER, PIPE, DEFINITION 225.0
WELL, DISTANCE FROM SEWAGE DISPOSAL ....721.1, Table 721.1, H 101.8, Table H 101.8
WET VENTS 908.0
Definition 225.0
Fixtures 908.2.3, 908.2.4
Horizontal ....908.2, B 101.2, C 601.6
Vent connection 908.2
Vertical 908.1
WHIRLPOOL BATHTUBS 225.0, 409.0
WINDOW, TERMINATION OF VENTS FROM 906.2
WOODEN SINKS AND TILE WASH TRAYS OR SINKS 405.3
WORK, CONCEALED ....105.0 – 105.2.1.2, 309.2, 402.10, 503.0 – 503.2, 509.7.3.2, 1004.1, 1203.2, 1203.3, 1210.4, 1212.3.1, 1212.5.3, 1213.2.1, 1303.2, 1323.6, 1406.1
WORK, EXISTING 102.2, 102.4, 207.0
WORKMANSHIP 309.0, 609.1, 1208.6.8
WRAPPING OF PIPE ....609.3, 1210.3, E 403.6
WROUGHT IRON (GALVANIZED) ....402.6.1, Table 604.1, 604.6, 604.7, 701.2, 707.1, 1101.4, 1208.6.3.1, 1208.6.10.5, E 403.5
WYE CONNECTIONS .... 414.3, 419.2, 706.2 – 706.4, 707.5, 707.6, 710.4, 719.4, 907.2, 909.1, 1101.16.2

– X –

X-RAY DARKROOMS 811.8

– Y –

YOKE VENTS 227.0, 907.0

– Z –

ZINC ALLOY COMPONENTS 406.4, 1208.6.10.5